INTERNATIONAL UNION OF CRYSTALLOGRAPHY
BOOK SERIES

Crystalline Molecular Complexes and Compounds

Structures and Principles

Volume 2

FRANK H. HERBSTEIN

Emeritus Professor of Chemistry, Technion-Israel
Institute of Technology, Israel

OXFORD

UNIVERSITY PRESS

OXFORD

UNIVERSITY PRESS

Great Clarendon Street, Oxford OX2 6DP

Oxford University Press is a department of the University of Oxford.
It furthers the University's objective of excellence in research, scholarship,
and education by publishing worldwide in

Oxford New York

Auckland Cape Town Dar es Salaam Hong Kong Karachi
Kuala Lumpur Madrid Melbourne Mexico City Nairobi
New Delhi Shanghai Taipei Toronto

With offices in

Argentina Austria Brazil Chile Czech Republic France Greece
Guatemala Hungary Italy Japan Poland Portugal Singapore
South Korea Switzerland Thailand Turkey Ukraine Vietnam

Oxford is a registered trade mark of Oxford University Press
in the UK and in certain other countries

Published in the United States
by Oxford University Press Inc., New York

British Library Cataloguing in Publication Data

Data available

Library of Congress Cataloging in Publication Data

Data available

Typeset by Newgen Imaging Systems (P) Ltd., Chennai, India
Printed in Great Britain
on acid-free paper by
Biddles Ltd., King's Lynn

ISBN 0–19–856893–2 (Vol 1) 978–0–19–856893–3
ISBN 0–19–856894–0 (Vol 2) 978–0–19–856894–0
ISBN 0–19–852660–1 (Set) 978–0–19–852660–5

10 9 8 7 6 5 4 3 2 1

TWENTY: Give up learning and put an end to your troubles.

From TAO TE CHING by Lao Tsu
(A new translation by Gia-fu Feng and Jane English
Wildwood House, London 1973)

OR

How charming is divine philosophy,
Not harsh and crabbed as dull fools suppose,
But musical as Apollo's lute,
And a perpetual feast of nectared sweets,
Where no crude surfeit reigns,

John Milton (1608–74): Comus, 476.

This book has been written in many places and over too many years. It would never have been completed without the help and support of my wife Any. It is dedicated to her and to four teachers and friends:

R. W. James, FRS, formerly Professor of Physics in the University of Cape Town
G. M. J. Schmidt, formerly Professor of Chemistry in the Weizmann Institute of Science
J. D. Dunitz FRS, emeritus Professor of Chemical Crystallography in the Eidgenössische Technische Hochschule, Zürich
Sir Aaron Klug P-PRS, OM, NL. MRC Laboratory of Molecular Biology, Cambridge, UK

Preface

My intention is to give an account of the structure and properties of crystalline binary adducts, perhaps better known as molecular compounds and complexes, which are a broad group of materials whose several members are of interest to chemists (e.g. separations *via* crown ethers and identifications *via* charge transfer compounds), physicists (e.g. high-conductivity organics), biologists (is not DNA an excellent example of a hydrogen-bonded molecular compound?) and technologists (zeolites for separations and as catalysts). I have tried to cater to them all; extensive inclusion of chemical formulae (for the nonchemists) and stereodiagrams (for the noncrystallographers) will hopefully make it easier to assimilate some of the unavoidable complexities. Most emphasis will be given to geometrical structures derived from crystal structure analyses for here lies the bulk of the available information. I refer to interactions between the components wherever this is possible, including both thermodynamic and electronic aspects. Consideration of the relation between structure and properties will be principally confined to the solid state and the implications of solid state results for understanding chemical reactivity in other phases will not be pursued.

I restrict myself to crystalline materials because the results, and their meanings, are least unequivocal for this state of matter. Interactions between components in the fluid states are undoubtedly important but I leave these aspects to others. The word "molecular" appears in the title because most of the relevant materials are indeed molecular, but many contain charged entities and I have licensed myself to include what seems relevant, regardless of the formal restrictions of the title. Most of the substances considered are organic, some are inorganic and many have both organic and inorganic parts.

I discuss representatives of many of the various types of molecular compound and complex, but I early realized that any attempt to cover all examples of all types would be self-defeating. The problems of choice beset us all, at all levels of our lives, and an author, struggling to compress the vast expanses of knowledge into a practical physical confine, is no exception. I have given preference to mature areas, where what is known presents a model for the treatment of those regions as yet unexplored, and I have emphasized the fundamentals – structure and thermodynamics. I guess (no other word seems realistic) that I have managed to include about 20% of what is available in the literature. The series *Inclusion Complexes*, dealing with less than half the topics covered here runs to five volumes and over two thousand pages. The series *Comprehensive Supramolecular Chemistry*, again with half the present coverage, stretches to eleven volumes and five thousand pages. Perhaps the most serious of my many acts of omission is exclusion of material on "Zeolites". This is not because of any lack of importance of the subject but because it is adequately covered in a number of books and an on-going journal. There is also little said about complexes between large biomolecules – this would have required a separate book.

An overall theory is hardly possible but some areas have had sophisticated theories applied to them, e.g. the quantum-mechanical treatments of charge transfer interactions and the statistical mechanical treatments of some phase diagrams. In order to provide some unifying factors, I have given special emphasis to structural relationships and the classification scheme used is structural rather than chemical in nature. The classification scheme proposed here should not be regarded as more than a convenient framework – Nature is too complex and subtle for the imposition of straitjackets.

Titles of books and review articles have been included in the (close to 4000) references, which are attached to each chapter for the convenience of the reader. Much of the information is conveyed through tables, some 200 in number. The tabulated material shows that considerable systematization is possible, but also that considerable variety remains as exceptions to those rules that I (and many others) have succeeded in developing. Crystal packing and other diagrams (some 600) are a challenge to author and reader alike. Colour would have helped but was ruled out as impractical.

I am most grateful to authors, editors and publishers for permission to use published material. Some more detailed acknowledgments are made in the text. My thanks go to many friends and to Caltech, Northwestern, Cambridge (U.K.), The Royal Institution, the Universities of Cape Town and Witwatersrand and, last but not least, Technion for help and facilities. Needless to say, the responsibility for the contents is entirely mine.

Haifa,   Frank H. Herbstein
November 2004

Contents

Volume 1

PART I SOME PRELIMINARIES

PART II MOIETIES WITHIN MOLECULES

Introduction to Part II

PART III   HOST–GUEST INCLUSION COMPLEXES

Introduction to Part III

Volume 2

PART V   MOLECULAR COMPOUNDS WITH LOCALIZED INTERACTIONS

PART VI   MOLECULAR COMPOUNDS WITH DELOCALIZED INTERACTIONS

(Note. The components in the ground states of these molecular
compounds are taken to be neutral unless explicitly stated otherwise).

Acknowledgements

The author wishes to thank the following for permission to reproduce published material.

Academic Press:

Inclusion Compounds. Figs. 6.39, 6.65, 7.1, 7.24(a), 7.24(b), 7.38, 7.39, 7.40, 7.42, 7.43, 9.19, 9.22.
Non-Stoichiometric Compounds: Fig. 8.4.

American Association for the Advancement of Science

Science. Fig. 10.24.

American Chemical Society

Accts. Chem. Res. Figs. 7.36, 10.15.
Biochemistry. Figs. 5.7, 5.8, 5.9, 5.12.
Chemistry of Materials: Figs. 6.3(b), 6.14, 6.20, 6.30, 8.6.
Cryst. Growth & Design: Figs. 12.35, 12.36.
Inorg. Chem. Figs. 9.18, 9.21, 10.19, 11.41, 11.58, 11.59, 11.71, 11.100, 11.119, 15.27, 17.38.
J. Am. Chem. Soc. Figs. 3.4(a), (b), 3.8, 3.9, 3.17, 3.19, 3.25, 3.26, 3.28, 4.4, 4.15, 6.55, 6.60, 6.61, 6.62, 6.69, 7.5, 7.6, 7.32, 8.8, 8.53, 10.13, 11.16, 11.31, 11.43, 11.60, 11.63, 11.116, 12.6, 12.12, 12.14, 12.16, 12.18, 12.21, 12.27, 15.13, 15.32, 15.33, 16.43, 16.44, 17.9, 17.15, 17.19, 17.23, 17.42, 17.53.
J. Chem. Educ. Figs. 4.1, 7.29.
J. Med. Chem. Fig. 5.21.
J. Org. Chem. Figs. 11.3, 11.72.
J. Phys. Chem. Figs. 6.19, 6.26, 6.27, 6.28, 7.26, 17.22(b), 17.37.
Organometallics: Figs. 11.113, 11.121.

American Crystallographic Association

Fig. 3.13.

American Institute of Physics

J. Chem. Phys. Figs. 6.17, 6.25, 7.7, 7.15, 7.34, 7.35, 7.37, 15.29, 16.5, 16.20, 16.23.
Sov. Phys. Crystallogr. Figs. 10.20, 10.22

American Physical Society

Phys. Rev. Letts.: Figs. 13.4, 16.36, 17.46.
Phys. Rev.: Fig. 16.38.

Elsevier

Adv. Organometall. Chem.: Fig. 9.6.
CALPHAD: Fig. 10.21.
Carbohydrate Research: Figs. 4.5(a), 4.11(a).

Carbon: Figs. 9.10, 9.11.
Chem. Phys. Letts: Fig. 6.11.
Comp. Rend. Acad. Sci. (Paris), Ser. C,: Fig. 17.18.
Coord. Chem. Revs. Fig. 10.40.
FEBS Letters: Fig. 5.23.
Inorg. Chim. Acta: Figs. 11.46, 11.53.
J. Chromatography: Fig. 15.9.
J. Mol. Biol. Figs. 5.6, 5.11, 5.17, 5.18, 5.20.
J. Organometall. Chem.: Figs. 10.34, 11.120.
J. Phys. Chem. Solids: Figs. 6.13, 17.52.
Mater. Sci.: Fig. 16.14.
Sol. State Comm. Fig. 16.39.
Synth. Mets. Fig. 16.42.

International Union of Crystallography

Acta Crystallographica, Figs. 6.31, 6.51, 10.14, 10.16, 11.1, 11.15.
Acta Crystallographica, B. Figs. 4.3, 4.16, 4.17, 4.19, 6.1, 6.3(a), 6.8, 6.50, 6.53, 6.58, 7.2,
    7.25, 7.31, 7.41,10.7,10.8, 10.26, 11.40, 12.2, 12.5, 12.9, 12.10, 12.30, 12.31, 15.2,
    15.3, 15.5, 15.7, 15.15, 15.18, 15.21, 16.4, 16.9, 16.10, 16.11, 16.12, 16.13, 16.22,
    16.32, 17.17, 17.20, 17.22(a), 17.24, 17.27, 17.29, 17.32, 17.33, 17.43(a), 17.48, 17.49.
Acta Crystallographica, C. Figs. 6.18, 8.7, 8.31, 8.37, 8.38, 10.4, 10.17, 10.17A, 10.32(a),
    10.32(b), 12.24, 12.28, 15.6, 17.6, 17.7.
Acta Crystallographica, D. Fig. 5.16.
IUCr Monographs on Crystallography: Figs. 15.34, 15.35.
J. Appl. Cryst. Fig. 7.27.

Kluwer Academic Publishers

J. Incl. Phenom. Figs. 3.11, 6.43, 7.13, 7.14, 7.24, 9.13.

Macmillan Publishers

Nature. Figs. 5.15, 7.16, 10.33.

National Academy of Sciences U. S. A.:

Proceedings. Figs. 5.13, 5.14, 5.22.

NRC Research Press (Canada)

Can. J. Chem.: Figs. 11.45, 11.52.

Oldenbourg Verlag

Z. Kristallogr. Fig. 6.21.

Pergamon

Comprehensive Supramolecular Chemistry. Figs. 4.2, 6.59 (Vol. 6).
Tetrahedron Letters: Figs. 3.12, 3.22.
Tetrahedron: Fig. 12.13.

Plenum

Water–a comprehensive treatise. Fig. 7.30.
J. Cryst. Spectroscop. Res.: Fig. 10.31.

Professor M. Le Cointe.

Ph.D. thesis, University of Rennes I: Figs. 16.35, 16.37, 16.40.

RIA-Novosti, Paris

La Recherche: Fig. 7.18 (I am grateful to Professor Rose Marx, Saclay, for her help in obtaining this figure).

Springer:

Monatshefte Chem.: Fig. 10.10.
Springer Series in Materials Science No 18: Figs. 9.7, 9.8, 9.9.
Topics in Current Chemistry. Figs. 6.41, 12.26.

Taylor and Francis:

Adv. Phys.: Fig. 9.2.
Contemp. Phys.: Fig. 17.57.
Mol. Cryst. Liq. Cryst. Figs. 6.12, 6.22, 6.23, 6.33, 10.2, 10.23, 11.25, 15.30, 16.29, 17.3.

The Chemical Society of Japan

Bull. Chem. Soc. Jpn. Figs. 4.5(b), 4.6(a), (b), 4.7, 4.9, 4.10, 4.13, 4.18, 11.42, 13.9, 13.10, 13.11, 16.33, 16.34, 17.10, 17.16.
Chem. Letts. Figs. 7.17, 10.46, 14.6, 17.44.

The Physical Society of Japan

J. Phys. Soc. Jpn.: Figs. 13.7, 17.54, 17.55.

The Royal Society of Chemistry

Chemical Communications: Figs. 3.3, 3.4(c), 3.5, 3.16, 3.23, 6.47, 6.48, 8.21, 8.22, 10.30, 10.35, 10.41, 12.19.
Chem. Soc. Revs.: Fig. 9.17.
Chemistry in Britain, Fig. 1.2.
J. Chem. Soc. A: Fig. 11.49.
J. Chem. Soc. B: Figs. 6.67, 6.68.
JCS Dalton. Figs. 8.15, 8.16, 8.17, 8.18, 9.23, 9.24, 11.12, 11.33, 11.39, 11.101, 17.11, 17.36.
JCS Perkin II: Figs. 3.15, 6.40, 6.42, 7.3, 12.15, 15.14.
JCS Trans. Farad. Soc. Figs. 6.4, 6.5, 6.6.
J. Mater. Chem.: Figs. 17.39, 17.40.
New J. Chem.: Fig. 12.3.

The Royal Society of London:

Proceedings, Ser. A: Figs. 8.19, 9.12, 15.4, 16.8, 16.30, 16.31.

Verlag Chemie-Wiley

Angew. Chem. Intl. Ed.: Figs. 3.27, 5.3, 10.36, 11.115, 14.2.
Chem. Ber. Figs. 10.28, 11.35, 14.8, 14.12, 14.13, 14,14, 16.45.
Chemistry Eur. J. Fig. 6.56.
Prog. Inorg. Chem. Figs. 7.21, 7.23.

Verlag Helvetica Chimica Acta:

Helvetica Chimica Acta. Figs. 5.19, 7.4, 10.11.

Verlag der Zeitschrift für Naturforschung

Z. Naturforsch. (b): Figs. 11.110, 11.114, 11.118.

Worth Publishers New York

Lehninger Biochemistry, 2nd edition. Figs. 5.1, 5.2.

Various

Acta Chem. Scand.: Figs. 11.7, 11.8, 11.9, 11.30.
Acta Chem. Scand A: Figs. 11.10. 11.11.
J. Phys. D: Fig. 9.1.
J. Struct. Chem. USSR: Figs. 6.2, 7.33.
Liebigs Annalen: Fig. 15.20.
Molecular Complexes: Fig. 13.8.
Phys. Chem. Low-dimens. Materials: Fig. 9.15.

I am grateful to Dr Moshe Kapon and Dr Mark Botoshansky for help of many kinds and to the staff of the Chemistry-Biology Library at Technion for their assistance in tracking down material.

Part V

Molecular compounds with localized interactions

Introduction to Part V

Molecular compounds with localized interactions

In the nineteenth century convincing experimental proof was collected of interactions between molecular systems generally regarded as chemically "saturated", leading to more or less stable complexes. In several cases stoichiometric solid compounds could even be isolated from such mixtures ... about the atomic arrangements in the complexes almost nothing was known until direct interferometric experiments were carried out ... Particular importance may be attributed to complexes in which direct bonding exists between one atom belonging to the donor molecule and another atom belonging to the acceptor molecule. Complexes of this kind are above all those formed by donor molecules containing atoms possessing "lone pair electrons" and halogen or halide molecules.

Odd Hassel, from Nobel Lecture, Chemistry, 1969.

Donor–acceptor charge-transfer interactions fall into two groups – those with localized and those with delocalized interactions between the component molecules. Part V deals with localized and Part VI with delocalized interactions. The localized interactions can be further divided into charge transfer interactions and hydrogen bonding interactions, here treated in two separate chapters. While charge transfer interactions may have received the greater attention – Nobel prizes in Chemistry to Odd Hassel for experimental studies and to R. S. Mulliken for development of theory – it would not be unfair to say that hydrogen bonding has had a greater impact on chemistry in the broader sense. Chapter 11 is the longest in the book and ranges from Hassel's compounds with iodine as acceptor through Menschutkin's compounds of aromatic hydrocarbons with antimony trihalides. Hydrogen bonding receives rather briefer treatment, partly because of the existence of a number of excellent texts (Jeffrey and Saenger, and Desiraju and Steiner being among the more recent) but also because we have concentrated on zero, one and two-dimensional systems and largely omitted the much more numerous three dimensional systems with their complicated and varied structures.

Chapter 11

Donor–acceptor molecular compounds (essentially localized interactions)

Electron donors D and acceptors A are here defined as all those entities such that, during interaction between a particular species of D and a particular species of A entities, transfer of negative charge from D to A takes place, with the formation as end-products either of additive combinations or new entities. The additive combinations may be $1:1$, $m:1$, $1:n$ or in general $m:n$ combinations.

R.S. Mulliken (1952)

Summary: The components discussed in this chapter are linked, in the main, by localized interactions. There is a variety of donor and acceptor types ranging from n-donors and σ^*-acceptors, through n-donors and p-acceptors, n-donors and p*-acceptors, π-donors and s,p-acceptors to p-donors and σ^*-acceptors. The first of these categories includes some self-complexes. A distinction has been made between "pure acceptors," discussed in the first part of this chapter, and "self-interacting acceptors," discussed in the second part. The "pure acceptors" are involved only in donor–acceptor interaction with the donor component of the molecular compound. In the "self-interacting acceptors" the principal acceptor atom (e.g. Ag^+ in $AgClO_4$ or Sb in $SbCl_3$), in addition to its donor–acceptor interaction, is also involved in interactions with other atoms of the acceptor moiety – e.g. the oxygens of $AgClO_4$ or the chlorines of $SbCl_3$. These additional interactions are essential to the stability of the crystalline molecular compounds. Despite widespread use of the term 'charge transfer' to describe these compounds, physical measurements (e.g. NQR, Mössbauer effect) suggest that there is very little actual transfer of charge in the ground state, as Mulliken often emphasized.

11.1   Introduction and classification

Many chemical species can be distinguished by their tendency either to donate electrons (i.e. to be nucleophilic, or to act as electron donors, or as Lewis bases, in various different terminologies) or to accept electrons (i.e. to be electrophilic, or to act as electron acceptors, or as Lewis acids). In the limit, complete transfer of an electron from donor to acceptor leads to formation of an ionic structure. Appreciable interaction, sufficient for the formation of stable crystalline molecular compounds, can occur even for partial transfer of charge and the substances considered in the following chapters are called (more or less interchangeably) charge-transfer (CT) or donor–acceptor (DA) molecular compounds.

Most modern theoretical treatments are based on proposals developed by Mulliken (1952a, b), which in turn rest on ideas of Weiss and Brackmann in the specific area of molecular compounds and on the more general approaches to chemical structure and reactivity introduced by many others, including G. N. Lewis and C. K. Ingold (for a review of theoretical approaches see Bender (1986)). Mulliken proposed that the wave functions of a donor–acceptor adduct in ground (G) and first excited (E) states could be represented by:

$$\Psi_G = a\psi_0(D, A) + b\psi_1(D^+, A^-)$$
$$\Psi_E = -b^*\psi_0(D, A) + a^*\psi_1(D^+, A^-)$$

where D, A represents the no-bond structure of the adduct and D^+, A^- the dative structure. It is assumed that $a \gg b$, $a^* \gg b^*$ and physical measurements (e.g. NQR, Mössbauer Effect; see later for details) confirm that the charge transfer in the ground state is small. This is not a contradiction of the basic premises. On the one hand, the major part of the lattice energy of crystalline CT molecular compounds comes, as Mulliken has insisted, from the contributions of dispersion and electrostatic forces of various kinds; on the other hand, the crystal structures to be described provide incontravertible geometrical evidence for specific donor–acceptor interactions and it is these interactions which determine the *details* of the crystal structures.

The classification used here derives from Mulliken's (1952b) detailed classification of donors and acceptors, but is biased towards the structural evidence obtained from diffraction analyses of crystalline molecular compounds. The structural evidence shows that it is convenient to distinguish between two types of acceptors, which we shall call "pure acceptors" and "self-interacting acceptors." The "pure acceptors" are exemplified by Br_2 molecules in the benzene···Br_2 molecular compound, where the only donor–acceptor interaction is between benzene π-orbitals and Br_2 σ^*-orbitals; this group of molecular compounds is described in the first part of this chapter. The "self-interacting acceptors" are exemplified by $AgClO_4$ in aromatic hydrocarbon···$AgClO_4$ molecular compounds, where the vacant s-orbitals of Ag^+ accept not only π-electrons from the aromatic molecule but also lone-pair donation from oxygen atoms of neighbouring ClO_4^- ions. Both interactions are necessary for the formation and stability of these molecular compounds but it is the aromatic moiety···Ag^+ interaction which identifies them as localized interaction donor–acceptor molecular compounds; this group of molecular compounds is

Table 11.1. Classification scheme for crystalline donor–acceptor molecular compounds

Pure acceptors		Self-interacting acceptors	
Interaction type	Example	Interaction type	Example
$n \rightarrow s$	dioxane···$AgClO_4$	$n \rightarrow s$	pyrazine···$AgNO_3$
$n \rightarrow \sigma^*$	dioxane···Br_2		
$n \rightarrow p$	benzoyl chloride···$AlCl_3$	$n \rightarrow p$	dioxane···$HgCl_2$
$\pi \rightarrow \sigma^*$	benzene···Br_2	$\pi \rightarrow s$	benzene···$AgClO_4$
$\pi \rightarrow \pi^*$	mesitylene··· $NO^+PF_6^-$	$\pi \rightarrow p$	naphthalene···$SbCl_3$
	(localized interaction)		
	anthracene···picric acid		
	(delocalized interaction)		

described in the second part of this chapter. Our classification is set out in general terms in Table 11.1 and in more detail in the tables of contents.

We use the following nomenclature conventions. For mnemonic reasons the donor moiety is always listed first in the name of the molecular compound which we write as {benzene···Br$_2$} or {benzene···AgClO$_4$}. The term 'secondary interactions' is used as a catch-all for all bonding interactions intermediate in strength between covalent and van der Waals bonding, and will usually be indicated by···; weaker interactions are denoted by . . . ; interatomic distances are indicated as d(D···A). Pioneering reviews are by Bent (1968) and Alcock (1972).

Although the divisions are not sharp, as features of one structure are often partially employed in another (Nature's principle of structural parsimony (Herbstein, 1987)), nevertheless each of these groups has sufficiently distinctive structural characteristics to justify separate treatment. Furthermore, there is so much information available about the *delocalized* $\pi \rightarrow \pi^*$ category that it is discussed, together with related systems, in separate chapters later.

One consequence of a structurally based classification scheme is that perhaps-expected resemblances are not always found in practice. For example, SbCl$_3$ interacts with unsubstituted aromatics through $\pi \rightarrow$ p interactions, but with acetylated aromatics such as p-diacetylbenzene through n$\rightarrow$p interactions without participation of the aromatic ring. SbCl$_3$ always behaves in some respects as a self-interacting acceptor but AsCl$_3$ is a pure acceptor in its (few known) molecular compounds with aromatics.

Donors can be divided into delocalized-orbital donors (the HOMOs of aromatic systems) and localized-orbital donors (the lone pair s or p electrons). A similar distinction can be applied to acceptors. The delocalized acceptors (generally substituted aromatics or heteroaromatics) employ LUMOs as acceptor orbitals. The localized acceptors are dihalogens or halogenated molecules, with σ^* or s acceptor orbitals, nitronium salts with localized π^* acceptor orbitals, or metal salts in low oxidation states. For this latter group our classification is based on foundations developed by Rundle (1957), Amma, Schmidbaur (1985) and their coworkers. The metal-ion acceptors are divided into groups with analogous acceptor orbitals (Table 11.2).

Many olefins and aromatics form molecular compounds with Ag(I) salts, and one example is known with a Cu(I) salt. A fair number of aromatics form molecular compounds with Ga(I), In(I), Tl(I), Sn(II) and Pb(II) salts, where the anion is a conjugate ion of a strong acid. Many crystal structures have been reported and it appears that the d^{10}s^0 ions have an η^2 or η^3 mode of interaction whereas the d^{10}s^2 ions have an η^6 mode of interaction. This was indeed predicted many years ago by Rundle and Corbett (1957):

Consider an M$^+$-benzene complex with the cation on the benzene axis and therefore of symmetry C$_{6v}$. The highest filled π-orbitals of the benzene ring belong to the irreducible representation e$_1$

Table 11.2. Outer electron configurations of metal ions in low oxidation states

Outer electron configuration of ion	Acceptor orbitals	Examples
(n − 1)d^{10}	ns	Cu^I Ag^I Hg^{II}
(n − 1)d^{10}ns^2	np^3	Ga^I Ge^{II} As^{III} In^I Sn^{II} Sb^{III} Tl^I Pb^{II} Bi^{III}

(doubly degenerate) and the orbital of the cation must belong to the same representation to allow bonding by electron transfer from benzene to the cation. For ions such as Ag^+, Hg^{2+} etc. the lowest acceptor orbital is an s-orbital belonging to a_1, orthogonal to the upper filled π-orbitals of the ring. Charge transfer is then impossible without considerable electronic promotion, so that movement of the cation to a position of lower symmetry is necessary for bonding. However, for Ga(I), In(I) and Tl(I), the lowest acceptor orbitals available are the p-orbitals, where the degenerate pair p_x, p_y belonging to e_1 can accept the highest energy p-electrons from the benzene ring. Hence a symmetry C_{6v} for the benzene-M^+ ion is by no means excluded in this case.

The results described below for the appropriate molecular compounds fit these predictions very well.

PART 1: Pure acceptors

11.2   n-Donors and s-acceptors

11.2.1   *N, O, S containing ligands as donors and Ag(I) salts as acceptors*

Silver salts form molecular compounds with molecules containing N, O, S as donor atoms and a number of crystal structures have been reported. As most of these materials appear to be on the border between metal coordination complexes and molecular compounds (for example, piperazine···2AgI (Ansell and Finnegan, 1969; PIPZAG) and morpholine···AgI (Ansell, 1976; MORAG10)), we mention them only briefly.

In {3(dioxane)···$AgClO_4$} (Prosen and Trueblood, 1956; AGPDOX) the Ag^+ ions at the corners of the cubic unit cell ($a = 7.67$ Å, space group $Pm3m$, $Z = 1$) are spanned by dioxane molecules along the cube edges, rotationally disordered about their O...O axes; thus each Ag^+ ion is surrounded by a regular octahedron of oxygens, with $d(O···Ag^+) = 2.46$ Å. The ClO_4^- ions are at the cube centres and are rotationally disordered; these oxygens do not interact with Ag^+. The compound is surprisingly stable and only starts to lose dioxane at 85 °C; however, it is hygroscopic and the Ag^+ is easily reduced even in the absence of direct sunlight. Powder photographs of compounds of formula {3(dioxane)···NaX} (X=ClO_4, DIOXSP01; BF_4; QQQHBJ; I, QQQHBG) indicate that these are isomorphous or isostructural with {3(dioxane)···$AgClO_4$} (Barnes and Duncan, 1972; AGPDOX01)·{3(dioxane)···$NaClO_4$} has a more ordered form stable below 317.6K (rhombohedral, $a = 7.62$ Å, $\alpha = 91.2°$, space group $R\bar{3}$) in which the dioxane molecules occupy fixed positions although the perchlorate ions are still disordered (Barnes and Weakley, 1978; DIOXSP); an analogous ordered form of the silver compound has not been found. It thus appears that metal-dioxane ion–dipole interactions are sufficient to explain the formation of this type of compound and there is no need to invoke any special Ag^+–oxygen interactions. The enthalpies of decomposition have been measured (Barnes, 1972) for {3(dioxane)···$AgClO_4$} and {dioxane···$AgClO_4$}.

In {bis(2-imino-4-oxo-1,3-thiazolidene)···$AgClO_4$} (Murthy and Murthy, 1977) the Ag^+ is three-coordinated with two N···Ag interactions to the amine nitrogens of the two ring molecules of the formula unit ($d = 2.20(2)$, 2.28(2) Å) and a weaker interaction to carbonyl oxygens of symmetry-related molecules ($d = 2.54(1)$ Å); these four atoms are coplanar, giving triangular coordination about Ag^+. A perchlorate oxygen is also weakly linked to Ag^+ ($d = 2.90(3)$ Å).

1,3,5-Trithian forms a $1:1$ molecular compound with $AgNO_3$, and monohydrates of the $1:1$ molecular compounds with $AgNO_3$, $AgClO_4$ and $AgBF_4$; (Dalziel and Hewitt, 1986) the structures of the first three have been determined (Ashworth, Prout, Domenicano and Vaciago, 1968; TRTAGN, TRTAGH, TRTAGP). Structures are also known for the $1:1$ compounds of dimethylbut-3-enyl methyl sulphide with $AgNO_3$ (NMBEAG10), $AgBF_4$ (MTBAGB10) and $AgClO_4$ (MTBCAG10) (Alyea et al., 1981); this ligand was chosen because it was thought possible that the double bond coordinated to Ag^+, but in fact the interaction is through S.

11.3   n-Donors and σ^*-acceptors

11.3.1   *N, O, S or Se containing donors and dihalogens or halogenated molecules as acceptors*

11.3.1.1   *Summary of available results*

There are many molecular compounds of I_2 where lone-pair electrons of a donor atom (N, O, S, Se) are partially transferred into an antibonding σ^*5p orbital of I_2; these are n...σ^* molecular compounds. Br_2 behaves similarly, as do ICl and IBr. Although some of these compounds were first prepared in the nineteenth century, major advances in understanding were not made until the theoretical studies of Mulliken (1952a, b) and Mulliken and Person (1969a, b) and the crystallographic studies of Hassel and coworkers (Hassel and Rømming, 1962; Hassel, 1970a, b). Parenthetically we note an intriguing curiosity connected with the 1969 Nobel Prize for Chemistry that Hassel shared with D. H. R. Barton "for their contributions to the development of the concept of conformation and its application in chemistry". Despite Hassel's classical contributions to conformational analysis, his Nobel lecture dealt exclusively with the structures of charge transfer molecular compounds. In addition to the molecular compounds with dihalogens, there are many analogs where halogenated molecules such as halomethanes serve as electron acceptors. Evidence for CT interaction is often based on a comparison of

Table 11.3. Values of covalent, ionic and van der Waals radii (Å) (for a summary of sources see Douglas, McDaniel and Alexander, 1983)

Moiety	Covalent	Ionic	van der Waals	Moiety	Covalent	Ionic	van der Waals
chlorine	0.99	1.67	$1.75 - 1.8$	Cu^+	–	0.74*	–
bromine	1.14	1.82	1.85	Ag^+	–	0.81*	–
iodine	1.33 (gas phase)	2.06	$2.1 - 2.15$	Hg^{2+}	–	1.10*	–
	1.37 (cryst. I_2)						
hydrogen	0.38	–	1.2	nitrogen	0.55	–	1.55
oxygen	0.60	1.26 (O^{2-})	1.52	sulphur	1.06 (in S_8)	1.70 (S^{2-})	1.80

* These values depend on coordination number.

experimentally determined interatomic distances with those predicted on the basis of a set of radii of various kinds, such as those in Table 11.3.

Perhaps the most striking feature of the crystal structures of the present group of molecular compounds is the occurrence of short distances between donor and acceptor atoms; also, when the acceptor is a dihalogen, the distance within the molecule increases. This is shown in Tables 11.4–11.6, where the principal classification is in terms of how the dihalogen interacts with the donor atoms. Three situations are distinguished:

a. dihalogen symmetrically bonded to donor atoms (Table 11.4);
b. dihalogen unsymmetrically bonded to donor atoms (Table 11.5);
c. dihalogen bonded through only one of its atoms to donor (Table 11.6).

There are further subdivisions according to the nature of the donor and acceptor atoms. The results for halogenated molecules (e.g. $CHCl_3$) interacting with various donors are collected in Table 11.7.

Table 11.4. Symmetrical interaction of dihalogens with donor atoms. Measurements at room temperature unless stated otherwise; distances in Å, angles in degrees. Donor . . . Acceptor (Charge Transfer) Molecular Compound is abbreviated to 'DA MC'

Interaction type	DA MC/Refcode	$d(D\cdots A)$	$d(A–A)$	Angle $D\cdots A–A$	Reference and remarks
N$\cdots$I	phenazine$\cdots I_2$ PHNAZI01	2.982(5)	2.726(1)	180	Uchida and Kimura, 1984. Chains of $\cdots$ph$\cdots I_2\cdots$ph$\cdots$
S$\cdots$I	(merocyanine)$_2\cdots I_2$	3.098(5)	2.750(1)	178.4 <C=S . . . I = 99.5(7)	Bois d'Enghien-Peteau *et al.*, 1968. Structural unit is mer$\cdots I_2\cdots$mer
N$\cdots$Br	MeCN$\cdots Br_2\cdots$MeCN; m·pt. 232K; measured at 202K; ACTNBM	2.84	2.328	179.4(1)	Marstokk and Strømme, 1968. <CN . . . Br = 171.6(4)
O$\cdots$Br	1,4-dioxane$\cdots Br_2$ DOXABR	2.71	2.31		Hassel and Hvoslef, 1954a. I_2 analogue isomorphous.
	$(CH_3OH)_2\cdots Br_2$; m.pt. 207K, measured at 183K. METHOB	2.78	2.29		Groth and Hassel, 1964a. H-bonding also present.
	$(CH_3)_2CO\cdots Br_2$; m.pt. 265K; measured at 243K. ACETBR	2.82	2.28	180	Preparation: Maass and McIntosh, 1912; Structure: Hassel and Strømme, 1959a. <Br . . . O . . . Br = 110.
O . . . Cl	1,4-dioxane . . . Cl_2	2.67	2.02		Hassel and Strømme, 1959b. Isomorphous with Br_2 analogue (DOXABR).

Note: Raman spectra of crystalline dioxane complexes with I_2, Br_2 and Cl_2 (all 1:1, samples at 77K) are compatible with the x-ray results (Anthonsen, 1976). NMR and NQR spectra of dioxane . . . Cl_2 show that there is no phase change down to 78K (Gordeev *et al.*, 1970).

Table 11.5. Unsymmetrical interactions of donor atoms with dihalogens. Other details as in legend of Table 11.4

Interaction type	DA MC/refcode	d(D···A)	d(A–A)	D···A–A angle	References and remarks
N···I	HMT*... I_2...NI_3 HMTNTI	2.474(8) 3.23(1)	2.808(1)		Pritzkow, 1974a.
Se···.I	1-Oxa-4-selenacyclohexane···I_2 OXSELI	2.755(4) 3.708(4)	2.956(3)	174.8(3) 77.4(3)	Maddox and McCullough, 1966. I_2 axial to Se
	Tetrahydro-selenophene···I_2 THSELI01	2.762(5)	2.914(4) 3.64	179.4(3)	Hope and McCullough, 1964.

* HMT is hexamethylenetetramine.

11.3.1.2 *Generalizations based on the experimental results*

(i) Stereochemistry of the D···A interactions The D···A–B (A, B halogens) arrangement is always linear or nearly so. The distance between donor and acceptor atoms varies from a small increase (*ca.* 10%) over the covalent bond distance (e.g. as in $(CH_3)_3N$···I–I) to a small reduction (also *ca.* 10%) below the sum of the van der Waals' radii (e.g. as in $3(S_8)$···ICH_3). The donor molecules do not show significant changes of dimensions on formation of the molecular compounds, nor do the trihalogenomethanes and similar acceptors; however, the dihalogens (I_2, Br_2, Cl_2, IBr, ICl) all show significant lengthening compared to their gas phase dimensions. There is an experimental caveat here – halogen positions will generally be rather precisely determined in x-ray diffraction studies of these molecular compounds, and carbon (or N, S etc) positions much less so, with the consequence that C–halogen distances will have, essentially, the standard uncertainties of the C atom. Neutron diffraction may provide a means of obtaining more precise C–halogen distances in appropriate crystals.

The coordination about the donor atom depends on its nature and situation:

(a) nitrogen – tetrahedral about 4-coordinated nitrogen, trigonal planar about 3-coordinated N and linear about 2-coordinated N.

(b) oxygen, sulphur, selenium – the arrangement about these atoms is generally tetrahedral when there is 4-coordination and pyramidal when there is 3-coordination. O and S in ring compounds link the A atom equatorially, irrespective of whether A is in a dihalogen or a halomethane. Se in a ring compound links axially to A except for 1,4-selenothiane···I_2, where the bonding to both S and Se is equatorial, and in 1,4-diselenane···$(I_3CH)_2$, where one link is equatorial and the other axial. Oxygen with two secondary linkages forms a two-dimensional cross-linked structure in acet-one···Br_2. The secondary linkage from sulphur forms "elbow" bonds (<X=S···A ca. 95–110°) in dithizone···I_2 and (merocyanine)$_2$···I_2.

Table 11.6. Dihalogen bonded to donor moiety through only one of its halogen atoms. Other details as in legend of Table 11.4. Some chemical formulae are given at the end of this Table

Interaction type	DA MC/refcode	$d(D{\cdots}A)$	$d(A{-}A)$	angle $D{\cdots}A{-}A$	References and remarks
N···I	(CH₃)₃N···I₂ at 253K. TMEAMI	2.27	2.83	179	Strømme, 1959. Tetrahedral coordination about N
	4-picoline···I₂; m.pt. 356K. PICOLI	2.31	2.83	180	Hassel, Rømming and Tufte, 1961. I₂ at 14° to donor plane
	HMCP*···I₂ HMCPZI	2.417(7)	2.823(1)	177.8(2)	Markila and Trotter, 1974. Trigonal planar coordination about N
	HMT···I₂ HXMTDI	2.439(8)	2.830(1)	173.1	Pritzkow, 1975. Tetrahedral coordination about N
	HMT···(I₂)₂ HXMIOD	2.496(5)	2.791(1)	173.8(1)	Pritzkow, 1975. Bromine analogue is isomorphous (Eia and Hassel, 1956); (ZZZHEM).
	HMT···(I₂)₃ YUYNUB	2.593(6) 2.520(3)	2.746(1) 2.764(1)	177.6(2) 177.1	Tebbe and Nagel, 1995.
	9-cyclohexyladenine···I₂ CHXADI10		2.54		Van der Helm, 1973. Approximate trigonal planar coordination about N
	(CH₃)₃N···ICl	2.30(1)			Hassel, 1958. Isostructural with I₂ compound (TMEAMI)
	Pyridine···ICl PYRIIC10	2.29(1)	2.513(4)	178.7(3)	Rømming, 1972. Approximate trigonal planar coordination about N
	Pyridine···IBr PYIOBR	2.26(4)	2.66(1)	180	Dahl, Hassel and Sky, 1967. Approximate trigonal planar coordination about N
	Pyridine···ICN PYCYNI	–	2.57(2)		Dahl, Hassel and Sky, 1967. Approximate trigonal planar coordination about N
	Pentamethylene-tetrazole···ICl PMTTIC01	2.34	2.44	177	Baenziger et al., 1967. (PMTTIC). I bonded to N(2); ICl coplanar with donor; see note to this Table.

Table 11.6. (*Continued*)

Interaction type	DA MC/refcode	d(D···A)	d(A–A)	angle D···A–A	References and remarks
O···I	Dioxane···(ICl)$_2$	ca. 2.6	ca. 2.3	linear	Hassel and Hvoslef, 1956. Not known whether O···I axial or equatorial
S···I	1,3-Dimethyl-2-thioimidazole···I$_2$	2.616	2.967	<C=S···I = 97.7	Freeman et al., 1998. α-polymorph (GEGNUB)
	do.	2.607	2.984	99.7	β-polymorph (GEGNUB01)
	Dithizone···I$_2$ BZHTIC10	2.664	2.918	178.4 Angle C=S···I 95.6	Herbstein and Schwotzer, 1984. See note to this Table.
	N-methylthio-caprolactam···I$_2$ TCAPLI	2.687	2.879	176.2 Angle C=S···I 109.9	Ahlsen and Strømme, 1974. See note to this Table.
	Benzyl sulphide···I$_2$ BENZSI	2.78(2)	2.82(1)	179. Angle C=S···I 100.	Preparation: Fromm, 1913; structure: Rømming, 1960. S pyramidal. See note to this Table.
	Co(py)$_4$(NCS)$_2$···(I$_2$)$_2$ PYTTCC	2.797(2)	2.804(1)	178.8(1) Angle C=S···I 95.2(3).	Hartl and Steidl, 1977.
	1,4-dithiane . . . (I$_2$)$_2$ DTHINI	2.867(2)	2.787(2)	177.9(4)	Preparation: Husemann, 1863; structure: Chao and McCullough, 1960. Structural moiety is I$_2$···donor···I$_2$; I equatorial to S
	dithia[3.3.1]-propellane···(I$_2$)$_2$ FAJPUB	2.806(2) 2.852(2)	2.797(1) 2.796(1)		Herbstein et al., 1986. One S···I interaction is axial and one equatorial (for C$_4$S rings) for both propellane compounds
	dithia[3.3.2]-propellane···(I$_2$)$_2$ FAJRAJ	2.803(2) 2.902(2)	2.794(1) 2.767(1)		Herbstein et al., 1986.
	1,4-dithiane··· (IBr)$_2$ DTHIBR10	2.687(2)	2.646(1)	178.2(4)	Knobler et al., 1971

Se···I	1,4-diselenane···I_2 DSEIOD	2.829(4)	2.870(3)	180.0(3)	Chao and McCullough, 1961. Moiety as above but I axial to Se.
	1-oxa-4-selenocyclohexane··· ICl OXSEIC	2.630(5)	2.73(1)	175.8(2)	Knobler and McCullough, 1968. Moiety is C_4H_8OSe···ICl; I axial to Se.
	1,4-selenothiane···$(I_2)_2$ ZZZLSI				Hope and McCullough, 1962. Disorder prevented detailed analysis.
N···Br	$(CH_3)_3N$···Br_2	1.98(3)	2.29(1)	112(3)	Shibata and Iwata, 1985. Electron diffraction
S···Br	C_4H_8S···Br_2 THINBR	2.724(2)	2.321(4)	178	Allegra et al., 1970. Pyramidal coordination about S.

Notes:

Chemical formulae:

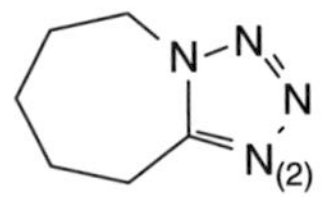

Pentamethylenetetrazole

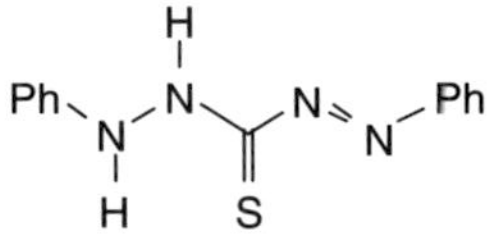

Dithizone

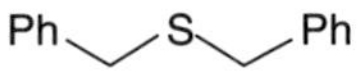

Benzyl sulphide

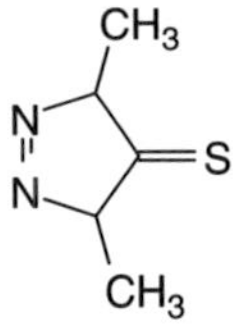

1,3-Dimethylimidazole

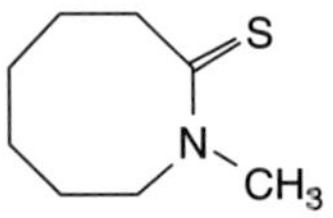

N-methylthiocaprolactam

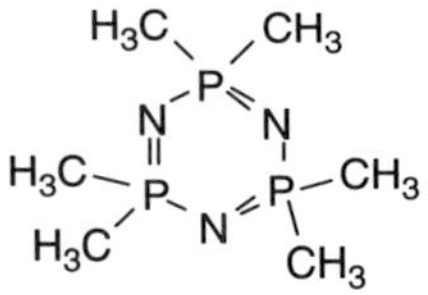

Hexamethylcyclotriphazene (HMCP)

Table 11.7. Donor$\cdots$halogen interactions in molecular compounds of N, O, S or Se-containing donors and halogenated molecule acceptors. Other details as in legend of Table 11.4

Interaction type	DA MC/refcode	$d(D\cdots A)$	Remarks
N$\cdots$I	$H_3N\cdots I_3N$	2.53	Hartl, Bärnighausen and Jander, 1968.
	pyridine$\cdots I_3N$ PYDTIN	2.58	Hartl and Ullrich, 1974.
	HMT$\cdots I_2\cdots NI_3$ HMTNTI	2.567(6) 2.582(8)	Pritzkow, 1974a. Both contacts are between I of NI_3 and N of HMT. Angles N-I$\cdots$N 175.4, 178.6
	HMT$\cdots I_3$CH HEXAIF10	2.93	Dahl and Hassel, 1970. C–H$\cdots$.N links also present ($d = 3.22$). Tetrahedral coordination of each molecule about the other. N$\cdots$I–C nearly linear.
	pyrazine$\cdots I_4C_2$ IETPYA10	2.98	Dahl and Hassel, 1970. Angle C–N$\cdots$I 175. Isomorphous with C_2Br_4 analogue
	(quinoline)$_3\cdots I_3$CH; m.pt. 336K. IFOQUI	3.05	Preparation: Rhoussopoulos, 1883; structure: Bjorvatten and Hassel, 1962. Angle C–N$\cdots$I 177
	4,4$'$-bipyridine$\cdots C_2I_4$ QIHBIS		Walsh *et al.*, 2001
	1,2-bis(4-pyridyl) ethylene$\cdots C_2I_4$ QIHCEP		Walsh *et al.*, 2001
	HMT$\cdots\ C_2I_4$. QIHCUF		Walsh *et al.*, 2001
O$\cdots$I	1,4-dioxane$\cdots$1,4-diiodoacetylene. DOXIAC	2.65	Gagnaux and Susz, 1960. I equatorial to O; chains in crystal.
S$\cdots$I	1,4-dithiane$\cdots$1,4-diiodoacetylene	3.27	Holmesland and Rømming, 1960. I equatorial to S.
	1,4-dithiane$\cdots I_3$CH. DTHIOF	3.32	Bjorvatten and Hassel, 1961. Poor crystals, I equatorial to S; chains in crystal.
	Co(py)$_4$(NCS)$_2\cdots$ 2(I_3CH). TCPYCO.	3.361(5) 3.460(5) 3.510(5)	Hartl and Steidel, 1980. Each S coordinated to three I of different CHI$_3$ molecules
	(S$_8$)$_3\cdots I_3$CH, IFOSUL	3.50	Bjorvatten, 1962. I equatorial to S; (S$_8$)$_3\cdot I_3$R (R=CH, P, As, Sb) are isomorphous.
	(S$_8$)$_3\cdots I_3$Sb	3.60	Bjorvatten, Hassel and Lindheim, 1963. Additional Sb$\cdots$I and S$\cdots$I interactions postulated to account for stability

Table 11.7. (*Continued*)

Interaction type	DA MC/refcode	d(D$\cdots$A)	Remarks
Se$\cdots$I	1,4-diselenane$\cdots$1,4-diiodoacetylene	3.34	Holmesland and Rømming, 1960. I axial to Se; not isomorphous with 1,4-dithiane analogue.
	1,4-diselenane$\cdots I_4C_2$ DISTIE	3.465	Dahl and Hassel, 1965.
	1,4-diselenane$\cdots 2(I_3CH)$ IFODSE	3.514	Bjorvatten, 1963. Angle C-I$\cdots$Se 178.2°. Se forms both equatorial and axial bonds to different iodines.
N$\cdots$Br	diazabicyclo[2.2.2]-octane*$\cdots$ (N-bromo-succinimide)$_2$ CEGVUF	2.332(4)	Crowston *et al.*, 1984. <N-Br$\cdots$N = 175.1(2)
	quinuclidine$\cdots Br_4C$ FEGYEV	2.53	Blackstock *et al.*, 1987. C-Br$\cdots$N linear
	diazabicyclo[2.2.2]-octane*$\cdots Br_4C$ FEGYAR	2.76, 2.88	Blackstock *et al.*, 1987.
	2,3-diazabicyclo[2.2.2]-octene[#]$\cdots Br_4C$ FEJFAB	2.91(1)	Blackstock and Kochi, 1987.
	HMT$\cdots(Br_3CH)_2$, at 238K. BFHMTA	3.00	Dahl and Hassel, 1971. Also N$\cdots$H–C (d = 3.13 Å) links.
	pyrazine$\cdots Br_4C_2$ BEPYRZ10	3.02	Dahl and Hassel, 1968b.
O$\cdots$Br	1,4-dioxane$\cdots$oxalyl bromide at 298K. OXBDOC	3.21	Damm *et al.*, 1965. Chains of alternating donor and acceptor molecules along [111].
O$\cdots$Cl	1,4-dioxane . . . oxalyl chloride at 253 to 233K. OXCDOC	3.18	Damm *et al.*, 1965. Isomorphous with bromo analogue.

Notes:
* DABCO;   # DBO, with formulae given below.

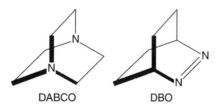

DABCO   DBO

(ii) Arrangement of the moieties in crystals

(a) Individual moieties linked by Van der Waals interactions

In many crystals of this kind, the structural (packing) units are donor$\cdots$acceptor groupings such as $(CH_3)_3N\cdots I_2$ or $I_2\cdots$1,4-dithiane$\cdots I_2$ (Fig.11.1); these structural units interact only through dispersion forces and thus the crystals are molecular crystals, the

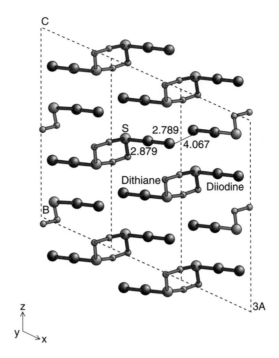

Fig. 11.1. Crystal structure of $\{I_2 \cdots (1,4\text{-dithiane}) \cdots I_2\}$ projected onto (010). The dark circles are iodines, the intermediate circles sulphur and the smallest circles carbon; hydrogens are not shown. Dithiane molecules are located about crystallographic centres of symmetry and are in the chair form; I_2 is equatorial with respect to dithiane. The I–I, I...S distances are shown, as well as a van der Waals contact between two iodines. (Data from Chao and McCullough, 1960.)

'molecule' being the D$\cdots$A structural unit. $\{1,5\text{-Dithiacyclooctane}\cdot 2I_2\}$ (Hope and Nicols, 1981; THOCIA) may well have a structure similar to that of $\{I_2 \cdots 1,4\text{-dithiane} \cdots I_2\}$ but atomic coordinates are not available.

The D$\cdots$A interactions seem to be stronger (using distances between donor and acceptor atoms as the criterion) within these quasi-isolated structural groupings than when extended interactions occur, leading to the formation of chains of the type $\cdots$D$\cdots$A$\cdots$D$\cdots$A$\cdots$, as described in the next section. Such a situation is exemplified by comparison of the $\{$quinuclidine$\cdots$CBr$_4\}$ (Blackstock, Lorand and Kochi, 1987; FEGYEV) (cubic, $a = 10.906(8)$ Å, space group $P2_13$, $Z = 4$) and $\{$diazabicyclo[2.2.2]octane$\cdots$CBr$_4\}$ (Blackstock, Lorand and Kochi, 1987; FEGYAR) structures (cell dimensions summarized in Table 11.8 below). The former has isolated D$\cdots$A units (Fig. 11.2) in its crystal structure, with rather strong and linear N$\cdots$Br–C links (d = 2.531(6) Å). One of the two C–Br distances reported in $\{$quinuclidine$\cdots$CBr$_4\}$ is 2.25 Å, which is much longer than the value found in Phase II of CBr$_4$ itself (1.91(4) Å) (More $et\ al.$, 1977; CTBROM). The discrepancy is due to inaccuracy in determining the carbon position, as noted above and by Blackstock $et\ al.$ The other C-Br distance is normal at 1.86(2) Å. $\{$Quinuclidine$\cdots$CBr$_4\}$ is unusual among CT molecular compounds in that it is enantiomorphic, although the absolute configuration was not reported. NaClO$_3$ is an example in the same space group,

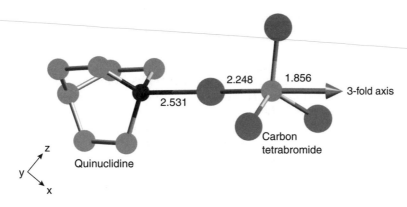

Fig. 11.2. Packing unit of the {quinuclidine···CBr₄} structure showing only the nonhydrogen atoms. Each donor–acceptor pair has three-fold symmetry about its C-Br···N axis and this takes up three orientations in the crystal. As noted in the text, the C–Br bond length of 2.248 Å is an artifact due to systematic error. (Data from Blackstock, Lorand and Kochi, 1987.)

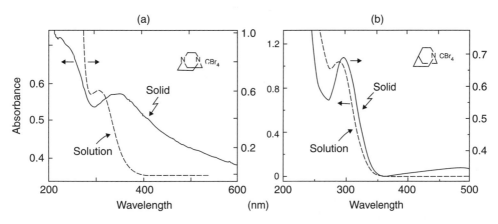

Fig. 11.3. Comparison of solution and solid-state absorption spectra of (a) {diazabicyclo[2.2.2]-octane···CBr₄} and (b) {quinuclidine···CBr₄}. (Reproduced from Blackstock, Lorand and Kochi, 1987.)

much studied in regard to the relation between crystal structure and optical activity (Devarajan and Glazer, 1986).

Solution and solid-state absorption spectra of {quinuclidine···CBr₄} are virtually identical (Fig. 11.3(b)) with the latter being shifted to the red by ≈5 nm with respect to the former; this close similarity suggests that the D···A units of the crystals are largely maintained in solution. The N···Br distances are 2.76, 2.88 Å in {diazabicyclo [2.2.2]octane···CBr₄} and solution and solid state spectra are rather different (Fig. 11.3(a)), suggesting only weak interactions in solution.

Symmetrical interactions of both halogen atoms (Table 11.4) appears to be rare for I₂; {merocyanine···I₂···merocyanine} is one example where the centrosymmetric arrangement has an iodine molecule interacting symmetrically with the sulphurs of two merocyanine molecules. Br_2 acts fairly commonly as a symmetrical acceptor (e.g. $\{CH_3CN\cdots Br_2\cdots NCCH_3\}$). IBr and ICl appear to act only as monofunctional acceptors, interaction always being through the iodine atom.

Caution must be exercised in inferring isomorphism – for example, both $\{I_2\cdots 1,4\text{-}$ dithiane$\cdots I_2\}$ and $\{BrI\cdots 1,4\text{-dithiane}\cdots IBr\}$ have $A\cdots D\cdots A$ moieties as structural units and also rather similar unit cells, but the structure analyses show that the moieties are arranged very differently. Comparison of the $I\ldots S$ distances in these two molecular compounds shows that IBr is a stronger acceptor than I_2.

(b) Unsymmetrical interactions of dihalogens

There are a number of examples where weaker secondary $D\ldots A$ interactions link the $D\cdots A$ moieties in the crystals. These can be considered as situations which are intermediate between isolated moieties ((a) above) and symmetrical linkings ((c) below). In $\{$tetrahydroselenophene$\cdots I_2\}$ (THSELI) approximately linear chains are found. The crystals are orthorhombic (12.804(3) 9.256(3) 7.625(3) Å) but there is some doubt about the space group. For convenience, we have shown in Fig. 11.4 an ordered structure in space group $Pna2_1$, Z = 4, but Hope and McCullough (1964) preferred the centrosymmetric $Pnma$, with disorder of the tetrahydroselenophene rings. There is very little difference between the two possibilities. The striking features of the structure are the lengthening of the iodine covalent bond (compared to 2.66 Å in the gas phase) and the shortening of one distance between I and Se to 2.764 Å, while the other distance is close to the

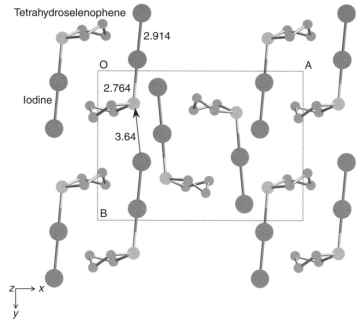

Fig. 11.4. A line of unsymmetrically interacting moieties in $\{$tetrahydroselenophene$\cdots I_2\}$. The Se $\ldots$ I(1) distance is 3.64 Å and the angles I(1)I(2)$\ldots$Se and I(2)$\ldots$Se$\ldots$I(1) are 179.4(3) and 167.2(4)° respectively. (Data from Hope and McCullough, 1964.)

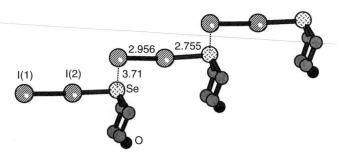

Fig. 11.5. Chain of unsymmetrically interacting moieties along [100] in {1-oxa-4-selenacyclo-hexane···I_2}; carbons are light-dotted, oxygens dark-dotted and iodines diagonally hatched. The angles I(1)-I(2)...Se, I(2)...Se...I(1) and Se...I(1)-I(2) are 174.8(3), 82.5(4) and 77.4(3)° respectively. (Adapted from Maddox and McCullough, 1966.)

sum of the van der Waals radii. Thus there are distinct bimolecular {C_4H_8Se...I_2} units in the crystal.

In {1-oxa-4-selenacyclohexane···I_2} (not in CSD) the arrangement is zigzag: In the examples shown in Figs. 11.4 and 11.5, the I–I distance is about 0.3 Å longer than the gas-phase value of 2.66 Å and requires checking.

In {1,4-diselenane···I_2} (DSEIOD) there are weaker secondary bonds of length 3.89 Å between Se of one DA moiety and I(1) of another, the molecules being linked in helical chains about alternate twofold screw axes. Unsymmetrical chains are also found in the ternary compound {I_3N···I_2···hexamethylenetetramine} (HMTNTI), with much stronger bonding between I_3N and I_2 than between I_2 and HMT; there is additional bonding between iodines of I_3N and nitrogens of HMT, leading to a three-dimensional arrangement of the three components (Fig. 11.6).

(c) Symmetrical arrangements:
When both donor and acceptor molecules are bifunctional then infinite arrangements can be formed, their shape depending on the coordination at the interacting atoms. Linear chains are formed in {phenazine···I_2} and {dioxane···Br_2} and zigzag chains in {12[ane]S_4·I_2}$_\infty$ (12[ane] is 1,4,7,10-tetrathiocyclododecane (Baker *et al.*, 1995; LINHEV) and {acet-one···Br_2} (Fig. 11.7). In {12[ane]S_4·I_2}$_\infty$ the I–I and S...I distances are 2.736(1) and 3.220(3) Å, indicative of rather weak interactions. Baker *et al.* carried out semi-empirical MO calculations on $(CH_3)_2S$...I_2 and $(CH_3)_2S$...I_2...$S(CH_3)_2$ to study the differences between terminal and bridged molecular compounds. These calculations confirmed that the principal interaction is between the thioether HOMO and the diiodine LUMO (the σ* antibonding orbital), with a consequent weakening of the I–I bond. A more complex arrangement, but based on similar structural principles, is found in {9[ane]S_3·(I_2)} (9[ane]S_3 is 1,4,7-trithiacyclononane) (Blake *et al.*, 1993; PELSUX). Similar extended structures are formed by the centrosymmetrical, bifunctional donor and acceptor molecules in {pyrazine···tetrabromoethylene} (BEPYRZ10) and {dioxane···diiodoacetylene} (DOX-IAC). Analogous structures are found in some trihalogenomethane molecular compounds, only two of the halogen atoms then interacting with donor atoms.

A zigzag arrangement of moieties linked by charge transfer I...S interactions (d(I...S) = 3.42 Å) is found in {dithiane···CHI_3} (Fig.11.8); {dioxane···CHI_3}, while

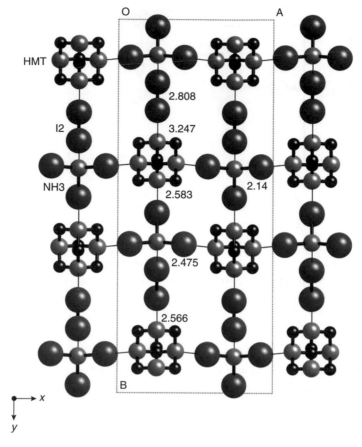

Fig. 11.6. The ternary molecular compound {I$_3$N$\cdots$I$_2$$\cdots$HMT}, shown in projection onto (001). The overall arrangement is three-dimensional as NI$_3$ is pyramidal and the nitrogens of HMT are arranged tetrahedrally. Iodines are shown as large filled circles and nitrogens as small filled circles; carbons are the small light circles. The secondary N . . . I interactions of different lengths are shown. (Data from Pritzkow, 1974a.)

not isomorphous with {dithiane$\cdots$iodoform}, appears to have an analogous structure, although details are lacking because of poor crystal quality. A similar chain arrangement is found in {diazabicyclo[2·2·2]octane$\cdots$CBr$_4$}, where the comparison of solution and solid-state absorption spectra (Fig. 11.3) shows a clear charge-transfer band in the solid; presumably there is only weak interaction between the moieties in solution. There is also a chain arrangement in {DBO$\cdots$CBr$_4$}. These molecular compounds are all isostructural, as is shown by comparison of cell dimensions and relationships among the projections (Table 11.8). These chain arrangements should be compared with the D$\cdots$A packing units found in {quinuclidine$\cdots$CBr$_4$} (Fig. 11.2).

^{14}N and ^{81}Br NQR spectra have been measured for {DABCO$\cdots$CBr$_4$} over the temperature range 77–320K (Okuda *et al.*, 1984) and provide, potentially, a means for measuring the strengths of the CT interactions. Unfortunately, this cannot yet be done as

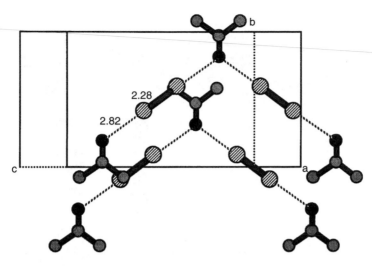

Fig. 11.7. Projection onto (100) of part of the {acetone∙∙∙Br$_2$} structure, showing two superposed planar layers of linked bromine and acetone molecules; the O . . . Br links are shown by dashed lines and oxygens are dark-dotted, bromines cross-hatched and carbons light-dotted. The outline of the unit cell ($a = 7.12$, $b = 7.48$, $c = 12.90$ Å, $\beta = 111.5°$, $C2/c$, $Z = 4$) is shown. (Reproduced from Hassel and Strømme (1959a).)

Table 11.8. Comparison of cell dimensions (Å, deg.) for three isostructural molecular compounds. The axes of analogous projections are emphasized

Molecular compound	a	b	c	β	Space group	Z	Molecular positions
Dithiane∙∙∙ CHI$_3$	**6.56**	**21.06**	4.47	104.4	$P2_1/m$	2	Dithiane at centers; CHI$_3$ on mirror.
DBO∙∙∙CBr$_4$	**8.509**	6.101	**12.006**	90	$Pmc2_1$	2	DBO and CBr$_4$ molecules on alternate mirror planes
DABCO∙∙∙ CBr$_4$	6.126	**16.80**	**12.634**	101.8	$P2_1/m$	4	Two disordered DABCO's on independent centers; two CBr$_4$ on independent mirrors.

there are some contradictions between diffraction and NQR results, perhaps because of the occurrence of a phase transition at ≈320K.

In {diselenane∙∙∙2(CHI$_3$)}, the donor Se atoms form two secondary bonds and the diselenane molecules link across zigzag chains (Fig. 11.9(a)). A formally similar arrangement is found in {(CH$_3$OH)$_2$∙∙∙Br$_2$} (Fig.11.9(b)), where there are hydrogen bonds as well as O∙∙∙Br interactions. A common feature of these two groups of structures is that only two of the three (or four) halogens are involved in charge transfer bonding.

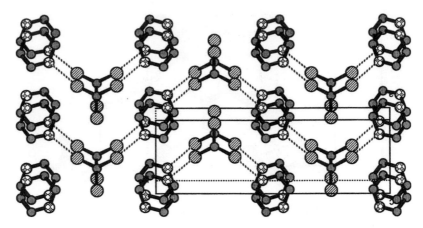

Fig. 11.8. Structure of the chains in {dithiane···CHI₃}; figured circles indicate S, cross-hatched circles I and small dotted circles C; one I does not participate in the bonding between the moieties. The dashed links are 3.42 Å. Two layers are shown, projected onto (001). The unit cell dimensions are: $a = 6.56$, $b = 21.06$, $c = 4.47$ Å, $\beta = 104.4°$, space group $P2_1/m$, $Z = 2$, dithianes are on centres of symmetry and iodoforms on mirror planes. (Adapted from Bjorvatten and Hassel, 1961.)

There is a three-dimensional tetrahedral arrangement in {HMT···CHI₃} of the four intermolecular bonds linking each molecule to its neighbours. Three of these links are $N \rightarrow I$, with equal lengths of 2.93 Å, while the fourth is a weak CH···N interaction (3.22 Å). {HMT···I₂} and {HMT···(I₂)₂} both form crystals without extended linkages, although there are some weak I...N interactions ($d(I...N) = 3.48$ Å) in the first of these and some weak I...I interactions in the second (Pritzkow, 1975). {HMT···(I₂)₃} is similar (Tebbe and Nagel, 1995; YUYNUB).

(d) Octasulphur as donor and halogenated molecules as acceptors:
S₈ forms 3:1 molecular compounds with CHI₃, PI₃, AsI₃ and SbI₃, first prepared (Auger, 1908) and examined crystallographically (Demassieux, 1909) before the First World War. The molecular compounds {3S₈···CHI₃} and {3S₈···AsI₃} are isomorphous and form an unbroken series of solid solutions (Hertel, 1931), although CHI₃ and AsI₃ have essentially no mutual solid solubility. The isomorphism was confirmed by West (1937) and later extended to the SbI₃ molecular compound (Table 11.9).

The crystal structures of {3S₈···CHI₃} (Bjorvatten, 1966; IFOSUL) and {3S₈··· SbI₃} (Bjorvatten, Hassel and Lindheim, 1963) have been determined and we describe them in terms of the latter (m.pt. 116–8°, sensitive to moist air). The sulphur molecules are located on mirror planes and the SbI₃ molecules on trigonal axes; dimensions are given in Table 11.9. Each iodine atom has a shorter link to one S atom of a particular S₈ molecule, and two longer links to two pairs of others; thus five of the eight S atoms of a sulphur molecule have close iodine neighbors. The situation is complicated by each Sb atom having three additional I...Sb interactions at distances of 3.85 Å, all from the nearest SbI₃ molecule situated along the same trigonal axis. We consider the S...I interactions to be more important for determining the overall crystal structure than the I...Sb interactions (or analogous CH...I interactions in {3S₈···CHI₃} and hence have classified SbI₃, and by extension the other RI₃ molecules, as "pure acceptors" in this group of molecular

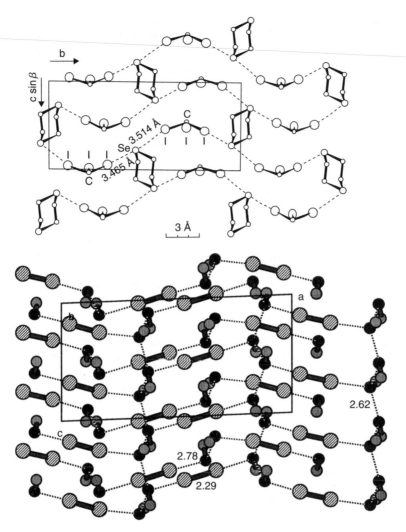

Fig. 11.9. (above) Structure of {diselenane···2(CHI₃)} projected down [100]; one I does not participate in intermoiety bonding. The dimensions of the unit cell are $a = 6.73$, $b = 17.27$, $c = 8.31$ Å, $\beta = 104.3°$, $P2_1/c$, $Z = 2$, m.pt. 93 °C. (Reproduced from Bjorvatten, 1963.) (below) Two layers of {(CH₃OH)₂···Br₂}, viewed down [010]. This compound melts at 207K and the intensity measurements were made at 183K. Because of the limited precision of the structure analysis, only average values of crystallographically independent interatomic distances are given. The charge transfer (Br...O) links extend horizontally across the figure, and the hydrogen bonds (hydrogen positions were not determined) vertically. Bromines are cross-hatched, oxygens dark-dotted and carbons light-dotted. (Reproduced from Groth and Hassel, 1964a.)

compounds. Fairly extensive NQR and Mössbauer effect studies have been made of this group of compounds (see Section 11.2.3.3). Second harmonic generation has been studied in {3S₈···CHI₃} and {3S₈···SbI₃} (Samoc *et al.*, 1992); powder SHG efficiency is about twice as large in the CHI₃ compound as in the SbI₃ compound.

Table 11.9. Crystallographic data (given for the triply primitive hexagonal cell) and some results for the isomorphous molecular compounds $\{3S_8\cdots RI_3\}$ where R=CH, As, Sb (rhombohedral, space group $R3m$, $Z=1$). Distances in Å, angles in degrees

R	a	c	$d(R–I)$	$<I–R–I$	$d(S\cdots I)$	$d(I\cdots R)$
CH	24.32	4.44	2.10		3.50(2)	3.84, 3.88
As	24.60	4.523				
Sb	24.817(7)	4.428(2)	2.747(2)	96.56(4)	3.60(1), 3.78, 3.88	3.85

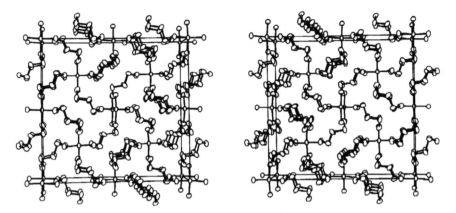

Fig. 11.10. Stereoview of the unit cell of $\{2S_8\cdot SnI_4\}$ viewed along [001]; [100] is vertical and [010] horizontal. The crystals are orthorhombic, $a=20.89$, $b=21.81$, $c=11.40$ Å, space group $Fdd2$, $Z=8$, m.pt. 101 °C. (Reproduced from Laitinen *et al.*, 1980.)

The above preparative results were confirmed by Feher and Linke (1966), who further showed that S_6 did not form corresponding molecular compounds; thus treatment of a sample of S_6 with a suitable RI_3 affords a method of removing unwanted S_8.

S_8 forms 4 : 1 molecular compounds with SnI_4 (Feher and Linke, 1966) and P_2I_4 (Feher *et al.*, 1962) (m.pt. 66–8°; very unstable in air) but crystal structures were not reported. 4 : 1 and 2 : 1 molecular compounds of S_8 with SnI_4 have also been prepared and studied by NQR (Ogawa, 1958) (see Section 11.2.3.2); the crystal structure of the 2 : 1 compound, and of a related compound, have been reported but not that of the 4 : 1 compound. Although these materials are probably better described as packing complexes, it is convenient to discuss them here and not in Chapter 10. The crystal structure of $\{2S_8\cdot SnI_4\}$ was determined by Hawes (1962) and refined by Laitinen *et al.* (1980) (Fig. 11.10). As no I...S distances of less than 3.8 Å were found, it was inferred that this was a packing complex without appreciable charge transfer interactions. There is some controversy about the structure of $\{2S_nSe_{8-n}\cdot SnI_4\}$ with Hawes (1963) claiming that the compound with $n=7$ is a distorted version of the sulphur complex, i.e. with similar cell dimensions and space group but with a displacement of the individual molecules from their positions in the parent complex, presumably because of the strain induced by replacement of S_8 by $S_nSe_{8-;n}$. Hawes also noted that 16% of the sulphur rings in the S_8 complex could be replaced by S_7Te. Laitinen *et al.* using a similar method of preparation, obtained a triclinic

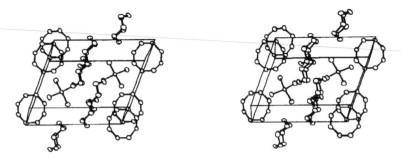

Fig. 11.11. Stereoview of the unit cell of $\{2S_nSe_{8-n}\cdot SnI_4\}$ seen along [001]; the [100] axis is vertical and the [010] axis horizontal. The crystals are triclinic, $a = 11.41$, $b = 16.14$, $c = 8.10$ Å, $\alpha = 92.3$, $\beta = 110.5$, $\gamma = 68.9°$, $Z = 2$, space group $P\bar{1}$. (Reproduced from Laitinen et al., 1980)

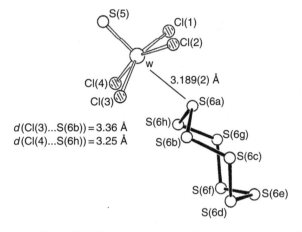

Fig. 11.12. View of the adduct $\{WCl_4S\cdot S_8\}$; the closer nonbonded distances are shown. The crystals are tetragonal, $a = 16.606$, $c = 10.546$ Å, $P\bar{4}\,2_1\,c$, $Z = 8$. (Reproduced from Hughes et al., 1991.)

complex (Fig.11.11), which appears to differ from that of Hawes, and contains (according to evidence from Raman spectroscopy), S_8 and S_nSe_{8-n}, with n probably mainly 6, although there was also a small $n = 7$ content.

The deep-red, air-sensitive crystals of WCl_4S_9 obtained by refluxing a suspension of $W(CO)_6$ with S_2Cl_2 in CH_2Cl_2 under tungsten light irradiation, has been shown, by x-ray crystallography, to be the adduct $\{WCl_4S\cdot S_8\}$ (Fig. 11.12) (Hughes et al., 1991). The W...S distance is close to the sum of the metallic radius of W and the van der Waals radius of S and is thus a normal packing interaction. The two S...Cl distances shown in the figure are significantly shorter than the sum of the van der Waals radii and suggest a possible charge transfer interaction. A spectroscopic study would be of interest.

(e) A diversion – an early C–H...O hydrogen bond
Diethyl ether forms 1:1 molecular compounds with $CHCl_3$ (m.pt. 182K) and $CHBrCl_2$ (m.pt. 179K) (Andersen and Thurmann-Moe, 1964). Two polymorphs were found for $\{(C_2H_5)_2O\cdot CHBrCl_2\}$ with appreciable differences in unit cell dimensions and volume

(1111 Å^3 (ETHBME01) and 959 Å^3 (ETHBME)) and hence presumably also in structure, although both had the same space group (*Pna2$_1$*, *Z* = 4). The crystal structure of ETHBME was determined, intensity measurements being made at 143K. It had been expected that there would be a link between O and one of the halogens but, in fact, the closest approach was between CH and O, with length 3.10 Å. This was called a hydrogen bond, well in advance of current usage (Desiraju and Steiner, 1999).

11.3.2   *S containing molecules as donors and iodine molecules as acceptors (the polyiodines)*

There are a number of molecular compounds in which a molecule containing S as donor atom binds to iodine molecules. We call these polyiodines; in contrast to the polyiodides,

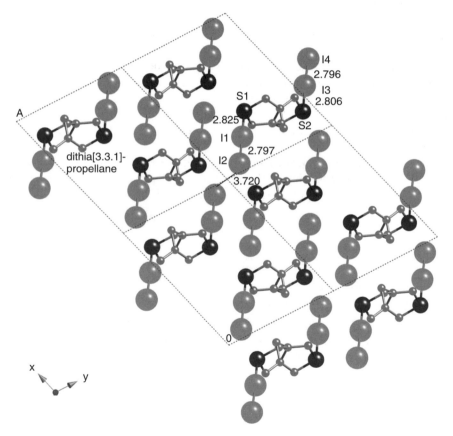

Fig. 11.13. The {I$_2$·dithia[3.3.1]propellane·I$_2$} structure (FAJPUB) viewed down [001]. The large spheres are iodine, the small dark spheres sulphur, and the small light spheres are carbon; hydrogens are not shown. The chains of moieties are linked by two kinds of weak interactions – iodine to iodine, 3.720 Å and iodine to sulphur 2.806 and 2.825 Å. The covalent bonds within the iodine molecule 2.796 and 2.797 Å are considerably extended above the gas phase length (2.66 Å). For comparison we note that in {(merocyanine)$_2$·I$_2$} (d'Enghien-Peteau *et al.*, 1968) *d*(I–I) = 2.750(1) Å and *d*(S...I) = 3.098(5) Å. (Data from Herbstein, Ashkenazi, *et al.*, 1986.)

which are salts, the polyiodines are composed of *formally* neutral entities. Two groups can be distinguished, both with a limited number of examples. In the first group the interactions are weak and there is no polarization towards the formation of quasi-charged structures. The first example to be reported (Herbstein *et al.*, 1986) was the bis(diiodine) adduct of dithia[3·3·1]propellane (Fig. 11.13) which is associated into pairs of adducts across crystallographic centers of symmetry, as shown in FAJPUB.

The distance of 3.720(1) Å between adjacent centrosymmetrically related iodines is markedly less than the van der Waals separation of ≈ 4.2–4.3 Å. The I–I...I angle is 154.2(1)° so there is an appreciable deviation from linearity. The other intra-adduct S...I and I–I distances are normal (Table 11.6). Thus there is no resemblance to the well-known I_4^{2-} ions (Herbstein *et al.*, 1983). The similar adduct {I_2·dithia[3·3·2]propellane·I_2} does not show analogous association in its crystals. (Herbstein *et al.*, 1986). The molecular compound {bis(morpholinothiocarbonyl)···bis(diiodine)} {[$OC_4H_8NC(S)C(S)NC_4H_8O$]···$(I_2)_2$} (Atzei *et al.*, 1988; GIGLOX) shows an intramolecular C=S...I_2 arrangement and intermolecular I...I distances (3.705, 3.758, 3.931 and 3.746(3) Å) similar to those found in {I_2·dithia[3.3.1]propellane·I_2}.

The second group is characterized by intra-adduct polarization towards an ionic, polyiodide-type structure, but without complete separation into the discrete ions of the true polyiodides. {Dithizone···I_2} (BZHTIC10) shows marked polarization towards the canonical form (dithizone...I)$^+$...I$^-$ (Table 11.6) and {(ethylenethiourea)···$(I_2)_2$} (CEWMIA) is polarized towards (ethylenethiourea...I)$^+$...$(I_3)^-$ (Herbstein and Schwotzer, 1984).

There are three bridged structures with very similar overall geometries, where subtle differences in bond lengths show different degrees of participation of the polarized canonical forms in the overall resonance structure. The compounds $R_2(I_2)_3$, where

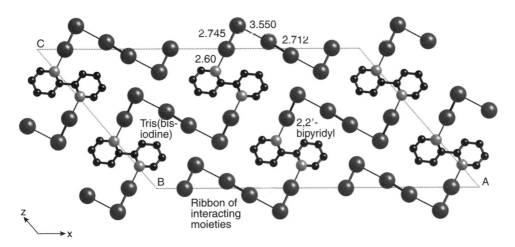

Fig. 11.14. Crystal structure of {$(2,2'$-bipyridyl)···$(I_2)_3$} projected down [010]. The crystallographic parameters are $a = 32.15$, $b = 4.399$, $c = 18.24$ Å, $\beta = 130.8°$, C2/c, Z = 4. The secondary interactions are shown by thin lines. (Data from Pohl, 1983.)

R = triphenylphosphine sulphide (Schweikert and Meyers, 1968; Bransford and Meyers, 1978) or ethylenethiourea (Herbstein and Schwotzer, 1984; CEWMOG), are polarized towards $(R \ldots I)^+ \ldots I_4^{2-} \ldots {}^+(I \ldots R)$, while $\{(2,2'\text{-bipyridyl}) \ldots (I_2)_3\}$ (Pohl, 1983; CECZAL) (Fig. 11.14) shows essentially no polarization as the bond lengths in the I_2 molecules are close to standard values. On the other hand, $\{(\text{dithizone})_2(I_2)_7\}$ is strongly polarized towards $(\text{dithizone} \ldots I)^+ \ldots I_5^- \ldots I_2 \ldots I_5^- \ldots {}^+(I \ldots \text{dithizone})$ (Herbstein and Schwotzer, 1984; BZHTID20).

11.3.3 Physical measurements on molecular compounds of the type discussed above

11.3.3.1 N and S containing donors and I_2 as acceptor

There have been a number of measurements by various physical techniques (principally Mössbauer (^{129}I) and NQR (^{127}I) spectroscopy and, to a lesser extent, Raman and IR spectroscopy) on some of the molecular compounds discussed earlier. Interpretation of the experimental results gives information about the interactions linking the components. Related compounds are treated together.

These results are in excellent qualitative agreement with the known crystal structures; in {phenazine···I$_2$} the iodine molecules bridge between nitrogens of successive phenazines and there is one independent iodine atom in the asymmetric unit, while in both {HMT···I$_2$} and {dithiane···(I$_2$)$_2$} there are two crystallographically independent iodine atoms in the asymmetric unit, one bridging (I$_B$) and linked to N (or S) and the second terminal (I$_T$); the I$_2$ molecules are pendant in both structures. The $(e^2Qq)_{obs}$ values for I$_B$ and I$_T$ in the two compounds of known structure are clearly different and there is similar differentiation for the other three molecular compounds. Thus one can infer that they also

Table 11.10. Experimental results from ^{129}I Mössbauer measurements. Where two results are given for a parameter, the first refers to the bridging iodine I$_B$ and the second to the terminal iodine I$_T$

Molecular compound	Quadrupole coupling constant $e^2Qq(^{129}I)$ (MHz)	Asymmetry parameter η	Isomer shift ∂ (mm/s)	$d(D \cdots I_B)$ (Å)	$d(I_B - I_T)$ (Å)
(a) Nitrogen-containing donors (measurements at 85K) (Ichiba, Sakai, Negita and Maeda, 1971)					
phenazine···I$_2$	-2230 ± 30	0.06	0.93(2)	2.982(5)	2.726(1)
acridine···I$_2$	-2840 ± 20	0.13	1.64(2)		
	-1308 ± 20	0.16	0.29(2)		
HMT···I$_2$	-2582 ± 20	0.19	1.51(2)	2.439(8)	2.830(1)
	-1272 ± 20	0.06	0.28(2)		
(b) Sulfur-containing donors (measurements at 16K) (Sakai et al., 1986)					
thiane···I$_2$	-2513 ± 13	0.02(4)	1.54(4)		
	-1310 ± 13	0.02(4)	0.32(4)		
1,4-oxathiane···I$_2$	-2512 ± 13	0.04(4)	1.47(4)		
	-1475 ± 13	0.09(4)	0.43(4)		
1,4-dithiane···(I$_2$)$_2$	-2526 ± 13	0.02(4)	1.47(4)	2.867(2)	2.787(2)
	-1553 ± 13	0.06(4)	0.50(4)		

Table 11.11. Bonding parameters derived from the experimental results summarised in Table 11.10 – h_p is the number of p electron holes; U_p is the number of unbalanced p electrons; N_x, N_y, N_z are the numbers of 5p electrons in these orbitals. Where two values are given for a parameter, the upper refers to I_B and the lower to I_T. The electron configuration is given as $5s^2\,5p^t$. The errors of h_p, N_x, N_y, N_z are estimated (Ichiba *et al.*, 1971) as $\pm 2\%$ while those of U_p are $\pm 1\%$

Molecular compound	h_p	U_p	N_x	N_y	N_z	t	Charge on I_B/I_T
(a) Nitrogen-containing donors (Ichiba, Sakai, Negita and Maeda, 1971)							
phenazine$\cdots$I$_2$	0.91	0.97	2.04	2.00	1.05	5.09	0.09
acridine$\cdots$I$_2$	1.45	1.24	1.98	1.88	0.69	4.55	+0.45
	0.55	0.57	2.03	1.97	1.44	5.44	−0.44
HMT$\cdots$I$_2$	1.37	1.13	1.99	1.85	0.79	4.63	+0.37
	0.55	0.55	2.01	1.99	1.45	5.45	−0.45
(b) Sulphur-containing donors (Sakai *et al.*, 1986)							
thiane$\cdots$I$_2$	1.39	1.10	1.91	2.10	0.80	4.81	+0.19
	0.57	0.57	2.00	2.00	1.43	5.43	−0.43
1,4-oxathiane$\cdots$I$_2$	1.34	1.10	1.94	2.03	0.82	4.86	+0.14
	0.65	0.63	2.01	2.10	1.36	5.40	−0.40
1,4-dithiane$\cdots$(I$_2$)$_2$	1.34	1.10	1.93	2.09	0.82	4.84	+0.16
	0.69	0.68	2.01	2.02	1.32	5.35	−0.35

have pendant I$_2$ molecules (note that the spectrum of {acridine$\cdots$I$_2$} contained two extra peaks that were not explained; the crystal structure is not known). One notes that the $(e^2Qq)_{obs}$ values for I_B are very similar and those for I_T cover only a slightly larger range. This holds also for the isomer shift (∂) values while the asymmetry parameters (η) are close to zero for the molecular compounds with S donors but have appreciable nonzero values for the N-donors. ^{127}I NQR measurements (Terao *et al.*, 1985) on the three compounds in Part (b) of Table 11.10 (and {1,3,5-trithiane$\cdots$I$_2$}) lead to similar, but less detailed, conclusions; it was inferred that {1,3,5-trithiane$\cdots$I$_2$} had a bridged structure analogous to that of {phenazine$\cdots$I$_2$}.

The bonding parameters can be analyzed quantitatively and the results are shown in Table 11.11. There are major similarities among the various derived parameters, as must indeed be the case because of the similarities among the experimental results from which they are obtained. Thus we shall treat them as a whole without attempting to account for minor differences. The picture that emerges from the 5p electron populations is that the $5p_x$ and $5p_y$ lone pair orbitals are essentially completely filled for both I_B and I_T in all the molecular compounds and thus the bonding orbital in the I$_2$ molecule is $5\sigma p_z$.

In {phenazine$\cdots$I$_2$} there is a charge transfer of ≈ 0.09 e$^-$ from nitrogen to iodine; this charge enters the $5\sigma^* p_z$ antibonding orbital leading to a small increase in the I–I bond length to 2.726 Å as compared to 2.66 Å in the gas phase. The charge transfers for the pendant I$_2$ molecules are larger and in consequence so are the changes in the bond lengths.

The approximate charge distributions can be written as follows (where D represents a donor atom such as N, O or S):

$$
\begin{array}{cc}
\text{+0.05}\diagdown\overset{\displaystyle -0.05}{\text{N}^{\text{iiiiiiii}}\text{I}\,\text{━━}\,\text{iiiiiiii}\text{N}}\diagup & \text{+0.2}\diagdown\overset{\displaystyle +0.2}{\text{D}^{\text{iiiiiiii}}\text{I}\,\text{━━}\,\text{I}}\;\;-0.4\\[4pt]
\text{Symmetrical diiodide} & \text{Unsymmetrical diiodide}\\
\text{binding to phenazine} & \text{binding to a donor D such}\\
& \text{as N, O or S.}
\end{array}
$$

The weakening of the I–I bond is also shown by a decrease in the bond stretching constant which, in $\{1,4\text{-dithiane}\cdots(I_2)_2\}$, has been calculated as $158\,\text{cm}^{-1}$ from the far IR spectrum (Hendra and Sadavisan, 1965), to be compared with $213\,\text{cm}^{-1}$ in free I_2 (Rosen *et al.*, 1971).

Theoretical analysis based on a four electron–three-center molecular orbital treatment can reproduce the experimental results with appropriate values of the Coulomb integral of donor atom D and the resonance integral of the D . . . I bond (Bowaker and Hacobian, 1969).

11.3.3.2 S_8 as donor and SnI_4 molecules as acceptors

$\{4S_8 \ldots SnI_4\}$, $\{2S_8 \ldots SnI_4\}$ and SnI_4 have been studied by ^{127}I NQR spectroscopy and the latter also by ^{129}I and ^{119}Sn Mössbauer spectroscopy (Bukshpan and Herber, 1967). There is no evidence for phase transformations in any of these materials in the temperature range 77–301K. The coupling constants and asymmetry parameters are summarized in Table 11.12. There are two independent NQR resonances for $\{2S_8\cdots SnI_4\}$ and SnI_4 and four for $4S_8 \ldots SnI_4$; this is in accord with the crystal structures of SnI_4 and $\{2S_8\cdots SnI_4\}$ while that of $\{4S_8\cdots SnI_4\}$ is not known.

The differences among the various ^{127}I coupling constants are small and the values of the asymmetry parameter are all close to zero; hence the Sn–I bonding in all of them is very similar. Analysis of the Mössbauer data for both ^{119}Sn and ^{129}I in SnI_4 (Ehrlich and Kaplan, 1969) indicates that the Sn–I bond has about 24% π-character and that $\approx5\%$ s-hybridization is involved in the iodine σ-bonding orbital; the atoms in SnI_4 are essentially neutral with charges of $-0.015e$ on each of the iodines and $+0.060e$ on Sn. It seems reasonable to infer also from these measurements, as from the crystal structures, that I . . . I and S . . . I interactions are very weak and that the two molecular compounds would be better classified as packing complexes (Chapter 10).

Table 11.12. Parameters derived from ^{127}I NQR measurements, which were made over the temperature range 77–301K; 77K values are given here unless noted otherwise

Compound	$(e^2Qq)_{obs}$ (Mhz)		η	
SnI_4 (301K)	1356.36	1363.28	0.008	0.000
$2S_8\cdots SnI_4$	1383.03	1402.68	0.002	0.039
$4S_8\cdots SnI_4$	1364.3; 1405.0	1368.3; (1)	0.0; 0.03	0.0; (1)

Notes:
(1) as ν_2 was not observed, it was not possible to calculate values of $(e^2Qq)_{obs}$ and η for the fourth iodine.
(2) the Mössbauer measurements for SnI_4 give a positive sign for $(e^2Qq)_{obs}$; the values of the isomer shift are $+1.55\,\text{mm/sec}$ for ^{119}Sn compared to a SnO_2 source and $+0.43\,\text{mm/sec}$ for ^{129}I compared to a ZnTe source.

11.3.3.3 S_8 as donor and RI_3 (R=As, Sb) as acceptor:

The NQR spectra of $AsCl_3$, $SbCl_3$, $SbBr_3$ (noted below), AsI_3, SbI_3 and of the molecular compounds $\{3S_8\dots AsI_3\}$ and $\{3S_8\dots SbI_3\}$ were measured more than 40 years ago (Ogawa, 1958) and also studied in a complementary Zeeman investigation (Abe, 1958) for AsI_3 and $\{3S_8\dots AsI_3\}$. Later a comprehensive Mössbauer study was made of AsI_3, SbI_3, $\{3S_8\dots AsI_3\}$ and $\{3S_8\dots SbI_3\}$ and of BiI_3 (Ogawa, 1958; Sakai, 1972). Pertinent experimental information is summarized in Table 11.13. Although the NQR measurements were made over the temperature range 300–77K we give only the 77K results; there is no evidence for phase changes in this range except, perhaps, for SbI_3; 45K results are quoted here and we assume these refer to the trigonal polymorph. The Zeeman measurements were made at room temperature and the Mössbauer study carried out at 85K.

The $(e^2Qq)_{obs}$ values and asymmetry parameters η for all the compounds in Table 11.13, except for SbI_3, fall within the range of values found for the terminal iodines I_T in the compounds discussed in Section 11.2.3.2. The electron configurations and a summary of relevant molecular dimensions are given in Table 11.14. We compare the conclusions drawn about the bonding from the physical measurements with changes found in bond lengths. In crystalline AsI_3 there is additional π-bonding due to transfer of $5p_y$ electrons from I to the As atom at 3.467 Å from I. This charge transfer is at the expense of the covalent component of the As_1–I bond, which has 54% covalent character, 36% ionic character and 10% π (intermolecular) character. There is only a small increase in As–I bond length (0.034 Å) between gas phase and crystalline AsI_3. In $3S_8\dots AsI_3$ the As–I bonding is $5\sigma p_z$ with transfer of $0.33e$ from As to I, the bond being 67% covalent and 33% ionic. The secondary S$\dots$I interactions appear to have a negligible effect on the nature of the As–I bonding. The crystal structure of $\{3S_8\dots AsI_3\}$ is not known.

Larger effects are found when trigonal SbI_3 is compared with the gas phase, where there is an increase of $\approx$0.15 Å in d(Sb–I) over the gas-phase value. However, the values of d(Sb–I) for the gas phase and monoclinic SbI_3 differ by only $\approx$0.05 Å. The effects of secondary interactions in the sulphur molecular compounds are smaller as is shown in $\{3S_8\dots SbI_3\}$, where the Sb–I bond now has 9% π-character, and the increase in length

Table 11.13. Summary of parameters derived from NQR and Mössbauer measurements (temperatures are 77K for NQR and 85K for Mössbauer measurements, except where stated otherwise)

Compound	^{75}As/^{121}Sb $(e^2Qq)_{obs}$ (MHz)	^{127}I $(e^2Qq)_{obs}$ (MHz)	η	∂ (mm/sec)
AsI_3	58.676(10)	1330.23–1350	0.1891 0.20	0.57
$3S_8\cdots AsI_3$	99.002(10)	1516.86–1544	0.007 < 0.05	0.47
SbI_3 (at 45K)	169.37	895.83–911	0.565 0.57	0.54
$3S_8\cdots SbI_3$	251.8 (250K)	1226.25–1209	0.03 0.11	0.49

Notes:
(1) The $(e^2Qq)_{obs}$ values from the ^{129}I Mössbauer measurements have been converted to the ^{127}I scale for comparison with the NQR measurements. The limits of error are given as $\pm$20MHz, while for η and ∂ the limits of error are $\pm$0.05.
(2) As ^{75}As and ^{121}Sb have nuclear spin 3/2, the values of (e^2Qq)obs and η cannot be determined separately. We follow Ogawa (1958) and assume, because of the crystal structure, that $\eta = 0$; then $(e^2Qq)_{obs} = 2\nu_{meas}$.
(3) ∂, the isomer shift, is given with respect to a ZnTe source.

Table 11.14. Electron configurations in iodine of RI_3 as deduced from NQR and Mössbauer measurements, together with a summary of relevant molecular dimensions in different environments; distances in Å, angles in degrees

Compound	h_p	U_p	N_x	N_y	N_z	$t(1)$	$d(R–I)$; $[d(R···I)]$ (Å) References	$< I–R–I$
AsI_3	0.74	0.59	1.99	1.91	1.36	5.26	Crystal (EG80) 2.591(1) [3.467(2)] gas (M66) 2.557(5)	99.67(5)
$\{3S_8···AsI_3\}$	0.67	0.66	2.00	2.00	1.33	5.38	Crystal structure not reported.	
SbI_3	0.72	0.39	1.96	1.82	1.50	5.28	gas (AB63) 2.719(2) Trigonal crystal (TZ66) 2.87(1); [3.32] Monoclinic crystal (PS84) 2.754(3), 2.768, 2.774 [3.540, 3.605, 3.880]	99.1 95.8(3) 95.1(1) 96.6(1) 100.2(1)
$\{3S_8···SbI_3\}$	0.69	0.53	1.95	1.94	1.42	5.31	Crystal (BHL63) 2.747(2); [3.85]	96.56(4)

Notes: (1) The electron configuration is given as $5s^2 5p^t$

References:
AB63 – Almenningen and Bjorvatten, 1963; BHL63 – Bjorvatten, Hassel and Lindheim, 1963; EG80 – Enjalbert and Galy, 1980; M66 – Morino *et al.*, 1966; TZ66 – Trotter and Zobel, 1966; PS84 – Pohl and Saak, 1984.

(0.028 Å) over the gas-phase value is small. The Sb–I bond in crystalline SbI_3 has been estimated to have 36% covalent character, 42% ionic character and 22% π character, but it is not clear which polymorph was used for the measurements (Pohl and Saak (1981) quote Cooke (1877/8) as reporting that monoclinic SbI_3 transforms into trigonal at 367K, suggesting that monoclinic, although rarer, is the stable phase at lower temperatures). Repetition of these studies using current techniques will surely yield information of much interest.

11.3.4 *Halogenated molecules as donors and dihalogens as acceptors*

The phase diagrams of a number of systems containing Cl_2 and various halogeno-methanes (CCl_4, $CHCl_3$, CH_2Cl_2, CH_3Cl) have been determined from freezing-point measurements (Wheat and Browne, 1936, 1938, 1940a, b). The binary compounds found are listed in Table 11.15. The $HCl–Cl_2$ system shows a 2:1 (m.pt. $-121°$) and a 1:1($-115°$) compound. No compounds were found in the $CHCl_3–Br_2$ system, which is a simple eutectic. No structures are known; it was suggested that the halogenomethanes act as donors and the Cl_2 molecules as acceptors. This seems reasonable as all the evidence summarized in this

Table 11.15. Reported compositions and melting points (°C) in various halogenomethane – Cl_2 systems

Halogenomethane	Ratio halogenomethane : Cl_2				Reference
	2:1	1:1	1:2	1:3	
CCl_4	−67	−90.5	−112.5	−115.5	Wheat and Browne, 1938.
$CHCl_3$	−80	−96.5	−112.5	−115	idem, 1936
CH_2Cl_2	–	−124.5	−126.5	−117.5	idem, 1940a
CH_3Cl	–	–	−120	−122	idem, 1940b.

chapter indicates that Cl_2 is a stronger acceptor than the chloromethanes. However, these results should perhaps be viewed with some caution as it has been reported that the CCl_4–Cl_2 system has only an eutectic at −114.4° (Abassalti and Michaud, 1975).

11.3.5 *Self-complexes – N, O, S, Se to halogen interactions in one-component systems*

Interactions similar to those important in the formation of n–σ* charge-transfer compounds in binary systems also operate in many one-component (unary) crystals, which can then be described as 'self-complexes' (there are also 'self-complexes' in π–π* systems, which are discussed later); according to our convention, these should be called 'self-compounds' but we defer here to common usage. Two excellent reviews are noted; Bent (1968) has discussed and compared donor–acceptor interactions in both unary and binary systems, while Alcock (1972) has concentrated on unary systems. Our discussion will be illustrative rather than comprehensive, and will be organized in terms of the donor-acceptor atom pairs involved in the secondary bonding.

11.3.5.1 *Nitrogen...halogen interactions*

In molecular compounds with nitrogen•••halogen interactions, the nitrogen is found in cyano groups or as the heteroatom in, for example, pyridine and hexamethylenetetramine. Similar combinations occur in self-complexes. The cyanogen halides (ClCN (Heiart and Carpenter, 1956), BrCN (Geller and Schawlow, 1955), ICN (Ketelaar and Zwartsenberg, 1939) and halocyanoacetylenes ($XC{\equiv}C{-}C{\equiv}N$; X=Cl, Br (Bjorvatten, 1968), I (Borgen et al., 1962)) have been studied crystallographically, but with varying degrees of completeness. All can be represented as rod-like molecules and it is not surprising that all should be arranged in their crystals as arrays of exactly or approximately close-packed cylinders. In ClCN, BrCN and the three halocyanoacetylenes, alternate rows of molecules are antiparallel and the crystals are nonpolar; cyanogen iodide and a low-temperature form of iodocyanoacetylene are reported to form polar arrays, but more detailed study seems desirable in both examples. For all these molecules there is head-to-tail packing within the rows and all N•••X distances (Table 11.16) are less than the sums of the appropriate van der Waals distances. The strength of the N•••X interaction increases as X changes from Cl through Br to I; this also happens among the molecular compounds. The shortening has been attributed to electrostatic interactions but the more generally accepted view is that

Table 11.16. Comparison of N···X distances (Å) in cyanogen halides and halogenoacetylenes with sums of van der Waals radii

Compound	X=Cl	X=Br	X=I
Cyanogen halides X–C≡N	3.01(1)	2.87 (estimated)	Not measured
Halocyanoacetylenes X–C≡C-C≡N	2.983(8)	2.98(2)	2.93
Sum of vdW radii of N and X	3.33	3.43	3.93

there is partial charge transfer from N to halogen. In the crystals of N, N'-dibromoethane-diimidoyl difluoride there are also rather strong, essentially linear N-Br···N interactions $(d(N \ldots Br) = 2.863(7), 2.891(6)$ Å, with $<$N-Br···N $= 171.1, 174.6°)$ (Waterfeld *et al.*, 1983; BONFIT).

Similar but weaker interactions are found in other crystals. Although full structures have not been reported for 2,4,6-tri-(p-chlorophenyl)-s–triazine (Belitskus and Jeffrey, 1965) and cyanuric chloride (Hoppe *et al.*, 1957), the overall arrangements in both crystals are such that linear C≡N···Cl–C arrangements appear probable for all three such groupings in each molecule. 2,4,6-Trichloro-1,3,5-tricyanobenzene (space group $P2_12_12_1$, $Z = 4$) (Britton, 1981; BAFJOH) is not isomorphous with cyanuric chloride and there is a linear arrangement only for one C≡N···Cl–C grouping $(d(N···Cl) = 3.03$ Å), with a much weaker interaction for a second such grouping and no interaction for the third potential donor–acceptor pair. Linear C≡N···Cl–C arrangements have also been found in 2,4,6-trichloro- (CLBECN01) and tribromobenzonitrile (BRBZNT; Carter and Britton, 1972) which have similar but not isomorphous structures. Pairs of molecules are formed in trichlorobenzonitrile with each C≡N and *ortho*-Cl coordinated to the corresponding groups of an adjacent molecule $[d(N···Cl) = 3.22$ Å] with an arrangement reminiscent of that found in carboxylic acid dimers. In tribromobenzonitrile polymeric chains are formed, each N being coordinated to two Br atoms related by a mirror plane $[d(N \ldots Br) = 3.06$ Å] and the C≡N···X angle is about 120° instead of the linear arrangement found in the cyanogen halides. ClC(C≡N)$_3$ is isomorphous with the methyl analog and there are no close approaches (Witt *et al.*, 1972). Among other crystals with appropriate N···X approaches (Britton, 1967) are p-chlorobenzonitrile (Britton *et al.*, 1979; CLBZNT) and p bromobenzonitrile (Britton *et al.*, 1977; BRBNIT) (which are not isomorphous but have related structures), 9-dicyanomethylene-2,7-dibromofluorene (Silverman *et al.*, 1973; CYBFME20) and 1,4-dimethoxy-2,3-dicyano-5,6-dichlorobenzene (Reddy, Panneerselvam *et al.*, 1993; LAGMOV).

11.3.5.2 *Oxygen···halogen interactions*

As with N···X interactions, the order of increasing strength for O···X interactions is Cl $<$ Br $<$ I; no O···F interaction is found in the only relevant fluorinated compound (2,3-difluoro-1,4-naphthoquinone) (Gaultier *et al.*, 1972; DFNAPQ) studied to date. Some typical O···X distances are given in Table 11.17.

A systematic and extended study of the secondary bonding in crystalline halogenated quinones (benzoquinones, naphthoquinones and anthraquinones) has been carried out, mainly by the Bordeaux school (Gaultier *et al.*, 1971a); the structural arrangement in

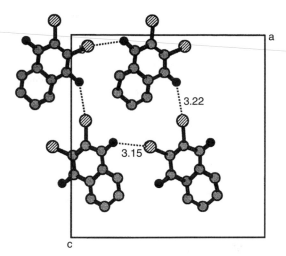

Fig. 11.15. Schematic diagram showing the O···Br interactions in the crystals of 2,3-dibromo-1,4-naphthoquinone; Br diagonally hatched, O small dark circles. (Reproduced from Breton-Lacombe, 1967.)

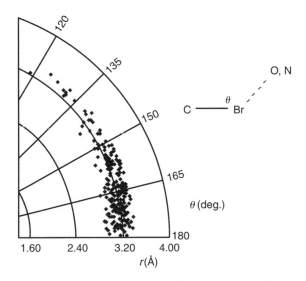

Fig. 11.16. Polar scatter plot (r vs. θ) for the nucleophiles O or N around Br. The average value of θ for these contacts are 158(13)° for Cl, 162(12)° for Br and 165(8)° for I. The crowding around 162° is especially pronounced for Br. (Reproduced from Ramasubbhu *et al.*, 1986.)

2,3-dibromo-1,4-naphthoquinone (Breton-Lacombe, 1967; DBRNPQ10) is shown as an example in Fig. 11.15. Halogenated benzoquinones and naphthoquinones generally show O···X interactions but the distances suggest that these are comparatively weak; thus it is not surprising that they do not appear when the structural arrangement is determined by hydrogen

Table 11.17. Some typical oxygen···halogen approach distances (Å) in unary crystals

Compound/refcode	$d(O···X)$	Compound	$d(O···X)$
Type of approach: O···I		**Type of approach: O···Cl**	
α-HIO$_3$; RH41	2.45, 2.70, 2.95	N-chlorosuccinimide[#] CSUCIM; B61	2.88
N,N-diiodoformamide IFORAM; P74b	2.565(7); 3.13(1)	γ-N-(p-tolyl)-tetrachloro-phthalimide TOCPIM11; HK81	2.91(1)
N-iodosuccinimide; P90	2.580(6)	α-N-(p- tolyl)-tetrachlorophthalimide; TOCPIM; K78	2.98(1)
p-chloro-iodoxy-benzene; CIOBEN; A48	2.72; 2.87	2,3,4,4-tetrachloro-1-oxodihydronaphthalene (h); VM71	2.973
KIO$_2$F$_2$; RH40	2.82; 2.88	3-chloro-1,2-naphthoquinone; CLNAPQ; C70	3.22
3-iodo-1,4-naphthoquinone; INAPQU; GHHS71b	3.21		
Type of approach: O···Br			
3-methoxy-5β,19-cyclo-5,10-secoandrosta-1(10),2,4-trien-17β-ol-p-bromobenzoate; CSANDR; HC68	2.93		
POBr$_3$*; OM69	3.08; 3.27		
oxalyl bromide (m.pt. $-19.5\,°C$) **; OXALYB; GH62			
3-bromo-1,4-naphthoquinone BRNAPQ; GaHa65	3.11		

\# N-bromosuccinimide is isomorphous
* POCl$_3$ is isomorphous
** Oxalyl chloride is not isomorphous and there are no close O···Cl approaches in its crystals.

References:
A48 – Archer, 1948; B61 – Brown, 1961; C70 – Courseille *et al.*, 1970; GAHA65 – Gaultier and Hauw, 1965; GH62 – Groth and Hassel, 1962; HHS71 – Gaultier, Hauw, Housty and Schvoerer, 1971; HC68 – Hope and Christensen, 1968; HK81 – Herbstein and Kaftory, 1981; K78 – Kaftory, 1978; OM69 – Olie and Mijlhoff, 1969;

bonding, as happens, for example, in 4-amino-3-bromo-1,2-naphthoquinone (Bechtel *et al.*, 1976; BANAQP10). O···X interactions have not been found among the halogenated anthraquinones. In the benzoquinone and naphthoquinone series the angle O···X–C does not deviate from 180° by more than about 20°, but < C=O···X lies between 120° and 180°.

Polar scatter plots of Y ... X-C interactions (where Y is the nucleophile O or N and X is Cl, Br or I) have been presented, using the Cambridge Structural Datafile as database (Ramasubbu *et al.*, 1986); 332 Cl ... (O, N) entries were used, 397 Br ... (O, N) and

60 I...(O, N). The shorter distances, i.e. those relevant to self-complexation, are found in the region of head-on approaches (Fig. 11.16, which shows C–Br...(O, N) interactions, those for Cl and I being qualitatively similar). Using the same methodology, it was shown that the distribution of halogens around a carbonyl oxygen ranges from 75 to 180°, but is concentrated in the 105–140° region (especially the shorter approaches) and agrees well with an angular distribution representative of the lone pair(s) in a conventional sp^2 orbital diagram.

11.4   n-Donors and p-acceptors

11.4.1   *N, O or S containing ligands as donors and Group VA metal halides as acceptors*

Earlier work on the binary adducts of MX_n [M=Al, Ga ($n = 3$); Ti ($n = 4$); P, As, Sb ($n = 5$); X=F, Cl, Br] has been summarized by Lindqvist (1963) while the Group V adducts have also been discussed by Webster (1966). In the group of compounds of present interest only an organic ligand, such as nitrobenzene or benzoyl chloride, and metal halide are present in the adduct. A second group contains HX as well but will not concern us here because these are salts.

(a) Some preparative studies of molecular compounds of $AlCl_3$
A common experimental problem in the study of these materials is their sensitivity to moisture ("even more readily hydrolyzed than $AlCl_3$ [itself]," according to Jones and Ward (1966); however, they are generally well-crystallized and thermodynamically stable if adequate precautions are taken to work under dry conditions. The earliest systematic studies appear to be those of Menschutkin before the First World War (see Thomas (1941) and Olah (1973) for summaries). Phase diagrams were determined and congruently melting 1:1 compounds obtained of $AlCl_3$ and $AlBr_3$ with, severally, nitrobenzene, benzoyl chloride, benzophenone (needles up to 8 cm long were reported for $C_6H_5COC_6H_5$···$AlCl_3$! (Menschutkin, 1910)) and the various isomers of $XC_6H_4NO_2$ (X = halogen). Later work showed that $2[C_6H_5NO_2]$···$AlCl_3$ could also be prepared. Among other crystalline compounds are $2CH_3CN$···$AlCl_3$, $2POCl_3$···$AlCl_3$, $POCl_3$···$AlCl_3$ and (perhaps) $6POCl_3$···$AlCl_3$ (Groeneveld and Zuur, 1958). Other ligands that have been reported to form 1:1 molecular compounds with $AlCl_3$ include acetone, diethyl ketone and nitromethane (Jones andWard, 1966). Preliminary single-crystal X-ray diffraction and polarized infrared absorption studies were made of a number of compounds in this group (Gagnaux and Susz, 1961) and many crystal structures have now been reported.
(b) Preparative studies of molecular compounds of Group VA halides
Many compounds of oxygen-containing molecules with $SbCl_5$ have been reported (Meerwein and Maier-Hüser, 1932); among those crystallizing above room temperature are (diethyl ether)···$SbCl_5$; dioxane···$2SbCl_5$; dimethylpyrone···$SbCl_5$; acetyl chloride ···$SbCl_5$; and 2(benzoyl chloride)·dimethylpyrone···$2SbCl_5$ (prepared by adding benzoyl chloride to the dimethylpyrone···$SbCl_5$ molecular compound).
(c) Structural studies of molecular compounds of Group III and V halides
There are two quite distinct structural types: ionic crystals without any special interaction between cation and anion other than ionic, and molecular compounds where the donor is

linked to the acceptor by an oxygen-metal bond of varying strength. Le Carpentier and Weiss (1972c) have suggested that an equilibrium is set up between molecular and ionic moieties in solution, as illustrated here for an acyl chloride and $AlCl_3$:

$$RCOCl + AlCl_3 \rightleftharpoons RCO^+ + AlCl_4^-$$

This equilibrium is a consequence of the bifunctionality of the acyl chloride, which has both an ionizable chlorine and an oxygen atom with a lone pair which can complete the outer electron shell of a metallic acceptor. The validity of this proposal has been strikingly demonstrated by the determination of the crystal structures of donor–acceptor and ionic forms of the adducts between p-toluoyl chloride and $SbCl_5$ (Chevrier et al., 1972a). The molecular compound was crystallized from CCl_4 and the salt from the more polar $CHCl_3$; the two materials were interconvertible by recrystallization from the appropriate solvent. An analogous influence of solvent polarity occurs among the π–π^* molecular compounds (see Chapter 17). Presumably ligands without ionizable groups can form only molecular compounds, although a relevant example has not been reported.

It is convenient also to list some of the ionic structures that have been determined (Le Carpentier and Weiss, 1972a, b) although they will not be discussed further here: $[CH_3CO]^+SbX_6^-$ (X = F,Cl); $[CH_3CO]^+AlCl_4^-$ (Ga analog isomorphous) (Le Carpentier and Weiss, 1972a); $[CH_3CH_2CO]^+GACl_4^-$, $[(CH_3)_2CHCO]^+SbCl_6^-$ (Le Carpentier and Weiss, 1972b); $[o\text{-}CH_3C_6H_4CO]^+SbCl_6^-$ (MPOCCA; Chevrier, Le Carpentier and Weiss (1972c)). $[p\text{-}CH_3C_6H_4CO]^+SbCl_6^-$ (MPOCSB; Chevrier, Le Carpentier and Weiss (1972d)).

The structures of a number of $AlCl_3$ and $SbCl_5$ molecular compounds are known (Le Carpentier and Weiss, 1972c, d; Chevrier, Le Carpentier and Weiss, 1972a, b) all of which can be represented schematically as

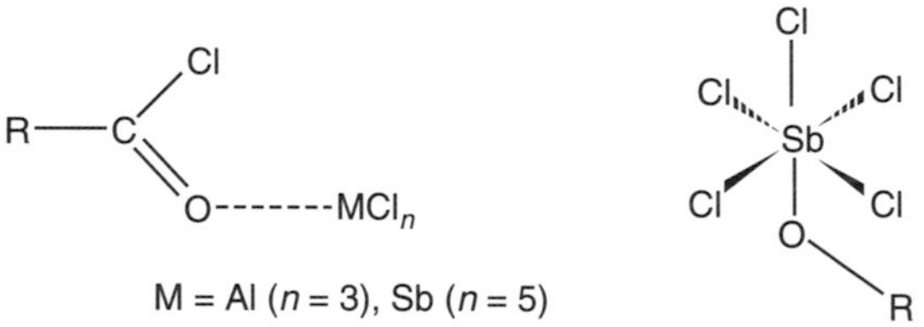

M = Al (n = 3), Sb (n = 5)

Values of $d(O\cdots M)$ and C=O$\cdots$M are summarized in Tables 11.18 and 11.19. In the $AlCl_3$ molecular compounds values of $d(O\cdots Al)$ lie in the range 1.83 ± 0.02 Å and the angle at O has a range of 4–5° about a mean of 143° the coordination about Al is distorted tetrahedral. In minerals the Al$\cdots$O distance is about 1.79 Å so the interaction in these molecular compounds is relatively strong. The Sb$\cdots$O distances in the $SbCl_5$ molecular compounds range over about 0.5 Å, although the >C=O$\cdots$Sb angle is rather constant around 145°, except for the dimethylformamide compound, where a lengthening of $d(C=O)$ to 1.30 Å has been reported (Brun and Branden, 1962). In all these compounds there is a square pyramid of Cl about Sb, with the octahedron completed by the oxygen of the donor moiety. A similar geometry is found in $S_4N_4\cdots SbCl_5$, which has $d(Sb\cdots N) =$ 2.17 Å (Neubauer and Weiss, 1960).

The molecular compound 1,4-dioxane$\cdots$2[Al(CH$_3$)$_3$] is an analogous example; it has overall symmetry C_{2h}-2/m, with dioxane in the chair form (Atwood and Stucky, 1968;

Table 11.18. Geometrical parameters for various 1:1 $AlCl_3$ and $SbCl_5$ molecular compounds. Distances in Å, angles in degrees

Molecular compound	$d(Al \cdots O)$	$< Al \cdots O=C$	Molecular compound	$d(Sb \cdots O)$	$<Sb \cdots O=C$
$CH_3CH_2COCl \cdots$ $AlCl_3$ PROPAL	1.847(6)	141.6	$ClCO(CH_2)_2COCl \cdots$ $2(SbCl_5)$ SUCCSB	2.428	143.6
$m\text{-}CH_3C_6H_4\text{-}$ $COCl \cdots AlCl_3$ ALCMTU	1.835(2)	138.3	$C_6H_5COCl \cdots SbCl_5$	2.317	145.9
$p\text{-}CH_3C_6H_4\text{-}$ $COCl \cdots AlCl_3$ ALTPCU	1.828(2)	146.6	$m\text{-}CH_3C_6H_4\text{-}$ $COCl \cdots SbCl_5$	2.295	144.7
$o\text{-}CH_3C_6H_4\text{-}$ $COCl \cdots AlCl_3$ ALCOTU	1.824(4)	146.2	$p\text{-}CH_3C_6H_4\text{-}COCl \cdots$ $SbCl_5$ TOLCSB	2.253	146.6
$C_5H_4COCl \cdots$ $AlCl_3$	1.819(5)	141.7	$Cl_3PO \cdots SbCl_5$[a]	2.17	145
			$(CH_3)_3PO \cdots SbCl_5$[a]	1.94	145
			$HCON(CH_3)_2 \cdots$ $SbCl_5$[b]	2.05	124.5
			$(C_6H_5)_2SO \cdots SbCl_5$[c]	–	–

Notes:
[a] these two compounds are isomorphous with $Cl_3PO \cdots NbCl_5$, where $d(Nb \cdots O) = 2.16$ Å, $< Nb \cdots O=C = 149°$; however, it should be noted that the interactions differ because Nb is a transition metal.
[b] N, N-dimethylformamide.
[c] Hannson and Vänngaed, 1961.

TMALOX) and the two $Al(CH_3)_3$ groups linked equatorially to the oxygens of the dioxane $(d(Al \cdots O) = 2.02$ Å); there is also an incompletely studied orthorhombic polymorph.

(d) $SnCl_4$ as an acceptor

The older literature contains reports of the formation of molecular compounds between various aromatic hydrocarbons or quinones and $SnCl_4$ and/or $SbCl_5$. The products vary considerably in stability and crystallinity; indeed many are perhaps best described as "ill-defined." 2,3,10,11-Dibenzoperylene was reported to form crystalline 1 : 1 adducts with both $SnCl_4$ and $SbCl_5$, and 2,3-dibenzoanthraquinone forms 'glittering violet needles' with $SbCl_5$ (composition 1 : 2) (Brass and Fanta, 1936). These, and perhaps other examples (Brass and Tengler, 1931a, b), may repay further study but no one seems yet to have taken up the gauntlet.

Cyclohexanol forms 2 : 1 and 4 : 1 adducts with $SnCl_4$. The crystal structure of the 4 : 1 adduct has been reported (Fournet and Theobald, 1981); there are [cyclohexanol $\cdots$ $SnCl_4 \cdots$ cyclohexanol] molecular groupings linked to cyclohexanols of crystallization by $OH \cdots O$ and $OH \cdots Cl$ hydrogen bonds. The Sn atoms are at crystallographic centres of symmetry and the first coordination shell about Sn has approximately D_{4h} symmetry. Presumably the same molecular grouping occurs in the 2 : 1 adduct, but without 'cyclo-hexanols of crystallization.'

Mössbauer effect studies have been made (using ^{119}Sn) of some molecular compounds of aliphatics with $SnCl_4$ (Ichiba et al., 1968). There is little evidence for charge transfer.

(e) AsI_3, $SbCl_3$ and SbI_3 as acceptors

The structures of {1,4-dithiane···$SbCl_3$} (Kiel and Engler, 1974; $P2_12_12_1$, $Z=4$; DTANSC) and {1,4-dithiane···$2SbI_3$} (Bjorvatten, 1966; SBIDTH) (have points of resemblance – there are weak Sb...S interactions in both, with considerably weaker secondary Sb...X interactions. There is also a 1 : 1 compound but its structure was not reported. Kiel and Engler described the {1,4-dithiane···$SbCl_3$} structure as built up by molecules of $SbCl_3$ and $S_2C_4H_8$, and these are arranged in chains along [100] (Fig. 11.17). The $SbCl_3$ molecules have their usual pyramidal shape and the dithianes have a chair conformation. The covalent Sb–Cl distances are 2.385, 2.419 and 2.438 Å. Each Sb is linked to two different sulphurs (3.065, 3.135 A). There are no other Sb...S limks shorter than 3.5 Å. The crystals are chiral.

The space group of {1,4-dithiane···$2SbI_3$} is $C2/c$ with $Z=4$; the dithiane molecules are chair-shaped and at centres of symmetry, and the SbI_3 molecules at general positions (Fig. 11.18). Each Sb forms donor–acceptor links with two sulphurs of different dithiane molecules (d(S···Sb) = 3.274(7) and 3.336(6) Å). Thus the coordination about Sb can be described as square pyramidal (three covalent Sb–I bonds (2.746, 2.767 and 2.774 Å) and two S···Sb links (3.275 and 3.337 A) or as distorted octahedral if the probable location of the lone pair is assumed to be opposite the axial covalent Sb–I bond. Each S forms *two* links to different Sb atoms, one in the equatorial and one in the axial direction. The Sb...S distances are similar to those in DTCHAC (Fig. 11.19). There are no short Sb...I interactions although there is a distance of 3.73 Å between a sulphur and an iodine atom. Thus the overall arrangement has chains along the [110] and [1$\bar{1}$0] directions in the crystal.

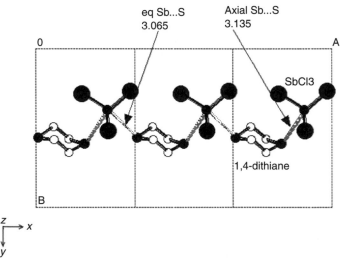

Fig. 11.17. Crystal structure of {1,4-dithiane···$SbCl_3$} showing the chain arrangement of moieties connected by secondary Sb...S interactions. (Data from Kiel and Engler, 1974.)

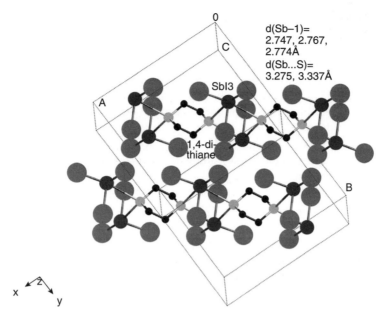

Fig. 11.18. The {1,4-dithiane$\cdots$2SbI$_3$} structure viewed approximately down [001] and showing the layer structure. The chains along {$\bar{1}$10} are shown in the figure. There are only van der Waals interactions between adjacent layers. (Data from Bjorvatten, 1996.)

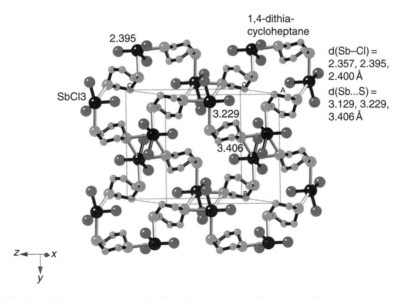

Fig. 11.19. {1,4-Dithiacycloheptane·SbCl$_3$} showing a two-dimensional layer in the *bc* plane. One notes that a rhombus is formed by equatorial S(2)...Sb secondary links to a pair of Sb atoms (3.129 (not shown) and 3.406 Å), while S1 forms an axial linkage to Sb (indicated by 3.229 Å in the diagram). A typical covalent Sb–Cl bond is shown by 2.395 Å. (Data from Schmidt *et al.*, 1979.)

{1,4-Dithiacycloheptane·SbCl$_3$; DTCHAC} crystallizes in space group $P2_1/c$ ($Z=4$); the molecules of the two components are linked by secondary Sb . . . S interactions. There are no secondary Sb . . . Cl interactions. Two dimensional slabs are formed of thickness a (Fig. 11.19). There are only van der Waals forces normal to the slabs. There is approximately octahedral coordination about Sb, with three covalent Sb–Cl bonds and three secondary Sb . . . S interactions, two from S2 and one from S1·{1,4-Dithiacycloheptane·2SbI$_3$} has also been reported but its structure is not known (DTANSC; Schmidt *et al.*, 1979).

{1,3,5-Trithiane·SbX$_3$} (X=Cl, TRTHAC; Br TRTHAB; the crystals are isomorphous) is another example of a molecular compound stabilized by weak S . . . Sb interactions (Fig. 11.20). The crystals are hexagonal (8.129(4) 9.141(1) Å, space group $P6_3mc$ (no. 186), $Z=2$ for the Cl compound) (Lindemann *et al.*, 1977). Changes in dimensions of the component molecules are not discernible but consequences of the interaction are shown by changes in the Raman spectrum (from $1000\,\text{cm}^{-1}$) and the far IR spectrum

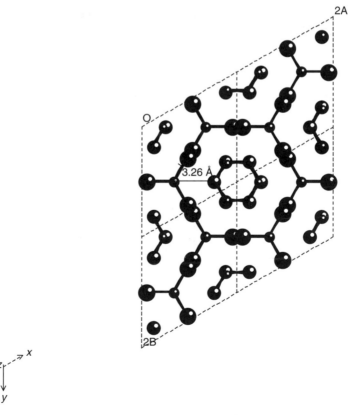

Fig. 11.20. Projection of {trithiane·SbCl$_3$} down the hexagonal c axis. This is a layer structure, with the trithiane molecules superimposed in columns (two, mutually rotated, are shown in the center of the diagram). The structure is polar, with the pyramidal SbCl$_3$ molecules oriented with the Sb atoms below the plane of the three chlorines. The Sb . . . S interaction of 3.26 Å is shown in the diagram; symmetry-related interactions are not shown for clarity. (Data from Lindemann *et al.*, 1977.)

(from $400 \, cm^{-1}$) (Schmidt *et al.*, 1977). The adduct shows a general shift (to higher wave numbers) of Raman frequencies by $5–20 \, cm^{-1}$, accompanied by a reduction in the number of absorptions; both effects were ascribed to the higher symmetry (hexagonal) of the adduct compared to trithiane (orthorhombic).

1,3,5,7-Tetramethyl-2,4,6,8,9,10-hexathia-adamantane (THA) forms molecular compounds with AsI_3 – yellow crystals (m.pt. 109°) of unreported composition and orange-red crystals of 1 : 1 composition whose structure has been reported ($P2_1/c$, $Z=4$; Kniep and Reski, 1982; BIBFOH). The coordination polyhedron of AsI_3 is distorted octahedral with contacts to three of the six S atoms of THA ($d(S\cdots As) = 3.274$ to $3.310 \, Å$); the As lone pair is assumed to be directed towards the six-membered ring containing the linked sulphurs. There do not appear to be any close $As\cdots I$ interactions. There are alternating layers of THA and AsI_3 molecules parallel to the (100) planes.

11.5 n-Donors and π*-acceptors

There are a number of molecular compounds with localized interactions between donor and acceptor moieties where the interaction appears to be of the $n\cdots π^*$ type – among these are the 2 : 1 compounds of 9,10-diazaphenanthrene with 2,3-dichloro-5,6-dicyanobenzoquinone (DDQ) (Shaanan *et al.*, 1982) and tetracyanoethylene (TCNE) (Shmueli and Majorzik, 1980), and the 1 : 2 compounds of TCNE with 2,3-diazabicyclo[2.2.2]octene (DBO) (Blackstock and Kochi, 1987), and DBO-oxide and DBO-dioxide (Blackstock, Poehling and Greer, 1995).

The first two molecular compounds are isostructural but not isomorphous and are characterized by mutually *perpendicular* arrangements of the planes of donor and acceptor molecules. The structure of (9,10-diazaphenanthrene)$_2\cdots$DDQ (red monoclinic crystals; ZPHCYQ10) is shown in Fig. 11.21; the arrangement in the donor layer is remarkably similar to that found in 9,10-diazaphenanthrene itself (Van der Meer, 1972) and thus this is a mimetic molecular compound in which there is a resemblance between the structure of one of the components and that of the molecular compound (cf. Section 10.1). The interaction *between* the two types of layer is due to localized charge transfer of lone pair electrons from the nitrogens of the 9,10-diazaphenanthrene to acceptor $π^*$ orbitals of DDQ or TCNE molecules.

The crystal structure (triclinic, $P\bar{1}$, $Z=1$; 6.676(2) 8.670(2) 9.355(2) Å, 67.05(5) 75.40(3) 69.84(4)°) of the dark red-brown 2 : 1 compound of 2,3-diazabicyclo-[2.2.2]octene (DBO) with TCNE (Blackstock and Kochi, 1987; FEJDUT) contains discrete packing units of the formula given (Fig. 11.22). The TCNE acceptor is sandwiched between the azo groups of the azo-alkane in a manner very similar to that found for DDQ with the aromatic donor 9,10-diazaphenanthrene (compare Figs. 11.21 and 11.22). Blackstock *et al.* (1995) further show that the N-oxide and N,N'-dioxide of DBO associate with TCNE in solution (CH_3CN, charge-transfer bands in the UV-vis spectra peaking at 330 and 472 nm respectively), while the crystals of (DBO-oxide)$_2$·TCNE (ZAGKOH) and (DBO-dioxide)$_2$. TCNE (ZAGKUN) have "D...A...D sandwich structures [similar to those of (DBO)$_2$.TCNE] in which the topology of the D...A...D interaction is effectively described as a pericyclic array, with local D...A cycles apparently influencing the geometry of the non-covalent interactions in the solid state."

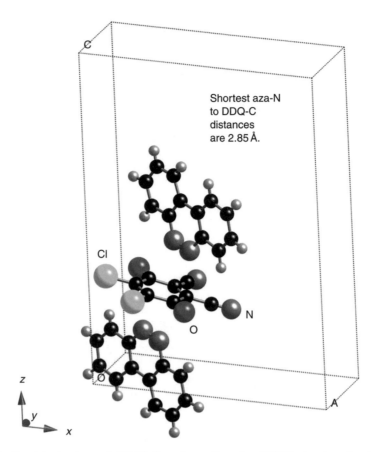

Fig. 11.21. Crystal structure of {(9,10-diazaphenanthrene)$_2$···DDQ} showing the termolecular packing unit consisting of a DDQ molecule sandwiched between 9,10-diazaphenanthrene molecules. The planes of the two moieties are almost mutually perpendicular. These packing units are arranged in herringbone fashion in the crystal. (Data from Shaanan *et al.*, 1982.)

The 9,10-diazaphenanthrene – TCNE system is noteworthy in that two types of 2:1 molecular compound are formed, triclinic and monoclinic. This is an example of an isomeric molecular compound where there are polymorphic forms of the same chemical composition but these have different structures (see Section 15.11; Herbstein, 2000). In the triclinic polymorph there is localized charge transfer of lone pair electrons from the nitrogens of the 9,10-diazaphenanthrene to acceptor π^* orbitals of TCNE, while there is delocalized π···π^* charge transfer in the monoclinic polymorph.

Another possible additional example is triethylphosphate···benzotrifurazan (Cameron and Prout, 1972; BTFZEP10), where mixed stacks of the two moieties along [100] are found (Figs. 11.23 and 11.24). The oxygen atom of the phosphoryl group (O(1)) is only 2.51 Å from the plane of the central ring of the benzotrifurazan molecule and 2.88 Å, on the average, from the six carbons of this ring; the average value of <P=O...C is 150°.

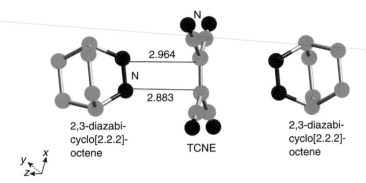

Fig. 11.22. The termolecular centrosymmetric packing unit of formula (2,3-diazabicyclo [2.2.2]octene)$_2$···TCNE shown with the TCNE acceptor sandwiched between the nitrogens (darkened circles) of two donor molecules. C...N distances less than 3 Å are shown. (Data from Blackstock and Kochi, 1987.)

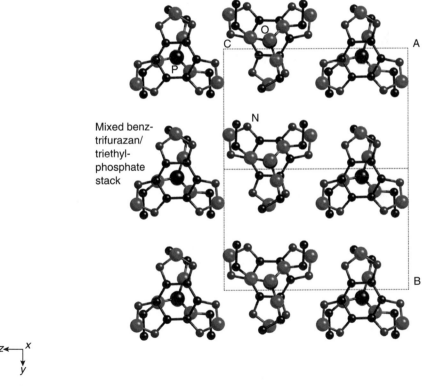

Fig. 11.23. The crystal structure of triethylphosphate···benzotrifurazan at 153K; the crystal data are 7.70(2) 8.87(2) 13.13(2) Å, $\gamma = 107.0°$, $Z = 2$, space group $P2_1$, with the [001] axis unique. The structure is seen in projection down [100], showing close packing of the mixed stacks. The P and O atoms of the triethylphosphate are superimposed. (Data from Cameron and Prout, 1972.)

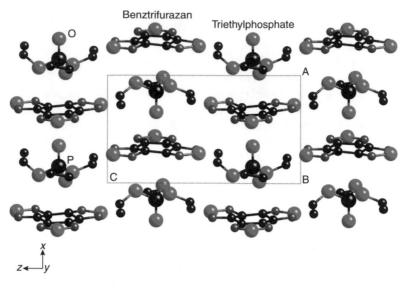

Fig. 11.24. The crystal structure of triethylphosphate···benzotrifurazan at 153K. The mixed stacks of alternating triethylphosphate and benzotrifurazan moieties are shown; the polar nature of the individual stacks is evident. (Data from Cameron and Prout, 1972.)

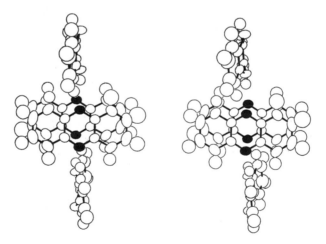

Fig. 11.25. Stereoview of the "dimeric" packing unit seen along a vector perpendicular to the plane of a phenazine molecule. d(N...H–N) = 3.115(2) Å. Nitrogens of phenazine emphasized. (Reproduced from Stezowski *et al.*, 1983.)

The phosphoryl oxygen could be considered as n-donor and the benzotrifurazan molecule as π^*-acceptor. However, there is no charge transfer absorption band in the solution spectrum of the two components and Cameron and Prout (1972) preferred to classify this molecular compound as polarization bonded with important dipole–dipole interactions; the O...ring interactions are then perhaps similar to those found in crystals such as alloxan (ALOXAN to ALOXAN13), barbituric acid (BARBAC), parabanic acid (PARBAC) and chloranil (TCBENQ02 to 06) (Bolton, 1964).

Another example where the planes of the two components are approximately mutually perpendicular is {phenazine···phenothiazine} (Stezowski et al., 1983; BUNRAD; no coordinates). The binary phase diagram of the system (obtained by DSC techniques) shows formation of a 1:1 molecular compound (m. pt. 435.9K); there is essentially no solid solubility of phenazine in phenothiazine, or conversely. Crystals were grown by high vacuum plate sublimation techniques; their dark red color was due to a broad, unstructured absorption edge from 660 to 580 nm; both components are light yellow in color. Cell parameters (at 297K) were $a = 9.072(2)$, $b = 8.872(2)$, $c = 25.934(5)$ Å, $\beta = 112.65(2)°$, $V = 1926.4$ Å^3, $P2_1/c$, $Z = 4$. The molecular volume of phenazine (in the α-phase) is 222.9 Å^3 and that of phenothiazine (in the orthorhombic phase) is 243.9 Å^3 (see Table 10.5). Thus there is a 3.2% expansion on formation of the molecular compound from these polymorphs of its components. The packing unit (Fig. 11.25) was found to consist of two phenazines in plane-to-plane arrangement, with approximately perpendicular phenothiazines linked by weak N...H–N hydrogen bonds. There is no potential donor to provide hydrogen bonding to the second nitrogen of phenazine, nor is there any evidence for interaction of some kind with the sulphur atom of phenothiazine. The packing arrangement does not indicate any π-orbital interaction beyond that between superposed phenazines.

One can argue whether {phenazine . . . phenothiazine} has been correctly located at this place in the text. Certainly {triethylphosphate···benzotrifurazan} and {phenazine··· phenothiazine} form an interesting contrast. In the first of these molecular compounds there are strong geometrical suggestions of charge transfer interaction but a colourless binary compound is formed from colorless components, while in the second there is spectroscopic evidence for charge transfer interaction, but no support for this from the crystal structure. These paradoxes remain to be resolved.

The molecular compounds {1,4-dioxane···dinitrogen tetroxide} (Groth and Hassel, 1965; NODIOX) and {1,4-dioxane···oxalyl fluoride} (Möller et al., 1987; FAYNOI) have isoelectronic acceptors and their triclinic crystals are isomorphous; there are chains of alternating donor and acceptor molecules along [111] with short O···N (2.90, 2.76 Å) and O···C (2.59, 2.61 Å) distances. The common feature (Möller et al., 1987) is the near-symmetric interaction between an electron-rich centre (O of dioxane) and the comparatively weakly linked central atoms of a 34 valence-electron system (N_2O_4 or $C_2F_2O_2$).

11.6   π-Donors and σ*-acceptors

11.6.1   Aromatic molecules as donors and dihalogens as acceptors

The color of an iodine solution depends on the nature of the solvent (Kleinberg and Davidson, 1948); for example, iodine–ethanol solutions are violet and their absorption spectra are similar to those of iodine vapour, while iodine–benzene solutions are brown. The implication is that iodine interacts with benzene in a manner that is more substantial than its negligible interaction with ethanol (Benesi and Hildebrand, 1949). Theoretical studies of this interaction were initiated by Mulliken and continue (e.g. Bruns, 1977). The geometry of the benzene···I_2 molecular compound formed in solution or in the vapour phase remains controversial; in the 'resting' model the halogen molecule is parallel to the plane of the benzene ring, but perpendicular to this plane in the 'axial' model.

There is a considerable body of publication, much of it contradictory, on the structures of aromatic molecules with the dihalogens Cl_2 and Br_2. However, rather recent studies lead to a resolution of many of the problems. The structures of the isomorphous $\{C_6H_6\cdots Cl_2\}$ and $\{C_6H_6\cdots Br_2\}$ molecular compounds were first reported by Hassel and Strømme (1958, 1959c); the melting points are 233 K and 259 K respectively and the temperatures of measurement were 183 K and 223–233 K. The halogen molecules were reported to lie along the sixfold axes of the benzene rings (i.e. a very symmetrical form of the 'axial' model) forming an infinite arrangement

$$\cdots X_2 \cdots C_6H_6 \cdots X_2 \cdots C_6H_6 \cdots X_2 \cdots C_6H_6 \cdots.$$

The intrahalogen distances are the same as those in the gas phase, testifying to rather weak interactions, while the distances between halogen atom and center of benzene ring are 3.28 and 3.36 Å for the Cl_2 and Br_2 compounds respectively. The crystals were originally reported to be monoclinic (space group $C2/m$) but re-examination (Herbstein and Marsh, 1982) strongly suggests that the crystals are rhombohedral ($a_H = 8.49$, $c_H = 8.54$ Å for the chlorine compound), the correct space group being $R\bar{3}m$. $\{C_6H_6\cdots Br_2\}$ has been studied further by Vasilyev, Lindeman and Kochi (2001, 2002; BENZBR01, BENZBR02), who found a phase transition at 203K. Above this temperature the diffraction pattern (i.e. that found by Hassel and Strømme) was very weak; slow cooling and careful manipulation led to a single crystal to single crystal transformation to trigonal crystals which gave diffraction patterns to high angles ($a_H = 8.721(2)$, $c_H = 8.701(2)$ Å, space group $P3_22_12$, $Z = 3$). The absolute configuration (of the crystal used) was checked via the Flack parameter. The structure at 123K deviated appreciably from the axial model (the 'A structure' of Vasilyev et al.) and had the Br_2 molecule located over a C–C bond with $d(C\ldots Br) = 3.18$ and 3.36 Å (the 'B structure' of Vasilyev et al.). The structure of $\{$toluene$\ldots Br_2\}$ was more complicated (triclinic, $a = 5.516(1)$, $b = 11.715(2)$, $c = 13.551(3)$ Å, $\alpha = 79.76(1)$, $\beta = 80.89(1)$, $\gamma = 85.56(1)°$, space group $P\bar{1}$, $Z = 4$). There are two crystallographically independent toluenes in the unit cell, two bromine molecules at centers of symmetry and one bromine molecule at a general position. The values of $d(C\ldots Br)$ lie in the range 3.01–3.17 Å, and the bromine molecules are again positioned over the rims of the aromatic ring.

Spectroscopic studies (Abe and Ito, 1978) showed that toluene and p-xylene form crystalline compounds with Cl_2 and Br_2 but only $\{$toluene$\cdots Br_2\}$ has been studied crystallographically. Despite the extensive IR and Raman studies (Anthonsen and Møller, 1977; Abe and Ito, 1978), there is no agreement among spectroscopists about the space groups; clearly 77K spectra of $\{C_6H_6\cdots Cl_2\}$ and $\{C_6H_6\cdots Br_2\}$ must be checked against noncentrosymmetric space groups, while $\{$toluene$\cdots Br_2\}$ could still be centrosymmetric at 77K.[1] Vasilyev et al. give a comprehensive list of references to spectroscopic and theoretical studies. Matrix isolation spectroscopy of $\{C_6H_6\cdots Cl_2\}$ and $\{C_6H_6\cdots Br_2\}$ at 20K was interpreted to give a geometry where the halogen molecules do not lie along the sixfold axis of the benzene ring but are inclined to it (Fredin and

---

[1] Two errors in the literature require correction. Hassel and Stromme reported atomic coordinates for $\{C_6H_6\cdots Cl_2\}$ (not noted in CSD) but not for $\{C_6H_6\cdots Br_2\}$ (BENZBR), as Vasilyev et al. appear to have assumed. The statement of Herbstein and Marsh (1982) that "the spectroscopic and diffraction studies were made on the same phases" also requires revision.

Table 11.19. Stretching force constants for Cl_2 and Br_2 in different environments (units of mdyn $\mathring{A}^{-1}$) (Abe and Ito, 1978).

Molecule	Gas phase	In benzene solution	In crystalline $\{C_6H_6 \cdots X_2\}$
Cl_2	3.208	2.974	2.705
Br_2	2.336	2.133	2.049

Nelander, 1974), thus supporting the off-axis model (structure B). The benzene$\cdots$$I_2$ compound does not appear to be known in crystalline form, but the off-axis model described above was supported by the matrix isolation study of Fredin and Nelander and by the 16K ^{129}I Mössbauer spectrum of iodine dissolved in benzene (Bukshpan et al., 1975; Sakai et al., 1983).

The degree of charge transfer in the benzene$\cdots$halogen compounds has been estimated from NQR spectroscopy. Hooper (1964) found that the NQR frequency of ^{81}Br was 321.83 Mhz in $\{C_6H_6 \cdots Br_2\}$ compared to 319.46 Mhz in crystalline Br_2 (both samples at 77K). These values indicate very little charge transfer, the small decrease in the resonance frequency in Br_2 being ascribed to intermolecular interactions present in the crystalline halogen but absent in the molecular compound. Later work (Kadaba et al., 1971) over an extended temperature range (15–150K) gave similar results; a break in the frequency–T curve was ascribed to a possible phase transition at 60K. A discontinuity at 120K in the curve of the ^{35}Cl NQR frequency against temperature in $\{C_6H_6 \cdots Cl_2\}$ was ascribed to the onset of precession of the Cl_2 molecules about the normal to the benzene-ring plane as the crystals warm up (Gordeev et al., 1974), but this could be evidence for a phase transition. Spectroscopic evidence for some charge transfer comes from determination of the force constants for intrahalogen stretching vibrations in different environments (Table 11.19). The weakening of the intramolecular halogen–halogen bond is ascribed to transfer of charge from benzene π-orbitals to antibonding σ^* orbitals of the halogen molecules.

The 1:1 molecular compound of benzene and acetylene is rhombohedral at 201 and 123K ($a_H = 8.504(1)$, $c_H = 8.206(2)$ at 123K, $Z = 3$, space group $R\bar{3}m$ (Boese, Clark and Gavezzotti, 2003) and is thus isomorphous with $\{C_6H_6 \cdots Br_2\}$ above 203K. Boese et al. remark that "there are no obvious peculiarities about the ADPs that might indicate static or dynamic disorder". However, anomalous values for the acetylene triple bond length led them to infer "an essentially dynamic disorder process in which the acetylene molecule moves along a double cone between the benzene molecules", also described as a wobbling or precession motion between the two rings so that the acetylene is perpendicular to them only in a time-averaged sense. Perhaps there is a phase change at lower temperatures.

There are many reports of formation of 'complexes' of aromatic hydrocarbons with halogens; most attention seems to have been paid to pyrene and perylene as donors and Br_2 and I_2 as acceptors. The nature of these materials remains in doubt, especially when ratios other than 1:1 are reported and we shall not consider them further. However, coronene does form a 1:1 molecular compound with I_2 and this has a structure (Shibuya, 1961; Pepinsky, 1962; Mitani, 1986; DUPCIA10) not much different from those of the $\{aromatic \cdots X_2\}$ compounds described above; the I_2 molecule lies between stacks of superimposed coronene molecules, with both moieties at centers of symmetry in space

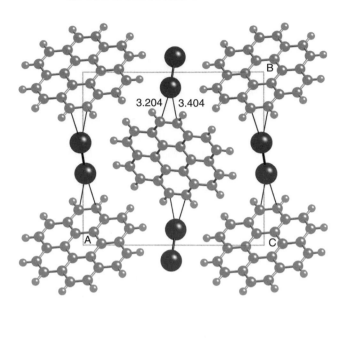

Fig. 11.26. A layer in the {coronene···I_2} structure showing the zigzag chains of interacting moieties. The two close I...C distances are 3.20 and 3.40 Å. The unit cell ($a = 14.025$, $b = 13.138$, $c = 4.819$ Å, $\beta = 105.05°$, $Z = 2$, $P2_1/a$) is outlined. (Coordinates from Mitani, 1987.)

group $P2_1/a$ (Figs. 11.26, 11.27). There are very similar stacks of coronene molecules here and in neat coronene (Robertson and White, 1945; CORONE01; Fawcett and Trotter, 1966, CORONE), and the difference between the two structures is expressed in the η^2 interaction between peripheral carbons of coronene and the iodines in {coronene···I_2}, leading to bridging between adjacent stacks and the formation of zigzag ..C...I–I...C...I–I...chains, where we use C as an abbreviation for coronene. Some measure of the interactions between the moieties can perhaps be inferred from the intramolecular I–I distance, which is increased to 2.710(2) Å, the value found in crystalline iodine, where there are strong intermolecular interactions, and is ≈0.05 Å longer than that found for I_2 in the gas phase. Although precise room temperature C–C bond lengths are now available for neat coronene (Krygowski *et al.*, 1996; COR-ONE02), it seems unlikely that those in {coronene···I_2} are precise enough to warrant comparison.

^{129}I Mössbauer measurements on crystalline {coronene···I_2} at 16K (Sakai, Matsuymam, Yamaoka and Maeda, 1983) are compatible with the crystal structure, and show that the I–I bond consists of a pure p_σ bond, the acceptor orbital being described as the σ_u antibonding molecular orbital of the iodine molecule. 0.06 electrons are transferred to this orbital.

The B structures found for {C_6H_6···Br_2} and {toluene···Br_2} fit neatly with the η^2 interaction found in {coronene···I_2}. We shall see below that an η^2 interaction is also found in some of the aromatic hydrocarbon-SbCl$_3$ compounds.

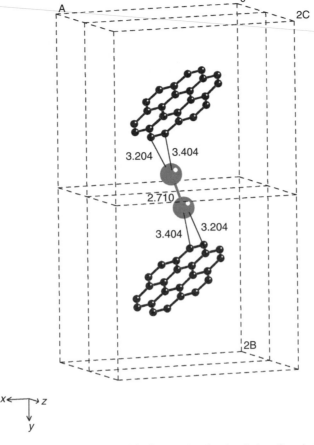

Fig. 11.27. Arrangement of coronene and iodine molecules in chain of moieties along [010]. (Coordinates from Mitani, 1987.)

11.6.2 *Aromatic molecules as donors and polyhalogenated methanes as acceptors*

The first molecular compounds in this group appear to have been discovered by Wyatt (1936) in his studies of the phase diagrams of a number of binary systems; relevant here are the incongruently melting 1:1 and 1:2 compounds found in the benzene – CCl_4 system, and the reported absence of compounds in the benzene-$CHCl_3$ system. These results were extended by Kapustinskii and Drakin (1947), who confirmed the occurrence of the 1:1 and 1:2 compounds in the benzene – CCl_4 system (melting points -25 and $-35°C$ respectively) and reported a 1:3 compound (m.pt. $-40°$ they also prepared $\{C_6H_6\cdots CBr_4\}$ (m.pt. $30.4°$) and reported incongruently-melting 1:1, 1:2 and 1:3 compounds in the benzene – CH_2Cl_2 system. Spectroscopic studies (IR and Raman) suggested that 2:1 (perhaps), 1:1 and 1:2 compounds do occur in the benzene – $CHCl_3$ system (Chantry *et al.*, 1967), in contradiction to Wyatt's results. A variety of methods (DTA, Raman and ^{35}Cl NQR spectroscopy) show the formation of loosely bound {benzene...CH_3Cl}, melting at 198K (Rupp and Lucken, 1986).

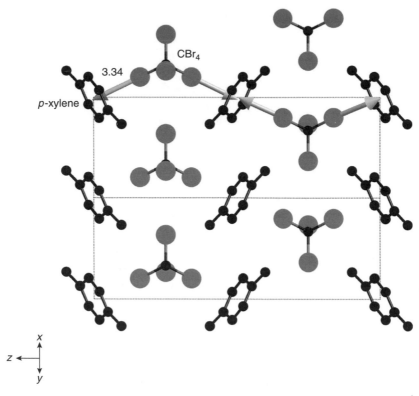

Fig. 11.28. Crystal structure of {p-xylene···CBr$_4$} ($a = 8.48(3)$, $b = 8.89(3)$, $c = 17.46(5)$ Å, space group *Cmcm*, $Z = 4$) in projection down [110]. The arrows show the interactions between the aromatic rings and the bromines, with distances of 3.34 Å between ring centres and Br. (Data from Strieter and Templeton, 1962.)

The formation of {C$_6$H$_6$···2CCl$_4$} has been confirmed twice (Viennot and Dumas, 1972; Ott and Goates, 1963); the 1:1 compound between 1,2,4-trimethylbenzene and CCl$_4$ melts at 229.85K (Goates *et al.*, 1987) and its enthalpy of formation has been measured (see below). Extensive spectroscopic studies have been made of a number of crystalline {p-xylene···CBr$_4$} compounds, where the p-xylene was deuterated in various ways (Lebas and Julian-Laferriere, 1972). There are only small differences between the spectra of the pure components and those of the molecular compounds, from which it was inferred that the interactions between the components in the molecular compounds are rather weak. Only one crystal structure has been reported – that of {p-xylene···CBr$_4$} (m.pt. 53 °C) (Strieter and Templeton, 1962; CTBRXY). A layer structure is formed (Fig. 11.28), with two of the four C–Br bonds directed towards the centers of the benzene rings; d(Br···ring centre) $= 3.34$ Å, close to that found in {C$_6$H$_6$···Br$_2$}.

The liquidus curves have been determined in the phase diagrams of 1,2,4-trimethylbenzene (Goates *et al.*, 1987) and of benzene, toluene and p-xylene with CCl$_4$ (Boerio-Goates *et al.*, 1985) and the detailed forms of these curves were then used to determine the standard

Table 11.20. Calculated enthalpies of formation of 1:1 aromatic hydrocarbon⋯CCl$_4$ molecular compounds

Aromatic hydrocarbon	$\Delta H°m$(kJ/mol)	Tm(K)
Benzene	−3.24	239.12
Toluene	−4.01	205.97
p-Xylene	−9.43	269.28
1,2,4-trimethylbenzene		229.85

enthalpies of formation of the molecular compounds at their melting points, a typical reaction being

$$C_6H_6(s, 239.12K) + CCl_4(s, 239.12K) \Rightarrow C_6H_6{\cdots}CCl_4(s, 239.12K)$$

The calculated enthalpies are summarized in Table 11.20.

A different method was used for {benzene⋯CBr$_4$}, where the vapor pressure of benzene over the molecular compound was measured in the temperature range 9–25 °C (Kapustinski and Drakin, 1950). Using the Arrhenius equation, this gives the enthalpy of the reaction

$$\{C_6H_6 \cdot CBr_4\}(s) \Rightarrow C_6H_6(g) + CBr_4(s)$$

as $\Delta H_{diss} = 50.6$ kJ mol^{-1}. The enthalpies of vaporization and fusion of benzene (at 278.66K) are $\Delta H_{vap} = 34.1$ kJ/mol and $\Delta H_{fus} = 9.87$ kJ/mol. Combining these values gives the (formal, as used above, not standard) enthalpy of formation of crystalline {C$_6$H$_6$·CBr$_4$} as −6.6 kJ/mol. Thus there is little difference in the stabilities of the two molecular compounds. A more useful comparison would be *via* the free energies of formation but entropy values are unfortunately not available.

11.7   π-Donors and p-acceptors

11.7.1   *Aluminum tribromide as an acceptor*

Aromatic hydrocarbons (ArH) form 1:1 and 1:2 molecular compounds with aluminum trichloride and aluminum tribromide. Relevant results are available only for the aluminum tribromide compounds. Compositions are generally given as ArH:AlBr$_3$ (i.e. 1:1) or ArH:Al$_2$Br$_6$ (i.e. 1:2) but there is direct evidence for the nature of the acceptor species only for benzene:Al$_2$Br$_6$ (see below). Only one crystal structure appears to have been reported, that of the easily hydrolyzed {benzene⋯Al$_2$Br$_6$} (Eley *et al.*, 1961; ALBROB) There are layers of Al$_2$Br$_6$ molecules in the structure with the benzenes in interstices. There was no evidence of interactions other than those due to van der Waals forces; however, the results are not very accurate because of the experimental difficulties involved in the structure determination and it would be desirable to have structures of some analogs before firm chemical conclusions are drawn. NQR measurements made on this system (Okuda *et al.*, 1972) at 301K indicate that there is very little charge transfer (≤0.01e) from benzene to Al$_2$Br$_6$.

Table 11.21. Thermodynamic parameters for the *dissociation* of 1:2 and 1:1 {aromatic hydrocarbon···aluminum tribromide} molecular compounds (for reactions see below); ΔG (at 0 °C) and ΔH values in kJ/mol and ΔS (at 0 °C) in J/mol deg.

Aromatic ligand	1:2 molecular compounds					
	$\Delta G(0°)$	ΔH	$\Delta S(0°)$	$\Delta G(0°)$	ΔH	$\Delta S(0°)$
benzene	9.5	44.3	125			
toluene	13.1	48.9	30			
p-xylene	16.7	52.7	130			
o-xylene	17.1	48.5	117			
m-xylene	15.2	52.7	138	7.2	20.4	46
mesitylene	18.0	51.9	125	8.9	22.3	50

The reactions are: $[\text{ArH}\cdots\text{AlBr}_3](s) \Rightarrow 1/2[\text{ArH}\cdots\text{Al}_2\text{Br}_6](s) + 1/2\text{ArH}(g)$
$[\text{ArH}\cdots\text{Al}_2\text{Br}_6](s) \Rightarrow \text{Al}_2\text{Br}_6](s) + \text{ArH}(g)$

Thermodynamic parameters (Table 11.21) were measured (Choi and Brown, 1966) for a number of compounds by determining dissociation pressures at five temperatures between 0° and –45°.

11.7.2 *Miscellany-mainly MX$_3$ (M = As, Sb; X = Cl, Br) as acceptors and aromatic molecules as donors*

Binary phase diagram studies (Shaw *et al.*, 1968) have shown the occurrence of congruently melting equimolar molecular compounds of hexamethylbenzene with triphenylarsene (m.pt. 144 °C), triphenylphosphine (145°) and triphenylstibine (137°). No structures have been reported.

The 1:2 molecular compounds of benzene with AsCl$_3$ (m.pt. –25.5°) and of hexaethylbenzene with AsBr$_3$ have been prepared. The first of these has been studied (Biedenkapp and Weiss, 1968) by ^{35}Cl NQR and it was deduced that the chlorines occur pairwise at independent sites in the unit cell but the overall structure is not known. The structure of {C$_6$(C$_2$H$_5$)$_6$···2AsBr$_3$} (rhombohedral, $R\bar{3}$, 13.356(2), 12.355(2) Å, $Z=3$) has been determined; this contains discrete Br$_3$As···C$_6$(C$_2$H$_5$)6···AsBr$_3$ moieties of S_6–3 symmetry (Schmidbaur, Bublak, Huber and Müller, 1987a; SALCIR); this was called an 'inverse sandwich' structure. The geometries of the components in the molecular compound differ only slightly from those in the individual neat crystals; this resemblance extends even to the conformations of the ethyl groups. Furthermore, there are no weak Br . . . As interactions as in crystalline AsBr$_3$ (Bartl, 1982). The large colourless rhombohedral crystals of {C$_6$(C$_2$H$_5$)$_6$···2AsBr$_3$} are only slightly sensitive to air and moisture. A number of other AsBr$_3$ molecular compounds were obtained but disorder prevented determination of the structures. Hexaethylbenzene also forms 1:1 molecular compounds with SbCl$_3$ and SbBr$_3$ (Schmidbaur, Nowak, Huber and Müller, 1987), with the structure of the former reported as orthorhombic, *Pnma* (no. 62), $Z=4$, 18.003 10.118 12.235 Å (FOSWUF). There are discrete C$_6$(C$_2$H$_5$)$_6$···SbCl$_3$ moieties of C_3-3 symmetry, the Sb being above the center of the benzene ring at a distance of 2.96 Å. The interaction is thus much weaker than that in, say, hexaethylbenzene-chromium tricarbonyl, where the

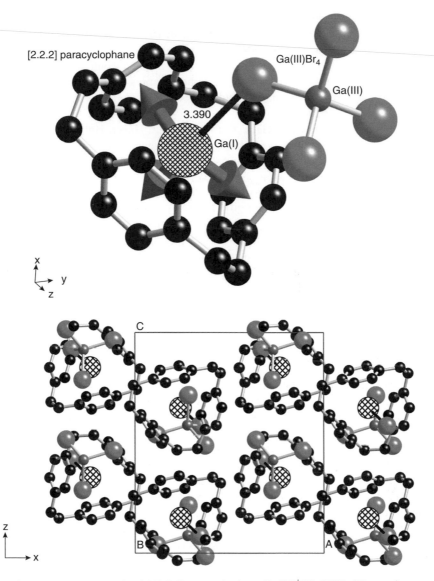

Fig. 11.29. (upper) The ion-pair of {[2.2.2]paracyclophane·Ga(I)}$^{+}$(Ga(III)Br$_4$)$^{-}$ seen in perspective; the covalent Ga(III)–Br distances range from 2.306 to 2.334 Å and the secondary Ga(I) to carbon distances from 2.91 to 3.1 Å. There is a weak link between Ga(I) and Br indicating the formation of ion pairs.
(lower) The packing arrangement of the ion pairs seen down [010]. (Data from Schmidbaur, Hager, Huber and Müller, 1987.)

metal-to-ring distance is 1.72 Å. A similar alternating up-down conformation of ethyl groups is found in both Br$_3$As•••$_6$(C$_2$H$_5$)$_6$•••AsBr$_3$ (SALCIR) and C$_6$(C$_2$H$_5$)$_6$•••SbCl$_3$ (FOSWUF) and also in (hexaethylbenzene)(toluene)Ga(I) tetrachloro-gallate(III)·hemi (hexaethylbenzene) (Section 11.12.2).

The triangular [2.2.2]paracyclophane molecule forms a $1:1$ complex with $AsCl_3$, where the $AsCl_3$ coordinates weakly through As to the ring system as a whole (Probst *et al.*, 1991; KISWUE); the crystals are orthorhombic, $a = 10.235(1)$, $b = 11.501(1)$, $c = 19.330(2)$ Å, $Z = 4$, space group $P2_12_12_1$, and the absolute configuration (of the crystal used in the analysis) was determined. The authors express their surprise at the lack of $Cl_3As \ldots$ (aromatic ring) coordination, and ascribe the formation of the complex to weak van der Waals type interactions.

$\{[2.2.2]Paracyclophane \cdot Ga\}^+(GaBr_4)^-$ (Schmidbaur, Hager, Huber and Müller, 1987; FIGKOV) is discussed at this point because of an overall, but incomplete, structural resemblance to $\{[2.2.2]paracyclophane \cdot AsCl_3\}$; alternatively it could be included, on grounds of chemical resemblance, in Chapter 11.12.2. The colourless, only slightly air-sensitive crystals are monoclinic (13.741(1) 11.856(1) 15.973(1) Å, 90.24(1)°, $Z = 4$, $P2_1/c$) and contain ion-pairs (Fig. 11.29). The Ga(I) atom is equidistant, within narrow limits, to all 18 C atoms of the arene; distances to the ring centers are all 2.65 Å. However, the Ga(I) is displaced by 0.43 Å above the plane containing the three ring centers; the authors ascribe this to interaction with Br2, distant 3.388 Å, and to Ga(I) being just too large to fit into the inter-ring space.

11.8 π-Donors and (localized) π^*-acceptors

Crystalline complexes have been prepared between benzene or polymethylbenzenes and $NO \cdot MX_6$, where M may be P, As or Sb and X may be F or Cl (Brownstein *et al.*, 1986; Kochi, 1990), and structures have been reported for $\{[C_6(CH_3)_6] \cdot NOSbCl_6\}$ (COMROL10), $\{2[C_6H_5CH_3] \cdot NOSbCl_6\}$ (FATFEL), $\{[C_6(CH_3)_6] \cdot NOAsF_6\}$ (FATDUZ) and $\{[1,3,5-C_6H_3(CH_3)_3] \cdot NOPF_6\}$. While the crystals all have different structures, there is a common feature in that much the same mutual disposition of aromatic hydrocarbon donor and the NO^+ acceptor appears to be found in all of them, as illustrated for the mesitylene compound (Fig. 11.30). The bond lengths are not precise enough for one to infer the degree of charge transfer in the molecular compounds.

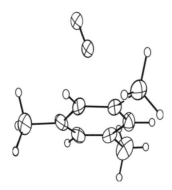

Fig. 11.30. View of the mutual disposition of mesitylene and NO^+ moieties in the $\{mesitylene \cdot NOPF_6\}$ molecular compound. The N atom is probably adjacent to the ring but this has not yet been established beyond doubt. (Reproduced from Kochi, 1990.)

PART 2: Self-interacting acceptors

11.9   n-Donors and s-acceptors

11.9.1   *N, O, and S containing ligands as donors and Ag(I) salts as acceptors*

Silver salts form molecular compounds with molecules containing N, O, S as donor atoms and a number of crystal structures have been reported. As noted in the corresponding section (Chapter 11.2) in Part 1, many of these materials are peripheral to the major theme of this book and are thus noted only briefly. {Pyrazine$\cdots$AgNO$_3$} (Vranka and Amma, 1966; AGPYRZ) and {HMT$\cdots$AgNO$_3$} (Michelet *et al.*, 1981; BARPUF) (HMT = hexamethylenetetramine) have somewhat similar structures. The first compound contains kinked chains of the type

$$\cdots Ag^{+}\cdots pyr\cdots Ag^{+}\cdots pyr\cdots Ag^{+}\cdots pyr\cdots$$

with $d(N\cdots Ag^{+}) = 2.21$ Å and $<N\cdots Ag^{+}\cdots N = 159(1)°$. The next nearest neighbors of the Ag^{+} ions are two oxygens at 2.72(2) Å and another two at 2.94(2) Å. The chains are held together by weak van der Waals forces. In HMT . . . AgNO$_3$ there are sheets of composition

$$\cdots Ag^{+}\cdots HMT\cdots Ag^{+}\cdots HMT\cdots Ag^{+}\cdots HMT\cdots$$

where Ag^{+} and HMT lie on mirror planes, and the nitrate ions are disordered across mirror planes. Each Ag^{+} is linked to three different HMT molecules ($d(N\cdots Ag^{+}) = 2.406(5)$ (twice) and 2.335(6) Å) and three oxygens of nitrate ions with $d(N\cdots O) = 2.62(1)$(twice) and 2.593(8) Å. Cohesion between the sheets is by van der Waals forces.

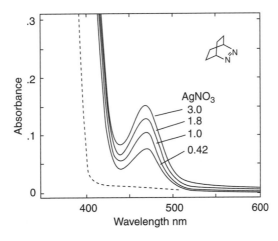

Fig. 11.31. The charge transfer absorption bands from solutions of DBO in acetonitrile to which varying molar concentrations of AgNO$_3$ have been added. The dashed line shows the spectrum of DBO alone; that of AgNO$_3$ alone is rather similar. (Reproduced from Blackstock and Kochi, 1987.)

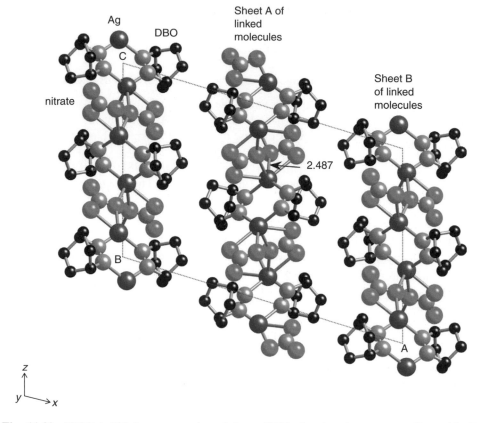

Fig. 11.32. {DBO·AgNO$_3$} structure viewed down [010], showing the two crystallographically independent sheets A and B. In the central sheet the distances are: d(Ag...N) = 2.298, 2.269 Å; d(Ag...O) = 2.487, 2.509, 2.754, 2.856 Å. The values in the second sheet are only slightly different. In each sheet only the shortest Ag...O distance is shown, for clarity. (Data from Blackstock and Kochi, 1987.)

A particularly detailed picture is obtained from the structure of {diazabicyclo [2.2.2]octene (DBO)·AgNO$_3$}, which shows a distinct charge transfer band in its spectrum (Fig. 11.31) (Blackstock and Kochi, 1987; FEJFEF). The canary yellow compound crystallizes in space group $P2_1/c$, with 8 formula units in the unit cell ($a = 19.80$, $b = 7.45$, $c = 13.12$ Å, $\beta = 106.9°$). The crystal is composed of sheets of interconnected DBO molecules, Ag$^+$ and NO$_3^-$ ions. The nitrogens of the diaza group each interact with a different Ag$^+$ cation. The oxy ligands serve as multiple bridges to link the silver(I)-azo chromophores. Each Ag$^+$ cation is six-coordinate (two DBO nitrogens at ≈2.3 Å, four oxygens at 2.4–2.9 Å). The sheets interact by van der Waals forces (Fig. 11.32). Although the layers shown are crystallographically independent, the interactions within them differ only slightly.

11.9.2 N, O, S containing ligands as donors and HgX_2 (X = Cl, Br, I) as acceptors

The lone pairs of O, N, S atoms in many molecules interact with $HgCl_2$ to form addition compounds; $HgBr_2$ forms some, while among the few HgI_2 analogs reported is dioxane···HgI_2 (Patterson *et al.*, 1973). In order to be able to assess the degree of ligand···Hg(II) interaction it is necessary to compare ligand···Hg(II) distances with those found in covalently-bonded compounds. The covalent radius of Hg(II) depends on the type of coordination (Grdenic, 1965); a value of 1.28 Å is used for linear bicovalent Hg(II). Thus distances in molecular compounds should lie within a range whose lower limit is the sum of the appropriate covalent radii and upper limit the sum of the appropriate van der Waals radii (Table 11.22). These limiting values themselves have uncertainties of ≈ 0.1 Å.

Our classification will be based on the coordination number of Hg(II), where we include among the nearest neighbours both covalently bonded ligands and those linked by secondary interactions. In geometrical terms this gives distorted octahedral arrangements for six-coordinate Hg(II), distorted square pyramidal and trigonal bipyramidal arrangements for five-coordinate Hg(II) and distorted tetrahedral arrangements for four-coordinate Hg(II). The distinctions are sometimes blurred because of the secondary interactions. The subject has been extensively reviewed (Branden, 1964c; Dean, 1978; Graddon, 1982).

HgX_2 molecules are linear in the gas phase, with dimensions given in Table 11.22. In their crystals $HgCl_2$, $HgBr_2$ and the yellow form of HgI_2 all have structures in which each Hg is surrounded by six halogens along orthogonal axes; two *trans* Hg-X bonds have lengths very similar to those in the gas phase (and thus the linear HgX_2 molecules are preserved), and four are appreciably longer (Table 11.22). Thus the HgX_2 molecules in the solid state are linked in arrays by secondary interactions, which are often also found in the six-coordinated HgX_2 molecular compounds considered in the next section.

We shall to some extent infringe on the requirement imposed in Chapter 1 that "the properties of the components be very largely conserved;" the linear HgX_2 molecule does not appear *as such* in all the molecular compounds to be discussed here but the advantages of considering the HgX_2 compounds as a group outweigh too strict an allegiance to self-imposed definitions.

Table 11.22. Range of donor–acceptor distances (Å) expected in molecular compounds of Hg(II) halides together with some limiting values

Type of interaction	Lower limit	Upper limit	HgX_2 in gas phase	HgX_2 in crystal phase
Hg···O	2.0	2.7		
Hg···N	2.1	2.8		
Hg···S	2.3	3.3		
Hg···Cl	2.3	3.3	2.29 (a)	2.25, 3.34, 3.63 (all × 2) (b)
Hg···Br	2.4	3.45	2.41	2.48 (×2), 3.23 (×4)
Hg···I	2.6	3.65	2.59	2.62 (×2), 3.51 (×4)

References:
(a) Kashiwabara *et al.*, 1973; (b) Subramanian and Seff, 1980.

11.9.2.1   *Six-coordinate Hg(II)*

(i) Structures based on HgCl$_2$ ribbons with weak Hg$\cdots$Cl interactions (bicoordinate Cl)   This first group of addition compounds contains those based on ribbons of HgCl$_2$ molecules, with secondary Hg$\cdots$Cl interactions as shown schematically below, and completion of the approximately octahedral coordination by two Hg$\cdots$A (A=O, N or S) interactions which run, up and down, approximately normal to the plane of the ribbon.

The HgCl$_2$ molecule participates in these compounds almost without change from its gas-phase geometry where d(Hg–Cl) $\approx$ 2.30 Å and <Cl–Hg–Cl $\approx$ 175°. The secondary Hg$\cdots$Cl distances are about 3.2 Å, only slightly smaller than the sum of their van der Waals radii; however, the consistent appearance of so many similar arrangements is strong evidence for the reality of the secondary Hg$\cdots$Cl interactions. The essential requirement for a structure to be classified in this group is that the Cl atoms of the HgCl$_2$ molecules should be bicoordinate, with one covalent bond to an Hg of the same HgCl$_2$ molecule and one secondary link to Hg in a neighbouring HgCl$_2$ molecule.

Rather similar arrangements, characteristic of this structural group, are found in {quinoline-N-oxide$\cdots$HgCl$_2$} (McPhail and Sim, 1966a; HGCQIN) (octahedrally coordinated by two Cl's at 2.30 Å, <Cl–Hg–Cl = 174°, two O's at 2.56, 2.61 Å and two Cl's of neighbouring HgCl$_2$ molecules at 3.12, 3.35 Å), and in {dihydrouracil$\cdots$HgCl$_2$} (DHURHG) and {uracil$\cdots$HgCl$_2$} (URILHG) (Carrabine and Sundaralingam, 1971); and {(CH$_3$OH)$_2$$\cdots$HgCl$_2$} (Brusset and Madaule-Aubry, 1966; HGCMEO10). In {(phenoxathiin)$_2$$\cdots$HgCl$_2$} (McEwen and Sim, 1969; HGCPHO10) which also belongs to this structural group, the link to Hg is through S and not O and it was argued that this is evidence for a charge-transfer rather than an ion-dipole interaction (S has a lower ionization potential than O but will carry a smaller (partial) negative charge).

A number of variations are found in the coordination about Hg. For example, in both {azoxyanisole$\cdots$HgCl$_2$} (not described in detail) and {coumarin$\cdots$HgCl$_2$} (Kitaigorodskii, 1961) the quasi-octahedral coordination polyhedron is made up of two covalently bonded chlorines, three chlorines at $\approx$3.2 Å and only one oxygen, at $\approx$2.6 Å. Another example is found in {tetrahydrofuran$\cdots$HgBr$_2$}, (Frey, Leligny and Lederset, 1971; THFHGB) where HgBr$_2$ ribbons of the familiar type are found (with d(Hg–Br) = 2.48 Å; d(Hg$\cdots$Br) = 3.15, 3.27 Å), the fifth site is occupied by the oxygen of tetrahydrofuran (d(Hg$\cdots$O) = 2.67 Å) and the sixth site is occupied by a Br from an adjacent ribbon, with d(Hg$\cdots$Br) = 3.47 Å. Quasi-octahedral coordination and the familiar HgCl$_2$ ribbons are found in {diphenyl sulphoxide$\cdots$HgCl$_2$} (Biscarini *et al.*, 1973; DPSOHG) but the sixth coordination site is occupied by a phenyl ring whose centre is 3.51 Å from the Hg atom (Fig. 11.33; the interaction is η^6). There is a somewhat similar (but η^2) benzene$\cdots$Hg interaction in {CoHg$_2$(SCN)$_6$·C$_6$H$_6$} (Grønbaek and Dunitz, 1964).

The overall arrangement in the crystal depends on a number of factors. Monofunctional ligands, as in {diphenyl sulphoxide$\cdots$HgCl$_2$} (Fig.11.33), give single ribbons in the crystal. If the ligand is bifunctional, as in {pyridine-N-oxide$\cdots$HgCl$_2$} (Sawitzki and

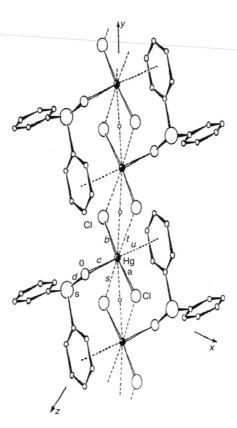

Fig. 11.33. Clinographic projection of chain in the structure of the polymeric adduct {diphenyl sulphoxide···HgCl₂} ($a = 2.291(4)$, $b = 2.289(4)$, $c = 2.58(1)$, $s = 3.230(6)$, $t = 3.284(5)$ Å). The Cl–Hg–Cl angle is 172.4(1)°. The phenyl ... Hg interactions are shown by broken lines ($u = 3.51(1)$ Å). (Reproduced from Biscarini *et al.*, 1973.)

von Schnering, 1974) then sheets are formed in the crystal (see below). The oxygen has a single secondary interaction in the first of these examples and two in the second. Analogous double oxygen interactions are also found in the polymeric structure of {2,2′-trimethylenedipyridine-1,1′-dioxide···(HgCl₂)₂} (Alshaikh-Kadir *et al.*, 1978) which has the familiar ribbons of chlorine-bridged HgCl₂ molecules.

2,2′-trimethylenedipyridine-1,1′-dioxide

In {cyclohexa-1,4-dione ... HgCl₂} (Groth and Hassel, 1964b; CHEXHG; $a = 7.57$, $b = 17.11$, $c = 7.56$ Å, $\beta = 108.5°$, $C2/c$, $Z = 4$)) the Hg atom lies on a crystallographic

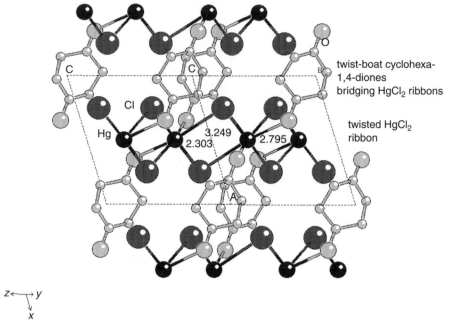

Fig. 11.34. View of a single sheet of the {cyclohexa-1,4-dione···HgCl$_2$} crystal structure projected approximately down [011]. The large dark spheres are Cl, medium dark spheres Hg, smaller lighter spheres O and C. The linked HgCl$_2$ moieties are sandwiched between the bridging cyclohexa-1,4-dione molecules, giving the sheets a hydropohobic exterior. (Data from Groth and Hassel, 1964b.)

twofold axis and is octahedrally coordinated by two Cl's at 2.30 Å, <Cl–Hg–Cl = 173°, two O's at 2.79 Å and two Cl's of neighbouring HgCl$_2$ molecules at 3.25 Å (note that Graddon's Fig. 2 has two oxygens instead of chlorines but the text is correct). The cyclohexadiones have a twist-boat conformation and are also located about two fold axes. The overall structure consists of a series of parallel layers along the [010] axis (Fig. 11.34) with juxtapositioning of the cyclohexa-1,4-dione molecules. The HgCl$_2$ ribbon, is not planar but twisted, and the two cyclohexa-1,4-dione molecules linked to a particular octahedrally coordinated Hg are *cis* to each other. The twist-boat cyclohexadiones bridge between the twisted HgCl$_2$ ribbons to form sheets in the (010) planes, with the cyclo-hexadiones on their outer surfaces.

{Pyridine-N-oxide···HgCl$_2$} (Sawitzki and von Schnering, 1974; PYNOHG) has an interesting structure (Fig.11.35) that is a variation on the theme of HgCl$_2$ ribbons. There are indeed such ribbons but they have a sawtooth rather than planar shape. The liganded oxygens of the pyridine-N-oxide molecules are bicoordinate and two such molecules link between adjacent sawtooth HgCl$_2$ ribbons. The consequence is a layer structure. The geometry of the coordination about Hg is approximately pentagonal bipyramidal with one equatorial site vacant (however, there is no lone pair available to occupy this site). {Pyridine-N-oxide···CdI$_2$} (PYNOCD) has an analogous structure

In the structures discussed until now the integrity of the HgX$_2$ molecule has been clearly preserved. However, this feature disappears gradually in a number of structures.

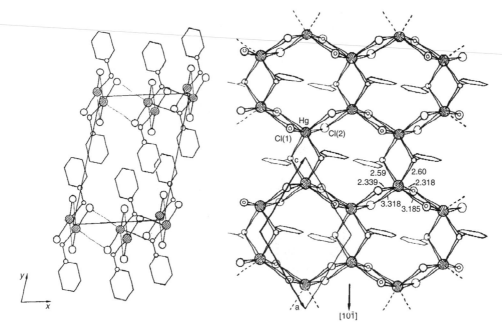

Fig. 11.35. The crystal structure of {pyridine-N-oxide···HgCl$_2$} (7.060 10.132 6.890 Å 105.28 117.04 75.35°, $P\bar{1}$, $Z = 2$; this is the reduced cell with a nonstandard choice of orgion). The projection on (001) is on the left and shows the layers in the (010) planes, while the projection on (010), shown on the right, illustrates the structure of a layer. Hg atoms above and below the mean planes have different shadings. Although the crystals are triclinic, the a and c directions are closely equivalent structurally. Tests show that the space group is correct. The angle Cl(1)-Hg-Cl(2) = 164(1)°. (Reproduced from Sawitzki and von Schnering, 1974.)

For example, in {bis(thiosemicarbazide)···HgBr$_2$} (Chieh, Lee and Chin, 1978; TSCHGB) there are HgBr$_2$ ribbons of the type shown above (at the beginning of Section 11.9.2(i)) but with a tendency towards equality of the Hg–Br links; complete equality is not yet attained as d(Hg–Br) = 2.860(4), 3.436(4) Å (cf. values of d(Hg–Br) in Table 11.22). There is a rather strong Hg–S interaction ($d = 2.45(1)$ Å) normal to the plane of the ribbon. The corresponding HgCl$_2$ compound (Chieh, 1977b; CTSCHG) is isomorphous and has a similar coordination arrangement (d(Hg–S) = 2.417(3), d(Hg–Cl) = 2.821(3), 3.250(3) Å). (Mono(thiosemicarbazide)···HgCl$_2$} has five-coordination and is discussed later; Graddon (1982, p. 246) has confused the coordinations in the mono-and bis-liganded HgCl$_2$ compounds. In {bis(pyridine)···HgCl$_2$} (Canty *et al*, 1982; ZZZLCI10) the two independent Hg–Cl distances are nearly equal at 2.754(2) and 2.765(2) Å, and there is strong Hg to N bonding (d(Hg . . . N) = 2.266(6) Å). The Hg–Cl distances are close to the mean of the short (≈ 2.3 Å) and long (≈ 3.2 Å) contacts encountered previously. The ribbons of Hg-centered octahedra remain. A similar structure is found in {bis(thioacetamide)···HgCl$_2$} (Rolies and De Ranter, 1977; TAMHGC) (Fig.11.36); there is a linear S-Hg-S arrangement, with stronger Hg to S interaction ($d = 2.39$ Å) than Hg to Cl (d = 3.07 Å).

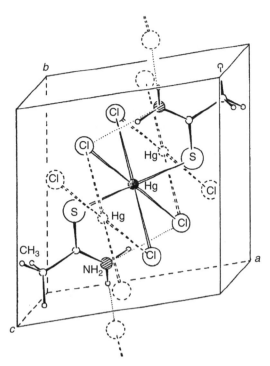

Fig. 11.36. {Bis(thioacetamide)···HgCl$_2$}. The crystals are triclinic, space group $P\bar{1}$. Note the N–H···Cl hydrogen bond in the upper and lower parts of the figure. (Reproduced from Rolies and de Ranter, 1977.)

HgCl$_2$ ribbons or sheets with $(2 + 4)$ or $(2 + 2 + 2)$ quasi-octahedral coordination are found in many crystal structures, not only those of addition compounds; for example ribbons are found in K$_2$HgCl$_4$·H$_2$O and sheets in NH$_4$HgCl$_3$ (Wells, 1975). Almost planar HgI$_2$ ribbons are found in {bis(thiourea)···HgI$_2$} (*Pnna*, $Z = 4$; Korczynski, 1968; HGITUR), with the thiourea molecules coordinated to Hg (d(Hg . . . S) = 2.46 Å) and extending above and below the planes of the ribbons. However, the angle between the two covalent Hg-I bonds is 102° and no semblance of a linear HgI$_2$ entity remains.

In contrast to these polymeric structures, the {bis(pyridine)···HgX$_2$} (X = Br (BEJLAD), I (BEJLEH)) compounds are monomeric with approximately tetrahedral coordination about Hg (Canty *et al.*, 1982), but not isomorphous.

(ii) Structures based on ribbons with weak Hg···O or Hg···S interactions   Bridging in ribbons sometimes occurs through O or S atoms rather than through Cl – an essential

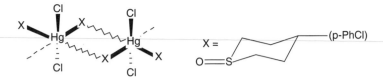

Fig. 11.37. Schematic structure of the centrosymmetric dimeric molecule {bis[(cis-4-*p*-chlorophenylthian oxide)···HgCl$_2$]. The moiety X is linked to Hg through the oxygen of the thione group, with d(Hg···O) = 2.48 and 2.97 Å in the bridge and 2.70 Å externally. The HgCl$_2$ molecule is only slightly distorted, with d(Hg–Cl) = 2.29 Å and <Cl-Hg-Cl = 164°. The quasi-octahedral coordination about the Hg atoms is completed (geometrically, if hardly electronically) by remote approaches from X moieties through Cl atoms, with d(Hg···Cl) = 3.88 Å, which is appreciably larger than the sum of the van der Waals radii.

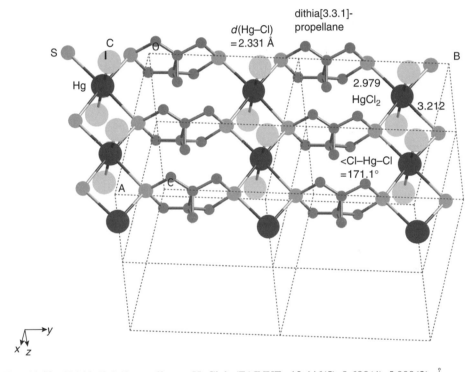

Fig. 11.38. {Dithia[3.3.1]propellane···HgCl$_2$} (FAJNUZ; 12.446(5) 8.680(4) 5.000(3) Å. space group *Pmmn*, Z=2), showing one sheet of HgCl$_2$ molecules loosely linked by Hg···S interactions. The Hg and Cl atoms lie in the (100) mirror plane, to which the Hg–Cl vectors are nearly normal. The largest spheres are Cl, with Hg and S intermediate in size. (Data from Herbstein, Ashkenazi *et al.*, 1986.)

requirement for inclusion of a structure in this group is that the O or S should be linked to two different Hg atoms, generally through secondary linkages which are not necessarily of equal strength; one consequence is that the Hg–Cl links will be normal to the plane of the ribbon. One of the relatively few examples studied is {bis(biuret)···HgCl$_2$(Birker *et al.*,

1977; BURHGC) (biuret is $NH_2CONHCONH_2$), where the $HgCl_2$ molecules are hardly perturbed and ribbons are formed containing quasi-octahedral Hg(II), with the bridging through amide oxygens of four biuret molecules. The biuret ligands are monodentate but each coordinating oxygen is linked to two Hg atoms $(d(Hg\cdots O) = 2.76, 2.95$ Å); the Hg–Cl distance is normal at 2.30 Å. The delicacy of the balance between different arrangements is shown by the occurrence of another isomer with chlorine bridging between Hg atoms (Nardelli and Chierci, 1960). Another example, but where discrete dimeric molecules are formed rather than infinite chains, is {bis[(cis-4-*p*-chlorophenylthian oxide)$\cdots HgCl_2$}, (McEwen *et al.*, 1967; HGCTOX) shown in Fig. 11.37.

Another example, this time involving bridging through sulfur, is found in dithia[3.3.1]propellane (Herbstein *et al.*, 1986; FAJNUZ) which forms sheets of composition $C_7H_{10}S_2\cdots HgCl_2$. The essential structural element is a ribbon where the sulfurs of the propellane moiety bridge between $HgCl_2$ molecules (with $d(Hg–Cl) = 2.330$ Å and $<Cl–Hg–Cl = 171.1°$), the interaction being rather weak as $d(Hg\cdots S) = 2.978, 3.212$ Å. As the propellane molecules contain two thiolane rings they bridge between ribbons to give a sheet structure (Fig. 11.38).

The structure of {2[$(CH_3)_2SO]\cdots 3[HgCl_2]$} (triclinic, $P\bar{1}$, Z = 1) (Biscarini *et al.*, 1974; DMSOMC) can be described in terms of $HgCl_2$ molecules and dimeric [$(CH_3)_2SO\cdots HgCl_2$] moieties (Fig. 11.39; Graddon (1982) incorrectly refers to this compound as having a 1:1 stoichiometry). The $HgCl_2$ molecule is at a crystallographic

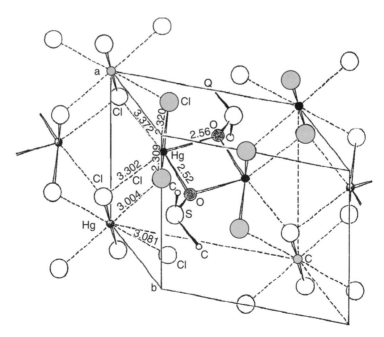

Fig. 11.39. Clinographic projection of the {2[$(CH_3)_2SO]\cdots 3[HgCl_2]$} structure ($a = 6.672$, $b = 9.286$, $c = 8.764$ Å, $\alpha = 60.0$, $\beta = 95.5$, $\gamma = 90.1°$, $P\bar{1}$, Z = 1; the reduced cell is 6.672 8.764 9.036 Å 62.87 84.77 84.50°). One of the $HgCl_2$ molecules and one of the oxygen-linked dimers is emphasized; both groupings are centrosymmetric. (Reproduced from Biscarini *et al.*, 1974.)

centre of symmetry and hence must be linear; d(Hg-Cl) = 2.30 Å. In addition there are two pairs of Cl atoms, at 3.00 Å, belonging to adjacent dimer moieties and these complete the oblate quasi-octahedral coordination which is of the frequently found 2(short) + 4(long) type. The second $HgCl_2$ molecule, in a general position, approaches an ideal arrangement (d(Hg-Cl) = 2.309, 2.320 Å and <Cl–Hg–Cl = 166°). However, there are also two close approaches from O atoms of the sulphoxides (d(Hg···O) = 2.52, 2.56 Å). The highly distorted octahedral coordination about this Hg is completed by two chlorines of adjacent $HgCl_2$ molecules (at centres) with d(Hg...Cl) = 3.302, 3.372 Å, distances close to the van der Waals' distances. There is a thiourea compound of $HgCl_2$ with the same unusual 2:3 stoichiometry, which is monoclinic with space group $P2_1/c$, $Z = 2$ (Brotherton and White, 1973; HGCLTU). There are some structural similarities as $[2\{[NH_2]_2CS\}\cdots 3(HgCl_2)]$ also has a linear $HgCl_2$ molecule at the origin of the unit cell. The thioureas and second $HgCl_2$ molecule are linked in a chain with a number of secondary interactions. Many features appearing singly in other structures are combined in these two compounds.

(iii) Ribbon structures based on the polyfunctionality of ligand molecules The prototype structure in this group is that of {bis(dioxane)···HgBr$_2$} (Frey and Monier, 1971; HGBDOX). Each $HgBr_2$ molecule is essentially unperturbed, the coordination about Hg being completed by four oxygens of four different dioxane molecules, each dioxane being bonded to two different $HgBr_2$ molecules. The Hg atoms are in the same plane so that sheets of composition $HgBr_2\cdots$(dioxane)$_2$ are formed (Fig. 11.40). There are no weak Hg···Br interactions in this structure and the oxygens are each linked to only one Hg atom. The structure of the $HgCl_2$ analog does not appear to have been reported.

The arrangement in {*trans*-naphthodioxane···HgCl$_2$} (Hassel and Rømming, 1956; only one projection was studied but a diagram was given later by Branden (1964c) appears to be similar in principle – each oxygen is linked to only one Hg atom. Here too the

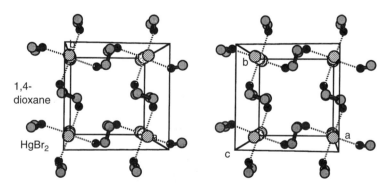

Fig. 11.40. Stereodiagram of one layer in the structure of {bis(dioxane)···HgBr$_2$} viewed down the [001] axis of the tetragonal crystals ($a = b = 7.454$, $c = 12.439$ Å, $I4/m$, $Z = 2$). d(Hg–Br) = 2.433 Å and d(Hg...O) = 2.83 Å, which is only slightly less than the sum of the van der Waals radii. The dioxanes are in the chair conformation. The layers, parallel to (001), are stacked one above the other along [001] but are mutually offset by $1/2(\mathbf{a} + \mathbf{b} + \mathbf{c})$. Hg – partly hidden open circles, Br diagonally hatched, C cross-hatched, O dark dotted circles. (Reproduced from Frey and Monier, 1971.)

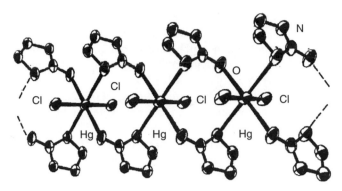

Fig. 11.41. Ribbon structure in {bis(2-imidazolidinone)$\cdots$HgCl$_2$}. (Reproduced from Majeste and Trefonas, 1972.)

oxygens of the donor molecule interact with different HgCl$_2$ molecules and ribbons are formed. A difference between these ribbons and those discussed in Section 11.6.2.1(ii) is that the latter are based on bicoordinate oxygens. A third example is given by the structure of {bis(2-imidazolidinone)$\cdots$HgCl$_2$} ((Majeste and Trefonas, 1972; ETURHG) (2-imidazolidinone is also known as ethylene urea), where there is quasi-octahedral coordination about Hg with essential preservation of the HgCl$_2$ molecules (d(Hg–Cl) $= 2.309(4)$ Å) and formation of ribbons with weak Hg$\cdots$O ($d = 2.67(1)$ Å) and Hg$\cdots$N (d $= 2.95(1)$ Å) interactions (Fig. 11.41); the coordination about Hg is 2Cl + 2O + 2NH, in descending order of interaction strength. Formation of ribbons is a consequence of the bifunctionality of the organic ligand.

A twisted chain structure is also found in {(S)-(−)-n-butyl-t-butyl sulfoxide HgCl$_2$} (Drabowicz *et al.*, 2001; BABNAU; absolute cofiguration determined). This has some unusual crystallographic features; the space group is $P1$ and the unit cell contains 4(C$_8$H$_{18}$OS) + 5 HgCl$_2$. Four HgCl$_2$ molecules are linked to four molecules of the sulfoxide by Hg–O links, Hg thus being 6-coordinate, with Hg–O distances varying from 2.366 to 2.533 Å. The remaining HgCl$_2$ molecule forms coordination bonds to Cl's of the neighbouring HgCl$_2$ molecules with Hg . . . Cl distances of 3.173 and 3.216 Å. The overall result is a twisted chain sheathed on both sides by hydrophobic sulfoxide molecules. These chains interact by van der Waals forces.

A stepped chain is found in {(S)$_S$(S)$_C$(−)$_{589}$-methyl 2-phenylbutyl thioether HgCl$_2$} (Biscarini and Pelizzi, 1988; GOKFAN) (space group $P2_1$, $Z = 2$; absolute configuration determined). The authors comment that there is "a polymeric structure in which one of the two crystallographically independent chlorine atoms unsymmetrically bridges three adjacent symmetry-related Hg atoms. The overall coordination around Hg is not simple to interpret, as the ligand arrangement cannot be described in terms of a regular polyhedron." Hg has two close neighbours, S at 2.433(4) Å and Cl(1) at 2.315(5) Å, with <S–Hg–Cl(1) $= 149.8(2)°$; thus no sign remains of the original linear HgCl$_2$ molecule. Also there are three Cl(2) neighbors to Hg at 2.677(5), 2.838(10) and 2.995(10) Å. The sixth coordination site is a weak link to the phenyl ring (≈ 3.6 Å). Overall, the stepped chain is sheathed by the phenyl and methyl groups.

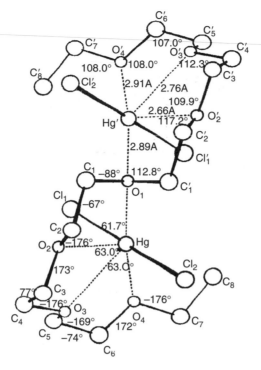

Fig. 11.42. Diagram of the molecular unit in {hexaethylene glycol diethyl ether·2(HgCl$_2$)} (compound III); d(Hg–Cl) = 2.298, 2.319 Å, <Cl–Hg–Cl = 175.9°. The crystals are monoclinic, $C2/c$, Z = 4. (Reproduced from Iwamoto, 1973c.)

(iv) Formation of discrete molecular structures   There are a number of examples where HgCl$_2$ interacts with a particular ligand to form a discrete molecular entity of composition nL·HgCl$_2$ (n a small integer) with quasi-octahedral coordination about Hg. In earlier work complexes of polyethylene oxide had been made with HgCl$_2$, compositions with CH$_2$CH$_2$O : HgCl$_2$ ratios of 1 : 1 and 1 : 4 being obtained. However, these results will not be discussed as the structural inferences drawn from their fibre diagrams appear less certain than later conclusions from analyses of single crystals of simpler analogs. Thus HgCl$_2$ forms 1 : 1 molecular compounds with the dimethyl and diethyl ethers of tetra-ethylene glycol, the compositions being RO(CH$_2$CH$_2$O)$_4$···HgCl$_2$ (R = CH$_3$ (compound I; MTEGMC) (Iwamoto, 1973a), C$_2$H$_5$ (compound II; ETEGMC)) (Iwamoto, 1973b), and with the diethyl ether of hexaethylene glycol, the composition being {[EtO (CH$_2$CH$_2$O)$_2$CH$_2$CH$_2$OCH$_2$CH$_2$(CH$_2$CH$_2$O)$_2$Oet]···2HgCl$_2$} (III) (Iwamoto, 1973c; EHXGMC). (II) was disordered but single-crystal structural results were obtained for (I) and (III), showing the Hg atoms of the HgCl$_2$ molecules to be encompassed by 5 and 4 oxygen atoms respectively, the oxygens being arranged in planar hexagons, with one and two sites vacant; details are given for III (Fig. 11.42).

   The structure of {1,15-bis(2-bromophenyl)-2,5,8,11,14-pentaoxapentadecane···HgCl$_2$} (Weber, 1980; BPOHGB) is very similar to that of (I) above as is dibenzo-18-crown-6···HgCl$_2$, where the linear HgCl$_2$ molecule in the centre of the crown ether is surrounded

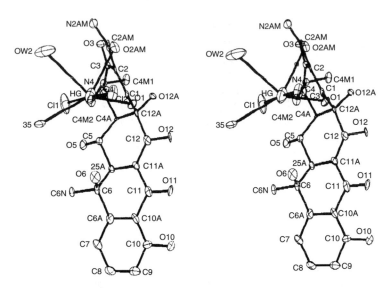

Fig. 11.43. Stereoview of the oxytetracycline···HgCl$_2$ interaction, viewed approximately normal to the molecular plane. (Reproduced from Jogun and Stezowski, 1976.)

by six oxygens in the equatorial plane (Kawasaki and Matsuura, 1984; COCGAC). A somewhat similar distorted octahedral arrangement of oxygens about Hg is found in the dihydrate compound of zwitterionic oxytetracycline with HgCl$_2$ ({C$_{22}$H$_{24}$N$_2$O$_9$···HgCl$_2$ ·2H$_2$O}) (Jogun and Stezowski, 1976; OXTETH10). The Hg atom is surrounded by two chlorines (d(Hg···Cl) = 2.80 Å; Cl···Hg···Cl = 169°) and four oxygens (distance of Hg to O(1) is 2.80 Å, to O(2am) 2.65 Å, to OW(2) 3.025 Å, and to O(3′) 3.22 Å (i.e. O(3) of a neighbouring molecule) Fig. 11.43). There are no weaker Hg···Cl interactions in any of these three structures.

11.9.2.2 *Five-coordinate Hg(II)*

Although only relatively few structures containing five-coordinate Hg(II) are known, it is possible to distinguish between those where the Hg has square pyramidal coordination and those where there is trigonal biyramidal coordination about Hg. The relation between octahedral and square pyramidal coordination is clearly shown in {3,5-dibromopyridine-N-oxide···HgCl$_2$} (Genet and Leguen, 1965; HGCOPO10) where ribbons of HgCl$_2$ molecules are found with dimensions rather similar to those encountered previously (d(Hg–Cl) = 2.29, 2.31 Å, <Cl–Hg–Cl = 173.5°. d(Hg···Cl) = 3.12, 3.21 Å). A fifth coordination site is occupied by oxygen of the organic ligand, with d(Hg . . . O) = 2.51 Å, which is somewhat shorter than values cited above. The next closest neighbor to Hg is at 3.45 Å so the coordination about Hg is distorted square pyramidal. Another weak stabilizing interaction in this compound is between O and Br; the approach distance is 3.06 Å, close to values found in some self-complexes (Section 11.2.4); the oxygen coordination is trigonal. Distorted square pyramidal coordination is also found in {di-*n*-butyl sulphoxide···HgCl$_2$} (Biscarini *et al.*, 1981; BSHGCL) where HgCl$_2$ ribbons

are again found (d(Hg–Cl) = 2.33, 2.35 Å, <Cl–Hg–Cl = 167.0(5)°) with the organic ligand linked through oxygen (d(Hg–O) = 2.59 Å) only to *one* side of the ribbon. In {2,4-dimethylpyridine···HgBr$_2$} (Bell *et al.*, 1980a; BDMPHG) there are rather unsymmetrical HgBr$_2$ ribbons (d(Hg–Br) = 2.49 (to Br(1)), 2.62 (to Br(2)), 2.91 and 3.55 Å, <Br(1)–Hg–Br(2) = 122.2(2)°) and distorted square pyramidal coordination about Hg, with the organic ligands directed alternately above and below the plane of the HgBr$_2$ ribbons. The HgCl$_2$ compound (CDMPHG) is isomorphous.

There is also trigonal bipyramidal coordination about Hg(II) in {collidine···HgCl$_2$} (collidine is 2,4,6-trimethylpyridine) (Fig. 11.44; Kulpe, 1967; MCOLID). The Hg–N distances are close to those characteristic of covalent interactions (Table 11.22) while the covalent Hg–Cl interactions are weaker, and the secondary interactions stronger than those found in the chains of octahedrally coordinated Hg. In particular there is no retention of a linear HgCl$_2$ molecule – indeed the Cl–Hg vectors are nearly mutually perpendicular. A rather similar arrangement is found in {triethylphosphine···HgCl$_2$} (Bell *et al.*, 1981; ETPHGC10).

There are discrete, centrosymmetric, bridged dimers in mono(thiosemicarbazide)dichloromercury(II) (Chieh and Cowell, 1977) based on similar structural principles (Fig. 11.45); the coordination about Hg is trigonal bipyramidal when all interactions are taken into account. Again there is no sign of a linear HgCl$_2$ molecule <Cl(1)–Hg–Cl(2) = 96.1°. The molecules are linked by weak N–H···Cl hydrogen bonds. There are similar discrete dimeric bridged moieties in {2,2′-bipyridine···HgBr$_2$} (Craig *et al.*, 1973) with trigonal bipyramidal coordination about Hg(d(Hg–Br) = 2.51 (terminal), 2.61, 3.14 Å (bridging); d(Hg–N) = 2.37, 2.40 Å).

In contrast to {triethylphosphine···HgCl$_2$} the corresponding methyl compound (Bell *et al.*, 1981; MPHGCL10) has been formulated as ionic [ClHgP(CH$_3$)$_3$]$^+$ Cl$^-$; cations and

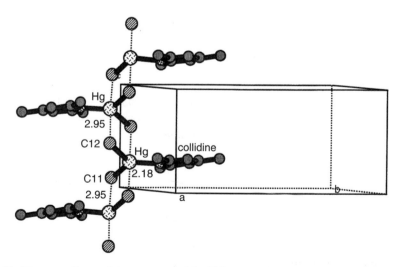

Fig. 11.44. Diagram of part of the structure of {collidine···HgCl$_2$} (a = 8.15, b = 17.81, c = 7.44 Å, β = 92.6°, $P2_1/c$, Z = 4). Some additional dimension (Å; deg.) are: Hg–Cl(1) = 2.455; Hg–Cl(2) = 2.542; Hg–N = 2.18; <Cl(2)–Hg–Cl(1) = 110.1. The overall structure consists of symmetry-related columns along [001]. (Data from Kulpe, 1967.)

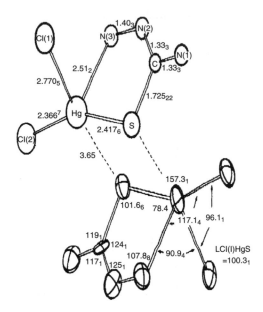

Fig. 11.45. View of the dimer in {mono(thiosemicarbazide)dichloromercury(II)}. (Reproduced from Chieh and Cowell, 1977.)

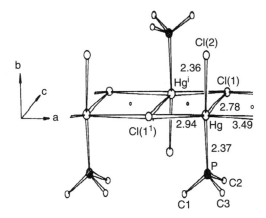

Fig. 11.46. The linear polymeric ladder structure of {trimethylphosphine···HgCl$_2$} (triclinic, $P\bar{1}$, $Z=1$; MPHGCL10). Atoms of the chain are coplanar to within 0.14 Å. (Reproduced from Bell *et al.*, 1981.)

anions are arranged in a polymeric ladder (Fig. 11.46) and the distinction between an ionic and charge-transfer formulation would appear to be arbitrary. Perhaps the compound could be more reasonably described as a linear array of weakly interacting dimers of composition [Cl$_2$HgP(CH$_3$)$_3$]$_2$, containing five-coordinate square pyramidal Hg, the weak Hg···Cl interaction (d(Hg···Cl) = 3.489(4) Å) being taken into account in this description. Br$_2$HgP(CH$_3$)$_3$ appears to be isostructural with the chloro compound, but the iodo compound is not. There is also a structural resemblance to {(tetrahydrothiophene)·HgCl₂} (Branden, 1964b; HGCTHS).

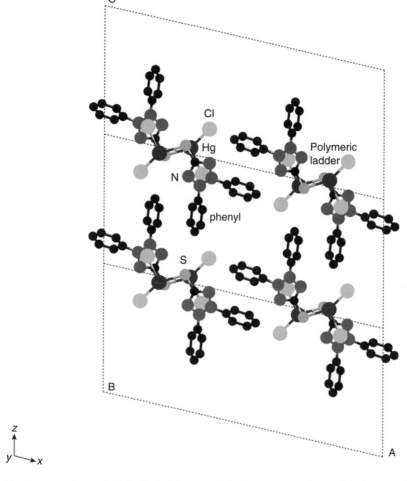

Fig. 11.47. The structure of {dehydrodithizone···HgCl₂}; projected down [010], the polymeric ladder direction. The crystals are monoclinic, $a = 25.717$, $b = 6.476$, $c = 11.476$ Å, $\beta = 102.79°$, corrected space group $C2/m$, $Z = 4$. Some important dimensions (Å; deg.) are: Hg–Cl(1) = 2.34; Hg–Cl(2) = 2.57; Hg···S′ = 2.40; Hg···S = 3.14, 3.42; S′-C(ring) = 1.71 Å; <S′–Hg-Cl(1) = 152; <Cl(2)–Hg–S′ = 96; <Cl(2)–Hg–Cl(1) = 112°. (Adapted from Kozarek and Fernando, 1973.)

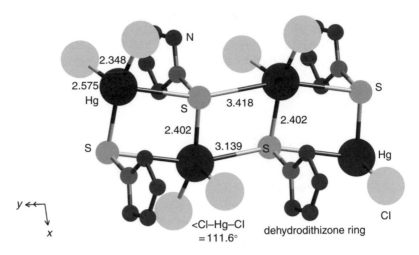

Fig. 11.48. The structure of {dehydrodithizone···HgCl$_2$}, showing detail of the polymeric ladders, projection down [01$\bar{1}$]. The two phenyl groups attached to each dehydrodithizone ring have been omitted for clarity. (Adapted from Kozarek and Fernando, 1973.)

Hg···Cl interactions are replaced by (rather weak) Hg···S interactions in the 1:1 adduct of dehydrodithizone with HgCl$_2$ (Kozarek and Fernando, 1973; DTIZHG10) (dehydrodithizone, C$_{13}$H$_{10}$N$_4$S, is the meso-ionic sydnone anhydro-5-mercapto-2,3-diphenyltetrazolium hydroxide). The original space group (C2) was corrected to C2/m by Marsh and Herbstein (1983) who also showed that there was residual solvent (identity not established) in the crystals.

The arrangement is shown in Figs. 11.47 and 11.48; the polymeric ladder structure is not quite planar as <S′-Hg···S is 162°. One Hg-Cl bond has lengthened appreciably to 2.57 Å while the other has a standard value; the HgCl$_2$ molecule is not linear, <Cl–Hg–Cl being 111.6°. There is no appreciable difference between the dimensions of the organic ligand before and after formation of the molecular compound. The Hg···S interactions are so weak that this material is perhaps in the border region between molecular compounds and chelate coordination complexes.

11.9.2.3 Four-coordinate Hg(II)

Four coordinate Hg(II) is found in a number of compounds of HgCl$_2$ and HgBr$_2$ with oxygen-containing and, especially, with sulphur-containing ligands. In the following it is convenient to distinguish between unbridged structures (i.e. with one Hg per molecule or structural unit, the latter phrase allowing for the possibility of polymeric arrangements in the solid state) and bridged structures, which range from discrete dimeric molecules to polymeric arrangements in the solid state. We first discuss the unbridged structures. Approximately square-planar coordination about Hg is found in {cyclononanone···HgCl$_2$} (Dahl and Groth, 1971; CNONHG). In this molecular compound, where the cyclononanone molecule is in a twisted chair boat conformation of approximately C$_2$–2 symmetry (both enantiomers are present in the centrosymmetric crystals), there are zigzag chains of composition C$_9$H$_{16}$O···HgCl$_2$ along [100]. The oxygen atoms are bicoordinate and linked to

Hg atoms of different $HgCl_2$ molecules with $<O \cdots Hg \cdots O \approx 83°$ and $d(Hg \ldots O) = 2.76$, 2.94 Å; the $HgCl_2$ molecules are essentially unchanged from their usual linear structure $(d(Hg-Cl) = 2.26(1), 2.28(1)$ Å; $<Cl-Hg-Cl = 177.5(5)°)$. There is similar approximately square-planar coordination about Hg in {ioxane $\cdots HgCl_2$} (Hassel and Hvoslef, 1954) (triclinic, $Z = 1$; only one projection was studied but a diagram was given by Branden (1964); the molecule is centrosymmetric and necessarily linear, $d(Hg-Cl) = 2.34$ Å; $d(Hg-O) = 2.66$ Å). This structure should be compared with that of {bis(dioxane) $\cdots HgBr_2$} (Fig. 11.40).

The arrangement about Hg is closer to tetrahedral (see diagram) in {(testosterone)$_2$ ... $HgCl_2$} (Cooper *et al.*, 1968; TESTHG) where discrete units of the molecular compound are formed. The Hg-Cl bond length (2.30 Å) is the same as in linear $HgCl_2$ but the deviations from an ideal tetrahedral arrangement are shown by the angles between the bonds), with $<HO \cdots Hg \cdots OH = 87°$, $Cl-Hg-Cl = 165°$, and $d(Hg \cdots O) = 2.48$ Å. The dihedral angle between the Cl-Hg-Cl and $O \cdots Hg \cdots O$ planes is 92°. There are no weak Hg $\cdots$ Cl interactions.

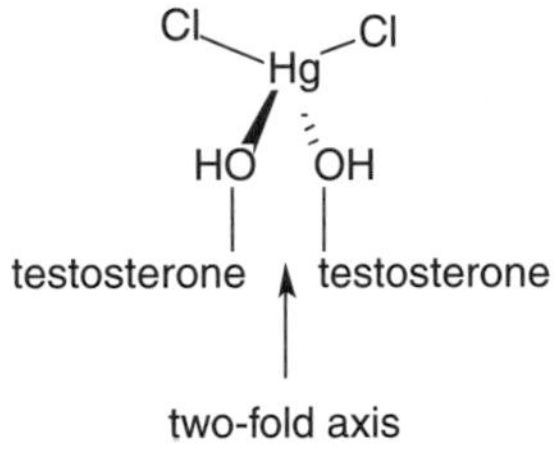

two-fold axis

A similar distorted tetrahedral arrangement is found in the isomorphous structures of {bis(triphenylarsine oxide) $\cdots HgCl_2$} (HGPARO) and {bis(triphenylphosphine oxide) $\cdots HgCl_2$} (Branden, 1963), where $d(Hg-Cl) = 2.32, 2.33$ Å, $d(Hg-O) = 2.32, 2.37$ Å, Cl-Hg-Cl = 146.6° and $<O-Hg-O = 92.5°$. Distorted tetrahedral coordination is also found in {1,10-phenanthroline $\cdots$ phenylmercury(II) cyanide} (Ruiz-Amil *et al.*, 1978; PCYPHG).

Much closer approximations to tetrahedral arrangements are found with sulphur-containing ligands; examples are {bis(1,4-thioxan) ... $HgCl_2$} (McEwen and Sim, 1967; HGCTOX) and the tetra-alkylthiuram compounds of $HgBr_2$ and HgI_2, (BETHSH) (Chieh, 1978; BTETHG; see also IHGTUR, IMTSHG), all of which crystallize as discrete molecules. A rather similar arrangement is found in {1,4,8,11-tetrathiacyclotetradecane $\cdots (HgCl_2)_2$} (Alcock *et al.*, 1978; TTPHGC10).

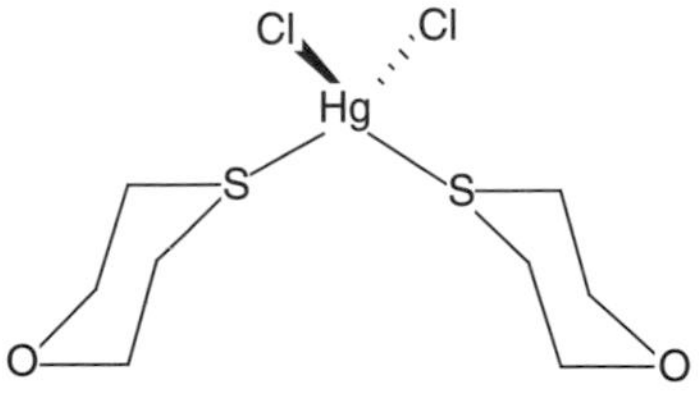

1,4-Dithioxan...$HgCl_2$
Cl-Hg-Cl = 114°; <S-Hg-S = 115°

The structure of {1,3,5-trithian $\cdots HgCl_2$} (first prepared in 1870 by Hofmann and Girard) is polymeric (Fig. 11.49; Costello *et al.*, 1966; HGTRIT) with tetrahedral coordination about Hg, the chain structure resulting from the bifunctionality of the trithian

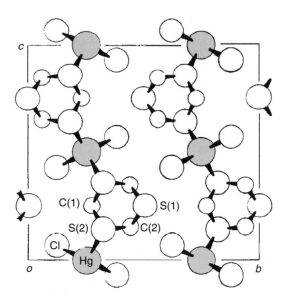

Fig. 11.49. Projection of the crystal structure of {1,3,5-trithian···HgCl$_2$} down [100] ($a=4.29$, $b=14.33$, $c=13.62$ Å, space group *Pbcm*, $Z=4$. There are zigzag polymeric chains along [001]. (Reproduced from Costello *et al.*, 1966.)

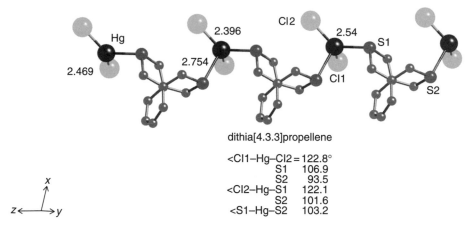

dithia[4.3.3]propellene

<Cl1–Hg–Cl2 = 122.8°
                      S1   106.9
                      S2   93.5
       <Cl2–Hg–S1   122.1
                      S2   101.6
       <S1–Hg–S2   103.2

Fig. 11.50. {Dithia[4.3.3]propellene···HgCl$_2$}. The structure is viewed down the conjoining (propellene) bond. The quasi-cylindrical chains are approximately close packed along [010]. The standard uncertainties of bond lengths were reported as ≈ 0.005 Å and of bond angles $\approx 0.2°$. (Data from Herbstein, Ashkenazi *et al.*, 1986.)

molecule (one S does not participate in intermoiety bonding). The trithian molecule is in the chair conformation and the S … Hg bonds are equatorial; d(Hg–Cl) $=2.44(1)$ Å, d(Hg-S) $=2.61(1)$Å, <Cl-Hg-Cl $=117(0.4)°$, <S···Hg···S $=113(0.4)°$ (Fig.11.49).

A linear polymeric structure analogous to that of {1,3,5-trithian···HgCl$_2$} is found in {dithia[4.3.3]propellene···HgCl$_2$} (FAJPAH; triclinic, $P\bar{1}$, $Z=2$, 11.765(5) 8.701(4)

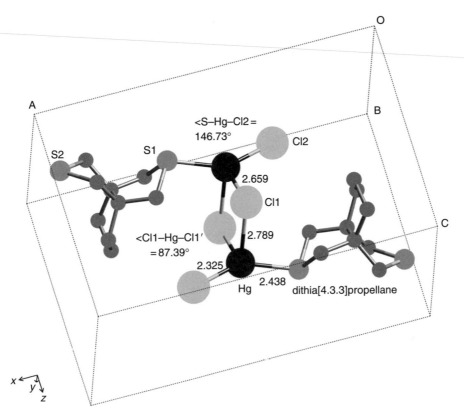

Fig. 11.51. The quasi-cylindrical dimeric centrosymmetric structural (packing) unit of {dithia[3.3.3]propellane···HgCl₂ is shown in the unit cell ($a = 14.320(6)$, $b = 9.762(4)$, $c = 9.058(4)$ Å, $\beta = 90.32(5)°$, $P2_1/c$, $Z = 4$). {Oxathia[4.3.3]-propellane···HgCl₂} ($a = 14.93$, $b = 9.42$, $c = 8.74$ Å, $\beta = 94.8°$, $P2_1/c$, $Z = 4$) is isostructural with {dithia[3.3.3]propellane···HgCl₂}; the two centrosymmetric dimers have similar shapes. Standard uncertainties are as given in the caption to Fig. 11.48. (Data from Herbstein, Ashkenazi *et al.*, 1986.)

7.552(40 Å, 115.33(5), 103.97(5) 83.18(5)°; Herbstein *et al.*, 1986; Fig. 11.50), where the chain nature follows from the bifunctionality of the propellene moiety; the interatomic distances in the two structures are similar but not identical.

A polymeric structure with a zigzag chain is found in {*catena*(μ-chloro)chloro (guanosine-N⁷) mercury(II),} where Hg is covalently bonded to Cl ($d = 2.339$ Å), and to N(7) of guanosine ($d = 2.16$ Å), and linked in chains by bicoordinate Cl (d(Hg ... Cl) $= 2.659$, 2.761 Å). This structure is enantiomorphic (space group $P2_12_12_1$, $Z = 4$) and the absolute configuration of the crystal used in the analysis was determined (Authier-Martin *et al.*, 1978; CLGUHG).

Bridging leading to formation of discrete dimers is found in the isomorphous 1 : 1 HgCl₂ molecular compounds with triphenylphosphine oxide (Branden, 1963; HGPARO) and triphenylarsine oxide ((Branden, 1964a; HGCPAO), where the bridging atoms are oxygen, in {1-methylcytosine···HgCl₂} (Authier-Martin and Beauchamp, 1977; MCYTHG) (where there are additional weak Hg···O interactions), and in {oxathia[4.3.3]propellane···HgCl₂}

(Herbstein *et al.*, 1986; FAJPIP) and {dithia[3.3.3]propellane$\cdots$HgCl$_2$} (FAJPEL; Herbstein *et al.*, 1986), where there are (nearly) symmetrical Cl bridges (Fig. 11.51). Similar structures are found in the 1 : 1 compounds of HgCl$_2$ with other ligands, variability being provided by the detailed dimensions and especially by the degree of asymmetry in the bridges. Four structures with symmetrical bridges are noted immediately above, while unsymmetrical bridges are found in the HgCl$_2$ compounds of 1,2,5-triphenylphosphole (2.54, 2.75(1) Å) (Bell *et al.*, 1980b; CLBHGC), triphenyphosphine selenide (2.60, 2.78 Å) and S-methylpyrrollidene-1-carbodithioate (2.57, 2.78 Å) (Brotherton *et al.*, 1974; CMPRTM). The enthalpy of formation of crystalline [Ph$_3$PO]$_2$ . . . HgCl$_2$ according to the following reaction

$$2Ph_3PO(s) + HgCl_2(s) \;=\; [(Ph_3PO)_2 \cdots HgCl_2](s)$$

has been measured by solution calorimetry as $-12.3{\pm}0.6$ kJ/mol at 298K (Jorge *et al.*, 1978).

A one-dimensional twisted-ribbon structure is found in the compound of composition {C$_{22}$H$_{31}$O$_2$S$\cdots$HgBr$_2$} formed between the thiosteroid spiroxazone (7α-(thio-acetyl)-(17R)-spiro[androst-4-ene-17,2(3H)-furan]) and HgBr$_2$ (Terzis *et al.*, 1980; SBHGAF), which was studied in connection with the possible use of thiosteroids as antidotes for mercury poisoning. The ribbon has both S and Br bridging, the distorted tetrahedral coordination about Hg giving rise to a twist of the ribbon. The dimensions of the steroid molecule are hardly affected by the interaction with HgBr$_2$ and it was inferred that it acts *via* direct complexation and removal of Hg rather than through a conformational change of the steroid which in turn was postulated to trigger some other defence mechanism.

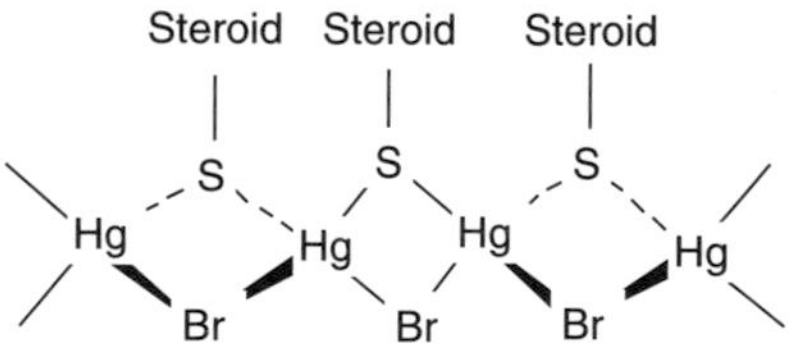

{Dibromo[N-(2-pyridyl)acetamide]mercury(II)} has a polymeric structure (Lechat *et al.*, 1980; BPACHG) in which the bidentate ligands are coordinated to Hg through pyridyl-N and acetyl-O of different molecules, forming a chain structure; d(Hg$\cdots$N) $=$ 2.49 Å, d(Hg–Br) $= 2.44$ Å, d(Hg$\cdots$O) $= 2.77$ Å, $<$O$\cdots$Hg$\cdots$N $= 85°$ and $<$Br–Hg–Br $= 158°$. There is also a weak secondary interaction ($d = 3.31$ Å) between Hg and Br atoms of different chains.

The free energy differences between different arrangements must be rather small, as is shown by the occurrence of a number of structures in which more than one stereochemical form appears. One example is the briefly-described 1,10-phenanthroline-trichloromercury chloride adduct (Redhouse, 1972; CMHGPN). This contains an ordered array of mono-mers and bridged dimers in 1 : 1 proportion and thus could also be classified as a packing complex (cf. Section 10.3.4.2). The monomer has a structure very similar to that of {1,10-phenanthroline.phenylmercury(II)cyanide} (d(Hg$\cdots$N) $= 2.53$, 2.61 Å), while the dimer is centrosymmetric and bridged through chlorine atoms as in triphenylarsine oxide$\cdots$HgCl$_2$; however, here the chlorine bridging is markedly unsymmetrical with d(Hg–Cl) $= 2.30$, 3.05 Å.

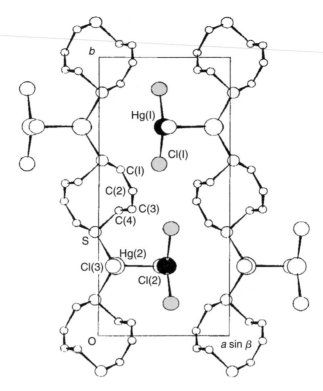

Fig. 11.52. Projection of crystal structure of {1,6-dithiacyclodeca-*cis*-3,*cis*-8-diene···(HgCl$_2$)$_2$} down [001] ($a = 7.29$, $b = 17.01$, $c = 6.20$ Å, $\beta = 92.7°$, space group $P2_1/m$, $Z = 2$). There are mirror planes perpendicular to [010] at $y = 1/4$, 3/4. Two of the HgCl$_2$ molecules are emphasized for clarity. (Reproduced from Cheung and Sim, 1965.)

Both tetrahedral Hg(II) and essentially undistorted HgCl$_2$ molecules are found in the polymeric structure of {1,6-dithiacyclodeca-*cis*-3,*cis*-8-diene···(HgCl$_2$)$_2$} (Cheung and Sim, 1965; HGCSCD10). The crystal structure (Fig. 11.52) shows that the tetrahedral Hg(2) has two Cl neighbors ($d = 2.50$, 2.51 Å) and two S neighbors ($d = 2.53$ Å). Hg(1) of the HgCl$_2$ molecule actually has five Cl neighbors, two close at 2.30 Å and three further away at 2.93, 3.02 and 3.21 Å. The moieties are arranged in ribbons along [010], with very weak interactions (d(Hg....Cl) = 4.44 Å) along [100] giving some degree of layer character about (001).

A combination of tetrahedral and octahedral coordination is found in {(iso-propylthio)HgCl$_2$} (Biscarini, Foresti and Pradella, 1984; CIDFAW). The space group is $C2$, with $Z = 2$; the absolute configuration of the crystal used does not appear to have been determined. The two independent Hg atoms are on two fold axes, one (Hg(1)) at Wyckoff position 2(a) (000) and the other (Hg(2)) at 2(b) (0,y,1/2); the other atoms are at general positions. Hg(1) has two S atoms covalently linked in a linear arrangement (d(Hg–S) = 2.378(6) Å, <S(1)–Hg(1)–S(1A) = 178.2(2)°; thus the linear HgCl$_2$ molecule has not been carried over to the solid state. Hg(1) forms four weaker coordinative bonds to chlorine (2.985, 3.342 Å), leading to (distorted) octahedral coordination for Hg(1)·Hg(2)

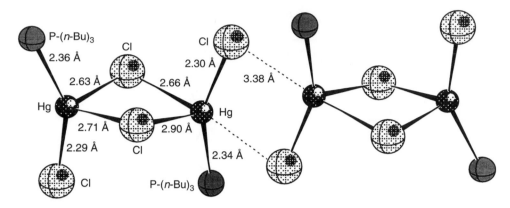

Fig. 11.53. The weakly linked dimer of the α-polymorph of {tri-n-butyl phosphine $HgCl_2$}, showing the dimensions. The n-butyl groups have been omitted for clarity. (Adapted from Bell *et al.*, 1980b.)

has tetrahedral arrangement of two S and two Cl atoms, covalently linked at 2.537(6) and 2.534(8) Å. Two dimensional networks are stacked about (100) planes, with the alkylic tails interspersed between the layers and perhaps determining the packing arrangement.

The discrete, centrosymmetric, molecules of $\{\alpha\text{-}[Bu_3P]\cdots HgCl_2\}$ (Bu $= n$-butyl (Bell *et al.*, 1980b; NBUPHG10 at 300K and NBUPHG11 at 100K) also show co-existence of 4- and 5-coordinate Hg(II) (Fig. 11.53).There is fivefold coordination about the central Hg and fourfold about the terminal Hg's; one notes also the combination of highly unsymmetrical and somewhat unsymmetrical Cl bridging. The central and outer rings are approximately mutually perpendicular. The tetrameric moieties are well separated in the crystals, which have the rather low melting point of 80.5°. There is also a β-polymorph (Kessler, 1977).

A comprehensive ^{35}Cl NQR study of $HgCl_2$ molecular compounds with organic donors has been carried out by Fichter and Weiss (1976) and some comparisons between NQR spectra and crystal structures are possible. Those compounds containing essentially unperturbed $HgCl_2$ molecules ({dioxane$\cdots HgCl_2$}; {bis(dioxane)$\cdots HgCl_2$}, assumed isomorphous with {bis(dioxane)$\cdots HgBr_2$}; {cyclohexa-1,4-dione$\cdots HgCl_2$}) all have ^{35}Cl NQR frequencies in the range 20–21.5 MHz. However, there are significant differences in regard to NQR frequency and its temperature dependence among the various compounds and it has not yet been possible to explain these effects in crystallographic or molecular terms. Tetrahydrothiophene$\cdots HgCl_2$, which has been assigned an ionic structure $[ClHgSC_4H_4]^+\cdot Cl^-$ by Branden (1964c), shows a single-line spectrum ($\nu(^{35}Cl)$ ≈ 20 MHz) due to the Cl bonded to Hg (d(Hg–Cl) $= 2.30$ Å), while the frequency due to the Cl^- ion was too low to be detected under the conditions of measurement. A noticeable difference was found for {1,3,5-trithiane$\cdots HgCl_2$} where there is quasi-tetrahedral coordination about Hg; $\nu(^{35}Cl) \approx 15$ MHz.

The adducts of $HgCl_2$ and analogs show a wide range of structural features, starting from retention of the linear $HgCl_2$ molecule, to the disappearance of this feature and its replacement by a variety of other structural features. Presence of linear $HgCl_2$ is an

Table 11.23. Thermodynamic quantities for the decomposition reactions of crystalline dioxane···HgX$_2$ molecular compounds at 298 K

Compound	$-\Delta G$ (kJ/mol)	ΔH (kJ/mol)	ΔS (J/mol deg)
dioxane···HgCl$_2$	22.7(4)	65.5(4)	297(5)
dioxane···HgBr$_2$	17.7	66.6	281
dioxane···HgI$_2$	11.4	58.3	233

essential requirement for conformation to our definition of molecular compounds, The other examples stray beyond our permitted limits to different extents; however, as we have noted, there are many advantages in discussing together the whole group of mercury(II) halide adducts.

Thermodynamic parameters (Table 11.23) have been determined (Barnes, 1972) for the decomposition reactions

$$\text{dioxane} \cdot \text{HgX}_2(\text{s}) \Rightarrow \text{HgX}_2(\text{s}) + \text{dioxane(g)} \quad (\text{X} = \text{Cl}, \text{Br}, \text{I}).$$

The structure of dioxane···HgI$_2$ is not known.

11.10   n-Donors and p-acceptors

11.10.1   *N, O, S containing ligands as donors and MX$_3$ (M = As, Sb; X = Cl, Br, I) as acceptors*

11.10.1.1   *SbX$_3$ (X = Cl, Br, I) as acceptor*

Although the first molecular compounds of aromatic hydrocarbons with SbCl$_3$ were made more than one hundred years ago (Smith, 1879; Smith and Davis, 1882),[2] the first systematic work was by Menschutkin (1910–1912) who studied, via binary phase diagrams, some 60 molecular compounds formed by SbCl$_3$ and SbBr$_3$ with alkylbenzenes, phenol, methoxy- and ethoxybenzenes, fluorobenzene, naphthalene, substituted naphthalenes and other aromatics. Later work has shown that there is a fairly clear distinction, both chemical and structural, between those molecular compounds where the interaction between ligand and SbX$_3$ is preferentially with the nitrogen or oxygen atoms of a substituted aromatic molecule (n-donor) and those where the interaction is preferentially with the aromatic ring of a ligand which is an unsubstituted aromatic hydrocarbon or one which has only alkyl substituents (π-donor). Thus we shall discuss these two types of interaction separately, with the p-donors discussed here and the π-donors in Section 11.11.1.

Menschutkin reported stable 1 : 1, 2 : 1, 3 : 1, 4 : 1 and metastable 6 : 1 molecular compounds of aniline with SbCl$_3$ and the structure of the first of these has been determined (ANISBC; Hulme and Scruton, 1968); there is also a brief report of the structure of the 2 : 1 complex

---

[2] "It was observed by one of us some time ago (this Journal, June 1879) that on melting a mixture of antimony trichloride and naphthalene, after removing the source of heat, a beautiful crystallisation commences in the still liquid mass, the minute but perfectly symmetrical clinirhombic crystals in their rapid growth performing during the process singular gyrations upon the liquid surface on which they float." The compositions given for the benzene and naphthalene molecular compounds with antimony trichloride were 2 : 3 instead of the 1 : 2 ratios of more recent work.

(ANILBC; Hulme, Mullen and Scruton, 1969). There are also 1 : 1 compounds with substituted anilines (May, 1911), *trans*-cinnamic acid and substituted cinnamic acids, bis(*p*-methylstyryl) ketone, terephthaldehyde, *p*-acetylbenzene, *o*-nitrocinnamaldehyde, α-methylcinnamaldehyde, *p*-phenyl-benzaldehyde, phenyl-benzoate, 5-methyl-1-phenyl-1-hex-3-one, 5-phenyl-2,4-pentadienophenone, *trans*-1,2-dibenzoylethylene, chalcone (Park, 1969); {(1,3,5-triacetylbenzene)···2SbCl$_3$} has also been reported (Park, 1969). These compounds are not particularly hygroscopic.

Six crystal structures have been reported where the donor atom is nitrogen or oxygen – these are the 1 : 1 compounds of SbCl$_3$ with aniline (noted above), terephthalaldehyde (TPHALD10) and *p*-diacetylbenzene (DACBZA10) (Baker and Wiiliams, 1972), and 2,2$'$-bipyridyl (Lipka and Wunderlich, 1980; BPYSBC10), the 1 : 2 compound of 1,3,5-triacetylbenzene with SbCl$_3$ (Baker, 1976) and the 3 : 2 compound of 4-phenylpyridine with SbCl$_3$ (Lipka, 1983; PYRSBA10). In aniline···SbCl$_3$ there is trigonal bipyramidal AX$_4$E coordination about Sb, with the lone pair (E) equatorial; d(Sb···N) = 2.53(4) Å, which is long compared to the Sb...N distance of 2.17 Å found in {S$_4$N$_4$...SbCl$_5$} (Neubauer and Weiss, 1960). In {terephthaldehyde···SbCl$_3$} the coordination about Sb is remarkably similar to that in SbCl$_3$ itself, except that the two shorter Sb...Cl secondary interactions in SbCl$_3$ are replaced by Sb···O interactions (two independent approaches, both 2.932 Å). The coordination type is thus 1 : 5:1 AX$_6$E, a distorted pentagonal bipyramid with the lone pair presumably located opposite one covalently-bonded Cl in an axial position. In {*p*-diacetylbenzene···SbCl$_3$} the coordination about Sb is distorted octahedral AX$_6$, with little surface area of the coordination polyhedron left over for the lone pair, which is thus probably not stereochemically active. The first neighbour atoms to Sb are three Cl (covalently bonded at 2.360, 2.385 and 2.372(2) Å), two O (at 2.668, 2.806(5) Å) and one Cl (at 3.213 Å). There is also quasi-octahedral coordination about the two Sb atoms in {1,3,5-triacetylbenzene···SbCl$_3$}, with average covalent Sb-Cl distance 2.38 Å; one Sb has three close oxygens (at 2.76, 3.04 and 3.07 Å) and the other has one O (at 2.84 Å) and two more remote chlorines (at 3.42, 3.44 Å). There is no significant

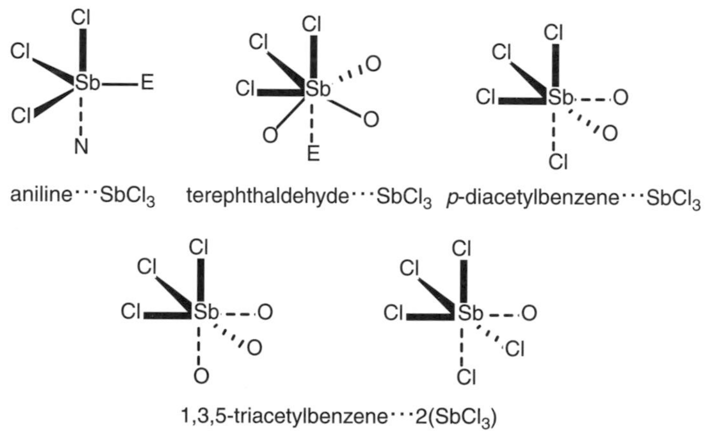

aniline···SbCl$_3$ terephthaldehyde···SbCl$_3$ *p*-diacetylbenzene···SbCl$_3$

1,3,5-triacetylbenzene···2(SbCl$_3$)

Fig. 11.54. Schematic diagrams of coordination arrangements in *p*-donor···SbCl$_3$ molecular compounds. Distances between atoms are given in the text.

interaction between Sb and aromatic ring in any of these compounds. These coordination polyhedra are summarized in Fig. 11.54.

The SbCl$_3$ moiety in {aniline$\cdots$SbCl$_3$} and in one of the crystallographically independent units of {1,3,5-triacetylbenzene$\cdots$2SbCl$_3$} behaves structurally as a 'pure acceptor' rather than as a 'self-interacting acceptor;' this again emphasizes that we introduce definitions and classifications for our own convenience and that Nature does not always follow the rules we may attempt to impose.

4-Phenylpyridine forms two molecular compounds with SbCl$_3$, with compositions 3:2 (melts congruently at 173–176.5°) and 2:1 (decomposes at 161°). Neither of these colorless materials is markedly hygroscopic. The structure of the 3:2 compound contains two structural units A and B, which are linked in alternating fashion in a chain by weak Sb$\cdots$Cl links (d(Sb$\cdots$Cl) = 3.59, 3.37 Å) and also laterally (d(Sb$\cdots$Cl) = 3.23 Å) to give a sheet-like structure. The structural units A and B can be represented as shown:

$$\text{A}: \quad \text{Cl}_3\text{Sb}\cdots\text{NC}_5\text{H}_4-\text{C}_6\text{H}_5 \quad d(\text{Sb}\ldots\text{N}) = 2.47\text{Å}$$

$$\text{B}: \quad \text{C}_6\text{H}_5-\text{C}_5\text{H}_4\text{N}\cdots\text{Sb}\cdots\text{NC}_5\text{H}_4-\text{C}_6\text{H}_5$$
$$\underset{\text{Cl}_3}{\overset{|||}{}}$$
$$d(\text{Sb}\cdots\text{N}) = 2.39\text{ Å}$$

The SbCl$_3$ moiety has its usual pyramidal shape in A, although the covalent Sb-Cl bond lengths (2.378, 2.399, 2.505 Å) suggest some change in the details of the bonding pattern; this is accentuated in B where the SbCl$_3$ moiety is now T-shaped and the covalent Sb–Cl bondlengths are even more different (2.361, 2.474, 2.736 Å). A similar situation occurs in 2,2'-bipyridyl$\cdots$SbCl$_3$ (Lipka and Wunderlich, 1980) (yellow, slightly hygroscopic crystals), which is a *chelate* coordination complex, the Sb$\ldots$N distances being 2.25(1) and 2.32(2) Å, with a T-shaped SbCl$_3$ moiety (Sb-Cl = 2.505, 2.549 and 2.588(6) Å). In our terms this material is not a molecular compound and would be better described as containing chelated molecules of Sb(C$_{10}$H$_8$N$_2$)Cl$_3$ linked by weak Sb$\ldots$Cl interactions (d = 3.34 Å). These structural differences are presumably a consequence of the small increments in basicity of the nitrogens as one passes from the anilines through 4-phenylpyridine to 2,2'-bipyridyl.

11.10.1.2 *AsCl₃ as an acceptor*

The structure of only one AsCl$_3$ compound has been determined, that with 2,2'-bipyridyl (Cameron and Killean, 1972; ASCDPY). The structure is ionic with (2,2'-bipyridyl–AsCl$_2$)$^+$ cations and chloride anions. There is no interaction between aromatic ring and As. This material is not a molecular compound in our present terms. Less basic ligands do not appear to have been studied.

11.10.1.3 *{S₈⋯SbCl₃}*

This molecular compound (Müller and Mohammed, 1983), prepared by the reaction of CS$_2$ with SbCl$_5$ at 5 °C, is considered separately as there are structural resemblances to some aromatic hydrocarbon$\cdots$SbX$_3$ molecular compounds discussed later (Section 11.11.1.2). The Raman spectrum of the molecular compound was found to be the superposition of the spectra of the individual components, indicating very small mutual

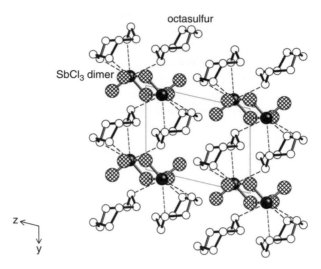

Fig. 11.55. Crystal structure of {$S_8\cdots SbCl_3$} at 268K projected down [100]. The secondary interactions within the $SbCl_3$ dimers are shown by broken lines, the $S\cdots Sb$ interactions (distances up to 4 Å) by dashed lines. The three covalent Sb–Cl distances are 2.359, 2.388 and 2.349 A. The intra-ring S–S distances range from 2.036 to 2.055(2) Å. (Data from Müller and Mohammed, 1983.)

perturbation. The yellow crystals are stable for some weeks at room temperature but decompose rapidly into the components on heating.

There are centrosymmetric Sb_2Cl_6 pairs to which crown-shaped S_8 molecules (ideal symmetry $\bar{8}\ 2m$) are linked by S . . . Sb distances of 3.33, 3.73 and 3.96 Å (to one S_8) and of 3.49 Å to another S_8 (Fig. 11.55). There is quasi-octahedral coordination about Sb if distances of less than 3.5 Å are taken into account. There are distinct resemblances to the {diphenyl$\cdots 2SbBr_3$} structure type shown below in Figs. 11.73 and 11.74. Although the Sb . . . S interactions are similar in {1,4-dithiane$\cdots SbCl_3$} (Section 11.4) and {$S_8\cdots SbCl_3$}, the important distinction is that the former has "pure acceptors" and the latter "self-interacting" acceptors.

11.11 π-Donors and s-acceptors

The molecular compounds treated in this section are between acetylenes, olefins or aromatics as electron donors and soft metal ions in low oxidation states as s-acceptors; specifically we consider Ag(I) and Cu(I) salts, with primacy given to Ag(I) because of the larger amount of information available about these systems. These molecular compounds exist essentially only in the solid state and dissociate appreciably in solution. Consideration of the structures suggests that the differences between aromatic and olefin adducts are smaller than their resemblances; we introduce them in sequence and then compare their crystal chemistries. The earliest studies of interactions between aromatic hydrocarbons and silver salts were studies of the phase diagrams of $AgClO_4$ (in the presence of water) with benzene (Hill, 1922), aniline (Hill and Macy, 1924), pyridine (Macy, 1925) and toluene (Hill and Miller, 1925). The following solid phases were isolated: benzene$\cdots AgClO_4$;

6(aniline)···$AgClO_4$; 4(aniline)···$AgClO_4$; 2(aniline)···$AgClO_4$; aniline···$AgClO_4$; 5(pyridine)···$AgClO_4$; 4(pyridine)···$AgClO_4$; 2(pyridine)···$AgClO_4$ and toluene··· $AgClO_4$. An early study by Winstein and Lucas (1938) of the coordination of unsaturated compounds with silver ions in solution led to the synthesis of crystalline 1:1 adducts of dicyclopentadiene with $AgNO_3$ and $AgClO_4$. The 2:1, 1:1 and 2:3 crystalline com- pounds of cyclotetratetraene with silver salts, and a 1:1 cyclo-octatriene···$AgNO_3$ adduct, were reported in 1950 by Cope and Hochstein and by 1962 some thirty crystalline compounds of $AgNO_3$ and $AgClO_4$ with various olefins had been reported (Bennett, 1962). These early results, for both aromatics and olefins, show that more than one stoichiometry is possible (perhaps even common) and thus a unique mode of interaction of Ag^+ with the π-system of the donor is not to be expected; fortunately a sufficient variety of stoichiometries has been studied to provide a comprehensive picture. More recent synthetic work with $AgClO_4$ compounds of aromatics is summarized in Table 11.24; crystal structures of most of these compounds have been reported. The results fall into two time periods. First, there is the early work of Rundle and Amma and coworkers over the period 1950–1975; then there was a fallow period for the next twenty years, with the field being resuscitated by Munakata and coworkers.

Most compounds (olefins and aromatics alike) have $AgNO_3$ and $AgClO_4$ as acceptor moieties but other anions of strong acids have been used, such as $AlCl_4^-$, BF_4^- (Quinn and Glew, 1962; Buffagni et al., 1968), F_3COO^- and $F_3CSO_3^-$ (trifluromethanesulphonate or triflate) (Gash et al., 1974). The crystalline adducts range from fairly stable in the atmosphere (decompose within a few days) to very unstable; rigorously dry conditions are often essential for preparation and preservation of the most reactive adducts. Most adducts are unstable to light (reduction of Ag^+ to Ag) and many to x-rays. Many of the crystalline adducts are disordered to some extent; thus details of molecular dimensions, especially in older work, should be treated with judicious reserve.

Formation of an adduct with a silver salt has become an accepted method of purifying olefins and aromatics, of stabilizing labile molecules and of preparing crystalline adducts suitable for crystal structure analysis (before the advent of direct methods this technique was important for determining structures of complicated molecules by heavy-atom methods; it is still important when the crystals of the parent hydrocarbon are disordered or liquid at ambient temperatures). Interactions between olefins and silver salts have been used in the petroleum industry since at least the 1930s.

11.11.1 *Aromatics as donors and Ag(I) salts as acceptors; also {benzene···CuAlCl$_4$}*

Most compounds of aromatic hydrocarbons with silver salts are with $AgClO_4$, especially those for which crystal structures have been determined (Table 11.24). We emphasize first the resemblances and then discuss the differences. In the $AgClO_4$ adducts of aromatic hydrocarbons the Ag^+ is four-coordinated, with a rather distorted tetrahedral arrangement of the four first neighbors. Usually these are two C–C aromatic bonds, either in the same or in two different molecules, and two oxygen atoms of different ClO_4^- ions; however, either one or two oxygens of a particular ClO_4^- ion can interact with Ag^+ ions. The closest Ag^+···C distances lie in the range 2.4–2.5 Å and the Ag^+···O distances in the 2.35–2.5 Å range, with occasional outliers. Many combinations of the individual features of this bonding pattern are possible (and also some deviations from it) and the interatomic

Table 11.24. Molecular compounds of aromatic hydrocarbons with $AgClO_4$ (ratio hydrocarbon to silver salt given). The unbracketed entries in the table were reported by Peyronel *et al.* (1958) and the bracketed entries by other authors; all references except the first are to crystal structure analyses. No attempt has been made to assign priorities or achieve complete coverage

2:1	1:1	3:4	2:3	1:2	1:4
phenanthrene	naphthalene; anthracene*	(Coronene) M98	Fluoranthene M99	anthracene	Rubrene· 4H$_2$O M98
o–xylene TA75a	(benzene) RG50; SR58			(pyrene) NOKNUW; M97	
p-xylene	toluene (dec.25°)			(perylene) NOKPAE; M97	
(cyclohexyl-benzene) HGA71	*m*-xylene			(toluene) FADPEF; M98	
(1,4-benzodioxan) BB85	1- and 2-methylnaphthalene (1,6-, 2,3- and 2,6-dimethylnaphthalenes) B60 (acenaphthene) RA72; (acenaphthylene) RA72 (indene) R72 (1,2-diphenylethane) TA75b (diphenyl, diphenylmethane) V60 [2.2]paracyclophane M99 (η^2-benzene) (η^2–9,10-diphenylanthracene) FADNUT M98				

* Needles and prisms were reported, so there are perhaps two polymorphs.

References:

B60 – Buffagni *et al.*, 1960; BB85 – Barnes and Blyth, 1985. HGA71 – Hall Griffith and Amma, 1971; M97 – Munakata *et al.*, 1997; M98 – Munakata *et al.*, 1998; M99 – Munakata *et al.*, 1999; R72 – Rodesiler *et al.*, 1972; RA72 – Rodesiler and Amma, 1972; RG50 – Rundle and Goring, 1950; SR58 – Smith and Rundle, 1958; T69 – Taylor *et al.*, 1969; TA75a – Taylor and Amma, 1975a; TA75b – Taylor and Amma, 1975b; V60 – Vezzosi *et al.*, 1960.

In addition, the following solvated compounds have been prepared: 9,10-diphenylanthracene· AgClO$_4$·2(benzene) (k; FADNUT); rubrene.4(AgClO$_4$).4H$_2$O (k); benzo[a]pyrene.2(AgClO$_4$).2(toluene) (k; FADPEF); 4(benzo[g,h,I]pyrene).4(AgClO$_4$).toluene (l); decacyclene.2 (AgClO$_4$).(benzene) (l).

distances are altered in a complicated way to give the lowest free-energy structures; disordered arrangements are not uncommon.

In the molecular compounds of benzene and substituted benzenes with AgClO$_4$, each aromatic molecule interacts only with a single Ag$^+$ ion while each Ag$^+$ ion interacts with two different aromatic molecules. For example, in bis(*o*-xylene)···AgClO$_4$ (Taylor and

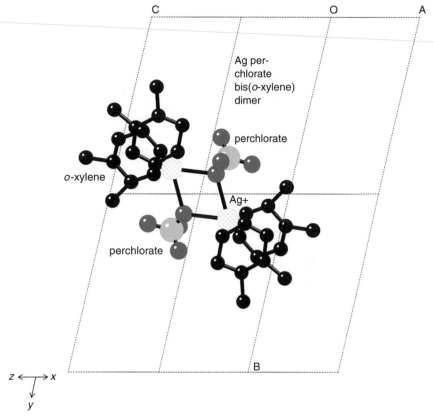

Fig. 11.56. An indivdual dimer of {bis(*o*-xylene)···AgClO₄}, viewed down [201] of the triclinic unit cell. The dimer is located about the centre of inversion at the origin of the unit cell; the dimer packing units shown interact through dispersion forces. The Ag···O distances are 2.56 and 2.60 Å, the Ag···C distances are 2.44 and 2.59 Å to adjacent carbons in one *o*-xylene, and 2.49 and 2.57 Å to the second *o*-xylene. Cl–O distances are 1.28, 1.28, 1.38 and 1.39 Å. The precision of these rather early measurements is not high – compare, for example, the Cl–O distances of 1.441(1) and 1.451(1) Å found in {benzene···AgClO₄} discussed below. (Data from Taylor and Amma, 1975b.)

Amma, 1975a; OXAGPC) discrete (*o*-xylene)···Ag⁺ units are bridged by two tri-coordinated oxygens of centrosymmetrically-related ClO₄⁻ ions to form discrete dimeric formula units (Fig. 11.56); each Ag⁺ interacts with a single aromatic bond in each of two first-neighbor *o*-xylenes.

However, the formation of infinite chains seems to be a more common arrangement and is illustrated by the bis(*m*-xylene)···AgClO₄ structure (Taylor, Hall and Amma, 1969;

MXAGCO) which can be described in terms of chains of the type shown in the scheme below, where Ar represents the m-xylene molecule.

The crystals are chiral (space group $P2_122_1$), which is unusual in this group of structures and is a consequence of the arrangement of the chains as the individual structural units are achiral. Rather similar chain arrangements are found in {bis(cyclohexylbenzene)$\cdots$AgClO$_4$} (Griffith and Amma, 1971; CHBSPC10), {benzene$\cdots$AgClO$_4$} (Rundle and Goring, 1950; Smith and Rundle, 1958) and (1,2-diphenylethane$\cdots$AgClO$_4$} (Taylor and Amma, 1975b; DPEAGP), where the two phenyl rings behave as independent aromatic entities. {Bis(1,4-benzenedioxan)silver(I) perchlorate} has a similar chain structure (Barnes and Blyth, 1985; DAVDIN), and is clearly a π-complex, there being no interaction between Ag$^+$ and oxygen atoms; the crystals are chiral (space group $C2$). This is a particularly interesting example as both the σ-complex of silver perchlorate with 3(1,4-dioxane) (Prosen and Trueblood, 1956; AGPDOX) and the π-complex with benzene are known.

The {benzene$\cdots$AgClO$_4$} structure has been re-determined at 18, 78 and 158K by neutron diffraction, giving roughly equal precision to all parameters (McMullan et $al.$, 1997; AGPCBE01, 02, 03). We shall discuss only the 18K structure (orthorhombic, space group $Cmcm$ (no. 63), 7.913(1) 7.837(2) 11.798(3) Å, $Z = 4$), which is the most precise structure considered in this chapter (and, perhaps, in this book). There is no phase change between 300 and 18K. The structure viewed down $[\bar{1}\bar{1}0]$ is shown in Fig. 11.57. Ag and Cl are at Wyckoff position 'c', mm symmetry, 0,y,1/4, O1 at 'f' (0yz) and O2 at 'g' (xy,1/4), both symmetry m, C1 and H1 at 'e', symmetry 2, and C2 and H2 at general positions. The benzene molecules are centered at 'a' (000), symmetry 2/m, and the C$_6$ rings are planar. Ag$^+$ ions are centered over C–C bonds and are slightly outside the C$_6$ ring; these two bonds are lengthened by 0.007(1) Å compared to the other four. The principal effect of Ag$^+$ on the benzene molecule is to push the two nearest H atoms 0.064(1) Å away from the C$_6$ plane. This minor distortion allows the p–π orbitals to point inward towards the metal ion. The ion is four coordinate (one to the C–C bond center, and three to perchlorate oxygens, 2.612(1) Å ($\times$2) and 2.785(1) Å ($\times$1). The differential effects on the Cl–O bond lengths (1.451(1) and 1.441(1) Å) are small but significant. The very small geometrical effects of complexation found in benzene$\cdots$AgClO$_4$ possibly provide a benchmark against which other reported distortions (generally measured at room temperature) should be compared. The considerable disorder (static or dynamic was not established) found in the 1958 room-temperature XRD structure is consistent with the NMR results, which show rapid benzene reorientation in the solid at room temperature (Gilson and McDowell, 1964).

Thermodynamic studies have been made on {benzene$\cdots$AgClO$_4$}. Vapor pressure measurements give $\Delta G_{298} = -26.9(4)$kJ/mol, $\Delta H_{298} = 92.1(1)$kJ/mol and $\Delta S_{298} = 399(12)$J/molK, while measurements of heat of solution give $\Delta H_{298} = 33.9$ kJ/mol (Tildesley and Sharpe, 1953). These results suggest that the crystals are entropy stabilized. However, the suggestion of Smith and Rundle of Ag$^+$ disorder was not confirmed by the much more precise study of McMullan et $al.$ A calorimetric study over the temperature range 8–300K (Clayton et $al.$, 1978) showed no transitions or major anomalies in the C_p–T curve, which accords with the results of the neutron diffraction study. It was estimated that the crystals had $\approx$10 J/mol K of configurational entropy at 298K.

The AgClO$_4$ molecular compounds of acenaphthene, acenaphthylene (Rodesiler and Amma, 1972) and indene (Rodesiler et $al.$, 1972) have been studied. In all three molecules

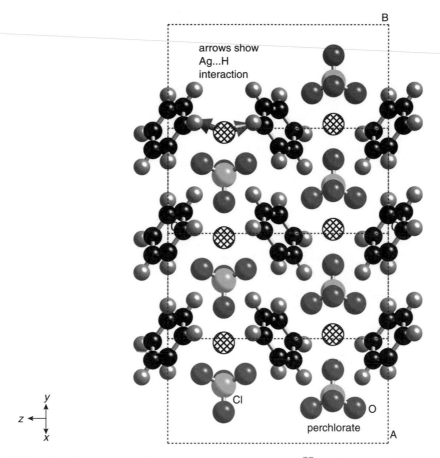

Fig. 11.57. The {benzene•••AgClO₄} structure viewed down [1̄1̄0]. Examples of the various interactions are shown. The distance of Ag from the C–C bond center is 2.467(1) Å, slightly shorter than d(Ag•••C). The neutron diffraction measurements permit meaningful inclusion of hydrogen atoms in the diagram. (Data from McMullan *et al.*, 1997.)

the two C=C bonds linked to Ag^+, as indicated by the arrows in the formulae, are separated by two other C–C bonds.

acenaphthene   acenaphthylene   indene

The mutual arrangement of aromatic molecules and Ag^+ is very similar in {acenaphthene•••AgClO₄} (ACENAG) (Fig. 11.58), and {indene•••AgClO₄} (INDAEP), although there are some differences in the disposition of the perchlorate ions.

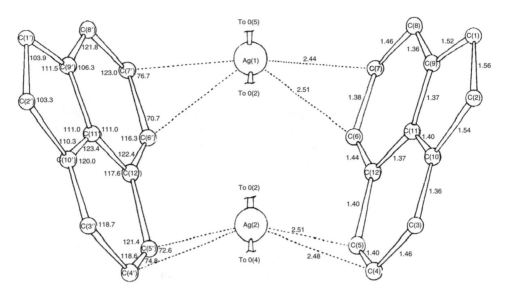

Fig. 11.58. View down [001] of the {acenaphthene···AgClO$_4$} structure showing only Ag...C interactions ($a = 18.531(2)$, $b = 15.586(5)$, $c = 7.877(3)$ Å, space group *Pmnb*, $Z = 8$. In this view a mirror plane normal to the page bisects the Ag$^+$ and perchlorate ions (latter not shown); [100] is horizontal and [010] vertical. There are two crystallographically independent Ag$^+$ and perchlorate ions in the asymmetric unit, both on mirror planes. (Reproduced from Rodesiler and Amma, 1972.)

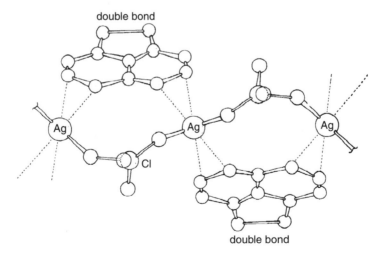

Fig. 11.59. Part of the {acenaphthylene···AgClO$_4$} structure as viewed down [100]. The crystals are orthorhombic, $a = 6.416(1)$, $b = 10.286(2)$, $c = 18.056(2)$ Å, $Z = 4$, space group $P22_12_1$. The silver atoms lie on the twofold axes normal to the page; the crystals are chiral but the absolute configuration was not determined. There is appreciable disorder, not shown, in the structure. (Reproduced from Rodesiler and Amma, 1972.)

In {acenaphthylene···AgClO$_4$} (ACYLAG), which is disordered, there is slippage along the chain (Fig. 11.59); this structure is very similar to that of {bis(m-xylene)··· AgClO$_4$} if one replaces the two m-xylene molecules by one acenaphthylene molecule.

In {naphthalene···4AgClO$_4$·4H$_2$O} (Griffith and Amma, 1974a; NAPAGP) and {anthracene···4AgClO$_4$·H$_2$O} (Griffith and Amma, 1974b; ANTAGP) the 1–2 and analogous bonds of the aromatic molecules provide four equivalent binding sites per molecule for Ag$^+$. However, the water molecules in these two hydrated crystals introduce appreciable changes from the bonding patterns described above for anhydrous compounds. In both the hydrated compounds Ag$^+$ interacts with one C–C bond of a particular aromatic molecule (thus each aromatic molecule interacts with four different Ag$^+$ ions) and completes its coordination requirements by Ag$^+$···O interactions with one water molecule (d(Ag$^+$...O) $= 2.35$ Å) and two oxygens of different perchlorate ions (d(Ag$^+$...O) ≈ 2.6 Å). In the naphthalene compound the water molecules each interact with one Ag$^+$ ion and are also hydrogen-bonded together to form a sheet; in the anthracene compound the water molecule interacts with two Ag$^+$ ions. The Ag$^+$...C distances are extended (≈ 2.6 Å) in the naphthalene compound and normal (≈ 2.45 Å) in the anthracene compound. Anhydrous naphthalene and anthracene molecular compounds with AgClO$_4$ have been prepared (Table 11.24) but structures have not been reported.

The structural chemistry of arene···AgClO$_4$ compounds attracted little further attention for about 25 years until its recent revival by Munakata and coworkers, who have prepared {pyrene·2AgClO$_4$}, {perylene·2AgClO$_4$}, {2(fluoranthene)··3AgClO$_4$}, {3(coronene)·4AgClO$_4$} (FADQAC), {[2.2]paracyclophane·AgClO$_4$}, {9,10-diphenylanthracene··AgClO$_4$}·2(benzene) (FADNUT), {rubrene·4(AgClO$_4$)·4H$_2$O} (FADPAB), {benzo[a]-pyrene·2(AgClO$_4$)}·2(toluene) (FADPEF), {4(benzo[ghi]pyrene)· 4(AgClO$_4$)}·toluene and {decacyclene·2(AgClO$_4$)}·(benzene) and determined their crystal structures. References are given in Table 11.24. The motivation for their work was described as follows: "Polycyclic hydrocarbons have been selected as potential donor molecules for preparing metal ion – aromatic π-donor – acceptor complexes with possible applications in electrical conductors and photosensitive devices." Only time can tell whether these goals will be achieved, but there is no doubt that many remarkable crystal structures have been reported.

Munakata and coworkers (1999) have classified the overall arrangements into five groups (Fig. 11.60). The linkage principles found in the simpler compounds are also encountered in these more intricate structures, but there can be additional features. A particular Ag$^+$ is linked to two C–C bonds in different arene molecules (distance ≈ 2.5 Å), and to two oxygens of different ClO$_4^-$ ions (distance ≈ 2.5 Å), and can thus be described either as 6-coordinate or as pseudo-tetrahedral. These give two-dimensional (crumpled) sheets which are then packed to give a three-dimensional supramolecular architecture. However, {9,10-diphenylanthracene.-AgClO$_4$}·2(benzene) (FADNUT) is unusual in having a monomeric structure. Some of the benzene (toluene) third components are present as space-filling solvent of crystallization (e.g. {4(benzo[ghi]-pyrene)·4(AgClO$_4$)}· toluene), but there are examples where these molecules interact with Ag$^+$ (e.g. {decacyclene·2(AgClO$_4$)}·(benzene)). We shall only describe {3(coronene)·4AgClO$_4$} (FADQAC; triclinic, $P\bar{1}$, $Z = 1$, 12.771(4) 11.224(2) 10.894(3) Å, 89.43(2) 65.25(2)

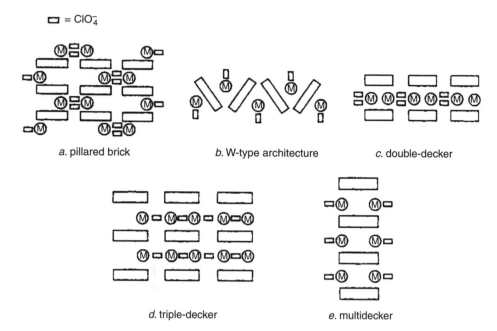

a. pillared brick   b. W-type architecture   c. double-decker

d. triple-decker   e. multidecker

Fig. 11.60. Classification of the structure types found by Munakata *et al.* (1999) for the metal sandwich systems derived from assembly of Ag(I) perchlorate with polycyclic hydrocarbons. (Reproduced from Munakata *et al.* (1999).)

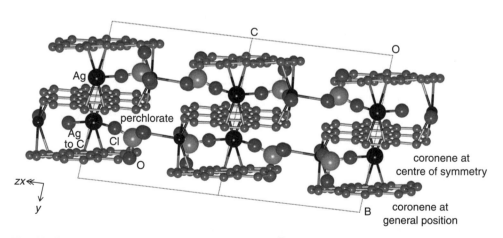

Fig. 11.61. {3(coronene)·4AgClO$_4$} viewed down [10$\bar{1}$]. The strips extend along [001] (see next figure) and repeat along [010], with π–π interactions between adjacent strips. The central coronene interacts with four Ag$^+$ cations, and the outer coronenes each with two Ag$^+$. (Data from Munakata *et al.*, 1998.)

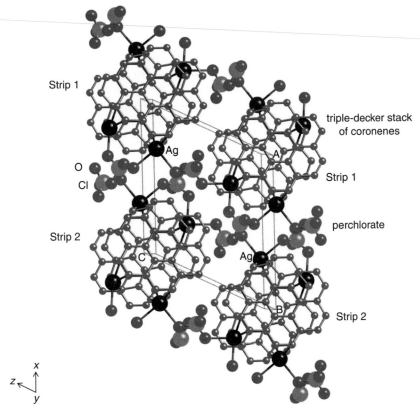

Fig. 11.62. {3(coronene)·4AgClO₄} viewed down [010]. Strips 1 and 2 extending along [001] only have van der Waals interactions. This view shows that each strip has an organic core and inorganic surfaces. (Data from Munakata *et al.*, 1998.)

79.57(2)°) in some detail. One coronene molecule is at a center of symmetry, all the other moieties being at general positions (Fig. 11.59). The central coronene is planar to within 0.018 Å but the outer coronene is bow shaped (convex towards the center), with the outer carbons (those linked to the Ag⁺ ions) deviating by ≈0.25 Å from the plane of the central hexagon. This is an unusual feature.

There are a number of interesting compounds of macro-rings of aromatic molecules with silver salts (Beverwijk *et al.*, 1970; Gmelin, 1975). For example, {([2](1,5) Naphthalino[2]paracyclophane)silver(I) perchlorate} crystallizes in space group $P2_1/n$, $Z = 4$ (12.495(2) 8.546(1) 16.196(2) Å, $\beta = 101.27(1)°$) (Schmidbaur, Bublak *et al.*, 1988; SACFOR). Each Ag is connected with the naphthalene rings of two paracyclophanes and with two perchlorate ions, two oxygens of which are chelating, one monodentate and the fourth noninteracting. The cation–anion chains are linked through the organic ligands in such a way that a sheet structure is formed. The naphthalene rings are η^3-bonded to the silver atoms, which are not ring-centered; the benzene rings of the cyclophanes are not involved in the metal coordination, while the silver atoms are η^3-linked to the naphthalene rings, and shifted from the centers of the rings.

[2$_3$](1,4)Cyclophane (called [2.2.2]paracyclophane by Cohen-Addad *et al.*, (1983)) forms 1 : 1 compounds with AgClO$_4$ (chiral crystals, space group $P2_12_12_1$; CAKDAT) and with silver triflate (AgCF$_3$SO$_3$) (space group $P2_1/a$, $Z = 16$) (Kang *et al.*, 1985; CUXTIY), in which the silver ions lie on the pseudo-threefold axis of the molecules at distances of 2.54–2.67(2) Å from the six nearest carbons for the AgClO$_4$ salt and of 2.40–2.69 Å for the triflate salt, and are also each coordinated to one oxygen (d(Ag–O) = 2.51(2) and 2.41 Å respectively).

[2$_3$](1,4)Cyclophane

Deltaphane ([2$_6$](1,2,4,5)cyclophane also forms a 1 : 1 compound with silver triflate (space group $P2_1/n$, $Z = 4$; Kang *et al.*, 1985; CUXTAQ) with rather similar coordination for the silver ion; the Ag$^+$ ions are 0.23 Å above the planes of the three nearest carbons (Fig. 11.63).

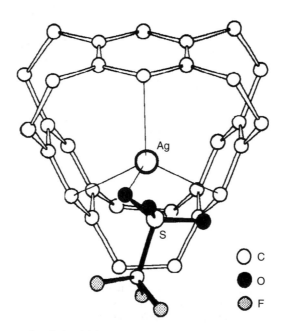

○ C
● O
◉ F

Fig. 11.63. The structural unit in {deltaphane···silver triflate}. (Reproduced from Kang *et al.*, 1985.)

The structure of {pentacyclo-[12.2.2.2^{2,5}·2^{6,9}·2^{10,13}]-1,5,9,13-tetracosa-tetraene···AgCF$_3$SO$_3$} has been briefly reported (McMurry *et al.*, 1986; DURDIF, no coordinates); the Ag$^+$ ion enters into the macro-ring and is coordinated to the four C=C bonds. This latter structure is noted here rather than in the next section, where it really belongs, because of the macro-ring nature of the ligand.

One (Ag(I)) structure has been reported with AlCl$_4^-$ as counterion – this is {C$_6$H$_6$···Ag(I)AlCl$_4$} (Turner and Amma, 1966b; SALBEN), which is compared here with {C$_6$H$_6$···Cu(I)AlCl$_4$} (Turner and Amma, 1966a; BZCATC10). The unit cells for {C$_6$H$_6$···Ag(I)AlCl$_4$} ($a = 9.09(3)$, $b = 10.22(3)$, $c = 12.73(3)$ Å, $\beta = 95.1(3)°$, $Z = 4$, space group $P2_1/c$) and {C$_6$H$_6$···Cu(I)AlCl$_4$} ($a = 8.59(1)$, $b = 21.59(3)$, $c = 6.07(1)$ Å, $\beta = 93.0(3)°$, $Z = 4$, space group $P2_1/n$) appear different at first sight but can be related. The important common feature is in the immediate surroundings of the metal atoms – the coordinations are very similar. Both structures are polymeric with η^2 coordination of metal to benzene, together with interactions to two Cl atoms of one AlCl$_4$ anion and one Cl of another anion. In {C$_6$H$_6$···Ag(I)AlCl$_4$} there are infinite sheets, held together by van der Waals forces, about the (100) planes (Fig. 11.64). The sheets are composed of AlCl$_4$ tetrahedra interconnected by Ag...Cl links; each Ag$^+$ is linked to chlorines of three different tetrahedra in such a way that every Cl of every AlCl$_4$ is involved in a link to

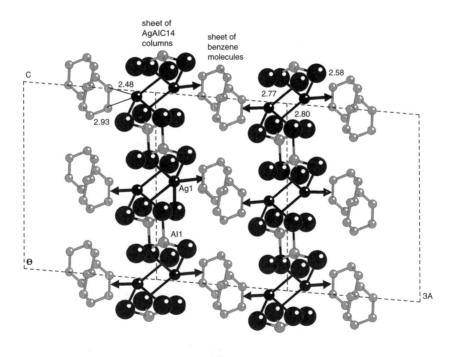

Fig. 11.64. Projection of {C$_6$H$_6$···Ag(I)AlCl$_4$} structure down [010]. The Al–Cl distances are not significantly different from 2.14 Å, with the Ag···Cl distances 2.58, 2.80 and 2.78 A. The Ag$^+$···C=C interaction is unsymmetrical η^2, with d(Ag···C) = 2.48, 2.93 Å. (Data from Turner and Amma, 1966b.)

Ag^+ ($d(Ag^+ \ldots Cl = 2.59, 2.77, 2.80$ and 3.04 Å). Each Ag^+ interacts with one bond of a benzene ring ($d(Ag^+ \ldots C = 2.47, 2.92$ Å, with the distance to the midpoint of the bond being 2.47 Å); such asymmetry is found in many of these compounds. The benzene ring extends out of the $AgAlCl_4$ sheets, with its plane approximately normal to the $Ag^+ \rightarrow \parallel$ vector.

The $\{C_6H_6 \cdots Cu(I)AlCl_4\}$ structure is based on pleated sheets lying about glide planes at $y = 1/4, 3/4$ (Fig. 11.65). The Cu^+ ion is placed asymmetrically with respect to one of the bonds of the benzene ring, and the coordination environment is completed by three chlorines of three different $AlCl_4^-$ ions. The Cu to 'midpoint of C–C bond' vector is nearly perpendicular to the plane of the benzene ring (angle $= 98°$). The overall structure is thus similar to those of the silver-salt molecular compounds, and the anion plays a correspondingly important role in stabilizing the structure. Despite the similarities between the two structures, the differences must not be disregarded. Comparing interatomic distances, one notes that the $AlCl_4$ tetrahedra in the two structures are indeed similar, with $d(Al - Cl) \approx 2.13$ Å, but other interatomic distances are appreciably shorter in $\{C_6H_6 \cdots Cu(I)AlCl_4\}$ than in $\{C_6H_6 \cdots Ag(I)AlCl_4\}$. For example, $d(Cu \ldots Cl)$ ranges from 2.36 to 2.56 Å, while $d(Ag \ldots Cl)$ is 2.59–3.04 Å; $d(Cu \ldots C)$ is 2.15 and 2.30 Å,

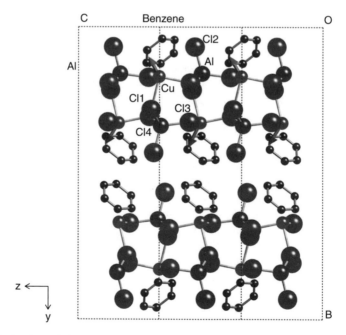

Fig. 11.65. Projection of the crystal structure of $\{C_6H_6 \cdots Cu(I)AlCl_4\}$ down [100], showing the pleated sheets lying about alternate (040) planes. The η^2 interactions between Cu and the benzene ring are shown by the arrows in the upper sheet in the cell. The values of $d(Cu \ldots C)$ are 2.15(3) and 2.30(3) Å respectively; the $Cu \ldots Cl$ distances are – to Cl_1 2.56(1), to Cl_3 2.40(1) and to Cl_4 2.37(1) Å. The Al–Cl distances are 2.14(1) Å except for Al–Cl_2 (not bonded to Cu), which is significantly shorter at 2.07(1) Å. (Data from Turner and Amma, 1966a.)

whereas $d(Ag \cdots C)$ is 2.47 and 2.98 Å. The bonding in the two compounds was discussed by Turner and Amma but we shall not elaborate.

$\{C_6H_6 \cdots Cu(I)AlCl_4\}$ is important in the Tenneco process for the purification of carbon monoxide by formation of weak complexes with Cu(I) salts (Sneedon, 1979):

$$nCO + Cu(I) = Cu(I)(CO)_n$$

The problem of poor solubility of Cu(I) salts is resolved by complexing them with aromatic solvents such as benzene or toluene. These complexes are not affected by CO_2, N_2 or H_2 but must be protected from O_2, H_2O, H_2S and SO_2.

Another situation where there are differences in the coordination around Ag and Cu are the $1:1$ coordination polymers of [2.2]cyclophane with $Ag(I)(GaCl_4)$ (FAPCAA) and of [3.3]cyclophane with $Cu(I)(GaCl_4)$ (FAPBUT) (Schmidbaur, Bublak, Huber, Reber and Müller, 1986). Both Ag and Cu are approximately η^2 bonded to aromatic rings of the cyclophanes but each Ag atom is linked to other silver atoms through two $(GaCl_4)^-$ anions while the coordination sphere of the Cu atom is saturated by a chelating $(GaCl_4)^-$ anion. Schmidbaur et al. comment "It remains an open question whether the different arene coordination of Cu(I) and Ag(I)... is due to the nature of the metal or to the difference in the cyclophane bridges."

Aromatic ... Ag^+ interactions can also be important in more complicated structures: an example is $Ag(I)[Cu(II)_2(C_8H_4O_4)_2OH] \cdot 5H_2O$ ($C_8H_4O_4$ is the o-phthalate ion) (Cingi et al., 1979; AGPHCU), where the silver ion is coordinated asymmetrically with a C–C bond of the benzene ring of the o-phthalate ion (distance of Ag^+ to midpoint of C–C bond is 2.43 Å), with two water molecules ($d(Ag...O) = 2.33(3)$ and 2.54(3) Å) and to an oxygen of the anion ($d(Ag...O) = 2.26(1)$ Å); the overall coordination about Ag^+ is distorted tetrahedral.

11.11.2 Olefins as donors and Ag(I) salts as acceptors

$AgNO_3$ is the silver salt most widely used in forming molecular compounds with olefins, although $AgBF_4$ has also been used. The $2:3$, $1:1$ and $2:1$ compounds of cyclo-octatetraene with $AgNO_3$ have already been mentioned (Cope and Hochstein, 1950); norbornadiene forms $1:1$ and $1:2$ compounds with $AgNO_3$ (Traynham and Olechowski, 1959; Traynham, 1961; Abel et al., 1959); linear mono-olefins form $3:1$ and $2:1$ compounds with $AgBF_4$ (and some other compositions as well), while butadiene forms $\{C_4H_6 \cdots AgBF_4\}$ and $\{3[C_4H_6] \cdots AgBF_4\}$ and 1,4-pentadiene forms a $1:1$ and 2-methyl-1,3-butadiene a $3:2$ compound (Quinn, 1967). Both anhydrous and hydrated compounds can be made: for example, norbornadiene forms anhydrous compounds as above while crystallization from aqueous solutions gives $\{C_7H_8 \cdots AgBF_4 \cdot xH_2O\}$ ($x < 2$), where the water appears to be zeolitic, and $\{3[C_7H_8] \cdots 2AgBF_4 \cdot yH_2O\}$ ($2 < y < 4$), where the water appears to be structural (Quinn, 1968). Solid state studies suggested that the stability of the compounds decreases in the order $SbF_6^- > BF_4^- > ClO_4^- > NO_3^-$. The stronger the acid, the more stable the compound containing the conjugate base of the acid as anion. Thus the most stable compounds would be expected to be those trifluoromethanesulphonic acid, the strongest monobasic acid known. This has been demonstrated in solution studies (Lewandos et al., 1976); these authors also made crystalline compounds of mono-enes, dienes and trienes with $AgCF_3SO_3$ (i.e. silver triflate compounds, as already noted above).

The crystal structures of the anhydrous and hydrated compounds compared below are quite complicated; in particular the olefin molecules have more complicated shapes than the aromatics discussed above, leading to greater variety in the arrangement of the moieties. As before, we emphasize overall resemblances. The feature common to the structures is the interaction of the silver ion with C=C double bonds, and structures will therefore be classified in terms of the numbers of such interactions, with Ag^+ . . . O interactions used as a secondary basis for classification. Most of the crystal structures can be described as multiple sandwiches, with sheets of $Ag^+NO_3^-$ chains forming one part of the sandwich (the ionic part) and sheets of organic molecules acting as insulation between the charged sheets. The $Ag^+\!\!\leftarrow\!\!\|$ interactions provide the "glue" between the two types of sheet.

Silver ions each interacting with only one double bond are found in the 1 : 1 compounds of $AgNO_3$ with *exo*-tricyclo[3.2.1.0^{2,4}]oct-6-ene (Gibbons and Trotter, 1971; TOEAGN) and with a *pseudo*-Diels–Alder dimer of norbornadiene (Caughlan *et al.*, 1976; HCY-TAG), in the 1 : 2 compounds of norbornadiene (Baenziger *et al.*, 1966; BORNAG) and humulene (McPhail and Sim, 1966b; HULAGN) with $AgNO_3$ and in the 1 : 3 compound of *cis,cis,cis*-1,4,7-cyclononatriene with $AgNO_3$ (Jackson and Streib, 1967; CNTAGN10). In {*trans*-cyclooctene$\cdots$AgNO$_3$} (*Pbca*, $Z=8$) (Rencken *et al.*, 1988; GOJFEQ), the silver ion is coordinated almost symmetrically to the double bond ($d(Ag\cdots C) = 2.341(4)$, 2.327(4) Å) and to two oxygens of different nitrate ions ($d(Ag\cdots O) = 2.295(3)$ Å), the third oxygen being involved only in a van der Waals interaction. The conformation of the cyclooctene ring is twist-chair-chair with approximate D_2 symmetry, which is also the minimum strain energy conformation of the free ring.

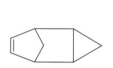 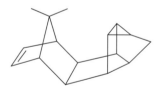

exo -tricyclo[3.2.1.0^{2,4}]oct-6-ene   *pseudo* -Diels-Alder dimer of
                                               norbornadiene

The {norbornadiene . . . 2[AgNO$_3$]} structure (space group $Cmc2_1$, $Z=4$) is rather symmetrical and can serve to show both the multiple sandwich arrangement of "charged" and "insulating" sheets and the arrangement of olefin double bonds and oxygen atoms around Ag^+ (Fig.11.66). The $Ag^+\cdots C$ distances are 2.31(5) and 2.41(4) Å and the shortest $Ag^+\cdots O$ distances are 2.27(4) and 2.34(4) Å. These distances are fairly typical of those found in these three compounds; in particular the $Ag^+\cdots O$ distances are short enough to indicate some degree of covalence. It has been shown (Marsh and Herbstein, 1983) that the space group $Cmcm$ is equally compatible with the atomic positions reported; the changes that would be required in the description of the structure are small and would make it more symmetrical and thus the space group change is not important in the present context.

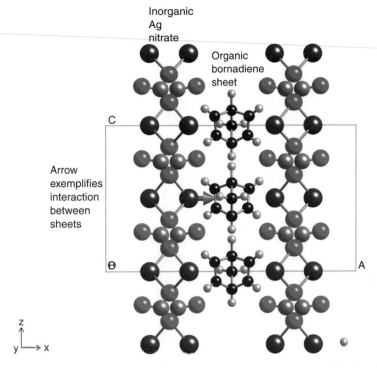

Inorganic
Ag
nitrate

Organic
bornadiene
sheet

Arrow
exemplifies
interaction
between
sheets

Fig. 11.66. The structure of {norbornadiene···2AgNO$_3$} projected onto (010). The interatomic distances are not accurate enough to distinguish among chemically similar interactions. (Data from Baenziger *et al.*, 1966.)

The silver ion is coordinated to two carbon double bonds in the same molecule in {β-gorgonene···AgNO$_3$} (Hossain and Van der Helm, 1968; GORAGN10) and in different molecules in {germacratriene···AgNO$_3$} (Allen and Rogers, 1971; GRMTAG10) {cyclo-octatetraene···AgNO$_3$}, (Mathews and Lipscomb, 1959; COCAGN), in the 1 : 1 compound of the cyclo-octatetraene dimer (m.pt. −35.5 °C) and AgNO$_3$ (Nyburg and Hilton, 1959; COCDSN), in {costunolide···2AgNO$_3$} (Linek and Nowak, 1976; COST-AG10) and in the four isostructural compounds of composition {2R···AgNO$_3$}, where R is *trans*-cyclodecene (Ganis and Dunitz, 1967; AGCDEC), *trans*-cyclododecene (Ganis *et al.*, 1971; AGCDEG), *cis*-cyclodecene (Ermer *et al.*, 1971; CDEAGN), and 1,1,4,4-tetra-methyl-*cis*-cyclodec-7-ene (Ermer *et al.*, 1971; TMCDAG). The structures of two hydrated compounds have been determined where Ag$^+$ interacts with two carbon double bonds; these are {pregeijerene···AgNO$_3$.H$_2$O} (Coggon *et al.*, 1966; PREJAG10) and {bullvalene···AgBF$_4$·H$_2$O} (McKechnie and Paul, 1968; BULAGF10). The overall structural features are the same as in the anhydrous compounds; the water molecules are not coordinated to Ag$^+$ in the first of these compounds (d(Ag$^+$ . . . OH$_2$) = 3.20 Å) but there is strong coordination in the second (d(Ag$^+$ . . . OH$_2$) = 2.32, 2.41 Å). The water molecules play an essential role in the cohesion of the crystals of the second compound as they both occupy one of the silver-ion coordination sites and also provide a link (by hydrogen bonding to BF_4^-) between cations and anions; the silver ions do not interact directly with the BF_4^- anions.

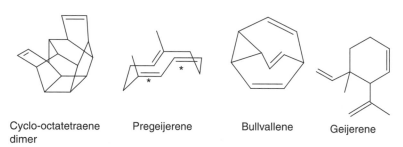

β-gorgonene　　Germacratriene　　Costunolide

Mixed situations are also found, where one Ag^+ interacts with one double bond and another with two double bonds. This shows that there is no great energetic advantage either to a single or a double interaction. In racemic {geijerene···$2AgNO_3$} (Robinson and Kennard, 1970; GEIJAG10; $P2_1/n$, $Z = 4$; geijerene is 3-isoprenyl-4-methyl-4-vinylcyclohexene) one Ag^+ interacts with the isopropenyl and vinyl groups of a particular geijerene molecule ($d(Ag^+ \ldots C) = 2.39(5)$, 2.59, 2.54, 2.54 Å) and completes its distorted tetrahedral coordination (each double bond being considered to occupy a single coordination site) by interacting with oxygens from two bridging nitrate groups ($d(Ag^+ \ldots O) = 2.42$, 2.49 Å). The other Ag^+ is trigonally coordinated to two bridging nitrate groups ($d(Ag^+ \ldots O) = 2.31$, 2.48 Å) and to the remaining double bond of the cyclohexene ring ($d(Ag^+ \ldots C) = 2.30$, 2.33 Å).

Cyclo-octatetraene　　Pregeijerene　　Bullvallene　　Geijerene
dimer

In the molecular compound {3(1,5-hexadiene)···$2AgClO_4$} (HDEAGC; Bassi and Fagherazi, 1965) two hexadiene molecules are in the *transoid-gauche-transoid* conformation and one in the centrosymmetric *gauche-transoid-gauche* conformation. The two double bonds of the *tgt* molecule are coordinated to one Ag^+ and one of the double bonds in the *gtg* molecule is coordinated to the other Ag^+ ion (Fig. 11.67). Thus the cation is the centrosymmetric moiety (hexadiene...Ag^+...hexadiene...Ag^+...hexadiene), with the outer and central hexadienes having different conformations. The perchlorate ions are disordered (rotating according to the authors) so nothing is known about how the Ag^+ coordination is completed.

In {3(bullvallene)...$AgBF_4$} (AGBULV10; McKechnie *et al*, 1967) double bonds of three *different* bullvallene molecules are coordinated to a single silver ion, giving discrete $Ag^+(C_{10}H_{10})_3$ cations (Fig. 11.68; note that the three double bonds in a bullvallene molecule are chemically equivalent). Bullvallene I has two double bonds more weakly linked to

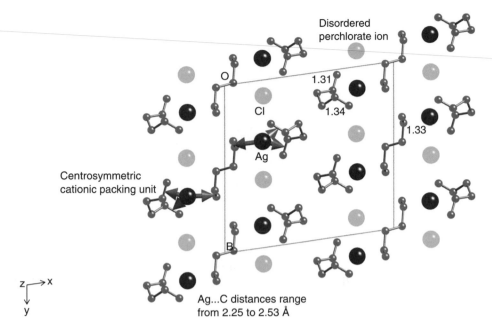

Fig. 11.67. The crystal structure of {3(1,5-hexadiene)···2AgClO₄} projected down [001]. The largest circles represent the disordered perchlorate ions. The crystals are monoclinic and the cell was described with [001] unique ($a = 13.39$, $b = 13.05$, $c = 7.22$ Å, $\gamma = 98°$, $Z = 4$, space group $P2_1/b$). (Data from Bassi and Fagherazi, 1965.)

Ag^+ while molecules II and III each have one double bond more strongly linked to this Ag^+. The BF_4^- ion is remote from the cation and does not interact directly with Ag^+. This molecular compound, which is sensitive to light and x-rays, really belongs in Part 1 of this chapter, as the tetrafluoroborate ions do not interact with Ag^+, as has already been noted for {bullvallene···AgBF₄·H₂O} (McKechnie and Paul, 1968). This is also probably the situation for {pentacyclo-[12.2.2.2²,⁵.2⁶,⁹.2¹⁰,¹³]1,5,9,13-tetracosatetraene . . . AgCF₃SO₃} (McMurry et al., 1986; DIRDUF, no coordinates in CSD) although for the different reason that the Ag^+ ion is here essentially contained within the macrocycle. However, the number of examples is still too small, and uncertain, to lose the convenience of grouping together chemically similar compounds.

The only example so far of AgClO₄ interacting with an olefin is provided by the allyl cyanide compound {CH₂ = CH-CH₂-CN···AgClO₄} (Zavalii and Prots, 1987; FOBZEB) where the Ag^+ ions are joined in a chain by bridging perchlorate groups ($d(O···Ag^+) =$ 2.52, 2.56 Å). The allyl cyanide molecules are coordinated to Ag^+ both through the double bond ($d(C . . . Ag^+) = 2.35$, 2.43 Å) and the N atom ($d(N . . . Ag^+) = $ 2.23 Å) and these interactions crosslink the chains into layers between which there are only van der Waals forces. This molecular compound can be considered to be a hybrid with both n-s and π-s interactions.

An apparent exception to the structures described above is the reported formation of a 1 : 1 compound of cyclobutadiene with AgNO₃ (Avram et al., 1959); however, later

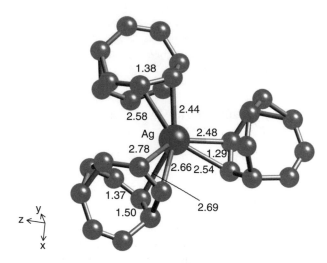

Fig. 11.68. The structure of the cation in {3(bullvallene)···AgBF$_4$}. The distances of Ag$^+$ from the centres of double bonds involved in secondary linkages are shown; the cutoff has been set at 2.70 Å. The lengths of the double bonds are also shown, indicating that the carbon positions are not very precise. (Data from McKechnie *et al.*, 1967.)

work (Avram *et al.*, 1961) showed that the correct formulation of the organic moiety was *syn*-tricyclo-[4.2.0.0^{2,5}]octadiene(3,7). Thus the compound should be formulated as {C$_8$H$_8$···2A$_g$NO$_3$}, which would be compatible with other structural results. A monomeric formulation had earlier been suggested (Fritz *et al.*, 1960) on the basis of IR spectra. A crystal structure analysis should be decisive but has not been performed.

Possible interaction between silver and the active methylene group of a chelate ring should also be noted – addition of silver salts to a solution of a metal acetylacetonate gives adducts with up to one silver per chelate ring. The crystal structures of tris(silver) dinitrate tris(acetylacetonato)nickelate(II) monohydrate (Watson and Lin, 1966; AGACNI) and of mono(silver) perchlorate iron(III) tris(acetylacetonate) monohydrate [AgClO$_4$· Fe(C$_5$H$_7$O$_2$)$_3$·H$_2$O] (Nassimbeni and Thackeray, 1974; FACAGP) have been reported; both show four coordinated silver with close aproaches to the methylene carbon ($\approx$2.3 Å) and to oxygens; the first compound has approximately tetrahedral coordination about silver and the second planar.

Finally, {C$_{60}$·5(AgNO$_3$)} is a striking inclusion complex of the fullerene in a silver nitrate network that incorporates some of the structural features described above (Olmstead, Maitra and Balch, 1999, HIFCOO; Balch and Olmstead, 1999). There is an extended network of AgNO$_3$ moieties in which each Ag$^+$ is in contact with six oxygens from the nitrate ions, with Ag...O distances ranging from 2.340(12) to 2.814(12) Å. There is fourfold disorder of the fullerene molecule within the cavity, and one of the Ag^+ ions is also disordered. There is η^2 interaction of Ag^+ to C (2.59(1) and 2.70(1) Å) and a rare, stronger, η^1 interaction $(d(Ag^+)\ldots C = 2.213(6)$ Å.

11.11.3   Some general structural principles emerging from Sections 11.11.1 and 11.11.2

Some general principles emerge from the preceding description of the available structural information; such differences as there are between the silver-salt compounds of aromatic and olefinic ligands seem to stem mainly from the greater variability of double-bond location in the olefinic molecules, and we continue to treat them together.

  (a) Coordination about Ag^+: in the molecular compounds of aromatics with $AgClO_4$: the silver ion interacts with one C–C bond of the aromatic molecule only in the hydrated naphthalene and anthracene compounds. In all the other compounds of this group the silver ion interacts with two different aromatic molecules, as in the compounds with benzene, o- and m-xylene, cyclohexylbenzene, 1,2-diphenylethane, acenaphthene, acenaphthylene and indene. There is always four-coordination about Ag^+, with a distorted tetrahedral arrangement of coordination sites; the coordination sites additional to the C–C bonds are provided by oxygens of the perchlorate ions in the anhydrous compounds while water oxygens play an important role in the two hydrated compounds. When Ag^+ interacts with two aromatic molecules, the planes of these molecules are approximately perpendicular. An idealized location for Ag^+ could be above the centre of the aromatic bond; such completely symmetrical positioning is not found in practice and the usual situation is for the two $Ag^+\ldots C$ distances to differ by 0.1–0.2 Å and for the vector from the midpoint of the double bond to Ag^+ to be displaced by about 20° from the normal to the plane of the aromatic molecule. Most authors suggest that bonding is responsible for these displacements from the idealized situation but the variety of steric arrangements is such that a clearcut explanation does not appear possible as yet.

  In the molecular compounds of olefins with $AgNO_3$, one finds Ag^+ interacting with one double bond or with two. Two crystals are known in which one Ag^+ interacts with one double bond while a second interacts with two; these are {geijerene···$AgNO_3$} and {3(1,5-hexadiene)···$2AgClO_4$}. Both three- and four-coordination have been found in the molecular compounds with olefins, the coordination sites other than double bonds generally being occupied by oxygen (an exception is {3(bullvalene)···$AgBF_4$}. The mode of interaction of NO_3– groups with Ag^+ ions varies from one molecular compound to the next; three oxygens of different NO_3– groups can be involved at one extreme, as in {cyclonona-triene···3($AgNO_3$)} or three NO_3–groups can be involved in bidentate fashion at the other extreme, as in {exo -tricyclo-oct-6-ene···$AgNO_3$}.

  (b) overall crystal structures: most of these silver-salt molecular compounds, albeit with aromatic or olefinic moieties, have layer structures composed of sheets of Ag^+ and anions, separated by sheets of the organic moieties. Anions bridging between Ag^+ ions make important contributions to the binding within the ionic sheets, while the organic sheets are bonded to the ionic sheets through the $Ag^+\leftarrow\|$ interactions.

In formal terms these substances are salt-molecule compounds, with the organic moiety neutral and the charge residing on Ag^+ and the anions. Clearly there is interaction among all three moieties but the essential points are

(a) that the type of interaction between Ag^+ and organic moieties is similar and consistent in this whole group of compounds, which justifies their treatment together, and
(b) the perturbation of the organic moiety is small.

The latter point follows from the observations that the aromatic molecules remain planar and that the changes in bond lengths and angles are small (indeed are hardly established in most of the crystal structure analyses). This situation should be contrasted with that found in the compounds formed by aromatic molecules with some alkali-metal coordination complexes; for example in bis{(tetramethyl-ethylenediamine) Li(I)} anthracenide (Rhine *et al.*, 1975) the tetramethylethylene-diamine ligands are neutral while the anthracene moiety appears as a dianion. This hardly meets our criterion (Chapter 1) that the properties of the individual components of a molecular compound should be "very largely conserved" and hence such compounds, despite their interest, are not considered further in this book. Some other salt-molecule molecular compounds that do meet our criterion are considered in Chapter 17.

11.11.4 *Acetylides as donors and Ag(I) salts as acceptors*

Silver nitrate forms two molecular compounds with silver acetylide – $\{Ag_2C_2\cdots AgNO_3\}$ and $\{Ag_2C_2\cdots 6AgNO_3\}$ (Shaw and Fisher, 1946) The first of these is tetragonal and the Debye–Scherrer pattern has been reported (Redhouse and Woodward, 1964) but not the structure. The structure of the second has been briefly reported (Osterlof, 1954) – the crystals are rhombohedral and the linear $Ag-C\equiv C-Ag$ molecules lie along the long diagonal of the cell, surrounded by six Ag^+ and NO_3- ions. Details of the interactions were not given but presumably the triple bonds act as electron donors to the Ag^+ ions, whose coordination spheres are completed by oxygens of the nitrate groups.

Detailed crystal structure analyses are available for the compounds of silver(I) triflate with 1,6-cyclodecadiyne (space group $P2_1/n$, $Z=4$; TACXIE), 1,7-cyclododecadiyne (space group $P2_12_12_1$, $Z=4$; TACXOK) and 1,8-cyclotetradecadiyne (space group $Pna2_1$, $Z=4$; TACYEB) (Gleiter *et al.*, 1990). Each silver atom has a distorted tetrahedral coordination, being linked to two triple bonds and two oxygens of the triflate ion. There is a two dimensional sheet arrangement in the first of these molecular compounds and three dimensional arrangements in the other two.

11.11.5 *Acetylides as donors and Cu(I) salts as acceptors*

1,2 : 5,6 : 9,10-tribenzocyclododeca-1,5,9-triene-3,7,11-triyne forms adducts with copper(I) triflate, which are in the border area between molecular and organometallic compounds. A pale-yellow, air-sensitive, polycrystalline adduct $Cu(C_{24}H_{12})$-(O_2SOCF_3) has been prepared, while single crystals were obtained of the related packing complex $Cu(C_{24}H_{12})(O_2SOCF_3)\cdot(C_{24}H_{12})$ (Ferrara *et al.*, 1987; FEHTER). The crystal structure (Fig. 11.69) showed that the copper atom was in the centre of one macrocycle (but 0.18 Å above its mean plane), coordinated to the three triple bonds, which were 0.04(1) Å longer than those in the uncoordinated ligand of the packing complex, which was unchanged

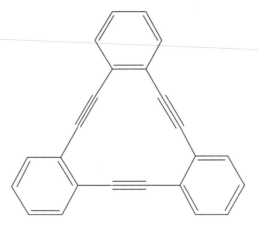

Fig. 11.69. The triacetylide ligand 1,2:5,6:9,10-tribenzocyclododeca-1,5,9-triene-3,7,11-triyne $C_{24}H_{12}$.

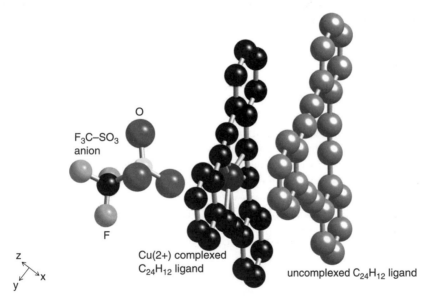

Fig. 11.70. Partial view of the crystal structure of the packing complex $\{Cu(C_{24}H_{12})(O_2SOCF_3)\cdot(C_{24}H_{12})\}$ [$a = 11.119(2)$, $b = 12.256(2)$, $c = 28.191(2)$ Å, $\beta = 105.39(1)°$, $Z = 4$, space group $P2_1/c$]. (Data from Ferrara *et al.*, 1987.)

from the neat ligand (Irngartinger *et al.*, 1970; TBZTDY10). The intraligand adduction is reminiscent of those found with adducts of silver salts in macrocycles containing double bonds. The crystal structure is characterized by a striking plane-to-plane juxtaposition of complexed and uncomplexed ligands (Fig. 11.70); there is a formal analogy to that of 1,4-diphenylbutadiene tricarbonyliron·1/2[1,4-diphenylbutadiene] (Section 10.3.4.3).

A more typical coordination occurs in the air-sensitive $\{Cu(C_{24}H_{12})\text{-}(O_2SOCF_3)\cdot C_6H_6\}$ (Ferrara *et al.*, 1988; GEKZOL) (Fig. 11.71), where each copper is linked to one triple bond of the macrocycle and to two oxygens of different triflate ions.

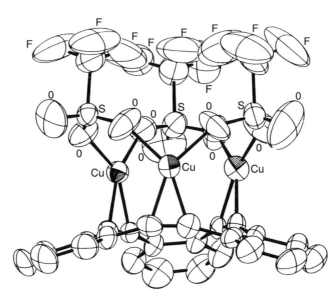

Fig. 11.71. ORTEP diagram of the $Cu(C_{24}H_{12})(O_2SOCF_3)$ moiety (crystals are triclinic, $P\bar{1}$, $Z=2$). (Reproduced from Ferrara *et al.*, 1988.)

A similar local coordination is found in the copper triflate compound of 1,7-cyclododecadiyne; here, however, the overall arrangement is that of a three dimensional polymer (Gleiter *et al.*, 1990; TACYAX). Although the 1,7-cyclododecadiyne could be expected to have some conformational flexibility, there is essentially no difference between its uncomplexed and complexed conformations. The space group of the crystals is the non-centrosymmetric *C*2, so enantiomorphous crystals are formed; however, the absolute configuration of the crystal used in the structure analysis was not determined.

11.11.6 *Aromatics as donors and Hg(II) salts as acceptors*

Under this heading we recall the phenyl···Hg interaction found in {diphenyl sulphoxide··· HgCl₂} (see Section 11.9.2).

Solid molecular compounds of various aromatics with HgX_2 (X = Cl, Br, I) have been reported (Vezzosi *et al.*, 1974) but no structures seem to have been determined. The toluene and ethylbenzene adducts of $HgCl_2$ decomposed too rapidly for satisfactory analyses to be obtained. However, the corresponding adducts with $HgBr_2$ and HgI_2 (Table 11.25) were very stable in air, soluble in ethanol and only slowly decomposed by HCl. The toluene and ethylbenzene adducts of HgI_2 showed the interesting effect of changing colour from yellow to red over a few days, apparently without change of crystal structure; no explanation has been tendered.

The compound C_6H_6···$2HgAlCl_4$ has been reported (Turner and Amma, 1966c) but its structure was not determined because of lack of suitable crystals. The crystal structure of {hexamethylbenzene···mercuric trifluoroacetate} $[Hg(O_2CCF_3)_2]$ at 108K has been reported ($P\bar{1}$, $Z=2$; Lau and Kochi, 1985; BODYEY10). There are dimeric structural units showing η^2 interaction of Hg with hexamethylbenzene (Fig. 11.72), in accordance with the predictions of Rundle and Corbett (1957).

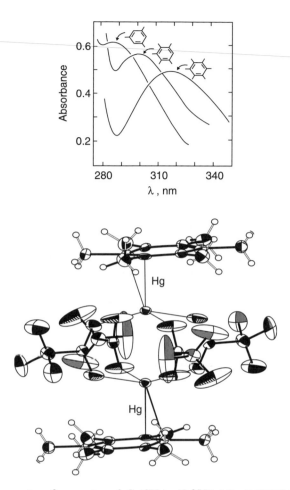

Fig. 11.72. (a) Charge transfer spectra of $C_6(CH_3)_6$ (0.0025 M), $C_6H(CH_3)_5$ (0.0025 M) and mesitylene (0.005 M) with equimolar amounts of $Hg(O_2CCF_3)_2$ in CH_2Cl_2 at 233K; (b) ORTEP diagram of the centrosymmetric dimeric $\{C_6(CH_3)_6 \cdots [Hg(O_2CCF_3)_2]\}$ species (crystals are triclinic, $P\bar{1}$, $Z = 1$). The Hg...arene carbon distances are 2.56, 2.58 Å. (Reproduced from Lau and Kochi, 1986.)

Table 11.25. Solid molecular compounds reported (Vezzosi *et al.*, 1974) between various aromatic hydrocarbons and $HgBr_2$ and HgI_2. The ratios given are for donor : HgX_2

Acceptor $HgBr_2$		Acceptor HgI_2		
1 : 1	1 : 2	2 : 1	1 : 1	1 : 2
acenaphthene	toluene	anthracene	pyrene	toluene
	ethylbenzene		pyrene:dioxane	
	acenaphthene			

also 2(ethylbenzene)$\cdots$3HgI$_2$.

11.12   π-Donors and p-acceptors

11.12.1   *Aromatics as donors and MX₃ (M = Sb, Bi; X = Cl, Br) as acceptors*

11.12.1.1   *Introduction*

We have already noted that SbX_3 acts as an acceptor for two types of donor, alkylaro-matics on the one hand and aromatics with N-, O- and S-containing substituents on the other. There are chemical and structural differences between the two groups and the latter group has already been discussed (Section 11.10). The requirement for inclusion here is aromatic ring···Sb (or Bi) interaction in the crystalline molecular compound, while lack of this feature leads to discussion of the molecular compound in Section 11.10. Considerable information is available for the SbX_3 group with aromatics and alkylaromatics as donor, together with an important extension to analogous BiX_3 molecular compounds. We shall first outline what is available, proceeding from isolated (M_2X_6) dimers to more compli-cated $(M_2X_6)_n$ oligomers. These Menshutkin molecular compounds exist only in the solid state. The first crystal structures were briefly reported in the early 1960s (Beagley and Edington, 1963); interest has continued over the years with many more (and more detailed) results reported by Mootz, Schmidbaur and coworkers, and others (see refer-ences in Tables 11.28 to 11.32). The crystal structures of SbX_3 molecular compounds that have been reported show both broadly-based common features, and differences in detail; this also appears to hold for the smaller number of structures known for BiX_3 molecular compounds. A wide variety of physico-chemical techniques are potentially available for investigation of these compounds, but relatively little has been reported.

11.12.1.2   *Structures of MX₃ (M = Sb, Bi; X = Cl, Br) crystals*

Interactions between the halogen donor atoms of MX_3 groups and the metal acceptor atoms (M = Sb, Bi) play an essential role in stabilizing the molecular compounds and so we summarize some relevant aspects of the crystal chemistry of the MX_3 compounds. A comprehensive review of the crystal chemistry of the VA element trihalides has been given by Galy and Enjalbert (1982), the structures belonging to a limited number of groups. For those relevant here (Table 11.26), it was found that β-$SbCl_3$, β-$SbBr_3$, and β-$BiCl_3$ belong to the YF_3 group, while the $AsBr_3$ group contains $AsCl_3$, $AsBr_3$, "α-$SbCl_3$" and α-$SbBr_3$. The AsI_3 group contains AsI_3, α-$AsBr_3$, α-SbI_3 and BiI_3. The structures of $SbCl_3$ (Lindqvist and Niggli, 1956; Lipka, 1979b), of two polymorphs of $SbBr_3$ (α (Cushen and Hulme, 1964) and β (Cushen and Hulme, 1962)), and of $BiCl_3$ (Nyburg, Ozin and Szymanski, 1971) are known. Another polymorph of $SbCl_3$ was reported by Groth (1906) but its structure is not known. Monoclinic β-SbI_3 is reported to be intermediate between α-SbI_3 and a layered molecular type.

The β-$SbCl_3$ structure has two equivalent covalent bonds at 2.368(2) Å and a third at 2.340(2) Å; there are secondary interactions of lengths 3.457 ($\times 2$), 3.609 ($\times 1$) and 3.736 Å($\times 2$). The coordination is best described as $3 + 2 + 3$ (bicapped trigonal prismatic), with a three-dimensional network of secondary linkages joining the $SbCl_3$ molecules. The covalent interactions are maintained virtually unchanged in the molecular compounds, while the pattern of secondary interactions is matched to the arene molecule. Two careful studies (references in Table 11.26) support the non-centrosymmetric space group for

Table 11.26. Crystal data (Å) for some MX_3 compounds (cf. Section 11.2.3.3)

Compound	a	b	c	Space group	Z	Reference
β-SbCl$_3$	8.111	9.419	6.313	$Pnma$	4	L79
β-SbBr$_3$	8.25(1)	9.96(1)	6.68(1)	$Pnma$	4	CH62
β-BiCl$_3$	7.641	9.172	6.291	$Pn2_1a$	4	NOS71; B82
α-SbBr$_3$	10.12(1)	12.30(1)	4.42(1)	$P2_12_12_1$	4	CH64
AsBr$_3$	10.17	12.09	4.32	$P2_12_12_1$	4	B35
α-SbI$_3$	7.48	7.48	20.90	$R\bar{3}$	6	TZ66
BiI$_3$	7.5117(3)	7.5117(3)	20.700(1)	$R\bar{3}$	6	C96
AsI$_3$	7.193(2)	7.193(2)	21.372(7)	$R\bar{3}$	6	EG80
β-SbI$_3$	7.281	10.902 109.93	8.946	$P2_1/c$	4	PS84

Notes:

α-SbCl$_3$ was reported by Lindqvist and Niggli (1956) but no x-ray work appears to have been done. There are cubic polymorphs of BiCl$_3$ and BiBr$_3$ (8.14 Å, 9.23 Å, both $P2_13$, $Z=4$) obtained as polycrystalline samples and of unknown structure (Wolten and Mayer, 1958).

References:

B35 – Braekken (1935, 1938), cited in CH62; EG80 Enjalbert and Galy, 1980; B82 – Bartl, 1982; C96 – Carmalt *et al.*, 1996; structure from polycrystalline sample; CH62 – Cushen and Hulme, 1962 (two-dimensional analysis); CH64 – Cushen and Hulme, 1964; L79 – Lipka, 1979; NOS71 – Nyburg, Ozin and Szymanski, 1971; PS84 – Pohl and Saak, 1984; TZ66 – Trotter and Zobel, 1966.

BiCl$_3$, but the deviations from the centrosymmetric structure are very small; Galy and Enjelbart (1982) prefer *Pnma*, citing a 1972 NMR study in support of their contention. Re-examination would seem desirable.

11.12.1.3   *Binary molecular compounds*

Phase diagrams have been determined for a number of aromatics with SbCl$_3$; the donors include benzene (Perkampus and Sondern, 1980); pyrene (Mootz and Händler, 1985); diphenylamine and triphenylamine (Lipka, 1979a); durene, pentamethylbenzene, hexamethylbenzene, naphthalene, 2,3-dimethylnaphthalene, diphenyl, anthracene, phenanthrene (Perkampus and Schönberger, 1976); benzo[b]thiophene (Korte, Lipka and Mootz, 1985). Mootz and Händler (1985) have reported the phenanthrene–SbBr$_3$ phase diagram. Noteworthy individual preparations include the 1 : 1 compounds of benzene and toluene with SbCl$_3$, SbBr$_3$ and BiCl$_3$; and 1 : 2 compounds of benzene and toluene with SbCl$_3$ and SbBr$_3$ (Peyronel *et al.*, 1968), and many other individual preparations by this group. Spectroscopic studies include applications of optical spectroscopy, nuclear quadrupole resonance and Mössbauer spectroscopy to many Menshutkin compounds, and these will be discussed below.

A wide variety of molecular compounds has been reported, with the predominant composition ratio 1 : 2 and this is also the most widely investigated group. There is a fair number of 1 : 1 compounds and a much smaller number with other ratios. Some of the latter are based only on the results of chemical analyses and confirmation by other techniques would be useful; this seems even more desirable for the ternary compounds (discussed separately below in Section 11.1.2.1.4). Melting points (Table 11.27) suggest that the 1 : 2 compounds are more stable than the 1 : 1 where both are present in a system,

Table 11.27 Melting points (K) of some molecular compounds of aromatics with SbCl₃. Other melting points in Table 11.32

Donor	1:1	1:2	Donor	1:2
benzene		352	naphthalene	359
toluene	289	316	anthracene	482
o-xylene	292	307	phenanthrene	378
m-xylene	281	311	pyrene	416
p-xylene	329	343	diphenyl	345
mesitylene	316	349		
durene		377		
pentamethylbenzene	422			
hexamethylbenzene	466			

and melting points generally increases with molecular weight of the donor, although benzene···2SbCl₃ is an exception.

11.12.1.4 Crystal structures of arene/MX₃ (M = Sb, Bi; X = Cl, Br) binary molecular compounds

There are two features of the MX₃ molecular compounds that hardly change from structure to structure: firstly, the MX₃ molecules all have a pyramidal shape with the M–X bond lengths and the X–M–X angles having essentially invariant values (e.g. d(Sb–Cl) $\approx$ 2.35 Å and <Cl–Sb–Cl $\approx$ 95°), which are the same as in the neat crystalline acceptors and in the gas phase; secondly, a ring of the aromatic molecule always has a planar MX₂ group approximately parallel to its plane with the M located above the ring (i.e. η^2, η^3 or η^6 bonding), with the third M–X bond pointing upwards away from the ring (Fig. 11.73). The MX₃ molecule always interacts with the aromatic molecule and with other MX₃ molecules and this gives a coordination number of 5, 6 or 7 for M, with one secondary link to an aromatic molecule and a number of secondary M···X links to neighbouring MX₃ molecules; more than one coordination number is often found in a particular adduct. The overall crystal structure then depends on whether the interaction between MX₃ molecules is pairwise or extended; if the latter then chain or layer structures are usually found. The description is not entirely clearcut because it depends on what interatomic distances are accepted as indicating 'secondary interaction.' We emphasize here the overall crystal structures; full details of dimensions are given in the original papers. Reports of changes in bond lengths in the aromatic molecules should be treated with caution. Our classification is based, in general, on the degree of interaction within the MX₃ oligomers, proceeding from isolated M₂X₆ dimers, and more complex *isolated* arrangements of (M₂X₆)ₙ oligomers, to extended arrangements based on interactions of various kinds among the dimers.

(a) Structures based on isolated M₂X₆ dimers   In the simpler situations there are centrosymmetric (or nearly so) M₂X₆ dimers, as in {diphenyl···2SbBr₃} (Fig. 11.73; PYRSBB10; Lipka and Mootz, 1982), {2,2'-dithienyl···2SbCl₃} (DITHSB10; Korte, Lipka and Mootz, 1985), {diphenylamine···SbCl₃} (Fig. 11.78; ZZZBAM10; Lipka,

Table 11.28. Crystal data (Å, deg) for binary {arene···n(MX$_3$)} molecular compounds ($n = 2$ except for the last four entries where $n = 1$). There are three isostructural groups (a pair, a triple and a group of four). The further entries are 'miscellaneous' in the sense that there do not appear to be close relationships among the several crystal structures

Formula/refcode	a/α	b/β	c/γ	Space group	Z	Reference
diphenyl···2SbBr$_3$ PYRSBB10	11.305	11.645 107.66	8.183	$P2_1/c$	2	LM82
2,2'-dithienyl···2SbCl$_3$ DITHSB10*	11.188	11.116 110.6	7.561	$P2_1/c$	2	KLM85
naphthalene···2SbCl$_3$ NAPSBC	9.154	9.368 122.7	11.989	$P2_1/c$	2	HS69
benzo[b]thiophene···2SbCl$_3$ DAZZIN	9.205	9.39 124.35	11.856	$P2_1/c$	2	KLM85
p-xylene···2SbCl$_3$ (at 163 K) PXYLSB	9.13	8.44 125.4	12.79	$P2_1/c$	2	HM76
9,10-dihydroanthracene ···2SbBr$_3$ KEFJAG	11.189	12.396 90.75	7.674	$P2_1/c$	2	S90
pyrene···2SbBr$_3$ PYRABR	10.699	13.001 90.00	7.634	$P2_1/c$	2	BPV72
pyrene·2BiCl$_3$ FUNJUT	10.564	12.564 90.33	7.220	$P2_1/c$	2	V88
pyrene·2 BiBr$_3$ ** JEJPET	10.589	13.21 91.60	7.533	$P2_1/c$	2	B90
[2.2]paracyclophane·2BiBr$_3$ JOLBAN	7.896	14.198 101.27	10.803	$P2_1/c$	2	VBC92
phenanthrene···2SbCl$_3$ PHENBS*	8.990 90.7	11.10 80.4	10.44 69.1	$P\bar{1}$	2	D72
phenanthrene···2SbBr$_3$ DEJYEW **	9.015 108.39	11.387 109.6	11.862 98.32	$P\bar{1}$	2	MH85
diphenyl···2SbCl$_3$ BIPHSB10*	13.498 86.4	7.884 110.1	9.341 91.41	$P\bar{1}$	2	LM78
pyrene···2SbCl$_3$ DEJYAS	18.423	9.599 107.09	11.789	$C2/c$	4	B96
diphenylamine···2SbCl$_3$ DPASBC10	7.802	9.415 91.08	26.037	$P2_1/c$	4	L78
triphenylamine···2SbCl$_3$ TPASBD	14.066	9.905 107.09	17.513	$C2/c$	4	KLM79; no coordinates
dibenzyl···2SbCl$_3$ DBESBC	7.46	11.54 96.7	23.44	$P2_1/c$	4	HH66
diphenylamine···SbCl$_3$ ZZZBAM10	11.585	7.750	32.005	$Pbca$	8	L80
1,2,3-trimethylbenzene·BiCl$_3$ PUVLOH	12.746	9.197 114.98	11.495	$P2_1/c$	4	FR98
Benzene···BiCl$_3$ LAHJAF	7.343	11.631	23.072	$P2_12_12_1$	8	F93
{[2.2.2.2]paracyclophane··· SbBr$_3$} GOMWAG	13.024	11.962 100.58	19.795	$P2_1/n$	2	F98

Table 11.28. (*Continued*)

* Reduced cell (unconventional setting); for PHENBS standard settings are 8.990 10.44 11.10 Å, 90.70 110.90 99.60°.
** reduced cell (standard setting).

References:
B90 – Battaglia *et al.*, 1990; B96 – Battaglia *et al.*, 1996; BPV72 – Bombieri, Peyronel and Vezzosi, 1972; D72 – Demalde *et al.*, 1972; F93 – Frank *et al.*, 1993; F98 – Ferrari *et al.*, 1998. FR98 – Frank and Reiland, 1998; HH66 – Hulme and Hursthouse, 1966; HM76 – Hulme and Mullen, 1976; HS69 – Hulme and Szymanski, 1969; KLM79 – Korte, Lipka and Mootz, 1979; KLM85 – Korte, Lipka and Mootz, 1985; L78 – Lipka, 1978; L80 – Lipka, 1980; LM78 – Lipka and Mootz, 1978; LM82 – Lipka and Mootz, 1982; MH85 – Mootz and Händler, 1985; S90 – Schmidbaur, Nowak, Steigelmann and Müller, 1990; V88 – Vezzosi *et al.*, 1988; VBC92 – Vezzosi, Battaglia and Corradi, 1992.

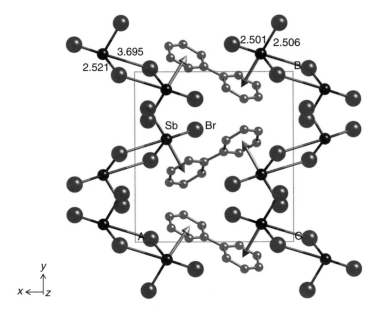

Fig. 11.73. Projection of {diphenyl···2SbBr$_3$} structure down [001]. A linear arrangement of diphenyl and Sb$_2$Br$_6$ groups extends along [101]; this shows as a horizontal (projected) band in the diagram. Both the diphenyl molecules and the Sb$_2$Br$_2$ parallelograms are centrosymmetric and planar; one edge of the parallelogram is a covalent Sb–Br bond (2.521 Å), and the other is a secondary Sb...Br interaction (3.695 A). A particular Sb$_2$Br$_6$ group does not interact with other Sb$_2$Br$_6$ groups but only with phenyl rings of different diphenyls. These interactions are indicated by arrows, the Sb...phenyl ring distance being 3.23 Å. The larger dark spheres are Br and the smaller Sb; the arene moieties are lightly shaded. Hydrogens are not included. (Data from Lipka and Mootz, 1982.)

1980) and {[2.2.2.2]paracyclophane···2SbBr$_3$} ((Fig. 11.79; GOMWAG; Ferrari *et al.*, 1998). In {diphenyl··· 2SbBr$_3$} linear chains of diphenyl and SbBr$_3$ are found with diphenyl and isolated Sb$_2$Br$_6$ moieties located about crystallographic centres of symmetry, as shown in Fig. 11.73, each Sb interacting with one phenyl ring. The Br···Sb distance

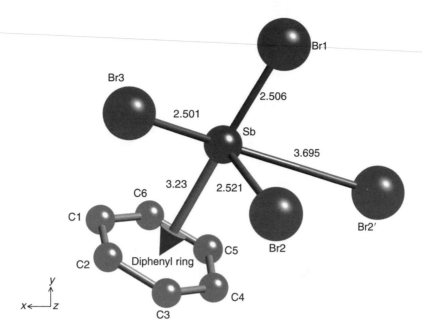

Fig. 11.74. The immediate environment of the Sb atom in {diphenyl···2SbBr$_3$}. Only atoms within 4.0 Å of the Sb are included. In this example there is five-fold 'square' pyramidal coordination about Sb. These dimers are termed 'isolated' because there is only one Sb...Br interaction of less than 4.0 Å (the actual value is 3.695 Å), and this is *within* the Sb$_2$Br$_6$ dimer; The Sb...C(ring) distances (from C1 to C6) are: 3.66, 3.48, 3.38, 3.45, 3.63 and 3.73 Å. The angles subtended at Sb are Br1/Br3 93.40, Br2/Br3 95.59, Br1/Br2 96.48, Br3/Br2′ 168.43°. (Data from Lipka and Mootz, 1982.)

within the Sb$_2$Br$_6$ dimer is 3.695(2) Å, with all other such distances greater than 4 Å, so description in terms of isolated dimers is an acceptable first approximation. The environment of the Sb atom is shown in Fig. 11.74, which also serves as a prototype for the other similar close environments encountered among these molecular compounds.

Although the overall arrangement in the isostructural {2,2′-dithienyl···2SbCl$_3$} is similar to that in {diphenyl·2SbBr$_3$}, there is not the same degree of isolation, as the intradimer Sb...Cl distance of 3.552(3) Å, is supplemented by other such contacts between dimers at 3.716 and 3.761 Å. Indeed, {2,2′-dithienyl···2SbCl$_3$} has ties to {pyrene···2SbBr$_3$} and related structures considered below. This serves as a warning that resemblance in cell dimensions and overall arrangement do not necessarily extend to the finer details of the structure. The Sb atom in {2,2′-dithienyl···2SbCl$_3$} is located almost directly above the S of the ring system, with d(S···Sb) = 3.31 Å; the S···Sb distances in molecular compounds of SbCl$_3$ with aliphatic hetero-compounds are in the range 3.06 to 3.3 Å and hence the coordinative S···Sb interaction is somewhat weaker, but not absent, in the dithienyl compound. In these compounds both diphenyl and dithienyl are at centres of symmetry, and hence are not twisted about the central C–C bond. A different situation occurs in {diphenyl···2SbCl$_3$} (Fig. 11.87), where the diphenyl rings are twisted.

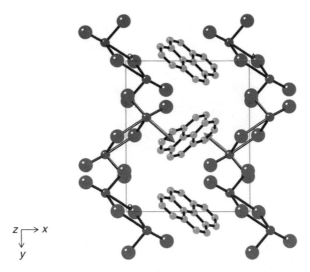

$z \rightarrow x$
$\downarrow$
y

Fig. 11.75. The {pyrene·2SbBr₃} and {pyrene·2BiCl₃} compounds are isostructural (see Table 11.28); the first of these is shown, viewed down [001]. The structure is also isostructural with that of {9,10-dihydroanthracene·2SbBr₃}. The arrows show examples of the Sb...arene interactions. (Data from Bombieri, Peyronel and Vezzosi, 1972.)

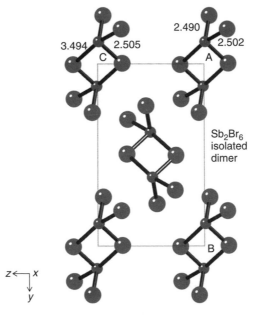

$z \leftarrow x$
$\downarrow$
y

Fig. 11.76. The arrangement of the isolated Sb_2Br_6 dimers in {pyrene·2SbBr₃}, showing the primary and secondary linkages between Sb and Br; pyrenes have been omitted for clarity. As the {9,10-dihydroanthracene·2SbBr₃} and {pyrene·2BiCl₃} compounds are isostructural, this diagram applies also to these compounds, with small dimensional differences. (Data from Bombieri, Peyronel and Vezzosi, 1972.)

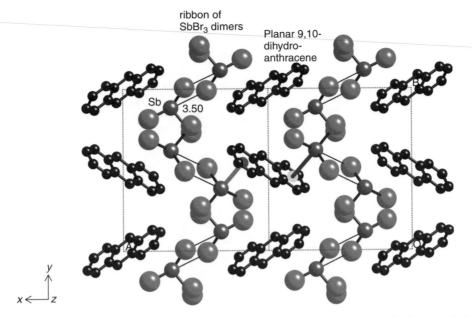

Fig. 11.77. Structure of {9,10-dihydroanthracene·2SbBr$_3$} viewed down [001]. The lengths of the primary covalent Sb–Br bonds are (to Br1) 2.538(1) Å, (to Br2) 2.502(1) Å, (to Br3) 2.532(1) Å. The secondary Sb...Br approaches are shown (there is also a longer approach of 4.035 Å which is not included). The two arrows show the interaction of the Sb atom with two of the rings of 9,10-dihydroanthracene. (Data from Schmidbaur, Nowak, Steigelmann and Muller *et al.*, 1990.)

The {pyrene···2SbBr$_3$} structure (Fig. 11.75) is a prototype for the four structures grouped together in Table 11.28. The Sb$_2$Br$_6$ dimers are indeed isolated, as is shown in Fig. 11.76, and this applies to the other isostructural compounds.

The {9,10-dihydroanthracene···2SbCl$_3$} structure (Fig. 11.77) resembles that of {pyrene···2SbBr$_3$} (cf. Table 11.28). The arene molecule is planar in the molecular compound, although it has a fold angle of ≈140° in its neat crystals (Herbstein, Kapon and Reisner, 1986; Reboul *et al.*, 1987). There is a similar occurrence of planar 9,10-dihydroanthracene in its molecular compound with 1,3,5-trinitrobenzene (Herbstein, Kapon and Reisner, 1986).

The phase diagram of the diphenylamine – SbCl$_3$ system (Lipka, 1979) shows two congruently melting compounds: 1:1 at 359–361 K and 1:2 at 354–358 K. Isolated Sb$_2$Cl$_6$ dimers are found in {diphenylamine.SbCl$_3$; ZZZBAM10}, where only one of the phenyl rings is involved in a Sb···ring interaction (Fig. 11.78). {Diphenylamine.2SbCl$_3$} (see below) has chains of interacting moieties, and both phenyl rings interact with Sb (Fig. 11.88).

The arrangement in {[2.2.2.2]paracyclophane·2SbBr$_3$} consists of Sb$_2$Br$_6$ dimers linked through Sb···arene interactions to one of the four benzene rings of the paracyclophane (Ferrari *et al.*, 1998; GOMWAG; Fig. 11.79). There are van der Waals interactions among these units which pack as in a normal molecular crystal. The [2.2.2.2]paracyclophane molecule shows some flexibility. It has no symmetry here but the

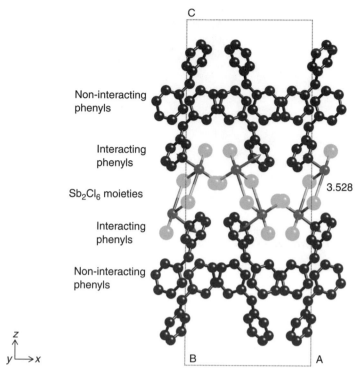

C

Non-interacting
phenyls

Interacting
phenyls

Sb$_2$Cl$_6$ moieties   3.528

Interacting
phenyls

Non-interacting
phenyls

z
y └→x   B   A

Fig. 11.78. Structure of {diphenylamine·SbCl$_3$} projected down [010]. The three covalent Sb–Cl distances (not shown) are 2.373, 2.374 and 2.390 Å, and the angles between the pairs of covalent bonds are 91.9, 94.3 and 92.6°. The secondary Sb...Cl interaction is shown, as well as representative Sb...phenyl ring interactions (arrows; Sb to ring distance 3.13 Å); only one of the two phenyls interacts with Sb. There is no Sb to nitrogen interaction. (Data from Lipka, 1980.)

gauche conformation is G$^+$G$^+$G$^+$G$^+$; it is centrosymmetric and G$^+$G$^+$G$^+$G$^+$ in its neat crystals (Cohen-Addad *et al.*, 1983), and centrosymmetric and G$^+$G$^-$G$^+$G$^-$ in its molecular compound with tetracyanoethylene (Cohen-Addad *et al.*, 1984) and in {[2.2.2.2]paracyclophane·2BiBr$_3$}(C$_7$H$_8$) (Ferrari *et al.*, 1998).

(b) Chain and layer structures   All the SbX$_3$ chain arrangements encountered among the present group of structures can be described in terms of two motifs. The first is the centrosymmetric Sb$_2$X$_6$ pair noted above, while the second is a distorted pentagonal bipyramidal arrangement, with three covalent Sb–X bonds and three secondary X···Sb links to neighbouring SbX$_3$ moieties. Addition of a secondary link to a ring of an aromatic molecule converts the array of chains into a layer. Sometimes only two X···Sb links are found and the coordination about Sb then becomes pseudo-octahedral. In {*p*-xylene···2SbCl$_3$} (PXYLSB; an early low-temperature structure, determined at 163 K) there are prominent layers parallel to (100) (Fig. 11.80), and a complicated arrangement of linked Sb$_2$Cl$_6$ pairs (Fig. 11.81). Two Sb atoms interact with a single *p*-xylene molecule

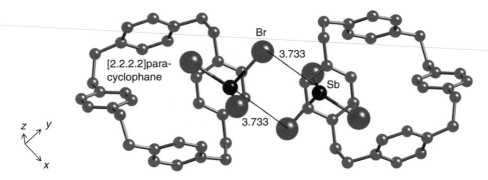

Fig. 11.79. The packing unit in {[2.2.2.2]paracyclophane···SbBr₃}. The covalent Sb–Br distances are 2.528(2), 2.539(2) and 2.522(2) Å; the Sb···Br secondary interaction is shown on the diagram. There are also secondary interactions normal to the page between the Sb atoms and the benzene rings below (on left) and above (on right). This is approximately η^2, with distance from Sb to ring centre 3.28 Å. (Data from Ferrari *et al.*, 1998.)

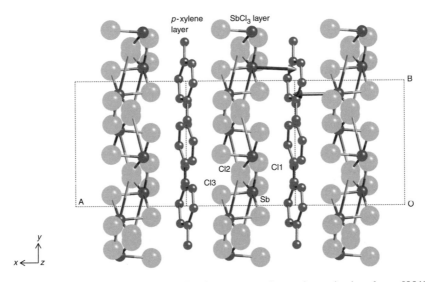

Fig. 11.80. The {p-xylene ... 2SbCl₃} molecular compound seen in projection down [001]. The segregation of organic and inorganic moieties into layers is clearly seen. The arrows represent the Sb ... arene interaction. (Data from Hulme and Mullen, 1979.)

from opposite sides; the distance from Sb to the ring plane is 3.09 Å. The closest Cl···Sb secondary interaction is remarkably short at 3.24 Å. Despite the quite close resemblance between the cell dimensions of {p-xylene···2SbCl₃} and those of {naphthalene···2SbCl₃} (Table 11.28), the structures are rather different. The {p-xylene···SbCl₃} molecular compound has also been studied, but details do not appear to have been published. Hulme and Mullen (1976) remark "The 1 : 1 SbCl₃ : C₈H₁₀ compound is very similar to the 2 : 1

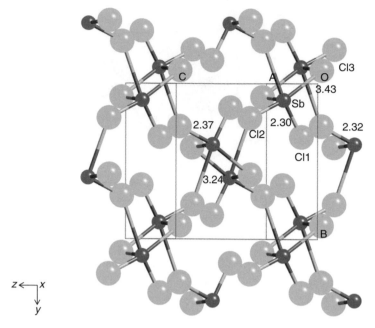

Fig. 11.81. The structure of the inorganic layer of the {p-xylene⋯2SbCl$_3$} molecular compound. (Data from Hulme and Mullen, 1979.)

compound in its layer-like nature, its bond distances and the antimony trichloride–aromatic interactions. Both compounds crystallize in the space group $P2_1/c$, with closely similar a, b and β values. To accommodate the additional p-xylene molecule the c axis is longer (17.65 Å) than in the 1 : 1 cell." Redetermination by current techniques seems desirable.

The {naphthalene⋯2SbCl$_3$} structure (NAPSBC; another early structure) differs from that of {p-xylene⋯2SbCl$_3$} as the Sb atoms interact with individual rings of different naphthalene molecules and from opposite sides (Figs. 11.82 and 11.83). The Cl...Sb distance within the Sb$_2$Cl$_6$ pair is 3.66 Å and the links between adjacent pairs forming a corrugated ribbon are 3.58 Å long; there is also a Sb...Cl distance of 3.83 Å. Thus decision whether the arrangement should be described as 'extended SbCl$_3$ ribbons' or as 'discrete Sb$_2$Cl$_6$ pairs' is somewhat arbitrary.

Layer structures are also found in the triclinic crystals of {pyrene⋯2SbCl$_3$}, {phenanthrene⋯2SbCl$_3$} and {phenanthrene⋯2SbBr$_3$} (Table 11.28); however, the latter two are not isostructural despite a resemblance in cell dimensions. The first of these has sheets of pyrene molecules and interacting SbCl$_3$ moieties parallel to (100) (Fig.11.84). Somewhat similar SbCl$_3$ sheets are found in {phenanthrene⋯2SbCl$_3$} but here the aromatic-molecule sheets contain two phenanthrenes rather than one as in the other structures (Fig. 11.85). A consequence is that the two SbCl$_3$ moieties interacting with the two outer rings of a particular phenanthrene molecule are both on the same side of its plane rather than on opposite sides as in most of the other structures. Single phenanthrene sheets are found in {phenanthrene⋯2SbBr$_3$} (Fig. 11.86) but here the SbBr$_3$ arrangement is rather

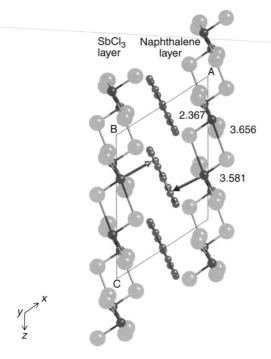

Fig. 11.82. The {naphthalene···2SbCl$_3$} structure viewed down [010]. The naphthalene molecules are at crystallographic centres of symmetry and are seen edge-on. The alternating naphthalene and SbCl$_3$ layers are clearly seen. Compare Fig. 11.84. (Data from Hulme and Szymanski, 1969.)

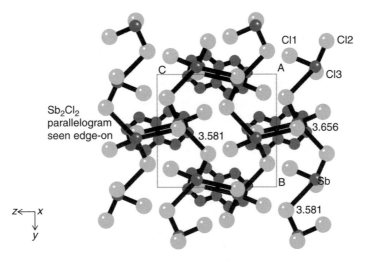

Fig. 11.83. The {naphthalene···2SbCl$_3$} structure viewed down [100]. The alternating naphthalene and SbCl$_3$ layers are clearly seen. The Sb$_2$Cl$_4$ parallelograms (sometimes described as Sb$_2$Cl$_2$ or Sb$_2$Cl$_6$) are cross-linked to form layers. The Sb atoms are shown by open circles. (Data from Hulme and Szymanski, 1969.)

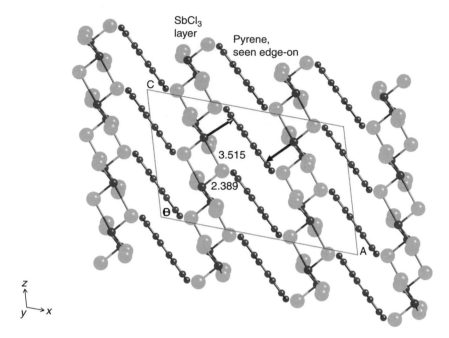

Fig. 11.84. Projection of the {pyrene···2SbCl₃} (DEJYAS) structure down [010] (*C2/c*, *Z* = 4). The layered structure is clearly visible. The pyrene molecules lie on crystallographic two-fold axes parallel to [010] while the SbCl₃ moieties are all crystallographically equivalent. The arrows show the Sb . . . ring interactions. The structure resembles that of {naphthalene···2SbCl₃} (Fig. 11.82) and differs from those of {pyrene···2SbBr₃} and {pyrene···2BiCl₃}, which are isomorphous. (Data from Mootz and Händler, 1985.)

more complicated than in the previously-mentioned SbCl₃ layers (Fig.11.84); again the SbBr₃ moieties interacting with a particular phenanthrene molecule are on the same side of the phenanthrene plane.

In {diphenyl···2SbCl₃} (BIPHSB10; Lipka and Mootz, 1978) the two phenyl groups and the two SbCl₃ groups are crystallographically independent (Fig. 11.87). One of the Sb atoms (Sb(1) participates in a centrosymmetric chain (d(Cl···Sb = 3.44, 3.46 Å), with the sixth coordination site occupied by one of the phenyl groups of the diphenyl molecule. The chlorine atoms above and below the plane of the chain form secondary links (d(Cl···Sb = 3.44 Å) to the second type of Sb atom (Sb(2)), the coordination of which is completed by three covalent Sb–Cl bonds and a fifth interaction to the second ring of a diphenyl molecule. Thus Sb(1) is six-coordinated and Sb(2) five-coordinated; the torsion angle about the central C–C bond in diphenyl is 40.5°, quite different from that in the SbBr₃ analog.

In {diphenylamine···2SbCl₃} there are dimers but these are folded (and hence not centrosymmetric) and linked in chains, with Sb atoms here interacting with both rings of the diphenylamine molecule (Fig.11.88).

(c) 'Half-sandwich' structures   The monoclinic crystals of {1,2,3-trimethylbenzene . . . BiCl₃} have been described as having a layer coordination structure, built up from

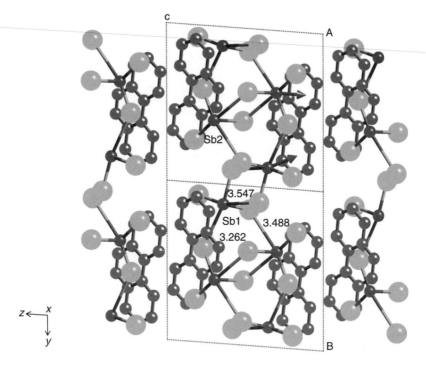

Fig. 11.85. Projection of the {phenanthrene⋯2SbCl$_3$} (PHENBS) structure along [100], showing the double layer of phenanthrene molecules related by centres of symmetry. There are two crystallographically independent SbCl$_3$ groups in the unit cell; the Sb interact with the two outer benzene rings of a phenanthrene molecule from the same side, as indicated by the arrows. Secondary Sb . . . Cl distances are shown. The Sb$_2$Cl$_2$ parallelograms are linked into chains along [010]. (Data from Demalde *et al.*, 1972.)

quasi-dimeric units of arene-coordinated BiCl$_3$ fragments (Fig. 11.89; Frank and Reiland, 1998; PUVLOH). Only non-bonding arene-arene contacts are observed between adjacent layers, showing that the structure has a backbone chain of linked Bi$_2$Cl$_6$ parallelograms sheathed on both sides by the arene moieties. The covalent Bi–Cl distances are 2.468, 2.523 and 2.528 a, while there are secondary interactions at 3.293 and 3.406 Å. The shorter Bi⋯C interaction is η^2 (distances are 3.17 and 3.24 Å to C5 and C4 respectively). Because the metal⋯arene interaction is only on one side of the aromatic ring, the arrangement was described as a 'half sandwich'.

Orthorhombic crystals of {benzene⋯BiCl$_3$} (Frank *et al.*, 1993; LAHJAF) that rapidly lose benzene *in vacuo*, have space group, $P2_12_12_1$ (Z = 8). This is the only chiral structure so far encountered in this family of compounds, and Frank *et al.* determined the absolute configuration of the crystal used in their structure analysis. There are two crystallographically independent BiCl$_3$ moieties in the crystal, and similarly for the benzenes, and the Bi⋯C interaction is weaker for one pair than for the other (Fig. 11.90). The structure was interpreted as that of a layer coordination polymer, built up from quasi-molecular BiCl$_3$ units linked by weak chlorine bridges. Alternatively, as the

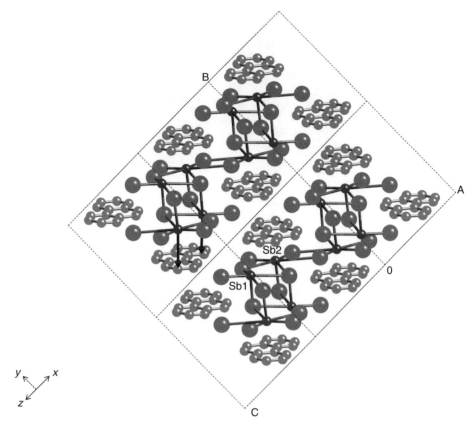

Fig. 11.86. Projection of {phenanthrene···2SbBr$_3$} (DEJYEW) down [$\bar{1}0\bar{1}$]. Sb(1) and Sb(2) are rystallographically independent and have distorted six- and seven-fold coordination respectively. The arrows show the Sb...ring interactions, both from the same side of the phenanthrene molecule. Not all the secondary Sb...Br interactions have been included. (Data from Mootz and Händler, 1985.)

metal ... arene interaction is only on one side of the arene, this can also be called a 'half-sandwich' structure (compare Figs. 11.89 and 11.90). The overall packing arrangements in these two structures do not differ in principle from those (based on chains and/or layers) found in the other arene ... MX$_3$ compounds discussed until now.

Mesitylene forms isomorphous triclinic 1 : 1 compounds with MX$_3$ (M = Sb, Bi; X = Cl, Br) (Schmidbaur, Wallis *et al.*, 1987a). Although these have also been described as 'half-sandwiches', there are important differences from the first group. The crystals of {mesitylene···SbBr$_3$} and {mesitylene···BiCl$_3$} were reported to be triclinic, with unit cells of similar dimensions containing eight formula units. From a crystallographic point of view, it is interesting to note that the reported unit cells show a greater resemblance than the reduced cells (Table 11.29). Indeed, the two structures are similar but not identical; there is no evidence for disorder. The MX$_3$ groups are in an approximate η^6 relationship to the mesitylene rings, with half the metal atoms having quasi-octahedral coordination and the other half having pentagonal bipyramidal coordination; as the MX$_3$ groups are only on one side of the mesitylene rings, these were also called 'half sandwich'

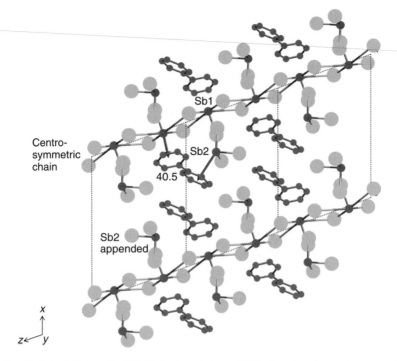

Fig. 11.87. Triclinic {diphenyl···2SbCl₃} viewed down [010], showing the centrosymmetric chain of Sb(1)Cl₃ groups, with appended Sb(2)Cl₃ groups. The arrows show the secondary interactions of Sb with the two phenyl rings of diphenyl. Dimensions in the chain are given in the text. (Data from Lipka and Mootz, 1978.)

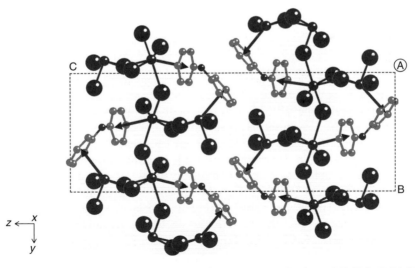

Fig. 11.88. Projection down [010] of the structure of {diphenylamine···2SbCl₃} (DPASBC10), showing the folded Sb₂Cl₆ dimers linked into zigzag chains, with both rings of the diphenylamine molecules interacting with Sb atoms. (Data from Lipka, 1978.)

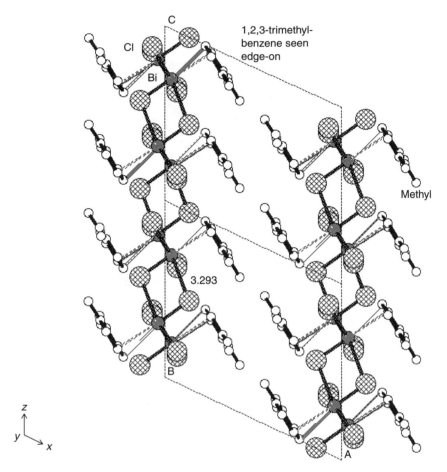

Fig. 11.89. {1,2,3-trimethylbenzene...BiCl₃} viewed down [010]. There is a striking resemblance to the analogous projections shown for {naphthalene.2SbCl₃} (Fig. 11.82) and {pyrene·2SbCl₃} (Fig. 11.84). (Data from Frank and Reiland, 1998.)

structures. In {mesitylene⋯BiCl₃} covalent d(Bi–Cl) are very similar at 2.465(5), 2.489(4) and 2.483(5) Å, with secondary d(Bi...Cl) = 3.275, 3.302, 3.180, 3.368 Å. The corresponding values in {mesitylene⋯SbBr₃} are 2.5 Å and 3.6 Å.

The overall packing in {mesitylene⋯BiCl₃} is quite different from that in the {benzene...BiCl₃} and {1,2,3-trimethylbenzene⋯BiCl₃} despite use of the same 'half-sandwich' name for both (compare Figs. 11. 89 and 11.90 on the one hand with that of the hexamers of {mesitylene⋯BiCl₃} (Fig. 11.91) on the other). The hexamers of 'half-sandwich' {mesitylene⋯BiCl₃} lead us on to the tetramers of 'full-sandwich' {hexamethylbenzene⋯2AsCl₃} (Fig. 11.96).

Sandwich structures   There is a remarkable group of four isomorphous structures (Table 11.30; Burford *et al.*, 1996) based on that of the (very hygroscopic) {benzene⋯ 2SbCl₃} (Mootz and Händler, 1986). Burford *et al.* entitled their paper "Tethered diarenes

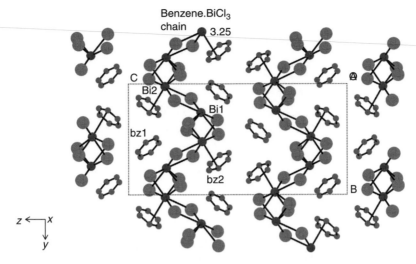

Fig. 11.90. Structure of {benzene•••BiCl₃} viewed down [100]. The two independent Bi atoms are designated Bi1 and Bi2, and the benzenes bz1 and bz2. The shortest Bi2...C distance is 3.25 Å (to bz2) and is shown in the diagram. The Bi1...C distances (to bz1) are all (slightly) greater than 3.5 A. The covalent Bi–Cl distances range from 2.444 to 2.482 Å, and the secondary Bi...Cl distances from 3.228 to 3.621 Å. (Data from Frank *et al.*, 1993.)

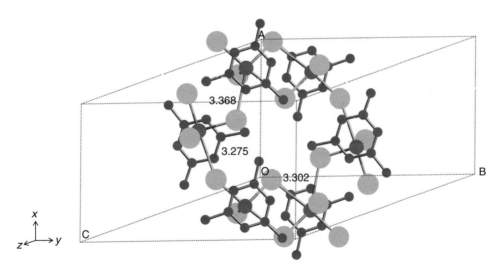

Fig. 11.91. One of the hexameric units in {mesitylene•••BiCl₃}; there is a second, crystallographically independent hexamer with slightly different dimensions. The mesitylene molecules alternate above and below the mean plane. The distances corresponding to the secondary interactions are for this particular hexamer. There are even longer M...X distances between hexamers. (Data from Schmidbaur, Wallis *et al.*, 1987.)

Table 11.29. Cell dimensions (Å, deg.) for {mesitylene···SbBr₃} and {mesitylene···BiCl₃}. Both compounds crystallize in space group $P\bar{1}$, Z = 8 (Schmidbaur, Wallis, Nowak, Huber and Müller, 1987)

Compound	Reported cells			Reduced cells		
	a/α	b/β	c/γ	a/α	b/β	c/γ
{mesitylene···SbBr₃}	9.401	14.200	22.759	9.401	14.200	22.501
FOWYAR (1)	109.27	102.94	87.94	107.29	101.76	92.06
{mesitylene···BiCl₃}	8.716	14.150	21.740	8.716	14.150	21.740
FAPYOK10 (2)	108.64	101.24	88.88	71.36	78.76	88.88

Notes:
(1) reported and reduced cells are different. The reduced cell is Type II (all angles obtuse).
(2) reported cell is a non-standard orientation of the reduced cell., which is Type I (all angles ≤ 90°).

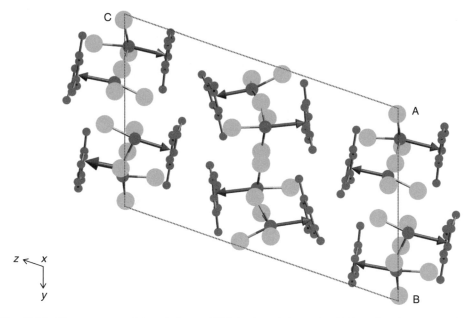

Fig. 11.92. The structure of {mesitylene···BiCl₃} viewed down the [100] axis. The arrows show the arene . . . Bi interactions. The central group of molecules is that shown in Fig. 11.91. This is a layer structure. (Data from Schmidbaur, Wallis, Nowak, Huber and Müller, 1987.)

as four-site donors to SbCl₃." The prototype {benzene···2SbCl₃} structure is of the layer type, with the feature that SbCl₃ moieties interact with the benzene molecule from both above and below with an η^6 arrangement. In this respect it resembles the {p-xylene···2SbCl₃} and {C₆(CH₃)₆···2MX₃} structures (for the latter, see below). We illustrate for the {tolane . . . 4SbCl₃}, the best determined of the four crystal structures (Figs. 11.93 and 11.94). Comparison of the four structures suggests that they are

Table 11.30. Four isomorphous structures. All crystallize in space group $P\bar{1}$; FAGFAU has 2 formula units in the unit cell and the others have 1. The triclinic cells are reduced but the ordering of the axes, where we follow reference B96, is not standard

Compound	a/α	b/β	c/γ	Reference
$C_6H_6\cdots2SbCl_3$ FAGFAU	8.211	11.833	8.165	MH86
	94.00	108.55	94.22	
$PhCH_2CH_2Ph\cdots4SbCl_3$	8.384	11.950	8.103	HH66, B96
DBDSBC01	94.81	108.64	97.42	
$PhCH{=}CHPh\cdots4SbCl_3$	8.363	11.950	8.099	HH66, B96
STISBC01	95.44	109.15	96.28	
$PhCCPh\cdots4SbCl_3$	8.379	11.965	8.122	B96
ZOQWUX	95.51	110.37	94.69	

References:
B96 – Burford *et al.*, 1996; HH66 – Hulme and Hursthouse, 1966; MH86 – Mootz and Händler, 1986.

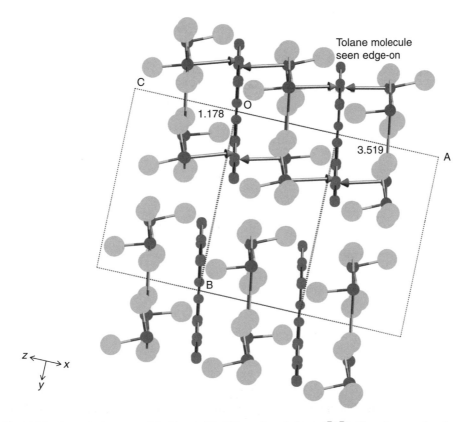

Fig. 11.93. Crystal structure of {tolane$\cdots$4SbCl$_3$} projected down [$\bar{1}0\bar{1}$]. The tolane molecules are at centres of symmetry, with the acetylene bond indicated by its length 1.178 Å. The arrows show the interactions between Sb and phenyls, the Sb to ring centroid distances being 3.21 and 3.35 Å. (Data from Burford *et al.*, 1996.)

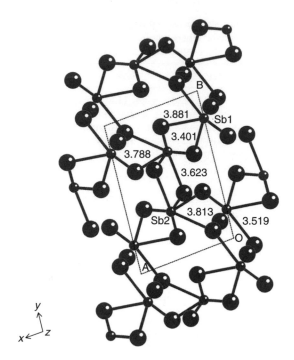

Fig. 11.94. Detail of the SbCl$_3$ network in the crystal structure of {tolane$\cdots$4SbCl$_3$} projected down [$\bar{1}0\bar{1}$]. A number of secondary interactions not listed by Burford *et al.* are included. (Data from Burford *et al.*, 1996.)

determined by the interaction between benzene ring and Sb, and that the groups linking the benzene rings (none; –CH$_2$–CH$_2$–; –CH=CH–; –C≅C–) have little structural importance. These are 'inverse sandwich' structures, with the benzene rings coordinated at both faces. The 1 : 4 compounds of dibenzyl, and stilbene with SbCl$_3$ were briefly reported more than thirty years ago (Hulme and Hursthouse, 1966).

The 1 : 2 molecular compounds of C$_6$(CH$_3$)$_6$ (HMB) with SbCl$_3$, SbBr$_3$, BiCl$_3$ and BiBr$_3$ (Schmidbaur, Nowak, Schier, Huber and Müller, 1987) all crystallize in isomorphous tetragonal unit cells, where the cell dimensions of {C$_6$(CH$_3$)$_6$$\cdots$2SbCl$_3$}, {C$_6$(CH$_3$)$_6$$\cdots$2BiCl$_3$} and {C$_6$(CH$_3$)$_6$$\cdots$2SbBr$_3$}, {C$_6$(CH$_3$)$_6$$\cdots$2BiBr$_3$} are pairwise closely similar (Table 11.31); {C$_6$(CH$_3$)$_6$$\cdots$2AsCl$_3$} (Schmidbaur, Nowak, Steigelmann and Müller, 1990) also fits into this group; they are discussed together, the minor differences being ignored. The projection down [100] for {C$_6$(CH$_3$)$_6$$\cdots$2AsCl$_3$} is shown in Fig. 11.95, and shows the hexamethylbenzene molecules sandwiched between groups of AsCl$_3$ molecules.

The η^6 AsCl$_3$$\cdots$C$_6$(CH$_3$)$_6$$\cdots$AsCl$_3$ moieties are arranged in tetramers (Fig. 11.95 and 11.96); there is no phase transition nor change in the disorder pattern on cooling to 123K. The structure as whole is built up by cross-linking of tetramers at the metal centres with other tetramers by double-sided arene coordination. The disorder makes some of the details of these structures suspect: for example, the unusual secondary Sb . . . Cl distance

Table 11.31. Crystal data for the similar ("isotypical") {HMB···2MX₃} molecular compounds (all crystallize in space group $P4_2/nnm$ with $Z = 4$)

Formula	a	c	Reference
HMB···2SbCl₃ At 300K FOWXAQ	12.683	12.379	SN87a
HMB···2SbCl₃ At 123K FOWXAQ01	12.581	12.315	HLK00
HMB···2BiCl₃ FOWXEU	12.599	12.315	SNS87
HMB···2SbBr₃ FAPYUQ	12.950	13.015	SNS87
HMB···2BiBr₃ FOWXOE	12.915	12.966	SNS87
HMB···2AsCl₃ (at 233K)·TADDUX	12.644	11.967	SNSM90

References:
HLK00 – Hubig, Lindemann and Kochi, 2000; SNS87 – Schmidbaur, Nowak, Schier, Wallis, Huber and Müller,1987; SNSM90 – Schmidbaur, Nowak, Steigelmann and Müller, 1990.

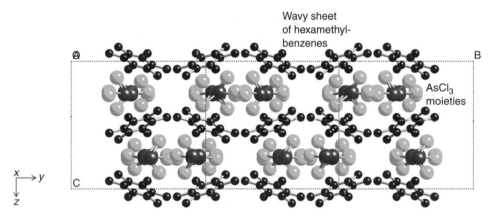

Fig. 11.95. Crystal structure of $C_6(CH_3)_6$···2AsCl₃ seen in projection down [100]. The disorder in the AsCl₃ moieties has not been removed. The $C_6(CH_3)_6$ molecules interact with AsCl₃ moieties above and below the benzene ring in an approximate η^6 arrangement. (Data from Schmidbaur, Nowak, Steigelmann and Müller, 1990.)

of 2.895(3) Å in {$C_6(CH_3)_6$···2SbCl₃} is not encountered elsewhere and appears to require independent confirmation.

The As atoms are 3.2 Å from the centers of the benzene rings; this is the longest arene-to-metal distance observed in the series of group 15 element/hexamethylbenzene adducts (As/Sb/Bi: 3.20/3.15/3.07 Å). There are other differences of detail between the analogous compounds. For example, the three covalent d(Bi–Cl) distances are respectively 2.404(8), 2.438(8) and 2.887(6) Å, considerably distorted from its gas phase structure of three similar Cl–Bi bonds; this does not appear to happen in the corresponding SbX₃ molecular compounds. No explanation has been offered. If one includes the secondary interactions, then the M atoms are all 7-coordinate and are excellent examples of "split-site octahedral coordination" (Herbstein *et al.*, 1989). Pentamethylbenzene gives similar adducts with

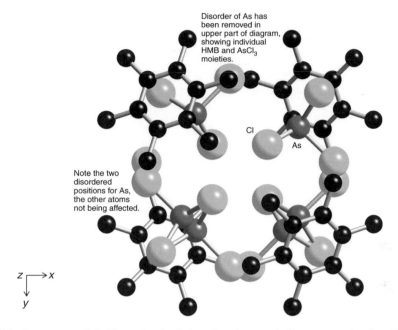

Fig. 11.96. A tetramer of $AsCl_3$ molecules below four hexamethylbenzene molecules. There is a similar layer of $AsCl_3$ molecules above the hexamethylbenzene molecules. The covalent As–Cl bonds have lengths 2.158(2), 2.195(2) and 2.186(2) Å, and are considerably shorter than the secondary As$\cdots$Cl interactions not shown in the diagram. Disorder is found in all the analogous crystals. (Data from Schmidbaur, Nowak, Steigelmann and Müller, 1990.)

MX_3; here there must be additional disorder of the arene molecule, and the structures were not determined. Durene also forms a $1:2$ adduct with $SbCl_3$ (tetragonal, 17.326 24.990 Å, $I4_1/acd$, $Z = 16$). This also has tetrameric chlorine-bridged Sb_4Cl_{12} units linked to the arene, and thus is structurally analogous to the adducts described above, but not isostructural (different cell dimensions and space group).

11.12.1.5   *Ternary molecular compounds*

A number of ternary molecular compounds have been reported; a few structures are known. The third component is presumably usually present as solvent of crystallization without interaction with the MX_3 group. Examples are $2(1,4$-diphenylbutadiene$)\cdots7(BiCl_3)\cdot$ $8(p$-xylene$)$; $3(1,4$-diphenylbutadiene$)\cdots4(BiBr_3);(p$-xylene$)$; $3(1,4$-diphenylbutadiene$)\cdots$ $4(BiBr_3)\cdot$-(ethylbenzene) (all Peyronel *et al.*, 1968); naphthalene$\cdots2(BiBr_3)\cdot1/3$(toluene); pyrene$\cdots2(BiBr_3)\cdot xS$, where S $= p$-xylene, CH_2Cl_2 or diethyl ether (Buffagni *et al.*, 1968); 4(fluorene)$\cdots7(BiBr_3)\cdot(p$-xylene$)$; fluoranthene$\cdots2(BiBr_3)\cdot x(p$-xylene$)$; acenaphthene$\cdots$ $2(BiBr_3)\cdot$; x$(p$-xylene$)$; acenaphthene$\cdots6(BiCl_3)\cdot3$(toluene); (all Vezzosi *et al.*, 1968) and 5(fluoranthene)$\cdots9(SbCl_3)\cdot$cyclohexane (Peyronel *et al.*, 1970). These earlier results should be viewed with some caution because more recent results for $1:2$ compounds of perylene, pyrene (isomorphous with the $BiCl_3$ compound), acenaphthene, phenanthrene and

Table 11.32. Crystal data for ternary arene·MX$_3$·solvent molecular compounds.

Formula	a/α	b/β	c/γ	Space group	Z	Reference
{[2.2.2.]paracyclophane··· 2SbCl$_3$·1/2(benzene)} KISXAL	14.371	8.716 101.05	24.384	P2$_1$/n	4	P91
{[2.2.2.2]paracyclophane··· 2SbBr$_3$}·toluene GOMWEK*	12.513 90.76	12.957 101.13	7.165 112.95	P1̄	1	F98
{[2.2.2]paracyclophane. 3BiCl$_3$·benzene} (KISXEP)	12.155(3)	20.523(3)	29.952(5)	Pbca	8	P91
Fluoranthene·2BiBr$_3$· 0.5(p-xylene) JEJPAP**	9.812 71.59	11.199 80.28	12.861 74.38	P1̄	2	B90
(perylene)$_3$·4BiBr$_3$ ZUVTAL**	10.250 101.44	10.605 91.33	14.418 106.23	P1̄	2	B96
{(pyrene) [2.2]paracyclophane··· 2BiBr$_3$} ZOCWOD**	7.871 92.77	10.440 102.67	10.853 104.61	P1̄	1	B95

* Reduced cell (unconventional setting).
** reduced cell (standard setting).

References:
B90 – Battaglia et al., 1990; B95 – Battaglia et al., 1995; B96 – Battaglia et al., 1996; F98 – Ferrari et al., 1998; P91 – Probst et al., 1991;

fluorene with BiBr$_3$ (Battaglia et al., 1990) show solvent of crystallization only in the fluoranthene compound. The crystal structures (Table 11.32) of the ternary compounds fall into two groups: in the first the third component fills the familiar 'included solvent of crystallization' role. However, in the second group, the arene component fills two different structural roles, in one of which it interacts with the metal atom while in the other it appears to behave only as a space-filler. There are two examples known; in ZUVTAL the same arene (perylene) fills two different structural roles, while in ZOCWOD there are two different arenes, [2.2]paracyclophane interacting with BiBr$_3$ and pyrene filling space.

The arrangement in {[2.2.2]paracyclophane···2SbCl$_3$·1/2(C$_6$H$_6$)} (KISXAL) consists of Sb$_4$Cl$_{12}$ tetramers linked through Sb . . . arene interactions to two of the three benzene rings of the paracyclophane, thus forming a set of zigzag chains along [001] (Figs. 11.97 and 11.98; crystal data in Table 11.32; Probst et al., 1991). This arrangement leaves a set of channels along [010], which contain the benzenes of crytsallization (seen edge-on in the diagram). Both tetramers and benzenes are located at crystallographic centres of symmetry. There is no Sb . . . benzene of crystallization interaction.

The structure of {fluoranthene·2BiBr$_3$}·0.5(p-xylene) (Battaglia et al., 1990; JEJPAP) is related to this group. The triclinic crystals have centrosymmetric tetramers of loosely linked BiBr$_3$ molecules interlayered by pairs of face-to-face fluoranthene molecules. The p-xylenes, at centres of symmetry, are solvent of crystallization. The two crystallographically independent BiBr$_3$ molecules each have three covalently bonded bromines with d(B$_i$–B$_r$) = 2.571, 2.679, 2.695 (Bi1) and 2.601, 2.628, 2.670 Å. There are two pairs of secondary Bi . . . Br distances (bridging bromines) at longer distances (3.165, 3.494 and 3.267, 3.375 Å). Two benzene rings of a fluoranthene molecule interact with Bi1 and Bi2 as shown in Fig. 11.99 (pairs of closest d(Bi . . . C) are in the range 3.23–3.38 Å, indicating

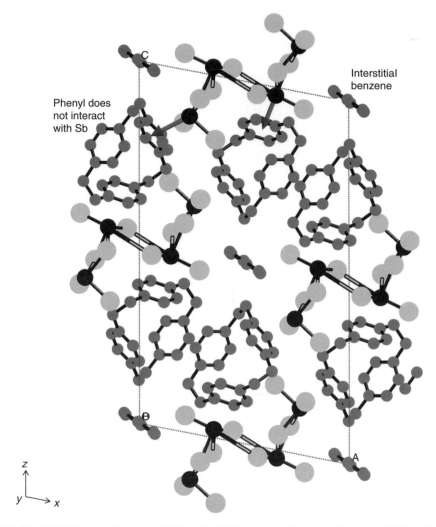

Fig. 11.97. {[2.2.2]paracyclophane·2SbCl$_3$·1/2(C$_6$H$_6$)} viewed in projection down [010]. The two arrows show examples of the Sb . . . benzene interaction, which is η^6 to a good approximation. The distances between Sb and benzene planes are 3.047 and 3.050 Å respectively. (Data from Probst *et al.*, 1991.)

η^2 interactions). The isolated BiBr$_3$ tetramers can be contrasted with the interlinked tetramers found in phenanthrene·2SbBr$_3$.

[2.2.2.2]paracyclophane·2BiBr$_3$·toluene (GOMWEK; Ferrari *et al.*, 1998) has centro-symmetrical paracyclophane molecules of G$^+$G$^-$G$^+$G$^-$ conformation interacting with Bi atoms of chains of linked BiBr$_3$ molecules (Fig. 11.100). The disordered. toluene molecules are at interstitial positions and do not interact with Bi. The Bi . . . arene interaction is η_2, the two shortest distances being 3.27(2) and 3.31(2) Å. The chains are made up of interacting BiBr$_3$ moieties with twisted Bi$_2$Br$_4$ parallelograms; although it is out-of-order, this is an appropriate place to note the similarity of these parallelograms to

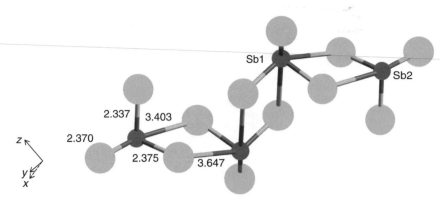

Fig. 11.98. Detail of the dimensions in the inorganic tetramer in {[2.2.2]paracyclophane·2SbCl₃· 1/2(C₆H₆)}. The coordination spheres about Sb are completed by interactions to two benzene rings of the [2.2.2]paracyclophane; thus Sb1 is 6-coordinate and Sb2 5-coordinate. (Data from Probst *et al.*, 1991.)

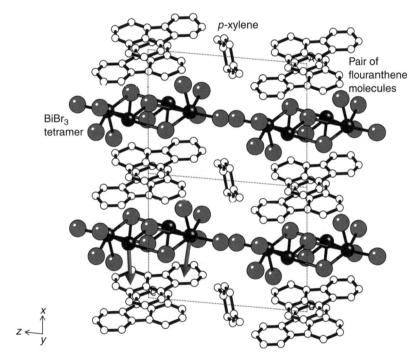

Fig. 11.99. The structure of {fluoranthene·2BiBr₃}·0.5(*p*-xylene) viewed in projection down [010]. (Data from Battaglia *et al.*, 1990.)

those in pyrene·[2.2]paracyclophane·2BiBr₃ (ZOCWOD; Fig. 11.106). The structures of GOMWEK and [2.2]paracyclophane·2BiBr₃ (JOLBAN; Vezzosi *et al.*, 1992) are compared in Figs. 11.100 and 11.101, emphasizing the similar structures of the inorganic chains and the modes of their interactions with benzene rings of the two cyclophanes.

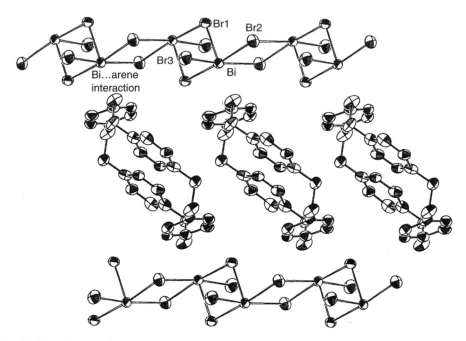

Fig. 11.100. View of the [2.2.2.2]paracyclophane·2BiBr₃·toluene (GOMWEK) structure, showing chains of BiBr₃ molecules linked by secondary interactions between themselves, and also to benzene rings of the cyclophanes. The covalent Bi–Br distances are 2.668(2), 2.659, 2.5982 Å to Br1, 2 and 3. The secondary distances are 3.221(2) (to Br1) and 3.419 Å (to Br2). The toluene solvate is not shown. (Reproduced from Ferrari *et al.*, 1998.)

The interatomic distances quoted in the captions to these figures are hardly significantly different. Thus this pair of structures is a model illustration of 'isostructurality.'

Perhaps the most complicated of the structures belonging to the present family is that of {[2.2.2]paracyclophane·3BiCl₃·benzene; KISXEP} (Probst *et al.*, 1991). The crystals are orthorhombic with the chemical moieties at general positions. The projection down [100] (Fig. 11.102) suggests a layer structure but this is misleading because of the secondary Bi . . . arene interactions to the benzene rings of the cyclophane. The BiCl₃ have a ladder structure of interlinked quadrilaterals extending (roughly) along [100], with cross-links along [010] (Fig. 11.103). The cross-linked sheet is made up of interacting elements that have been encountered in many of the examples discussed above.

{(Perylene)₃·Bi₄Br₁₂} (ZUVTAL) has perhaps the most intricate arrangement of moieties thus far encountered among the Menshutkin compounds with *isolated* (BiBr₃)ₙ oligomers (Battagalia, Bellito *et al.*, 1996). The preparation was noteworthy "The compound was obtained by mixing a *p*-xylene 1 : 1 solution of perylene and pyrene and a solution of BiBr₃ . . . *In the absence of pyrene,* the title compound was not obtained" (italics added). No explanation was offered. The asymmetric unit of the black triclinic crystals (Table 11.32) contains one perylene (a) at a general position, and half a perylene (b) centred at inversion centers. The Bi₄Br₁₂ moiety is also centred at inversion centers (Fig. 11.104). Perylene

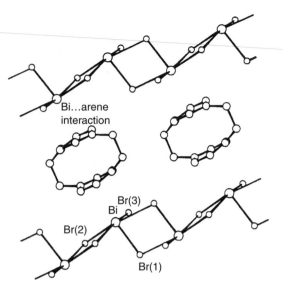

Fig. 11.101. View of the [2.2]paracyclophane·2BiBr₃ (JOLBAN) structure, showing chains of BiBr₃ molecules linked by secondary interactions between themselves, and also to benzene rings of the cyclophanes. The covalent Bi–Br distances are 2.676(3), 2.663, 2.595 Å to Br1, 2 and 3. The secondary distances are 3.325(2) (to Br1) and 3.402 Å (to Br2). (Reproduced from Vezzosi *et al.*, 1992.)

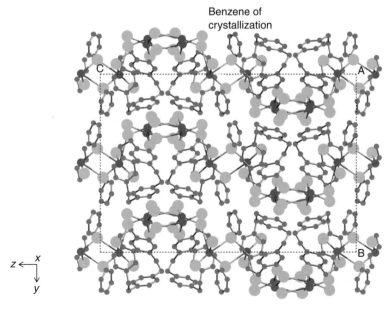

Fig. 11.102. Structure of {[2.2.2]paracyclophane·3BiCl₃·benzene} (KISXEP) seen in projection down [100]. The benzenes of crystallization are located in the (002) planes, and do not interact with the Bi atoms. The secondary Bi . . . arene interaction is not shown. (Data from Probst *et al.*, 1991.)

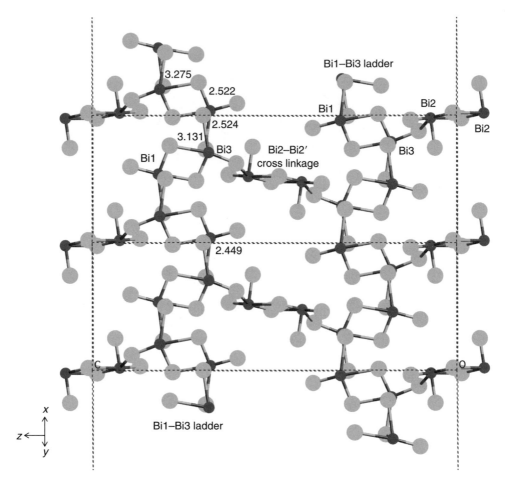

Fig. 11.103. The $(BiCl_3)_n$ cross-linked two-dimensional polymeric sheet shown after removal of the organic moieties. (Data from Probst *et al.*, 1991.)

interacts with Bi_4Br_{12} in a way that is similar to the other examples described here, including interaction between two Bi atoms above perylene (a) and two below, thus interacting with the four outer rings of perylene (a) (Fig. 11.105). Perylene (b) does not show any special interaction with Bi_4Br_{12} and appears to act just as a filler of space. There are a number of examples of similar behaviour in other molecular compounds.

In {pyrene·[2.2]paracylophane···$2BiBr_3$} (ZOCWOD) there are $BiBr_3$ chains which are layered with stacks of alternating pyrene and [2.2]paracytclophane molecules (Battaglia, Cramarossa and Vezzosi, 1995). There is a secondary interaction between Bi and the benzene rings of the [2.2]paracytclophane molecules, while pyrene, which showed high atomic displacement parameters, appears to be present as an inert "filler" (Figs. 11.106 and 11.107). The ternary compound melts without decomposition at 443K. The unusual stability of this compound compared to analogous compounds was ascribed to "the

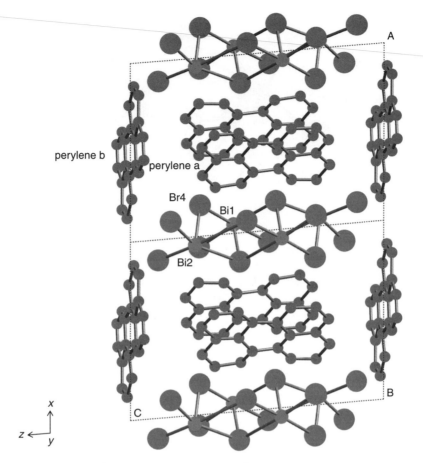

Fig. 11.104. {(Perylene)$_3$·Bi$_4$Br$_{12}$} viewed down the [010] axis of the triclinic cell. (Data from Battaglia, Bellito *et al.*, 1996.)

remarkable effects of the transannular π-electronic interactions of one benzene with another."

11.12.1.6 *Application of various physico-chemical methods*

(a) Thermodynamics The enthalpy and entropy of formation of {benzene···2SbCl$_3$} from the liquid components have been measured by Perkampus and Sondern (1980) by three independent methods – melting equilibrium, vapour pressure and calorimetrically (enthalpy) – with good agreement. The mean values are $\Delta H^0_{complex} = -55.4$ kJ/mol, $\Delta S^0_{complex} = -138.9$ J/K mol. Values of $\Delta H^0_{complex}$ have also been measured for a number of polymethylbenzene···SbCl$_3$ compounds (Table 11.33; Sondern and Perkampus, 1983).

(b) Nuclear quadrupole resonance studies and other spectroscopies NQR spectroscopy provides a rich alternative for study of the {arene·nSbX$_3$} molecular compounds

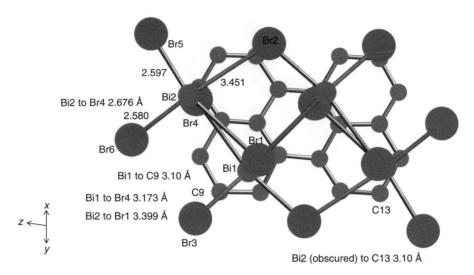

Fig. 11.105. {(Perylene)$_3$·Bi$_4$Br$_{12}$} showing detail of interaction between perylene (a) molecule and Bi$_4$Br$_{12}$ moiety. The perylene molecule above the inorganic moiety has been omitted for clarity. The sandwich [perylene (a)·Bi$_4$Br$_{12}$·perylene (a)] is the packing unit of the structure, to which perylene (b) must be added. (Data from Battaglia, Bellito *et al.*, 1996.)

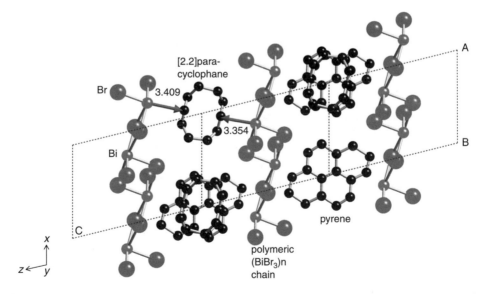

Fig. 11.106. Structure of {pyrene·[2.2]paracylophane·2BiBr$_3$} viewed in projection down [010]. A superposed pyrene has been removed for clarity in the upper left part of the diagram, and a [2.2]paracylophane in the lower right-hand corner. The arrows show the secondary interactions between Sb and the benzene rings of [2.2]paracylophane. Battaglia *et al.*, describe the structure as follows: "linear (BiBr$_3$)$_n$ polymeric chains are cross-linked in a two-dimensional network in which pyrene is included; these layers are packed by means of van der Waals contacts." (Data from Battaglia, Cramarosa and Vezzosi, 1995.)

Fig. 11.107. Structure of {pyrene·[2.2]paracylophane·2BiBr$_3$} viewed approximately down [001]. Details of the polymeric (BiBr$_3$)$_n$ chain. The interaction with the benzene rings, which completes the Bi coordination sphere, is not shown. (Data from Battaglia, Cramarosa and Vezzosi 1995.)

Table 11.33. Enthalpies of formation of crystalline polymethylbenzene/2SbCl$_3$ molecular compounds from the liquid components at the melting points. Values from Sondern and Perkampus, 1983. Other melting points in Table 11.27

Polymethylbenzene	M.Pt. (K)	ΔH^0 complex	Polymethylbenzene	M.Pt. (K)	ΔH^0 complex
benzene	352	−55.0	1,3,5-trimethylbenzene	348	−53.5
toluene	316	−45.1	1,2,3,4-tetramethylbenzene	367	−59.0
o-xylene	307	−48.0	1,2,3,5-tetramethylbenzene	372	−59.6
m-xylene	311	−51.6	1,2,4,5-tetramethylbenzene	375	−61.0
p-xylene	343	−52.7	Pentamethylbenzene	422	−74.8
1,2,3-trimethylbenzene	326	−47.5	Hexamethylbenzene	467	−80.4
1,2,4-trimethylbenzene	331	−49.1			

because ^{35}Cl, ^{121}Sb, ^{123}Sb, and ^{81}Br (also ^{79}Br) resonances can be obtained. Each crystallographically independent ^{35}Cl and ^{81}Br quadrupole nucleus gives a single line spectrum in the ranges 18–21 MHz and 130–145 MHz respectively, while NQR spectra of Sb contain a single line for each of the five ^{121}Sb and ^{123}Sb quadrupole transitions, where generally only one line for each isotope is studied. Although a wide ranging review of the combination of diffraction and NQR methods in crystal chemistry (Weiss, 1995) does not include any discussion of the {arene·nSbX$_3$} molecular compounds, these have been considered in the present context by Herbstein (2004).

The number of crystallographically independent SbX$_3$ molecules in the unit cell is shown up well both by Raman spectroscopy and by NQR. The Raman spectra in the $300 - 400$ cm^{-1} region (the SbCl$_3$ region) are very similar for SbCl$_3$ and {naphthalene···2SbCl$_3$}, but there is a doubling of the lines for {diphenyl···2SbCl$_3$} and {phenanthrene···2SbCl$_3$}. We have already noted that the first of these has one SbCl$_3$ moiety in the asymmetric unit and the others two, and a similar correspondence is obtained (Kozulin, 1971) for other examples such as {p-xylene···SbCl$_3$}, {p-xylene···2SbCl$_3$} and {benzene···2SbCl$_3$}.

The advantages of NQR spectroscopy appear when one compares the sensitivity of crystallographic and spectroscopic methods. Covalent d(Sb–Cl) covers a range of 2.3 Å to 2.4 Å, with s.u.'s of 0.01 Å (in early work) and 0.002 Å later; bond angles are in the range

92° to 96° with s.u.'s of 1° (early work) to 0.1° later. d(C–C) in arene groups will have s.u.'s of 0.01 A or more; caution suggest that little credence should be placed on deviations of C–C bond lengths in the arene moieties from standard values. Frequencies in NQR spectra are measured with a precision of a few kHz and so the halogen NQR spectrum of a particular compound gives detailed information about small differences in chemical state among these atoms *provided that frequencies can be assigned to atoms*. For this a Zeeman analysis of the single crystal NQR spectrum is needed; spectra from polycrystalline samples do not carry enough information. Problems of assignment for Sb spectra arise only if there is more than one Sb atom in the asymmetric unit.

Early work (Okuda, Nakao, Shiroyama and Negita, 1968; Okuda, 1971; Okuda *et al.*, 1972) used the Zeeman effect to match particular atoms to measured frequencies, and there was also some study of the temperature dependence of the resonance frequencies; in later work (Kyuntsel and Mokeeva (2002) who give references to earlier studies) primary attention was given to the thermal behavior but correlation of crystal structure and spectra was not neglected. Parallel crystallographic and NQR spectroscopic results are available only for three {(arene)·nSbCl$_3$} molecular compounds: {naphthalene·2SbCl$_3$} (NAPSBC; Zeeman analysis by Okuda, 1971; nuclear spin–lattice relaxation times T_1 over the range 77–293 K by Kyuntsel, 1998); {p-xylene·2SbCl$_3$} (PXYLSB; Zeeman by Ishihara, 1981; nuclear spin-lattice relaxation times T_1 over the range 77–293 K by Kyuntsel and Mokeeva, 2001; {benzene·2SbCl$_3$}(FAGFAU; Zeeman analysis by Okuda, Nakao, Shiroyama and Negita, 1968.)

There are also some partial results. The ^{35}Cl and ^{121}Sb NQR spectra have been measured for {ethylbenzene·2SbCl$_3$} and {diphenylmethane·2SbCl$_3$} (Kyuntsel, 2000) but crystal structures and Zeeman analyses are lacking. A Zeeman analysis has been carried out for {benzene·2SbBr$_3$} (Okuda, Terao, Ege and Negita, 1970) but the crystal structure is lacking; {benzene·2SbBr$_3$} is not isomorphous with {benzene·2SbCl$_3$}. The ^{35}Cl NQR spectra have been reported (Gordeev and Kyuntsel, 1992) for {diphenylamine·SbCl$_3$} (ZZZBAM10) and {diphenylamine·2SbCl$_3$} (DPASBC10) but Zeeman analyses are lacking.

The fundamental structural unit in all the {(arene)·nSbX$_3$} molecular compounds reported until now is the centrosymmetric (or approximately so) (arene)$_2$·Sb$_2$Cl$_6$ moiety. There are three differently linked Cl atoms, which give a three-line ^{35}Cl NQR spectrum. Zeeman analysis of single crystal NQR spectra enables assignment of frequencies to particular Cl atoms. On this basis the lowest frequency ν_L derives from the Cl in the Sb$_2$Cl$_2$ parallelogram (in a few examples, this is a bent quadrilateral), the middle frequency ν_M from the Cl with a secondary link to another moiety, and the highest frequency ν_H to the third Cl, without secondary linkages. There is a fair correlation between ν_L values and d(Sb...Cl) of the Sb$_2$Cl$_2$ parallelogram but hardly for ν_M and ν_H. The interaction of the π-system of the arene with Sb is shown by Sb NQR frequencies; for ^{121}Sb the frequencies for the first group of six molecular compounds listed in the previous paragraph are similar, although there are internal differences. The frequencies for the two diphenylamine compounds are in a second, lower, group.

Despite the limited database, it is possible to draw some conclusions. Firstly, the qualitative correlation of the ν_H, ν_M and ν_L values with the structural features shown in Fig. 5 seems to be well founded and one may risk predicting that it will hold for most of these molecular compounds. Quantitative correlation of ν_L with d(Sb...Cl) seems plausible but requires critical testing. Correlations for ν_H and ν_M are less convincing. The situation for [121]Sb . . . arene is intriguing; it is unfortunate that the value for {benzene·2SbBr$_3$} is lacking. Clearly there is a wealth of chemical information presently locked away, and waiting to be unravelled – the many crystal structures available need to be complemented by Zeeman analyses of strategically chosen single crystal [35]Cl and other NQR spectra.

Some early reports indicated that the present group of molecular compounds formed highly colored crystals, possibly suggesting a high degree of charge transfer. However, measurements of the UV and visible spectra of carefully prepared crystalline arene···2SbCl$_3$ compounds have shown (Perkampus and Schönberger, 1976) that the colours are due to impurities, principally the radical ions of the aromatic components; the pure molecular compounds are colourless or slightly yellow. This agrees well with conclusions drawn from NQR and Mössbauer spectroscopy (Usanovich *et al.*, 1974) that there is little change in electron density at the metal atoms or arene rings on molecular-compound formation.

It has been suggested that there are three criteria based on NQR spectra which can be used to distinguish terminal and bridging halogens in unsymmetrically bridged molecular compounds in the absence of information from crystal structure analysis, and these have been applied to HgX$_2$ and SbX$_3$ molecular compounds (Wulfsberg and Weiss, 1980):

1. If a series of NQR signals is split by more than the crystal field effect (Weiss, 1972) then the lower-frequency set of resonances will belong to the more bridging halogen atoms; this is supported by the behaviour of ν_L noted above.
2. The asymmetry parameters (η) will be larger for bridging than for terminal halogens; the η values are smaller for unsymmetrical than for symmetrical bridging and tend to merge with the values for terminal atoms.
3. The NQR frequencies of terminal halogens decrease with increasing temperature while those of bridging halogens are less temperature dependent and may even increase slightly with increasing temperature. This has been illustrated for {4-picoline-N-oxide···HgCl$_2$} (Ramakrishnan *et al.*, 1972).

This set of criteria is likely to be more immediately applicable to the HgCl$_2$ molecular compounds than to those of MX$_3$ because the former show a much more definite distinction between bridging and terminal halogens than the latter. Appropriate test examples in the latter group could be {diphenyl···2SbCl$_3$} and {p-xylene···2SbCl$_3$}.

11.12.1.7   *Summary*

There are small differences in the detailed structures (bond lengths and angles) of the SbX$_3$ molecules as they are found in the various molecular compounds and even in the same molecular compound when crystallographically independent. However, there is no obvious regularity in these effects, suggesting that the interactions involved are all about equally strong (or weak). It has sometimes been claimed that the dimensions of aromatic molecules in some molecular compounds differ from those in the corresponding neat crystals; more precise measurements appear necessary before expenditure of effort in explaining such putative differences. Conformational effects are, however, incontrovertible, e.g. in {diphenyl···2SbCl$_3$} the torsion angle about the central C–C bond is 40.5°, whereas the diphenyl molecule is planar in {diphenyl···2SbBr$_3$} and (on the average) in neat diphenyl, while 9,10-dihydroanthracene is planar in its molecular compounds with

SbBr$_3$ and 1,3,5-trinitrobenzene, but folded in its neat crystals. [2.2.2.2]paracyclophane also has different conformations in its neat crystals and in its molecular compounds with SbBr$_3$ and BiBr$_3$.

A particular aromatic moiety can, in principle, form four different Menshutkin compounds with SbCl$_3$, SbBr$_3$, BiCl$_3$ and BiBr$_3$, and different arene/MX$_3$ ratios are also possible. Thus, from comparisons of groups of compounds, one could perhaps hope to predict the effects of replacing Sb by Bi, and/or Cl by Br. For example, cell dimensions for the HMB/MX$_3$ quartet are given in Table 11.31; here replacement of Sb by Bi produces a smaller perturbation than replacement of Cl by Br. Caution is needed before generalization: for example, {phenanthrene$\cdots$2SbCl$_3$} and {phenanthrene$\cdots$-2SbBr$_3$} are not isomorphous, neither are {benzene$\cdots$2SbCl$_3$} and {benzene$\cdots$2SbBr$_3$}. However, the three pyrene compounds with SbBr$_3$, BiCl$_3$ and BiBr$_3$ are isomorphous while {pyrene$\cdots$2SbCl$_3$} is different (data in Table 11.28).

The Menshutkin molecular compounds of aromatics with MX$_3$ (M $=$ Sb, Bi; X $=$ Cl, Br) are relatively easily prepared and handled as single crystals despite their sensitivity to moisture. This family of compounds has the great advantage, whose potential has not yet been entirely realized, of being amenable to study by NQR spectroscopy (^{35}Cl, ^{79}Br, ^{81}Br, ^{121}Sb and ^{123}Sb) and Mössbauer spectroscopy (^{121}Sb), in addition to the techniques of optical spectroscopy and crystal structure analysis. The aromatic moiety acts as an electron donor towards the MX$_3$ moieties, which have similar mutual arrangements in all these molecular compounds. Various physical measurements agree in indicating that the actual transfer of charge is very small. The MX$_3$ moieties form arrangements based on the interaction of six- and seven-coordinated metal atoms, the ligands being the three covalently bound halogen atoms, a ring of the aromatic molecule and halogens of other MX$_3$ moieties linked by secondary bonding. The mutual arrangement of MX$_3$ moieties varies from one molecular compound to the next.

The arene and inorganic components of the Menshutkin compounds themselves undergo only subtle and hardly discernible changes on formation of the adducts, but the structural nature of their interaction is demonstrated by the crystal structure analyses. The arene . . . metal interaction is remarkably invariant, even to the extent of being η^2 in nature in the vast majority of the examples; only a few high-symmetry donors show η^6 interactions. These interactions have been interpreted by qualitative molecular orbital methods. Much less attention has been paid to the 'self interactions' between the MX$_3$ moieties. Why are these sometimes localized, and sometimes extended? And what governs the type of extension that is found? How should one interpret the meager thermodynamic information available? These questions can be asked also of the compounds of silver and mercury salts discussed earlier, and of those of gallium (and its congeners) discussed in the final section of this chapter.

11.12.2   *Aromatics as donors and np^3 metal ions (Ga^I, In^I, Tl^I, Sn^{II}, Pb^{II}) as acceptors*

11.12.2.1   *Introduction*

The outer electron configurations of metal ions in low oxidation states have been given in Table 11.2, and the complexes of d^{10}s^0 ions (Cu^I, Ag^I and Hg^{II}) and of some d^{10}s^2 ions (As^{III}, Sb^{III} and Bi^{III}) have been discussed in earlier parts of this Chapter. Rundle and

Corbett (1957) predicted that $d^{10}s^2$ ions would give η^6 complexes with benzene, and this prediction has proved remarkably resilient for the pure acceptors considered in Section 11.7.2. Although the benzene, *sym*-tri- and hexa-alkylbenzene complexes (alkyl = methyl, ethyl) show η^6 interactions of the metal ions with the benzene rings, the *self-interacting* complexes of MX_3 (M = As^{III}, Sb^{III} and Bi^{III}; X = Cl, Br) with (multi-ring) aromatic hydrocarbons show η^2 interactions. The interactions among the oligomeric $(MX_3)_n$ groups and the lower symmetry of the multiring aromatics presumably introduce complications not satisfactorily taken into account by simple theory. What of the remaining groups in Table 11.2 – Ga^I, In^I, Tl^I and Ge^{II}, Sn^{II} and Pb^{II}? Ge does not form complexes with benzene and its simple derivatives, but an appreciable number of complexes are known with the other elements. M^I and M^{II} arene complexes have many features in common but there are also differences, and this is reflected in our discussion. At present, 'arene' is limited to benzene or polyalkyl-benzenes, increasing alkyl substitution in benzene strongly favouring formation of complexes, while complexes with other aromatic hydrocarbons (apart from some cyclophanes) do not appear to have been reported, and perhaps do not occur.

The chemistry (preparative, NMR, crystallography) of the arene complexes of mono-valent gallium, indium and thallium has been comprehensively reviewed by Schmidbaur (1985). Schmidbaur, Thewalt and Zafiropoulos (1984) have remarked "Over a hundred years ago [in 1881], the discoverer of the element gallium [Lecoq de Boisbaudran] reported the astonishing observation that the preparations which he termed gallium(II) halides of empirical formula GaX_2 (X = Cl, Br) are freely soluble in anhydrous benzene and can be precipitated from solution with 'crystal benzene'." One milestone on the path to understanding this observation was the demonstration that crystalline $GaCl_2$ was in fact composed of Ga^+ and $Ga^{III}Cl_4^-$ ions (Garton and Powell, 1957), and this has been sup-plemented by analogous results for $GaBr_2$ (Schmidbaur, Nowak, Bublak, Burkert, Huber and Müller, 1987; Hönle, Simon and Gerlach, 1987). The complexes studied crystal-lographically can be classified into a number of groups. The invariant feature is that the M^I ion always interacts in η^6 fashion with a (possibly substituted) benzene ring[3]. Further-more, a particular metal ion may interact with one or two benzene rings. The fundamental difference between ferrocene-type compounds on the one hand and those of the present group (and of Ag^I and Cu^I) with aromatics, on the other, is the essential role played by the partially filled d-orbitals in the bonding in the ferrocene group, whereas these orbitals are filled in the second group and hence cannot participate. All the M^I (M = Ga, In, Tl) complexes have $[(A^{III})X_4]^-$ counterions, where A is Al, Ga or In. The counterions form a number of complicated arrangements, including interactions with the M^I moieties, which are somewhat reminiscent of the situations found in the self-interacting $\{(arene)M^{III}X_3\}$ complexes described in Section 11.12.1. An important chemical feature is the high sensitivity to moisture of most of the molecular compounds of aromatics with post-transition metal salts. Thus almost all of the compounds described below were prepared under pure and dry nitrogen; many of the compounds decompose under reduced pressures.

The benzene complexes of Sn(II) and Pb(II) were quite extensively studied in the 1970s by Amma and coworkers, and since then further structures have been reported by

[3] Bis[1,5-ditolylpentaazadienido-thallium(I)] (FAMPIS) and bis[1,3-diphenyltriazenido-thallium(I)] (FAMPOY) are examples of unary crystals where η^6 Tl(I)...benzene interactions are important in formation of polymeric chains in the crystal structures (Beck and Strähle, 1986).

Schmidbaur and coworkers. One group of Sn(II) and Pb(II) complexes has $Al^{III}Cl_4^-$ as counterion, while a second group of Sn(II) complexes has $(Sn_2Cl_2)^{2+}$ moieties with $Al^{III}Cl_4^-$ as counterions, coordinations being completed in both groups by secondary interactions including metal . . . benzene.

Among other complexes reported (but not yet studied crystallographically) are $C_6H_6 \cdots 2TlAlCl_4$ and $2(C_6H_6) \cdots 2TlAlCl_4$ (Auel and Amma, 1968); $C_6H_6 \cdots Ga(AlCl_4)$ and $C_6H_6 \cdots Ga(GaCl_4)$ (Rundle and Corbett, 1957); $C_6H_6 \cdots BiCl_2 \cdot AlCl_4$ (Peyronel *et al.*, 1968).

11.12.2.2 *Ga, In or Tl to One benzene ring*

1. (Hexamethylbenzene)gallium(I) tetrachlorogallate(III) tetramer·$[\{((CH_3)_6C_6)Ga^I\}^+$ $[Ga^{III}Cl_4]^-]_4$. Thewalt, Zafiropoulos and Schmidbaur, 1984a; CUTTUG.

The very soft rhombohedral crystals ($a = 12.449$ Å, $\alpha = 91.01°$, $Z = 4$, $R3m$) melt at 441 K (dec.); the structure was determined at 233(5)K. The overall structure (Fig. 11.108) may be described as consisting of four $\{[((CH_3)_6C_6)Ga^I]^+$ cations and four $[Ga^{III}Cl_4]^-$ anions, generating a hydrocarbon surface around an inorganic skeleton. The cation and

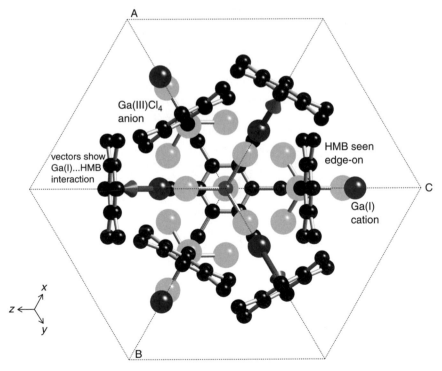

Fig. 11.108. (Hexamethylbenzene)gallium(I) tetrachlorogallate(III) tetramer viewed down the [111] rhombohedral axis. The interaction between Ga(I) (open ellipsoid) and (edge-on) hexam-ethylbenzene is shown by the arrow; the corresponding interaction along the view axis is partially obscured. The covalent Ga(III)–Cl distances in the tetrachlorogallate anions are all close to 2.17 Å, while the secondary Ga(I) . . . Cl interactions are 3.224 Å in length. (Adapted from Thewalt, Zafiropoulos and Schmidbaur, 1984.)

anion on [111] (in the center of the diagram) have crystallographic $3m$ symmetry, while those on the edges of the diagram have m symmetry. Each arene can be assigned to a discrete Ga$^+$ centre as an η^6 ligand. The distances of Ga^I to the centers of the hexamethylbenzene rings are 2.43 and 2.51 Å. Thewalt *et al.* note that "The four hexamethylbenzene rings form an efficient cover around the inorganic [Ga$^+$GaCl$_4$]$_4$ framework. There are no Ga...Cl contacts between neighbouring "clusters" and the contents of a unit is to be regarded as a tetrameric species of composition (C$_6$Me$_6$)$_4$Ga$_8$Cl$_{16}$."

2. (Hexamethylbenzene)gallium(I)   Tetrabromogallate(III)   Network.   CEFSEL. [{(CH$_3$)$_6$C$_6$)Ga(I)$^+$} {Ga^{III}Br$_4$}$^-$]. Schmidbaur, Thewalt and Zafiropoulos, 1984a.

This material (monoclinic, 10.906(3) 13.994(5) 12.415(4) Å, 103.90(2)°, Z = 4·$P2_1/c$), which is stable in a vacuum at room temperature, melts at 419K. There is a complicated network of interacting {(CH$_3$)$_6$C$_6$)Ga(I)$^+$} cations and {Ga(III)Br$_4$}$^-$ anions. Each arene can be assigned to a discrete Ga$^+$ center as an η^6 ligand, with the Ga$^+$ cations paired through bidentate bromines, the coordination sphere about Ga$^+$ being completed by

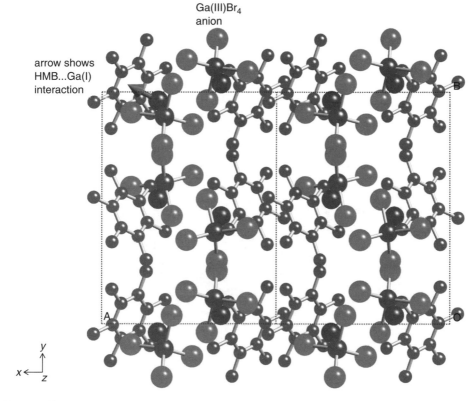

Fig. 11.109. (Hexamethylbenzene)gallium(I) Tetrabromogallate(III) Network viewed down [001]. In this figure the overall structure is emphasized; there are inorganic sheets viewed edge-on, to which arene moieties are attached on both sides. The Ga(I) atoms are identified by the associated arrows showing interactions with the hexamethylbenzenes. The covalent Ga(III)–Br distances in the tetrabromogallate anions are all close to 2.30 Å, while the secondary Ga(I)...Br interactions range up to 3.5 Å in length. (Data from Schmidbaur, Thewalt and Zafiropoulos, 1984a.)

monodentate bromines (Fig. 11.109). The distance of Ga(I) to the centers of the hexa-methylbenzene rings are 2.55 Å, with the average Ga(I)...C distance is 2.89(2) Å. The five close bromines come from one monodentate and two bidentate $\{Ga(III)Br_4\}^-$ tetrahedra.

The benzene ring of the complexed hexamethylbenzene is planar to within 0.02 Å but the methyl groups deviate from this plane by up to 0.14 Å. The crystal structures of the tetrachloro-and tetrabromogallates are quite different; Thewalt et al.(1984) consider "that the arrangement of halogens in these compounds around Ga^I centers is governed largely by packing effects originating in the relative size of the halogen atoms."

11.12.2.3   Ga, In or Tl to Two benzene rings

1.   and   2.   {Bis(benzene)gallium(I)   Tetrachlorogallate(III)}   Dimer·3(benzene). $[\{(C_6H_6)_2Ga^+\}[Ga(III)Cl_4]^-]_2\cdot3(C_6H_6)$. Schmidbaur, Thewalt and Zafiropoulos, (1983) (CEFPAE),   and   the   isomorphous   {Bis(benzene)gallium(I)   Tetrabromogallate(III)} Dimer·3(benzene). $[\{(C_6H_6)_2Ga^+\}[Ga(III)Br_4]^-]_2\cdot3(C_6H_6)$. Uson-Finkenzeller et al., 1986 (DIVNOL).

These two (colorless) compounds crystallize in isomorphous triclinic cells. We concen-trate on the bromo compound, determined at 233K (9.390(2) 10.847(2) 13.118(2) Å,

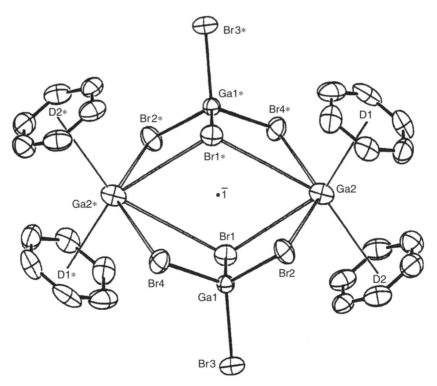

Fig. 11.110. The centrosymmetric $[\{(C_6H_6)_2Ga^+\}[Ga(III)Br_4]^-]_2$ molecule; ORTEP diagram, 50% probability ellipsoids. The atom labellled Ga2 is Ga^I while Ga1 is Ga^{III}. The four independent covalent Ga–Br distances range from 2.300(1) to 2.337(1) Å; the secondary Ga...Br distances range from 3.205 to 3.803 Å. (Reproduced from Uson-Finkenzeller et al., 1986.)

85.54(1) 102.91(1) 105.62(10°, $Z = P\bar{1}$). The structure consists of centrosymmetric dimers of composition $[\{(C_6H_6)_2Ga^+\}[Ga(III)Br_4]^-]_2$ (Fig. 11.110), with three benzene solvate molecules not coordinated to Ga. The distances of Ga(I) to the centers of the two coordinated benzene rings are 2.78 Å (Ga to C 3.08 to 3.16 Å) and 3.00 Å (Ga to C 3.24 to 3.36 Å) (both have η^6 arrangements), appreciably longer than for Ga linked to a single benzene. The angle between the vectors from Ga to the ring centers is 124.5°·Ga(I) is 6-coordinate.

These crystallographic results enable one to make some comments on earlier work in this area. The $(Ga_2Br_4)/(C_6H_6)$ pressure-composition phase diagram shows 1 : 1 and 1 : 3 phases (Oliver and Worrall, 1967), and the latter is presumably to be identified with the 1 : 3.5 compound described above. A 1 : 1 (Ga_2Cl_4)-(C_6H_6) phase giving a satisfactory chemical analysis was reported by Carlston, Griswold and Kleinberg, (1958), while Rundle and Corbett (1957) reported a 1 : 1 pseudo-hexagonal phase $a = 11.89$, $c = 30.05$ Å) but could not obtain a structure. Schmidbaur, Thewalt and Zafiropoulos (1983) remark that crystals of $[\{(C_6H_6)_2Ga^+\}[Ga(III)Cl_4]^-]_2 \cdot 3(C_6H_6)$ must not be dried by lowering pressure "as any loss of benzene in vacuo leads to phase changes and deterioration of crystal quality. It should be noted that the alleged 1 : 1 complex $Ga_2Cl_4 \cdot C_6H_6$ is reported to be obtained from $GaCl_2$ solutions in benzene upon removal of solvent at –5°C." No preparative details were given by Rundle and Corbett. It would seem that existence of a crystalline 1 : 1 phase has yet to be demonstrated.

3. Bis(mesitylene)gallium(I) Tetrachlorogallate(III) Chain Polymer. $[\{1,3,5\text{-}(CH_3)_3 H_3C_6]_2Ga(I) (Ga(III)Cl_4)]_n$. Schmidbaur, Thewalt and Zafiropoulos, 1984b; CUPJUS.

This material is stable in a vacuum at room temperature; the crystals are monoclinic (17.717(3) 10.856(2) 12.959(2) Å, 110.21(1)° at 233K, $Z = 4$, space group Cc). There are bent-sandwich moieties similar to those found in the benzene analog, but here these are linked through $(Ga(III)Cl_4)$ tetrahedra into polymeric chains running along [001] (Fig. 11.111). The Ga to mesitylene ring-centre distances are equal at 2.673 Å; the angle between the vectors from Ga2 to the two ring centers is 136.5°. The Ga–Cl distances in the anion range from 2.156(3) to 2.178(3) Å. Ga(I) is 4-coordinate.

4. Bis(mesitylene)indium(I) Tetrabromoindate(III) chain polymer. $[1,3,5\text{-}(CH_3)_3 H_3C_6]_2In(I) (In(III)Cl_4)$. Ebenhöch et al. (1984); COCFEF.

This chain polymer structure is closely related to that of the bis(mesitylene)gallium(I) tetrachlorogallate(III) chain polymer, but the two structures are not isomorphous. The colourless monoclinic crystals (at 233K $a = 10.624(2)$, $b = 13.384(3)$, $c = 17.697(4)$ Å, $\beta = 94.56(2)°$, $Z = 4$, space group $P2_1/n$) are sensitive to air and light, and readily lose mesitylene on reduction of ambient pressure. Helical chains extend along [010]; complex cations and anions are coupled so that the tetrabromoindate(III) tetrahedra are bidentate chelate-forming on the one side and monodentate-bridging on the other. Each In(I) forms two η^6 links to separate mesitylenes, with distances of 2.83 and 2.89 Å from In(I) to ring center and an angle of 47.3° between the mesitylene planes. The coordination about In(I) is completed by two links to the chelated bromines with distances of 3.446(1) and 3.503(1) Å. In(I) is 5-coordinate. Crystals of $[1,3,5\text{-}(CH_3)_3H_3C_6]_2In(I) (Al(III)Br_4)$ and $[(CH_3)_6C_6]_2In(I) (In(III)Br_4)$ were also obtained but structures have not yet been reported.

5. (Hexaethylbenzene)(toluene)Ga(I) Tetrachlorogallate(III)·hemi((hexaethylbenzene) Monomer·$[\{(C_6Et_6)(C_6H_5Me)Ga\}^+GaCl_4^-]\cdot1/2(C_6Et_6)$. Schmidbaur, Nowak, Huber and Müller, 1988; SARKEB.

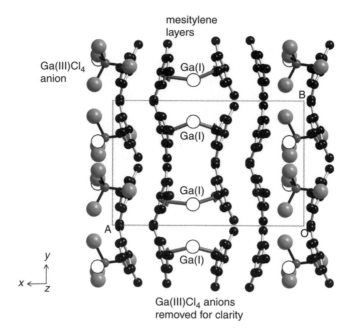

mesitylene
layers

Ga(III)Cl₄
anion

Ga(I)

B

Ga(I)

Ga(I)

A

Ga(I)

y ↑

x ← ↓ z

Ga(III)Cl₄ anions
removed for clarity

Fig. 11.111. Projection down [010] of the bis(mesitylene)Ga(I) tetrachlorogallate(III) structure (CUPJUS). The anions have been removed from the vertical band in the centre of the figure in order to show the secondary interaction of Ga(I) with bis(mesitylene). Other secondary interactions have not been included here but details are available from the original paper. (Data from Schmidbaur, Thewalt and Zafiropoulos, 1984b.)

In structural terms this is perhaps the simplest of the present group of complexes, and the only one in which Ga(I) interacts with two chemically different arenes. The colourless transparent monoclinic crystals (9.760(1), 21.447(2), 17.754(2) Å, β not given, space group $P2_1/n$; $Z=4$, measurements at 238K) melt sharply at 387K, are unstable in a vacuum and moist air but stable under a protective atmosphere. The molecules of the complex are at general positions with the uncoordinated solvation hexaethylbenzenes at centers of symmetry. Although the Ga2 to benzene ring interactions are both η^6, the linkage from Ga2 to hexaethylbenzene is markedly stronger than that to toluene (Fig. 11.112). Ga(I) is 4-coordinate.

Both coordinated and solvent hexaethylbenzene molecules have the alternating up . . . down conformation of the ethyl groups shown in Fig. 11.112, which is also found in other crystals containing hexaethylbenzene.

6. {[2.2]paracyclophane-gallium(I)   tetrabromogallate(III)·DUMVAI·[(1,4-$C_6H_4CH_2$ $CH_2)_2Ga$]⁺$(GaBr_4)^-$. Schmidbaur, Bublak, Huber and Müller, 1986a.

Mixing of the components in benzene gives the complex (both tetrachloro- and tetrabromogallate complexes were obtained but only the latter was examined crystallographically). The complexes were insoluble in most solvents so NMR spectra could not be obtained. In(I) and Tl(I) salts did not give complexes on treatment with [2.2]para-cyclophane. The bromo complex crystallized in space group *Pnma*, $Z=4$, $a=12.836(4)$, $b=10.004(3)$, $c=14.943(4)$ Å (note: CSD interchanges b and c but not y and z). There are chains of interacting moieties along [010] (Fig. 11.113). Each Ga(I) center is

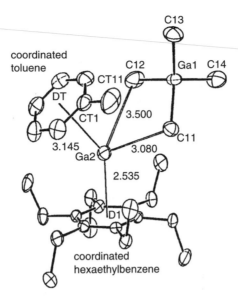

Fig. 11.112. ORTEP diagram of the monomeric molecule of (hexaethylbenzene)(toluene)Ga(I) Tetrachlorogallate(III)·hemi((hexaethylbenzene); SARKEB. Ga1 is Ga(III) and Ga2 is Ga(I). The angle DT–Ga2–D1 is 137.2(1)°, and that between the hexaethylbenzene and toluene planes is 38.8°. The Ga1–Cl distances range from 2.152(1) to 2.192(1) Å. (Adapted from Schmidbaur, Nowak, Huber and Muller, 1988.)

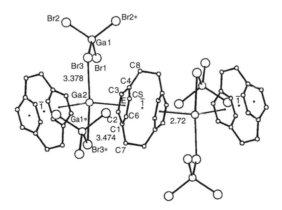

Fig. 11.113. Detail of the folded Ga(I)...paracyclophane...Ga(I) chain running along [010]. Atoms Ga1, Ga2, Br1 and Br3 are situated on crystallographic mirror planes, and the paracyclophanes are at centers of symmetry. (Reproduced from Schmidbaur, Bublak, Huber and Müller, 1986a).

η^6–coordinated to two different cyclophane molecules ('external' coordination); the crystal symmetry requires that the Ga(I)–arene distances (2.72 A) are equal. The para-cyclophanes are at centers of symmetry, and the angle subtended at Ga2 by the centers of adjacent coordinated benzene rings is 131.1(8)°. This folding is similar to that found in other bis(arene)metal complexes. The paracyclophane geometry hardly differs from that found in crystals of the neat compound. When all interactions are considered, it becomes clear that the overall structure is best described as a three-dimensional network. Ga(I) is 4-coordinate.

7. {[3.3]paracyclophane-gallium(I)   tetrabromogallate(III).   [(1,4-C$_6$H$_4$CH$_2$CH$_2$CH$_2$)$_2$Ga]$^+$(GaBr$_4$)$^-$. Schmidbaur, Bublak, Huber and Müller, 1986b; FEFWUI.

[3.3]paracyclophane forms 1 : 1 complexes with both Ga$^+$(GaCl$_4$)$^-$.and Ga$^+$(GaBr$_4$)$^-$, and the structure of the latter (monoclinic, $P2_1/m$, $Z = 2$, 8.082(1) 15.960(3) 8.818(1) Å, 107.18(1)°) has been reported. The crystals are less sensitive to air and moisture than other arene complexes of Ga(I). The structure has resemblances to that of the [2.2]para-cyclophane analog, but has two-dimensional sheets rather than a three-dimensional network. However, the immediate environment of Ga(I) is similar to that shown in Fig. 11.113, the angle subtended at Ga2 by the centers of adjacent coordinated benzene rings being 134.2°. Crystal symmetry requires that the Ga(I)–arene distances (2.76 Å) are equal.

8. {[2](1,4)Naphthalino[2]paracyclophane)gallium(I)   tetrabromogallate(III)}. [(C$_{20}$H$_{18}$)$_2$Ga]$^+$(GaBr$_4$)$^-$. Schmidbaur, Bublak, Huber and Müller, 1987b; GANPOA.

[2](1,4)Naphthalino[2]paracyclophane

The monoclinic crystals (12.353(2) 15.035(2) 12.368(2) Å, 97.98(1)°, $Z = 4$, $P2_1/n$) have a layer structure based on the interaction of Ga(I) with the linked benzene rings of the cyclophane (those on the left hand side of the structural formula shown above); the distance from Ga(I) to the center of the benzene ring is 2.67 Å (d(Ga(I) to C range from 2.89 to 3.15(2) Å), and that to the naphthalene ring is 2.85 Å (d(Ga(I) to C range from 3.06 to 3.35(2) Å), suggesting a stronger complexation with the unperturbed π-system of the former. There are also important secondary interactions between Ga(I) and two bromines of the GaBr$_4$ ions. The covalent bond lengths in the GaBr$_4$ ions are (twice) 2.306(3) Å for the two bromines that do not interact and 2.324 and 2.338(3) Å for the bromines involved in secondary interactions (distances of 3.310, 3.331 and 3.523(4) Å). The angle between the vectors from Ga(I) to the two ring centers is 124.8°. Ga(I) is 5-coordinate (Fig. 11.114).

11.12.2.4 *Mixed mono- and bis(arene) complexes*

1. {Hexakis[mesitylene]tetrathalliuml(I) Tetrakis(tetrabromogallate(III)) {(1,3,5-(CH$_3$)$_3$H$_3$C$_6$]$_6$(Tl(I))$_4$ [Ga(III)Br$_4$]$_4$}. Schmidbaur, Bublak, Riede and Müller, 1985; DEXLOH.

This is the only example (so far) of a mixed complex in which both mono- and bis(arene)M(I) moieties are present in the same cluster (Fig. 11.115). The triclinic crystals have centrosymmetric tetrameric units. In the monoarene cations the Tl(I) is located 2.94 Å above the center of the mesitylene, while in the bis(arene) cation the distances are 3.00 and 3.02 Å, with a mutual inclination of 60.5°. The cationic units are linked via tetrahedral [Ga^{III}Br$_4$]$^-$ anions into a complicated network of thallium–bromine contacts.

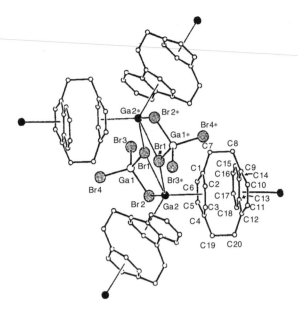

Fig. 11.114. {[2](1,4)Naphthalino[2]paracyclophane)gallium(I) tetrabromogallate(III)}, showing the centrosymmetric $[(C_{20}H_{18})_2Ga]^+(GaBr_4)^-$ grouping, which repeats to form the layer structure. Ga(I) is shown by filled spheres and Ga(III) by open spheres. Br1 and Br2 of the tetrabromogallate(III) ions are linked to Ga(I) while Br3 and Br4 do not have secondary interactions. (Reproduced from Schmidbaur, Bublak, Huber and Müller, 1987b.)

11.12.2.5   *Sn or Pb to one benzene ring*

1. The polymeric chain complex $[(\pi\text{-}C_6H_6)\cdot Sn^{2+}(AlCl_4)_2]_n \cdot nC_6H_6 \cdot BENZSN10$. Rodesiler, Auel and Amma, 1975.

The crystals are monoclinic (11.282(3) 17.317(8) 12.646(6) Å, 110.08(1)°, $Z = 4$, space group $P2_1/n$); the Sn(II) environment and chain structure are shown in Figs.11.116 and 11.117. The Al–Cl distances range from 2.068(3) to 2.199(3) Å, and the Sn...Cl distances from 2.766(2) to 3.280(2) Å. The Sn-benzene ring interaction is η^6, with the distance to the ring center 2.74(1) Å while the average Sn...C distance is 3.06 Å. A simple molecular orbital scheme (which we do not detail) was put forward to rationalize the present structure, and its relationship to $(C_6H_6)U(AlCl_4)_3 \cdot Sn(I)$ is 7-coordinate.

2. The polymeric chain complex $[(\pi\text{-}C_6H_6)\cdot Pb^{2+}(AlCl_4)_2]_n \cdot nC_6H_6$. Gash *et al.*, (1974; BZACLB).

This structure is isomorphous with the Sn(II) analog described in the previous paragraph (11.365(1) 17.307(2) 12.707(2) Å, 110.14(1)°, $Z = 4$, space group $P2_1/n$). The Al–Cl distances range from 2.05 to 2.21 Å, and the Pb...Cl distances from 2.85 to 3.22 Å. The Pb-benzene ring interaction is η^6, with the distance to the ring center 2.77 Å while the average Pb...C distance is 3.11 Å. There is approximate pentagonal bipyramidal coordination about Pb(II), with Cl(1), Cl(2), Cl(6), Cl(3′) and Cl(4′) in the equatorial plane and Cl(7) and the centre of the benzene ring in the axial sites; Pb is displaced by ≈0.6 Å out of the equatorial plane towards the benzene. The primed atoms are related

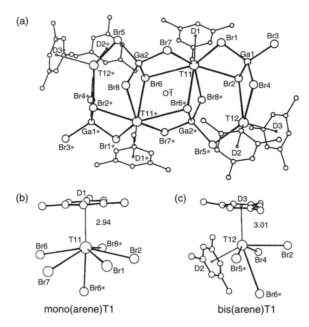

Fig. 11.115. (above) A general view of the mixed arene {hexakis[mesitylene]tetrathalliuml(I) tetrakis(tetrabromogallate(III)) structure. (below) Detail of the mono- and bis(arene)Tl parts of the structure. In the bis(arene)Tl part, Tl is 7-coordinate and the two distances from Tl to the midpoint of the arene ring are essentially equal, and a little longer than the distance for 6-coordinate mono(arene)Tl. (Adapted from Schmidbaur, Bublak, Riede and Müller. 1985.)

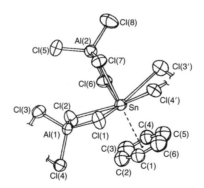

Fig. 11.116. (π-C_6H_6)·Sn^{2+} ($AlCl_4$)$_2$·C_6H_6 structure: ORTEP plot (50% probability ellipsoids) showing the sevenfold coordination about Sn(II) (two axial chlorines, four equatorial chlorines, and the center of the benzene ring). Two chlorines are not coordinated; the equatorial $AlCl_4$ groups provide the propagating element for the chains. (Reproduced from Rodesiler, Auel and Amma, 1975.)

by symmetry elements to the unprimed and their repetition forms the Pb($AlCl_4$)$_2$ chain. The second benzene is intercalated in the structure between the $AlCl_4$ groups. The overall structure can be described as a chain where one group of $AlCl_4$ tetrahedra is bridged by Pb ions and those of the second group act as chelating ligands forming axial and equatorial

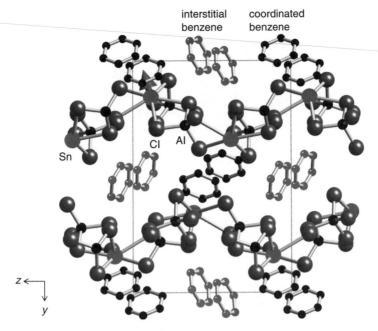

Fig. 11.117. Projection of the $(\pi\text{-}C_6H_6)\cdot Sn^{2+}$ $(AlCl_4)_2\cdot C_6H_6$ structure down [100], showing zigzag chains approximately along [001]. Most vectors linking the centers of the coordinated benzene rings (filled atoms) to Sn have been omitted for clarity. Two chlorines on one $AlCl_4$ moiety are not linked to Sn but all other chlorines form secondary links to Sn. (Data from Rodesiler, Auel and Amma, 1975.)

Pb...Cl links. Thus both benzene···metal ion and cation...anion interactions are important in stabilizing the structure. There are analogies to the structural arrangements in $C_6H_6\cdots U^{III}(AlCl_4)_3$ (Cesari *et al.*, 1971a; BNZUAL) and $\{[U^{IV}(\eta^6\text{-}C_6Me_6)Cl_2]_2(\mu\text{-}Cl_3)\}$ $AlCl_4$ (Cotton and Schwotzer, 1985; DACRII), where the rings are π-bonded to the uranium atoms.

3. The dimeric coordination complex of hexamethylbenzene with $Sn^{2+}(AlCl_4)_2$]. $[(\eta^6 - C_6(CH_3)_6)\cdot Sn^{2+}(AlCl_4)_2]_2\cdot 3C_6H_6$. Schmidbaur, Probst, Stiegelmann and Müller, (1989); JAVJIZ.+

The triclinic crystals (11.852(1) 11.839(1) 12.031(1) Å, 85.94(1) 115.59(1) 98.30(1)°, $Z=1, P\bar{1}$; reduced cell, non-standard choice of origin) contain centrosymmetric $[(\eta^6 - C_6(CH_3)_6)\cdot Sn^{2+}(AlCl_4)_2]_2$ dimers (Fig. 11.118). The difference from the polymeric chain structure $[(\pi\text{-}C_6H_6)Sn^{2+}(AlCl_4)_2\cdot C_6H_6]_n$ appears in the mode of interaction between Sn(II) and $AlCl_4$ moieties. Here the Sn(II) atoms are linked by bidentate bridging $AlCl_4$ counterions, with the Sn coordination completed by η^6 interaction to benzene, and bidentate interaction to another (nonbridging) $AlCl_4$ counterion, which has no other Cl contacts. The Al–Cl distances range from 2.092(1) to 2.178(2) Å, and the Sn...Cl distances from 2.920(1) to 3.097(1) Å. The Sn-benzene ring center, distance is 2.45 Å while the Sn...C distances range from 2.75 to 2.88 Å; these are the shortest such distances so far recorded.

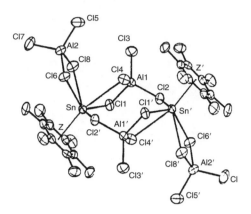

Fig. 11.118. Environment of the Sn(II) atoms in the dimeric coordination complex of hexamethylbenzene with $Sn^{2+}(AlCl_4)_2$]. (Reproduced from Schmidbaur, Probst, Stiegelmann and Müller, (1989).)

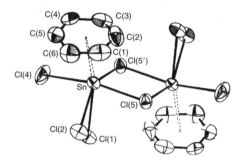

Fig. 11.119. ORTEP diagram (50% probability ellipsoids) showing the local environment of the centrosymmetric $(Sn_2Cl_2)^{2+}$ dimer in $((\pi\text{-}C_6H_6)\cdot Sn^{2+}Cl^-(AlCl_4)$. The two independent Sn–Cl bonds in the dimer parallelogram have lengths 2.614 and 2.659(4) Å. (Reproduced from Weininger, Rodesiler and Amma, 1979.)

11.12.2.6 *Sn to one benzene ring, $(Sn_2Cl_2)^{2+}$ moiety being present*

1. $\{((\pi\text{-}C_6H_6)\cdot Sn^{2+}\ Cl^-(AlCl_4)^-\}$ dimer. Weininger, Rodesiler and Amma, 1979; CBZSNA10.

The monoclinic crystals (19.624(6) 9.531(1) 7.099(1) Å, 93.65(10°, $Z=4$ formula units, space group $P2_1/n$) have centrosymmetric (and hence planar) $(Sn_2Cl_2)^{2+}$ moieties linked to $(AlCl_4)^-$ moieties (d(Al–Cl) 2.08 to 2.19 Å) to form chains along [001], with the coordination of each Sn being completed by three longer Sn . . . Cl links (2.84 to 3.32 Å) and an approximately η^6 link to benzene, with the distance to the ring center 2.90(2) Å while the Sn . . . C distances range from 3.05 to 3.39 Å (Fig. 11.119).

2. $\{(p\text{-xylene})\ Sn^{2+}\ Cl((AlCl_4)_2^-\}$ dimer. Weininger, Rodesiler and Amma, 1979; CPXSNA10.

The monoclinic crystals were described in the non-standard $I2/c$ space group ($Z=8$, 18.970 10.903 15.470 Å, 107.33°); the local environment of the planar centrosymmetric

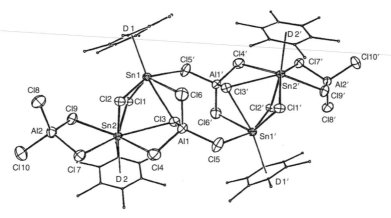

Fig. 11.120. The centrosymmetric tetrameric $[(\eta^6 - (C_6(CH_3)_6)_2Sn^{2+}Cl(AlCl_4)]_4$ molecule. (Reproduced from Schmidbaur, Probst, Huber, Müller and Krüger, 1989.)

$(Sn_2Cl_2)^{2+}$ dimer (d(Sn–Cl) = 2.616(3) and 2.684(3) Å) is much the same as shown in Fig. 11.119 although there is evidence for somewhat stronger Sn . . . *p*-xylene interaction as the Sn to ring carbon distances range from 2.92(1) to 3.27 Å.

3. Tetrameric $[(\eta^6\text{-}(C_6(CH_3)_6)_2Sn^{2+}Cl(AlCl_4)]_4 \cdot 3C_6H_5Cl$ at 238K. Schmidbaur, Probst, Huber, Müller and Krüger, 1989; SANMUP.

Colourless triclinic crystals (12.481(2) 14.749(2) 12.899(2) Å, 100.77(1) 93.62(1) 100.98(1)°, $Z = 1$ (solvated tetramer), space group $P\bar{1}$; reduced cell, nonstandard choice of origin) were obtained from chlorobenzene. The structure is composed of dimeric $[(\eta^6\text{-}C_6(CH_3)_6)_2Sn^{2+}Cl(AlCl_4)]_2$ units containing crystallographically non-equivalent monomers. These dimers are cross-linked to form centrosymmetric tetramers (Fig. 11.120). The hexamethylbenzene molecules are each η^6-coordinated to a particular Sn atom, the Sn to ring center distances being 2.60 Å (d(Sn . . . C) range from 2.831(3) to 3.110(5) Å) and 2.73 Å (d(Sn . . . C) range from 3.014(5) to 3.106(5) Å). The chlorobenzene molecules, one of which is disordered about a center of symmetry, do not make any metal contacts. The covalent Al–Cl distances (8 independent values) range from 2.104(2) to 2.191(2) Å, the Sn–Cl distances (4 independent values) within the dimer range from 2.586(1) to 2.709(1) Å, whereas the secondary Sn . . . Cl distances (7 independent values) range from 2.808(1) to 3.625(2) Å,

11.12.2.7 *Sn to two benzene rings, $(Sn_2Cl_2)^{2+}$ moiety being present*

1. Dimeric $[(\eta^6 - (C_6H_6)_2Sn^{2+}Cl(AlCl_4)]_2 \cdot C_6H_6$. Schmidbaur, Probst, Huber, Stiegelmann and Müller, 1989; VAWCAX.

The crystals are orthorhombic (19.869(1) 15.941(1) 25.954(2) Å, $Z = 8$, space group $Pbca$). Two $[(\eta^6\text{-}(C_6H_6)_2Sn^{2+}$ cations are bridged by two 1,3-bidentate tetrahedral $(AlCl_4)^-$ anions and by two bicoordinate Cl^- anions (Fig. 11.121). The angles between the planes of the pairs of coordinated benzenes are 101.9° and 99.7°, and the distances for Sn(II) to the ring centers range from 3.11 to 3.26 Å, somewhat longer than found in analogous complexes.

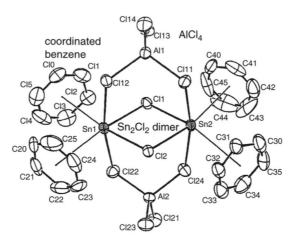

Fig. 11.121. ORTEP diagram (50% probability ellipsoids) showing the local environment of the $(Sn_2Cl_2)^{2+}$ dimer in $[(\eta^6\text{-}C_6H_6)\cdot Sn^{2+}Cl^-(AlCl_4)]_2$. The four independent Sn–Cl bonds in the dimer parallelogram range from 2.607(2) to 2.633(2) Å, with angles of 78.5° at Sn and 101.5° at Cl. The molecule of crystal benzene has been omitted. (Reproduced from Schmidbaur, Probst, Huber, Stiegelmann and Müller, 1989.)

11.13 Summary

Four of the groups of molecular compounds considered in this Chapter have similar overall features but differences in detail. The MX_3 compounds with arenes contain only neutral moieties, the AgX compounds with arenes contain a single type of charged species, $M(I)A(III)X_4$ compounds with arenes contain metal species in two oxidation states, and the HgX_2 compounds with arenes, which show the most structural variability, fall between the limits of charged and uncharged species. Of course, the extensive networks of secondary interactions will modify formal charge distributions. The inorganic part of the AgX compounds with arenes shows little variability, and the same is true for the $A(III)X_4$ moieties. The SbX_3 moieties also maintain their pristine shapes, but BiX_3 shows departures. As already noted, HgX_2 shows wide structural variety. Invariance holds rather strictly for the arene … metal interactions $-\eta^2$ for the AgX and MX_3 compounds and η^6 in the $M(I)A(III)X_4$ compounds. Isomorphism is rare in this family of molecular compounds – there are examples of replacement of Cl by Br, or Sb by Bi, or even Sn by Pb, that do not lead to changes of crystal structure – but it is more usual for the overall structure to change, even though recognizable structural features are carried over from one compound to the next.

References

Abassalti, M. and Michaud, M. (1975). *Rev. Chim. Mineral.*, **12**, 134–138.
Abe, H. and Ito, M. (1978). *J. Raman Spectroscopy*, **7**, 35–40.
Abe, Y. (1958). *J. Phys. Soc Japn.*, **13**, 918–927.

Abel, E. A., Bennett, M. A. and Wilkinson, G. (1959). *J. Chem. Soc.*, pp. 3178–3182.

Ahlsen, E.N. and Strømme, K.O. (1974). *Acta Chem. Scand.*, A**28**, 175–184.

Alcock, N. W. (1972). *Adv. Inorg. Chem. Radiochem.*, **15**, 1–58.

Alcock, N. W., Herron, N. and Moore, P. (1978). *J. Chem. Soc. Dalton Trans.*, pp. 394–399.

Allegra, G., Wilson, G. E. Jr., Benedetti, E., Pedone, C. and Albert, R. (1970). *J. Am. Chem. Soc.*, **92**, 4002–4007.

Allen, F. H. and Rogers, D. (1971). *J. Chem. Soc.* (B), pp. 257–262.

Almenningen, A. and Bjorvatten, T. (1963). *Acta Chem. Scand.*, **17**, 2573–2574.

Alshaikh-Kadir, K., Drew, M. G. B. and Holt, P. (1978). *Acta Cryst.*, B**34**, 939–941.

Alyea, E. G., Ferguson, G., McAlees, A., McCrindle, A., Myers, R., Siew, P. Y. and Dias, S. A. (1981). *J. Chem. Soc. Dalton Trans.*, pp. 481–490.

Andersen, P. and Thurmann-Moe, T. (1964). *Acta Chem. Scand.*, **18**, 433–440.

Ansell, G. B. and Finnegan, W. G. (1969). *J. Chem. Soc D*, p. 1300.

Ansell, G. B. (1976). *J. Chem. Soc. Perkin II*, pp. 104–106.

Anthonsen, J. W. (1976). *Spectrochim. Acta*, **32A**, 987–993.

Anthonsen, J. W. and Møller, C. K. (1977). *Spectrochim. Acta*, **32A**, 963–970.

Archer, E. M. (1948). *Acta Cryst.*, **1**, 64–69.

Ashworth, R. S., Prout, C. K., Domenicano, A. and Vaciago, A. (1968). *J. Chem. Soc.* (A) pp. 93–104.

Atwood, J. L. and Stucky, G. D. (1967). *J. Am. Chem. Soc.*, **89**, 5362–5366.

Atzei, D., Deplano, R., Trogu, E. F., Bigoli, F., Pellinghelli, M. A. and Vacca, A. (1988). *Can. J. Chem.*, **66**, 1483–1489.

Auel, Th. and Amma, E.L. (1968). *J. Am. Chem. Soc.*, **90**, 5941–5942.

Auger, V. (1908). *C. R. Acad. Sci. Paris*, **146**, 477–479.

Authier-Martin, M. and Beauchamp, A. L. (1977). *Can. J. Chem.*, **55**, 1213–1217.

Authier-Martin, M., Hubert, J., Rivest, R. and Beauchamp, A. L. (1978). *Acta Cryst.*, B**34**, 273–276.

Avram, M., Dinulescu, I. G., Marica, E., Mateescu, G., Sliam, E. and Nenitzescu, C. D. *Tetr. Letts.*, (1961). pp. 21–26.

Avram, M., Marica, E. and Nenitzescu, C. D. (1959). *Chem. Ber.*, **92**, 1088–1091.

Baenziger, N. C., Haight, H. L., Alexander, R. and Doyle, J. R. (1966). *Inorg. Chem.*, **5**, 1399–1401.

Baenziger, N. C., Nelson, A. D., Tulinsky, A., Bloor, J. H. and Popov, A. I. (1967). *J. Am. Chem. Soc.*, **89**, 6463–6465.

Baker, P. K., Harris, S. D., Durrant, M. C., Hughes, D. L. and Richards, R. L. (1995). *Acta Cryst.*, C**51**, 697–700.

Baker, W. A. and Williams, D. E. (1978). *Acta Cryst.*, B**34**, 1111–1116.

Baker, W. A., Ph. D. Dissertation, Univ. of Kentucky, Louisville, Ky. (1976); Xerox University Microfilms, 76–26, 197. 129, pp. From Diss. Abstr. Int. B, 37(5), 2267. (1976).

Balch, A. L. and Olmstead, M. M. (1999). *Coord. Chem. Revs.*, **185–186**, 601–617.

Barnes, J. C. (1972). *Inorg. Chem.*, **11**, 2267–2268.

Barnes, J. C. and Blyth, C. S. (1985). *Inorg. Chim. Acta*, **98**, 181–184.

Barnes, J. C. and Duncan, C. S. (1972). *J. Chem. Soc., Dalton Trans.*, pp. 1732–1734.

Barnes, J. C. and Weakley, T. J. R. (1978). *Acta Cryst.*, B**34**, 1984–1985.

Bartl, H. (1982). *Fresenius' Z. Anal. Chem.*, **312**, 17–18.

Bassi, I. W. and Fagherazzi, G. (1965). *J. Organometall. Chem.*, **13**, 533–538.

Battaglia, L. P., Bellitto, C., Cramarossa, M. R. and Vezzosi, I. M. (1996). *Inorg. Chim.*, **35**, 2390–2392.

Battaglia, L. P., Corradi, A. B., Vessozi, I. M. and Zanoli, F. A. (1990). *J. Chem. Soc., Dalton Trans.*, pp. 1675–1678.

Battaglia, L. P., Cramarossa, M. R. and Vezzosi, I. M. (1995). *Inorg. Chim. Acta*, **237**, 169–172.

Beagley, B. and Edington, J. W. (1963). *Brit. J. Appl. Phys.*, **14**, 609.

Bechtel, F., Chasseau, D., Gaultier, J. and Hauw, C. (1976). *Acta Cryst.*, **B32**, 1738–1748.

Beck, J. and Strähle, J. (1986). *Z. Naturforsch.*, **41b**, 1381–1386.

Belitskus, D. and Jeffrey, G. A. (1965). *Spectrochim. Acta*, **21**, 1563–1567.

Bell, N. A., Goldstein, M., Jones, T. and Nowell, I. W. (1980a). *Acta Cryst.*, **B36**, 710–712.

Bell, N. A., Goldstein, M., Jones, T. and Nowell, I. W. (1980b). *Inorg. Chim. Acta*, **43**, 87–93.

Bell, N. A., Goldstein, M., Jones, T. and Nowell, I. W. (1981). *Inorg. Chim. Acta*, **48**, 185–189.

Bender, C. J. (1986). *Chem. Soc. Revs.*, **15**, 475–502.

Benesi, H. A. and Hildebrand, J. H. (1949). *J. Am. Chem. Soc.*, **71**, 2703–2707.

Bennett, G. M. (1962). *Chem. Revs.*, **62**, 611–652 (see especially pp. 612–616).

Bent, H. A. (1968). *Chem. Revs.*, **68**, 587–648.

Beverwijk, C. D. M., Van der Kerk, G. J. M., Lensink, A. J. and Noltes, J. G. (1970). Organosilver Chemistry, *Organomet. Chem. Rev.*, Sect. A, **5**, 215–280.

Biedenkapp, D. and Weiss, Al. (1964). *Z. Naturforsch.*, **19a**, 1518–1521.

Biedenkapp, D. and Weiss, Al. (1968). *Z. Naturforsch.*, **23b**, 172–175.

Birker, P. J. M. W. L., Freeman, H. C., Guss, J. M. and Watson, A. D. (1977). *Acta Cryst.*, **B33**, 182–184.

Biscarini, P., Foresti, E. and Pradella, G., (1984). *J. Chem. Soc. Dalton Trans.*, pp. 953–957.

Biscarini, P., Fusina, L., Mangia, A., Nivellini, G. D. and Pelizzi, G. (1974). *J. Chem. Soc. Dalton Trans.*, pp. 1846–1849.

Biscarini, P., Fusina, L., Nivellini, G. and Pelizzi, G. (1981). *J. Chem. Soc. Dalton Trans.*, pp. 1024–1027.

Biscarini, P., Fusina, L., Nivellini, G. D., Mangia, A. and Pelizzi, G. (1973). *J. Chem. Soc. Dalton Trans.*, pp. 159–161.

Biscarini, P.and Pelizzi, G. (1988). *J. Chem. Soc. Dalton Trans.*, pp. 2915–2919.

Bjorvatten, T. (1962). *Acta Chem. Scand.*, **16**, 749–754.

Bjorvatten, T. (1963). *Acta Chem. Scand.*, **17**, 2292–2300.

Bjorvatten, T. (1966). *Acta Chem. Scand.*, **20**, 1863–1873.

Bjorvatten, T. (1968). *Acta Chem. Scand.*, **22**, 410–420.

Bjorvatten, T. and Hassel, O. (1961). *Acta Chem. Scand.*, **15**, 1429–1436.

Bjorvatten, T. and Hassel, O. (1962). *Acta Chem. Scand.*, **16**, 249–255.

Bjorvatten, T., Hassel, O. and Lindheim, A. (1963). *Acta Chem. Scand.*, **17**, 689–702.

Blackstock, S. C. and Kochi, J. K. (1987). *J. Am. Chem. Soc.*, **109**, 2484–2496.

Blackstock, S. C., Lorand, J. P. and Kochi, J. K. (1987). *J. Org. Chem.*, **52**, 1451–1460.

Blackstock, S. C., Poehling, K. and Greer, M. L. (1995). *J. Am. Chem. Soc.*, **117**, 6617–6618.

Blake, A. J., Gould, R. O., Radek, C. and Schroder, M. (1993). *Chem. Commun.*, pp. 1191–1193.

Boerio-Goates, J., Goates, S. R., Ott, J. B. and Goates, J. R. (1985). *J. Chem. Thermodyn.*, **17**, 665–670.

Boese, R., Clark, T. and Gavezzotti, A. (2003). *Helv. Chim. Acta*, **86**, 1085–1100.

Bois d'Enghien-Peteau, M., Meunier-Piret, J. and Meerssche, M. van. (1968). *J. Chim. Phys.*, **65**, 1221–1226.

Bolton, W. (1964). *Nature, Lond.*, **201**, 987–989.

Bombieri, G., Peyronel, G. and Vezzosi, I. M. (1972). *Inorg. Chim. Acta*, **6**, 349–354.

Borgen, B., Hassel, O. and Rømming, C. (1962). *Acta Chem. Scand.*, **16**, 2469–2470.

Bowaker, G. A. and Hacobian, S. (1969). *Austral. J. Chem.*, **22**, 2047–2059.

Branden, C.-I. (1963). *Acta Chem. Scand.*, **17**, 1363–1374.

Branden, C.-I. (1964a). *Ark. Kemi.*, **22**, 485–493.

Bränden, C.-I. (1964b). *Ark. Kemi.*, **22**, 495–500.

Branden, C.-I. (1964c). *Ark. Kemi.*, **22**, 501–516.

Bransford, J. W. and Meyers, Edward A., *Cryst. Struct. Commun.*, **7**, 697–702.

Brass, K. and Fanta, K. (1936). *Ber.*, **69B**, 1–11.

Brass, K. and Tengler, E. (1931a). *Ber.*, **64B**, 1650–1653.

Brass, K. and Tengler, E. (1931b). *Ber.*, **64B**, 1654–1664.

Breton–Lacombe, M. (1967). *Acta Cryst.*, **23**, 1031–1037.

Britton, D. (1967). *Persp. Struct. Chem.*, **1**, 109–171.

Britton, D. (1981). *Cryst. Struct. Commun.*, **10**, 1061–1064.

Britton, D., Konnert, J. and Lam, S. D. (1977). *Cryst. Struct. Commun.*, **6**, 45–48.

Britton, D., Konnert, J. and Lam, S. D. (1979). *Cryst. Struct. Commun.*, **8**, 913–916.

Brotherton, P. D. and White, A. H. (1973). *J. Chem. Soc. Dalton Trans.*, pp. 2698–2700.

Brotherton, P. D., Epstein, J. M., White, A. H. and Willis, A. C. (1974). *J. Chem. Soc. Dalton*, pp. 2341–2343.

Brown, R. N. (1961). *Acta Cryst.*, **14**, 711–715.

Brownstein, S., Gabe, E., Lee, F. and Piotrowski, A. (1986). *Can. J. Chem.*, **64**, 1661–1667.

Brun, L. and Branden, C.-I. (1964). *Ark. Kemi.*, **20**, 485–493.

Bruns, R. E. (1977). *J. Molec. Struct.*, **31**, 121–126.

Brusset, H. and Madaule-Aubry, F. (1966). *Bull. Soc. Chim. Franc.*, pp. 3121–3127.

Buffagni, S., Vezzosi, I. M. and Peyronel, G. (1960); *Ann. Chim. (Roma)*, **50**, 343, 348–351 *Chem. Abstr.*, **56**, 15145i (1960).

Buffagni, S., Vezzosi, I. M. and Peyronel, G. (1968). *Gazz. Chim. Ital.*, **98**, 156–161.

Bukshpan, S. and Herber, R. H. (1967). *J. Chem. Phys.*, **46**, 3375–3378.

Bukshpan, S., Goldstein, C., Sonnino, T., May, L. and Pasternak, M. (1975). *J. Chem. Phys.*, **62**, 2606–2609.

Burford, N., Clyburne, J. A. C., Wiles, J. A., Cameron, T. S. and Robertson, K. N. (1996). *Organometallics*, **15**, 361–364.

Cameron, J. U. and Killean, R. G. C. (1972). *Cryst. Struct. Commun.*, **1**, 31–33.

Cameron, T. S. and Prout, C. K. (1972). *Acta Cryst.*, **B28**, 447–452.

Canty, A. J., Raston, C. L., Skelton, B. W. and White, A. H. (1982). *J. Chem. Soc. Dalton*, pp. 15–18.

Carlston, R. C., Griswold, E. and Kleinberg, J. (1958). *J. Am. Chem. Soc.*, **80**, 1532–1534.

Carmalt, C. J., Clegg, W., Elsegood, R. J., Havelock, J., Lightfoot, P., Norman, N. C. and Scott, A. J. (1996). *Inorg. Chem.*, **35**, 3709–3712.

Carrabine, J. A. and Sundaralingam, M. (1971). *Biochemistry*, **10**, 292–299.

Carter, V. B. and Britton, D. (1972). *Acta Cryst.*, **B28**, 945–950

Caughlan, C. N., Smith, G. D., Jennings, P. W. and Voecks, G. E. (1976). *Acta Cryst.*, **B32**, 1390–1393.

Cesari, M., Pedretti, U., Zazzetta, A., Lugli, G. and Marconi, W. (1971). *Inorg. Chim. Acta*, **5**, 439–444.

Chantry, G. W., Gebbie, H. A. and Mirza, H. N. (1967). *Spectrochim. Acta*, **23**, 2749–2752.

Chao, G. Y. and McCullough, J. D. (1960). *Acta Cryst.*, **13**, 727–732.

Chao, G. Y. and McCullough, J. D. (1961). *Acta Cryst.*, **14**, 940–945.

Cheung, K. K. and Sim, G. A. (1965). *J. Chem. Soc.*, pp. 5988–6004.

Cheung, K. K., McEwen, R. S. and Sim, G. A. (1965). *Nature*, **205**, 383–384.

Chevrier, B., Le Carpentier, J.-M. and Weiss, R. (1972a). *Acta Cryst.*, **B28**, 2659–2666.

Chevrier, B., Le Carpentier, J.-M. and Weiss, R. (1972b). *Acta Cryst.*, **B28**, 2666–2672.

Chevrier, B., Le Carpentier, J.-M. and Weiss, R. (1972c). *Acta Cryst.*, **B28**, 2673–2677.

Chevrier, B., Le Carpentier, J.-M. and Weiss, R. (1972d). *J. Am. Chem. Soc.*, **94**, 5718–5723.

Chieh, C. (1977a). *Can. J. Chem.*, **55**, 1115–1119.

Chieh, C. (1977b). *Can. J. Chem.*, **55**, 1583–1587.

Chieh, C. (1978). *Can. J. Chem.*, **56**, 974–975.

Chieh, C. and Cowell, D. H. (1977). *Can. J. Chem.*, **55**, 3898–3900.

Chieh, C., Lee, L. P. C., and Chiu, C. (1978). *Can. J. Chem.*, **56**, 2526–2529.

Choi, Sang Up and Brown, H. C. (1966). *J. Am. Chem. Soc.*, **88**, 903–909.

Cingi, M. B., Lanfredi, A. M. M., Tiripicchio, A. and Camellini, M. T. (1979). *Acta Cryst.*, B**35**, 312–316.

Clayton, P. R., Cassell, S. and Staveley, L. A. K. (1978). *J. Chem. Thermodyn.*, **10**, 387–394.

Coggon, P., McPhail, A. T. and Sim, G. A. (1966). *J. Chem. Soc.* (B), pp. 1024–1028.

Cohen-Addad, C., Baret, P., Chautemps, P. and Pierre, J.-L. (1983). *Acta Cryst.*, C**39**, 1346–1349.

Cohen-Addad, C., Lebars, M., Renault, A. and Baret, P. (1984). *Acta Cryst.*, C**40**, 1927–1931.

Cooper, A., Gopalkrishna, E. M. and Norton, D. A. (1968). *Acta Cryst.*, B**24**, 935–941.

Cope, A. C. and Hochstein, F. A. (1950). *J. Am. Chem. Soc.*, **72**, 2515–2520.

Costello, W. R., McPhail, A. T. and Sim, G. A. (1966). *J. Chem. Soc.* (A), pp. 1190–1193.

Cotton, F. A. and Schwotzer, W. (1985). *Organometallics*, **4**, 942–943.

Courseille, C., Gaultier, J., Hauw, C. and Schvoerer, M. (1970). *Compt. Rend. Acad. Sci., Paris*, Ser.C, **270**, 687–689.

Crowston, E. H., Lobo, A. M., Prabhakar, S., Rzepa, H. S. and Williams, D. J. (1984). *J. Chem. Soc. Chem. Commun.*, pp. 276–278.

Cushen, D. W. and Hulme, R. (1962). *J. Chem. Soc.*, pp. 2218–2222.

Cushen, D. W. and Hulme, R. (1964). *J. Chem. Soc.(A)*, pp. 4162–4166.

Dahl, S. and Groth, P. (1971). *Acta Chem. Scand.*, **25**, 1114–1124.

Dahl, T. and Hassel, O. (1965). *Acta Chem. Scand.*, **19**, 2000–2001.

Dahl, T. and Hassel, O. (1968b). *Acta Chem. Scand.*, **22**, 2851–2866.

Dahl, T. and Hassel, O. (1970). *Acta Chem. Scand.*, **24**, 377–383.

Dahl, T. and Hassel, O. (1971). *Acta Chem. Scand.*, **25**, 2168–2174

Dahl, T., Hassel, O. and Sky, K. (1967). *Acta Chem. Scand.*, **21**, 592–593.

Dalziel, J. A. W. and Hewitt, T. G. (1966). *J. Chem. Soc.* (A) pp. 233–235.

Damm, E., Hassel, O. and Rømming, C. (1965). *Acta Chem. Scand.*, **19**, 1159–1165.

Dean, P. A. W. (1978). *Prog. Inorg. Chem.*, **24**, 109–178.

Demaldé, A., Mangia, A., Nardelli, M., Pelizzi, G. and Vidoni Tani, M. E. (1972). *Acta Cryst.*, B**28**, 147–150.

Demassieux, (1909). *Bull. Soc. Franc. Miner.*, **32**, 387–396.

Desiraju, G. R. and Steiner, Th. (1999). *The Weak Hydrogen Bond in Structural Chemistry and Biology*, IUCr and Oxford University Press.

Devarajan, V. and Glazer, A. M. (1986). *Acta Cryst.*, A**42**, 560–569.

Douglas, B., McDaniel, D. H. and Alexander, J. J. (1983). 2nd Edition, Wiley, New York etc.

Drabowicz, J., Dudzinski, B., Mikolajczyk, M., Wang, F., Dehlavi, A., Goring, J., Park, M. Rizzo, C. J., Polavarapu, P. L., Biscarini, P., Wieczorek, M. W. and Majzner, W. R. (2001). *J. Org. Chem.*, **66**, 1122–1129.

Ebenhöch, J., Müller, G., Riede, J., and Schmidbaur, H. (1984). *Angew. Chem. Int. Ed. Engl.*, **23**, 386–388.

Ehrlich, B. S. and Kaplan, M. (1969). *Chem. Phys. Letts.*, **3**, 161–163.

Eia, G. and Hassel, O. (1956). *Acta Chem Scand.*, **10**, 139–141.

Eley, D. D., Taylor, J. H. and Wallwork, S. C. (1961). *J. Chem. Soc.*, pp. 3867–3873.

Enjalbert, R. and Galy, J. (1980). *Acta Cryst.*, B**36**, 914–916.

Ermer, O., Eser, H. and Dunitz, J. D. (1971). *Helv. Chim. Acta*, **54**, 2469–2475.

Fawcett, J. K. and Trotter, J. (1966). *Proc. Roy. Soc. Lond.*, Ser., A, **289**, 366–376.

Feher, F. and Linke, K.-H. (1966). *Z. Naturforsch.*, B**21**, 1237–1238.

Feher, F., Hirschfeld, D. and Linke, K.-H. (1962). *Acta Cryst.*, **15**, 1182–1183.

Ferrara, J. D., Tessier-Youngs, C. and Youngs, W. J. (1987). *Organometallics*, **6**, 676–678.

Ferrara, J. D., Tessier-Youngs, C. and Youngs, W. J. (1988). *Inorg. Chem.*, **27**, 2201–2202.

Ferrari, M. B., Cramarossa, M. R., Iarossi, D. and Pelosi, G. (1998). *Inorg. Chem.*, **37**, 5681–5685.

Fichter, W. and Weiss, A. (1976). *Z. Naturforsch.*, **31b**, 1626–1634.

Fournet, F. and Theobald, F. (1981). *Inorg. Chim. Acta*, **52**, 15–21.

Frank, W. and Reiland, V. (1998). *Acta Cryst.*, **C54**, 1626–1628.

Frank, W., Schneider, J. and Müller-Becker, S. (1993). *J. Chem. Soc., Chem. Commun.*, pp.799–800.

Fredin, L. and Nelander, B. (1974). *Mol. Phys.*, **77**, 885–898.

Freeman, F., Zioller, J. W., Po, H. N. and Keindl, M. C. (1988). *J. Am. Chem. Soc.*, **110**, 2586–2591.

Frey, H., Leligny, H. and Ledesert, M. (1971). *Bull. Soc. Franc. Miner. Crist.*, **94**, 467–470.

Frey, M. and Monier, J.-C. (1971). *Acta Cryst.*, **B27**, 2487–2490.

Fritz, H. P., McOmie, J. F. W. and Sheppard, N. (1960). *Tetr. Letts.*, pp. 35–41.

Fromm, E. (1913). *Ann.*, **396**, 75–103.

Gagnaux, P. and Susz, B.-P. (1960). *Helv. Chim. Acta*, **43**, 948–956.

Gagnaux, P. and Susz, B.-P. (1961). *Helv. Chim. Acta*, **44**, 1128–1131.

Galy, J. and Enjalbert, R. (1982). *J. Sol. State Chem.*, **44**, 1–23.

Ganis, P. and Dunitz, J. D. (1967). *Helv. Chim. Acta*, **50**, 2379–2386.

Ganis, P., Giuliano, V. and Lapore, U. (1971). *Tetr. Letts.*, pp. 765–768.

Garton, G. and Powell, H. M. (1957). *J. Inorg. Nucl. Chem.*, **4**, 84–89.

Gash, A. G., Rodesiler, P. F. and Amma, E. L. (1974). *Inorg. Chem.*, **13**, 2429–2434.

Gaultier, J. and Hauw, C. (1965). *Acta Cryst.*, **18**, 604–608.

Gaultier, J., Hauw, C. and Schvoerer, M. (1971). *Acta Cryst.*, **B27**, 2199–2204.

Gaultier, J., Hauw, C., Housty, J. and Schvoerer, M. (1971b). *Compt. Rend. Acad. Sci. Paris*, Ser. C, **273**, 956–958.

Gaultier, J., Hauw, C., Housty, J. and Schvoerer, M. (1972). *Compt. Rend. Acad. Sci. Paris*, Ser. C, **275**, 1403–1406.

Geller, S. and Schawlow, A. L. (1964). *J. Chem. Phys.*, **40**, 2413.

Genet, F. and Leguen, J.-C. (1965). *Acta Cryst.*, **B25**, 2029–2033.

Gibbons, C. S. and Trotter, J. (1971). *J. Chem. Soc.* (A), pp. 2058–2062.

Gilson, D. F. R. and McDowell, C. A. (1955). *J. Chem. Phys.*, **23**, 779–783.

Glasser, L. S. D., Ingram, L., King, M. G. and McQuillan, G. P. (1969). *J. Chem. Soc.* (A), pp. 2501–2504.

Gleiter, R., Karcher, M., Kratz, D., Ziegler, M. L. and Nuber, B. (1990). *Chem. Ber.*, **123**, 1461–1468.

Gmelin, (1975). B5, System No. 61, pp. 90–119.

Goates, J. R., Boerio-Goates, J., Goates, S. R. and Ott, J. B. (1987). *J. Chem. Thermodynam.*, **19**, 103–107.

Gordeev, A. D., Grechishkin, V. S., Kyuntsel, I. A. and Rozenberg, Yu. I. (1970). *J. Struct. Chem. USSR*, **11**, 717–719.

Gordeev, A. D., Grechishkin, V. S., Kyuntsel, I. A. and Rozenberg, Yu. I. (1974). *J. Struct. Chem. USSR*, **11**, 827–829.

Gordeev, A. D. and Kyuntsel, I. A. (1992). *Russ. J. Coord. Chem.*, **18**, 840–844.

Graddon, D. P. (1982). *Revs. Inorg. Chem.*, **4**, 211–282.

Grdenic, D. (1965). *Quart. Revs.*, **19**, 303–328.

Griffith, E. A. H. and Amma, E. L. (1971). *J. Am. Chem. Soc.*, **93**, 3167–3172.

Griffith, E. A. H. and Amma, E. L. (1974a). *J. Am. Chem. Soc.*, **96**, 743–749.

Griffith, E. A. H. and Amma, E. L. (1974b). *J. Am. Chem. Soc.*, **96**, 5407–5413.

Groeneveld, W. L. and Zuur, A. P. (1958). *J. Inorg. Nucl. Chem.*, **8**, 241–244.

Grønbaek, R. and Dunitz, J. D. (1964). *Helv. Chim. Acta*, **47**, 1889–1897.

Groth, P. (1906). *Chem. Krist.* (Leipzig), **I**, 227.

Groth, P. and Hassel, O. (1962). *Acta Chem. Scand.*, **16**, 2311–2317.

Groth, P. and Hassel, O. (1964a). *Acta Chem. Scand.*, **18**, 402–408.

Groth, P. and Hassel, O. (1964b). *Acta Chem. Scand.*, **18**, 1327–1332.

Groth, P. and Hassel, O. (1965). *Acta Chem. Scand.*, **19**, 120–128.

Hannson, A. and Vänngard, M. (1961). U. S. Dept of Commerce, Office Tech. Serv., AD 264819, 2 pp. *Chem.Abstr.*, **58**, 84767 (1963).

Hartl, H. and Steidl, S. (1977). *Z. Naturforsch.*, **32b**, 6–10.

Hartl, H. and Ullrich, D. (1974). *Z. anorg. allgem. Chem.*, **409**, 228–236.

Hartl, H., Barnighausen, H. and Jander, J. (1968). *Z. anorg. allgem. Chem.*, **357**, 225–237.

Hassel, O. (1958). *Mol. Phys.*, **1**, 241–246.

Hassel, O. (1970). *Science*, **170**, 497–502. Chemistry 1963–1970 (Nobel Lectures), Elsevier, Amsterdam (lecture 1969, published 1970).

Hassel, O. and Hvoslef, J. (1954a). *Acta Chem. Scand.*, **8**, 873.

Hassel, O. and Hvoslef, J. (1954b). *Acta Chem. Scand.*, **8**, 1953.

Hassel, O. and Hvoslef, J. (1956). *Acta Chem. Scand.*, **10**, 138–139.

Hassel, O. and Rømming, C. (1962). *Quart. Revs.*, **16**, 1–18.

Hassel, O. and Romming, C. (1956). *Acta Chem. Scand.*, **10**, 136–137.

Hassel, O. and Strømme, K. O. (1958). *Acta Chem. Scand.*, **12**, 1146–1147.

Hassel, O. and Strømme, K. O. (1959a). *Acta Chem. Scand.*, **13**, 275–280.

Hassel, O. and Strømme, K. O. (1959b). *Acta Chem. Scand.*, **13**, 1775–1780.

Hassel, O. and Strømme, K. O. (1959c). *Acta Chem. Scand.*, **13**, 1781–1786.

Hassel, O., Rømming, C. and Tufte, T. (1961). *Acta Chem. Scand.*, **15**, 967–974.

Hawes, L. L. (1962). *Nature, Lond.*, **196**, 766–767.

Hawes, L. L. (1963). *Nature, Lond.*, **198**, 1267–1270.

Heiart, R. B. and Carpenter, G. B. (1956). *Acta Cryst.*, **9**, 889–895.

Hendra, P. J. and Sadavisan, N. (1965). *Spectrochim. Acta*, A**21**, 1127–1132.

Herbstein, F. H. (1987). *Top. Curr. Chem.*, **140**, 107–139.

Herbstein, F. H. (2000). *Advanced Methods in Structure Analysis* (published by Czech and Slovak Crystallographic Association, Praha), pp.114–154.

Herbstein, F. H. (2004). "Correlation of crystal structures and NQR spectra of {arene.nSbX$_3$} molecular compounds". In preparation.

Herbstein, F. H. and Kaftory, M. (1981). *Z. Kristallogr.*, **157**, 1–25.

Herbstein, F. H. and Marsh, R. E. (1982). *Acta Cryst.*, B**38**, 1051–1055.

Herbstein, F. H. and Schwotzer, W. (1984). *J. Am. Chem. Soc.*, **106**, 2367–2373.

Herbstein, F. H., Ashkenazi, P., Kaftory, M., Kapon, M., Reisner, G. M. and Ginsburg, D. (1986). *Acta Cryst.*, B**42**, 575–601.

Herbstein, F. H., Kapon, M. and Reisner, G.M. (1986). *Acta Cryst.*, B**42**, 181–187.

Herbstein, F. H., Kapon, M. and Reisner, G.M. (1989). *Z. Kristallogr.*, **187**, 25–38.

Herbstein, F. H., Kapon, M. and Schwotzer, W. (1983). *Helv. Chim. Acta*, **66**, 35–43.

Hertel, E. (1931). *Z. physik. Chem.* (B), **15**, 51–57.

Hill, A. E. (1922). *J. Am. Chem. Soc.*, **44**, 1163–1193.

Hill, A. E. and Macy, R. (1924). *J. Am. Chem. Soc.*, **46**, 1132–1150.

Hill, A. E. and Miller, F. W., Jr. (1925). *J. Am. Chem. Soc.*, **47**, 2702–2712.

Hofmann, A. W. and Girard, A. (1870). *Jahr. Ber.*, pp. 591.

Holmesland, O. and Rømming, C. (1960). *Acta Chem. Scand.*, **20**, 2601–2610.

Hönle, W., Simon, A. and Gerlach, G. (1987). *Z. Naturforsch.*, **42b**, 546–552.

Hooper, H. (1964). *J. Chem. Phys.*, **41**, 599–601.

Hope, H. and Christensen, A. T. (1968). *Acta Cryst.*, B**24**, 375–382.

Hope, H. and McCullough, J. D. (1962). *Acta Cryst.*, **15**, 806–807.

Hope, H. and McCullough, J. D. (1964). *Acta Cryst.*, **17**, 712–718.

Hope, H. and Nichols, B. G. (1981). *Acta Cryst.*, B**37**, 158–161.

Hoppe, W., Lenne, H. U. and Morandi, G. (1957). *Z. Kristallogr.*, **108**, 321–327.

Hossain, M. B. and Van der Helm, D. (1968). *J. Am. Chem. Soc.*, **90**, 6607–6611.

Hubig, S. M., Lindeman, S. V. and Kochi, J. K. (2000). *Coord. Chem. Rev.*, **200–202**, 831–873.

Hughes, D. L., Lane, J. D. and Richards, R. L. (1991). *J. Chem. Soc., Dalton Trans.*, pp. 1627–1629.

Hulme, R. and Hursthouse, M. B. (1966). *Acta Cryst.*, **21**, A143.

Hulme, R. and Mullen, D. J. E. (1976). *J. Chem. Soc., Dalton Trans.*, pp. 802–804.

Hulme, R. and Scruton, J. C. (1968). *J. Chem. Soc.* (A), pp. 2448–2452.

Hulme, R. and Szymanski, J. T. (1969). *Acta Cryst.*, B**25**, 753–761.

Hulme, R., Mullen, D. J. E. and Scruton, J. C. (1969). *Acta Cryst.*, **25**, S171.

Husemann, A. (1863). *Liebig's Annalen*, **126**, 269–298.

Ichiba, S., Mishima, M., Sakai, H. and Negita, H. (1968). *Bull. Chem. Soc. Jpn.*, **41**, 49–52.

Ichiba, S., Sakai, H., Negita, H. and Maeda, Y. (1971). *J. Chem. Phys.*, **54**, 1627–1629.

Irngartinger, H., Leiserowitz, L. and Schmidt, G. M. J. (1970). *Chem. Ber.*, **103**, 1119–1131.

Ishihara, H. (1981). *J. Sci. Hiroshima Univ.*, Ser. A, **45**, 319–XXX.

Iwamoto, R. (1973a). *Bull. Chem. Soc. Jpn.*, **46**, 1114–1118.

Iwamoto, R. (1973b). *Bull. Chem. Soc. Jpn.*, **46**, 1118–1123.

Iwamoto, R. (1973c). *Bull. Chem. Soc. Jpn.*, **46**, 1123–1127.

Jackson, R. B. and Streib, W. E. (1967). *J. Am. Chem. Soc.*, **89**, 2539–2543.

Jogun, K. H. and Stekowski, J. J. (1976). *J. Am. Chem. Soc.*, **98**, 6018–6026.

Jones, D. E. H. and Ward, J. L. (1966). *J. Chem. Soc.(A)*, pp. 1448–1453.

Jorge, R. A., Airoldi, A. and Chagas, A. E. (1978). *J. Chem. Soc. Dalton Trans.*, pp. 1102–1104.

Kadaba, P. K., O'Reilly, D. E., Peterson, E. M. and Scheie, C. E. (1971). *J. Chem. Phys.*, **55**, 5289–5291.

Kaftory, M. (1978). *Acta Cryst.*, B**34**, 471–475.

Kang, H. C., Hanson, A. W., Eaton, B. and Boekelheide, V. (1985). *J. Am. Chem. Soc.*, **107**, 1979–1985.

Kapustinskii, A. F. and Drakin, S. I. (1947). *Bull. acad. sci. URSS, Classe Sci. Chim.*, pp. 435–442; *Chem. Abstr.*, **42**, 1902e (1948).

Kapustinskii, A. F. and Drakin, S. I. (1950). *Bull. acad. sci. URSS, Classe Sci. Chim.*, pp. 233–236; *Chem. Abstr.*, **44**, 7788b (1950).

Karl, N., Heym, H. and Stezowski, J. J. (1983). *Mol. Cryst. Liq. Cryst.*, **131**, 163–192.

Kashiwabara, K., Konaka, S., and Kimura, M. (1973). *Bull. Chem. Soc. Jpn.*, **46**, 410–413.

Kawasaki, Y. and Matsuura, Y. (1984). *Chem. Letts.*, pp. 155–158.

Kessler, K. (1977). .Ph. D. thesis, University of Bristol.

Ketelaar, J. A. A. and Zwartsenberg, J. W. (1939). *Rec. Trav. Chim. Pays-bas*, **58**, 448–452.

Kiel, G. and Engler, R. (1974). *Chem. Ber.*, **107**, 3444–3450.

Kitaigorodsky, A. I. (1961). Consultants Bureau, New York, pp. 442–448.

Kleinberg, J. and Davidson, A. W. (1948). *Chem. Revs.*, **42**, 601–609.

Kniep, R. and Reski, H. D. (1982). *Inorg. Chim. Acta*, **64**, L83–87.

Knobler, C. and McCullough, J. D. (1968). *Inorg. Chem.*, **7**, 365–369.

Knobler, C., Baker, C., Hope, H. and McCullogh, J. D. (1971). *Inorg. Chem.*, **10**, 697–700.

Kochi, J. K. (1990). *Acta Chem. Scand.*, **44**, 400–432.

Korczynski, A. (1968). *Rocz. Chem.*, **42**, 393–406.

Korte, L., Lipka, A. and Mootz, D. (1979). Abstracts of 5th Europen Crystallography Meeting, Copenhagen.

Korte, L., Lipka, A. and Mootz, D. (1985). *Z. anorg. allgem. Chem.*, **524**, 157–167.

Kozarek, W. J. and Fernando, Q. (1973). *Inorg. Chem.*, **12**, 2129–2131.

Kozulin, A. T. (1971). *Chem. Abstr.*, **74**, 17791j.

Krygowski, T. M., Cyranski, M., Ciesielski, A., Swirska, B. and Leszczynski, P. (1996). *J. Chem. Inf. Comput. Sci.*, **36**, 1135–1141.

Kulpe, S. (1967). *Z. anorg. allgem. Chem.*, **349**, 314–323.

Kyuntsel, I. A. (1998). *Russ. J. Coord. Chem.*, **24**, 537–541.

Kyuntsel, I. A., (2001). *Russ. J. Coord. Chem.*, **27**, 835–837.

Kyuntsel, I. A. and Mokeeva, V. A. (2002). *J. Struct. Chem.*, **43**, 772–776.

Laitinen, R., Steidel, J. and Steudel, R. (1980). *Acta Chem. Scand.*, A**34**, 687–693.

Lau, W. and Kochi, J. K. (1986). *J. Org. Chem.*, **51**, 1801–1811.

Lebas, J. M. and Julien-Laferriere, S. (1972). *J. Phys. Chim. Physicochim. Biol.*, **69**, 503–512.

Le Carpentier, J.-M. and Weiss, R. (1972a). *Acta Cryst.*, B**28**, 1421–1429

Le Carpentier, J.-M. and Weiss, R. (1972b). *Acta Cryst.*, B**28**, 1430–1437.

Le Carpentier, J.-M. and Weiss, R. (1972c). *Acta Cryst.*, B**28**, 1437–1442.

Le Carpentier, J.-M. and Weiss, R. (1972d). *Acta Cryst.*, B**28**, 1442–1448.

Lechat, J. R., Francisco, R. H. P. and Airoldi, C. (1980). *Acta Cryst.*, B**36**, 930–931.

Lewandos, G. S., Gregston, D. K. and Nelson, F. R. (1976). *J. Organometall. Chem.*, **118**, 363–374.

Lindemann, W., Wögerbauer, R. and Berger, P. (1977). *Z. anorg. allgem. Chem.*, **437**, 155–158.

Lindqvist, I. (1963). "Inorganic Adduct Molecules of Oxo-Compounds," Springer-Verlag, Berlin.

Lindqvist, I. and Niggli, A. (1956). *J. Inorg. Nucl. Chem.*, **2**, 345–348.

Linek, A. and Novak, C. (1978). *Acta Cryst.*, B**34**, 3369–3372.

Lipka, A. (1978). *Z. anorg. allgem. Chem.*, **440**, 224–230.

Lipka, A. (1979a). *Thermochim. Acta*, **29**, 269–272.

Lipka, A. (1979b). *Acta Cryst.*, B**35**, 3020–3022.

Lipka, A. (1980). *Z. anorg. allgem. Chem.*, **466**, 195–202.

Lipka, A. (1983). *Z. Naturforsch.*, **38b**, 341–346.

Lipka, A. and Mootz, D. (1978). *Z. anorg. allgem. Chem.*, **440**, 217–223.

Lipka, A. and Mootz, D. (1982). *Z. Naturforsch.*, **37B**, 695–698.

Lipka, A. and Wunderlich, H. (1980). *Z. Naturforsch.*, **35b**, 1548–1551.

Maass, O. and McIntosh, D. (1912). *J. Am. Chem. Soc.*, **34**, 1273–1290.

Maddox, H. and McCullough, J. D. (1966). *Inorg. Chem.*, **5**, 522–526.

Majeste, R. J. and Trefonas, L. M. (1972). *Inorg. Chem.*, **11**, 1834–1836.

Markila, P. L. and Trotter, J. (1974). *Can. J. Chem.*, **52**, 2197–2200.

Marsh, R. E. and Herbstein, F. H. (1983). *Acta Cryst.*, B**39**, 280–287.

Marstokk, K. M. and Strømme, K. O. (1968). *Acta Cryst.*, B**24**, 713–720.

Mathews, F. S. and Lipscomb, W. N. (1959). *J. Phys. Chem.*, **63**, 845–850.

May, P. (1911). *J. Chem. Soc.*, **99**, 1382–1386.

McEwen, R. S. and Sim, G. A. (1967). *J. Chem. Soc.* (A), pp. 271–275.

McEwen, R. S. and Sim, G. A. (1969). *J. Chem. Soc.* (A), pp. 1897–1899.

McEwen, R. S., Sim, G. A. and Johnson, C. R. (1967). *J. Chem. Soc., Chem. Comm.*, pp. 885–886.

McKechnie, J. S. and Paul, I. C. (1968). *J. Chem. Soc.* (B), pp. 1445–1452.

McKechnie, J. S., Newton, M. G. and Paul, I. C. (1967). *J. Am. Chem. Soc.*, **89**, 4819–4825.

McMullan, R. C., Koetzle, T. F. and Fritchie, C. J, Jr. (1997). *Acta Cryst.*, B**53**, 645–653.

McMurry, J. E., Haley, G. J., Matz, J. R., Clardy, J. C. and Mitchell, J. (1986). *J. Am. Chem. Soc.*, **108**, 515–516.

McPhail, A. T. and Sim, G. A. (1966a). *J. Chem. Soc., Chem. Comm.*, pp. 21–22.

McPhail, A. T. and Sim, G. A. (1966b). *J. Chem. Soc.* (B), pp. 112–120.

Meerwein, H. and Maier-Hüser, H. (1932). *J. prakt. Chem.*, **134**, 51–81.

Menshutkin, B. (1910). *Izv. Petersburg Polytechnikum*, **13**, 1–16; *Chem. Zentralblat*, II, 154.

Menshutkin, B. N. (1911) *Chem. Zentr.* II, 751; ibid, I, 408, 807; II, 1436 (1912); *Zh. Russ. Fiz. Khim. Ova*, **44**, 1079–1145 (1912); *Chem. Abstr.*, **6**, 3403–3404 (1912).

Michelet, A., Viosset, B., Khodadad, P. and Rodier, N. (1981). *Acta Cryst.*, B**37**, 2171–2175.

Mitani, S. (1986) *Ann. Rep. Res. React. Inst. Kyoto Univ.*, **19**, 1–10; *Chem. Abstr.*, **107**, 15864 (1987).

Möller, G. M., Olmstead, M. M. and Tinti, D. S. (1987). *J. Am. Chem. Soc.*, **109**, 95–98.

Mootz, D. and Händler, V. (1985). *Z. anorg. allgem. Chem.*, **521**, 122–134.

Mootz, D. and Händler, V. (1986). *Z. anorg. allgem. Chem.*, **533**, 23–29.

More, M., Baert, F. and Lefebvre, J. (1977). *Acta Cryst.*, B**33**, 3681–3684.

Morino, Y., Ukaji, T. and Ito, T. (1966). *Bull. Chem. Soc. Jpn.*, **39**, 71–77.

Müller, U. and Mohammed, A. T. (1983). *Z. anorg. allgem. Chem.*, **506**, 110–114.

Mulliken, R. S. (1952a). *J. Am. Chem. Soc.*, **74**, 811–824.

Mulliken, R. S. (1952b). *J. Phys. Chem.*, **56**, 801–822.

Mulliken, R. S. and Person, W. B. (1969a). *Molecular Complexes – a Lecture and Reprint Volume*, Wiley, New York.

Mulliken, R. S. and Person, W. B. (1969b), "Electron Donor-Acceptor Complexes and Charge Transfer Spectra" in *Physical Chemistry* (An Advanced Treatise), edited by D. Henderson, Academic press, London.

Munakata, M., Wu, L. P., Kuroda-Sowa, T., Maekawa, M., Suenaga, Y. and Sugimoto, K. (1997). *Inorg. Chem.*, **36**, 4903–4905.

Munakata, M., Wu, L. P., Kuroda-Sowa, T., Maekawa, M., Suenaga, Y., Ning, G. L. and Kojima, T. (1998). *J. Am. Chem. Soc.*, **120**, 8610–8618.

Munakata, M., Wu, L. P., Ning, G. L., Kuroda-Sowa, T., Maekawa, M., Suenaga, Y. and Maeno, N. K. (1999). *J. Am. Chem. Soc.*, **121**, 4968–4976.

Murthy, R. V. A. and Murthy, B. V. R. (1977). *Z. Kristallogr.*, **144**, 259–273.

Nardelli, M. and Chierci, I. (1960). *J. Chem. Soc.*, pp. 1952–1953.

Nassimbeni, L. R. and Thackeray, M. M. (1974). *Acta Cryst.*, B**30**, 1072–1076.

Neubauer, D. and Weiss, J. (1960). *Z. anorg. allgem. Chem.*, **303**, 28–38.

Nyburg, S. C. and Hilton, J. (1959). *Acta Cryst.*, **12**, 116–121.

Nyburg, S. C., Ozin, G. A. and Szymanski, J. T. (1971). *Acta Cryst.*, B**27**, 2298–2304; Erratum ibid., B**28**, 2885 (1972).

Ogawa, S. (1958). *J. Phys. Soc Jpn.*, **13**, 618–628.

Okuda, T., (1971). *J. Sci. Hiroshima Univ.*, Ser. A, **35**, 213–229.

Okuda, T., Furukawa, Y. and Negita, H. (1972). *Bull. Chem. Soc. Jpn.*, **45**, 2245–2247.

Okuda, T., Suzuki, T. and Negita, H. (1984). *J. Mol. Struct.*, **111**, 177–182.

Okuda, T., Terao, H., Ege, O. and Negita, H. (1970). *Bull. Chem. Soc Jpn.*, **43**, 2398–2403.

Okuda,T., Nakao, A., Shiroyama, M. and Negita, H. (1968). *Bull. Chem. Soc. Jpn.*, **41**, 61–64.

Olah, G. A. (1973). "Friedel-Crafts Chemistry," Wiley, New York.

Olie, K. and Mijlhoff, F. C. (1969). *Acta Cryst.*, B**25**, 974–977.

Oliver, J. G. and Worrall, I, (1967). *Inorg. Nucl. Chem. Letts.*, **3**, 575–577.

Olmstead, M. M., Maitra, K. and Balch, A. L. (1999). *Angew. Chem. Int. Ed. Engl.*, **38**, 231–233.

Osterlof, J. (1954). *Acta Cryst.*, **7**, 637.

Ott, J. B. and Goates, J. R. (1983). *J. Chem. Thermodynam.*, **15**, 267–278.

Padmanabhan, K., Paul, I. C. and Curtin, D. Y. (1990). *Acta Cryst.*, C**46**, 88–92.

Park, W. C. (1969). Ph. D. Dissertation, Univ. of Kentucky, Louisville, Ky.; *Diss. Abstr. Int.*, B **31**, 103 (1970).

Patterson, D. B., Peterson, G. E. and Carnevale, A. (1973). *Inorg. Chem.*, **12**, 1282–1286.

Pepinsky, R. (1962). U. S. Dept. of Commerce, Office Technical Services AD285512; *Chem. Abstr.*, **60**, 10024c.

Perkampus, H. H. and Schönberger, E. (1976). *Z. Naturforsch.*, **31B**, 475–479.

Perkampus, H. H. and Sondern, Chr. (1980). *Ber. Bunsenges. Phys. Chem.*, **84**, 1231–1235.

Peyronel, G., Belmondi, G. and Vezzosi, I. M. (1958). *J. Inorg. Nucl. Chem.*, **8**, 577–581.

Peyronel, G., Buffagni, S. and Vezzosi, I. M. (1968). *Gazz. Chim. Ital.*, **98**, 147–155.

Peyronel, G., Vezzosi, I. M. and Buffagni, S. (1970). *Inorg. Chim. Acta*, **4**, 605–609.

Pohl, S. (1983). *Z. Naturforsch.*, **38b**, 1535–1538.

Pohl, S. and Saak, W. (1984). *Z. Kristallogr.*, **169**, 177–184.

Pritzkow, H. (1974a). *Z. anorg. allgem. Chem.*, **409**, 237–247.

Pritzkow, H. (1974b). *Monatsh.*, **105**, 621–624.

Pritzkow, H. (1975). *Acta Cryst.*, B**31**, 1589–1593.

Probst, T., Steigelmann, O., Riede, J. and Schmidbaur, H. (1991). *Chem. Ber.*, **124**, 1089–1093.

Prosen, R. J. and Trueblood, K. N. (1956). *Acta Cryst.*, **9**, 741–746.

Quinn, H. W. (1967). *Can. J. Chem.*, **45**, 1329–1336.

Quinn, H. W. (1968). *Can. J. Chem.*, **46**, 117–124.

Quinn, H. W. and Glew, D. N. (1962). *Can. J. Chem.*, **40**, 1103–1112.

Ramakrishnan, L., Soundararajan, S., Sastry, V. S. S. and Ramakrishna, J. (1979). *Aust. J. Chem.*, **32**, 931–932.

Ramasubbu, N., Parthasarathy, R. and Murray-Rust, P. (1986). *J. Am. Chem. Soc.*, **108**, 4308–4314.

Reboul, J. P., Oddon, Y., Caranoni, C., Soyfer, J. C., Barbe, J. and Pepe, G. (1987). *Acta Cryst.*, C**43**, 537–539.

Reddy, D. S., Panneerselvam, K., Pilati, T. and Desiraju, G. R. (1993). *J. Chem. Soc., Chem. Commun.*, pp. 661–662.

Redhouse, A. D. (1972). *J. Chem. Soc. Chem. Comm.*, pp. 1119–1120.

Redhouse, A. D. and Woodward, P. (1964). *Acta Cryst.*, **17**, 616–617.

Rencken, I., Boeyens, J. C. A. and Orchard, S. W. (1988). *J. Crystallogr. Spectroscop. Res.*, **18**, 293–306.

Rhine, W. E., Davis, J. and Stucky, G. D. (1975). *J. Am. Chem. Soc.*, **97**, 2079–2085.

Rhoussopoulos, O. (1883). *Chem. Ber.*, **16**, 202–203.

Robertson, J. M. and White, J. G. (1945). *J. Chem. Soc.*, pp. 607–617.

Robinson, D. J. and Kennard, C. H. L. (1970). *J. Chem. Soc.* (B), pp. 965–970.

Rodesiler, P. F. and Amma, E. L. (1972). *Inorg. Chem.*, **11**, 388–395.

Rodesiler, P. F., Auel, Th. and Amma, E. L. (1975). *J. Am. Chem. Soc.*, **97**, 7405–7410.

Rodesiler, P. F., Hall Griffith, E. A. and Amma, E. L. (1972). *J. Am. Chem. Soc.*, **94**, 761–766.

Rogers, M. T. and Helmholtz, L. (1940). *J. Am. Chem. Soc.*, **62**, 1537–1542.

Rogers, M. T. and Helmholtz, L. (1941). *J. Am. Chem. Soc.*, **63**, 278–284.

Rolies, M. and De Ranter, C. J. (1977). *Cryst. Struct. Comm.*, **6**, 157–161.

Rømming, C. (1960). *Acta Chem. Scand.*, **14**, 2145–2151.

Rømming, C. (1972). *Acta Chem. Scand.*, **26**, 1555–1560.

Rosen, H., Shen, Y. R. and Stenman, F. (1971). *Mol. Phys.*, **22**, 33–47.

Ruiz-Amil, A., Martinez-Carrera, S. and Garcia-Blanca, S. (1978). *Acta Cryst.*, B**34**, 2711–2714.

Rundle, R. E. and Corbett, J. D. (1957).*J. Am. Chem. Soc.*, **79**, 757–758.

Rundle, R. E and Goring, J. H. (1950). *J. Am. Chem. Soc.*, **72**, 5337.

Rupp, D. and Lucken, E. A. C. (1986). *Z. Naturforsch.*, **41a**, 208–210.

Sakai, H. (1972). *J. Sci. Hiroshima Univ.* Ser.A-2, **36**, 47–65.

Sakai, H., Matsuyama, T. and Maeda, Y. (1986). *Bull. Chem. Soc. Jpn.*, **59**, 981–986.

Sakai, H., Matsuymam, T., Yamaoka, H. and Maeda, Y. (1983). *Bull. Chem. Soc. Jpn.*, **56**, 1016–1020.

Samoc, A., Samoc, M. and Prasad, P. N. (1992). *J. Opt. Soc. Am.*, B**9**, 1819–1824.

Sawitzki, G. and Schnering, H. G. von, (1974). *Chem. Ber.*, **107**, 3266–3274.

Schmidbaur, H. (1985). *Angew. Chem. Int. Ed. Engl.*, **24**, 893–904.

Schmidbaur, H., Bublak, W., Haenel, M. W., Huber, B. and Müller, G. (1988). *Z. Naturforsch.*, B**43**, 702–706.

Schmidbaur, H., Bublak, W., Huber, B. and Müller, G. (1986a). *Organometallics*, **5**, 1647–1651.

Schmidbaur, H., Bublak, W., Huber, B. and Müller, G. (1986b). *Helv. Chim. Acta*, **69**, 1742–1747.

Schmidbaur, H., Bublak, W., Huber, B. and Müller, G. (1987a). *Angew. Chem., Int. Ed. Engl.*, **26**, 234–237.

Schmidbaur, H., Bublak, W., Huber, B. and Müller, G. (1987b). *Z. Naturforsch.*, **42b** 147–150.

Schmidbaur, H., Bublak, W., Huber, B., Reber, G. and Müller, G. *Angew. Chem., Int. Ed. Engl.*, **25**, 1089–1090 (1986).

Schmidbaur, H., Bublak, W., Riede, J. and Müller, G. (1985). *Angew. Chem. Int. Ed. Engl.*, **24**, 414–415.

Schmidbaur, H., Hager, R., Huber, B. and Müller, G. (1987). *Angew. Chem. Int. Ed. Engl.*, **26**, 338–340.

Schmidbaur, H., Nowak, R., Bublak, W., Burkert, P., Huber, B. and Müller, G. (1987). *Z. Naturforsch.*, **42b**, 553–556.

Schmidbaur, H., Nowak, R., Huber, B. and Müller, G. (1987). *Organometallics.* **6**, 2266–2267.

Schmidbaur, H., Nowak, R., Huber, B., and Müller, G. (1988). *Z. Naturforsch.*, **43b**, 1447–1452.

Schmidbaur, H., Nowak, R., Schier, A., Wallis, J. M., Huber, B. and Müller, G. (1987). *Chem. Ber.*, **120**, 1829–1835.

Schmidbaur, H., Nowak, R., Stiegelmann, O. and Müller, G. (1990). *Chem. Ber.*, **123**, 19–22.

Schmidbaur, H., Probst, T., Huber, B., Müller, G. and Krüger, C. (1989). *J. Organometall. Chem.*, **365**, 53–60.

Schmidbaur, H., Probst, T., Huber, B., Stiegelmann, O. and Müller, G. (1989). *Organometallics.* **8**, 1567–1569.

Schmidbaur, H., Probst, T., Stiegelmann, O. and Müller, G. (1989). *Z. Naturforsch.*, **44b**, 1175–1178.

Schmidbaur, H., Thewalt, U. and Zafiropoulos, T. (1983). *Organometallics*, **2**, 1550–1554.

Schmidbaur, H., Thewalt, U. and Zafiropoulos, T. (1984a). *Angew. Chem. Int. Ed. Engl.*, **23**, 76–77.

Schmidbaur, H., Thewalt, U. and Zafiropoulos, T. (1984b). *Chem. Ber.*, **117**, 3381–3387.

Schmidbaur, H., Wallis, J. M., Nowak, R., Huber, B. and Müller, G. (1987b). *Chem. Ber.*, **120**, 1837–1843.

Schmidt, M., Bender, and Burschka, C. (1979). *Z. anorg. allgem. Chem.*, **454**, 160–166.

Schmidt, M., Bender, R., Ellermann, J. and Gäbelein, H. (1977). *Z. anorg. allgem. Chem.*, **437**, 149–154.

Schwcikert, W. W. and Meyers, Edward A. (1968). *J. Phys. Chem.*, **72**, 1561–1565.

Shaanan, B., Shmueli, U. and Colapietro, M. (1982). *Acta Cryst.*, **B38**, 818–824.

Shaw, J. A. and Fisher, E. (1946). *J. Am. Chem. Soc.*, **68**, 2745.

Shaw, R. A., Smith, B. C. and Thakur, Ch. P. (1968). *Ann. Chem.*, **713**, 30–39.

Shibata, S. and Iwata, J. (1985). *J. Chem. Soc. Dalton*, pp. 9–10.

Shibuya, I. (1961). *Bussei*, **2**, 636.

Shmueli, U. and Mayorzik, H. (1980). Abstract 1-A-37, ECM-6, Barcelona.

Silverman, J., Krukonis, A. P. and Yannoni, N. F. (1973). *Acta Cryst.*, **B29**, 2022–2024.

Smith, H. G. and Rundle, R. E. (1958). *J. Am. Chem. Soc.*, **80**, 5075–5080.

Smith, W. (1879). *J. Chem. Soc.*, **35**, 309–311.

Smith, W. and Davis, G. W. (1882). *J. Chem. Soc.*, **41**, 411–412.

Sneedon, R. P. A. (1979). *L'actualite Chimique*, pp. 31–45.

Sondern, Chr. and Perkampus, H. H. (1983). *Ber. Bunsenges. Phys. Chem.*, **874**, 432–435.

Stezowski, J. J., Stigler, R.-D., Karl, N. and Schuller, K. (1983). *Z. Kristallogr.* **162**, 213–215.

Strieter, F. J. and Templeton, D. H. (1962). *J. Chem. Phys.*, **37**, 161–164.

Strømme, K. O. (1959). *Acta Chem. Scand.*, **13**, 268–274.

Subramanian, V. and Seff, K. (1980). *Acta Cryst.*, **B36**, 2132–2135.

Taylor, I. F., Jr. and Amma, E. L. (1975a). *J. Cryst. Mol. Struct.*, **5**, 129–135.

Taylor, I. F., Jr. and Amma, E. L. (1975b). *Acta Cryst.*, **B31**, 598–601.

Taylor, I. F., Jr., Hall, E. A. and Amma, E. L. (1969). *J. Am. Chem. Soc.*, **91**, 5745–5749.

Tebbe, K.-F. and Nagel, K. (1995). *Acta Cryst.*, **C51**, 1388–1390.

Terao, H., Sakai, H. and Negita, H. (1985). *Bull. Chem. Soc. Jpn.*, **58**, 2318–2322.

Terzis, A., Faught, J. B. and Pouskoulelis, G. (1980). *Inorg. Chem.*, **19**, 1060–1063.

Thewalt, U., Zafiropoulos, T. and Schmidbaur, H. (1984). *Z. Naturforsch.*, **39b**, 1642–1646.

Thomas, C. A. (1941). "Anhydrous Aluminum Chloride in Organic Chemistry." A. C. S. Monograph No. 87, Reinhold, New York. See particularly Chapter 3.

Tildesley, B. D. and Sharpe, A. G. (1953). *Research*, **6**, 51S–52S.

Traynham, J. G. (1961). *J. Org. Chem.*, **26**, 4694–4696.

Traynham, J. G. and Olechowski, J. R. (1959). *J. Am. Chem. Soc.*, **81**, 571–574.

Trotter, J. and Zobel, T. (1966). *Z. Kristallogr.*, **123**, 67–72.

Turner, R. W. and Amma, E. L. (1966a). *J. Am. Chem. Soc.*, **88**, 1877–1882.

Turner, R. W. and Amma, E. L. (1966b). *J. Am. Chem. Soc.*, **88**, 3243–3247.

Turner, R. W. and Amma, E. L. (1966c). *J. Inorg. Nucl. Chem.*, **28**, 2411–2413.

Uchida, T. and Kimura, K. (1984). *Acta Cryst.*, **C40**, 139–140.

Usanovich, M. I., Makarov, E. F., Kamysbaev, D. Kh., Aleksandrov, A. Yu., Sumarakova, T. N. and Amelin, I. I. (1974). *Dokl. Akad. Nauk. SSSR*, **217**, 151–153; *Chem. Abstr.*, **81**, 113273g (1975).

Uson-Finkenzeller, M., Bublak, W., Huber, B., Müller, G. and Schmidbaur, H. (1986). *Z. Naturforsch.*, **41b**, 346–350.

Van der Helm, D. (1973). *J. Cryst. Mol. Struct.*, **3**, 249–258.

Van der Meer, H. (1972). *Acta Cryst.*, **B28**, 367–370.

Vasilyev, A., Lindeman, S. V. and Kochi, J. K. (2001). *Chem. Commun.*, pp. 909–910.

Vasilyev, A., Lindeman, S. V. and Kochi, J. K. (2002). *New J. Chem.*, **26**, 582–592.

Veenvliet, H. and Migchelsen, R. (1971). *Z. Kristallogr.*, **134**, 291–304.

Vezzosi, I. M., Battaglia, L. P. and Corradi, A. B. (1992). *J. Chem. Soc. Dalton*, pp. 375–377.

Vezzosi, I. M., Buffagni, S. and Peyronel, G. (1960). *Ann. Chim.* (Roma), **50**, 343–347.

Vezzosi, I. M., Peyronel, G. and Buffagni, S. (1968). *Gazz. Chim. Ital.*, **98**, 162–166.

Vezzosi, I. M., Peyronel, G. and Zanoli, A. F. (1974). *Inorg. Chim. Acta*, **8**, 229–232.

Vezzosi, I. M., Zanoli, A. F., Battaglia, L. P. and Corradi, A. B. (1988). *J. Chem. Soc. Dalton*, pp. 191–193.

Viennot, J. P. and Dumas, G. G. (1972). *Compt. Rend. Acad. Sci., Paris, Ser. B*, **275**, 489–492.

Vranka, R. G. and Amma, E. L. (1966). *Inorg. Chem.*, **5**, 1020–1025.

Walsh, R. B., Padgett, C. W., Metrangolo, P., Resnati, G., Hanks, T. W. and Pennington, W. T. (2001). *Cryst. Growth Des.*, **1**, 165–175.

Waterfeld, A., Isenberg, W., Mews, R., Clegg, W. and Sheldrick, G. M. (1983). *Chem. Ber.*, **116**, 724–731.

Watson, W. H., Jr. and Lin, C.-T. (1966). *Inorg. Chem.*, **5**, 1074–1077.

Weber, G. (1980). *Acta Cryst.*, **B36**, 2779–2781.

Webster, M. (1966). *Chem. Revs.*, **66**, 87–118.

Weininger, M. S., Rodesiler, P. F. and Amma, E. L. (1979). *Inorg. Chem.*, **18**, 751–755.

Weininger, M. S., Rodesiler, P. F., Gash, A. G. and Amma, E. L. (1972). *J. Am. Chem. Soc.*, **94**, 2135–2136.

Weiss, Alarich. (1972). *Fortschr. Chem. Forsch.*, **30**, 1–76.

Weiss, Alarich. (1995). *Acta Cryst.*, **B51**, 523–539.

Wells, A. F. (1975). Clarendon Press, Oxford, 4th Edition, see, pp. 918–926.

West, C. D. (1937). *Z. Kristallogr.*, **96**, 459–465.

Wheat, J. A., II and Browne, A. W. (1936). *J. Am. Chem. Soc.*, **58**, 2410–2413.

Wheat, J. A., II and Browne, A. W. (1938). *J. Am. Chem. Soc.*, **60**, 371–372.

Wheat, J. A., II and Browne, A. W. (1940a). *J. Am. Chem. Soc.*, **62**, 1575–1577.

Wheat, J. A., II and Browne, A. W. (1940b). *J. Am. Chem. Soc.*, **62**, 1577–1578.

Winstein, S. and Lucas, H. J. (1938). *J. Am. Chem. Soc.*, **60**, 836–847.

Witt, J. R., Britton, D. and Mahon, C. (1972). *Acta Cryst.*, B**28**, 950–955.

Wolten, G. M. and Mayer, S. W. (1958). *Acta Cryst.*, **11**, 739–742.

Wulfsberg, G. and Weiss, Al. (1980). *Ber. Bunsenges. Phys. Chem.*, **84**, 474–484.

Wyatt, W. F. (1936). *Trans. Farad. Soc.*, **25**, 43–48.

Zavalii, P. Yu. and Prots', Yu. M. (1987). *Kristallografiya*. (1987). **32**, 343–346; Sov. *Phys. Cryst.*, **32**, 199–201.

Chapter 12

Hydrogen bonded molecular complexes and compounds

Although the hydrogen bond is not a strong bond (its bond energy ... lying in most cases in the range 2 to 10 kcal/mole), it has great significance in determining the properties of substances. Because of its small bond energy and the small activation energy involved in its formation and rupture, the hydrogen bond is especially suited to play a role in reactions occurring at normal temperatures. It has been recognized that hydrogen bonds restrain protein molecules to their native configurations, and I believe that as the methods of structural chemistry are further applied to physiological problems it will be found that the significance of the hydrogen bond for physiology is greater than that of any other structural feature.

Linus Pauling: *The Nature of the Chemical Bond*
(3rd Edition, pp. 449–450, 1963).

Summary: Hydrogen bonding, one of the most important secondary interactions between molecules of the same kind, is no less important among the binary adducts, and most of the same principles apply. Among the most important hydrogen bond donors are –OH, >NH, –NH$_2$ while >C=O, >O, –Cl and >S are important acceptors, together with charged analogs. The major structural distinction is between "appendage structures", where one component forms a framework to which the second component is hydrogen bonded, and "mixed framework structures", where both components form part of an alternating framework. The latter group can have the two components hydrogen bonded in pairs or in larger discrete groupings, the crystals being molecular crystals from a structural point of view, or in linear chains, or in layers, or in three-dimensional frameworks, where many complex arrangements are possible, especially if a third component such as water is present. Finally, the question of the circumstances under which proton transfer takes place (the formation of ions) is considered.

12.1   Introductory survey

12.1.1   *Introduction*

Hydrogen bonding is undoubtedly the most widespread of the *specific* interactions linking molecules with suitable functional groups together in the solid state (and, of course, also in the gas and liquid phases, which lie beyond the boundaries of this book). The molecules can be similar, as in ice, or different, as in the molecular compounds to be discussed in this chapter. The phrase 'hydrogen-bonded molecular *complex*' is widely used; in terms of our classification (Chapter 1) this is appropriate only when the interactions between one of the components predominates (A . . . A), as, for example, in the gas hydrates (Chapter 7, where we prefer to emphasize the clathrate properties rather than the hydrogen bonding), while '*compound*' is to be used when A . . . B interactions predominate. Our structural classification, based on the nature of the hydrogen-bonded unit formed, distinguishes between two major groupings. First are the appendage structures, where the A component forms a hydrogen-bonded framework to which the pendent B component is linked by a further hydrogen bond; only a relatively few examples of these molecular *complexes* are known. Ferguson and coworkers (e.g. Lavender *et al.*, 1998) have used the graphic phrase "stem and leaves motif" for this group[1]. Next are the framework structures, with the two components in alternating array; many such molecular *compounds* have been studied. The arrays may extend in zero, one, two or three dimensions. We use the term 'hydrogen bonded adduct', when these distinctions are not important. Graph theory provides a convenient way of describing hydrogen bonded arrangements and we shall introduce some of these concepts in Section 12.2.

We first apply these principles of classification to adducts in which hydrogen bonding is the major cohesive interaction, accompanied as always by van der Waals forces. Here we choose illustrative examples for discussion and then list (but certainly not completely) other examples of the same category. There are, however, many adducts in which hydrogen bonding accompanies other types of interaction, such as covalent linkages (as in the oligomers of nucleic acids), or ion–dipole or other types of electrostatic interaction. These are usually adducts with more than two components, the structures are complicated and it is not always easy to disentangle the relative importance of the various interactions. We give only a few illustrative examples.

The partners in hydrogen-bonded adducts may be neutral or charged; in the latter situation the material is formally a salt, with cation and anion linked by hydrogen bonding.

---

[1] The same phrase was used with a different meaning by Ferguson *et al.* (1998). We prefer to retain the unambiguous 'appendage structures.'

There do not appear to be clearcut structural differences between the neutral adducts and the salts and we treat them together. Apart from a few isolated examples, we do not include salts of simple inorganic cations or anions nor those in which there is no cation–anion hydrogen bonding. Hydrates are also generally excluded for reasons of space, despite their undoubted importance. We have found it convenient to distinguish between compounds where the two components have only donor or acceptor functions and those in which each component behaves both as donor and acceptor, i.e. is amphoteric. Pyridinium picrate is a simple example of the first group and the carboxylic acid dimer an example of the more numerous second group.

The last section of this chapter is concerned with the factors which lead to complete transfer of hydrogen from one component to the other, and hence to formation of a salt.

There are a number of classical treatments of the broad aspects of hydrogen bonding (Pimentel and McClellan, 1960; Hamilton and Ibers, 1968; Vinogradov and Linnell, 1971; Joesten and Schaad, 1974; Schuster et al., 1976; Jeffrey and Saenger, 1991; Jeffrey, 1997) dealing mainly with unary systems. Theoretical studies have been summarized by Scheiner (1997) and Del Bene (1998). Most emphasis has been placed on structure and physical properties and their inter-relation. Infrared spectra and dielectric properties, which can be used to identify structural types, have received particular attention. These properties, in turn, can be explained in terms of structure.

The most recent monograph is "The Weak Hydrogen Bond in Structural Chemistry and Biology" by Desiraju and Steiner, 1999. A modern approach to strong and weak hydrogen bonding is outlined in carefully argued introductory chapters; a later chapter on "The weak hydrogen bond in supramolecular chemistry" is particularly relevant to our present purposes, as discussed here and in Chapter 15.

There are many older records of binary molecular compounds, some predating the concept of hydrogen bonding, which we would now recognize as hydrogen bonded. For example, Plotnikow (1909) prepared well crystallized $1:2$, $1:1$ and $2:1$ addition compounds of dimethylpyrone (not defined more precisely) with trichloro- and tri-bromoacetic acids, their solutions being conducting. Bramley (1916) reported freezing point diagrams for the systems phenol–pyridine ($1:1$ and $2:1$ compounds) and phenol–quinoline ($2:3$ and $2:1$ compounds), and for some other systems. Similar methods showed that there was a $2:1$ compound of camphor and salicylic acid (Le Fevre and Tideman, 1931), which had indeed been reported earlier by Leger in 1890, and of 1,8-epoxymenthane with phenols (see Section 12.5.1.2) (Baeyer and Villiger, 1902).

12.1.2 The characteristic features of hydrogen bonds

The most important hydrogen bonds are formed between the following groups of atoms:

$$O–H\ldots O;\ N–H\ldots N;\ O–H\ldots N;\ N–H\ldots O;\ N–H\ldots Cl;\ N–H\ldots S,$$

with the donor atom mentioned first and then the acceptor atom. The same atom can act both as donor and acceptor. These are *strong* hydrogen bonds. The participating atoms can be neutral or charged. Appreciably weaker hydrogen bonds occur for $C–H\ldots O$, $C–H\ldots N$ and $C–H\ldots Cl$ (and some other) groups (Desiraju and Steiner, 1999).

A hydrogen bond is said to be formed when the distance between the proton and the acceptor atom $(d(X–H\ldots Y))^2$ is less than the sum of their van der Waals radii, i.e. Δ_{HB}

---

[2] X is used for the donor atom to avoid possible confusion with deuterium (D).

$\{=d(H \ldots Y)-[r(H)+r(Y)]\}$ should be negative. When $Y=O$, then $[r(H)+r(O)] \approx 2.5$ Å. When the hydrogen position is not known accurately, then the comparison has to be made between the measured value of $d(X \ldots Y)$ and $[r(X)+r(Y)]$; for $O-H \ldots O$ hydrogen bonds this means comparing $d(O_d \ldots O_a)$ and 3.1 Å; much the same values can be used for the N–H donor. Effective values (Table 12.1) for van der Waals radii of a number of important atoms, all singly bonded to C (H (Nyburg *et al.*, 1987), F to Se (Nyburg and Faerman, 1985)) have been derived from statistical analyses of the structural results given in the Cambridge Structural Database (CSD). For most of these atoms it is found that the van der Waals radius in the direction of the bond (a head-on contact) is shorter than that in the perpendicular direction (a side-on contact), i.e. there is polar flattening, and the atom must thus be described by a minor and a major radius. There is no polar flattening in N and O and relatively little in F. The averages of the major and minor radii are quite close to the standard (spherical) values of Pauling (1963) and Bondi (1968). For H bonded to O a minimum side-on contact of 1.03 Å was deduced by Steiner and Saenger (1991) from analysis of ND results for 23 carbohydrate structures. Baur (1992) has pointed out that "side-on [H ... H] contacts in inorganic crystals can be shorter than head-on contacts, while in organic crystals the opposite is true."

The *operational* definition of a strong hydrogen bond in terms of the van der Waals radius has advantages of convenience and applicability to most of the situations discussed in this chapter. but is not in tune with the more sophisticated approaches reviewed by Desiraju and Steiner in their Chapter 1, and has been much criticized (Jeffrey and Saenger, 1991; Desiraju and Steiner, 1999). We shall use it widely as, for our purposes, its advantages outweigh its defects, which are most obvious in borderline cases, and for the weak hydrogen bonds.

Hydrogen bonds can be two-centered ("linear"), three-centered ("bifurcated") or four-centered ("trifurcated") (Fig. 12.1). A three-centred bond is one in which the proton forms two contacts to hydrogen-bond acceptor atoms, such that both are in the forward direction (i.e. both α_1 and α_2 are greater than 90°) and both are shorter than the sum of the van der Waals radii of the atoms involved. The second acceptor atom (X) is approximately in the same plane as the donor (D), H and first acceptor (O) atoms, and in the general direction of the external bisector of the D–H ... O angle.

Table 12.1. Effective van der Waals radii (Å) of atoms X, singly bonded to C

Atom X	Spherical	Major	Minor
H, bonded to sp^3 C	1.20	1.26	1.01
H, bonded to sp^2 C*	1.20	1.38	1.35
F	1.47	1.38	1.30
Cl	1.76	1.78	1.58
Br	1.85	1.84	1.54
I	1.98	2.13	1.76
N	1.70	1.60	1.60
O	1.50	1.54	1.54
S	1.74	2.03	1.60
Se	2.00	2.15	1.70

* Centre of spheroid 0.42 Å away from proton towards C.

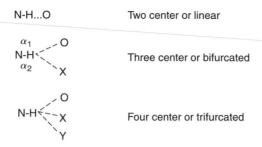

N-H...O   Two center or linear

α_1 O
N-H   Three center or bifurcated
α_2 X

            O
N-H   X   Four center or trifurcated
            Y

Fig. 12.1. Geometrical descriptions of the different multicenter hydrogen bonds. In the diagram the specific example illustrated has N as donor atom and O as the primary acceptor, with X and Y as subsidiary acceptors. Other donor–acceptor combinations are possible.

Structural details of hydrogen bonded systems are determined by x-ray and neutron diffraction; the latter method is to be preferred as the information obtained about the location of hydrogen (or deuterium) atoms is then as precise as that for the heavier atoms. Detailed statistical surveys have been made for O–H...O systems where the acceptor oxgen is from an ether or alcohol [XRD by Kroon *et al.* (1975) (196 bonds in 45 crystals of polyalcohols, saccharides and related compounds) and ND (100 bonds in 24 crystals of carbohydrates) by Ceccarelli *et al.* (1981)]; the samples of Kroon *et al.* and Ceccarelli *et al.* hardly overlap and thus can be combined. Such surveys have also been made for N–H...O=C systems, where the acceptor oxygen is part of a carbonyl group or a carboxylate ion [combined XRD and ND data (1509 bonds in 889 organic crystals, 1357 of which were intermolecular; 1982 CSD release, ≈ 27000 entries)] (Taylor, Kennard and Versichel, 1983, 1984a, b). A parallel study (Murray-Rust and Glusker, 1984) concerned X–H...O< interactions, where X=N, O and the oxygen acceptors were contained in a number of different chemical frameworks; similar analyses have been carried out for $-NH_2/>N-H...N(sp^2)$ (Llamas-Saiz and Foces-Foces, 1990), $R-O-H...N(sp^2)$ (1990 CSD release, 79 000 entries, ≈ 400 hits) (Llamas-Saiz *et al.*, 1992) and N–H...S (Taylor and Kennard, 1984c) systems. Almost all the crystals considered in these surveys were homomolecular. There were significant differences between the geometries of intramolecular and intermolecular hydrogen bonds and the results given here are mainly for the latter group. Detailed studies have not yet been reported for N–H...Cl. Hydrogen bonding of the weaker C–H...O, C–H...N and C–H...Cl types (Taylor and Kennard, 1982; Berkovitch-Yellin and Leiserowitz, 1984) is discussed below.

One-quarter of the H-bonds examined in the Ceccarelli study were found to be three centred. Among the 1354 intermolecular bonds in the Taylor sample, 1112 (82%) were two-centered and 242 three-centered. Among the 149 intramolecular bonds in the sample, 87 (58%) were two centred and 62 three centred. Four centre bonds were very uncommon, there being only six in the total sample of 1509; 2 of these were entirely intramolecular, while all involved positively charged nitrogen atoms. Thus two centre hydrogen bonds predominate but bifurcation is quite common, especially among intramolecular hydrogen bonds. Geometrical parameters of hydrogen bonds are defined in Fig. 12.2.

We first consider hydrogen bond parameters for $O_d-H...O_a$ hydrogen bonds. The distribution of $d(O_d...O_a)$ values has a mean of 2.77 Å and a standard deviation of

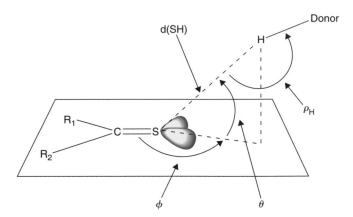

Fig. 12.2. The geometry of hydrogen bonding illustrated for a thiocarbonyl acceptor, with S to be replaced by O for carbonyls, and appropriate changes to be made for other situations. The lone pair lobes are shown for the S (O) atom, together with the angle at the hydrogen atom (the bond is linear when this is 180°). The H...lone pair directionality parameters are given by the angles θ (angle of elevation from the sp^2 lone pair plane) and ϕ (angle of rotation from the C=S vector in that plane). (Reproduced from Allen, Bird, Rowland and Raithby, 1997a.)

0.07 Å (Fig. 12.3). The distribution of $<(O_d–H...O_a)$ values has a mean value at 165°, with a standard deviation of about 8° (Fig. 12.4); this is not corrected for the Ramachadran-Kroon conical factor (Balasubramanian *et al.*, 1970), which states that, from purely geometrical considerations and excluding energy factors, the number of D–H...A bonds whose bending angle lies in the range θ to $\theta + \partial\theta$ is proportional to $2\pi \sin\theta \ \partial\theta$. There is a definite tendency for the H-bonds to be close to linear, and this is accentuated by application of the conical correction. Analysis of other distributions, not reproduced here, shows that the oxygen and hydrogen atoms of the donor moiety, particularly the hydrogens, prefer to lie in or near the lone pair plane of the acceptor moiety. When the O_a atom is a single hydrogen bond acceptor, there is no tendency for the hydrogen atom to adopt any particular position within the lone pair plane; in other words, a distinct preference for hydrogen bonding in the direction of one of the equivalent acceptor lone pairs was not observed. In bonds involving alcohol type acceptors, there is a tendency for the hydrogen bond to occur on the hydrogen side of the C–O_a–H plane. Although the observed parameters cover a fairly large range, Kroon *et al.* (1975) suggest that the extreme configurations have energies only about 4 kJ/mol higher than that of the optimum arrangement $(d(O_d...O_a)=2.77$ Å$;<(O_d–H...O_a)=165°)$.

The histogram of N...O distances is also shown in Fig. 12.3; the distribution is positively skewed (coefficient of skewness is 0.84, as against 0 for a normal distribution) and exhibits positive kurtosis (actual value 3.49, value for normal distribution 3). The mean distance is 2.89 Å, with a standard deviation of 0.11 Å, the extreme values being 2.6 and 3.4 Å. The mean value of the observed N–H...O angles is 164° (Fig. 12.3). The R–O–H...N(sp^2) hydrogen bonds have $<d(O...N)>=2.80(6)$ Å and are close to linear (at hydrogen); the strength of the interactions depends on R, increasing as $C(sp^3) < N < C(sp^2)$. The R–NH$_2$ and R, R$' >$ N–H...N(sp^2) hydrogen bonds were studied by Llamas Saiz and Foces-Foces (1990) using the January, 1989 CSD release with ≈ 70000 entries; about 300 hits were obtained for each of these hydrogen-bond types

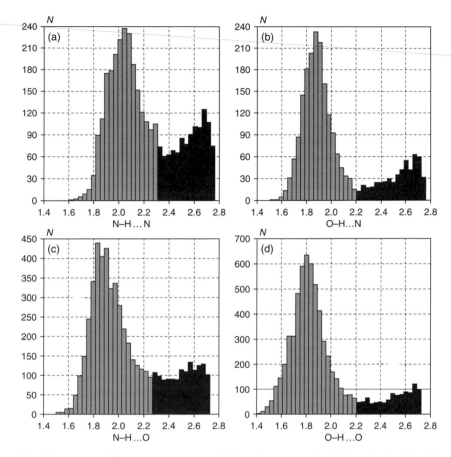

Fig. 12.3. Histograms of $d(\text{H}\ldots\text{A})$ for various $(\text{X–H}\ldots\text{A})$ systems: (a) $(\text{N–H}\ldots\text{N})$, (b) $(d(\text{O–H}\ldots\text{N})$, (c) $(\text{N–H}\ldots\text{O})$, (d) $(d(\text{O–H}\ldots\text{O})$, no distinction being made for different chemical roles of the elements. (Reproduced from Allen, Motherwell, Raithby, Shields and Taylor, 1999.)

and $<d(\text{N}\ldots\text{N})>$ was found be 3.04 and 2.95 Å for the two types, with minimum and maximum limits of 2.81 and 3.20 Å, and 2.70 and 3.20 Å. The means depend somewhat on the limiting values used for the distributions. The H bonds are close to linear, with the highest concentration of interactions in the directions of the $\text{N}(sp^2)$ lone pairs; the strength of the interactions depends on R, and increases as $\text{C}(sp^3) < \text{N} < \text{C}(sp^2)$.

The earlier distance/angle surveys for strong hydrogen bonds have been supplemented by a survey by Allen *et al.* (1999), based on about 160 000 structures given in the CSD release of October, 1996. More than 10 000 N–H$\ldots$O and O–H$\ldots$O fragments were found, and histograms were plotted of $d(\text{H}\ldots\text{O})$ and $d(\text{H}\ldots\text{N})$ in O–H$\ldots$O, N–H$\ldots$O, O–H$\ldots$N and N–H$\ldots$N hydrogen bonds. Values for the peaks of the distributions $d(\text{O–H}\ldots\text{O})$, $d(\text{N–H}\ldots\text{O})$, $d(\text{O–H}\ldots\text{N})$ and $d(\text{N–H}\ldots\text{N})$ were obtained here by adding 1.0 Å to the Allen values, giving 2.80 [2.77], 2.91[2.89], 2.87 and 3.05 [3.04 and 2.95] Å, with earlier mean values in square brackets. The limits were obtained from visual inspection of the minima in the histograms and were given as 3.20 (O–H$\ldots$O, O–H$\ldots$N), 3.25 (N–H$\ldots$O) and 3.30 Å (N–H$\ldots$N). Allen *et al.* note that the H-bond

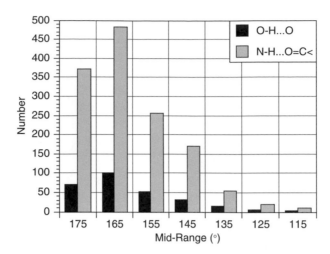

Fig. 12.4. Histogram of $<(O_d\text{--}H \ldots O_a)$, where O_a is an ether (or alcohol) oxygen, and of $<(N\text{--}H \ldots O_a)$, where O_a is a carbonyl oxygen. The angular values have not been corrected for the Ramachandran-Kroon conical factor.

distances are dependent on chemical environment, for example ether and carbonyl oxygens are not equally good acceptors; however, their survey, in contrast to the earlier work, does not make such distinctions, i.e. their histograms represent the superposition of severally chemically-independent distributions. Both means and limits should be considered as fuzzy values, useful but not to be taken too literally.

An earlier investigation of hydrogen bonding between N–H and thiocarbonyl was based on 650 intermolecular interactions, and showed a range of contacts from 3.2 to 3.7 Å, with a mean of 3.4 Å; the values for O–H are only slightly shorter. The H is approaching S in directions which are likely to be occupied by lone pairs and the mean C=S ... H angle is 106° (Taylor and Kennard, 1984c). These earlier results have been confirmed and supplemented by more extensive studies of $R_1R_2C{=}S$ and $R_1CS_2^-$ acceptor systems (Allen, Bird, Rowland and Raithby, 1997a) and of divalent S systems (Y–S–Z acceptors and R–S–H donors) (Allen, Bird, Rowland and Raithby, 1997b); these studies were based on 1994 and 1995 issues of the CSD, with some 150 000 entries. They were derived mostly from homomolecular systems but there is no obvious reason why they should not also be relevant to the heteromolecular systems of primary interest here.

An important feature was comparison of analogous C=O and C=S systems, with distinctions made between H–O– and H–N< donors. There are about ten times as many >N–H as –O–H donors in the sample. There are some fifteen times as many C=O as C=S systems in Table 12.2, but the ratio in the sample is about 45 (for subgroup (b) analysis was limited to about 1/3 of the total entries in the sample, see p. 691 of Allen *et al.*, 1997a). The fraction of acceptors forming hydrogen bonds is 2/3rds for both types (subgroups (a) and (b)); the advantage of C=O over C=S is shown in the competitive situation (subgroup (c)).

One important goal of these surveys was to answer the question whether there is a preference for bonding in the directions of the conventionally viewed O and S sp^2 lone pairs. Some of the results are summarized in Fig. 12.5. The distributions have clear

Table 12.2. Comparison of hydrogen-bond acceptor abilities of >C=S and >C=O groups (taken from Table 9 of Allen *et al.*, 1997a) n_P is the number of potential hydrogen bond acceptors (S or O) in each subgroup; n_b is the number of acceptors that actually form intermolecular hydrogen bonds. D–H is the average number of N, O–H donors in each structure of each subgroup

Subgroup	n_P	n_b(C=S) (%)	n_b(C=O) (%)	D–H
(a) One C=S only	202	133 (66)	–	1.97
(b) One C=O only	3549	–	2342 (66)	1.62
(c) One C=S plus one C=O	81	34 (42)	58 (72)	3.37

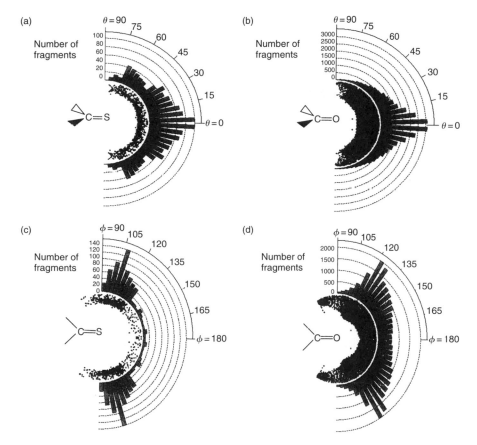

Fig. 12.5. Symmetrized polar scattergrams for the directionality parameters for S=O and C=O to H–(N or O) hydrogen bonds. θ above and ϕ below. (Reproduced from Allen *et al.*, 1997a.)

maxima ($\theta \approx 0°$ for both S and O (i.e. in the lone pair plane) and $\phi \approx 107°$ (S) and 120° (O)) (i.e. in the conventional lone pair directions). The peaks are not especially sharp and deviations of up to ±20° can be expected from the lone pair directions both in, and above and below, the lone pair mean plane. There are also qualitative differences between the N and O scattergrams (not reproduced here but shown by Allen *et al.* (1997a)).

The principal conclusions, based both on the CSD survey and *ab initio* computations, were that S in $R_1R_2C=S$ is an effective acceptor only when R_1, R_2 can form an extended delocalized system with C=S (one criterion for H-bond formation is $d(C=S)>1.65$ Å). The C=S...H–N, O arrangement tends strongly to linearity. Univalent =S in a conjugative environment acts as an effective, but not potent, acceptor of hydrogen bonds, appreciably weaker than =O. Among the reasons are that the σ and π components of the C=S bond are weaker than those of the analogous C=O bond; furthermore the electronegativity of S (≈ 2.6) is similar to that of C(≈ 2.6) and considerably lower than that of O (≈ 3.4). Thus the C=S bond spans a wider distance range than C=O, and will be less polar than $>C^{\delta+}=O^{\delta-}$. Hence =S is a weaker hydrogen bond acceptor than =O. Longer C=S bonds (>1.70 Å) are more polar than shorter (1.58 Å); the frequency of H-bond formation rises from 5% for shorter bonds to $>70\%$ for longer bonds. Thioethers (YSZ) are weak acceptors; of 1223 such sub-structures that co-occur with N(O)–H donors in the 1995 CSD release, only about 5% form hydrogen bonds ($d(S...H)<2.9$ Å). The corresponding values for dialkyloxy-ethers (YOZ) are 1900 and 30% (Allen *et al.*, 1997b).

Only 43 thiol –S–H groups were found in the 1995 CSD release. Of these 70% formed hydrogen bonds with a variety of acceptors (O, S, N, Cl, F), but the numbers were too small for a meaningful statistical analysis.

The mean value for $d(>Nsp^2–H...Cl^-)$ bonds is 3.18 Å (Steiner, 1998); the shortest value found is about 3.0 Å.

The role of C–H...X (X=O, N, S, Cl) interactions in crystals has been considered by Taylor and Kennard (1982) from a statistical point of view and by Berkovitch-Yellin and Leiserowitz (1984) by detailed examination of moiety packing in a variety of representative molecular crystals. Taylor and Kennard considered 113 structures deter-mined by neutron diffraction, in which there were 661 nearest-neighbour contacts; the mean value of Δ_{HB} (as defined above) was essentially zero, confirming the correctness of the van der Waals radii used in calculating Δ_{HB}. However, 46 contacts were found with $\Delta_{HB}<-0.3$ Å, of which 42 were C–H...O contacts. The proton in the majority of these short contacts lies within 30° of the plane containing the oxygen lone pair orbitals. The geometrical features of these short C–H...O contacts were similar to those of O–H...O hydrogen bonds and the interactions were attractive. Thus it was concluded that it would be reasonable to use the terminology "C–H...O hydrogen bond" in crystallographic studies. There were far fewer examples of shortened C–H...X (X=N, S, Cl) contacts but it was suggested that these too came into the same category. Supporting evidence for C–H...O and C–H...N hydrogen bonds has been marshalled by analysis of various crystal structures, including those of the 1:1 molecular compounds of thymine and chlorophenol with benzoquinone, and of barbital with urea and acetamide (Berkovitch-Yellin and Leiserowitz, 1984). A data set consisting of 30 carbohydrate structures analyzed by neutron diffraction has also been considered (Steiner and Saenger, 1992); about one-fifth of the hydrogens were found to be engaged in C–H...O interactions with H...O separations of <2.5 Å; the mean value of $d(C–H...O)$ was ≈ 3.5 Å for inter-molecular interactions. These conclusions match those of Desiraju (1991), the C–H...O hydrogen bond being described as a largely electrostatic long range interaction, the length of which (lying in the range 3.1 to 3.8 Å) depends on the acidity of the hydrogen. The energy is in the 0.5–1.0 kJ/mol range and the C–H group points towards the lone pairs of the oxygen. This suggests that many, but weak, C–H...O interactions will contribute to the cohesion of the crystal; this situation differs from that found with the stronger hydrogen bonds where a few can have a strong influence in determining the overall crystal structure. Theoretical justification, initially for unperturbed dimers in the gas phase, but extended to the solid state has been provided by Legon and Millen (1987). Earlier reviews by Desiraju (1991) and Steiner (1996) have been expanded into their definitive treatment (Desiraju and Steiner, 1999).

The π-system of a benzene ring can also act as an electron acceptor. The phase diagrams of benzene, toluene and mesitylene – HCl have been reported; incongruently melting benzene–HCl, toluene–2HCl and mesitylene–3HCl being found, together with congruently-melting toluene–HCl (180K) and mesitylene–HCl (209K). Crystal structures of toluene·2HCl and mesitylene·HCl were determined, with the HCl molecules normal to the benzene planes, above and below for the toluene complex and only above for the mesitylene complex (Deeg and Mootz, 1993). In toluene–2HCl, $d(\mathrm{Cl}\ldots\pi) = 3.504(1)$ Å, $d(\mathrm{Cl–H}) = 1.01(7)$ Å and $d(\mathrm{H}\ldots\pi) = 2.51(8)$ Å; similar values were found in the mesitylene complex. The mean value for $\mathrm{Cl_3C–H}\ldots\mathrm{Cl^-}$ bonds is 3.42 Å (Steiner, 1998).

We add that there is conclusive crystallographic and spectroscopic evidence for C–H$\ldots$A interactions where A is a strong proton acceptor and C–H forms part of a haloform molecule, HCN or an acetylene derivative (Sim, 1967).

12.2 Application of graph theory to the description of hydrogen bond patterns

The patterns found in the hydrogen bonded molecular compounds and complexes are conveniently described, and sometimes analyzed, by an adaptation of the mathematical theory of graphs (Harary, 1967) to hydrogen bonding in organic crystals (Etter, Macdonald and Bernstein, 1990). Our account closely follows that of Etter *et al.* More detailed expositions are by Grell *et al.* (1999; 2002). The set of molecules to be analyzed is called an *array*; some or all of these molecules are hydrogen bonded. A *network* is a subset of an array in which each molecule in the network is connected to every other molecule by at least one hydrogen bond pathway. A *motif* is a special type of network, a hydrogen bonded set in which only one type of hydrogen bond is present. Graph sets are assigned first to motifs and then to networks.

A graph set is specified by a pattern designator (G), its degree (r), and the number of donors (d) and acceptors (a); 'a' and 'd' were originally presented as superscripts and subscripts but it is more convenient to write them as G(a,d)(r). The pattern designator G has four different assignments depending on whether the hydrogen bonds are intra- or intermolecular:

> S(self) denotes an intramolecular hydrogen bond; C(chain) refers to infinite hydrogen bonded chains; R refers to rings, a cyclic carboxylic acid dimer being an example of a typical ring pattern; D (discrete) refers to non-cyclic dimers and other finite hydrogen-bonded sets, such as phenol hydrogen bonded to acetone.

The parameter 'r' refers to the degree, being either the number of atoms in a ring or the repeat length of a chain. When there is only one hydrogen bond in a D motif, then the degree is 2 and this is taken as the default value and not specified. The parameters 'd' and 'a' refer to the number of different kinds of donors (d) and acceptors (a) used in the hydrogen bond pattern. The default values are 1.

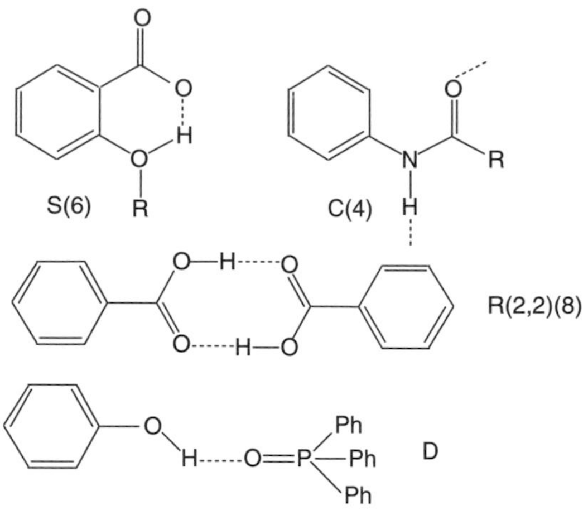

Scheme 1

Graph sets are assigned to motifs by first identifying the different types of hydrogen bond, one motif being generated for each type of hydrogen bond. The hydrogen bonds are then ranked by chemical priority, defined by an extended version of the Cahn–Ingold–Prelog rules. The highest priority hydrogen bond, H(1), is then identified and its motif is generated by finding all occurrences of this bond in the array. A graph set is then assigned to this motif; the process is repeated until graph sets have been assigned to all the motifs. A first-order network is then set up by listing all the motifs in order of increasing priority:

$$N_1 = M_j M_i \ldots M_3 M_2 M_1.$$

Rules have also been given for setting up higher order networks, which arise by combination of motifs from two different kinds of hydrogen bonds.

When there is only a single hydrogen bond type in an array which consists of a discrete pair of molecules, then $N_1 = D(2)$, or simply D. If one finds a centrosymmetric B–A–B discrete triple, then $N_1 = 2D$; if the two hydrogen bonds are not equivalent then $N_1 = DD$.

12.3   Statistics of hydrogen bond patterns

Allen *et al.* (1999) have developed a methodology for extracting from the CSD statistical data on the occurrence of hydrogen-bond patterns of various kinds between two organic molecules without any prior knowledge of the topology or chemical constitution of the motifs. This has enabled them to assess the strengths of various *supramolecular synthons* (the structure directing motifs involving non-covalent bonds) as a guide towards the synthesis of supramolecular structures. We use their data to assess the frequency of occurrence of the various patterns of hydrogen bonding that are found among the hydrogen-bonded molecular complexes and compounds. It should be emphasized that the methodology and results are restricted to the strong hydrogen bonds. Results are available for the graph-set group of rings (R).

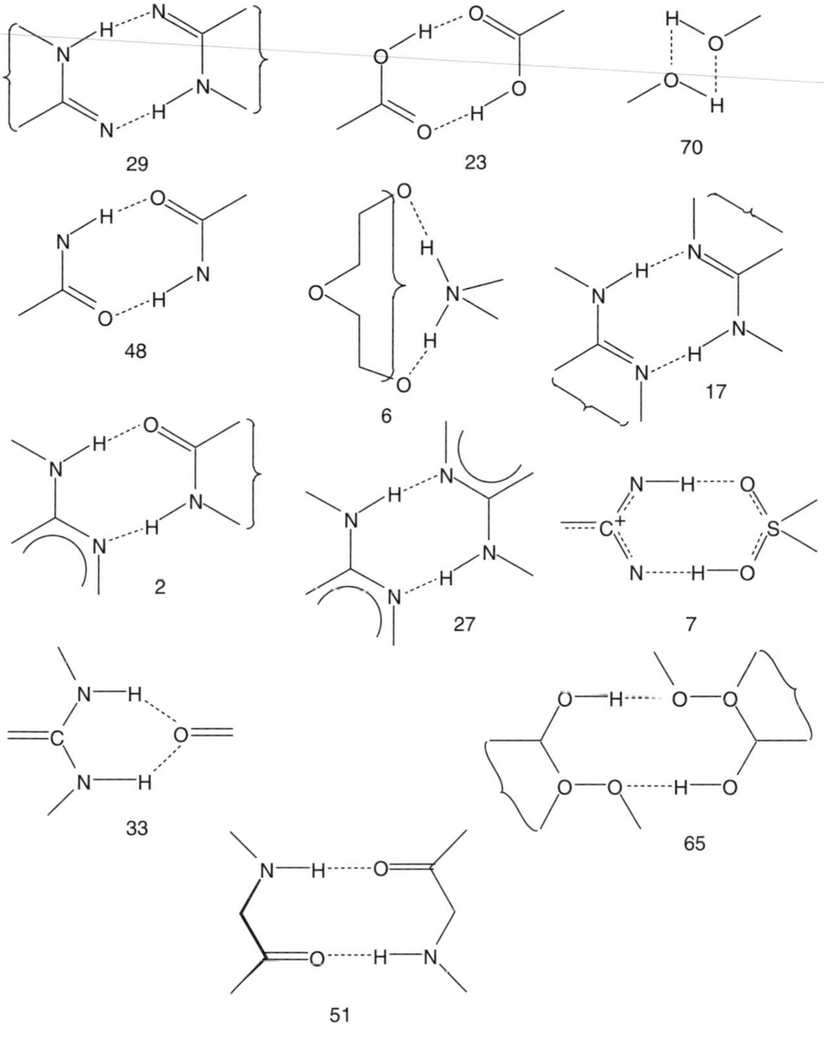

Scheme II

12.3.1 Methodology

We give a simplified account without technical detail. The first step is to determine the number of occurrences N_{obs} (in the CSD) of a particular motif, e.g. the carboxylic acid cyclic dimer. However, the carboxylic dimer may have a large N_{obs} value simply because there are many carboxylic acids in the CSD, not all of which participate in dimer formation. Thus one also needs the total number of carboxylic acids N_{poss} appearing in the CSD. The structural probability of carboxylic dimer occurrence is then given by $P_{m}=N_{obs}/N_{poss}$. This process is then repeated for all desired motifs and the results tabulated. The graph set symbols of the motifs were generated as part of the process.

Table 12.3. Statistical parameters for the most frequent *ring* motifs found in hydrogen-bonded molecular complexes and compounds. Note that Allen *et al.* ordered motifs in terms of decreasing P_{m} values. The numbers refer to the October, 1996 release of the CSD

Motif	Graph set	N_{obs}	N_{poss}	P_{m} (%)
29	R(2,2) (8)	876	3687	24
23	R(2,2) (8)	847	2541	33
70	R(2,2) (4)	408	21824	2
48	R(2,2) (8)	361	4310	8
6	R(2,2) (10)	206	354	58
17	R(2,2) (8)	204	556	37
2	R(2,2) (8)	199	218	91
27	R(2,2) (8)	172	660	26
7	R(2,2) (8)	158	290	54
33	R(1,2) (6)	153	904	17
65	R(2,2) (10)	135	3702	4
51	R(2,2) (10)	59	488	11

12.3.2 *Statistics of ring formation*

75 ring motifs were obtained, with associated parameters; rings were limited to 20 atoms and $N_{\mathrm{obs}} \geq 12$ was required for inclusion. We have abstracted (Table 12.3) the motifs with $N_{\mathrm{obs}} \geq 100$ (with one exception), retaining (to avoid confusion) the original motif numbering of Allen *et al.*; diagrams of the motifs are given in Scheme II.

12.4 Appendage structures (one component forms a hydrogen-bonded framework, to which the second component is appended by hydrogen bonding)

One component forms a hydrogen-bonded framework, to which the second component is appended by hydrogen bonding without being part of the framework. Perhaps the simplest example is the formic acid ... hydrogen fluoride complex HCOOH ... HF (NEWXAO; congruent melting point 242K) (Wiechert, Mootz and Dahlems, 1997). The structure (*Pnma*, Z = 4) was determined at 123K, and shows chains of hydrogen-bonded formic acid molecules to which HF molecules are appended by F–H ... O bonds. The authors point out that the formic acid chains here, with antiplanar carboxyl groups, are isomers of those found in the neat formic acid structure, where all the hydrogen bonds are reversed and the carboxyl groups are thus synplanar. Calculations suggest that the synplanar formic acid structure is some 30 kJ/mol more stable than the antiplanar structure. Weaker C-H ... O and C–H ... F interactions are also shown in Fig. 12.6. The description of the packing will change if these are given equal status with the strong hydrogen bonds. As the energy of a strong hydrogen bond is some thirty times as large as that of a weak hydrogen bond, it seems clear that the former are structure determining unless the number of weak H bonds is preponderant. This is the point of view we have adopted in describing moiety arrangements in this book.

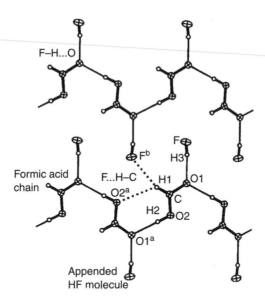

Fig. 12.6. Structure of formic acid...hydrogen fluoride. The ellipsoids correspond to 50% probability. The two C–O bond distances are 1.232(2) and 1.298(2) Å, showing localization of the proton of the carboxyl group. d(O–H...O)=2.649(2) and d(F–H...O)=2.556(2) Å. The C–H...O and C–H...F interactions are shown by dotted lines; the distances are 3.240(2) and 3.238(2) Å. (Adapted from Wiechert *et al.*, 1997.)

Another relatively clearcut example is provided by the 1:1 complex of *trans*-9, 10-dihydroxy-9,10-diphenyl-9,10-dihydroanthracene with ethanol (CENMOX; Toda *et al.*, 1984b). The crystals are triclinic, space group $P\bar{1}$ with two formula units per cell. The substituted DHA molecules are situated at two independent centres of symmetry. These two crystallographically distinct molecules, I and II, are alternately linked by O_I...H–O_{II} hydrogen bonds (2.893(8) Å) to form zigzag chains along [001]. The ethanol molecule is an *acceptor* in a hydrogen bond to O_I (2.767(8) Å) and is not linked to O_{II} (Fig. 12.7). A somewhat similar example is provided by 1,1,6,6-tetraphenylhexa-2,4-diyne-1, 6-diol:benzophenone (A...B; SOGHAX), where the linear framework is made up of a combination of covalent bonds within the bifunctional diol molecules and hydrogen bonds between them (Bond *et al.*, 1991). The pendent benzophenone molecule is linked to a hydroxyl of the diol by a hydrogen bond of length (2.670(2) Å), which is somewhat shorter than that between hydroxyls (d(O(1A)...O(1B) = 2.786(3) Å). Thus a more appropriate description for both these examples could be of A...B units linked by A...A hydrogen bonds.

We have encountered relatively few other examples in this category; among these are glucitol:pyridine (SORBPY20; Kim, Jeffrey and Rosenstein, 1971), diethylstilboestrol: dimethyl sulphoxide (DESTDM10; Busetta *et al.*, 1973), 3,5-diiodotyrosine: N-methylacetamide (where there is a short I...O=C< contact of 3.03 Å in addition to the hydrogen bonding) (ITYRMA10; Cody *et al.*, 1972) and trimethylammonium hydrogen oxalate (TMAHOX; Thomas and Renne, 1975). The first named component forms the framework in the first three examples and the hydrogen oxalate ions in the

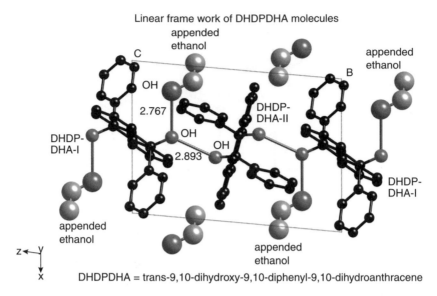

Linear frame work of DHDPDHA molecules

DHDPDHA = trans-9,10-dihydroxy-9,10-diphenyl-9,10-dihydroanthracene

Fig. 12.7. Projection down [010] of the 1:1 appendage complex of *trans*-9,10-dihydroxy-9,10-diphenyl-9,10-dihydroanthracene with ethanol. Hydrogen bonds are indicated by dark lines between oxygens (hydroxyls). Note that the two DHDPDHA molecules are at independent centers of symmetry. (Data from Toda *et al.*, 1984.)

fourth. The 1 : 1 compound between testosterone and *p*-bromophenol also falls into this category (space group $P2_12_12_1$, $Z=4$) (HOTEST10; Cooper *et al.*, 1969). Steroid molecules related by the two-fold screw axes parallel to [010] are hydrogen bonded end-to-end by O_{17}–H...O_3=C< links ($d = 2.71$ Å), while O_{17} also acts as an hydrogen bond acceptor to the pendent phenolic hydroxyl group ($d = 2.62$ Å); here too a more appropriate description could be of A...B units linked by A...A hydrogen bonds. Another example is the 1-(2,3-dideoxy-*erythro*-β-D-hexa-pyranosyl)-thymine – dioxane molecular complex (KINHIY) where there is an extensive network of hydrogen bonds between the thymine molecules and the dioxane is linked through one of its oxygens to a thymine C–H group through what may be a C–H...O hydrogen bond ($d = 3.182(5)$ Å), giving a pendent structure (De Winter *et al.*, 1991). A similar arrangement is found in the Grossularine II – tetrahydrofuran complex (DOGJIS), where there are layers of hydrogen-bonded Grossularine molecules, with pendent THF linked to them through weak N–H...O(THF) bonds ($d = 3.08$ Å) (Carré *et al.*, 1986).

2,2′-Bipyridyl (in its *trans* conformation) is pendent in its 1,3,5-trihydroxybenzene–2 (bipyridyl) (PUVMIC) and 4,4′-sulfonyldiphenol–bipyridyl (PUVMOI) molecular complexes (Lavender *et al.*, (1998a, b), who here introduced the phrase "stem and leaves motif"). In the first of these ($P2_1/c$, $Z=4$), two hydroxyls of phloroglucinol are linked to the two nitrogens of one bipyridyl molecule to form a chain along [010], with the second bipyridyl pendently linked to the third phloroglucinol hydroxyl group, while the remaining nitrogen has only a weak C–H...N link (3.49 Å). This connectivity is identical to that in {1,1,1-tris(4-hydroxyphenyl)ethane–2(hexamethylenetetramine)}{H_3C–$C(C_6H_4$ $(OH)_3$·(HMTA)} (RAWDEY), where half the HMTA molecules act as double acceptors within chains and half act as single acceptors pendent from the chains (Coupar *et al.*, 1997). In the 1 : 1 {4,4′-sulfonyldiphenol–bipyridyl} molecular complex, the bis-phenol molecules are linked into C(8) chains by means of O–H . . . O=S hydrogen bonds, and the 2,2′-bipyridyl moieties, which act as single acceptors only, are pendent from these chains through O–H . . . N hydrogen bonds. Pairs of these chains, related by centres of inversion, are mutually interlocked *via* their pendent arms (a "zip-fastener architecture"). The pendent moiety is the acceptor in all these examples.

12.5 Alternating framework structures (the components, in hydrogen bonded alternating array, form a mixed framework)

In this group of molecular compounds, the components, hydrogen-bonded together in alternating array, form a Mixed Framework. When the components are hydrogen bonded into a discrete group, then the framework has zero dimensions and the crystal is a molecular crystal, irrespective of whether the components are present as neutral molecules or ions. Other possibilities are that the mixed array is linear (one-dimensional), or has hydrogen bonding in two dimensions, forming layer structures, or has three dimensional hydrogen bonding. Which of these possibilities occurs depends on the number of functional groups in each component capable of hydrogen bonding. An important principle appears to be maximal formation of hydrogen bonds but there are examples where potential hydrogen bond donors or acceptors are left unengaged.

12.5.1 *Zero-dimensional frameworks*

Groups of two, three, four and even five linked molecules have been found in reported crystal structures. Pairs are by far the most frequent. We give some examples to illustrate each situation.

12.5.1.1 *Structures with discrete pairs (A–B) of components*

We start by choosing examples of a hydrogen-bonded neutral molecule pair and a salt (hydrogen-bonded cation–anion pair). The salt nicotinyl salicylate (NICSAL) (Kim and Jeffrey, 1971), and the neutral-molecule compound between antipyrine and salicylic acid (salipyrine; APYSAL) (Singh and Vijayan, 1974) are viewed normal to the benzene rings (Fig. 12.8).

Molecular compounds of triphenylphosphine oxide[3] provide interesting comparisons of proton positioning (Haupt *et al.*, 1977). The HCl adduct was prepared from Ph_3PO and HCl in benzene (HXTPPL). The space group is $P2_1/c$ and $Z = 4$. Discrete molecule pairs were found but it was inferred from the hydrogen position in the P=O . . . HCl bridge that the compound was better described as hydroxotriphenylphosphonium chloride, i.e. $Ph_3P^+OH . . . Cl^-$. The HF compound was also prepared in benzene and molecular pairs were found (TPOXHF; Thierbach and Huber, 1979). Although the space group was also $P2_1/c$, $Z = 4$, the two compounds were not isomorphous. Here the hydrogen was attached to the fluorine in an asymmetrical hydrogen bond. In the trichloroacetic acid compound

---

[3] Etter and Baures (1988) note that Ph_3PO (also abbreviated as TPPO) forms high quality crystals with many organic compounds, acting as a "crystallization aid."

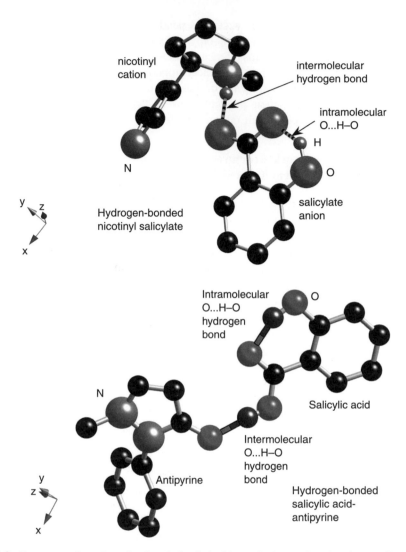

Fig. 12.8. Two examples of a pair of moieties linked by an hydrogen bond to form a discrete pair; the salt nicotinyl salicylate (NICSAL) is shown above and salipyrine (APYSAL), composed of the neutral molecules antipyrine and salicylic acid, below. For clarity only hydrogens involved in hydrogen bonding are shown. There is hydrogen transfer within the molecular compound in the first example but not in the second. In both examples there is a single donor–acceptor interaction between the components. H-bond distances are: intramolecular O–H...O NICSAL 2.550 Å, APYSAL 2.582 Å; intermolecular NICSAL N^+...O^- 2.628 Å; APYSAL O–H...O 2.534 Å. (Data from Kim and Jeffrey (1971) and Singh and Vijayan (1974).)

with triphenylphosphine oxide (TPOTCA) the structural units are discrete hydrogen bonded pairs; although the hydrogen bond is rather short (2.496(3) Å), it is unsymmetrical with the hydrogen remaining on the carboxyl group (Golic and Kaucic, 1976). It would be useful to verify these descriptions by neutron diffraction, preferably at low temperatures.

(a) Neutral molecule pairs (single intermoiety donor–acceptor interactions) Single O–H...N hydrogen bonds link pentachlorophenol in neutral molecule pairs with 4-acetylpyridine (PABBAV; Mayerz *et al.*, 1991), 3-pyridinecarbonitrile (FOMWIN; Malarski *et al.*, 1987a), 4-methylpyridine (with singly deuterated pentachlorophenol) (GADGUN01; Majerz *et al.*, 1990b), 4-methylpyridine (with singly hydrogenated penta-chlorophenol; GADGUN02) [these two compounds are not isomorphous] (Malarski *et al.*, 1987b), 3-oxo-azabicyclo[2.2.2]octane (Majerz *et al.*, 1990b) and 5,6,7,8-tetrahydroquinoline (SOGXUH; Malarski *et al.*, 1991). Single O–H...O hydrogen bonds are found in the compounds of *p*-nitrophenol with triphenylarsine oxide (DOYDAW; Lariucci *et al.*, 1986a), diphenylmethanol with triphenylphosphine oxide (FAXRAX; Lariucci *et al.*, 1986b), (2,4-dichlorophenoxy)acetic acid with triphenylphosphine oxide (d(O...O) = 2.58, 2.61 Å) (YACNUL; Lynch *et al.*, 1992), 1-(2′-hydroxy-3′,6′-diethylphenyl)-2-bromomethyl-naphthalene with triphenylphosphine oxide (PATROR; Peters *et al.*, 1993), triphe-nylmethanol with dioxane (JODXUV; $P\bar{1}$, $Z = 2$, d(O...O) = 2.839(3) Å) and tri-1-naphthylsilanol with dioxane (JODYEG; $P\bar{1}$, $Z = 2$, d(O...O) = 2.736(3) Å (Bourne, Johnson *et al.*, 1991); *p*-carboxy-phenylazoxycyanide with dimethyl sulphoxide (CPZXCY; Viterbo *et al.*, 1975), 2-aminophenol with 4-nitropyridine-N-oxide (NPOAPL; Cc, $Z = 4$; d(O...O) = 2.696(3) Å; Lechat *et al.*, 1981). 10,11-Dihydro-5-phenyl-5H-dibenzo[a,d] cyclo-hepten-5-ol forms a 1 : 1 compound with acetone (JIDCUU; $P2_1/c$, $Z = 4$), which is isomorphous with the 2 : 1 compound formed by the same donor with dioxane (JIDCOO) (Caira *et al.*, 1990). The second of these compounds could also be classified as a cen-trosymmetric D–A–D compound belonging in the following section.

Single N–H...O=C< hydrogen bonds link diphenylamine in neutral molecule pairs with benzophenone (BZPPAM); the space group is $P2_1/c$ and $Z = 4$. Although there are two polymorphs, only the structure of Polymorph I has been reported (Brassy and Murnou, 1972). Single P=O...H–N hydrogen bonds link triphenylphosphine oxide with N-acetylbenzamide (VIWBAE; 2.839(3) Å), N-propionylbenzamide (VIWBEI; 2.836(3) Å) and dibenzamide (VIWBIM; 2.897(3) Å; Etter and Reutzel, 1991) and with N-acetyl-*p*-toluenesulphonamide ($P2_1/a$, $Z = 4$; 2.705(2) Å) (GACCES; Etter and Baures, 1988). All these examples have graph set first order networks $N_1 = D$.

More complicated arrangements are also found. 1,8-Biphenylenediol forms two strong O–H...O hydrogen bonds with molecules such as O=P(NMe$_2$)$_3$ (COXKUV10) 1,2,6-trimethyl-4-pyridone (CEWYIM30) and 2,6-dimethyl-γ-pyrone (COXLEG; Hine *et al.*, 1990) which contain R=O groups; indeed the linkage in the latter molecular compound is so strong that it persists in solution, as shown by vapour pressure osmometry measurements (Hine *et al.*, 1984). 1,3-Bis(*m*-nitrophenyl)urea forms 1 : 1 complexes with tetrahydrofuran (THF; GIMROJ10; 2.995(2), 2.918(2) Å; graph set R(1,2)(6)), triphenyl-phosphine oxide (TPPO; GIMSEA10; 2.860(3), 2.969(2) Å; graph set R(1,2)(6)) N,N-dimethyl-*p*-nitroaniline (GIMRUP10; 3.003 Å; graph set R(2,2)(8)) where the two N–H groups of the urea molecule are both linked unsymmetrically (except for the third of these examples) to the oxygen of the acceptor molecule; a similar arrangement is found in the 1 : 1 complex of (1-*m*-nitrophenyl)-(3-*p*-nitrophenyl)urea with DMSO (SILVOY; 2.880(2), 3.100(2) Å; graph set R(1,2)(6), (Fig. 12.9). The authors (Etter, Urbanczyk-Lipkowska *et al.*, 1990) suggest that 1,3-bis(*m*-nitrophenyl)urea behaves primarily as a proton donor because very weak intramolecular C–H...O interactions inhibit its proton-accepting properties.

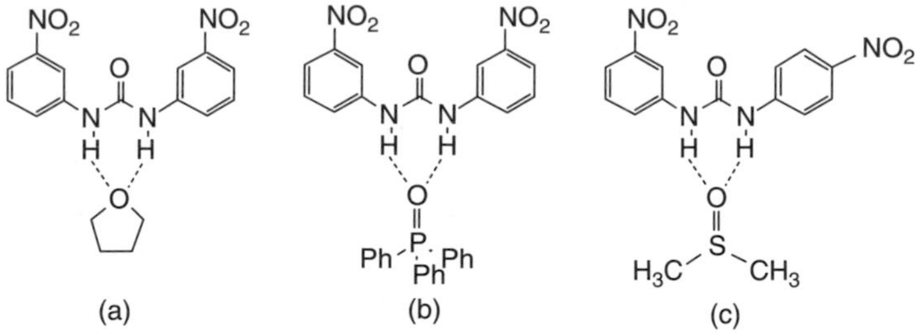

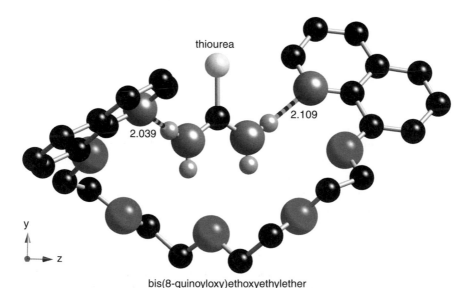

Fig. 12.9. Similar hydrogen bonding patterns are found in the discrete 1 : 1 complexes of (a) 1,3-bis(*m*-nitrophenyl)urea and THF; (b) 1,3-bis(*m*-nitrophenyl)urea and triphenylphosphine oxide; (c) (1-*m*-nitrophenyl)-(3-*p*-nitrophenyl)urea and DMSO. These are examples of ring motif 41 in the Allen *et al.* (1999) statistical analysis; it is not included in Scheme II because there are only 31 occurrences in the data set. (Adapted with permission from Etter *et al.*, 1990).

Finally we mention two remarkable examples, where discrete molecule pairs are formed with all possible intercomponent hydrogen bonds – the 1 : 1 compounds between bis[2-(*o*-methoxyphenoxy)ethoxyethyl]ether ($C_{22}H_{30}O_7$) and thiourea (MPEX-TU; Suh and Saenger, 1978), and between bis[(8-quinoyloxy)ethoxyethyl]ether ($C_{26}H_{28}N_2O_5$) and thiourea (QUETHU; Weber and Saenger, 1980, Fig. 12.10). Each thiourea interacts with only one ether molecule, and there are no stacking interactions between quinoline groups in the second example.

Fig. 12.10. View of the mutual arrangement of the two components in the ($C_{26}H_{28}N_2O_5$)–thiourea molecular compound showing how the polyether (the hydrogen bond acceptor) winds itself around the thiourea molecule (the hydrogen bond donor). The N–H...N hydrogen bonds are shown by broken lines. (Data from Weber and Saenger, 1980.)

(b) Salts (cation–anion pairs)   Mearsinium picrate $(C_9H_{14}NO^+ \cdot C_6H_2N_3O_7^-)$ has hydrogen bonded ion pairs packed as units in the crystal (CIJXOI10; Robertson and Tooptakong, 1985), as do fenpropimorph picrate (SENBUI; Jensen and Jensen, 1990), acridinium pentachlorophenolate (PABBEZ; Wozniak *et al.*, 1991) and pyridinium picrate (PYRPIC03; Botoshansky *et al.*, 1994), which is dimorphic (Kofler, 1944). Both phases of the latter contain hydrogen bonded ion pairs as the structural unit, with pyridinium nitrogen forming a bifurcated hydrogen bond with phenolic oxygen and an oxygen of an *ortho* nitro group. In Phase I (stable at room temperature) the pairs are stacked in parallel array while the antiparallel arrangement of Phase II (stable at higher temperatures) is somewhat similar to that found in mearsinium picrate. Note that the structure reported originally (Talukdar and Chaudhuri, 1976) for Phase I is wrong. Hydrogen bonded ion pairs are also found in N(6),N(6)-dimethyladeninium picrate and promethazine picrate (PROPIC10; Shastry *et al.*, 1987). A bifurcated hydrogen bond, somewhat different from that in pyridinium picrate, is found in the neutral molecule pair 2,3-dichloro-5-hydroxy-6-cyanobenzoquinone and 2-benzamidopyridine (GESSAY; Bruni *et al.*, 1988). A crystalline 1 : 1 molecular compound of N,N-dimethyl-*p*-toluidine and 1,2-benzisothiazol-3(2H)-one 1,1-dioxide (*o*-sulfobenzimide, Saccharin) can be prepared (with difficulty) and has been found to be composed of discrete, hydrogen bonded cation–anion pairs, in which a hydrogen has been transferred from the imide N of saccharin to the dimethylamino group of the toluidine moiety, followed by formation of a N^+–H . . . O=C< hydrogen bond $(d = 2.676(3)$ Å, $P2_1/c$, $Z = 4)$ (KIFYAZ; Courseille *et al.*, 1990). An antiparallel arrangement of hydrogen bonded cation–anion pairs is found in *trans*-4′-hydroxy-N-methyl-4-stilbazolium (+)-camphor-10-sulphonate (BOJWAY; Ziolo *et al.*, 1982) (these crystals are enantiomorphic, space group $P2_1$, with 4 ion pairs per unit cell).

Saccharin

In contrast to the above examples we give two examples of true salts. The first is ethyl 8-dimethylamino-1-naphthalenecarboxylate picrate (VIGJEA; Parvez and Schuster, 1991), in which the transferred hydrogen is located in an *intramolecular* hydrogen bond between protonated dimethylamino and carbonyl groups of the cation. The second, 1,8-bis(dimethylamino)naphthalene pentachlorophenolate. 2(pentachlorophenol), has an analogous proton sponge type cation. There is no hydrogen bonding between cation and anion in either of these salts but in the second the pentachlorophenolate anion and the two neutral pentachlorophenol molecules form a discrete hydrogen bonded system (TAPCES; Kanters *et al.*, 1992). The hydroxyl donors of these hydrogen bonds are each involved in an asymmetric three center hydrogen bond with a strong O–H . . . O$^-$ intermolecular branch and a weak OH . . . Cl intramolecular branch.

(c) Neutral molecule pairs (amphoteric intermoiety donor–acceptor interactions)   There are a number of examples where both components have donor and acceptor functions which are used in the formation of the molecule pair, i.e. both components are *amphoteric*.

This occurs in the molecular compound of 2-pyridone and 6-chloro-2-hydroxypyridine (left hand formula below) (PYOCHP; Almöf *et al.*, 1972). Other examples involving donor and acceptor functions from both components are the 1 : 1 molecular compounds between 1-chloro-2-hydroxynaphthalene-3-carboxylic acid and dimethylformamide (right hand formula below) (JIWNIM; Czugler *et al.*, 1991), where both intra- and intercomponent hydrogen bonds are shown in the diagram, and the 1 : 1 neutral-molecule molecular compounds between (E)-acetophenone oxime and benzoic acid (JUKJII) and (E)-benzaldehyde oxime and benzamide (JUKJOO; Maurin *et al.*, 1993).

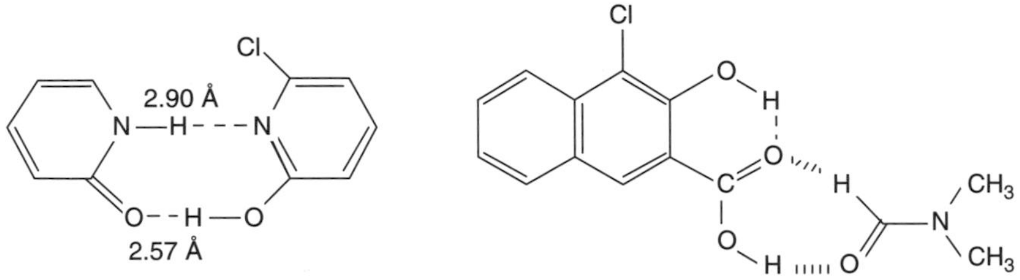

Scheme III

In the 1 : 1 compound of 3,4-dimethoxy- and 2,4-dinitrocinnamic acids (orange crystals) the carboxylic acid groups form the familiar hydrogen bonded pairs; furthemore, super-imposed antiparallel pairs interact by a π–π^* charge transfer mechanism, thus accounting for the colour (BOGGUZ10; Sarma and Desiraju, 1985). These crystals belong both here and in Chapter 15.

(d) Intermediate situations   The examples given above include structural units composed of neutral molecule pairs and of ion pairs; intermediate situations are also found, as in trichloroacetic acid – pyridine-N-oxide (Golic and Lazarini, 1974), where the IR spectrum shows the same features as are found in Type A acid salts (Golic *et al.*, 1971). The structure was first determined by x-ray diffraction at room temperature and has been reinvestigated by neutron diffraction with the crystal at 120K (PYOTCA10; Eichhorn, 1991). The hydrogen bond is nearly symmetrical, with $d(O \ldots H) = 1.148$ and 1.284 Å respectively, the hydrogen being closer to the pyridine-N-oxide oxygen.

(e) Some examples of molecular recognition   There is a simple demonstration of molecular recognition in the molecular compound of 2-phenyl-4-(4,6-dimethyl-2-pyrimidyl)-aminomethylene-5(4H)-oxazolone and acetic acid (Fig. 12.11) (TALHIX; Leban *et al.*, 1991) while a more complicated, but particularly neat example, is 1,4-bis[[(6-methylpyrid-2-yl)-amino]carbonyl]benzene–adipic acid (Fig. 12.12) (JEWNUU10; Garcia-Tellado *et al.*, 1991); both involve >N–H . . . O=C< and –O–H . . . N< hydrogen bonds and the interactions are amphoteric. C–H . . . O interactions occur in the second example, leading to formation of a seven-membered ring involving secondary interactions.

A classical example of molecular recognition *in solution* is of pyrazine, quinoxaline and phenazine by Rebek's diacid $C_{39}H_{43}N_3O_8$ (Rebek *et al.*, 1987). The association constants were interpreted to show two-point hydrogen bond binding, but this was later questioned by Jorgensen *et al.* (1993), who suggested one-point binding from Monte Carlo

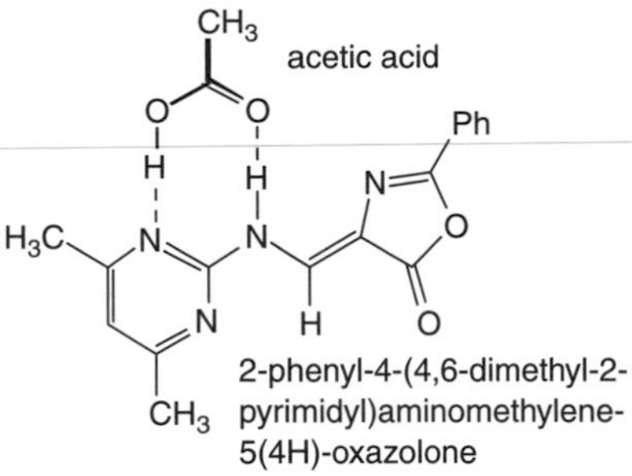

acetic acid

Ph

H₃C

2-phenyl-4-(4,6-dimethyl-2-
pyrimidyl)aminomethylene-
5(4H)-oxazolone

Fig. 12.11. Schematic diagram of the molecular compound of 2-phenyl-4-(4,6-dimethyl-2-pyrimidyl)-aminomethylene-5(4H)-oxazolone and acetic acid, demonstrating molecular recognition. After Leban *et al.*, 1991.

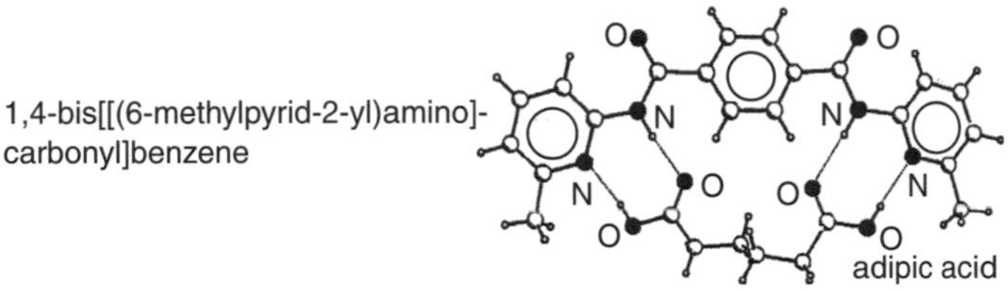

1,4-bis[[(6-methylpyrid-2-yl)amino]-
carbonyl]benzene

adipic acid

Fig. 12.12. Projection onto plane of the benzene ring of the discrete binary molecular unit formed by 1,4-bis[[(6-methylpyrid-2-yl)amino]carbonyl]benzene and adipic acid. Hydrogen bonds are shown by thinner lines. The space group is $P2_12_12_1$ with $Z=4$. The C–H...O interactions in the center of the diagram are not shown explicitly. (Reproduced with permission from Garcia-Tellado *et al.*, 1991.)

solution studies. An unambiguous demonstration of two-point binding was provided by the structure determinations of $\{C_{39}H_{43}N_3O_8.pyrazine.2.5(pyrazine)\}$ (POLFUR) and $\{C_{39}H_{43}N_3O_8.quinoxaline.0.5-(ethanol)\}$ (YAWJIP10; Fig. 12.13) by Pascal and Ho (1994); both crystals are triclinic, $P\bar{1}$, $Z=4$. It was shown that the Rebek-acid molecule "is quite flexible, and that small changes in critical framework bond angles may be amplified into substantial displacements at the diamine binding site".

The above examples demonstrate the formation of discrete polymolecular units from mutually compatible components, with evidence for the existence of such units in solution and demonstration of their structure in the solid state. Most of this book is concerned with the structures of binary adducts of crystallizable mutually-compatible components, generally encountered by chance. But what of design, especially with a view to controlling solid-state properties? One example of a brave attempt is given by Garcia-Tellado *et al.*, 1991.

The naphthalene compound (Scheme IV) forms triclinic 1:1 adducts with 1,8-octanedicarboxylic acid (JOHRAZ; $P1$, $Z=2$, thus there are two crystallographically-independent units of each component in the unit cell, without crystallographically imposed symmetry), and with 1,12-dodecanedicarboxylic acid (JOHPUR; $P\bar{1}$, $Z=1$, thus both components are located at independent centres of symmetry); the biphenyl compounds behaves similarly with 1,12-dodecanedicarboxylic acid (JOHPOL; $P\bar{1}$, $Z=1$,

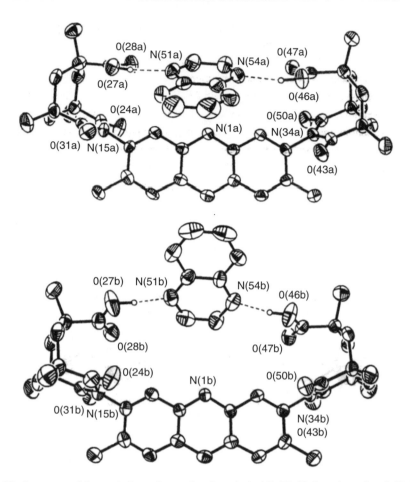

Fig. 12.13. Structures of the two independent molecular units in {C$_{39}$H$_{43}$N$_3$O$_8$·quinoxaline·0.5(ethanol)} showing the differences in the details of their structures. The N...O distances are 2.761 and 2.790 Å in the upper diagram and 2.691 and 2.661 Å (all s.u.'s 0.007 Å) in the lower diagram. The quinoxaline molecules are held by the crab-like claws of Rebek's diacid. (Reproduced with permission from Pascal and Ho, 1994.)

thus both components are located at independent centres of symmetry). There is amphoteric >N–H...O=C< and –O–H...N< hydrogen bonding, linking the components in head-to-tail fashion in chains (Fig. 12.14); the hydrogen bonding motif is similar to that shown in Fig. 12.12 (but without C–H...O interaction). Varying mutual dispositions of the components allow more than one possibility of recognition. It is not clear whether there are interactions in solution, but this seems unlikely, The authors comment "The development of molecules that can self-assemble into new solid-state materials is an area of recognized importance. The ability to control both the formation and details of the structure of these materials offers an interesting approach to fine tuning electrical or optical properties in the crystal."

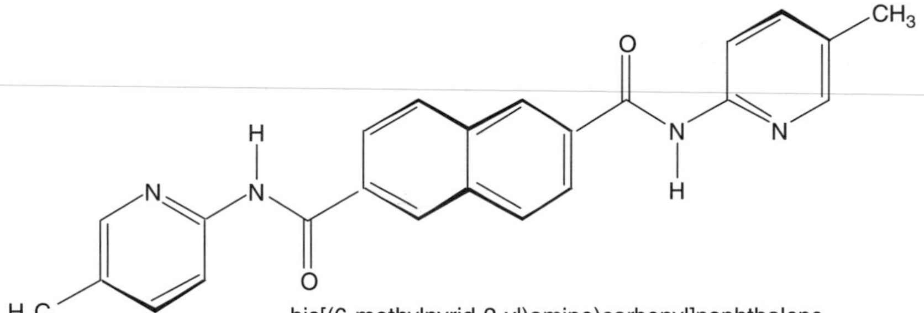

bis[(6-methylpyrid-2-yl)amino)carbonyl]naphthalene

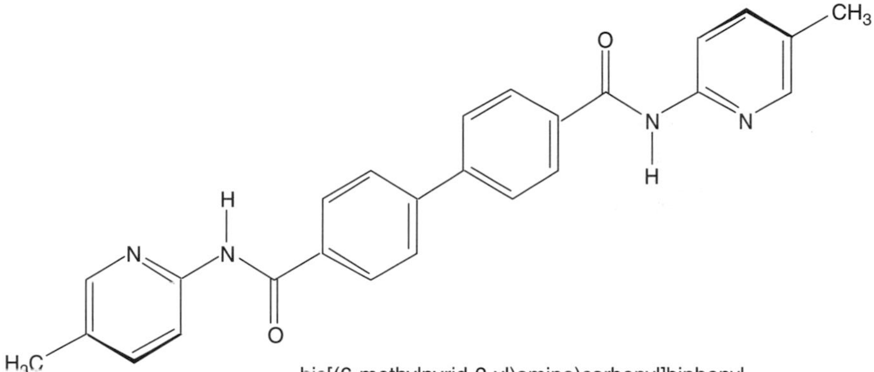

bis[(6-methylpyrid-2-yl)amino)carbonyl]biphenyl

Scheme IV

(f) Separation of enantiomers by formation of diastereoisomeric compunds There are now many examples where formation of discrete 1:1 hydrogen bonded compounds has been utilized to separate enantiomers (reviewed by Toda, 1987). For example, brucine forms a compound with *o*-bromophenyl-phenylacetylene-hydroxymethane which undergoes spontaneous resolution on crystallization (BAHLAX; space group $P2_12_12_1$, $Z=4$; Toda, Tanaka and Ueda, 1981). Another example is the preferential formation of the diastereomeric compound between (S)-N-(3,5-dinitrobenzoyl)-1-phenylethylamine and (R)-methyl-*p*-tolyl sulphoxide. The space group is $P2_1$ and $Z=2$; an $>N-H\ldots O=S<$ hydrogen bond is formed between the paired components, which form the packing units in the molecular crystal (Charpin *et al.*, 1981).

Bicyclic enones can be resolved by formation of specific diastereomers with the chiral donor[4] molecule (−)-(R,R)-*trans*-4,5-bis(hydroxydiphenylmethyl)-2,2-dimethyl-1,3-dioxacyclopentane; the adducting acceptor molecules tested were (−)-(R)-6-methylbicyclo-[4.4.0]dec-1-ene-3-one (KODWEF) and the corresponding 3,7-dione (KODWIJ) (Scheme V). Only the (−)...(−) hydrogen bonded compounds are formed, and these are isomorphous (space group $P2_12_12_1$, $Z=4$). Only the oxygen in the 3-position of the acceptors behaves as an hydrogen bond acceptor (Nassimbeni *et al.*, 1991).

---

[4] Although a host–guest terminology was used by Nassimbeni *et al.*, we prefer to distinguish between H-bond donors and acceptors as inclusion is not involved.

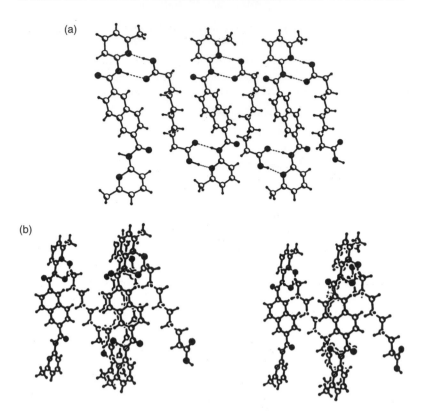

Fig. 12.14. Hydrogen bonding scheme in 2,6-bis[(6-methylpyrid-2-yl)amino)carbonyl]-naphthalene − 1,8-octanedicarboxylic acid (*P*1, *Z*=2); the upper part of the diagram (a) shows a projection onto the planes of the pyridine rings and the lower part (b) a stereoview in an approximately perpendicular direction. (Reproduced with permission from Garcia-Tellado *et al.*, 1991.)

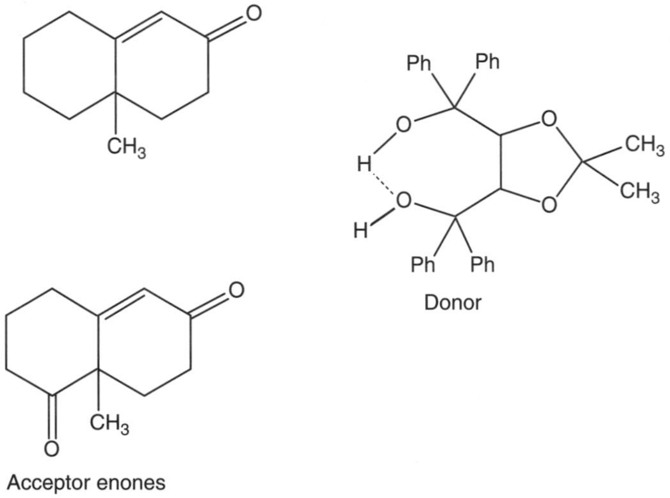

Scheme V

The unresolved donor forms discrete-pair molecular compounds with some amines, such as Donor : $(n-C_3H_7)_3N$ (SADFEI; $P\bar{1}$, $Z=2$) and another with the unusual composition 2(Donor) : $3(n-C_3H_7)_2NH$ (SADFAE; $P2_1/n$, $Z=2$), where one of the $(n-C_3H_7)_2NH$ molecules is disordered about a centre of symmetry and is not involved in the hydrogen bonding (Weber, Dörpinghaus and Goldberg, 1988).

12.5.1.2   Structures with discrete triples (B–A–B) of components

When one of the components has two functional groups which can act as acceptors and/or donors in hydrogen bond formation, and the other is monofunctional, then triples can be formed which pack in the crystal as discrete trimolecular units. It is convenient to distinguish the arrangements D – A – D and A – D – A, at least for purposes of classification. A simple example of the first group is pentafluorophenol – dioxane – pentafluorophenol ($P2_1/c$, $Z=2$, crystal structure determined at 138K) (DEVFOZ; Gramstad et al., 1985). The oxygens of the dioxane are hydrogen bonded, in roughly equatorial directions, to hydroxyls of the pentafluorophenols ($d(O-H...O) = 2.647(1)$ Å, $<O-H...O = 170(2)°$); the triples are centrosymmetric. There are rather similar centrosymmetric trimolecular units in (2-iodophenol)$_2$ – p-benzoquinone (GAPWEZ; Prout, Fail et al., 1988) and (phenol)$_2$ – p-benzoquinone (see Section 15.7.1), which are isostructural. Among representatives of the A – D – A group are the 1 : 2 molecular compounds of trans-9,10-dihydroxy-9,10-diphenyl-9,10-dihydroanthracene with acetophenone (VEKKEB) and 3-methylcyclopentanone (VEKKIF) (Bond et al., 1989a), and with 4-methylcyclohexanone (SEFXAC) and 2-methylcyclohexanone (SEFWUV) (Bond et al., 1989b), of fumaric acid with β–picoline-N-oxide (PICFUM; Gorres et al., 1975), of α-truxilic acid (2c,4t-diphenylcyclobutane-1r,3t-carboxylic acid) with dimethylacetamide (JOBXUT; Csöregh et al., 1991), of oxalic acid with triphenylphosphine oxide (space group $P2_1/c$, $Z=2$, hydrogen bond length 2.550(4) Å; TPOXLC; Thierback and Huber, 1981) and of quinol with 1,8-epoxymenthane (FUXYEC; Barnes, 1988) and 4-nitropyridine-N-oxide (NPOHQO; Shiro and Kubota, 1972).

In (2,5-piperazinedione)·2(salicylic acid,) the components are amphoteric, with the carboxylic acid group of the salicylic acid molecule (which also has intramolecular hydrogen bonding) forming hydrogen bonds with amide and carbonyl groups of the cis-peptide (DKPSAL10; Varughese and Kartha, 1982).

The termolecular ADA units are not centrosymmetric in the 1 : 2 compounds of 1,1,2,2-tetraphenylethane-1,2-diol with dimethyl sulfoxide (VENGUK; $P\bar{1}$, $Z=4$; Bond, Bourne, Nassimbeni and Toda, 1989), 3,5-lutidine (KOCHUF; $P\bar{1}$, $Z=4$), and 3,4-lutidine (KOCJAN; $P\bar{1}$, $Z=2$) (Bourne, Nassimbeni and Toda, 1991), nor in those of dimethyl-malonic acid with triphenylphosphine oxide (MALTPO; Declerq et al., 1974) and of tris(dimethylamino)phosphine oxide and heptasulphur imide (two independent N–H...O links of length 2.71(1) Å) (HMPASN; Steudel et al., 1977). The termolecular DAD units are not centrosymmetric in the 2 : 1 compounds of triphenylmethanol with dimethyl sulfoxide (JARROJ; Weber, Skobridis and Goldberg, 1989), pentachlorophenol with N,N,N′,N′-tetraethylphthaldiamide (SEHBAI; Grech et al., 1989), 2-methylphenol with p-benzoquinone (GAPWID; $P\bar{1}$, $Z=2$; Prout, Fail et al., 1988), and 2,4,6-triiodophenol with 1,2,4,5-tetramethylpyrazine (GAPWOJ; $C2/c$, $Z=8$; Prout, Fail et al., 1988). The three lutidine compounds with 1,1,2,2-tetraphenylethane-1,2-diol show interesting structural contrasts. As noted above, 3,4-lutidine and 3,5-lutidine form non-centrosymmetric ADA units but 2,6-lutidine forms a 1 : 1 compound with centrosymmetric ADDA units in its triclinic cell (KOCHOZ; $P\bar{1}$, $Z = 2$; Bourne, Nassimbeni and Toda, 1991). Competition experiments showed that 3,5-lutidine was more strongly adducted to 1,1,2,2-tetraphenylethane-1,2-diol than 2,6-lutidine despite the similar shapes of the two acceptor molecules.

A particularly interesting situation is provided by the 1:2 molecular complexes of 3,5-dimethylpyridine (A) with 3,5-dinitrobenzoic acid (B) (HAXFER) and its mono-deuterated analog (HAXFER01) (Jerzykiewicz et al., 1993). The all-hydrogen compound crystallizes in the triclinic space group $P\bar{1}$, with $Z = 2$ ($a = 11.680$, $b = 8.451$, $c = 24.382$ Å, $\alpha = 95.75$, $\beta = 108.17$, $\gamma = 91.48°$, volume per formula unit $= 590.9$ Å^3). There is proton transfer from the carboxyl group of one molecule to pyridine nitrogen, thus leading to formulation as {3,5-dimethylpyridinium 3,5-dinitrobenzoate 3,5-dinitrobenzoic acid} or {AH$^+$·B$^-$·B}. There is >N$^+$–H...O$^-$ hydrogen bonding between AH$^+$ and B$^-$(2.644(4) Å), and O$^-$...H–O– hydrogen bonding between B$^-$ and B (2.499(3) Å). When the carboxyl of B is deuterated, the crystals examined were found to be monoclinic, $P2_1/c$, $Z = 4$ ($a = 9.409$, $b = 10.813$, $c = 12.310$ Å, $\beta = 102.94°$, volume per formula unit $= 586.4$ Å^3) and the analogous hydrogen bond distances (but with deuterium) were lengthened at 2.739(5) and 2.526(5) Å. While the overall arrangements of {AH$^+$·B$^-$·B} and {AD$^+$·B$^-$·B} moieties are similar, there are differences of detail and this led to the suggestion that the difference in H(D)-bond distances is due to the difference in crystal structure (perhaps this phrase implies polymorphism) rather than to isotopic substitution. The difference in formula-unit volumes is suggestive. There are also 1:1 hydrogenated (PUHROZ; formula-unit volume 351.04 Å^3) and monodeuterated (PUHROZ2; formula-unit volume 350.65 Å^3) molecular complexes of 3,5-dimethylpyridine with 3,5-dinitrobenzoic acid (Jerzykiewicz et al., 1996; structures at 80K). These two compounds are isomorphous, with the acid hydrogen reported to be shared between carboxyl oxygen and pyridine nitrogen.

Resolution of enantiomers is obtained in some of the 1:2 compounds of chiral 1-(o-chlorophenyl)-1-phenyl-2-propyn-1-ol (CPPOH) (Scheme VI) with achiral 1,4-diazobicyclo[2.2.2]octane (DABCO) and achiral N,N'-dimethylpiperazine (DMP) (Yasui et al., 1989). For example, {[(−)-(R)-CPPOH]...DABCO...[(−)-(R)-CPPOH]} (chiral, $P2_1$, $Z = 2$; VAWLAG) is formed selectively from a racemic solution of CPPOH and thus molecular compound formation with achiral DABCO effects resolution of CPPOH. However, an achiral centrosymmetric molecular compound is formed from DMP {[(+)-(S)-CPPOH]...DMP...[(−)-(R)-CPPOH] (achiral, $P\bar{1}$, $Z = 1$) GERWEF10}. Racemic [(+)-(S)-CPPOH]...DABCO...[(−)-(R)-CPPOH] ($C2/c$, $Z = 8$; VAWLEK) and chiral [(−)-(R)-CPPOH]...DMP...[(−)-(R)-CPPOH] ($P2_1$, $Z = 2$; GERWIJ10) were obtained under (unspecified) "appropriate but severe" conditions.

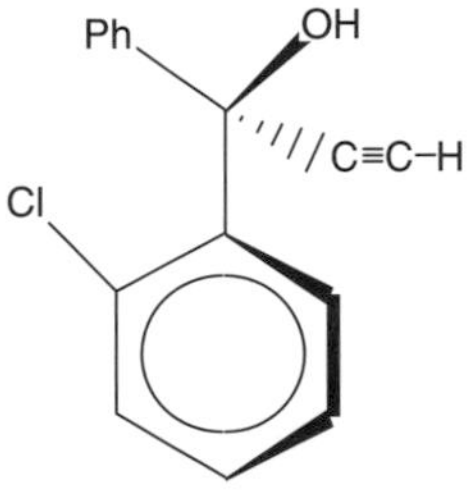

(−)-(R)-CPPOH

Scheme VI

1,4,7,10,13,16-Hexaoxacyclo-octadecane (18-crown-6, 18C6O) and 1,4,10,13-tetraoxa-7,16-diazacyclo-octadecane (diaza-18-crown-6, 18C(4O2N)) form 1:2 hydrogen bonded molecular compounds with a variety of ligands, showing interesting variations in the conformation of the macrocycle and the modes of hydrogen bonding. We consider first some of the proposed conformations for the macrocycle. Strain energy calculations by molecular mechanics (Bovill *et al.*, 1980) show that the most stable conformation ('c' of Fig. 12.15) is that described by the (a, a, a)(a, g$^+$, a)(a, g$^-$, g$^+$) sequence of partial conformations for the three successive non-equivalent –O–CH$_2$–CH$_2$–O– units of the (here assumed) centrosymmetric macrocycle (a=anti, g$^+$ and g$^-$ are the two gauche conformations) (Fig. 12.15); this biangular conformation is found in the crystalline uncomplexed macrocycle (Maverick *et al.*, 1980). The conformation with (ideal) D_{3d} symmetry ('b' of Fig. 12.15) has partial conformations (a, a, g$^-$) (a, a, g$^+$)(a, a, g$^-$) and a relative strain energy of 33 kJ/mol, due mostly to unfavourable interactions between the oxygen atoms separated by $\approx$2.8 Å. Another conformation ('e' of Fig. 12.15) with partial conformational sequence (a, g, a)(a, g$^-$, a) (g$^-$, g$^-$, a) has a relative strain energy of 21 kJ/mol. The 1:2 compounds are centrosymmetric (from the space groups) and have the ligands inclined with dihedral angles of $\approx$120° 'above' and 'below' the planes of the six oxygens of 18C6O, to some or all of which the ligand donor groups are hydrogen bonded. In the 18C(4O2N) compounds the hydrogen bonding is exclusively to the nitrogens of the macrocycle.

The macrocycle has approximately D_{3d} symmetry in its 1:2 compounds with malonitrile (HODMLN; Kaufmann *et al.*, 1980), 2,4-dinitrophenylhydrazine (CRHYDZ; Hilgenfeld and Saenger, 1981) and formamide (CEHGEB; Watson, Galloy, Grossie, Vögtle and Müller, 1984) and the (a, g$^-$, a)(a, g$^-$, a)(g$^-$, g$^-$, a) sequence when the ligand is dithiooxamide (CEHGOL; Watson, Galloy *et al.*, 1984) or 2,4-dinitroaniline (BAKHIE; Weber and Sheldrick, 1981). The sequence (g$^+$, a, a)(g$^+$, a, g$^+$)(g$^+$, a, a), the strain energy of which has not yet been reported, is found in the compounds of 18C6O with benzene-sulphonamide (Knöchel *et al.*, 1976) and 4-nitro-1,2-benzenediamine (BECVEK; Weber, 1982), and in the compounds of 18C(4O2N) with *p*-nitrobenzaldehyde oxime and N-hydroxybenzamide (Watson, Nagl and Kashyap, 1991).

The 2:1 molecular compounds of N-*m*-chlorophenylurea (DIWMOL10; Nastopoulos and Weiler, 1988) and N-*m*-bromophenylurea (KIVKAB; Nastopoulos *et al.*, 1991) with 18-crown-6 are not isomorphous although both contain discrete centrosymmetric [ligand–18C6O–ligand] arrangements, in which there is one medium ($d = 2.94$ (2.91) Å) and one weak ($d = 3.27$ (3.34) Å) N–H . . . O hydrogen bond. The macrocycle has the biangular, minimum strain energy conformation in both molecular compounds.

Interactions between hydrogens of methyl groups and the trigonally arranged oxygens in the two faces of 18C6O provide an opportunity for studying the formation (or not) of C–H . . . O hydrogen bonds. The crystal structure of the {O$_2$NCH$_3$ (18C6O) CH$_3$NO$_2$} complex has been determined at 298K (Rogers and Green, 1986) and at 123K (BIJWAS02; Rogers and Richards, 1987), the crystals being monoclinic ($P2_1/n$, $Z = 2$) at both temperatures. The methyl hydrogens appear to be ordered and linked by hydrogen bonds to the oxygens of the crown ether. The {NCCH$_3$ (18C6O) CH$_3$CN} complex has been studied at 298K (GEFREO02; Garrell *et al.*, 1988; Weller *et al.*, 1989) and at 123K (GEFREO01; Rogers, Richards and Voss, 1988). The acetonitrile complex at 298K is monoclinic and isomorphous with the nitromethane complexes, but there appears to be a phase transformation on cooling as the complex is triclinic at 123K ($P\bar{1}$, $Z = 2$,

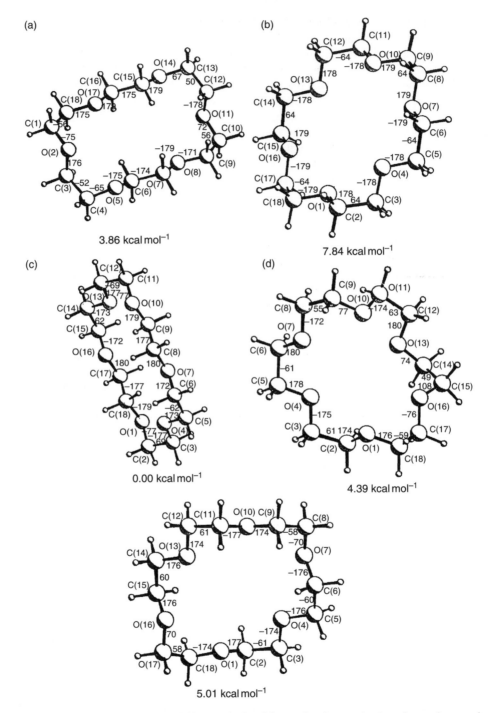

Fig. 12.15. Conformations of 18C6O as calculated by molecular mechanics; the strain energies appended are given with respect to conformation (c), the lowest energy conformation, as zero. (Reproduced with permission from Bovill *et al.*, 1980.)

termolecular moieties located about two independent centres of symmetry). The 18C6O molecule has (noncrystallographic) D_{3d} symmetry in all these structures. At room temperature the methyl hydrogens are rotationally disordered about the acetonitrile axis and H-bonding in the classic sense does not occur; this is corroborated by the lack of a shift in the C–H stretching frequency of the acetonitrile on complexation. At low temperature two of the three hydrogens of the acetonitrile are hydrogen bonded to 18C6O oxygens ($d(\text{C} \ldots \text{O}) = 3.19$, 3.27; 3.27, 3.46 Å). The differences between the two complexes are perhaps due to the greater acidity of the hydrogens of nitromethane compared to those of acetonitrile, or to the lower symmetry of nitromethane.

Diaza-18-crown-6 [18C(4O2N)] forms centrosymmetric 1:2 'discrete-molecule' compounds with p-nitrophenol (GOBYOL; $P2_1/n$, $Z = 2$), 2,4-dinitroaniline (GOBYUR; $P2_1/n$, $Z = 2$) and salicylaldoxime (GOBZEC; $P2_1/c$, $Z = 2$) (Watson et al., 1988b). In all these molecular compounds there are hydrogen bonds between amine or hydroxyl hydrogens and the nitrogens of the host and the crown ether has the D_{3d} conformation.

18C(4O2N) forms ionic as well as neutral molecular compounds. For example, both tropolone (FUYSOH) and 4-hydroxy-3-methoxybenzaldehyde (FUYSUN) form 1:2 ionic molecular compounds in which a proton from each ligand molecule is transferred to a nitrogen of the macrocycle (Watson et al., 1988a). In the first of these the macrocycle has the biangular, minimum strain energy conformation but the second shows a novel conformation where we list the torsion angles [−73, 162, 174, 66, −96, 164, −78, 168, 117] because of their appreciable deviations from the standard values of ±60, 180°.

18C6O also forms diaqua compounds in which a water molecule is hydrogen bonded, on each side of the macrocycle, to oxygens; an organic ligand is then hydrogen bonded to each water molecule. There is no direct link between organic ligand and macrocycle. Examples are {[18C6O·2H$_2$O].(3-nitrophenol)$_2$} (CEHGUR) and {[18C6O·2H$_2$O]·(p-nitrobenzaldehyde)$_2$} (CEHHAY); both have D_{3d} conformations (Watson, Galloy, Grossie et al., 1984). Somewhat analogous, at least from a formal point of view, is {bis(trichloroacetato-O,O')-Pb[18C6O]}(trichloroacetic acid), where each trichloroacetic acid molecule is hydrogen bonded to a trichloroacetate ion and not linked directly to the macrocycle (SOPRIY; Malinovski et al., 1990).

12.5.1.3 *Larger discrete groupings of components*

There are relatively few examples. Centrosymmetric quasi-linear tetramers are formed by linking p-nitrophenol molecules to diacetamide (VIVYUU) and N-butyrylbenzamide (VIVZAB) dimers (Etter and Reutzel, 1991; Fig. 12.16), and formally-similar arrangements are found in the 1:1 compound of 2-aminobenzothiazole with hexamethylphosphoramide (O=P(N(CH$_3$)$_2$)$_3$) (JOLJID; Armstrong, Bennett et al., 1992) and that of 2-aminobenzothiazole with dimethylpropyleneurea (PASYEN; Armstrong, Davidson et al., 1992). All three have solid-state structures which can be described as containing B...A...A...B units, with the three-dot linkages representing hydrogen bonds of possibly different kinds. There may also be differences in the ways these units are arranged in their respective crystals.

1,1,2,2-Tetraphenylethane-1,2-diol (T) forms a 1:1 compound with 2,6-lutidine (L) (KOCHOZ; Bourne, Nassimbeni and Toda, 1991; $P\bar{1}$, $Z = 2$) which resembles the arrangements shown in Fig. 12.16 in the sense that the packing unit is L...T...T...L, and differs from the T compounds with the other lutidine isomers noted above.

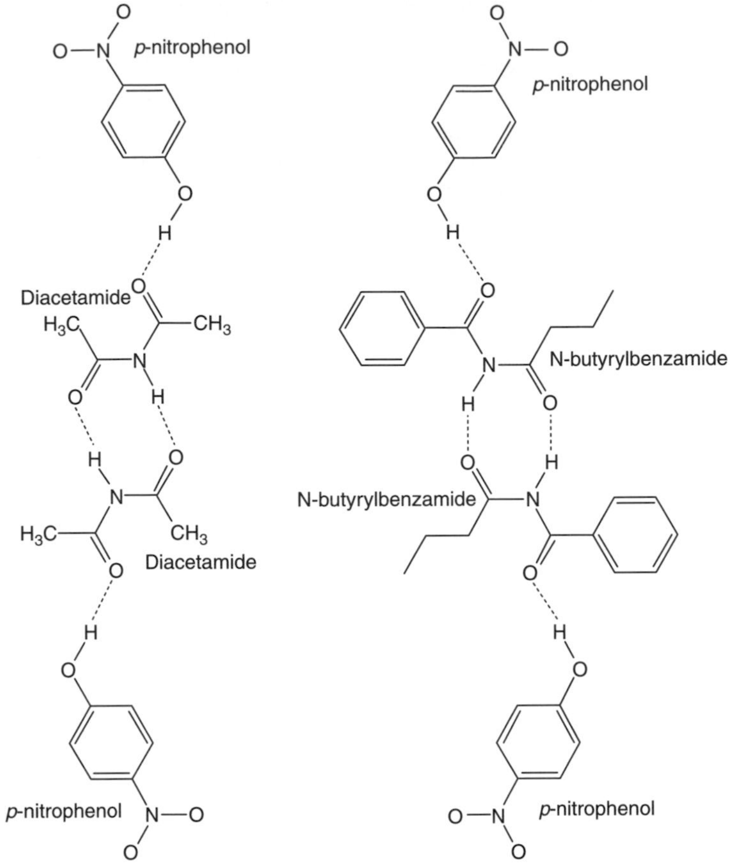

Fig. 12.16. Centrosymmetric tetramers of *p*-nitrophenol molecules linked to diacetamide (left) and N-butyrylbenzamide dimers (right). (Reproduced with permission from Etter and Reutzel,1991.)

In 8-dimethylamino-1-dimethylammonionaphthalene hydrogen squarate (2-hydroxy-3, 4-dioxocyclobut-1-en-1-olate) (KINKEX; *Pbca, Z* = 8), the two squarate ions are linked across a centre of symmetry to form hydrogen bonded 'quasi-carboxylic acid' dimers. Each squarate transfers an hydrogen to the proton sponge to form an ordered intramolecular hydrogen bond between dimethylamino groups, the hydrogen being additionally bonded in bifurcated fashion to a carbonyl oxygen of the squarate ion; there are a number of analogous structures (Kanters *et al.*, 1991). Somewhat similar arrangements are found in kinetin picrate (kinetin is 6-furfurylpurine) (VICPOM; Soriano-Garcia, Toscano *et al.*, 1985) and N(6)-methyladeninium picrate (VATCEY; Dahl and Riise, 1989), with the formation of discrete groups of four hydrogen-bonded moieties as illustrated below:

$$(PIC)^- \ldots (CAT)^+ \ldots (CAT)^+ \ldots (PIC)^-$$

The arrangement in 4(triphenylsilanol) – dioxane is formally rather similar (JODYAC; Bourne, Johnson *et al.*, 1991). This can be represented as B . . . B . . . [dioxane] . . . B . . . B, with both hetero and homo hydrogen bonds (Scheme VII).

In hexamethylenetetramine : tris(phenol) (HMTTPO10; Jordan and Mak, 1970) three phenols are linked to each HMT through O–H . . . N hydrogen bonds in the form of a three-bladed propeller, and the crystal structure (trigonal, $a = 14.88$, $c = 6.007$ Å, space group $P\bar{3}$, two units of HMT : (phenol)$_3$ per unit cell) is based on a packing of such units. The two tautomers of 3(5)-methylpyrazole form termolecular units with 1,1-bis (2,4-dimethylphenyl)but-2-yn-1-ol ($P2_1/n$, $Z = 4$) (GERYIL; Toda, Tanaka, Elguero, Stein and Goldberg, 1988) linked in a closed loop by N–H . . . O, O–H . . . N and N–H . . . N hydrogen bonds (Scheme VIII).

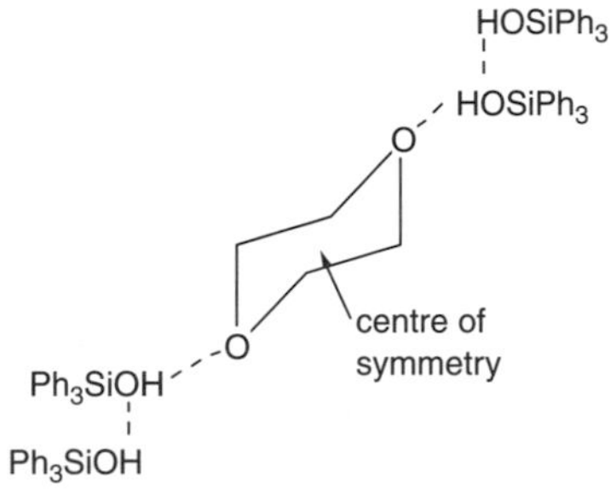

HOSiPh$_3$

HOSiPh$_3$

centre of symmetry

Ph$_3$SiOH

Ph$_3$SiOH

Scheme VII

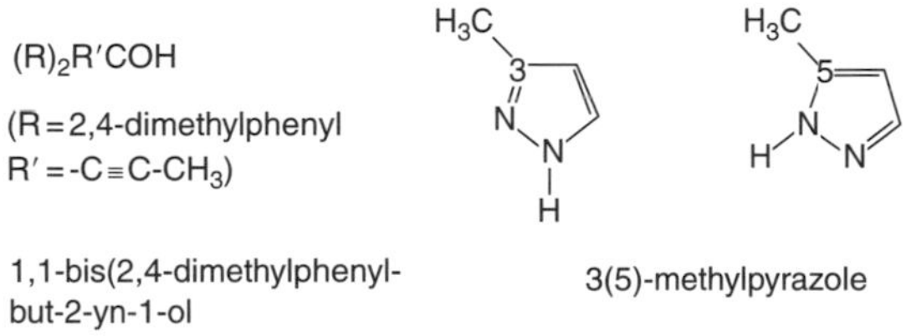

(R)$_2$R'COH

(R = 2,4-dimethylphenyl
R' = -C≡C-CH$_3$)

H$_3$C

H$_3$C

3(5)-methylpyrazole

1,1-bis(2,4-dimethylphenyl-
but-2-yn-1-ol

Scheme VIII

1,1,6,6-tetraphenylhexa-2,4-diyne-1,6-diol forms a 1:2 dioxane compound in which the diol hydrogen bonds to one dioxane to form an alternating . . . A . . . B . . . A . . . B . . . chain, while the second dioxane is not hydrogen bonded at all (JODXOP; Bourne, Johnson *et al.*, 1991).

We have encountered a number of examples containing parallelograms (some almost squares) of alternating N and O (or N and N, or O and O) atoms linked in different ways by hydrogen bonds. Neutral tetramolecular units are found in aniline – trichlorophenol (d(N–H . . . O) = 3.13 Å, d(O–H . . . N) = 2.78 Å) (ANCPOL; Bellingen *et al.*, 1970), and in triphenylmethanol – methanol (d(O–H . . . O) = 2.71, 2.71 Å) (JARRID; Weber, Skobridis and Goldberg, 1989). Discrete centrosymmetric 2 + 2 units with a roughly square arrangement of hydrogen bonds are formed in *trans*-9,10-dihydroxy-9,10-diphenyl-9,10-dihydro-anthracene – 1,4-butanediol (DEPTAT; Toda *et al.*, 1985). In thiomorpholine – 3, 5-dinitroimidazole a hydrogen is transferred from the pyrrole nitrogen to the nitrogen of the thiomorpholine, forming a >N$^+$H$_2$ ammonium cation. The ions form a four-membered hydrogen bonded ring lying on the two fold axis of the cell (space group $C2/c$, $Z = 8$), with d(N$^+$–H . . . N) = 2.936(4), 2.924(4) Å (JILWUW; Gzella and Wrzeciono, 1991).

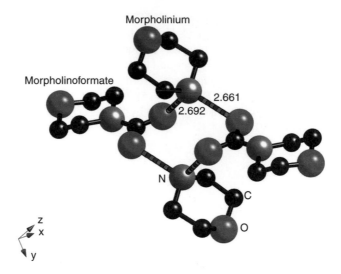

Fig. 12.17. The eight-membered ring composed of morpholinium cations and morpholinoformate anions. The neutral moiety packs as "molecular" units in the crystal. (Data from Brown and Gray, 1982.)

Larger rings are also found, often constructed of a combination of covalent and hydrogen bonds. Crystalline morpholinium morpholinoformate was isolated by morpholine extraction of the yellow-orange product formed during the corrosion of metallic lead by white mineral oil; its structure (Fig. 12.17) was deduced entirely by crystallographic methods ($P\bar{1}$, $Z = 2$; MORPLN; Brown and Gray, 1982). It was later realized that the compound had first been prepared in 1898 by bubbling CO_2 gas through morpholine (Knorr, 1898).

In the 4 : 1 compound of triphenylsilanol with ethanol, four triphenylsilanol molecules and one ethanol are linked together by homodromic hydrogen bonds (a circular pattern of hydroxyl groups all pointing in the same direction) in a five membered ring (SITKEL; Bourne, Nassimbeni, Skobridis and Weber, 1991). A rosette of trigonal symmetry has been constructed from three molecules each of N,N'-bis(p-butylphenyl)melamine and 5,5-diethylbarbituric acid (Fig. 12.18) (KUFPIK; Zerkowski et al., 1992), and linear tape and crinkled tape motifs (see below) have been obtained from 1 : 1 combinations of other substituted melamines and 5,5-diethylbarbituric acid. These constitute particularly beautiful examples of molecular recognition via hydrogen bonding between amphoteric moieties each having both donor and acceptor functions. The bulky p-butylphenyl groups encourage the formation of isolated rosettes rather than the extended sheet structure found in the melamine–cyanuric acid molecular compound (QACSUI; Ranganathan et al., 1999; see below).

A novel 3 : 2 structure has been found for the hydrogen-bonded molecular compound formed between [Ph₂Si(OH)OSi(OH)Ph₂] and pyridazine (1,2-diazine) (Fig. 12.19; KIRTOU; Ruud et al., 1991).

12.5.2   One-dimensional frameworks (linear chains of alternating components)

A 'ribbon' is defined in the Concise Oxford Dictionary as a "Long narrow chain of anything". In structural chemistry suitable ribbons can interact laterally to form sheets

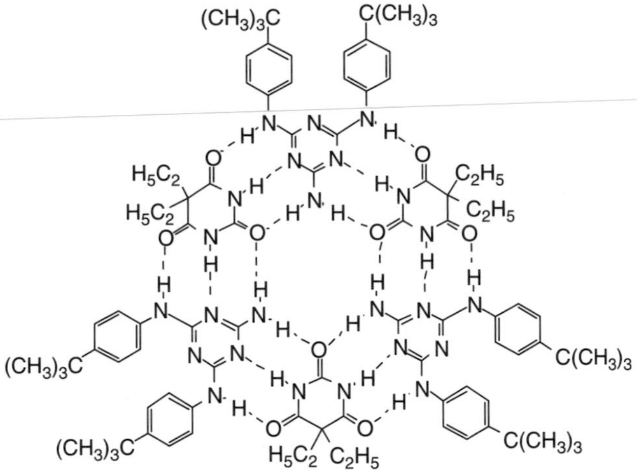

Fig. 12.18. Schematic diagram of the rosette formed between N,N'-bis(4-t-butylphenyl)melamine and 5,5-diethylbarbituric acid. The crystals are triclinic ($P\bar{1}$, two rosettes (i.e. six pairs) in the unit cell). (Adapted with permission from Zerkowski *et al.*, 1992.)

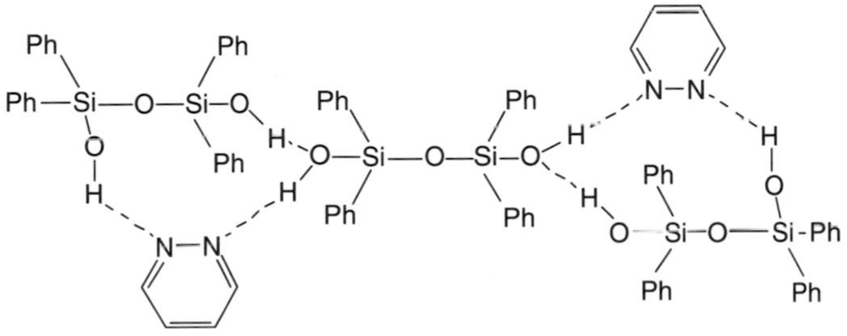

Fig. 12.19. The centrosymmetric 3:2 adduct between [Ph$_2$Si(OH)OSi(OH)Ph$_2$] and pyridazine (1,2-diazine) (space group $P\bar{1}$, $Z=1$). The central [Ph$_2$Si(OH)OSi(OH)Ph$_2$] molecule in the schematic diagram is centrosymmetric, with <Si–O–Si=180° (by symmetry) but not the others, which have <Si–O–Si=144.5°. (Reproduced with permission from Ruud *et al.*, 1991.)

(or layers). Many different kinds of ribbons are encountered among unary systems and also among the binary systems of interest here; their role in crystal engineering has been reviewed (Meléndez and Hamilton, 1998). The requirement for forming ribbons (linear chains) is that each component should be able to form two hydrogen bonds. This can be achieved in a number of ways.

12.5.2.1   *Component A has two donor groups and the single acceptor of component B can accept two hydrogen bonds*

An example is the quinol (A) – acetone (B) molecular compound, which was first reported more than one hundred years ago (Haberman, 1884; Schmidlin and Lang, 1910); other ketones do not form molecular compounds with quinol. The crystal structure (*C2/c*, $Z=4$) (QLNAC; Lee and Wallwork, 1959) has infinite chains of alternating quinol and acetone molecules joined by O–H . . . O hydrogen bonds ($d = 2.74$ Å). Each oxygen of acetone is linked to hydroxyls of two different quinols, with the hydrogen bonds in the directions of the oxygen lone pairs. The specificity to acetone was explained in terms of matching the dimensions of the two components. A formally similar example is trans-1,2-cyclohexanediol – N-methylmorpholine-N-oxide; hydroxyl groups of two different cyclohexanediols are hydrogen bonded to the exocyclic oxygen of the N-methylmorpholine-N-oxide molecule (d(O–H . . . O) = 2.677(4), 2.691(4) Å), forming ribbons along [100], with only dispersion interactions between neighbouring ribbons (BADCOY10; Chanzy et al., 1982). A one-dimensional chain but with different types of hydrogen bonding is found in 4-nitropyridine-N-oxide – 2-aminobenzoic acid (Cc, $Z = 4$) (SOJPEM; Fuquen et al., 1991). The hydroxyl group of the carboxylic acid acts as a donor to the oxygen of the N-oxide (d(O . . . O) = 2.629(5) Å), with a second link formed by O to H–N of the amino group (d(N . . . O) = 2.999(6) Å), thus forming a zigzag chain structure. There is an intramolecular hydrogen bond between N–H and carbonyl oxygen (d(N–H . . . O) = 2.675(7) Å). The components are stacked alternately face-to-face, leading to a charge transfer interaction as shown by the red color of the compound compared to the yellow colors of the individual components (cf. the molecular compound of 3,4-dimethoxy- and 2,4-dinitrocinnamic acid (Desiraju and Sarma, 1983) discussed earlier). The crown ether, 18-crown-6, can be considered as having the triangles of three oxygens on opposite sides of the macrocycle as its two acceptor groups, with analogous behaviour for diaza-18-crown-6. Linear chains of alternating diaza-crown ether and 1,4-dihydroxybut-2-yne molecules are found in their 1:1 complex (GOBZIG; Watson, Vögtle and Müller, 1988b) but the N . . . O distance (3.63 Å) is too long for hydrogen bonding. The authors note that the low density (measured 1.20, calculated 1.18 g cm^{-3}) could, perhaps, be ascribed to the difficulty of packing long linear and short cylindrical molecules closely together; there may also have been undetected solvent molecules, but this appears negated by the good agreement between measured and calculated densities. The last word is not yet in.

A somewhat more complicated example is provided by the 1:1 molecular compound of 5-bromouridine and DMSO (space group $P2_1$, $Z = 2$; BURDMS) (Iball et al., 1968), where the oxygen of DMSO acts as a double acceptor to two different hydroxyl donors of 5-bromouridine (d(O . . . H–O) = 2.66, 2.74 Å) while one of its carbonyls acts as acceptor to an N–H group of another 5-bromouridine molecule (d(C=O . . . H–N) = 2.86 Å), thus forming a linear chain (. . . 5-BU . . . 5-BU . . . (DMSO)$_2$ . . . 5-BU . . . 5-BU . . .) along the [10$\bar{1}$] direction (Fig. 12.20).

Similar packing arrangements, accompanied by optical resolution, are found in a number of examples. One of these is the molecular compound of (R,R)-(–)-1,6-bis (o-chlorophenyl)-1,6-diphenylhexa-2,4-diyne-1,6-diol and (–)-(R)-(Z)-benzyl-idene-2-(3-methylbutyl)azane oxide, which crystallizes in space group $P2_12_12_1$ ($Z = 4$). The diol and nitrone molecules alternate along [100], forming an infinite zigzag spiral chain, where different hydroxyls of the diol are involved in the two hydrogen bonds; d(O–H . . . O) = 2.788, 2.780 Å (VAXFUV; Toda et al., 1989). Mutual optical resolution is also obtained in the molecular compound of (R)-(+)-bis(β-naphthol) and (m-tolyl)(methyl)sulfoxide (space group $P1$, $Z = 1$), where infinite zigzag chains are formed along [100]; d(O–H . . . O) = 2.758(2), 2.782(1) Å (CONPAW; Toda et al., 1984); the bis (β-napthol) has a cisoid conformation with a torsion angle of $-78°$ about the C–C

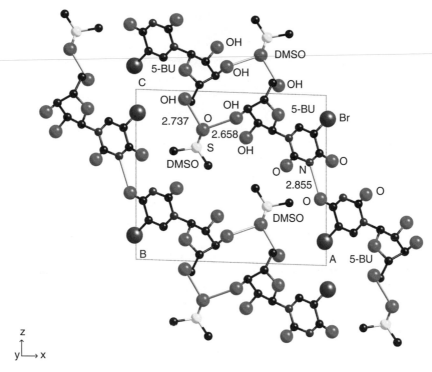

Fig. 12.20. Projection down [010] of the unit cell of 5-bromouridine – DMSO, showing the chains of hydrogen bonded molecules along the [10$\bar{1}$] direction. The hydrogen bonds are emphasized. (Data from Iball *et al.*, 1968.)

bond joining the aromatic rings, and both hydroxyls participate in the hydrogen bonding. Mutual optical resolution also occurs in the molecular compound of (S)-(−)-9, 9′-biphenanthryl-10,10′-diol and (+)-4-acetoxyazetidine-2-one (space group $P2_12_12_1$, $Z=4$), where infinite zigzag chains are formed along [010]; both hydroxyls and both carbonyls participate in the two nonequivalent hydrogen bonds, which nevertheless have essentially the same length (d(O–H . . . O=C) = 2.91(1) Å) (JEFRER; Toda *et al.*, 1990). Helical chains of alternating 1,1′-diphenyl-2,3-dicarboxylcyclopropane and *tert*-butanol molecules, linked by hydrogen bonds, are found in their 1 : 1 molecular compound ($P2_12_12_1$, $Z=4$) (KAXTUY; Weber, Hecker *et al.*, 1989). The two carboxyl groups of the cyclopropane derivative are linked by strong intramolecular hydrogen bonds (d(O–H . . . O=C) both in the neat compound (2.513(5) Å space group $P2_1$, $Z=2$) and 2.552(5) Å in the molecular compound with *t*-butanol. Spontaneous resolution of enantiomers has taken place on crystallization of both the neat cyclopropane derivative and the *t*-butanol molecular compound.

Chains are also found in the 1 : 2 molecular compounds of 1,1′-binapthyl-2,2′-dicarboxylic acid with methanol (CILLUE: TANDAN; Czugler and Weber, 1991; $P2_1/n$, $Z=4$; Fig. 12.21a) and with ethanol (CILMAN) and 2-propanol (CILMEP) (Fig. 12.21b, the 2-propanol compound being shown) and 2-butanol (CILMIT) (Weber, Csöregh, Stensland and Czugler, 1984). In the first three compounds both hydroxyl groups of the

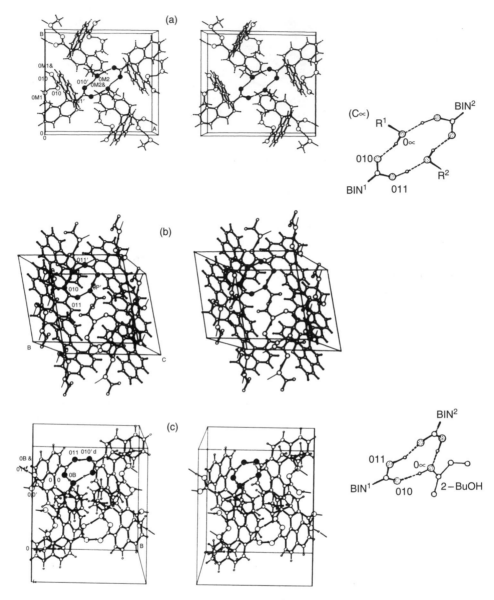

Fig. 12.21. Stereodiagrams of (a) 1,1′-binaphthyl-2,2′-dicarboxylic acid (BIN) – (methanol)$_2$; (b) BIN – (2-propanol)$_2$; (c) BIN – 2-butanol ($P2_1/n$, Z=4). Crystals of (b) with ethanol and 2-propanol are isomorphous, $C2/c$, Z=4. Two types of interruption of carboxylic acid dimers are shown on the right hand sides of the diagrams. (Reproduced from Weber, Csöregh, et al., 1984.)

alcohols are inserted into the two sides of the carboxylic acid dimer, giving what we may call a doubly interrupted dimer. In the 1:1 compound with 2-butanol ($P2_1/n$, Z = 4; Fig. 12.21c) only one side of the carboxylic acid dimer is so interrupted. The preferred bonding pattern of carboxylic acids is the formation of dimers, although some chain structures are known. Interruption of the dimers, as shown in these structures, is rare but not unprecedented and Weber *et al.* (1984) give other examples. Some of the trimesic acid channel inclusion complexes (Section 10.2) also fall into this group.

Trans-9,10-dihydro-9,10-diethanoanthracene-11,12-dicarboxylic acid (Fig. 12.22) has been shown to form molecular compounds (all compositions 1:1 except where noted otherwise) with 27 different moieties of different chemical types (Weber, Csöregh, Ahrendt, Finge and Czugler, 1988) (1-butanol, *t*-butanol, 1-pentanol, 1-octanol, ethylene glycol (1:2), 2-methoxyethanol, formic acid (1:2), acetic acid, propionic acid, 2-chloropropionic acid, valeric acid, lactic acid, tartaric acid (2:1), mercaptoacetic acid, thioacetic acid (2:1), propionaldehyde, acetone, DMF, acetonitrile, benzyl cyanide, DMSO, THF (1:2), dioxane (2:1), *o*-dichlorobenzene, 2,6-dimethylnitrobenzene, 1-propanol and 2-nitrophenol (1:2)). Crystal structures have been reported for the neat host (SABNAK) and the 1:1 compounds with DMF (SABNIS; $P2_1/c$, $Z=4$) and 1-butanol (SABNEO; $P\bar{1}$, $Z=2$). Both have unusual arrangements in which the components are linked in hydrogen bonded chains, one pair of carboxylic acid dimers having the usual uninterrupted arrangement while the second pair is interrupted by two hydroxyl groups of butanol, or two aldehyde groups, with consequent formation of C=O . . . HO and C–H . . . O=C hydrogen bonds (Fig. 12.22).

Pentachlorophenol ($pK_a = 4.5$) is a stronger acid than 2,4,5-trichlorophenol ($pK_a = 6$). Whereas the molecular compound aniline – 2,4,5-trichlorophenol (ANCPOL; Bellingen *et al.*, 1971a) appears to be composed of neutral molecules, that between aniline and pentachlorophenol has been reported to be the salt anilinium pentachlorophenolate (ANLPCP; Bellingen *et al.*, 1971b; there are infinite chains of alternating cations and anions linked by N^+–H . . . O^- hydrogen bonds (2.73, 2.60 Å). Three grounds were given for preferring the ionic formulation (i) the stronger hydrogen bonds compared to those in aniline – 2,4,5-trichlorophenol, (ii) the higher acidity of pentachlorophenol, (iii) the nature of the infrared spectrum (Zeegers-Huyskens, 1967). This conclusion is at variance with the ΔpK_a value quoted below (Section 12.5) and it would be useful to confirm these conclusions by determining hydrogen positions, which were not found in the 1971 structure analysis. In imidazolium picrate (SEZREU; Soriano-Garcia *et al.*, 1990; Herbstein and Kapon, 1991)), there are zigzag chains of hydrogen bonds, each 'picrate' oxygen being linked to H–N groups of two different imidazolium cations, which have effective symmmetry $mm2$ ($d(O^-$ . . . H–N) = 2.710(4), 2.825(5) Å); presumably the positive charge is preferentially localized on the N involved in the shorter of these two hydrogen bonds.

12.5.2.2 *Component A has two hydrogen bond donor groups and component B two acceptor groups*

The simplest examples are perhaps quinol – dioxane (SENYOK; Barnes *et al.*, 1990) and 1,4-diazabicyclo[2.2.2]octane (DABCO) and 4,4'-biphenol (NISLOQ; Ferguson *et al.*, 1998). [DABCO . . . biphenol] crystallizes in space group $C2/c$, $Z=4$, with DABCO disordered across centers of inversion and diphenol lying across two fold axes. Linear $C(2,2)(16)$ chains of alternating DABCO and (non-planar) biphenol molecules run parallel to [201] (Fig. 12.23). Because of very limited conformational flexibility there is no coiling of the chains.

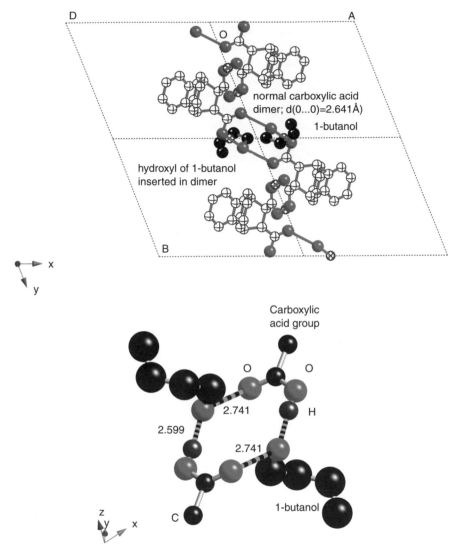

Fig. 12.22. *Trans*-9,10-dihydro-9,10-diethanoanthracene-11,12-dicarboxylic acid is noted above in the text for its prolixity in forming 27 molecular compounds. The overall arrangement in the molecular compound with 1-butanol, projected down [001], is shown in the upper part of the diagram, with the two independent carboxyl groups (both located about centres of symmetry) interacting in different ways. Details of the interrupted hydrogen bonding scheme found for one of the carboxyl pairs (the other is a standard dimer) are shown below. Both carboxyl groups are ordered (shown by different bond lengths for C–OH (1.307 Å) and C=O (1.207 Å)), hydrogens (other than those participating in H-bonds) have been omitted for clarity, while the butanol hydroxyl H was not found but is presumably located in the interaction denoted 2.741 (Å); the butanol molecule has been emphasized. (Data from Weber *et al.*, 1988.)

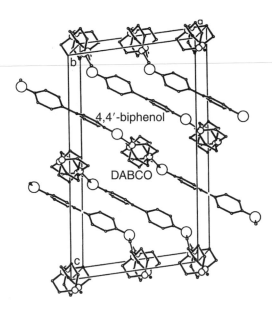

Fig. 12.23. [DABCO...biphenol] viewed approximately along [010], showing the alternating DABCO...biphenol chains running parallel to [201]. The oxygens of the hydroxyl groups have been emphasized. (Reproduced from Ferguson *et al.*,1998.)

Resorcinol – progesterone has chains of alternating donor and acceptor molecules (space group $P2_12_12_1$, $Z=4$, $d(\text{O–H}\ldots\text{O=C}) = 2.799$, 2.818 Å) (PRORES; Diderberg *et al.*, 1975). Resorcinol also forms a 1:1 compound with 2,9-dimethyl-1, 10-phenanthroline ($P\bar{1}$ $Z=2$; CECBOJ; Watson, Galloy, Vögtle and Muller, 1984), in which one of the hydroxyls of one resorcinol is linked in a bifurcated manner from '*above*' to the nitrogens of one phenanthroline molecule while the second hydroxyl binds similarly from '*below*' to the nitrogens of this phenanthroline, thus forming rows of alternating components. The planes of the two components are approximately mutually perpendicular. In 1,1,2,2-tetraphenylethane-1,2-diol – dioxane ($C2/c$, $Z=4$; JODXOP; Bourne, Johnson *et al.*, 1991) the first component is located on a diad axis and the dioxane at a centre of symmetry, with $d(\text{O–H}\ldots\text{O}) = 2.865(4)$ Å in the linear chains of alternating moieties. Analogous linear chains are found in 1,1,6,6-tetraphenylhexa-2,4-diyne-1, 6-diol – 2(dioxane) ($P\bar{1}$, $Z=1$), with the second dioxane in the asymmetric unit enclathrated between the chains and not hydrogen bonded in any way. A helical chain structure is found in *cis*-1-{[4-(1-imidazolylmethyl)cyclohexyl]methyl}-imididazole – succinic acid, shown schematically below (TAJVOP; Van Roey *et al.*, 1991; Scheme IX).

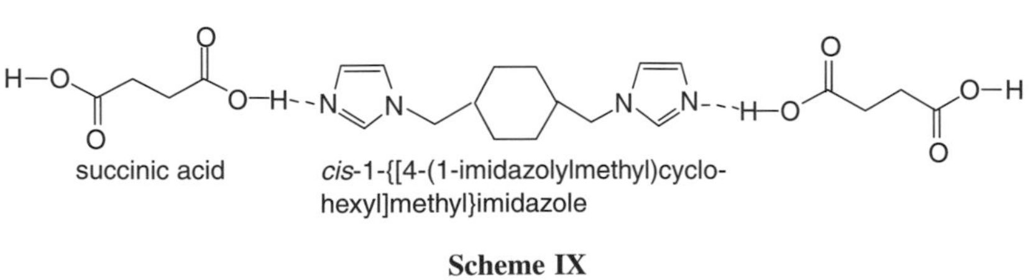

succinic acid   *cis*-1-{[4-(1-imidazolylmethyl)cyclo-
                          hexyl]methyl}imidazole

Scheme IX

The piperidine salts of *p*-chlorobenzoic (PIPCBZ), *p*-bromobenzoic (PIPBBZ) (Kashino *et al.*, 1972) and *p*-toluic acids (PIPTAC) (Kashino, 1973) and pyrrolidinium *p*-chlorobenzoate (PYMCBZ) and *p*-bromobenzoate (ZZZBVP) are isomorphous (*Pbca*, $Z = 8$) and pyrrolidinium *p*-toluate (PYMMBZ) is isostructural (*P2₁/c*, $Z = 4$; Kashino *et al.*, 1978). The moieties are linked by two kinds of N^+–H...O^- hydrogen bond to form ribbons along the two fold screw axes along [001] (the first group of salts) or [010] (the last example); the values of $d(N^+–H...O^-)$ range from 2.67 to 2.77 Å.

2-Aminopyridinium salicylate (SLCADB10; Gellert and Hsu, 1988) (Fig. 12.24) can also be classified in this group if some elasticity of definition is permitted. The two ionized moieties are linked by a pair of donor–acceptor hydrogen bonds and this unit is then joined in chains to other similar units by N–H...O=C< hydrogen bonds, employing the second N–H of the amino group and the second acceptor function of the carbonyl oxygen of the carboxylate.

As noted above, the two triangles of three oxygens on the opposite sides of the macrocyclic crown ether 18-crown-6 can behave as its two acceptor groups. Thus molecular compounds of composition 2(donor) – 18C6O, with molecules containing two hydrogen bond donors, can be classified in the present group; not all of the (potential) acceptor oxygens necessarily interact with the donors. Among the donors investigated in this

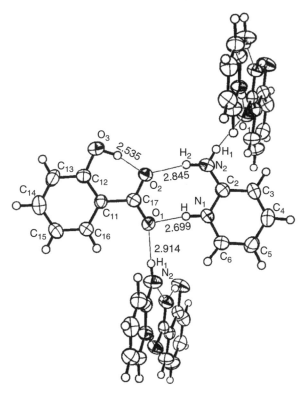

Fig. 12.24. The hydrogen bonding scheme in 2-aminopyridinium salicylate. The expected intramolecular hydrogen bond is found in the salicylate anion. (Reproduced with permission from Gellert and Hsu, 1988.)

context are N,N'-dimethylthiourea (($P2_1/c$, $Z=2$; 18C6O in D_{3d} conformation) (BUPRAF; Weber, 1983) and N,N'-diformohydrazide ($P2_1/c$, $Z=2$; 18C6O in D_{3d} conformation) (CEJVIW; Caira, Watson, Vögtle and Müller, 1984). These two molecular compounds are isostructural but not isomorphous. The N,N-dimethylthiourea molecular compound forms a chain of the following schematic type

... DMTU ... 18C6 ... DMTU ... DMTU ... 18C6 ... DMTU ... DMTU ... 18C6 ...

in which each dimethylthiourea forms an aminic hydrogen bond (d(N–H ... O = 2.955(6) Å) to one oxygen of the macrocycle, with C–H ... O interactions to the other two oxygens of the triangles on each side. The dimethylthioureas are linked by N–H ... S hydrogen bonds (d(N–H ... S = 3.394(6) Å). Similar chains are found in the N,N'-diformohydrazide molecular compound, with one N–H ... O hydrogen bond to one oxygen of the triangle and two between adjacent diformohydrazide molecules in the chain, and in the 1 : 2 18C(4N2O) complex with 5,5-diethylbarbituric acid (GOBZAY; Watson *et al.*, 1988b), where there are N–H ... N and N–H ... O links between barbituric acid and crown ether molecules, and similar links between adjacent barbituric acid molecules. The crown ether deviates from the D_{3d} conformation as there are two successive gauche interactions with torsion angles of 66 and 55°.

Optical resolution has been obtained with suitable combinations of donor and acceptor molecules. Thus, when chiral (R,R)-(–)-*trans*-2,3-bis(diphenyl-hydroxy-methyl)-1, 4-dioxaspiro[4,5]-decane and racemic dihydro-3-hydroxy-4H-dimethyl-2(3H)-furanone are crystallized together, optical resolution occurs because the crystalline hydrogen bonded molecular compound formed (space group $P2_1$, $Z=2$) contains only the (S)-(–)-furanone (VAXFOP; Toda *et al.*, 1989), the carbonyl and hydroxyl of which participate in the formation of spiral columns of alternating components along [010]. The two hydroxyl groups of the spiro compound are linked by a surprisingly short intramolecular hydrogen bond of length 2.365 Å; this should be checked.

12.5.2.3 *Both components have both hydrogen bond donor and acceptor functions*

For example, the phase diagram of 3,5-dichlorophenol (DCP) and 2,6-dimethylphenol (DMP) shows formation of a 1 : 1 molecular compound (Bavoux, 1975). Structure analysis shows that there are chains of alternating DCP and DMP molecules along [010], the O–H ... O hydrogen bonds having lengths of 2.83 and 2.67 Å respectively (EDCTMP10; Bavoux and Thozet, 1980). This can be considered to be a mimetic structure as both DCP (DCLPHE; Bavoux and Thozet, 1973) and DMP (DMEPOL10; Antona *et al.*, 1973) have similar one-dimensional chain structures. More complicated ribbon arrangements are found in the following molecular compounds

(1) *syn*-5-nitro-2-furaldehyde oxime – urea (URNFRO; Mathew and Palenik, 1972);

(2) 1-(4-bromophenyl)-4-dimethylamino-2,3-dimethyl-3-pyrazolin-5-one – 5,5-diphenyl-hydantoin (DPHPZL; Uno and Shimizu, 1980);

(3) 2,4-diamino-5-(3,4,5-trimethoxybenzyl)pyrimidine – 5,5-diethylbarbituric acid (BIG-CUP; Shimizu *et al.*, 1982);

(4) 2-amino-3,5-dibromo-N-cyclo-hexyl-N-methylbenzenemethanamine – 1,2-benzis-othiazol-3(2H)-one 1,1-dioxide, in which the components are ionized (BOYMEH; Shimizu and Nishigaki, 1983).

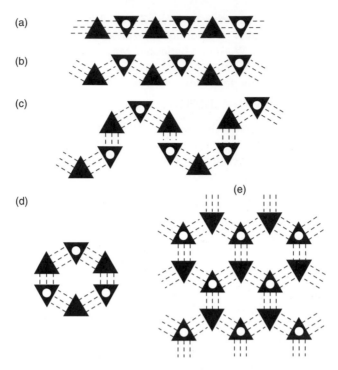

Fig. 12.25. Alternating arrangements of two components in (a) linear ribbon, (b) and (c) crinkled ribbons, (d) rosette (cf. Fig. 12.16) and (e) sheet aggregate. This diagram has been adapted from Fig. 2 of Meléndez and Hamilton (1998), which was in turn based on Fig. 1 of Zerkowski, Seto and Whitesides, 1992.

A particularly interesting group of "linear" and "crinkled" hydrogen-bonded tapes (Fig. 12.25) has been prepared from 1:1 combinations of suitable derivatives of melamines and cyanuric acid, where there are three hydrogen bonds between each pair of components, thus imposing rather stringent geometrical limitations on the possible structures.

The strategy (Whitesides, Mathias and Seto, 1991) for preparing the self-assembling motifs shown schematically in Fig. 12.25 was developed from "the pattern of hydrogen bonds present in the 1:1 complex formed from cyanuric acid and melamine." This pattern, shown in Figs. 12.25(e) and 12.26, was proposed by Finkel'shtein and Rukevich (1983) on the basis of an infra-red study. Diffraction-quality crystals of cyanuric acid–melamine (CA.M) were prepared only in 1999 by a hydrothermal technique (Ranganathan *et al.*, 1999); CA and M are both high-melting solids with limited solubility in most organic solvents. The XRD CA.M structure (QACSUI; 14.853(3) 9.641(2) 3.581(1) Å, 92.26(1)°, C2/m, Z = 2), and that of the isomorphous trithiocyanuric acid-melamine compound (QACTAP), have provided full confirmation of the postulated structure shown in Fig. 12.25(e). The molecules are arranged in planar layers, super-imposed in such a manner as to leave channels of diameter ≈4 Å along [001]. There

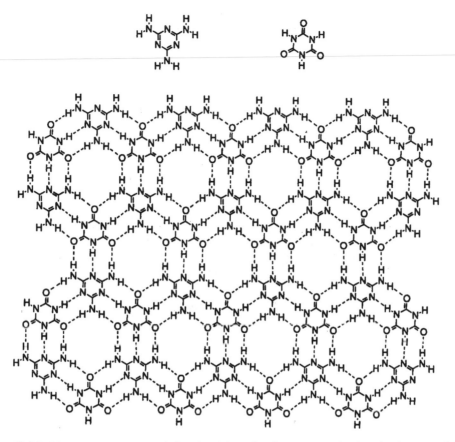

Fig. 12.26. The structure proposed for the 1:1 molecular compound of melamine (top left) and cyanuric acid (top right) on the basis of infra-red spectroscopy (Finkel'shtein & Rukevich, 1983). A representative portion of the infinite sheet is shown. This structure has been confirmed by XRD (see text); N–H...O and N–H...N distances are in the standard ranges of 2.94–2.98 and 2.85–2.88 Å respectively. (Reproduced with permission from Meléndez and Hamilton, 1998.)

has been interest in polymeric melamine–cyanuric acid complexes because of their potentialities as fire-retardants and lubricants; the layer structure may explain the latter property.

The success of the proposed strategy has been fully proved, as is shown by the structures of 1 : 1 combinations of suitable derivatives of melamines and 5,5-diethylbarbituric acid (Fig. 12.27; an analogous isolated rosette of trigonal symmetry (Zerkowski *et al.*, 1992) has been noted earlier in Fig. 12.18), and many other examples. Linear tapes have been found (Zerkowski, Seto, Wierda and Whitesides, 1990) in the 1 : 1 molecular compounds of 5,5-diethylbarbituric acid with N,N′-di-*p*-tolylylmelamine (JICVEW) and N,N′-diphenylmelamine (JICTIY) (for the latter, see Fig. 12.27 (left)), and a crinkled tape in the 1 : 1 molecular compound of 3,5-bis(N-4′-methoxycarbonylphenyl)melamine with 5,5-diethylbarbituric acid (KUFPIC; Zerkowski *et al.*, 1992; Fig. 12.27 (right)).

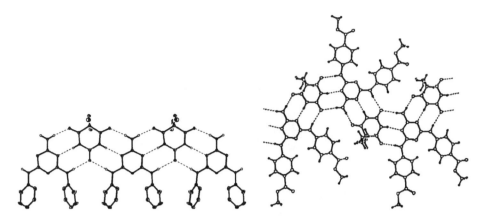

Fig. 12.27. Left: the linear triply hydrogen-bonded tape found in the 1:1 molecular compound of N,N'-diphenylmelamine and 5,5-diethylbarbituric acid (Zerkowski, Seto, Wierda and Whitesides,1990). Right: the crinkled triply hydrogen-bonded tape found in the 1:1 molecular compound of 3,5-bis(N-4'-methoxycarbonylphenyl)melamine and 5,5-diethylbarbituric acid (Zerkowski *et al.*, 1992). (Reproduced with permission from Zerkowski, Seto, Wierda and Whitesides (1990) and Zerkowski *et al.*, (1992).)

12.5.3 *Two-dimensional frameworks (layer arrangements of alternating components)*

We distinguish between molecular compounds of neutral components and those where hydrogen transfer has occurred between the components and are thus formally salts. A highly symmetrical 1:1 neutral-component molecular compound is formed between tris(4-hydroxy-3,5-dimethylbenzyl)amine ('TP') and hexamethylenetetramine ('HMTA') $(14.109(1)\ 26.697(5)$ Å, $R\bar{3}$, $Z = 6$; Fig. 12.28; HIFZIF; Bruyn *et al.*, 1996). We quote (with some abbreviation) "The central N of TP and one N of HMTA lie on sites of threefold symmetry, and thus one third of the two molecules forms the asymmetric unit. The phenol rings are perpendicular to (00.1). Each TP is H-bonded to three HMTA molecules, and each HMTA to three TPs; all molecules lie on threefold axes. Within each layer, N–H . . . O linkages give rise to large pear-shaped rings, with three HMTA and three TP molecules. Packing of these layers occurs in a space-filling way, which precludes the formation of any channels. There are no unusually short intermolecular distances between the layers." Comparison can be made with the close-packed layers found in CCP and HCP metals.

Not all layer structures are as symmetrical as HMTA–TP. Planar layers are found in {thiourea – parabanic acid} (structure at 93K (TUPRBN01; Weber and Craven, 1978), space group $P2_1/m$, $Z = 2$; no phase change between 93 and 298K); the N–H . . . S hydrogen bonds $(d(\text{H} . . . \text{S}) = 2.25$ Å) are shorter by 0.15 Å than any of those in thiourea itself. The layers are separated by 3.15 Å. The crystal structure of the 1:1 complex of acetamide with the E, Z (*cis, trans*) isomer of diacetamide has been determined at 123K $(a = 7.695(9)$, $b = 6.443(8)$, $c = 8.918(9)$, $\beta = 108.90(9)°$, $P2_1/m$, $Z = 2)$. The hydrogen-bonded molecules are in sheets in (010) planes, the sheets interacting by dispersion forces (DUVHOR10; Matias *et al.*, 1988). These two crystals are isostructural. The 1:1

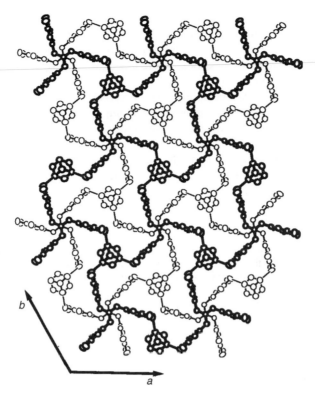

Fig. 12.28. HMTA – TP: projection down [001] showing the packing of two adjacent layers centrosymmetrically related through the origin. (Reproduced with permission from Bruyn *et al.* (1996).)

molecular compound between urea and 2,6-lutidine (2,6-dimethylpyridine) is an interesting example of molecular recognition in the solid state, providing a convenient method of separating 2,6-lutidine from other coal tar bases (Riethof, 1944, 1945). In the crystal (space group *C2/c, Z*=4, each molecule on a two fold axis) (LUTDUR; Lee and Wallwork, 1965) the planes of the two components are mutually perpendicular; there are strips of urea molecules linked by N–H . . . O=C< bonds, with the pyridine nitrogen acting as acceptor to the second N–H group of each urea. All the hydrogen bonding possibilities are used. The layers are packed one above the other, interacting by dispersion forces. There is also a 2 : 1 compound but its structure does not appear to have been reported.

In 2-thiohydantoin – 9-methyladenine (*C2/c, Z*=8) (BIFYOE; Cassady and Hawkinson, 1982) there are ribbons of alternating hydantoin (donor) and adenine (acceptor) molecules linked via N–H . . . N hydrogen bonds, with the ribbons linked laterally by N–H . . . N hydrogen bonds between adenine molecules, filling both donor and acceptor roles. The S and O atoms of the 2-thiohydantoin molecules do not appear to act as acceptors in strong hydrogen bonds. In 5-fluorouracil – 9-ethylhypoxanthine (*P2₁/c, Z*=4) (FUREHX; Kim and Rich, 1967) there are centrosymmetric pairs of 5-fluorouracils linked by two N–H . . . O hydrogen bonds, with each 5-fluorouracil linked to two different 9-ethylhypoxanthines.

Thus sheets of molecules are formed in the (100) plane. Puckered layers of the components are found in 9-ethyladenine – parabanic acid, with hydrogen bonding links between para-banic acid and adenine molecules and between pairs of adenine molecules; there are no links between parabanic acid molecules nor are there any unusual carbonyl-carbonyl interactions (EADPBA; Shieh and Voet, 1976).

In thiourea – hexamethylenetetramine oxide (HMTO) (*Pbca*, $Z = 8$) (HMTATV; Yu and Mak, 1978) three of the four N–H groups of thiourea act as hydrogen bond donors (while the fourth is inactive) towards the formally-negative oxygen of HMTO, which is the sole acceptor atom in this molecule. This leads to the formation of corrugated layers parallel to (001). The HMT portions of the HMTO molecules project from both sides of each layer to fill the space between adjacent layers; this is their only structural role. A somewhat similar arrangement of corrugated layers occurs in HMTO·H$_2$O$_2$·H$_2$O, where again the HMTO oxygen atom is the sole acceptor of hydrogen bonds to HMTO, and the HMT portion of the molecule has only a space-filling role (HMTOXH; Mak and Lam, 1978).

1,1-Di(*p*-hydroxyphenyl)cyclohexane forms isomorphous 1:1 triclinic layer com-plexes with *m*-cresol (GEFROY), *p*-cresol (GEFRUE), phenol (GEFSAL) and *o*-cresol (GEFRIS) (Goldberg *et al.*, 1988) (Fig. 12.29); although isomorphous, the complexes have slightly different cell dimensions and volumes (1029, 1012, 1014 and 964 Å^3). Selectivity studies show the order: *m*-cresol > *p*-cresol > phenol > *o*-cresol, and the

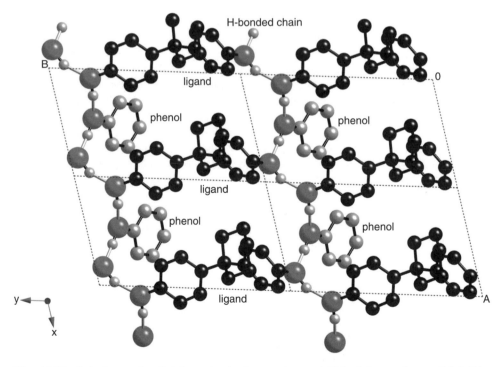

Fig. 12.29. A hydrogen bonded layer in the isomorphous triclinic 1:1 complexes of 1,1-di(*p*-hydroxyphenyl)cyclohexane with *m*-cresol, *p*-cresol, phenol and *o*-cresol, illustrated for the phenol compound. (Data with permission from Goldberg *et al.*, 1988.)

separation of *m*-cresol (b.pt. 202.0°) and *p*-cresol (b.pt. 201.8°) has been demonstrated. Every hydroxyl group is involved in two hydrogen bonds.

There are a number of $1:2$ molecular compounds forming sheet structures which have the formal similarity of crystallizing in space group $P2_1/c$ with $Z=2$, the first component thus being located at a crystallographic centre of symmetry. Among the many examples are oxalic acid – (urea)$_2$ (by XRD at 298K (UROXAL; Harkema *et al.*, 1972); also by ND at 100K), succinic acid – (benzamide)$_2$ (BZASUC) [oxalic acid – (furamide)$_2$ (FURAOX) is a triclinic variant of this structure type] (Huang *et al.*, 1973), 2,5-bis (2,4-dimethylphenyl)-hydroquinone – (ethanol)$_2$ (CAMMIN; Toda *et al.*, 1983) and α-truxillic acid – (methanol)$_2$ (JOBXON; Csöregh, Czugler *et al.*, 1991). A somewhat similar arrangement is found in N_2H_4 – $(C_2H_5OH)_2$ (m. pt. 242K), this being the only intermediate compound in the hydrazine–ethanol freezing point diagram (Corcoran *et al.*, 1953). The crystal structure (determined at 85K, space group *Pbcn*, $Z=4$, hydrazine on a two fold axis parallel to [010]) shows that hydrazine and ethanol molecules are hydrogen bonded to one another forming infinite layers parallel to (100). The hydroxyl of each ethanol forms one donor bond to hydrazine N, and accepts two N–H bonds. Each N accepts one hydroxyl H-bond and is the donor in two N–H bonds; the hydrogen bond lengths are N–H...O 3.041, 3.060 and O–H...N 2.730 Å (HYDETH; Liminga, 1967). The crystal structures of both phases of oxalic acid – bis(N-methylurea) have been determined at 295K (MUROXA; Harkema *et al.*, 1979). The phase stable up to 182K is orthorhombic (space group *Pnma*, $Z=4$); it seems that this phase can be (super)heated to at least 300K. The monoclinic phase stable at room temperature can be cooled down to 182K, where it changes to orthorhombic; it is not clear whether these are thermodynamic (equilibrium) or kinetic temperatures. Very similar mixed-component layer arrangements are found in both phases, with the layers in the mirror planes at $y=1/4$, 3/4 in the orthorhombic phase, and the oxalic acid molecules at inversion centres in the monoclinic phase (Fig. 12.30). The phase change is stated to be irreversible and exothermic (presumably this means the phase change 'monoclinic to orthorhombic on cooling'). All possible hydrogen bonds appear to be formed in both phases, although there are some differences in the details of their geometries. The volume per formula unit is 1.3% greater in the monoclinic than in the orthorhombic phase, with 0.9% ascribed to the difference in distance between parallel layers (3.21 and 3.24 Å), and 0.4 % to small differences in atomic positions in the planar layers.

There are some resemblances between the $1:1$ salt of piperazine and phenylbutazone (4-*n*-butyl-1,2-diphenyl-3,5-dioxo-1,2-pyrazolidine) (space group $P2_1/c$, $Z=2$; PBZPAZ10; Singh and Vijayan, 1977) and the $1:2$ salt of N-methylpiperazine and phenylbutazone (space group $P2_1/c$, $Z=2$) (Toussaint *et al.*, 1974). In the first of these salts there are two-dimensional sheets in (100) planes linked by N–H...O bonds of length 2.627(8) and 2.650(8) Å; all possible H-bonds are formed. There is a rather similar arrangement in the second salt, with N–H...O links of length 2.670(5) and 2.753(5) Å. Puckered layers are found in dimethylammonium hydrogen oxalate ($P2_1/c$, $Z=4$) (DMAHOX; Thomas and Pramatus, 1974). The $H_2C_2O_4^-$ anions are linked *via* strong O–H...O hydrogen bonds ($d=2.533(1)$ Å) to form chains along [100], which are linked transversely through the $(CH_3)_2NH_2^+$ cations by N–H...O hydrogen bonds. Histidinium trimesate·1/3(acetone) forms hydrogen bonded layers about $z \approx 0$, 1/2 in which the acetone molecules are enclosed by the cations and anions of the hydrogen bonded framework structure but not

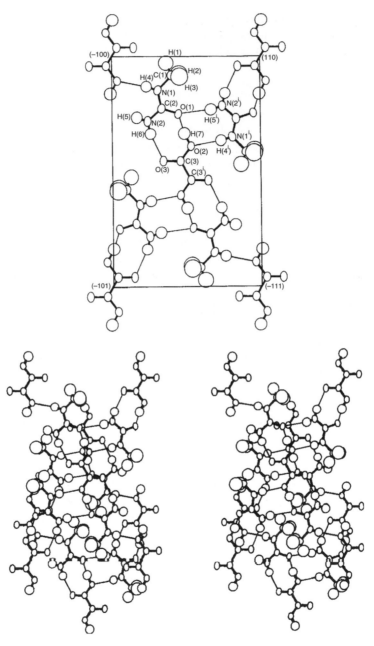

Fig. 12.30. Hydrogen bonding in the crystal structure of monoclinic bis(N-methylurea) – oxalic acid. The arrangement in the layer at $y=1/4$ is shown above, and a stereoview looking down [010] below. The thermal ellipsoids of the atoms in both diagrams are scaled to include 50% probability and hence the hydrogen atoms are largest. (Reproduced with permission from Harkema *et al.*, 1979.)

linked to them. The L- and DL-histidinium salts are isostructural, with the former having space group $P2_12_12_1$ (LHISTM) and the latter space group $Pna2_1$ (DLHTMS), with $Z=4$ in both examples (Herbstein and Kapon, 1979).

12.5.4 Three-dimensional frameworks (arrangements of alternating components in space)

The spatial arrangements in these molecular compounds can be quite complicated and we shall describe only a few examples in which the components are small molecules. The first of these is 'hyperol', urea – hydrogen peroxide, the room temperature structure of which was first determined in 1941 by what was then a crystallographic *tour de force*, (Lu *et al.*, 1941), and later redetermined by neutron diffraction, with crystals at 81K, thus enabling positions of all the hydrogens to be found (UREXPO11; Fritchie and McMullan, 1981). The 298 and 81K structures are essentially the same, and we shall describe only the latter. The space group is (nonstandard) *Pnca*, with $Z=4$ (Fig. 12.31). The urea molecule lies along and the H_2O_2 molecule across the crystallographic two fold axis along [001]. The hydrogen bonding of first-neighbor H_2O_2 molecules to urea is shown in Fig. 12.32. The peroxide oxygens act both as donors (to carbonyl O) and acceptors (from N–H of the amine groups). All possible hydrogen bonds are formed.

In monomethylammonium hydrogen oxalate (monoclinic, $P2_1/n$, $Z-4$) the ions are linked by O–H...O hydrogen bonds ($d = 2.515(2)$ Å) to form infinite chains along [101]. Transverse linkage between these chains is by a complex network of N–H...O bonds to the $CH_3NH_3^+$ cation (MAMHOX; Thomas, 1975). The hydrazine-methanol melting point diagram shows a 1:1 compound with an incongruent melting point at 226K, a 1:2 compound that melts at 215K and a 1:4 compound that melts at 203K; only the crystal structure of the latter has been reported (HYDTML; Liminga and Sørensen, 1967). Unusually, the

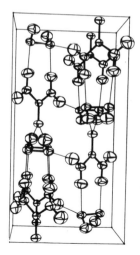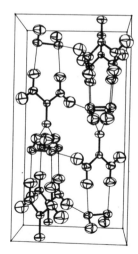

Fig. 12.31. Stereodiagram showing the hydrogen bonding in the crystal structure of hyperol at 81K. The [001] direction is vertical with [100] horizontal. The thermal ellipsoids are scaled to enclose 80% probability surfaces. (Reproduced with permission from Fritchie and McMullan, 1981.)

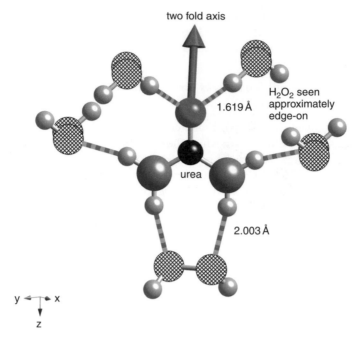

Fig. 12.32. Intercomponent hydrogen bonding in hyperol, showing the immediate surroundings of a urea molecule. The lengths (esds 0.001 Å) between H donor atoms and O acceptors are taken from the 81K neutron diffraction measurements. The crystallographic two fold axis along [010] is shown. (Adapted from Fritchie and McMullan, 1981.)

crystals are tetragonal and chiral ($P4_2$, $Z=2$). The methanol molecules are bonded to one another and to the hydrazines in a three-dimensional network ($d(\text{N–H}\ldots\text{O})=2.96, 3.03$ Å; $d(\text{O–H}\ldots\text{N})=2.68$ Å; $d(\text{O–H}\ldots\text{O})=2.74$ Å). In estradiol – urea ($P2_12_12_1$, $Z=4$) (ESOURE10; Duax, 1972) the steroid molecules are hydrogen bonded head-to-tail along [100] *via* hydroxyl groups; the urea molecules are interleaved, each being linked to four estradiol molecules and to two other ureas, forming infinite chains along [001], and in fact giving a three-dimensional arrangement of hydrogen bonds. As Duax notes, the ureas serve to bind the sterol molecules, not to enclose them. In salicylic acid – urea ($C2/c$, $Z=8$) (SLCADC10; Hsu and Gellert, 1983) the urea molecules form centrosymmetric dimers, while salicylic acid forms intermolecular hydrogen bonds exclusively to urea; one N–H group of each urea does not participate in a hydrogen bond. The strong O–H...O hydrogen bond ($d=2.54$ Å) found in this structure indicates that the carbonyl oxygen of urea is a stronger hydrogen bond acceptor than that of salicylic acid. In quinol – urea ($a=17.180(2)$, $b=6.601(1)$, $c=7.341(1)$, $\beta=94.4(1)°$, $P2_1/c$, $Z=4$) (QUOLUR; Mahmoud and Wallwork, 1975), the quinol molecules are situated at independent centres of symmetry. There are alternate sheets of mutually hydrogen bonded ureas and quinols parallel to (100); each O atom of urea receives two hydrogen bonds of length 2.683(2) and 2.702(2) Å from the hydroxyl groups of quinol molecules on either side of it in the [100] direction. In addition, two of the N–H groups of urea link to O atoms of two quinol molecules in the chain along [010] at distances of 2.980(2) and 3.049 (3) Å, and a third N–H links to a quinol

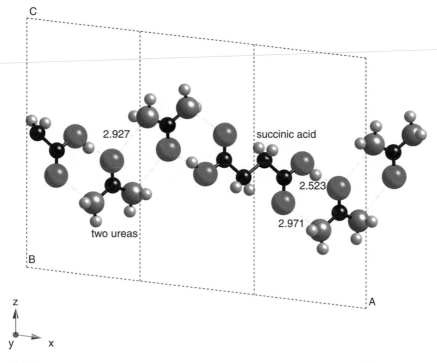

C

2.927

succinic acid

2.523

2.971

two ureas

B

z

y   x

A

Fig. 12.33. The primary hydrogen bonded chain in {2(succinic acid):urea}, showing the C=O...H–O and C=O...H–N hydrogen bonds, the two bonds of the latter kind are between urea and succinic acid (2.971 Å) and urea...urea (2.927 Å). The NH groups of urea *syn* to the oxygen are involved in the H-bonds within the primary chain. (Adapted from Videnova-Adrabinska (1996).)

O in an adjacent chain along [001] at a distance of 3.048(2) Å. The fourth N–H does not participate in the hydrogen bonding. Thus this is a three-dimensional structure, describable in terms of layers about (001) linked by hydrogen bonds roughly along [001].

The structure of (perdeuterated) parabanic acid – urea ($P2_1/c$, $Z = 4$; m. pt. 456K; determined by neutron diffraction at 116K) consists of alternating tilted stacks of nearly planar hydrogen-bonded ribbons which are symmetry-related and parallel to the crystal planes (211) and ($2\bar{1}1$) (URPRBN01; Weber, Ruble *et al.*, 1980). Close C...O approaches between the ribbons provide cohesion in addition to the hydrogen bonding. Such interactions, which have been found in a number of different crystals, are not well understood.

There are three-dimensional arrangements of hydrogen bonds in 18C6O–(urea)$_5$ (CRWNUR; Harkema, Hummel *et al.*, 1981) and 18C4O2N – (thiourea)$_4$ (BITYUY; Weber, 1982). In the first, two adjacent oxygens of the polyether bind to two urea molecules by N–H...O hydrogen bonds; the remaining urea molecules form hydrogen bonded layers, which alternate with the mixed 18C6O-urea layers. In the second, there are two intracyclic N–H...O hydrogen bonds, giving a biangular conformation to the macrocycle, and a three-dimensional arrangement of intermolecular hydrogen bonds.

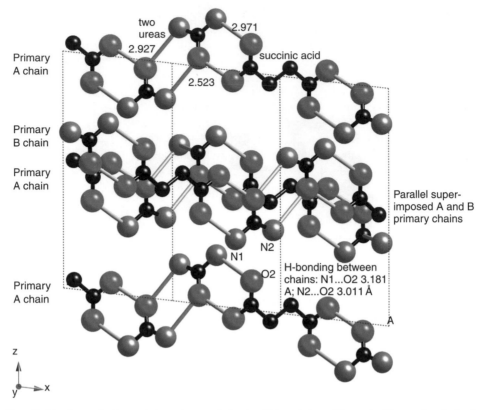

Fig. 12.34. Cross linking of the primary chains in {2(succinic acid):urea}. Shorter bonds in the primary chain are shown. The NH groups of urea *anti* to the oxygen are involved in the H-bonds between the primary chains. (Data from Videnova–Adrabinska, 1996).

12.5.5 *Accounting for formation of a molecular compound*

It is not often that one can assemble the facts (or most of them) needed to account for formation of a molecular compound from its components. The succinic acid–urea system meets *most* of the requirements. Two molecular compounds are formed in this system: 1:1 and 2:1. The crystal structures of succinic acid (neutron diffraction, 77K; SUCACB10; Leviel, Auvert and Savariault, 1981), urea (Swaminathan, Craven and McMullan, 1984; Section 6.2.1.1, Fig. 6.1) and {2(succinic acid):urea} (Wiedenfeld and Knoch (1990), VEJXAJ; Videnova-Adrabinska (1996), VEJXAJ10; 5.637(4) 8.243(3) 12.258(3) Å 96.80(5)°, $Z = 2$, $P2_1/c$) have been reported. The two structure determinations for {2(succinic acid):urea} agree but the interpretations differ; we follow Videnova-Adrabinska. No structure has been given in the CSD for the 1:1 compound although Videnova-Adrabinska does give a postulated one-dimensional self-assembled chain (her Fig. 3). The structures of {2(succinic acid):urea} and {2(fumaric acid):urea} (TIPWIY) are isomorphous (Videnova-Adrabinska, 1996), but there does not appear to be a 1:1 {(fumaric acid):urea} molecular compound.

{2(Succinic acid) : urea} has a three-dimensional hydrogen-bonded structure in which primary chains can be identified (Fig. 12.33). Comparison of hydrogen bond lengths in the molecular compound with those in the components shows that the only substantial difference is shortening of the distance from urea oxygen to succinic acid hydroxyl (2.532(3) Å) compared to that between carbonyl oxygen and hydroxyl in the succinic acid dimer (2.678(2) Å). The shortening (strengthening) of this H-bond is the decisive effect leading to formation of the heteromolecular cocrystal rather than separate crystals of the two components. An analogous H-bond shortening is found in {2(fumaric acid) : urea}, with a urea oxygen to fumaric acid hydroxyl distance of 2.499(2) Å. However, the reason why this shortening occurs is not known. The primary H-bond chains are crosslinked by weaker hydrogen bonds as shown in Fig. 12.34; these hydrogen bonds have close to standard distance values. Videnova-Adrabinska notes that "unlike the case of the urea-glutaric acid cocrystal, the three-dimensional networks of the present crystals do not contain independent two-dimensional hydrogen-bonded substructures (layers)."

12.6 Crystal engineering with hydrogen bonds

Crystal engineering is the process of designing a crystal with specific properties and then actually synthesizing the material, generally by self-assembly of components linked in specific ways. These linkages are usually provided in purely organic crystals by strong hydrogen bonds, which can be very usefully supplemented by coordinative interactions in inorganic-organic hybrid materials (Aakeröy and Beatty, 2001). Rather than attempting a comprehensive review of this rapidly expanding field, we give an evolving case history of a small but well-defined area "Engineering crystals for excited state diffraction studies". The purpose of such studies is to determine the geometrical changes that occur on excitation of a particular molecule from ground to excited state, the lifetime of the excited state being in the range of milliseconds to nanoseconds. To avoid possible crystal damage because of geometrical changes on excitation, it is essential to encapsulate the targeted molecules in some suitable framework rather than irradiate neat crystals. Ideally, one would wish to design a framework containing a single molecule in each of its cavities; the inclusion complex would be studied (at low temperature) in a third-generation synchrotron with synchronization of a pulsed laser beam (the pump beam) and the time-varying x-ray source (the probe beam) to achieve a sufficient concentration of molecules in the excited state (Zhang et al., 1999).

We limit ourselves to discussion of benzophenone as target molecule, and a framework composed of a hydrogen bonded *combination* of C-methylcalix[4]resorcinarene (CMCR) and the 'extender' 4,4'-bipyridine (bipy) (there are other extenders (MacGillivray and Atwood, 2000) not discussed here; 'spacer' and 'pillar' have also been used). It will be seen that there are many twists and turns on the road between conception and ultimate success.

C-methylcalix[4]resorcinarene has been found in four conformations in its various molecular complexes (Fig. 12.35). Many crystalline CMCR–bipy combinations have been prepared, generally by hydrothermal methods. Some do not include benzophenone at all (or benzil, as an alternative target molecule) even though these potential guests were present in the reaction mixture; examples are the 'supramolecular' compounds {CMCR·2bipy}, {CMCR·3bipy}, {CMCR·2bipy·2H_2O}, {2CMCR}·1.5bipy·5H_2O and

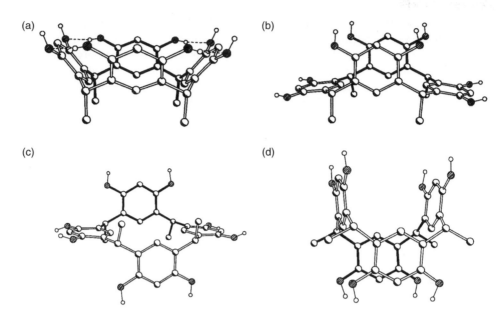

Fig. 12.35. The four conformations of C-methylcalix[4]resorcinarene: (a) crown or bowl, ideal symmetry $C_{2v} - mm2$; (b) boat or flattened cone; (c) chair; (d) saddle. (Reproduced with permission from Ma, Zhang and Coppens, 2001b.)

{CMCR·bipy}·3.5ethanol·H$_2$O·bipy, where the parentheses enclose framework components (Ma, Zhang and Coppens, 2002). As these are of no *direct* interest in the present context, they have not been included in Table 12.4, where only host-guest inclusion complexes are listed.

However, essentially identical hydrothermal conditions (4 mL water, 0.025 mol CMCR, 0.05 mol bipy, 0.05 mol benzophenone (or benzil); heated at 140°C for 1 day, cooled to room temperature at 20° per day) gave, in another set of experiments, the desired inclusion complexes of composition {CMCR·3bipy·2H$_2$O}·[benzophenone], {[CMCR·bipy·H$_2$O}· [benzophenone] and {(CMCR·bipy)·[benzophenone]} (Ma, Zhang and Coppens, 2001b), where the framework components are included in parentheses and the guest in square brackets. The first and second of these complexes, which are both of the framework type (Fig. 12.36), were considered suitable for photochemical experiments; the complexes differ in that the first has a single guest molecule in a cavity while the second has two. The third of these inclusion complexes has a carcerand type structure (cf. Chapter 3), and disordered guest in the cavities and appeared to be less suitable than the other two. It seems clear from these experiments that the type of complex formed (both those containing guests and those that are guest-free) depend in a subtle and not-yet-understood way on the crystallization conditions. A proposed design has yielded the desired product, but the road between beginning and end has yet to be mapped out convincingly.

One can consider the CMCR-bipy-H$_2$O-guest system in terms of phase diagrams. For example, the simplest of the ternary compounds have the composition {CMCR·2(bipy)}·[guest]. There are three groups of isostructural crystals; in the first the several guests are *p*-chlorotoluene, ferrocene adamantone and [2.2]paracyclophane,

Table 12.4. Crystal data for inclusion complexes in the CMCR-bipy-H_2O-guest system. The classification is based on chemical composition, CMCR conformation in the complex and its crystal structure. Data at room temperature unless stated otherwise. The REFCODES are given where available

Guest (reference)	a/α	b/β	c/γ	Space group	Z	Volume
CMCR·2(bipy)·n(guest): bowl; one-dimensional wave-like polymer						
p-chlorotoluene (MRR99) QAHNAO	10.025	23.982 112.71	11.375	$P2_1/m$	2	1261
Ferrocene (at 173K) (MSRR00)	9.730	25.440 110.71	10.825	$P2_1/m$	2	1253
Adamantanone (MRR99) QAHNAS	9.903	24.813 114.17	10.972	$P2_1/m$	2	1230
[2.2]paracyclophane (MRR99) QAHNIW	9.846	24.871 105.71	11.338	$P2_1/m$	2	1336
CH_3CN (MHA98)	7.893	29.243 99.67	10.054	$P2_1/m$	2	1144
CH_3NO_2 (MHA98) at 173K	7.709	29.307 100.55	9.982	$P2_1/m$	2	1120
Acetylferrocene (at 173K) (MSRR00)	18.477	11.742 107.794	26.086	$P2_1/n$	4*	1347
1,1'-diacetylferrocene (at 173K) (MSRR00)	18.808	11.948 108.306	25.595	$P2_1/n$	4*	1365
CMCR·2(bipy)·n(guest): bowl; zero-dimensional carcerand-like structure						
2(Nitrobenzene) (MDRR00)	10.731 95.727	15.075 94.860	16.882 93.915	$P\bar{1}$	2	1350
2(butanol)(MZC02)						
Benzophenone (MZC01)	22.824	25.799	36.354	$Fddd$	16	1338
CMCR·(bipy)·H_2O n(guest): flattened-cone; two-dimensional brick wall structure						
Benzophenone·(MZC01)	10.072 75.98	13.735 79.80	17.338 68.86	$P\bar{1}$	2	1080
CMCR·3(bipy)·$2H_2O$·n(guest): chair conformation; three-dimensional stepped network						
Benzophenone (MZC01a)						

* not 2, as given in paper.

References:
MHA98 – MacGillivray, Holman and Atwood, 1998; MRR99 – MacGillivray, Reid and Ripmeester, 1999; MDRR00 – MacGillivray, Diamente, Reid and Ripmeester, 2000; MSRR00 – MacGillivary, Spinney, Reid and Ripmeester, 2000; MZC01a – Ma et al., 2001a; MZC01b – Ma et al., 2001b; MZC02 – Ma et al., unpublished.

in the second acetonitrile and nitromethane, and in the third acetylferrocene and 1,1'-diacetylferrocene. In all three two guest molecules are included in a carcerand-like capsule, but these capsules are packed differently, as shown by the crystal data given in Table 12.4.

Thus the structural situation is indeed complicated. The CMCR·2(bipy) and {CMCR·2(bipy)·H_2O} combinations are versatile hosts in the sense of Chapter 8 and both

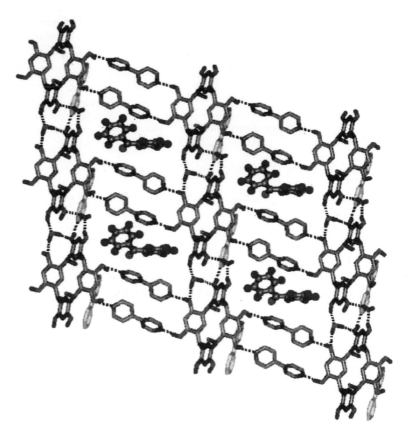

Fig. 12.36. Three-dimensional stepped sheet hydrogen bonded network of {[CMCR·3bipy·2H$_2$O]· benzophenone}. The horizontal layers contain CMCR and water molecules, while the bipy act as spacers, and the benzophenone guests are contained within the channels. (Reproduced with permission from Ma *et al.*, 2001b.)

crystallization conditions and nature of the guest determine the final structure; one of the tasks of the future is determine the relative roles of each factor.

12.7   Charged or neutral moieties – when is there hydrogen transfer between the components?

An early attempt to define conditions for the occurrence of neutral or charged components in hydrogen-bonded molecular compounds was made by Johnson and Rumon (1965). These authors prepared 18 *crystalline* molecular compounds of assorted combinations of pyridine and substituted pyridines, as hydrogen bond acceptors, and benzoic acid and substituted benzoic acids, as hydrogen bond donors. The infrared spectra of the solids were then determined, with emphasis placed on the N–H, O–H and C=O bands, most weight being given to the latter because the benzoic acid absorptions are much stronger than those of the pyridines. Detailed analysis of the spectra also permitted distinction between single-minimum and double-minimum potential wells. The pK_a value of the component in water was used as the best available measure of acidity, although this is not necessarily directly applicable to the situation in the solid. Their conclusion was that neutral component B...H–A compounds were formed when $\Delta pK_a < 3.8$ [$\Delta pK_a = pK_a$ (base) $- pK_a$(acid)], while for $\Delta pK_a > 3.8$ salts of the type B^+–H...A^- were formed. We summarize their results for the salts (ΔpK_a in brackets) – collidinium 3,4-dichlorobenzoate (3.99), collidinium 2,4-dichlorobenzoate (4.92), 2,6-lutidinium 3,5-dinitrobenzoate (6.47, erroneously printed as 3.88 in the original), collidinium 3,5-dinitrobenzoate (7.37, erroneously printed as 4.78 in the original), pyridinium 2,4-dinitrobenzoate (3.81), 3,5-lutidinium 2,4-dinitrobenzoate (4.92), 2,6-lutidinium 2,4-dinitrobenzoate (5.28), collidinium 2,4-dinitrobenzoate (6.18), pyridinium trichloroacetate (4.34), pyridinium trifluoroacetate (5.00). Crystal structures do not appear to have been reported for any of these molecular compounds. However, their conclusion is in accord with the following *crystalline* systems where both ΔpK_a values and structural results are available – pyridinium picrate (two polymorphs (Botoshansky *et al.*, 1994), 4.85), histidinium trimesate (3.98) (Herbstein and Kapon, 1979), nicotinyl salicylate (4.85) (Kim and Jeffrey, 1971), 2-aminopyridinium salicylate (3.86) (Gellert and Hsu, 1988), imidazolium picrate (6.57) (Soriano-Garcia *et al.*, 1990; Herbstein and Kapon, 1991), the isomorphous piperidinium salts of *p*-chlorobenzoic (7.14), *p*-bromobenzoic and *p*-toluic acids (6.76) (Kashino *et al.*, 1972; Kashino, 1973), pyrrolidinium *p*-chlorobenzoate (7.19) and pyrrolidinium *p*-toluate (6.81) (Kashino *et al.*, 1978) and 2,6-pyrido-2,7-crown-9 picrate (≈ 3.4) (Uiterwyk *et al.*, 1986).

Some exceptions should be noted – 9-methyladeninium salicylate has a ΔpK_a value of 0.9 but is nevertheless a salt; the pK_a of 9-methyladeninium (6-amino-9-methylpurine) is given as 3.9 in the tables but is labelled 'uncertain'. Acridinium pentachlorophenolate crystallizes as an ion pair (Wozniak *et al.*, 1991), although $\Delta pK_a = 5.58 - 4.5 = 1.1$; anilinium pentachlorophenolate is said to crystallize as an ion pair, although $\Delta pK_a = 4.63 - 4.5 = 0.13$. A better established surprise is found in dipyridinium oxalate oxalic acid ($\Delta pK_a = 5.23 - 1.23 = 4.00$) where "one oxalic acid donates two protons to [the two] pyridine bases rather than two oxalic acids giving up one proton each!" (Newkome *et al.*, 1985).

A similar approach, but applied to *solutions*, was taken by Brezinski *et al.* (1991) in their IR study of the proton potential as a function of ΔpK_a in hydrogen-bonded molecular compounds of various phenols (R–C_6H_4OH) with trimethylamine-N-oxide ($pK_a = 4.65$). For the most basic 4-methoxyphenol ($pK_a = 10.21$; $\Delta pK_a = -5.56$) there are two broad bands with maxima at about $2550\,cm^{-1}$ ($\nu(OH)$ vibration) and $1800\,cm^{-1}$ ($2\delta(OH)$ vibration). There was no continuous IR absorption. These results show that the proton is localized on the phenolic group in the $-OH...ON\equiv$ bond. With 4-chlorophenol ($pK_a = 9.37$; $\Delta pK_a = -4.72$), there are indications of a double minimum but the well is much deeper at the phenol and the weight of the polar structure $-O^-...H^+ON\equiv$ is very small. With R=$COOC_2H_5$ ($pK_a = 8.50$; $\Delta pK_a = -3.85$), the minimum of the proton potential at the phenol is less deep than for the 4-chlorophenol molecular compound, and the weight of the polar structure slightly larger. These changes are accentuated for R=NO_2 ($pK_a = 7.15$, $\Delta pK_a = -2.50$). The spectrum changes completely when R = 3,4-dinitro– and shows that there is a very strong hydrogen bond in which the proton fluctuates in a double minimum with a very low barrier, or in a broad flat single minimum potential. In the pentachlorophenol compound ($pK_a = 4.74$; $\Delta pK_a = -0.09$) the proton fluctuates in an $-OH\ldots ON\equiv\Leftrightarrow -O^-\ldots H^+ON\equiv$ equilibrium, as indicated by the continuum absorption. With 2,6-dichloro-4-nitrophenol ($pK_a = 3.70$; $\Delta pK_a = 0.95$) the polar structure has gained more weight. However, it was only with the picrate ($pK_a = 0.38$; $\Delta pK_a = 4.27$) that the polar structure was completely realized, the hydrogen bond no longer showing proton polarizability. A parallel study has been made with carboxylic acid/trimethylamine-N-oxide compounds in acetonitrile solutions (Böhner and Zundel, 1986).

Various attempts have been made to correlate ΔpK_a with hydrogen bond distances. Although some authors reported disappointing results, Lechat (1984) claimed that the equation $d(O\ldots O)$ Å $= 2.374 + 0.1754 \ln (0.405 \Delta pK_a + 1)$ reproduced the measured distances to within less than 0.02 Å for a number of molecular compounds where ΔpK_a ranges from -17.95 to 0. In order that the pK_a values measured in solution should be applicable to the situation in the solid, it seems desirable to restrict the sample to those molecular compounds which contain discrete pairs. Approximately linear relationships between $d(X-H\ldots Y)$ and ΔpK_a are found for the available examples in both $N-H\ldots O$ and $O-H\ldots O$ systems (Fig. 12.37), but it remains to be seen whether such a simple approach will survive the addition of more experimental values. The present situation is that pK_a values are lacking for relevant molecular compounds of known crystal structure, while crystal structures are not known for systems with known pK_a values. We note some

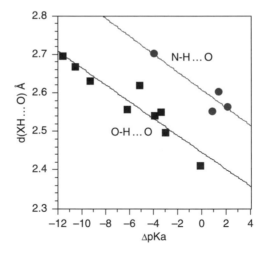

Fig. 12.37. Approximately linear plots are obtained for both $d(O-H\ldots O)$ and $d(N-H\ldots O)$ against ΔpK_a for hydrogen bonded, discrete pair molecular compounds. The equations are $d(O-H\ldots O) = -0.0219 \Delta pK_a + 2.45$ Å ($R^2 = 0.88$) and $d(N-H\ldots O) = -0.0232 \Delta pK_a + 2.61$ Å ($R^2 = 0.86$). The data on the figure are for the following molecular compounds: $O-H\ldots O$ (from left to right): 3-aminophenol/4-nitropyridine-N-oxide (NPNO); 3-chlorophenol/NPNO; 4-nitrophenol/ triphenylphosphine oxide (TPPO) (Moreno Fuquen & Lechat, 1992); 4-nitrophenol/triphenylarsine oxide; 4-aminobenzoic acid/NPNO; 2-carboxyl-1,3-xylyl-30-crown-9/urea (Aarts *et al.*, 1986); oxalic acid/TPPO; trichloroacetic acid/TPPO; salicylic acid/antipyrine; trichloroacetic acid/ pyridine-N-oxide. $N-H\ldots O$ (from left to right): All have pentachlorophenol as hydrogen bond donor and 3-pyridinecarbonitrile; 4-methylpyridine; 5,6,7,8-tetrahydroquinoline; N-methylmorpholine respectively as acceptors.

compilations of pK_a values (Serjeant and Dempsey, 1978; Smith and Martell, 1982, 1989; Klofutar *et al.*, 1967).

References

Aakeröy, C. B. and Beatty, A. M. (2001). *Aust. J. Chem.*, **54**, 409–421.

Aarts, V. M. L. J., Staveren, C. J. van, Grootenhuis, P. D. J., Eerden, J. van, Kruise, L., Harkema, S. and Reinhoudt, D. N. (1986). *J. Am. Chem. Soc.*, **108**, 5035–5036.

Allen, F. H., Bird, C. M., Rowland, R. S. and Raithby, P. R. (1997a). *Acta Cryst.*, B**53**, 680–695.

Allen, F. H., Bird, C. M., Rowland, R. S. and Raithby, P. R. (1997b). *Acta Cryst.*, B**53**, 696–701.

Allen, F. H., Motherwell, W. S. D., Raithby, P. R., Shields, G. P. and Taylor, R. (1999). *New J. Chem.*, pp. 25–34.

Almlöf, J., Kvick, Å. and Olovsson, I. (1971). *Acta Cryst.*, B**27**, 1201–1208.

Antona, D., Longchambon, F., Vandenborre, M. T. and Becker, P. (1973). *Acta Cryst.*, B**29**, 1372–1376.

Armstrong, D. R., Bennett, S., Davidson, M. G., Snaith, R., Stalke, D. and Wright, D. S. (1992). *J. Chem. Soc., Chem. Comm.*, pp. 262–264.

Armstrong, D. R., Davidson, M. G., Martin, A., Raithby, P. R., Snaith, R. and Stalke, D. (1992). *Angew. Chem. Int. Ed. Engl.*, **31**, 1634–1636.

Baeyer, A. and Villiger, V. (1902). *Chem. Ber.*, **35**, 1201–1212.

Balasubramanian, R., Chidambaram, R. and Ramachandran, G. N. (1970). *Biochim. Biophys. Acta*, **221**, 196–206.

Barnes, J. C. (1988). *Acta Cryst.*, C**44**, 118–120.

Barnes, J. C., Paton, J. D. and Blyth, C. S. (1990). *Acta Cryst.*, C**46**, 1183–1184.

Baur, W. H. (1992). *Acta Cryst.*, B**48**, 745–746.

Bavoux, C. (1975). *C. R. Acad. Sci. Paris*, Ser. C, **280**, 121–122.

Bavoux, C. and Thozet, A. (1973). *Acta Cryst.*, B**29**, 2603–2605.

Bavoux, C. and Thozet, A. (1980). *Cryst. Struct. Comm.*, **9**, 1115–1120.

Bellingen, I. van, Germain, G., Piret, P. and Meerssche, M. van, (1971a). *Acta Cryst.*, B**27**, 553–559.

Bellingen, I. van, Germain, G., Piret, P. and Meerssche, M. van, (1971b). *Acta Cryst.*, B**27**, 560–564.

Berkovitch-Yellin, Z. and Leiserowitz, L. (1984). *Acta Cryst.*, B**40**, 159–165.

Böhner, U. and Zundel, G. (1986). *J. Phys. Chem.*, **90**, 964–973.

Bond, D. R., Bourne, S. A., Nassimbeni, L. R. and Toda, F. (1989). *J. Cryst. Spectr. Res.*, **19**, 809–822.

Bond, D. R., Johnson, L., Nassimbeni, L. R. and Toda, F. (1991). *J. Sol. State Chem.*, **92**, 68–79.

Bond, D. R., Nassimbeni, L. R. and Toda, F. (1989a). *J. Cryst. Spectr. Res.*, **19**, 847–859.

Bond, D. R., Nassimbeni, L. R. and Toda, F. (1989b). *J. Incl. Phenom.*, **7**, 623–635.

Bondi, A. (1964). *J. Phys. Chem.*, **68**, 441–451.

Bondi, A. (1968). *Physical Properties of Molecular Crystals, Liquids and Glasses*, pp. 450 ff., Wiley, New York.

Botoshansky, M., Herbstein, F. H. and Kapon, M. (1994). *Acta Cryst.*, B**50**, 191–200.

Bourne, S. A., Johnson, L., Marais, C., Nassimbeni, L. R., Weber, E., Skobridis, K. and Toda, F. (1991). *J. Chem. Soc., Perkin II*, pp. 1707–1713.

Bourne, S. A., Nassimbeni, L. R. and Toda, F. (1991). *J. Chem. Soc., Perkin II*, pp. 1335–1341.

Bourne, S. A., Nassimbeni, L. R., Skobridis, K. and Weber, E. (1991). *J. Chem. Soc., Chem. Commun.*, pp. 282–283.

Bovill, M. J., Chadwick, D. J., Sutherland, I. O. and Watkin, D. (1980). *J. Chem. Soc., Perkin II*, pp. 1529–1543.

Bramley, A. (1916). *J. Chem. Soc.*, **109**, 469–496.

Brassy, C. and Mornou, J.-P. (1972). *C. R. Acad. Sci. Paris*, Ser. C, **274**, 1728–1730.

Brezinski, B., Brycki, B., Zundel, G. and Keil, T. (1991). *J. Phys. Chem.*, **95**, 8598–8600.

Brown, C. J. and Gray, L. R. (1982). *Acta Cryst.*, B**38**, 2307–2308.

Bruni, P., Tosi, G. and Valle, G. (1988). *J. Chem. Soc., Chem. Comm.*, pp. 1022–1023.

Bruyn, P. J. de, Gable, R. W., Potter, A. C. and Solomon, D. H. (1996). *Acta Cryst.*, C**52**, 466–468.

Busetta, B., Courseille, C. and Hospital, M. (1973). *Acta Cryst.*, B**29**, 2456–2462.

Caira, M. R., Nassimbeni, L. R., Niven, M. L., Schubert, W.-D., Weber, E. and Dörpinghaus, N. (1990). *J. Chem. Soc., Perkin II*, pp. 2129–2133.

Caira, M. R., Watson, W, H., Vögtle, F. and Müller, W. M. (1984). *Acta Cryst.*, C**40**, 136–138.

Carré, D., Moquin, C. and Guyot, M. (1986). *Acta Cryst.*, C**42**, 483–485.

Cassady, R. E. and Hawkinson, S. W. (1982). *Acta Cryst.*, B**38**, 2206–2209.

Ceccarelli, C., Jeffrey, G. A. and Taylor, R. (1986). *J. Mol. Struct.*, **70**, 255–271.

Chanzy, H., Maia, E. and Perez, S. (1982). *Acta Cryst.*, B**38**, 852–855.

Charpin, P., Dunach, E., Kagan, H. and Theobald, F. R. (1981). *Tetr. Lett.*, **27**, 2989–2992.

Cody, V., Duax, W. L. and Norton, D. A. (1972). *Acta Cryst.*, B**28**, 2244–2252.

Cooper, A., Kartha, G., Gopalakrishna, E. M. and Norton, D. A. (1969). *Acta Cryst.*, B**53**, 521–533.

Corcoran, J. M., Kruse, H. W. and Skolnik, S. (1953). *J. Phys. Chem.*, **57**, 435–437.

Coupar, P. I., Glidewell, C. and Ferguson, G. (1997). *Acta Cryst.*, B**25**, 2409–2411.

Courseille, C., Meresse, A., Dournel, P., Duparc, L. H. and Villenave, J.-J. (1991). *Acta Cryst.*, C**47**, 100–102.

Csöregh, I., Czugler, M., Kálmán, A., Weber, E. and Hecker, M. (1991). *Bull. Chem. Soc. Jpn.*, **64**, 2539–2543.

Czugler, M., Czerzö, M., Weber, E. and Ahrendt, J. (1991). *J. Cryst. Spectr. Res.*, **21**, 501–505.

Czugler, M. and Weber, E. (1991). *J. Incl. Phenom.*, **10**, 355–366.

Dahl, T. and Riise, B. (1989). *Acta Chem. Scand.*, **43**, 493–495.

Dahl, T. (1986). *Acta Chem. Scand.*, B**40**, 226–229.

De Winter, H. L., Blaton, N. M., Peeters, O. M., De Ranter, C. J., Van Aerschot, A. and Herdewijn, P. (1991). *Acta Cryst.*, C**47**, 838–842.

Declercq, J. P., Germain, G., Putzeys, J. P., Rona, S. and Meerssche, M. van, (1974). *Cryst. Struct. Commun.*, **3**, 579–582.

Deeg, A. and Mootz, D. (1993). *Z. Naturforsch.*, **48b**, 571–576.

Del Bene, J. E. (1998), "Hydrogen bonding I" in *Encyclopedia of Computational Chemistry*, edited by P. v. R. Schleyer, Wiley, Chichester, Vol. 2, 1263–1271.

Desiraju, G. R. and Sarma, J. A. R. P. (1983). *J. Chem. Soc., Chem. Comm.*, pp. 45–46.

Desiraju, G. R. and Steiner, Th. (1999). *The Weak Hydrogen Bond in Structural Chemistry and Biology*, IUCr and Oxford University Press.

Desiraju, G. R. (1991). *Acc. Chem. Res.*, **24**, 290–296.

Dideberg, O., Dupont, L. and Campsteyn, H. (1975). *Acta Cryst.*, B**31**, 637–640.

Duax, W. L. (1972). *Acta Cryst.*, B**28**, 1864–1871.

Eichhorn, K. D. (1991). *Z. Kristallogr.*, **195**, 205–220.

Etter, M. C. and Baures, P. W. (1988). *J. Am. Chem. Soc.*, **110**, 639–640.

Etter, M. C. and Reutzel, S. M. (1991). *J. Am. Chem. Soc.*, **113**, 2586–2598.

Etter, M. C., Macdonald, J. D. and Bernstein, J. (1990). *Acta Cryst.*, B**46**, 256–262.

Etter, M. C., Urbanczyk-Lipkowska, Z., Zia-Ebrahimi, M. and Panunto, T. W. (1990). *J. Am. Chem. Soc.*, **112**, 8415–8426.

Ferguson, G., Glidewell, C., Gregson, R. M., Meehan, P. R. and Patterson, I. L. J. (1998). *Acta Cryst.*, B**54**, 151–159.

Ferguson, G., Glidewell, C., Gregson, R. M. and Lavender. J. (1999). *Acta Cryst.*, B**55**, 573–590.

Finkel'shtein, A. I. and Rukevich, O. S. (1983). *Zh. Prikl. Spectroscop.*, **38**, 327–330.

Fritchie, C. J., Jr. and McMullan, R. K. (1981). *Acta Cryst.*, B**37**, 1086–1091.

Fuquen, R. M., Lechat, J. R. and Santos, R. H. de A. (1991). *Acta Cryst.*, C**47**, 2388–2391.

Fuquen, R. M. and Lechat, J. R. (1992). *Acta Cryst.*, C**48**, 1690–1692.

Garcia-Tellado, F., Geib, S. J., Goswani, S. and Hamilton, A. D. (1991). *J. Am. Chem. Soc.*, **113**, 9265–9269.

Garrell, R. L., Smyth, J. C., Fronczek, F. C. and Ganour, R. D. (1988). *J. Incl. Phenom.*, **6**, 73–78.

Gellert, R. W. and Hsu, I.-N. (1988). *Acta Cryst.*, C**44**, 311–313.

Goldberg, I., Stein, Z., Tanaka, K. and Toda, F. (1988). *J. Incl. Phenom.*, **6**, 15–30.

Golic, L. and Lazarini, F. (1974). *Vestn. Slov. Kem. Drus.*, **21**, 17–XX.

Golic, L. and Kaucic, V. (1976). *Cryst. Struct. Comm.*, **5**, 319–324.

Golic, L., Hadzi, D. and Lazarini, F. (1971). *J. Chem. Soc. Chem. Comm.*, p. 860.

Gorres, B. T., McAfee, E. R. and Jacobson, R. A. (1975). *Acta Cryst.*, B**31**, 158–161.

Gramstad, T., Husebye, S. and Maartman-Moe, K. (1985). *Acta Chem. Scand.*, B**39**, 767–771.

Grech, E., Nowicka-Scheibe, J., Lis, T. and Malarski, Z. (1989). *J. Mol. Struct.*, **195**, 1–10.

Grell, J., Bernstein, J. and Tinhofer, G. (1999). *Acta Cryst.*, B**55**, 1030–1043.

Grell, J., Bernstein, J. and Tinhofer, G. (2002). *Cryst. Rev.*, **8**, 1–56.

Gzella, A. and Wrzeciono, U. (1991). *Acta Cryst.*, C**47**, 980–982.

Habermann, J. (1884). *Monatsh. Chem.*, **5**, 329.

Hamilton, W. C. and Ibers, J. A. (1968). *Hydrogen Bonding in Solids*, Benjamin, New York.

Harary, F. (1967). *Graph Theory and Theoretical Physics*. Academic Press, New York.

Harkema, S., Bats, J. W., Weyenberg, A. M. and Feil, D. (1972). *Acta Cryst.*, B**28**, 1646–1648.

Harkema, S., Hummel, G. J. van, Daasvatu, K. and Reinhoudt, D. N. (1981). *J. Chem. Soc., Chem. Comm.*, pp. 368–369.

Harkema, S., Ter Brake, J. H. M. and Meutstege, H. J. G. (1979). *Acta Cryst.*, B**35**, 2087–2093.

Haupt, H. J., Huber, F., Krueger, C., Preut, H. and Thierback, D. (1977). *Z. anorg. allg. Chem.*, **436**, 229–236.

Herbstein, F. H. and Kapon, M. (1979). *Acta Cryst.*, B**35**, 1614–1619.

Herbstein, F. H. and Kapon, M. (1991). *Acta Cryst.*, C**47**, 1131–1132.

Hilgenfeld, R. and Saenger, W. (1981). *Z. Naturforsch.*, **36b**, 242–247.

Hine, J., Ahn, K., Gallucci, J. C. and Linden, S.-M. (1984). *J. Am. Chem. Soc.*, **106**, 7980–7981.

Hine, J., Ahn, K., Gallucci, J. C. and Linden, S.-M. (1990). *Acta Cryst.*, C**46**, 2136–2146.

Hsu, I.-N. and Gellert, R. W. (1983). *J. Cryst. Spectr. Res.*, **13**, 43–48.

Huang, C.-M., Leiserowitz, L. and Schmidt, G. M. J. (1973). *J. Chem. Soc., Perkin II*, pp. 503–508.

Hummel, G. J. and Helmholdt, R. B. (1991). *Acta Cryst.*, C**47**, 213–215.

Iball, J., Morgan, C. H. and Wilson, C. H. (1968). *Proc. Roy. Soc. Lond., Ser.* A, **302**, 225–236.

Jeffrey, G. A. and Saenger, W. (1991). *Hydrogen Bonding in Biological Structures*, Springer-Verlag, Berlin.

Jeffrey, G. A. (1995). *Cryst. Revs.*, **4**, 213–259.

Jeffrey, G. A. (1997). *An Introduction to Hydrogen Bonding*. Oxford University Press, New York.

Jensen, J. S. and Jensen, B. (1990). *Acta Cryst.*, C**46**, 779–781.

Jerzykiewicz, L. B., Lis, T., Malarski, Z. and Grech, E. (1993). *J. Cryst. Spectr. Res.*, **23**, 805–812.

Jerzykiewicz, L. B., Sobczyk, L., Lis, T., Malarski, Z. and Grech, E. (1998). *J. Mol. Struct.*, **440**, 175–185.

Joesten, M. D. and Schaad, L. J. (1974). Hydrogen bonding, Dekker, New York.

Johnson, S. L. and Rumon, K. A. (1965). *J. Phys. Chem.*, **69**, 74–86.

Jordan, T. H. and Mak, T. C. W. (1970). *J. Chem. Phys.*, **52**, 3790–3794.

Jorgensen, W. L., Boudon, S. and Nguyen, T. B. (1993). *J. Am. Chem. Soc.*, **111**, 755–757.

Kanters, J. A., Schouten, A., Kroon, J. and Grech, E. (1991). *Acta Cryst.*, C**47**, 807–810.

Kanters, J. A., Ter Horst, E. T. and Kroon, J. (1992). *Acta Cryst.*, C**48**, 328–332.

Kashino, S. (1973). *Acta Cryst.*, B**29**, 1836–1842.

Kashino, S., Sumida, Y. and Haisa, M. (1972). *Acta Cryst.*, **B28**, 1374–1383.

Kashino, S., Sumida, Y. and Haisa, M. (1978). *Bull. Chem. Soc. Jpn.*, **51**, 1717–1722.

Kaufmann, R., Knöchel, A., Kopf, J., Oehler, J. and Rudolf, G. (1977). *Chem. Ber.*, **110**, 2249–2253.

Kim, H. S. and Jeffrey, G. A. (1971). *Acta Cryst.*, **B27**, 1123–1131.

Kim, H. S., Jeffrey, G. A. and Rosenstein, R. D. (1971). *Acta Cryst.*, **B27**, 307–314.

Kim, S.-H. and Rich, A. (1967). *Science*, **158**, 1046–1048.

Klofutar, C., Krasovec, F. and Kusar, M. (1967). *Croat. Chem. Acta*, **39**, 23–28.

Knöchel, A., Kopf, J., Oehler, J. and Rudolf, G. (1978). *J. Chem. Soc., Chem. Comm.*, pp. 595–596.

Knorr, L. (1898). *Liebigs Ann. Chem.*, **301**, 4–18.

Kofler, A. (1944). *Z. Elektrochem.*, **50**, 200–207.

Kortüm, G., Vogel, W. K. and Andrussow, K. (1965). *Dissociation Constants of Organic Acids in Aqueous Solutions*, Butterworths, Supplement (1972), London.

Kroon, J., Kanters, J. A., van Duijneveldt-van de Rijdt, J. G. C. M., van Duijneveldt, F. B. and Vliegenthart, J. A. (1975). *J. Mol. Struct.*, **24**, 109–129.

Lariucci, C., Santos, R. H. de A. and Lechat, J. R. (1986a). *Acta Cryst.*, **C42**, 731–733.

Lariucci, C., Santos, R. H. de A. and Lechat, J. R. (1986b). *Acta Cryst.*, **C42**, 1825–1828.

Lavender, E. S., Glidewell, C. and Ferguson, G. (1998a). *Acta Cryst.*, **C54**, 1637–1639.

Lavender, E. S., Glidewell, C. and Ferguson, G. (1998b). *Acta Cryst.*, **C54**, 1639–1642.

Le Fevre, R. J. W. and Tideman, C. G. (1931). *J. Chem. Soc.*, pp. 1729–1732.

Leban, I., Svete, J., Stanovnik, B. and Tisler, M. (1991). *Acta Cryst.*, **C47**, 1552–1554.

Lechat, J. (1984). *Acta Cryst.* **A40**, C-264.

Lechat, J. R., Almeida Santos, R. H. de, and Bueno, W. A. (1981). *Acta Cryst.*, **B37**, 1468–1470.

Lee, J. D. and Wallwork, S. C. (1959). *Acta Cryst.*, **12**, 210–216.

Lee, J. D. and Wallwork, S. C. (1965). *Acta Cryst.*, **19**, 311–313.

Legon, A. C. and Millen, D. J. (1987). *Chem. Soc. Revs.*, **16**, 467–498.

Leviel, J.-L., Auvert, G. and Savariault, J.-M. (1981). *Acta Cryst.*, **B37**, 2185–2189.

Lii, J.-H. (1998), "Hydrogen bonding II" in *Encyclopedia of Computational Chemistry*, edited by P. V. R. Schleyer, Wiley, Chichester, Vol. 2, 1271–1283.

Liminga, R. and Sørensen, A. M. (1967). *Acta Chem. Scand.*, **21**, 2669–2678.

Liminga, R. (1967). *Acta Chem. Scand.*, **21**, 1206–1216.

Llamas-Saiz, A. L. and Foces-Foces, C. (1990). *J. Mol. Struct.*, **238**, 367–382.

Llamas-Saiz, A. L., Foces-Foces, C., Mo, O., Yanez, M. and Elguero, J. (1992). *Acta Cryst.*, **B48**, 700–713.

Lu, C.-S., Hughes, E. W. and Giguere, P. A. (1941). *J. Am. Chem. Soc.*, **63**, 1507–1513.

Lynch, D. E., Smith, G., Byriel, K. A. and Kennard, C. H. L. (1992). *Z. Kristallogr.*, **200**, 73–82.

Ma, B-Q., Zhang, Y. and Coppens, P. (2001a). *Cryst Eng Comm*, **20**, 1–3.

Ma, B-Q., Zhang, Y. and Coppens, P. (2001b). *Cryst. Growth Des.*, **1**, 271–275.

Ma, B-Q., Zhang, Y. and Coppens, P. (2002). *Cryst. Growth Des.*, **2**, 7–13.

MacGillivray, L. R. and Atwood, J. L. (2000). *J. Sol. State Chem.*, **152**, 199–210.

MacGillivray, L. R., Holman, K. T. and Atwood, J. L. (1998). *Cryst. Eng.*, **1**, 87–96.

MacGillivray, L. R., Reid, J. L. and Ripmeester, J. A. (1999). *Cryst Eng Comm*, 1–4.

MacGillivray, L. R., Diamente, P. R., Reid, J. L. and Ripmeester, J. A. (2000). *Chem. Commun.*, pp. 359–360.

MacGillavray, L. R., Spinney, H. A., Reid, J. L. and Ripmeester, J. A. (2000). *Chem. Commun.*, pp. 517–518.

Mahmoud, M. M. and Wallwork, S. C. (1975). *Acta Cryst.*, **B31**, 338–342.

Majerz, I., Malarski, Z. and Lis, T. (1990a). *J. Mol. Struct.*, **213**, 161–168.

Majerz, I., Malarski, Z. and Lis, T. (1990b). *J. Mol. Struct.*, **240**, 47–58.

Majerz, I., Malarski, Z. and Sawka-Dombrowska, W. (1991). *J. Mol. Struct.*, **249**, 109–116.

Mak, T. C. W. and Lam, Y.-S. (1978) *Acta Cryst.*, **B34**, 1732–1735.

Malarski, Z., Lis, T. and Grech, E. (1991). *J. Cryst. Spectr. Res.*, **21**, 255–259.

Malarski, Z., Majerz, I. and Lis, T. (1987). *Acta Cryst.*, **C43**, 1766–1769.

Malarski, Z., Majerz, I. and Lis, T. (1996). *J. Mol. Struct.*, **380**, 249–256.

Malinovskii, S. T., Simonov, Yu. A. and Nazarenko, A. Yu. (1990). *Sov. Phys. Cryst.*, **35**, 833–836.

Mathew, M. and Palenik, G. J. (1972). *J. Chem. Soc., Perkin II*, pp. 1033–1036.

Matias, P. M., Jeffrey, G. A. and Ruble, J. R. (1988). *Acta Cryst.*, **B44**, 516–522.

Maurin, J. K., Winnicka-Maurin, M., Paul, I. C. and Curtin, D. Y. (1993). *Acta Cryst.*, **B49**, 90–96.

Maverick, E., Seiler, P., Schweitzer, W. B. and Dunitz, J. D. (1980). *Acta Cryst.*, **B36**, 615–620.

Meléndez, R. and Hamilton, A. D. (1998). "Hydrogen-bonded ribbons, tapes and sheets as motifs for crystal engineering," *Top. Curr. Chem.*, **198**, 97–129.

Murray-Rust, P. and Glusker, J. P. (1984). *J. Am. Chem. Soc.*, **106**, 1018–1025.

Nassimbeni, L. R., Niven, M. L., Tanaka, K. and Toda, F. (1991). *J. Cryst. Spectr. Res.*, **21**, 451–457.

Nastopoulos, V. and Weiler, J. (1988). *Acta Cryst.*, **C44**, 500–503.

Nastopoulos, V., Weiler, J. and Germain, G. (1991). *Acta Cryst.*, **C47**, 1546–1548.

Newkome, G. R., Theriot, K. J. and Fronczek, F. R. (1985). *Acta Cryst.*, **C41**, 1642–1644.

Nyburg, S. C. and Faerman, C. H. (1985). *Acta Cryst.*, **B41**, 274–279.

Nyburg, S. C., Faerman, C. H. and Prasad, L. (1987). *Acta Cryst.*, **B43**, 106–110.

Parvez, M. and Schuster, I. S. (1991). *Acta Cryst.*, **C47**, 446–448.

Pascal, R. A., Jr. and Ho, D. M. (1994). *Tetrahedron*, **50**, 8559–8568.

Pauling, L. (1963). *The Nature of the Chemical Bond*, Third Edition, Cornell University Press, Ithaca, see p. 260.

Perrin, D. D. (1965). "Dissociation constants of organic bases" in *Aqueous Solutions*, Butterworths, Supplement (1972), London.

Perrin, M., Mahdar, E., Lecocq, S. and Bavoux, C. (1990). *J. Incl. Phenom. Mol. Recognit. Chem.*, **9**, 153–160.

Peters, K., Peters, E-A., von Schnering, H. G., Bringmann, H. G. and Göbel, L. (1993). *Z. Kristallogr.*, **203**, 265–268.

Pimentel, G. C. and McClellan, A. L. (1960). *The Hydrogen Bond*, Freeman, San Francisco and London.

Plotnikow, W. A. (1909). *Chem. Ber.*, **42**, 1154–1159.

Prout, K., Fail, J., Jones, R. M., Warner, R. E. and Emmett, J. C. (1988). *J. Chem. Soc., Perkin II*, pp. 265–284.

Ranganathan, A., Pedireddi, V. R. and Rao, C. N. R. (1999). *J. Am. Chem. Soc.*, **121**, 1752–1753.

Rebek, J., Jr., Askew, B., Killoran, M., Nemeth, D., and Lin, F.-T. (1987). *J. Am. Chem. Soc.*, **109**, 2426–2431.

Riethof, G. *U. S. Patents* 2,295,606 (1944); 2,376,008 (1945).

Robertson, G. B. and Tooptakong, U. (1985). *Acta Cryst.*, **C41**, 1332–1335.

Rogers, R. D. and Green, L. M. (1986). *J. Incl. Phenom.*, **4**, 77–84.

Rogers, R. D. and Richards, P. D. (1987). *J. Incl. Phenom.*, **5**, 631–638.

Rogers, R. D., Richards, P. D. and Voss, E. J. (1988). *J. Incl. Phenom.*, **6**, 65–71.

Ruud, K. A., Sepeda, J. S., Tibbals, F. A. and Hrncir, D. C. (1991). *J. Chem. Soc., Chem. Commun.*, p. 629–630.

Sarma, J. A. R. P. and Desiraju, G. R. (1985). *J. Chem. Soc., Perkin II*, pp. 1905–1912.

Scheiner, S. (1997). *Hydrogen Bonding: A Theoretical Perspective*, Oxford University Press, New York.

Schmidlin, J. and Lang, R. (1910). *Ber. dtsch. chem. Ges.*, **43**, 2806–2820.

Schuster, P., Zundel, G. and C. Sandorfy, C. (1976). (Eds). The Hydrogen Bond (Recent Developments in Theory and Experiments), Vol. I Theory, Vol. II Structure and Spectroscopy, Vol. III Dynamics, Thermodynamics and Special Systems. North-Holland, Amsterdam, New York and Oxford.

Serjeant, E. P. and Dempsey, B. (1978). "Ionization constants of organic acids" in *Aqueous Solution*, IUPAC Data Series No. 23, Pergamon Press, Oxford.

Shastry, C. I. V., Seshadri, T. P., Prasad, J. S. and Achar, B. N. (1987). *Z. Kristallogr.*, **178**. 283–288.

Shieh, H.-S. and Voet, D. (1976). *Acta Cryst.*, B**32**, 2361–2367.

Shimizu, N. and Nishigaki, S. (1983). *Acta Cryst.*, C**39**, 502–504.

Shimizu, N., Nishigaki, S., Nakai, N. and Osaki, K. (1982). *Acta Cryst.*, B**38**, 2309–2311.

Shiro, M. and Kubota, T. (1972). *Chem. Lett.*, pp. 1151–1152.

Sim, G. A. (1967). *Annu. Rev. Phys. Chem.*, **18**, 57–80.

Singh, T. P. and Vijayan, M. (1974). *Acta Cryst.*, B**30**, 557–562.

Singh, T. P. and Vijayan, M. (1977). *J. Chem. Soc., Perkin II*, pp. 693–699.

Smith, R. M. and Martell, A. E. (1982). *Critical Stability Constants, Plenum*, Vols. 1–4, Vol. 5 (First Supplement), Vol. 6 (Second Supplement), 1989, London.

Soriano-Garcia, M., Schatz-Levine, M., Toscano, R. A and Iribe, R. V. (1990). *Acta Cryst.*, C**46**, 1556–1558.

Soriano-Garcia, M., Toscano, R. A. and Espinosa, G. (1985). *J. Cryst. Spectroscop. Res.*, **15**, 651–662.

Steiner, T. (1996). *Cryst. Revs.*, **6**, 1–57.

Steiner, T. (1998). *Acta Cryst.*, B**54**, 456–463.

Steiner, T. and Saenger, W. (1991). *Acta Cryst.*, B**47**, 1022–1023.

Steiner, T. and Saenger, W. (1992). *J. Am. Chem. Soc.*, **114**, 10146–10154.

Steudel, R., Rose, F. and Pickardt, J. (1977). *Z. anorg. allgem. Chem.*, **434,** 99–109.

Suh, I.-H. and Saenger, W. (1978). *Angew. Chem. Int. Ed. Engl.*, **17**, 534.

Talukdar, A. N. and Chaudhuri, B. (1976). *Acta Cryst.*, B**32**, 803–808.

Taylor, R. and Kennard, O. (1982). *J. Am. Chem. Soc.*, **104**, 5063–5070.

Taylor, R. and Kennard, O. (1984). *Acc. Chem. Res.*, **17**, 320–326.

Taylor, R., Kennard, O. and Versichel, W. (1983). *J. Am. Chem. Soc.*, **105**, 5761–5766.

Taylor, R., Kennard, O. and Versichel, W. (1984a). *Acta Cryst.*, B**40**, 280–288.

Taylor, R., Kennard, O. and Versichel, W. (1984b). *J. Am. Chem. Soc.*, **106**, 244–248.

Thierback, D. and Huber, F. (1979). *Z. anorg. allg. Chem.*, **451**, 137–142.

Thierback, D. and Huber, F. (1981). *Z. anorg. allg. Chem.*, **477**, 101–107.

Thomas, J. O. and Pramatus, S. (1975). *Acta Cryst.*, B**31**, 2159–2161.

Thomas, J. O. and Renne, N. (1975). *Acta Cryst.*, B**31**, 2161–2163.

Thomas, J. O. (1975). *Acta Cryst.*, B**31**, 2156–2158.

Toda, F. (1987). *Top. Curr. Chem.*, **140**, 43–69.

Toda, F., Sato, A., Tanaka, K. and Mak, T. C. W. (1989). *Chem. Lett.*, pp. 873–876.

Toda, F., Tanaka, K, and Ueda, H. (1981). *Tetr. Lett.*, **22**, 4669–4672.

Toda, F., Tanaka, K. and Mak, T. C. W. (1983). *Chem. Lett.*, pp. 1699–1702.

Toda, F., Tanaka, K. and Mak, T. C. W. (1984a). *Chem. Lett.*, pp. 2085–2088.

Toda, F., Tanaka, K. and Mak, T. C. W. (1984b). *Tetrahedron Lett.*, **25**, 1359–1362.

Toda, F., Tanaka, K. and Mak, T. C. W. (1985). *J. Incl. Phenom.*, **3**, 225–233.

Toda, F., Tanaka, K. and Mak, T. C. W. (1989). *Chem. Lett.*, pp. 1329–1330.

Toda, F., Tanaka, K., Elguero, J., Stein, Z. and Goldberg, I. (1988). *Chem. Lett.*, pp. 1061–1064.

Toda, F., Tanaka, K., Yagi, M., Stein, Z. and Goldberg, I. (1990). *J. Chem. Soc., Perkin I*, pp. 1215–1216.

Toussaint, J., Diderberg, O. and Dupont, L. (1974). *Acta Cryst.*, B**30**, 590–596.

Uiterwyk, J. W. H. M., Staveren, C. J. van, Reinhoudt, D. N., Hertog, H. J. den, Jr., Kruise, L. and Harkema, S. (1986). *J. Org. Chem.*, **51**, 1575–1587.

Uno, T. and Shimizu, N. (1980). *Acta Cryst.*, **B36**, 2794–2796.

Van Roey, P., Bullion, K. A., Osawa. Y., Bowman, R. M. and Braun, D. G. (1991). *Acta Cryst.*, **C47**, 1015–1018.

Varughese, K. I. and Kartha, G. (1982). *Acta Cryst.*, **B38**, 301–302.

Venkatramana Shastry, C. I., Shashadri, T. P., Shashidhara Prasad, J. and Narayana Acher, B. (1987). *Z. Kristallogr.*, **178**, 283–288.

Vudenova-Adrabinska, V. (1996). *Acta Cryst.*, **B52**, 1048–1056.

Vinogradov, S. N. and Linnell, R. H. (1971). *Hydrogen Bonding*, Van Nostrand-Reinhold: New York.

Viterbo, D., Gasco, A., Serafino, A. and Mortarini, V. (1975). *Acta Cryst.*, **B31**, 2151–2153.

Wang, Y., Wei, B. and Wang, Q. (1990). *J. Cryst. Spectroscop. Res.*, **20**, 79–84.

Watson, W, H., Galloy, J., Vögtle, F. and Müller, W. M. (1984). *Acta Cryst.*, **C40**, 200–202.

Watson, W. H., Galloy, J., Grossie, D. A., Vögtle, F. and Müller, W. M. (1984). *J. Org. Chem.*, **49**, 347–353.

Watson, W. H., Nagl, A. and Kashyap, R. P. (1991). *Acta Cryst.*, **C47**, 800–803.

Watson, W. H., Vögtle, F. and Müller, W. M. (1988a). *Acta Cryst.*, **C44**, 141–145.

Watson, W. H., Vögtle, F. and Müller, W. M. (1988b). *J. Incl. Phenom.*, **6**, 491–505.

Weber, G. (1982a). *Acta Cryst.*, **B38**, 628–632.

Weber, G. (1982b). *Acta Cryst.*, **B38**, 2712–2715.

Weber, G. (1983). *Acta Cryst.*, **C39**, 896–899.

Weber, G. and Saenger, W. (1980). *Acta Cryst.*, **B36**, 424–428.

Weber, G. and Sheldrick, G. M. (1981). *Acta Cryst.*, **B37**, 2108–2111.

Weber, E., Csöregh, I., Ahrendt, J., Finge, S. and Czugler, M. (1988). *J. Org. Chem.*, **53**, 5831–5839.

Weber, E., Csöregh, I., Stensland, B. and Czugler, M. (1984). *J. Am. Chem. Soc.*, **106**, 3297–3306.

Weber, E., Dörpinghaus, N. and Goldberg, I. (1988). *J. Chem. Soc., Chem. Comm.*, pp. 1566–1568.

Weber, E., Hecker, M., Csöregh, I. and Czugler, M. (1989). *J. Am. Chem. Soc.*, **111**, 7866–7872.

Weber, E., Skobridis, K. and Goldberg, I. (1989). *J. Chem. Soc., Chem. Commun.*, pp. 1195–1197.

Weber, H.-P. and Craven, B. M. (1987). *Acta Cryst.*, **B43**, 202–209.

Weber, H.-P., Ruble, J. R., Craven, B. M. and McMullan, R. K. (1980). *Acta Cryst.*, **B36**, 1121–1126.

Weller, F., Borgholte, H., Stenger, H., Vogler, S. and Dehnicke, K. (1989). *Z. Naturforsch.*, **B44**, 1524–1530.

Whitesides, G. M., Mathias, J. P. and Seto, C. T. (1991). *Science*, **254**, 1312–1319.

Wiechert, D., Mootz, D. and Dahlems, T. (1997). *J. Am. Chem. Soc.*, **119**, 12665–12666.

Wiedenfeld, H. and Knoch, F. (1990). *Acta Cryst.*, **C46**, 1038–1040.

Wozniak, K., Krygowski, T. M., Kariuki, B. and Jones, W. (1991). *J. Mol. Struct.*, **248**, 331–343.

Yasui, M., Yabuki, T., Takama, M., Harada, S., Kasai, N., Tanaka, K. and Toda, F. (1989). *Bull. Chem. Soc. Jpn.*, **62**, 1436–1445.

Yu, P.-Y. and Mak, T. C. W. (1978). *Acta Cryst.*, **B34**, 3053–3056.

Zeegers-Huyskens, T. (1967). *Spectrochim. Acta*, **A23**, 855–866.

Zerkowski, J. A., Seto, C. T. and Whitesides, G. M. (1992). *J. Am. Chem. Soc.*, **114**, 5473–5475.

Zerkowski, J. A., Seto, C. T., Wierda, D. A. and Whitesides, G. M. (1990). *J. Am. Chem. Soc.*, **112**, 9025–9026.

Zhang, Y., Wu, G., Wenner, B. R., Bright, F. V. and Coppens, P. (1999). *Cryst. Eng.*, **2**, 1–8.

Ziolo, R. F., Günther, W. H. H., Meredith, G. R., Williams, D. J. and Troup, J. M. (1982). *Acta Cryst.*, **B38**, 341–343.

REFERENCES

Part VI

Molecular compounds with delocalized interactions

Introduction to Part VI

Molecular compounds with delocalized interactions

In this group of five related chapters we discuss principally molecular compounds with delocalized interaction between donor and acceptor molecules, due to electron transfer, in ground or excited states, from the highest occupied (HOMO) π orbitals of the donor to the lowest unoccupied (LUMO) π^* orbitals of the acceptor. Thus these are π–π^* molecular compounds (the π^* is often dropped for brevity). The molecular compounds are generally highly colored even though the parent molecules are colourless; this striking feature was already apparent in the first groups of molecular compounds to have been prepared (quinhydrone by Wohler (1844), picric acid molecular compounds of benzene, naphthalene and anthracene by Fritzsche (1858)). The early stages of research on π-molecular compounds were summarized by Pfeiffer (1928), whose book, despite the passage of the years, still contains much useful and interesting information. At that time there were two rival theories about the structures of π-molecular compounds; one, put forward by Lowry (1924) and Bennett and Willis (1929), envisaged the formation of covalent bonds between the two partners, whereas the other, due essentially to Pfeiffer, ascribed the cohesion to secondary valence interactions (Rest-affinitätskrafte). The first direct evidence for the Pfeiffer proposal was provided by the finding (Powell, Huse and Cooke, 1943) that all intermolecular distances in the crystal structure of {p-iodoaniline ⋯ TNB} were far too long to be ascribed to covalent bonding. Attention thus necessarily shifted to secondary interactions between the components, with major emphasis on finding an explanation for the intense colours developed in crystals and, especially, in solutions.

A major advance was made by Mulliken (1952a, 1952b, 1956), following earlier suggestions by Weiss (1942) and Brackmann (1949). The essential idea was that the ground state of the molecular compound can be described by combining a (major) contribution from a no-bond wave function of the two uncharged molecules and a (rather minor) contribution from a dative structure involving transfer of an electron from donor to acceptor. In the excited state, on the other hand, the major contribution comes from the dative or charge-transfer structure. The additional, frequently intense, charge-transfer band in the spectra of solutions or suitably-oriented crystals is ascribed to excitation of the system from its predominantly no-bond ground state to its predominantly dative excited state. Mulliken's theoretical contributions (Mulliken and Person, 1969) were primarily directed towards 1 : 1 'loose complexes' in solution but donor–acceptor interactions occur also in the gas (Tamres, 1973), liquid (melt) and solid phases. In solution there will be interactions with solvent molecules while in the solid longer-range interactions must be taken into account. These additional interactions may be sufficient to stabilize the ionic form as the ground state of the molecular compound.

The overall subject has been extensively discussed (Briegleb, 1949, 1961; Andrews and Kieffer, 1964; Rose, 1967; Foster, 1969; Slifkin, 1971; Soos and Klein, 1976) and the crystallographic aspects reviewed in detail (Prout and Wright, 1969; Herbstein, 1971;

Prout and Kamenar, 1973; Soos, 1974; Bleidelis, Shvets and Freimanis, 1976; Andrè, Bieber and Gautier, 1976; Ponte Goncalves, 1980). A major impetus was given in 1973 by the discovery (Coleman et al., 1973) that $\{[TTF][TCNQ]\}$ had a peak at $\approx 60K$ in its conductivity-temperature curve, with a maximum conductivity of $\approx 10^4 S/cm^{-1}$ (for comparison the conductivity of copper at 293K is $\approx 6 \times 10^5 S/cm^{-1}$). This lead to intense activity for a number of years in the field of "organic metals", many of which are not π-molecular compounds in our present sense.

There is so much material available that we have found it necessary to spread our discussion over five chapters. The first of these, Chapter 13, serves as an overall introduction, with many of the topics dealt with more extensively later. Chapter 14 deals with layered molecules with intramolecular charge transfer interactions, Chapters 15 and 16 with the crystal chemistry and physics of mixed-stack compounds and Chapter 17 with segregated-stack compounds.

In Chapters 13 to 16 the structure determining interaction is A···B leading to mixed-stack arrangements of donors and acceptors. This fits our original definition of molecular compounds (Chapter 1). The structures in Chapter 17 are also stacked but there is a striking difference – donors and acceptors are in separate stacks, hence the term 'segregated' – the structure determining interactions are A–A and B–B, and the materials are not molecular compounds in a strict sense. Nevertheless, it is desirable to keep both groups together for historical reasons and because delocalized charge transfer is important in both.

References

Andrè, J. J., Bieber, A. and Gautier, F. (1976). Ann. Phys., [15] 1, 145–256.

Andrews, L. J. and Keefer, R. M. (1964). Molecular Complexes in Organic Chemistry. Holden-Day, San Francisco.

Bennett, G. M. and Willis, G. H. (1929). J. Chem. Soc., pp. 256–268.

Bleidelis, J. J., Shvets, A. E. and Freimanis, J. F. (1976). J. Struct. Chem., 17, 930–944.

Brackmann, W. (1949). Rec. Trav. Chim. Pays-Bas, 68, 147–159.

Briegleb, G. (1961). Elektronen-Donator-Acceptor-Komplexen. Springer, Berlin-Göttingen-Heidelberg.

Briegleb, G. (1949). Zwischenmolekulkrafte, G. Brown, Karlsruhe.

Coleman, L. B., Cohen, M. J., Sandman, D. J., Yamagichi, F. G., Garito, A. F. and Heeger, A. J. (1973). Solid State Commun., 12, 1125–1132.

Foster, R. (1969). Organic Molecular Complexes. Academic Press, London.

Fritzsche, J. von, (1858). J. Prakt. Chem., 73, 282–292.

Herbstein, F. H. (1971). "Crystalline π-molecular compounds: chemistry, spectroscopy and crystallography" in Perspectives in Structural Chemistry, edited by J. D. Dunitz and J. A. Ibers, Wiley, New York etc., Vol. IV, pp. 166–395.

Lowry, T. M. (1924). Chem. and Ind. (London), 43, 218–221.

Mulliken, R. S. and Person, W. B. Molecular Complexes – a Lecture and Reprint Volume, Wiley, New York, 1969; also in Physical Chemistry (An Advanced Treatise), Vol. III, edited by D. Henderson, Academic Press, New York and London, 1969.

Mulliken, R. S. (1952a). J. Am. Chem. Soc., 72, 600–608.

Mulliken, R. S. (1952b). J. Phys. Chem., 56, 801–822.

Mulliken, R. S. (1956). Rec. Trav. Chim. Pays-Bas, 75, 845–852.

Pfeiffer, P. (1928). Organische Molekülverbindungen, 2nd Edition, Stuttgart: Enke.

Ponte Goncalves, A. M. (1980). *Prog. Solid State Chem.*, **13**, 1–88.

Powell, H. M., Huse, G. and Cooke, P. W. (1943). *J. Chem. Soc.*, pp. 153–157.

Prout, C. K. and Kamenar, B. (1973). *Molecular Complexes*, **1**, 151–207.

Prout, C. K. and Wright, J. D. (1968). *Angew. Chem. Int. Ed. Engl.*, **7**, 659–667.

Rose, J. (1967). Molecular Complexes. Pergamon Press, Oxford.

Slifkin, R. M. (1971). Charge-Transfer Interactions of Bio-Molecules. Academic Press, London.

Soos, Z. G. (1974). *Ann. Rev. Phys.* Chem., **27**, 121–153.

Soos, Z. G. and Klein, D. J. (1976). In Treatise on Solid State Chemistry, edited by N. B. Hannay, Plenum, New York, Vol. 3, 679–767.

Tamres, M. (1973). *Molecular Complexes*, **1**, 49–116.

Weiss, J. (1942). *J. Chem. Soc.*, pp. 245–252.

Wohler, F. (1844). *Ann. Chem.*, **51**, 145–163.

Chapter 13

Charge transfer molecular compounds with delocalized π–π* interaction – introduction and general survey

I now identified the Benesi–Hildebrand spectrum as an "intermolecular charge-transfer spectrum" of a benzene-iodine molecular complex resulting from partial transfer of an electron from the benzene to the iodine partner within the complex, which I called a "Charge-transfer Complex".... my use of the term "charge-transfer complexes" was not really appropriate because in the ground state there is usually only a little electron transfer. On the other hand, the name "charge-transfer state" is highly appropriate for the excited state, in which nearly a complete transfer of an electron has occurred.

R. S. Mulliken, 1985.

Summary: About one hundred years elapsed between the preparation of the first crystalline charge transfer molecular compound with delocalized π–π* interaction (anthracene•••picric acid) and the development of a generally accepted theory, based, in turn, on the experimental demonstration that these compounds crystallized in mixed stacks with alternation of donor and acceptor moieties. The conditions under which the moieties occur as neutral molecules (the predominant situation) or as radical ions have been defined. The properties of a donor–acceptor π-compound depend on the individual properties of the donor and acceptor participants, usually expressed in terms of donor ionization potential and acceptor electron affinity as modified for the difference between the gas and crystalline phases. These crystalline binary sytems should be characterized in terms of their binary phase diagrams, which will show the appearance of D_nA_m compounds, where 'n' and 'm' are usually but not always 1. These principles can be extended to self-complexes, where donor and acceptor portions are situated within the same molecule.

13.1   Introduction and historical development

The purpose of Chapter 13 is to provide physical and chemical background in advance of the following three more detailed chapters. We start with an overall classification that establishes chemical and structural foundations for these three chapters. This is followed by a description of the families of donor and acceptor structures, showing the special features underlying the general description. Among the acceptors we distinguish between the classical group of acceptors, exemplified by picric acid – a choice based on historical considerations – and quasi-acceptors, the majority of which appear to be fluorinated hydrocarbons. In the donor–quasi-acceptor molecular compounds the interaction between components is based on quadrupolar interactions rather than charge transfer interactions. An important consequence is that these molecular compounds are not coloured.

Returning to the classical charge transfer molecular compounds, measures of the strengths of the donor and acceptor interactions are given by the ionization potentials and electron affinities of donors and acceptors respectively; modification of the standard definitions applicable in the gas phase to the solid state are next discussed. We then consider the assessment of the degree of charge transfer encountered in various charge transfer molecular compounds, especially referring to the segregated stack structures. One of our constant themes has been that it is unsatisfactory to base discussion of particular donor–acceptor systems on the crystal structure of the pair obtained by crystallization from solution. Instead, it is desirable to report the phase diagram (at least a melting point diagram, but preferably a more elaborate version) of the system. This is illustrated towards the end of Chapter 13, followed by a brief discursion on ternary systems.

How important have charge transfer compounds been in the standard lexicon of the contemporary organic chemist? We have tried to obtain a rough answer to this question by examining a number of standard post-Second World War organic chemistry texts, all of which have some thousand pages or more. A text that bridges between pre- and post-War periods is that of Fieser and Fieser (1961). These authors allocate just over one page (pp. 694–695) to the topic under the title "Hydrocarbon–polynitro compound complexes". The main features are nicely summarized – the formation of colour on crystallization of the two-component system, exemplified first for picric acid as acceptor and then extended to trinitrobenzene, trinitrotoluene, trinitroresorcinol, 2,4,7-trinitrofluorenone, picryl chloride and picramide. The Fiesers point out the value of some of the 'complexes' "in identification and purification of hydrocarbons and in isolation of hydrocarbons from reaction mixtures because of their superior crystallizing tendency, sparing solubility, and relatively high melting points. The derivatives often have identifying colors . . . " Roberts and Caserio (1964), Hendrickson, Cram and Hammond (1970) and Streitwieser, Heathcock and Kosower (1992) cover much the same ground with similar allocations of space. However, the topic seems to have vanished from two more recent books (Solomons and Fryhle (2000) and Clayden, Greeves, Warren and Wothers (2001)).

A word about nomenclature – for mnemonic reasons we always place the donor molecule first. For mixed stack compounds we join donor and acceptor by the symbol$\cdots$, indicating a π–π* charge transfer interaction, but denote segregated stack complexes as {[donor][acceptor]}, dropping the $\cdots$ symbol to emphasize the difference between the two types.

13.2   Classification

13.2.1   *General considerations*

For the broad group of molecular compounds where the donor–acceptor interaction appears to be structure-determining, a classification can be set up both in structural terms and in terms of the nature of the ground state. However, there is overlap rather than exclusion between these two parts of the definition. Structurally it is possible to distinguish three types: those in which the two components are arrayed in mixed stacks, those where the components are segregated in separate stacks and those where there is an intermediate situation, either quite without stacks or with stackings of limited length. The terms 'heterosoric' have been suggested for mixed-stack structures, 'homosoric' for segregated-stack structures, and 'nonsoric' for those without stacks (Dahm *et al.*, 1975); however, most authors use the English nomenclature, and so shall we. Now the ground state can be neutral or ionic (when the material is formally a salt) and one can further differentiate between salt formation due to electron or to proton transfer. We distinguish among three possible types of charge transfer; which type occurs depends on the nature of the components involved (Fig. 13.1). The first group (Fig. 13.2), where only electron transfer is possible, can be further subdivided.

It seems useful to add three further subgroups for those π-compounds with neutral ground states. The first subgroup includes most π-compounds with a quinone as acceptor, including the quinhydrones; there appears to be a characteristic mode of interaction between aromatic hydrocarbons and quinones (supplemented by hydrogen bonding for the quinhydrones) which is worth emphasizing. The second subgroup is based on all other acceptors except for the quasi-acceptors, which are the basis for the third subgroup. A molecular compound based on a quasi-acceptor is structurally a mixed-stack molecular compound but lacks the characteristic charge-transfer band in its UV-visible spectrum and is not colored. The three subgroups merge into one another without sharply defined boundaries.

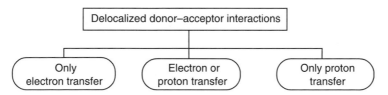

Fig. 13.1. Schematic description of the various types of donor–acceptor compounds with delocalized interactions.

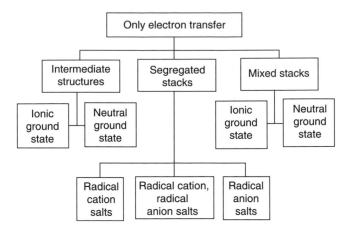

Fig. 13.2. The types of molecular compound found when only electron transfer occurs.

For π-compounds with ionic ground states we distinguish between proton transfer and electron transfer (Fig. 13.3). Only proton transfer occurs when the components are relatively strong acids and bases (e.g. aniline, $pK_b=9.51$; picric acid, $pK_a=2.80$) and here the binary compounds should be classified as organic salts rather than as molecular compounds. However, when the acid and/or base strength is reduced, a particular base–acid (donor–acceptor) pair may form a π-molecular compound (with neutral ground state) at higher temperatures and a salt at lower temperatures. The terms 'base/acid' are to be interpreted in the Lewis sense for the molecular compounds and in the Lowry–Brønsted sense for the salts. For example, 1-bromo-2-naphthylamine and picric acid form red 1 : 1 crystals that melt at 180 °C; these transform on cooling to a yellow compound stable below 117 °C. It has been shown (Carsten-Oeser *et al.*, 1968) that the red crystals are of a π-compound while the yellow compound is presumed to be a salt (π-molecular compounds are denoted '{donor$\cdots$picric acid}' while salts are denoted 'cation' picrate, and similarly for analogous situations). Many other examples were studied in the 1930s by Hertel and his coworkers and have been discussed by Herbstein (1971) (see pp. 187–192). The crystal structures of π-molecular compound and salt are likely to be appreciably different and so most phase transformations between them are likely to be (single crystal) $\rightarrow$ polycrystal (see Section 16.11.3).

Qualitative considerations based on spectral studies (Davis and Symons, 1965) and approximate calculations (McConnell *et al.*, 1965) suggest that crystalline 1 : 1 mixed stack π-molecular compounds are quite sharply divided into two classes, with nominally neutral or ionic ground states respectively. More elaborate calculations (Strebel and Soos, 1970) substantiate these conclusions; the situation in {TTF$\cdots$chloranil} near the neutral-ionic interface, has been considered in detail (Soos *et al.*, 1986) and will be discussed later. The energies of the lowest charge transfer excitations for neutral (N) and ionic (I) compounds are (McConnell *et al.*, 1965):

$$h\nu_{\mathrm{CT}}(N) = (I - A) - \langle e^2/a \rangle$$
$$h\nu_{\mathrm{CT}}(I) = -(I - A) + (2\alpha - 1)\langle e^2/a \rangle$$

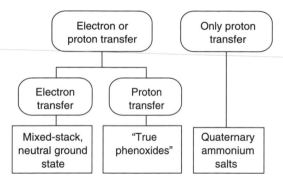

Fig. 13.3. The possible types of molecular compounds that are formed when there is either electron or proton transfer, or only proton transfer. Only those 'proton transfer salts' that have a 'π-compound' polymorph are included here. If, for example, there is hydrogen bonding between cation and anion, then the 'proton transfer salt' would belong more logically in Chapter 12.

where I and A are ionization potential and electron affinity of donor and acceptor respectively (see Section 13.3.5 for discussion of these quantities), $\langle e^2/a \rangle$ is the averaged electrostatic attraction between neighboring donor and acceptor moieties (defined to be positive) and α is the Madelung constant for the structure. These formulae have been tested (Torrance *et al.*, 1981) using the approximation that $(I - A)$, values of which are not always available, may be replaced by $\Delta E_{\mathrm{redox}}$, the difference between the oxidation potential of the donor and the reduction potential of the acceptor. The experimental values of $h\nu_{\mathrm{CT}}$ have been plotted against $\Delta E_{\mathrm{redox}}$ (Fig. 13.4) and lie close to two separate straight lines as predicted. Thus mixed-stack molecular compounds form a class of materials which are either neutral or ionic, but which range from being near to the neutral–ionic boundary to being far away. McConnell *et al.* (1965) suggested that for a particular π-compound the ionic ground state would be favored at high pressures and/or low temperatures. Indeed application of pressure does cause neutral → ionic transitions (Torrance *et al.*, 1981); critical pressures are given in Table 13.1. Furthermore {TTF···chloranil} has been shown to undergo a neutral → ionic phase change at 84K (Batail *et al.*, 1981). General forms can be suggested for the temperature–pressure phase diagrams of mixed-stack molecular compounds, based on the results reported for both electron-transfer and proton-transfer systems (Fig. 13.5). The crystal chemistry of mixed stack compounds, irrespective of the nature of the ground state, is described in Chapter 15 and some of their physical properties in Chapter 16.

The segregated stack compounds can be divided into two chemical groups. In those with radical cations and closed-shell anions there is complete transfer of an electron from cation to anion; this is dictated by the requirement that the discrete closed-shell anions bear an integral charge. The same argument applies to closed-shell cation, radical anion salts. The radicals are often arranged in stacks in both types of salt. In the so-called simple salts (e.g. K·TCNQ; Würster's blue perchlorate TMPD·ClO₄, both of which have stacked structures) each radical ion accepts (or loses) one electron; many different types of interaction between these electrons (holes) are possible along the stacks. In the so-called complex salts (e.g. $Cs_2(TCNQ)_3$) the stacks contain both radical ions and neutral molecules, which can sometimes but not always be distinguished in the crystal structures, there

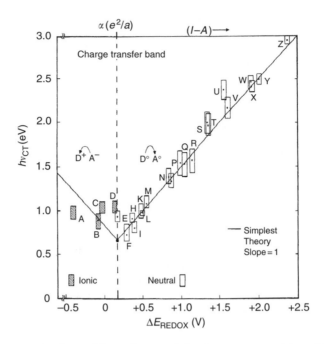

Fig. 13.4. Experimental values of $h\nu_{CT}$ (measured for the π-compound in a KBr disk) plotted against ΔE_{redox} (defined in the text). The scale of ΔE_{redox} is shifted with respect to that of $(I - A)$ in order to take solvation energy into account. The dashed line is the neutral–ionic boundary. The compounds are identified in Table 13.1. (Reproduced from Torrance *et al.*, 1981.)

being generally extensive delocalization of the electrons or holes along the stacks at moderate temperatures. A new feature enters with radical cation, radical anion salts of equimolar composition, which contain segregated stacks of radical cations and of radical anions very similar to those found individually in the radical-ion closed-shell ion salts. The requirement for transfer of charge in *discrete* overall amounts from radical-cation stack to radical-anion stack no longer holds. This is because of the possibility of delocalization of charge between radical ions and neutral molecules within the stacks. This confers varied and interesting properties on these compounds including, as noted earlier, some remarkably high electrical conductivities. The characteristic feature is a marked anisotropy and these materials are often described as quasi-one dimensional. Structures and physical properties are discussed in Chapter 17.

Finally there is a miscellaneous group of compounds, as yet limited in number, where the mixed stack or segregated stack description does not fit very well. There are often incipient stacks, either mixed or segregated, but the anisotropy associated with essentially-infinite stacks is lost. These examples are considered at appropriate points within the coming chapters.

13.2.2 *Intramolecular π-compounds and self-complexes*

Up to this point we have considered binary systems where one component is a donor and the second an acceptor. However, there is a group of compounds where donor and

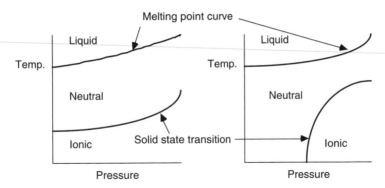

Fig. 13.5. Schematic representations of the *T–P* phase diagrams showing possible fields for the occurrence of neutral and ionic ground states for both "isomeric complexes" (proton transfer) and π-compounds (electron transfer). In the type of phase diagram on the left there will be an enantiotropic[1] I → N transition, while for the type shown on the right there will be a I → N transition only at higher pressures and the relationship between the two phases will be monotropic, with possible transitions above atmospheric pressure. In a third type of phase diagram (not shown) the I–N phase boundary occurs above the melting point for a range of pressures and the material is ionic at all temperatures in this pressure range; the form of the phase diagram may be more complicated at other pressures.

acceptor properties are located in different parts of the *same* molecule. These are called self-complexes (the term 'autocomplexes' is sometimes used, especially in translations from the Russian literature). Two types may be distinguished. The first are molecules based on a cyclophane skeleton where one of the interacting units is a donor and one an acceptor. The advantage of working with molecules of this type is that donor and acceptor can be brought into propinquity in the same molecule in a fixed (and determinable) orientation relationship; different mutual orientations of the same donor–acceptor pair can be obtained by synthesis of suitable cyclophanes. The interaction is thus intramolecular, although intermolecular interactions may also occur between different molecules in a crystal. This controlled and defined intramolecular interaction should be contrasted with that normally occurring in solution, where it is probable that a range of orientations exists (Matsuo and Higuchi, 1968), or in the solid, where the donor–acceptor orientation depends on the overall crystal structure and where the interaction is generally within stacks rather than between pairs. These 'intramolecular π-compounds' are discussed in detail in Chapter 14 under the title "Layered molecules".

The second group comprises those molecules made up of linked donor and acceptor portions, the molecule as a whole being approximately planar (Bleidelis *et al.*, 1976). Intramolecular charge transfer is possible when donor and acceptor portions are linked by

---

[1] In a pair of enantiotropically related polymorphs one is stable over a particular temperature range (at a particular pressure, e.g. 1 atm.) and transforms into the other outside this temperature range but below the melting point; the transformation will be reversible (in theory, if not necessarily in practice). The details of the relationship will depend on pressure. In a pair of monotropically related polymorphs one is stable (at a particular pressure, e.g. 1 atm.) over the whole temperature range up to its melting point, and the other metastable. However, the metastable polymorph can often be obtained (cf. Ostwald's rule) and may transform to the stable form below the melting point; the transformation is irreversible. The details of the relationship will depend on pressure.

Table 13.1. Auxiliary information about the molecular compounds included in Fig. 13.4 (for acronyms see Table 13.2)

Symbol	Compound	Neutral	ΔE_{redox}/ Ionic	$h\nu_{CT}$(eV)	P_c (kbar)	Stacking
A	TMPD···TCNQF$_4$	I	−0.42	0.96		?
B	Dimethylphenazine ···TCNQ	I	−0.08	0.89		Mixed
C	TMPD···TCNQ	I	−0.05	1.04		Mixed
D	TMPD···chloranil	I	0.13	1.03		Mixed
E	TMDAP···TCNQ	N	0.17	0.93	12	Mixed
F	TTF···chloranil	N	0.29	0.73	8	Mixed
G	TTF···fluoranil	N	0.34		9	Mixed
H	DibenzTTF···TCNQ	N	0.36	0.91	13	?
I	DEDTMSeF··· diethylTCNQ	N	0.39	0.80	5	?
J	TMDAP···fluoranil	N	0.4		19	?
K	TTF··· dichlorobenzoquinone	N	0.48	0.99	21	Mixed
L	perylene···TCNQF$_4$	N	0.5	0.97	11	?
M	perylene···DDQ	N	0.55	1.08		?
N	perylene···TCNE	N	0.84	1.38	18	Mixed
O	perylene···TCNQ	N	0.87	1.33		Mixed
P	TTF···dinitrobenzene	N	0.99	1.55		Mixed
Q	perylene···chloranil	N	1.05	1.53	35	?
R	pyrene···TCNE	N	1.14	1.57		Mixed
S	pyrene···chloranil	N	1.35	2.00		?
T	anthracene···chloranil	N	1.36	1.97		?
U	HMB···chloranil	N	1.57	2.36		Mixed
V	naphthalene···TCNE	N	1.62	2.17		Mixed
W	pyrene···PMDA	N	1.91	2.47		Mixed
X	anthracene···PMDA	N	1.92	2.42		Mixed
Y	anthracene···TCNB	N	2.01	2.50		?
Z	phenanthrene···PMDA	N	2.38	2.95		Mixed

a conjugated system. Intermolecular donor–acceptor interaction is possible in the crystal when the donor and acceptor portions of the self-complex are suitably oriented; for example in 1-phenyl-2-(methylthio)vinyl-2,4,6-trinitro-benzenesulphonate the interaction is pairwise (Meyers and Trueblood, 1969) while in 2-methyl-3-(N-methyl-anilinomethyl)-1,4-naphthoquinone donor–acceptor stacking occurs (Prout and Castellano, 1970). This area is discussed in more detail in Chapter 15.

13.3   Chemical nature of donors and acceptors

13.3.1   *Introduction*

Which molecules act as electron donors and which as electron acceptors in π-molecular compounds? How is donor or acceptor quality measured and how is it enhanced or

Table 13.2. Acronyms in common use

Electron donors		Electron acceptors	
TDAE	tetrakis(dimethylamino)-ethylene	BAQ	2,5-bis(methylamino)-p-benzoquinone
DMA	N,N-dimethylamine	BTF	benzotrifuroxan
HMB	hexamethylbenzene	DDQ	2, 3-dichloro-5,6-dicyano-p-benzoquinone
PD	p-phenylenediamine	DEQ	2,5-diethoxy-p-benzoquinone
DAD	durenediamine	HCBD	hexacyanobutadiene
TAB	1,3,5-benzenetriamine	PMDA	pyromellitic dianhydride
TMPD	N,N,N′,N′-tetramethyl-p-phenylenediamine	TCNB	1,2,4,5-tetracyanobenzene
TTF	tetrathiafulvalene	TCNE	tetracyanoethylene
TTT	tetrathiotetracene	TCNQ	7,7,8,8-tetracyanoquinodimethane
TSF	tetraselenofulvalene	TCPA	tetrachlorophthalic anhydride
TMTSF	tetramethyltetraselenofulvalene	TNB	1,3,5-trinitrobenzene
BP	3,4-benzopyrene	TNT	2,4,6-trinitrotoluene
DBA	1,2,5,6-dibenzanthracene (dibenz[a,h]anthracene)	TNP (PA)	2,4,6-trinitrophenol (picric acid)
		TMU	1,3,7,9-tetramethyluric acid
		TENF	2,4,5,7-tetranitro-9-fluorenone
		TNF	2,4,6-trinitro-9-fluorenone
		TNAP	11,11,12,12-tetracyanonaphtho-quinodimethane

diminished by various substituents? Theory suggests that a good donor has a low ionization potential while a good acceptor has a high electron affinity. Planarity of both donor and acceptor would be expected to enhance effective electron transfer. The broad effects of substituents have been known for many years (Martinet and Bornand, 1925); electron-donating substituents increase donor strength, and electron-withdrawing substituents diminish it, and conversely for electron acceptors. Nature, number and location of substituents are all important factors. There have been a number of qualitative studies of these questions and some results are summarized below; a quantitative approach would require measurement of thermodynamic parameters for the various molecular compounds but not enough information is yet available to permit a quantitative discussion.

Donor and acceptor are really relative terms since a particular compound can act as a donor in some circumstances and as an acceptor in others. An extreme example is provided by aromatic hydrocarbons which usually behave as reasonably strong electron donors, but act as electron acceptors in the presence of the very powerful donor tetrakis (dimethyl-amino)ethylene (Wiberg, 1968).

Acronyms for some donors and acceptors are given in Table 13.2.

13.3.2 *Donors*

Various types are shown in Fig. 13.4. Aromatic hydrocarbons are the classic donor type; indeed formation of molecular compounds between aromatic hydrocarbons and picric acid or *sym*-trinitrobenzene provided for a long time the most widely used laboratory method for identification and purification of aromatics. Shinomiya (1940a) found the following order for the ability of aromatic hydrocarbons to form molecular compounds with polynitrobenzenes:

> naphthalene, pyrene, fluoranthene > acenaphthene > benzene, phenanthrene > fluorene > anthracene, chrysene.

Complete planarity of the hydrocarbon is not essential as the acceptor can usually find a reasonably planar region with which to interact; many π-molecular compounds are formed by helicenes and substituted helicenes, e.g. hexahelicene···4-bromo-2,5,7-trinitrofluorenone (Mackay *et al.*, 1969); trithia[5]heterohelicene···TCNQ (Konno *et al.*, 1980).

Electron donating substituents such as alkyl groups (if not too bulky), methoxy groups and, particularly, amino and dimethylamino groups considerably enhance the donor strengths of aromatic hydrocarbons. Tetramethylphenylenediamine (TMPD) is a powerful donor, forming molecular compounds with ionic ground states with acceptors such as chloranil and TCNQ. 2,7-Diaminopyrene is also a strong donor with a tendency to give molecular compounds with ionic ground states. Some qualitative studies have been made of the effects of various substituents in naphthalene on compound formation with various polynitrobenzenes (Shinomiya, 1940b). The following sequence was found for the effect of a substituent (in a given position in the naphthalene ring) on molecular compound formation:

$$NH_2 > CH_3 > OH > C_2H_5 > OCH_3 > Cl > Br > OC_2H_5 > H > COOH$$
$$> COOCH_3 > OC_6H_5 > CN > COOC_6H_5 > NO_2.$$

Substitution in the 1-position tended to favor molecular compound formation while substitution in the 2-position sometimes hindered. Somewhat similar comparisons have been made for the picric acid compounds of polymethylbenzenes and of poly-hydroxybenzenes and naphthalenes, and of some aromatic hydrocarbons (Baril and Hauber, 1951). For benzene derivatives the tendency to compound formation is greater the larger the number of methyl substituents and the more symmetrical their disposition in the ring. Methyl groups were found to be even more effective in side chains than in the ring, a perhaps surprising result. Unsaturation in the side chains produced very explosive picrates! However, it should be appreciated that these considerations are intrinsically qualitative as they rest on the formation of *crystalline* molecular compounds, without taking into account their structures, which may be very varied.

Polycyclic hydrocarbons forming molecular compounds can be divided into the following groups:

(i)   Planar aromatic hydrocarbons, e.g. benzene, naphthalene, anthracene, phenanthrene, pyrene, perylene.
(ii) Nonplanar aromatic hydrocarbons, e.g. the helicenes, heterohelicenes, 9,10-dihydroanthracene.
(iii) Substituted aromatic hydrocarbons, e.g.

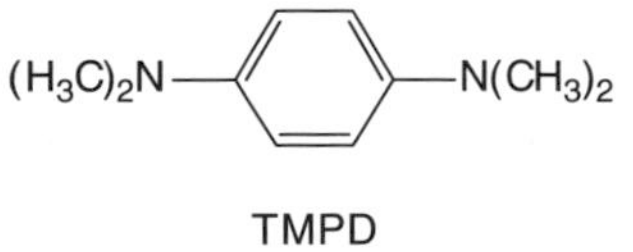

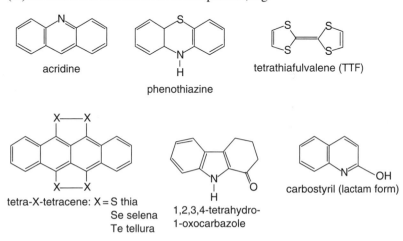

TMPD

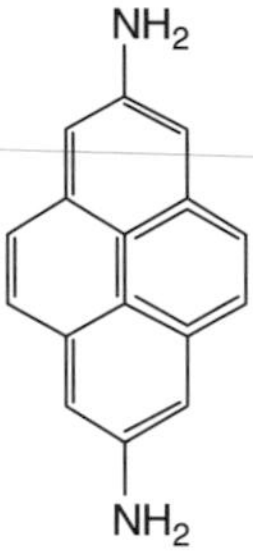

2,7-diaminopyrene

(iv) Heteroaromatics and related compounds, e.g.

acridine

phenothiazine

tetrathiafulvalene (TTF)

tetra-X-tetracene: X = S thia
                    Se selena
                    Te tellura

1,2,3,4-tetrahydro-
1-oxocarbazole

carbostyril (lactam form)

Heteroaromatic donors, mainly with nitrogen and oxygen substitution, have been known for many years while sulphur-containing molecules have become very important recently. Tetrathiafulavalene (TTF) and derivatives act as donors in many radical-cation salts and, of course, in {[TTF][TCNQ]} (see Chapter 17 for formulation), arguably the most-widely studied of the 'organic metals'. The effects of chemical modification of TTF will be illustrated later (Chapter 17). Another important donor containing sulphur is tetrathiatetracene; Se and Te analogs have also been prepared (Sandman *et al.*, 1982).

Among other donors are various metallocenes, including ferrocene, dibenzene chromium and tricarbonylchromium anisole, porphyrins and coordination complexes (e.g. metal oxinates with Cu(II), Pd(II) and Ni(II) as the metal atoms).

13.3.3 *Acceptors*

Acceptors are conveniently grouped on the basis of fundamental structures to which substituents have been appended in various ways. The fundamental structures are ethylene, *p*- and *o*-benzoquinone, cyclopentadienone, benzene, naphthalene and fluorene. The most important electron-withdrawing substituents are nitro, cyano, halo, anhydride, furoxan and furazan groups. We give below a selection of proven and potential acceptors.

Synergistic interaction of various substituents seems to be an important factor in the formation of electron acceptors. For example, maleic anhydride, phthalic anhydride and hexachlorobenzene do not form stable π-compounds with aromatic hydrocarbons while tetrachlorophthalic anhydride (TCPA) is a powerful electron acceptor. Unsymmetrical substitution, as in DDQ, enhances the effectiveness of electron acceptors, presumably because dipole and polarization interactions also contribute to the stability of the molecular compounds, in addition to the charge transfer interactions.

(a) Substituted olefins

Examples are TCNE, TCNQ and hexacyanobutadiene (HCBD). Of these TCNQ is undoubtedly the most prominent and occurs not only as the radical anion in many charge transfer salts and organic metals but also as a neutral acceptor in many mixed stack π-molecular compounds (for some other formulae see Section 17.2.3).

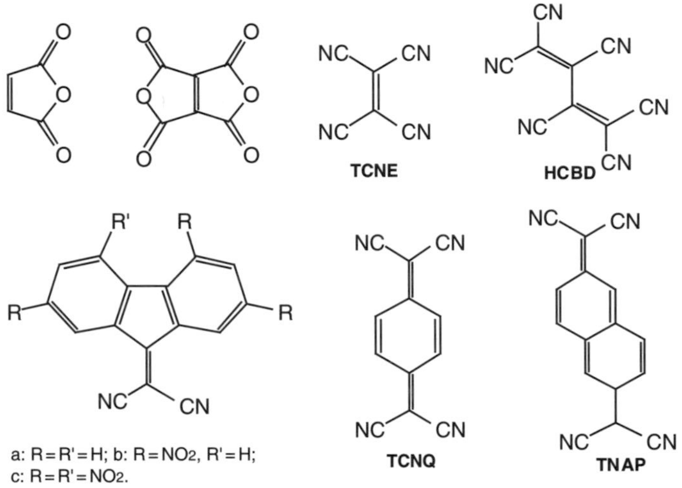

TCNE   HCBD

a: R = R' = H; b: R = NO2, R' = H;
c: R = R' = NO2.   TCNQ   TNAP

(b) Substituted aromatic hydrocarbons

Substituted polynitrobenzenes (e.g. TNB, picric acid) are the best known members of this group but polynitro naphthalenes and fluorenes also form molecular compounds. Anhydrides of benzene and naphthalene polycarboxylic acids are important acceptors, as well as polycyanobenzenes.

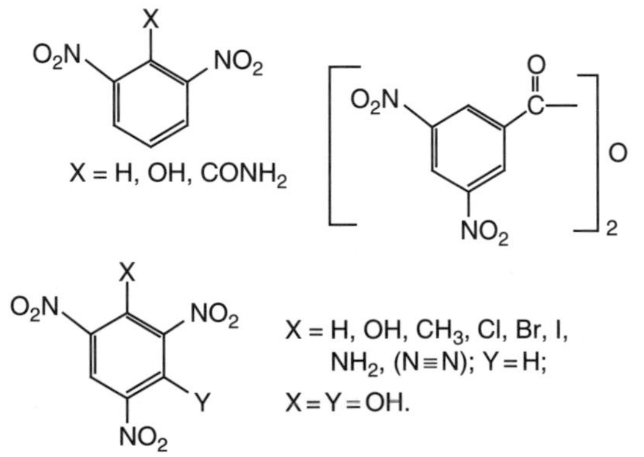

X = H, OH, CONH2

X = H, OH, CH3, Cl, Br, I,
    NH2, (N≡N); Y = H;
X = Y = OH.

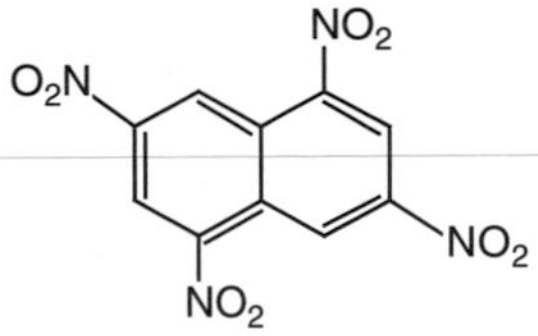

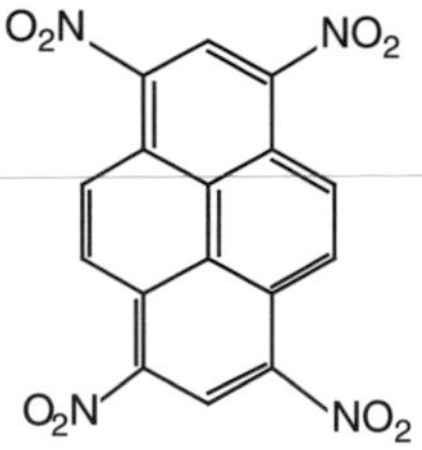

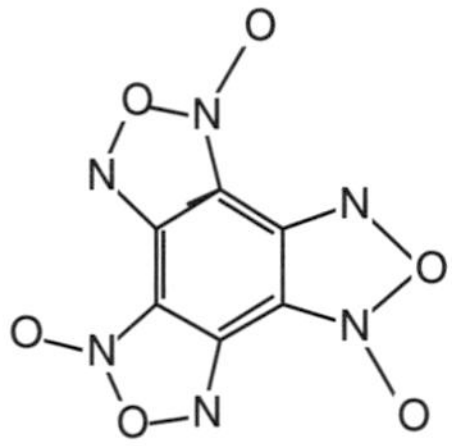

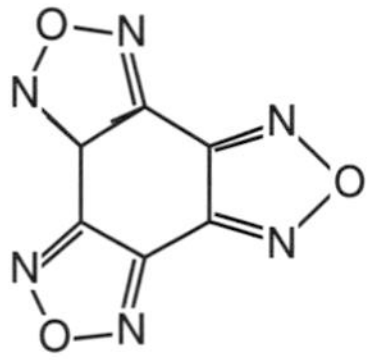

benzotrifuroxan (BTF).

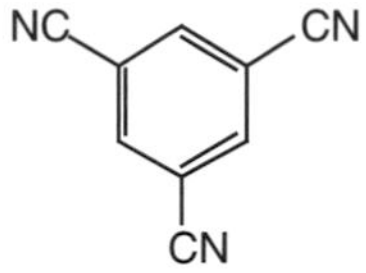

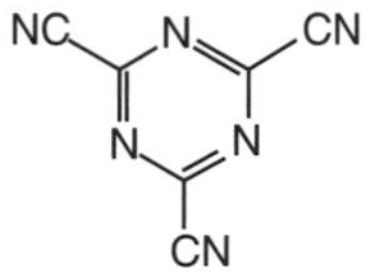

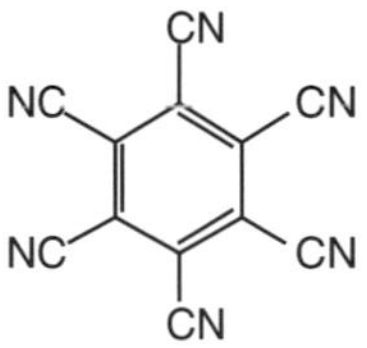

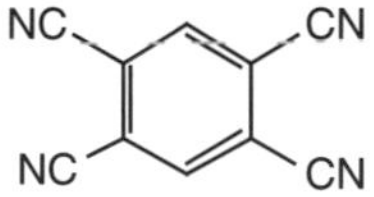

TCNB

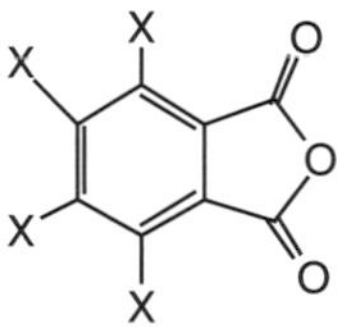

X=H, F, Cl, Br or I
X=Cl: TCPA

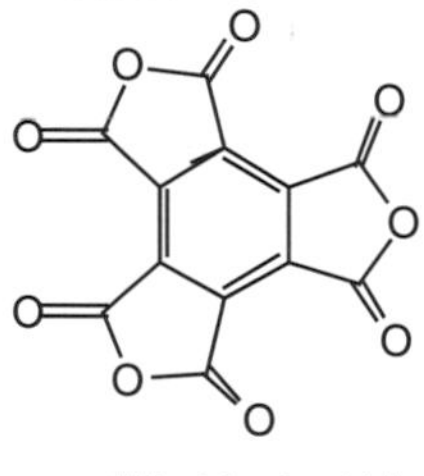

mellitic trianhydride

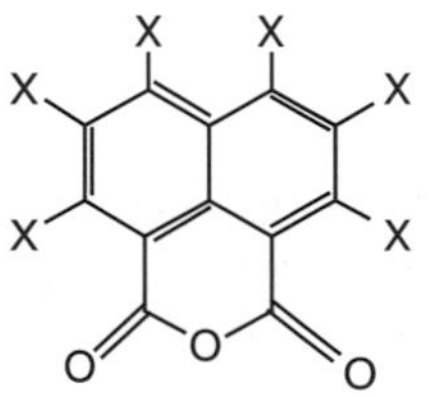

various anhydrides

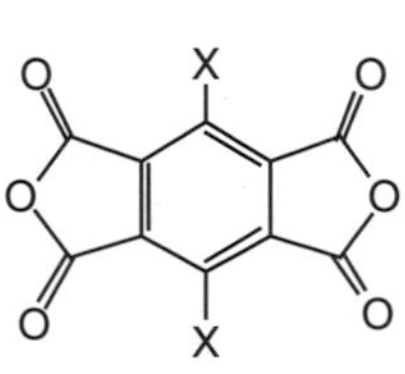

X=H: PMDA

(c) Unsubstituted and substituted quinones

Most attention has been paid to the tetra-substituted p-benzoquinones (see also Section 17.2.3 (v)), the less stable o-benzoquinones having been largely ignored; fluorenones, naphthoquinones, phenanthraquinones and anthraquinones have also been studied. Two vicinal triones are relatively new, powerful acceptors and have been shown to form 1:1 molecular compounds with pyrene (Gleiter and Schanz, 1980). We have already noted that the quinhydrones (Section 15.7.1), and {aromatic hydrocarbon···quinone} molecular compounds (Section 15.6) are best treated as separate groups.

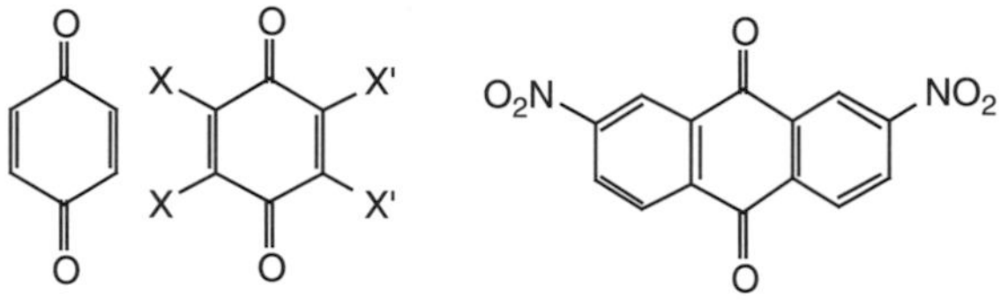

a: X=X'=F, Cl, Br, I, CN or CH$_3$.
b: X=Cl; X'=CN.

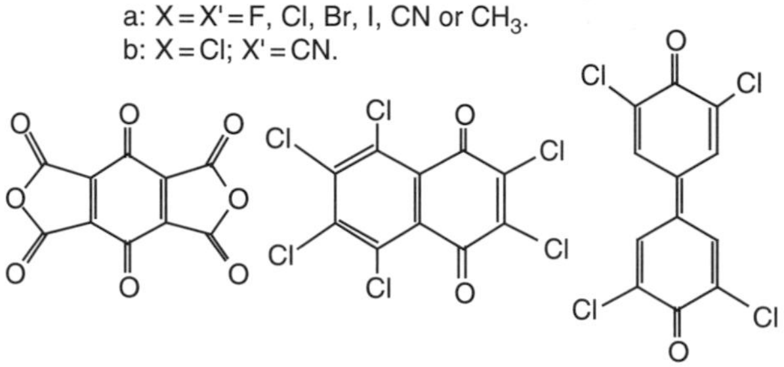

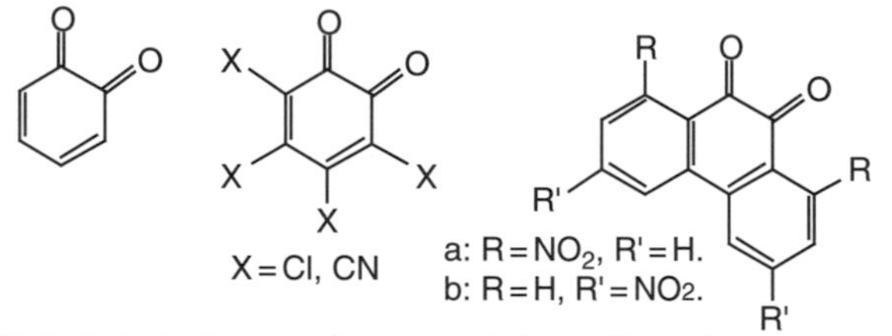

a: R=NO$_2$, R'=H.
b: R=H, R'=NO$_2$.

X=Cl, CN

Substituted o-benzoquinones and phenanthraquinones

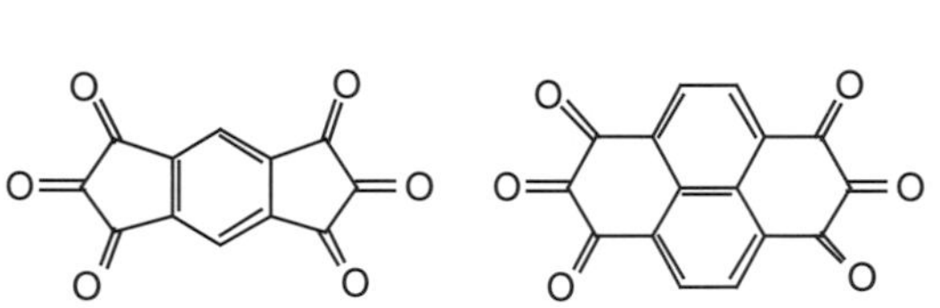

1,2,3,5,6,7-s-hydrinacenehexone   1,2,3,6,7,8-pyrenehexone

(d) Cyclopentadienone systems

The last of the molecules shown in this section is of particular interest because it contains an asymmetric carbon atom (asterisked); the presence of the carboxyl group allows resolution into enantiomers which can be used for the formation of diastereoisomeric π-complexes with nonplanar overcrowded aromatic hydrocarbons. The resolved aromatics can then be recovered (Newman and Lutz, 1956).

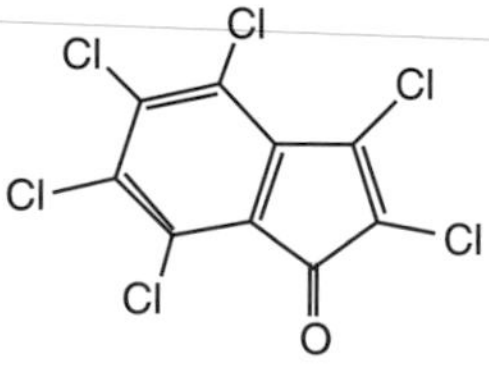

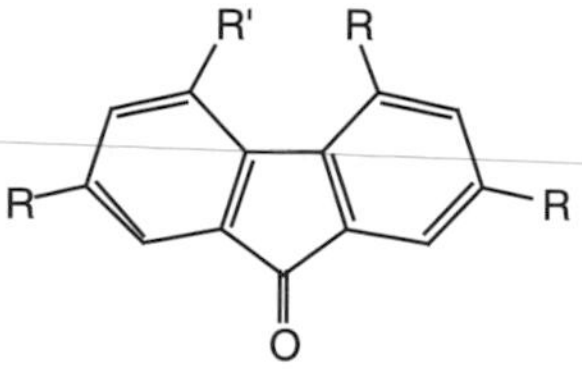

a: R=R'=H;
b: R=NO$_2$, R'=H;
c: R=R'=NO$_2$.

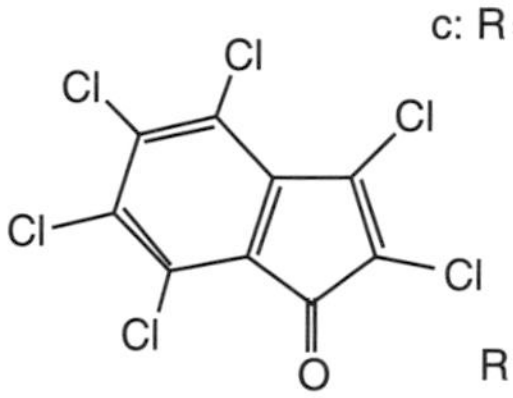

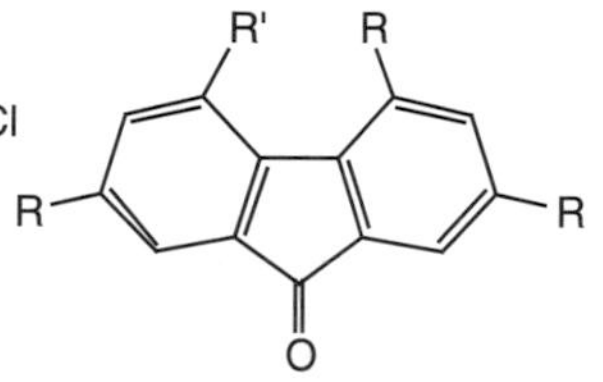

a: R=R'=H;
b: R=NO$_2$, R'=H;
c: R=R'=NO$_2$.

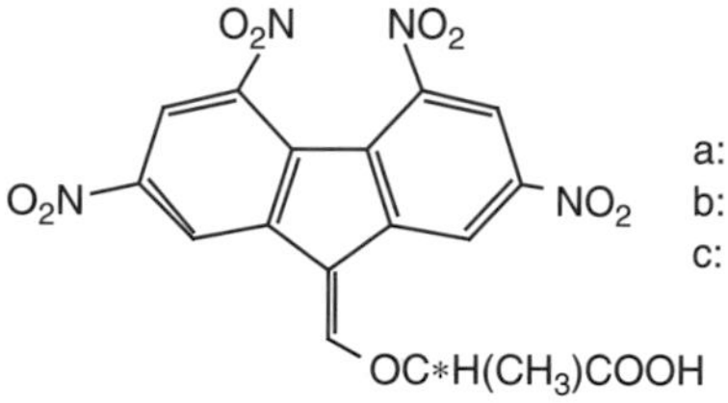

OC*H(CH$_3$)COOH

(e) Chelates

Chelates act as acceptors in some π-molecular compounds; examples are bis(di-fluoroborondimethylglyoximato)Ni(II) (on the left), metal bis(dithiolene) (on the right, where M = Ni, Pd, Pt) and bis(cis-1,2-trifluoro- methylethylene-1,2-dithiolato)Ni(II). The chemistry of the dithiolenes and related species and their metal complexes has been comprehensively reviewed (Mueller-Westerhoff and Vance, 1987).

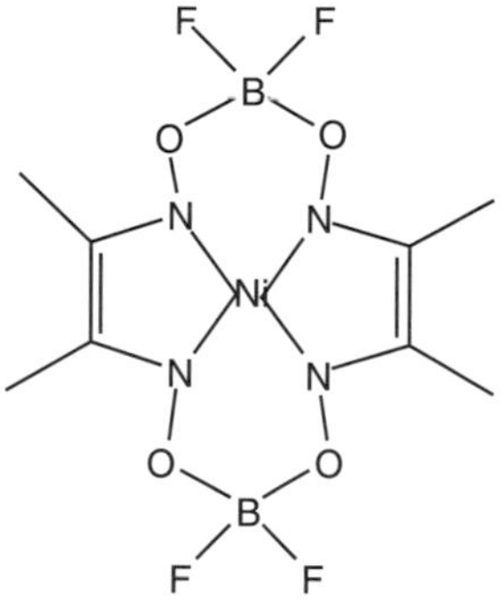

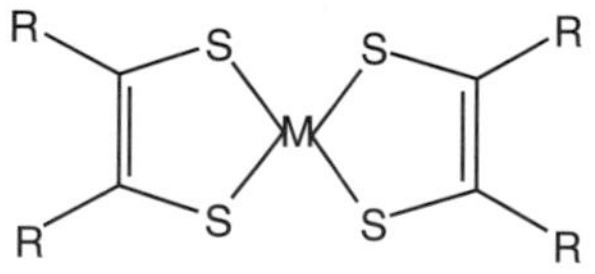

13.3.4 *Quasi-acceptors*

Three groups of molecular compounds are known which are structurally similar to the mixed stack π-donor–acceptor molecular compounds but do not show charge transfer bands in their absorption spectra. The donors are regular donors of the π-compound series and thus any difference in properties must derive from the second component, which we call a quasi-acceptor.

The first quasi-acceptor type is a halogenated aromatic hydrocarbon; the perfluorinated molecules hexafluorobenzene, octafluoronaphthalene and decafluoro-biphenyl are the best known examples, especially C_6F_6, and the structures of their molecular compounds will be described later (Section 15.9.1). Decachloropyrene perhaps behaves in a similar way but other examples are not known.

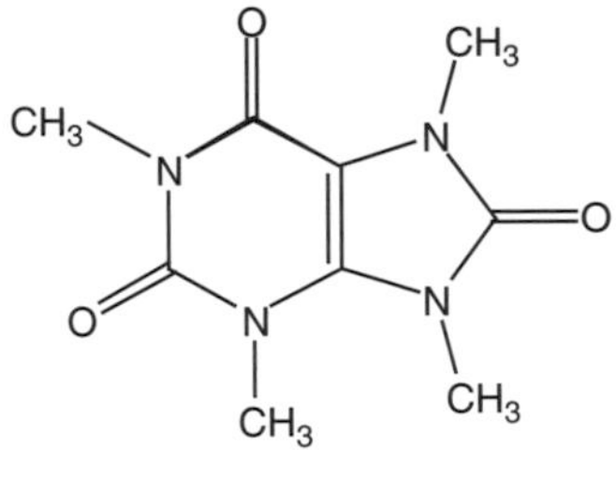

1,3,5,7-tetramethyluric acid (TMU) 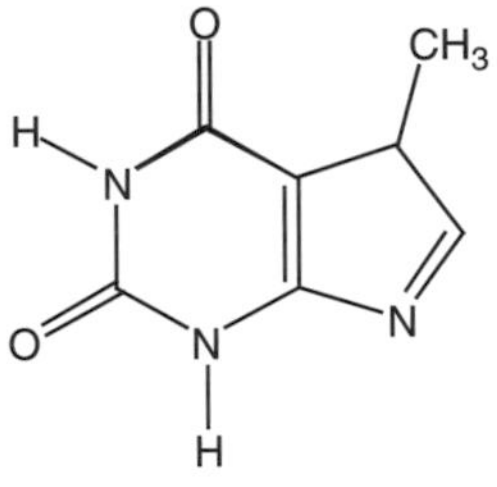   caffeine

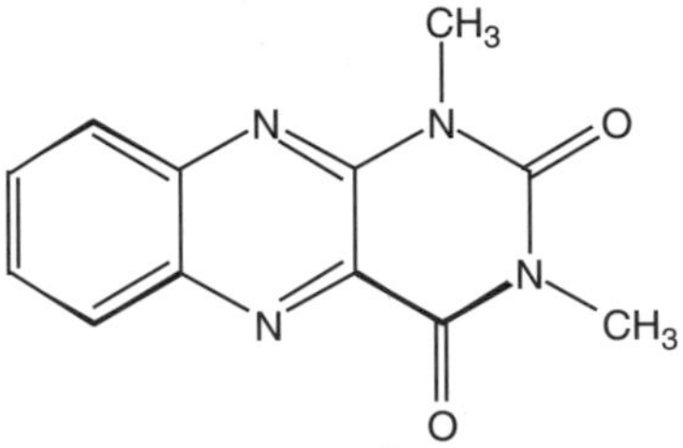

1,3-dimethylalloxazine

The second type is based on the purines 1,3,7,9-tetramethyluric acid and caffeine; some pyrimidines may play a similar role. Mixed stack molecular compounds of aromatic hydrocarbons and heteroaromatics with TMU and caffeine have been prepared (Weil-Malherbe, 1946; Booth and Boyland, 1953) and some crystal structures are known (Section 15.9.2). These have mixed stack structures but there is no charge transfer band in solution or solid state spectra. Pyrene, TCNE and DDQ form 1 : 2 molecular compounds with 1,3-dimethylalloxazine but the varying nature of the first component suggests that further study is needed before these can be reliably classified.

The third group is based on the flavins as quasi-acceptors. Various iosalloxazine derivatives have been reported to form molecular compounds of different types with electron donors such as phenols. The chemistry and structures of these molecular compounds are discussed in Section 15.7.2.

13.3.5 *Ionization potentials of donors and electron affinities of acceptors*

The first ionization potential (or energy) of a molecule is defined as the energy which is required to remove an electron from the highest occupied molecular orbital (HOMO) of the neutral molecule in its ground state and is denoted I_1. It is thus the enthalpy of the reaction

$$D \rightarrow D^+ + e \text{ (infinitely separated)}.$$

The electron affinity (EA) of a molecule is defined as the difference in energy between the ground state of the neutral molecule plus an electron at rest at infinity, and that of the negative ion. It is thus the enthalpy of the reaction

$$A + e \rightarrow A^-$$

The electron affinity of a stable negative ion is defined as positive, although this contradicts the usual thermochemical sign convention.

The difference in energy between the lowest vibrational level of the ground state of the neutral molecule and the corresponding level of the cation or anion is termed the 'adiabatic ionization energy', or 'adiabatic electron affinity', respectively. The energy difference between the energy level of the ground state and that part of the potential curve to which, on applying the Franck–Condon principle, transition is most likely to occur is denoted the 'vertical ionization energy' or the 'vertical electron affinity' as the case may be. Photoelectron spectroscopy, for example, gives the vertical ionization energy.

Experimental and calculated values of ionization potentials and electron affinities have been tabulated (Blaunstein and Christopherou, 1971; Rosenstock et al., 1980). Reasonably accurate values of ionization potentials have been available for many materials for some time; in the last few years many new values have been obtained and there has been a marked increase in accuracy stemming from widespread application of photoelectron spectroscopy (Eilfeld and Schmidt, 1981). Selected values for donors of interest here are given in Table 13.3.

There has been considerable confusion about the correct absolute values of electron affinities and two rather different sets of values were proposed at one point (Briegleb, 1964; Batley and Lyons, 1962). However, these can be correlated and put on an absolute scale (Chen and Wentworth, 1975) and absolute electron affinity values have been calculated for about 150 electron acceptors from measurements of charge transfer spectra and half-wave potentials (Table 13.4, which also includes some more recent values).

It is illuminating to compare typical values of I_1 and EA for organic donors and acceptors with those for alkali metals and halogens. For a typical organic donor $I_1 \approx 7$ eV, whereas for Li $I_1 = 5.4$ eV and for Cs $I_1 = 3.4$ eV; for the best organic acceptors EA≈ 2.8 eV, while EA for fluorine is 3.4 eV and for iodine 3.2 eV. Thus it is somewhat more difficult to form cations from organic donors than from alkali metal atoms, and also somewhat less advantageous to form anions from organic acceptors than from halogen atoms.

The values of ionization potential and electron affinity given in Tables 13.3 and 13.4 refer to isolated gaseous molecules and we shall denote them as I_G and EA_G respectively. However, the appropriate ionization potentials for molecules in crystals (I_C) will be less than those for isolated molecules because of polarization effects in the crystals. Similarly the electron affinity of a molecule in a crystal (EA_C) will be greater than that of a molecule in the gas phase because of polarization. Thus we can write $I_C = I_G - P$, where P is the polarization energy; similarly $EA_C = EA_G + P^*$, where both P and P^* are positive quantities. P and P^* depend on the polarizabilities of the molecules involved as well as on the crystal structures. Gutmann and Lyons (1981) have given a detailed discussion; the following values (Table 13.5) for P (for a single charge) are quoted from them and other sources. A comprehensive set of values of P has been determined by ultraviolet photoelectron spectroscopy (examination of line widths of the spectra) for 44 organic

Table 13.3. First π-ionization potentials (eV) for some aromatic hydrocarbons and other molecules. Values from Schmidt (1977) unless stated otherwise. Method is photoelectron spectroscopy (PES) unless stated otherwise

Molecule	I_1	Molecule	I_1	Molecule	I_1
benzene	9.24	Perylene (CS77)	6.97	Coronene (BL62)	7.29
naphthalene	8.15	Ovalene (BL62)	6.71	Diphenyl (R81)	7.95
	8.12 (adiabatic) (SSI81)				
anthracene	7.41	Biphenylene (BCS74)	7.61	hexamethyl-benzene (R80)	7.8 (CTS)
	7.36 (adiabatic) (CBS72)				
tetracene	6.97	Tetrathiofulvalene (R80)	6.83	TDAE	6.5
pentacene	6.61	Graphite (BL62)	4.9	N,N-dimethyl-p-phenylenediamine (BCS74)	6.28 (ETR)
phenanthrene	7.86	chrysene	7.59	1,2,7,8-dibenzochrysene	7.20
triphenylene	7.88	benzo[c]phen-anthrene	7.60	Pyrene (CS79)	7.41
pheno-thiazine[1]	6.54 (ETR)	N,N,N′,N′-tetramethyl-benzidine (CS77)	6.48 (ETR)		

Notes: PES photoelectron spectroscopy (molecule in gas phase); accuracy ± 0.02 eV for $I < 10$ eV; vertical ionization potential is measured. CTS charge transfer spectrum (molecule in solution). ETR electron transfer reactions (molecule in solution); adiabatic ionization potential is measured.

References:
B74 – Boschi *et al.*, 1974; BL62 – Batley and Lyons, 1962; C72 – Clark *et al.*, 1972; CS77 – Clar and Schmidt, 1977; CS79 – Clar and Schmidt, 1979; R80 – Rosenstock *et al.*, 1980; SSI81 – Sato *et al.*, 1981.

solids (Sato, Seki *et al.,* 1981); the values of P ranged from 0.9 to 3.0 eV. Planar condensed aromatic hydrocarbons appeared to have a common value of 1.7 eV. These values refer to the crystals of the pure components and somewhat different values would be expected to apply to these molecules in the crystals of the charge transfer molecular compounds. Measurements of photoelectric emission thresholds for neutral and ionic charge transfer compounds give values of P for anthracene and perylene in some of their molecular compounds as well as in their neat crystals (Batley and Lyons, 1968) and these values are also included in Table 13.5. Taken at face value these numbers would appear to mean that the donor qualities of anthracene are enhanced (by 0.2–0.4 eV) in its PMDA and TCNQ charge transfer molecular compounds compared to its behaviour in its neat crystals while those of perylene are diminished in its chloranil and PMDA π-compounds.

Analogous values for electron acceptors (P*) are hardly available as measurement of the solid state electron affinity is very difficult. Nevertheless the electron affinity of PMDA in PMDA molecular compounds has been estimated as 3.6(3) eV (Stezowski *et al.*, 1986).

Table 13.4. Electron affinities (eV). Most of the values have been taken from the more extensive collection in Table III of Chen and Wentworth (1975); these are charge transfer spectrum (CTS) values except where noted otherwise (PDS photodetachment spectra, $E_{1/2}$ half wave potentials)

Molecule	EA	Molecule	EA	Molecule	EA
p-benzoquinone	1.83	Hexafluorobenzene (SH80)	1.09 (PDS)	2,4,7-trinitro-fluoren-9-one	2.17
o-bromanil	2.62	o-chloranil	2.57	2,4,5,7-tetranitro-fluoren-9-one	2.24 ($E_{1/2}$)
p-bromanil	2.48	TCNB	2.00	2,4,5,7-tetranitro-fluorenemalononitrile	2.56
DDQ (CS77)	3.13	hexacyanobenzene	2.56	TCNE	2.77
p-chloranil	2.48	p-fluoranil	2.45	TCNQ	2.84
p-iodanil	2.43 ($E_{1/2}$)	hexacyanobutadiene	3.09	tetrafluoro-TCNQ	3.22 ($E_{1/2}$)
duroquinone	1.67	2,3-dicyano-5,6-dicyano-7-nitro-1,4-naphthoquinone	2.82	2,5-difluoro-TCNQ (SF79)	3.02
tetranitromethane	1.63	TNB	1.73	tetrabromophthalic anhydride	1.72
TNT	1.67	TCPA	1.72	PMDA	2.04
1,4,5,8-naphthalene tetracarboxylic acid anhydride	2.28	mellitic trianhydride	2.38	dibromo-PMDA	2.23

References:
CS77 – Clar and Schmidt, 1977; SF79 – Saito and Ferraris, 1979; SH80 – Sowada and Holroyd, 1980.

Table 13.5. Values of polarization energy P (in eV) for aromatic hydrocarbon molecules in their neat crystals (values from Gutmann and Lyons (1981), except where noted otherwise) and in some π-molecular compounds (values from Batley and Lyons (1962), except where noted otherwise)

Molecule in its neat crystals	P	Molecule in some π-molecular compounds	P
naphthalene	1.3, 1.7 (SSI81)		
phenanthrene	1.4		
tetracene	2.1, 1.57 (S81)		
chrysene	2.1, 1.7 (SSI81)		
pyrene	1.7, 1.6 (SSI81)		
coronene	2.4, 1.7 (SSI81)		
anthracene	1.8, 1.61 (KSS82)	anthracene···PMDA	2.16
		anthracene···TCNQ	2.39
perylene	1.9, 1.6 (SSI81)	perylene···PMDA	1.81
		perylene···chloranil	1.57

References:
KSS82 – Karl, Sato, Seki and Inokuchi, 1982; S81 – Silinski, 1981; SSI81 – Sato, Seki and Inokuchi, 1981.

13.3.6   *Determination of degree of charge transfer*

The degree of charge transfer (Z) from donor to acceptor, giving compositions $D^{Z+}A^{Z-}$ (where $0 \leq Z \leq 1$), is a quantity of prime importance in the discussion of delocalized charge transfer molecular compounds. The lattice energy, the conductivity, the nature of a possible Peierls transition, the type of ESR spectrum, are all heavily dependent on this parameter. A number of methods of general applicability have been used for determining Z. One method relies on diffraction studies and uses the differences in bond lengths between neutral and ionic forms of donor and acceptor moieties; two other methods rely on spectroscopic studies, one using the dependence of selected stretching frequencies on moiety charge, while the other uses the dependence of the oscillator strength of the charge transfer band on moiety charge. Photoelectron spectroscopy has been used to demonstrate the presence of two charge states but is not suitable for quantitative determination of Z. The most accurate method is based on analysis of diffuse x-ray (or neutron) scattering and will be discussed in Chapter 17.

   Dependence of moiety bond lengths on charge is shown in Table 13.6 for some donor and acceptor moieties. Clearly TMPD becomes more quinonoid and TCNQ more benzenoid on passing from neutral molecule to ion; this is in accordance with theoretical calculations for TMPD (Haddon, 1975) and TCNQ (Johanson, 1975).

   Four equations have been proposed for estimation of Z from changes in various bond lengths in TCNQ ((1) Flandrois and Chasseau, 1977; (2) Umland *et al.*, 1988 extending an earlier proposal of Coppens and Guru Row, 1978; (3) Kistenmacher *et al.*, 1982; (4) derived here from values in Table 13.6):

$$Z_{TCNQ} = 7.25(b - c) - 8.07(c - d) - 1 \tag{1}$$

$$Z_{TCNQ} = 26.24 - 29.92[(a + c)/(b + d)] \tag{2}$$

$$Z_{TCNQ} = 19.83 - 41.67c/(b + d) \tag{3}$$

$$Z_{TCNQ} = -1.374 + 8.13\{(b + d) - (a + c)\} \tag{4}$$

There seems little reason to prefer one or other of these equations at the available levels of precision of bond length measurements. One should note that all the equations assume a linear dependence of Z on dimensions, which is surely an over simplification. The values of the coefficients depend on the bond lengths used for neutral and ionic moieties, only one set being available for the neutral molecule but a number for the ionic moiety as it appears in different closed-shell salts; presumably it would be best to take a weighted average of the latter but such sophistication seems premature at the available levels of precision. It is possible to identify the most sensitive parameters ('discriminators') from the dimensions given in Table 13.6. We illustrate some of the problems by using {[TTF][TCNQ]} as an example, its crystal structure having been determined at a number of temperatures. We have used bonds (a) and (b) of TTF as standards although it has been suggested by Mayerle *et al.*, (1979) that the ring double bond (d) is more sensitive, based on values of 1.323(4) Å in {[TTF][TCNQ]} and 1.336 Å in TTF·$I_{0.71}$. Parenthetically, we had considered TTF·ClO$_4$ for use as our reference cation because four independent values are available for each bond length; however, using the ring double bonds as an example, we find values of 1.292(20), 1.305(15), 1.306(17) and 1.321(13) Å, with a clearly unacceptable spread.

Table 13.6. Dimensions of neutral and ionic forms of representative donor and acceptor moieties. The ionic forms are for salts where the counterion is a closed shell ion with an assumed integral single charge. The bond lengths (Å) are not corrected for thermal motion and have been averaged assuming D_{2h} symmetry. Values in brackets are experimental standard uncertainties divided by $(n-1)^{\frac{1}{2}}$ where n independent values have been averaged. $\Delta l = l_{ion} - l_{neutral}$. Equations showing the dependence of moiety charge (Z) on bond lengths (a, b, c, d) are given; the constants have dimensions such that Z is dimensionless. The values of a etc. in the equation for Z are those for the species under consideration

Moiety	Bond	l (neutral molecule)	l (ion)	Δl
		A (IK79)	B (dBV72)	
TMPD	a	1.390(5)	1.363(2)	−0.026(5)
	b	1.397(5)	1.428(2)	+0.031(5)
	c	1.418(6)	1.345(2)	−0.073(6)

$Z_{TMPD} = 10.85 - 7.69(c + a - b)$

Moiety	Bond	l (neutral molecule)	l (ion)	Δl
		C (CK71)	D (YN80)	
TTF	a	1.349(3)	1.404(10)	+0.055
	b	1.757(2)	1.713(3)	−0.042
	c	1.730(2)	1.725(4)	−0.005
	d	1.314(3)	1.306(8)	−0.008

$Z_{TTF} = 3.835 - 9.174(b - a)$

Moiety	Bond	l (neutral molecule)	l (ion)	Δl
		E (LST65)	F (HSV72)	
TCNQ	a	1.346(3)	1.373(1)	−0.027(3)
	b	1.448(2)	1.423(3)	+0.025(3)
	c	1.374(3)	1.420(0)	−0.046(2)
	d	1.441(3)	1.416(8)	+0.025(6)
	C≡N	1.140(2)	1.153(2)	+0.013(3)

See text for equations for Z_{TCNQ}.

Notes:
A. $TMPD^0$ at 298K; B. TMPD·ClO$_4$ at 110K; C. TTF^0; D. TTF·ClO$_4$; E. $TCNQ^0$ at 300K, libration corrected; F. Rb TCNQ at 113K.

References:
CK71 – Cooper, Kenny, Edmonds, Nagel, Wudl and Coppens, 1971; dBV72 – de Boer and Vos, 1972; HSV72 – Hoekstra, Spoelder and Vos, 1972; I79 – Ikemoto, Katagiri, Nishimura, Yakushi and Kuroda, 1979; LST65 – Long, Sparks and Trueblood, 1965; YN80 – Yakushi et al. 1980.

We illustrate possible use of discriminator bonds (a) and (b) of TTF and (c) and (d) of TCNQ by plotting these in Fig. 13.6, together with 'standard' values (300K) for the neutral and ionic moieties. The differences between paired bond lengths in the neutral and ionic moieties are only a few times the standard uncertainty of the measurements. If we make the approximation (permissible in terms of the available precision) that bond lengths are not dependent on temperature, then using mean values we calculate that $|Z_{TCNQ}| = 0.41$ (eq. 1), 0.37 (eq. 2), 0.44 (eq. 3) and 0.39 (eq. 4). Although the internal agreement is satisfactory, the accepted value of $|Z|$ (from X-ray diffuse scattering

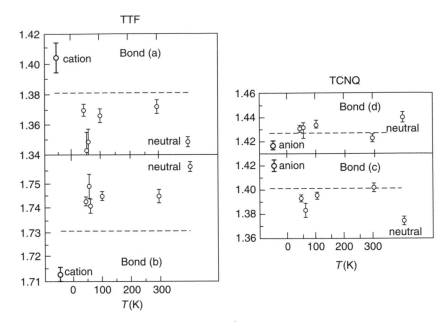

Fig. 13.6. The ordinates show measured values (Å) of two discriminator bond lengths in each of the TTF and TCNQ moieties of {[TTF][TCNQ]} at 300K (Kistenmacher *et al.*, 1974) and at 100, 60, 53 and 45K (Schulz *et al.*, 1976). The standard values deduced for these bonds in the anionic and neutral species are shown at the left and right extremes of the temperature scale and the measured values at the appropriate temperatures. The horizontal broken lines show the bond lengths corresponding to $Z = 0.59$ deduced from x-ray diffuse scattering measurements. See text for discussion.

(Section 17.7)) is 0.59. It is clear that much higher precision is required for all bond length measurements if meaningful conclusions are to be extracted.

The spectroscopic methods have been applied mainly to TTF and TCNQ, and we summarize the conclusions in Table 13.7. Again a linear dependence of Z on frequency is assumed. For example, resonance Raman measurements of the frequency of the exocyclic C=C bond stretch (ν_4) show this to be 1454 cm^{-1} for neutral TCNQ (Van Duyne *et al.*, 1979) and 1379 cm^{-1} for TCNQ$^-$ (Matsuzaki, Kuwata and Toyoda, 1980). Linear interpolation gives $Z_{TCNQ} = -0.01333\nu_{C=C} + 19.39$.

As the resolution of the spectroscopic methods is $\approx$1–3 cm^{-1}, a precision of $\approx$0.01 in Z can be expected. However, solid-state (crystal field) effects can alter vibration frequencies by $\approx$10 cm^{-1}. We also note that $\nu(C\equiv N)$ is very sensitive to the degree of charge transfer, in contrast to $d(C\equiv N)$.

Photoelectron spectroscopy using deconvolution of N-1s and S-2p spectra demonstrates the presence of neutral and charged moieties in segregated stack molecular compounds. This is illustrated for {[TMTTF][TCNQ]} in Fig. 13.7 (Tokumoto, Koshizuka, Murata, Kinoshita, Anzai, Ishiguro and Mori, 1982; Tokumoto, Koshizuka, Anzai and Ishiguro, 1982). Unfortunately the method does not give reliable quantitative results (Ritsko *et al.*, 1978).

Table 13.7. Determination of degree of charge transfer by spectroscopic techniques

ν(neutral species) cm^{-1}	ν(ionic species) cm^{-1}	(Charge transfer)/frequency relationship
TTF Raman, a_g ν_3, predominantly C=C stretch		
1515 (S80)	1413 (MMT80)	$Z_{TTF} = -0.00980\nu_{C=C} + 14.85.$
TCNQ Resonance Raman, ν_4 exocyclic C=C stretch		
1454 (VD79)	1379 (MKT80)	$Z_{TCNQ} = -0.01333\nu_{C=C} + 19.39$
TCNQ IR Absorption C≡N stretch		
2227 (CB81)	2183 (CB81)	$Z_{TCNQ} = -0.02273\nu_{C≡N} + 50.61$

References:
CB81 – Chappell *et al.*, 1981; MKT80 – Matsuzaki, Kuwata and Toyoda, 1980; MMT80 – Matsuzaki, Moriyama and Toyoda, 1980; S80 – Siedle *et al.*, 1980; VD79 – Van Duyne *et al.*, 1979.

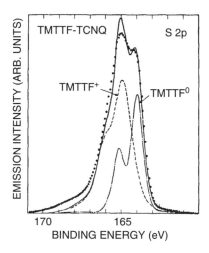 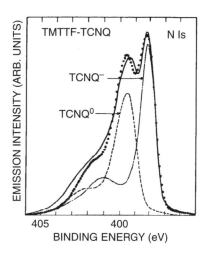

Fig. 13.7. Core level XPS spectra for {[TMTTF][TCNQ]} after data processing. Observed spectra are given by dotted lines, simulated spectra by full lines, and the individual moiety spectra by chain lines. The latter are obtained from TMTTF, {[TMTTF][(DDQ)2]}, TCNQ and K-TCNQ. (Reproduced from Tokumoto *et al.*, 1982.)

It has also been suggested that the value of Z can be obtained from measurement of the oscillator strength (f) of the charge transfer band (Jacobsen and Torrance, 1983). The relation between them is

$$f = \{8\pi^2 mc\nu_{CT} | R_{DA}|^2 Z\}/h = 3.254 \times 10^{-5}\nu CT | R_{DA}|^2 Z$$

where R_{DA} is the distance (Å) between the centers of the donor and acceptor molecules and ν_{CT} is the excitation energy of the charge transfer band in cm^{-1}. The oscillator strength is measured from

$$f = \frac{4\pi mc}{Ne^2} \int_{\nu_1}^{\nu_2} \sigma(\nu) \cos^2\theta \, d\nu,$$

where N is the number of DA pairs per unit volume and θ is the angle between the direction of light polarization and the transition dipole moment (Wooten, 1972). Results obtained by this method are given in Section 17.4 to demonstrate that a neutral to ionic phase transition occurs in {TTF···chloranil}.

Some care is needed in using the values of Z derived by these methods as systematic errors can occur. Cross checking of one method against another is always desirable.

13.4   Binary and quasi-binary donor–acceptor systems

13.4.1   *Phase diagrams*

Most charge transfer molecular compounds have been prepared by what are essentially hit-or-miss techniques – mix the components in a particular ratio in a suitable solvent and hope that the binary compound will crystallize. However, careful studies, involving determination of the phase diagram of the system, are now becoming more common. Ideally both the binary phase diagram, which covers a wide range of temperatures, and the ternary (D-A-solvent) phase diagram, at a particular or limited number of temperatures, should be determined, but this has rarely been done. The ternary phase diagram will give more reliable results in the lower temperature regions where solid–solid transitions are likely to be slow. In older work melting point diagrams were determined, perhaps with the addition of thaw points; in modern work use of differential scanning calorimetry (DSC) has greatly improved both the precision and the reliability of the results. Quite complicated binary diagrams can sometimes be obtained, as exemplified by that of carbazole/2,4,7-trinitrofluoren-9-one (Fig. 13.8; Krajewska and Pigon, 1980); we maintain our D/A nomenclature despite the converse usage in Fig. 13.8. This shows such typical features as a 1 : 1 molecular compound with congruent melting point and a 1 : 2 molecular compound

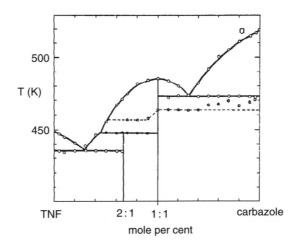

Fig. 13.8. Phase diagram for the carbazole-TNF system. ○ Liquidus and eutectic lines; ● peritectic lines; --ø--ø-- phase transition line. (Reproduced from Krajewska and Wasilewska, 1981.)

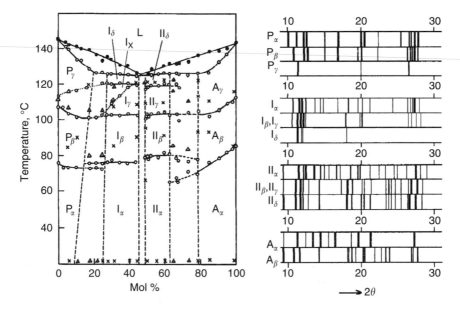

Fig. 13.9. (a) on left: Phase diagram for the mixed donor system (phenanthrene/anthracene)–picric acid: x single phase from Debye–Scherrer patterns; Δ two phase from Debye–Scherrer patterns. Mol.% anthracene runs from zero on the left to 100% on the right. Greek letters (α, β, γ) indicate phases stable at increasing temperatures; (b) on right: Debye–Scherrer patterns (Cu Kα) from the phases found in the phase diagram. In order from top to bottom these show

    (i)   the three phases of phenanthrene-picric acid and solid solutions

    (ii) the three patterns from the phases of intermediate solid solution I

    (iii) the three patterns from the phases of intermediate solid solution II

    (iv) the two patterns from anthracene-picric acid and solid solutions.

The very simple patterns of Pγ and I$_\delta$ have been interpreted as indicating that these are plastic phases (see Section 16.5). (Reproduced from Koizumi and Matsunaga, 1974.)

with incongruent melting point (in some systems ratios other than $1:1$ or $1:2$ are found but these are relatively rare). The $1:1$ compound is more stable than the $1:2$ compound, which is the usual situation. The composition ranges of both compounds appear negligibly small (although it should be remembered that the methods used are generally not sensitive to composition differences of less than $\approx 1\%$). The $1:1$ compound shows a solid state transition at about 460K; transitions below room temperature are also often found (see Chapter 16). Despite the improvements in methodology, controversy has not been entirely eliminated; for example, a phase diagram has been reported for the pyrene/picryl chloride system which shows DA, D_4A_3, D_2A and D_4A compounds (only DA melts congruently) (Bando and Matsunaga, 1976) while other workers could find only DA and D_3A_2 compounds (Krajewska and Wasilewska, 1981). Sluggishness in the attainment of equilibrium and problems of identification of incongruently melting compounds from small breaks in the liquidus curves can give rise to many difficulties.

    The binary phase diagram summarizes an appreciable amount of information about a system, and is always a desirable preliminary to a detailed study of the structure and/or

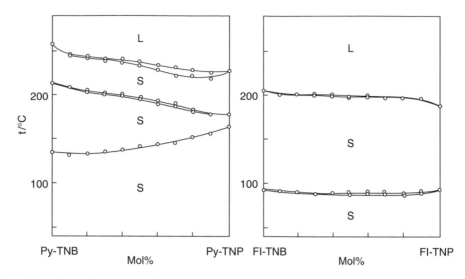

Fig. 13.10. Phase diagrams for the mixed acceptor systems pyrene–(TNB, TNP) and fluoranthene–(TNB, TNP) (TNP = 2,4,6-trinitrophenol or picric acid). The crystal structure of pyrene···TNB is known. (Reproduced from Inabe *et al.*, 1981.)

properties of a particular charge transfer molecular compound. Many such diagrams are available for mixed stack systems (for examples, see Herbstein, 1971; D'Ans and Kaufmann, 1956; Kofler, 1956; Radomska and Radomski, 1980a, b). There do not appear to be any phase diagrams for segregated stack systems but there are some for "complex isomers" e.g. 1-bromo-2-naphthylamine/picric acid (Hertel, 1926); 4-bromo-1-naphthylamine/2,6-dinitrophenol (Hertel and Van Cleef, 1928); o-bromoaniline/picric acid (Komorowski *et al.*, 1976).

Although the donor : acceptor ratio appears to be fixed at a ratio of small integers in the charge transfer molecular compounds, it is possible to replace to a considerable extent one donor by another, and the same holds for suitable pairs of acceptors. Thus about 60% of the phenanthrene molecules can be replaced by anthracene in the {phenanthrene···TNB} compound, and about 20% of the anthracene molecules in {anthracene···TNB} can be replaced by phenanthrene (Lower, 1977). Similar extensive mutual solid solubility of anthracene and phenanthrene has been found in the (anthracene–phenanthrene)/TCNB (Wright *et al.*, 1976) and (anthracene–phenanthrene)/picric acid (Koizumi and Matsunaga, 1974) systems. In the first of these there is complete mutual solid solubility of the two donors and the crystal structures of {anthracene···TCNB} and {phenanthrene···TCNB} are isostructural at room temperature (there is disorder of the phenanthrene molecule). In the second system (Fig. 13.9) the situation appears to be more complicated; in addition to limited solid solubility of anthracene in {phenanthrene···picric acid} and of phenanthrene in {anthracene···picric acid} two intermediate phases with fairly wide composition ranges appear. Only the crystal structure of {anthracene···picric acid} is known (Herbstein and Kaftory, 1976). Among the mixed acceptor systems there are some (Fig. 13.10) with a full range of solid solubility at lower temperatures as well as at higher temperatures; in accordance with these phase diagrams the {pyrene···picric

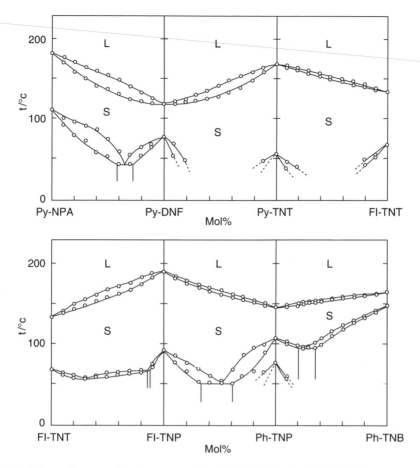

Fig. 13.11. Phase diagrams for the systems Py-(NPA, DNF); Py-(DNF, TNT); (Py, Fl)-TNT; Fl-(TNT, TNP); (Fl, Ph)-TNP; and Ph-(TNP, TNB). Py = pyrene; Fl = fluoranthene; Ph = phenanthrene; TNP = 2,4,6-trinitrophenol (picric acid); NPA = 2-nitrophthalic anhydride; DNF = 2,4-dinitrofluorobenzene. (Reproduced from Inabe *et al.*, 1981.)

acid} and {pyrene···TNB} compounds were reported to be isomorphous at room temperature, and also the pair {fluoranthene···picric acid} and {fluoranthene···TNB} (Herbstein and Kaftory, 1975a). Other systems (Fig. 13.11) show complete miscibility only at high temperatures and complicated diagrams at lower temperatures. There are many indications of solid state transitions. Some information about relevant crystal structures is noted in the captions to these diagrams; references are given in Chapter 15.

   The phase diagrams for the mixed acceptor systems (Fig. 13.10 and 13.11) are compatible with the more limited information available from crystallographic studies despite the different temperature ranges of the two types of study. It is not known whether there is ordered or random substitution of one donor (or acceptor) for another in the mixed stacks of these molecular compounds.

One ternary mixed-donor system has been studied – (anthracene, acridine, phenazine)/ PMDA (Karl *et al.*, 1982). The three individual molecular compounds ({anthracene··· PMDA}, {acridine···PMDA}, {phenazine···PMDA}) are isomorphous (triclinic, *P* $\bar{1}$, *Z* = 1; Table 15.2, Group 1a). Complete miscibility was found for all donor ratios over the temperature range 500–120K, the molecular compounds melting around 500K.

13.4.2   Component ratios in binary donor–acceptor systems

Most crystalline π-molecular compounds, be they mixed stack or segregated stack in structure, have a 1 : 1 donor : acceptor ratio and this has been the most extensively studied group, both in regard to crystal structures and physical properties. Many 1 : 2 and 2 : 1 compositions have been reported, and most of these occur as incongruently-melting compounds in systems where the 1 : 1 composition is the most stable. However, there are examples where the 1 : 2 (or 2 : 1) molecular compound appears to be the more stable e.g. (benzo[c]pyrene)$_2$···TMU (Weil-Malherbe, 1946); relative stabilities of the different compositions in such systems do not appear to have been investigated.

There are also reports of compositions other than 1 : 1, 1 : 2 or 2:1 (Table 13.8); these should be viewed with caution if based on chemical analyses alone. However, there are structural explanations for some unusual compositions. For example, in both

Table 13.8. Some examples of compositions other than 1 : 1, 1 : 2 or 2 : 1 reported for π-molecular compounds

Donor (D)	Acceptor (A)	D : A	Reference
Bromodurene	BTF	3 : 2	B60
Tetralin	Nitrobenzodifuroxan	1 : 3	BC58
Pyrene	PMDA	1 : 3	HS69
N,N-dibenzyl-*m*-toluidine	TNB	3 : 2	KMC39
Triphenylmethanol	TNB	3 : 2	KHM12
Phenanthrene	*p*-Dinitrobenzene	3 : 1	K08
Fluorene	TNB	3 : 4	HKR76
1-Naphthylamine	2,3-Dinitrophenol	3 : 2	S40b
2-Naphthylamine	2,3-Dinitrophenol	3 : 2	S40b
Phenanthrene	1,2,4,6-Tetranitrobenzene	2 : 3	S40a
Naphthalene	1,2,4,6-Tetranitrobenzene	3 : 2	S40c
1-Naphthol	Tetryl	3 : 2	S40c
p-Phenylenediamine	*p*-Benzoquinone	2 : 5	S09
Dibenzo[c,d]phenothiazine	DDQ	3 : 2	M64
Benzene	*o*-Chloranil	3 : 1	PJF17
1,4-Diphenylbutadiene	TNF	3 : 1	OW46
Tetrathiotetracene(TTT)	*o*-Chloranil	3 : 1	M65
TTT	*o*-Bromanil	3 : 1	M64
TTT	TCNE	3 : 1	M64

References:
B60 – Bailey, 1960; BC58 – Bailey and Case, 1958; HKR76 – Herbstein, Kaftory and Regev, 1976; HS69 – Herbstein and Snyman, 1969; K08 – Kremann, 1908; KHM12 – Kremann, Hohl and Muller, 1912; KMC39 – Kent, McNeil and Cowper, 1939; M64 – Matsunaga, 1964; M65 – Matsunaga, 1965; OW46 – Orchin and Woolfolk, 1946; PFJ17 – Pfeiffer, Jowleff, Fischer, Monti and Muuly, 1917; S09 – Schlenk, 1909; S40a – Shinomiya, 1940a; S40b – Shinomiya, 1940b; S40c – Shinomiya, 1940c.

(pyrene)$_3$···(picryl bromide)$_2$ (Herbstein and Kaftory, 1975b) and (TMTTF)$_{1.8}$···TCNQ (Kistenmacher et al., 1976) the molecules present over and above the 1 : 1 composition are included in the structure without participating in the charge transfer interaction, i.e. they are present as "molecules of crystallization". The remarkable structure of {(fluorene)$_3$···(TNB)$_4$} is discussed in Section 15.4.

The segregated-stack radical-cation, radical-anion salts seem all to have equimolar compositions but the M$_n$(TCNQ)$_m$ salts, where M is a closed-shell cation, have a wide range of compositions depending on the nature of the cation. There are good structural explanations for compositions such as Cs$_2$(TCNQ)$_3$ (see Section 17.4.6) or N-(n-butylpyridinium)$_4$(TCNQ)$_7$ (see Section 17.4.2.3).

13.5 Ternary π-molecular compounds

There are a number of molecular compounds which contain three components and we can distinguish three situations:

(i) Noninteracting third component, where the third component appears to be present essentially as "molecules of crystallization" and does not participate in the charge transfer interaction. However, no relevant structures have been reported so some reserve must be maintained about examples such as azulene···BTF·(propionic acid)$_{0.5}$ (Bailey, 1960); 2,2′,4,4′hexamethylstilbene···(picric acid)$_2$·benzene (Elbs, 1893); 1-naphthol···hexachloro-1-indenone·1/2X where X = benzene or acetic acid (Pfeiffer et al., 1924); phenanthrene···TNB·x(benzene) and tetra-benznaphthalene···picric acid. ethanol, where the third component may be present zeolitically in both examples (Herbstein, Kaftory and Regev, 1976).

(ii) Interacting third component: one example where all three components interact is Kofler's (1944) 1 : 1 : 1 ternary compound {1-naphthylamine···pyridine···picric acid}, the structure of which has been determined (Bernstein et al., 1980; see Section 15.11.4). A second example is the ordered ternary compound {3,3′-dimethylthio-azolinocarbocyanine···TCNQ···2,4,7-trinitrofluorenone} (Kaminskii et al., 1974), with mixed stacks of alternating TCNQ and TNF moieties. The pseudo-binary systems with mixed donors or acceptors presumably should be classified in this group.

Table 13.9. Examples of TNB molecular compounds formed by a particular donor and by its potassium salt. Melting points in °C

Donor	TNB molecular compound		TNB molecular compound of K salt	
	Crystals	M. Pt.	Crystals	M. Pt.
o-Aminobenzoic Acid	orange needles	192	deep-red needles	114
m-Aminobenzoic Acid			red-brown needles	118
p-Aminobenzoic Acid	red	151	red needles	115(dec)
1-Anthrol	red-brown plates	161	black needles	275

Notes: All the compounds are 1 : 1 (Sudborough and Beard, 1910) except for the K salt of anthrol. TNB which has composition $C_{14}H_9OK$···{$C_6H_3(NO_2)_3$}$_2$ (Cadre and Sudborough, 1916).

(iii)   There is another group of molecular compounds where the third component, although not participating in the charge transfer system, exerts a profound influence on the structure and its properties. These are the benzidine···TCNQ.solvent compounds discussed in Section 15.7.3.3, which have both charge transfer and inclusion characteristics.

(iv)   Finally we draw attention to a group where both a particular donor, and its potassium salt, form molecular compounds with TNB (Table 13.9). As both neutral compounds and the salts are highly colored, it seems probable that charge transfer interactions occur in both. These pairs of compounds could provide an opportunity to compare the donor characteristics of molecules and their anions.

References

Andrè, J. J., Bieber, A. and Gautier, F. (1977). *Ann. Phys.* (Paris), **1**, 145–256.

Andrews, L. J. and Keefer, R. M. (1964). Molecular Complexes in Organic Chemistry, Holden-Day, San Francisco.

Bailey, A. S. and Case, J. R. (1958). *Tetrahedron*, **3**, 113–131.

Bailey, A. S. (1960). *J. Chem. Soc.*, pp. 4710–4712.

Bando, M. and Matsunaga, Y. (1976). *Bull. Chem. Soc. Jpn.*, **49**, 3345–3346.

Baril, O. L. and Hauber, E. S. (1931). *J. Am. Chem. Soc.*, **53**, 1087–1091.

Batail, P., La Placa, S. J., Mayerle, J. J. and Torrance, J. B. (1981). *J. Am. Chem. Soc.*, **103**, 951–953.

Batley, M. and Lyons, L. E. (1962). *Nature (London)*, **196**, 573–574.

Batley, M. and Lyons, L. E. (1968). *Mol. Cryst. Liq. Cryst.*, **3**, 357–374.

Bennett, G. M. and Willis, G. H. (1929). *J. Chem. Soc.*, pp. 256–268.

Bernstein, J., Herbstein, F. H. and Regev, H. (1980). *Acta Cryst.*, **B36**, 1170–1175.

Blaunstein, R. P. and Christopherou, L. G. (1971). *Rad. Res. Rev.*, **3**, 69–118.

Bleidelis, J., Shvets, A. E. and Freimanis, J. (1976). *Zh. Strukt. Khim.*, **17**, 1096–1110.

Boer, J. L. de and Vos, A. (1972). *Acta Cryst.*, **B28**, 839–848.

Booth, J. and Boyland, E. (1953). *Biochim. Biophys. Acta*, **12**, 75–87.

Boschi, R., Clar, E. and Schmidt, W. (1974). *J. Chem. Phys.*, **60**, 4406–4418.

Brackman, W. (1949). *Rec. Trav. Chim.*, **68**, 147–159.

Briegleb, G., Förster, Th., Friedrich-Freksa, H., Jordan, R., Kortüm, G., Münster, A., Scheibe, G. and Wirtz, K. (1948). Zwischenmolekulare Krafte, Karlsruhe, G. Braun, 142 pp.

Briegleb, G. (1961). Elektronen-Donator-Acceptor-Komplexe. Springer, Berlin-Göttingen-Heidelberg.

Briegleb, G. (1964). *Angew. Chem. Int. Ed. Engl.*, **3**, 617–632.

Cadre, S.T. and Sudborough, J. J. (1916). *J. Chem. Soc.*, **109**, 1349–1354.

Carstensen-Oeser, E., Göttlicher, S. and Habermehl, G. (1968). *Chem. Ber.*, **101**, 1648–1655.

Chappell, J. S., Bloch, A. N., Bryden, W. A., Maxfield, M., Poehler, T. O. and Cowan, D. O. (1981). *J. Am. Chem. Soc.*, **103**, 2442–2443.

Chen, E. C. M. and Wentworth, W. E. (1975). *J. Chem. Phys.*, **63**, 3183–3191.

Clar, E. and Schmidt, W. (1977). *Tetrahedron*, **33**, 2093–2097.

Clar, E. and Schmidt, W. (1979). *Tetrahedron*, **35**, 2673–2680.

Clark, P. A., Brogli, F. and Heilbronner, E. (1972). *Helv. Chim. Acta*, **55**, 1415–1428.

Clayden, J., Greeves, N., Warren, S. and Wothers, P. (2001). Organic Chemistry. Oxford University Press, Oxford. 1512 pp.

Coleman, L. B., Cohen, M. J., Sandman, D. J., Yamagishi, F. G., Garito, A. F. and Heeger, A. J. (1973). *Sol. State Commun.*, **12**, 1125–1132.

Cooper, W. F., Kenny, N. C., Edmonds, J. W., Nagel, A., Wudl, F. and Coppens, P. (1971). *J. Chem. Soc., Chem. Commun.*, pp. 889–890.

Coppens, P. and Guru Row, T. N. (1978). *Ann. N.Y. Acad. Sci.*, **313**, 244–255.

D'Ans, J. and Kaufmann, E. (1956). "Solubility equilibria of organic substances in organic substances" in *Landolt-Bernstein Tables*, Springer-Verlag, Berlin-Göttingen-Heidelberg, Sixth Edition, Vol. II, Part 2, pp. 1–500.

Dahm, D. J., Horn, P., Johnson, G. R., Miles, M. G. and Wilson, J. D. (1975). *J. Cryst. Mol. Struct.*, **5**, 27–34.

Davis, K. M. C. and Symons, M. C. R. (1965). *J. Chem. Soc.*, pp. 2079–2083.

Eilfeld, P. and Schmidt, W. (1981). *J. Electr. Spectroscop. Related Phenom.*, **24**, 101–120.

Elbs, K. (1893). *J. prakt. Chem. (N.F.)*, **47**, 44–79.

Fieser, L. F and Fieser, Mary. (1961). Advanced Organic Chemistry. Reinhold, New York. 1155 pp.

Flandrois, S. and Chasseau, D. (1977). *Acta Cryst.*, **B33**, 2744–2750.

Fritzsche, J. von (1858). *J. prakt. Chem.*, **73**, 282–292.

Gleiter, R. and Schanz, P. (1980). *Angew. Chem. Int. Ed. Engl.*, **19**, 715–716.

Gutmann, F. and Lyons, L. E. (1981). Organic Semiconductors, Part A, original edition published by Wiley, New York and reprinted, with corrections, by R. E. Kriejer, Malabar, Florida.

Haddon, R. C. (1975). *Austr. J. Chem.*, **28**, 2333–2342.

Hendrickson, J. B., Cram, D. J. and Hammond, G. S. (1970). Organic Chemistry (Third edition). McGraw-Hill Kogakusha. Tokyo. 1279 pp.

Herbstein, F. H. (1971). "Crystalline π-molecular compounds: chemistry, spectroscopy and crystallography" in *Perspectives in Structural Chemistry*, edited by J. D. Dunitz and J. A. Ibers, Wiley, New York etc., Vol. IV, pp. 166–395.

Herbstein, F. H. and Kaftory, M. (1975a). *Acta Cryst.*, **B31**, 60–67.

Herbstein, F. H. and Kaftory, M. (1975b). *Acta Cryst.*, **B31**, 68–75.

Herbstein, F. H. and Kaftory, M. (1976). *Acta Cryst.*, **B32**, 387–396.

Herbstein, F. H. and Snyman, J. A. (1969). *Phil. Trans. Roy. Soc. Lond.*, **A264**, 635–666.

Herbstein, F. H., Kaftory, M. and Regev, H. (1976). *J. Appl. Cryst.*, **9**, 361–364.

Hertel, E. and van Cleef, J. (1928). *Ber.*, **61**, 1545–1549.

Hertel, E. (1926). *Ann. Chem.*, **451**, 179–208.

Hoekstra, A., Spoelder, T. and Vos, A. (1972). *Acta Cryst.*, **B28**, 14–25.

Ikemoto, I., Katagiri, G., Nishimura, S., Yakushi, K. and Kuroda, H. (1979). *Acta Cryst.*, **B35**, 2264–2265.

Inabe, T., Matsunaga, Y. and Nanba, M. (1981). *Bull. Chem. Soc. Jpn.*, **54**, 2557–2564.

Jacobsen, C. S. and Torrance, J. B. (1983). *J. Chem. Phys.*, **78**, 112–115.

Johanson, H. (1975). *Int. J. Quantum Chem.*, **9**, 459–471.

Kaminskii, V. F., Shibaeva, R. P. and Atovmyan, L. O. (1974). *J. Struct. Chem.*, **15**, 434–440.

Karl, N., Ketterer, W. and Stezowski, J. J. (1982). *Acta Cryst.*, **B38**, 2917–2919.

Karl, N., Sato, N., Seki, K. and Inokuchi, H. (1982). *J. Chem. Phys.*, **77**, 4870–4878.

Kent, A., McNeil, D. and Cowper, R. M. (1939). *J. Chem. Soc.*, pp. 1858–1862.

Kistenmacher, T. J., Emge, T. J., Bloch, A. N. and Cowan, D. O. (1982). *Acta Cryst.*, **B38**, 1193–1199.

Kistenmacher, T. J., Phillips, T. E. and Cowan, D. O. (1974). *Acta Cryst.*, **B30**, 763–768.

Kistenmacher, T. J., Phillips, T. E., Cowan, D. O., Ferraris, J. P., Bloch, A. N. and Poehler, T. O. (1976). *Acta Cryst.*, **B32**, 539–547.

Kofler, A. (1956). "Melting-point Equilibria in Organic Systems" in *Landolt-Bernstein Tables*, Springer-Verlag, Berlin-Göttingen-Heidelberg, Sixth Edition, Vol. II, Part 3, pp. 350–403.

Kofler, A. (1944). *Z. Elektrochem.*, **50**, 200–207.

Koizumi, S. and Matsunaga, Y. (1974). *Bull. Chem. Soc. Jpn.*, **47**, 9–13.

Komorowski, L., Krajewska, A. and Pigon, K. (1976). *Mol. Cryst. Liq. Cryst.*, **36**, 337–348.

Konno, M., Saito, Y., Yamada, K. and Kawazura, H. (1980). *Acta Cryst.*, **B36**, 1680–1683.

Krajewska, A. and Pigon, K. (1980). *Thermochim. Acta*, **41**, 187–197.

Krajewska, A. and Wasilewska, quoted by Pigon, K. and Chojnocki, H. (1981). "Electrical conductivity of solid molecular complexes" in *Molecular Complexes*, ed. by H. Ratajczak and W.J. Orville-Thomas, Vol. **2**, pp. 451–492.

Kremann, R. (1908). *Monatsh.*, **29**, 863–890.

Kremann, R., Hohl, H. and Muller, R., II. (1912). *Monatsh.*, **42**, 199–220.

Long, R. E., Sparks, R. A. and Trueblood, K. N. (1965). *Acta Cryst.*, **18**, 932–939.

Lower, S. K. (1968). *Mol. Cryst. Liq. Cryst.*, **5**, 363–368.

Lowry, T. M. (1924). *Chem. and Ind.*, **48**, 218–221.

Mackay, I. R., Robertson, J. M. and Sime, J. G. (1969). *J. Chem. Soc., Chem. Commun.*, pp. 1470–1471.

Martinet, J. and Bornand, L. (1925). *Rev. gen. sci.*, **36**, 569–577; *Chem. Abstr.*, **20**, 861 (1926).

Matsunaga, Y. (1964). *J. Chem. Phys.*, **41**, 1609–1613.

Matsunaga, Y. (1965). *J. Chem. Phys.*, **42**, 2248–2249.

Matsuo, T. and Higuchi, O. (1968). *Bull. Chem. Soc. Jpn.*, **41**, 518–519.

Matsuzaki, S., Kuwata, R. and Toyoda, K. (1980). *Sol. State Commun.*, **33**, 403–405.

Matsuzaki, S., Moriyama, T. and Toyoda, K. (1980). *Sol. State Commun.*, **33**, 857–859.

Mayerle, J. J., Torrance, J. B. and Crowley, J. I. (1979). *Acta Cryst.*, **B35**, 2988–2995.

McConnell, H. M., Hoffman, B. M. and Metzger, R. M. (1965). *Proc. Natl. Acad. Sci. U. S.*, **53**, 46–50.

Meyers, M. and Trueblood, K. N. (1969). *Acta Cryst.*, **B25**, 2588–2599.

Mulliken, R. S. (1952a). *J. Am. Chem. Soc.*, **72**, 600–608.

Mulliken, R. S. (1952b). *J. Phys. Chem.*, **56**, 801–822.

Mulliken, R. S. (1954). *J. chim. Phys.*, **51**, 341–344.

Mulliken, R. S. and Person, W. B. (1969a). *J. Am. Chem. Soc.*, **91**, 3409–3413.

Mulliken, R. S. and Person, W. B. (1969b). Molecular Complexes. Wiley-Interscience, New York, 498 pp.

Mueller-Westerhoff, U. T. and Vance, B. (1987). Dithiolenes and Related Species, Chapter 16.5 in Comprehensive Coordination Chemistry, Vol. **2**, 595–631, edited by G. Wilkinson, R. D. Gilard and J. A. McCleverty, Pergamon Press, Oxford etc.

Newman, M. S. and Lutz, W. B. (1956). *J. Am. Chem. Soc.*, **78**, 2469–2473.

Orchin, M. and Woolfolk, E. O. (1946). *J. Am. Chem. Soc.*, **68**, 1727–1729.

Pfeiffer, P., Goebel, F. and Angern, O. (1924). *Ann. Chem.*, **440**, 241–264.

Pfeiffer, P., Jowleff, W., Fischer, P., Monti, P. and Mully, H. (1917). *Ann. Chem.*, **412**, 253–335.

Pfeiffer, P. (1927). OrganischeMolekulverbindungen. 2nd Edition. F. Enke, Stuttgart.

Ponte Goncalez, A. M. (1980). *Prog. Sol. State Chem.*, **13**, 1–88.

Powell, H. M., Huse, G. and Cooke, P. W. (1943). *J. Chem. Soc.*, pp. 153–157.

Prout, C. K. and Castellano, E. E. (1970). *J. Chem. Soc. (A)*, pp. 2775–2778.

Prout, C. K. and Wright, J. D. (1968). *Angew. Chem., Int. Ed. Engl.*, **7**, 659–667.

Prout, C. K. and Kamenar, B. (1973). In *Molecular Complexes* **1**, 151–207. Elek: London.

Radomska, M. and Radomski, R. (1980a). *Thermochim. Acta*, **40**, 405–414.

Radomska, M. and Radomski, R. (1980b). *Thermochim. Acta*, **40**, 415–425.

Ritsko, J. J., Epstein, A. J., Salaneck, W. R. and Sandman, D. J. (1978). *Phys. Rev.*, **B17**, 1506–1509.

Roberts, J. D. and Caserio, M. C. (1964). Basic Principles of Organic Chemistry. Benjamin, New York. 1315 pp.

Rose, J. (1967). Molecular Complexes. Pergamon Press, Oxford.

Rosenstock, H. M., Sims, D., Shroyer, S. S. and Webb, W. (1980). J. National Standard Reference Data Series (U.S. Nat. Bur. Stands.) part I, no. 66.

Saito, G. and Ferraris, J. P. (1979). *J. Chem. Soc.Chem. Commun.*, pp. 1027–1029.

Sandman, D. J., Stark, J. C., Hamill, G. P., Burke, W. A. and Foxman, B. M. (1982). *Mol. Cryst. Liq. Cryst.*, **86**, 1819–1825.

Sato, N., Seki, K. and Inokuchi, H. (1981). *J. Chem. Soc., Faraday Trans. 2*, **77**, 1621–1633.

Schlenk, W. (1909). *Ann. Chem.*, **368**, 277–395.

Schmidt, W. (1977).*J. Chem. Phys.*, **66**, 828–845.

Schulz, A. J., Stucky, G. D., Blessing, R. H. and Coppens, P. (1976). *J. Am. Chem. Soc.*, **98**, 3194–3201.

Shinomiya, C. (1940a). *Bull. Chem. Soc. Jpn.*, **15**, 92–103.

Shinomiya, C. (1940b). *Bull. Chem. Soc. Jpn.*, **15**, 137–147.

Shinomiya, C. (1940c). *Bull. Chem. Soc. Jpn.*, **15**, 259–270.

Siedle, A. R., Kistenmacher, T. J., Metzger, R. M., Kuo, C.-S., Van Duyne, R. P. and Cope, T. (1980). *Inorg. Chem.*, **19**, 2048–2051.

Silinski, E. A. (1981). Organic Molecular Crystals, Springer Verlag, Berlin.

Solomons, T. W. G. and Fryhle, C. B. (2000). Organic Chemistry (Seventh edition). Wiley, New York. 1258 pp.

Soos, Z. G. (1974). *Ann. Rev. Phys. Chem.*, **25**, 121–153.

Soos, Z. G., Kuwajima, S. and Harding, R. H. (1986). *J. Chem. Phys.*, **85**, 601–610.

Sowada, U. and Holroyd, R. A. (1980). *J. Phys. Chem.*, **84**, 1150–1154.

Stezowski, J. J., Stigler, R.-D. and Karl, N. (1986). *J. Chem. Phys.*, **84**, 5162–5170.

Strebel, P. J. and Soos, Z. G. (1970). *J. Chem. Phys.*, **53**, 4077–4090.

Streitweiser, A., Heathcock, C. H. and Kosower, E. M. (1992). Introduction to Organic Chemistry (Fourth edition). Macmillan, New York. 1256 pp.

Sudborough, J.J. and Beard, S.H. (1910). *J. Chem. Soc.*, **97**, 773–798.

Tamres, M. L. and Strong, R. L. (1979). *Mol. Assoc.*, **2**, 331–456.

Tokumoto, M., Koshizuka, N., Anzai, H. and Ishiguro, T. (1982). *J. Phys. Soc. Jpn.*, **51**, 332–338.

Tokumoto, M., Koshizuka, N., Murata, K., Kinoshita, N., Anzai, H., Ishiguro, T. and Mori, N. (1982). *Mol. Cryst. Liq. Cryst.*, **85**, 195–202.

Torrance, J. B., Vazquez, J. E., Mayerle, J. J. and Lee, V. Y. (1981). *Phys. Rev. Lett.*, **46**, 253–257.

Umland, T. C., Allie, S., Kuhlmann, T. and Coppens, P. (1988). *J. Phys. Chem.*, **92**, 6456–6460.

Van Duyne, R. P., Suchanski, M. R., Lakovits, J. M., Siedle, A. R., Parks, K. D. and Cotton, T. M. (1979). *J. Am. Chem. Soc.*, **101**, 2832–2837.

Weil-Malherbe, H. (1946). *Biochem. J.*, **40**, 351–363.

Weiss, J. (1942). *J. Chem. Soc.*, pp. 245–252.

Wiberg, N. (1968). *Angew. Chem. Int. Ed. Engl.*, **7**, 766–779.

Wöhler, F. (1844). *Ann. Chem.*, **51**, 145–163.

Wooten, F. (1972). Optical Properties of Solids, Academic Press, New York, p. 72.

Wright, J. D., Ohta, T. and Kuroda, H. (1976). *Bull. Chem. Soc. Jpn.*, **49**, 2961–2966.

Yakushi, K., Nishimura, S., Sugano, T., Kuroda, H. and Ikemoto, I. (1980). *Acta Cryst.*, **B36**, 358–363.

Chapter 14

Layered molecules with intra-molecular donor–acceptor interactions

... lock'd in fierce embrace ...

Unknown.

Summary: Layered molecules (in the present context) are donor–acceptor cyclophanes in which the mutual orientation and/or location of the donor and acceptor ring systems can be altered in order to investigate the conditions for maximum overlap and hence charge transfer. The enhancement of charge transfer in pseudogeminal as opposed to pseudo-ortho cyclophanes has been demonstrated by comparison of UV-visible spectra for many diastereoisomeric pairs. Crystal structure determinations show that there is superposition of donor and acceptor moieties in the paracyclophane series, with appreciably more distortion in [2.2] than in [3.3] paracyclophanes. Charge transfer interactions persist in triply and quadruply layered donor–acceptor paracyclophanes but the orientation effects are lost. The charge transfer spectra of *syn* and *anti* diastereoisomers in the metacyclophane series are remarkably similar; crystal structure analyses show considerable intramolecular distortion and mutual displacement and nonparallelism of the donor and acceptor moieties. The ideas developed in the cyclophane studies are beginning to be applied to other areas of chemistry.

14.1 Introduction

We have already emphasized that the more stable π–π^* charge-transfer molecular compounds have equimolar ratios and that the most striking feature of their crystal structures is the alternating arrangement of donor and acceptor molecules, planes essentially parallel, in mixed stacks. However, the mutual donor–acceptor arrangements in the crystals, under which heading we include the aspects of D$\cdots$A interplanar distance, orientation and overlap, is not free of influence from neighbouring stacks and separation of the various contributory factors can be troublesome. Thus study of charge transfer in an *intra-molecular* situation has many advantages because the donor–acceptor arrangement is well-defined. This approach appears to have been suggested first by Cram and Day (1966), who synthesized **14.1** (the [2.2] compound); the corresponding [3.3] (Shinmyozu, Inazu and Yoshino, 1977) and [4.4] (Cram and Day, 1966) compounds have also been reported. Cram and Day (1966) anticipated that **14.2** and **14.3** would differ because of the possibility of interannular hydrogen bonding in **14.3** but not in **14.2** (Fig. 14.1). Such hydrogen bonding has not yet been encountered. However, in the early 1970s Staab pointed out a more important feature – that **14.2** and **14.3** differ in the mutual orientation of the π-systems of donor and acceptor moieties and thus permit a direct test of the validity of Mulliken's "overlap and orientation" principle for a D$\cdots$A pair.

Many cyclophanes suitable for this purpose have been synthesized in the last decade, principally by Staab and coworkers in Germany but also in Japan and elsewhere (Schroff, van der Weerdt *et al.*, 1973; Schroff, Zsom *et al.*, 1976) and we survey these results, which have been reviewed (Schwartz, 1990). The cyclophane systems needed were obtained by extensive synthetic programmes which will not be discussed here despite their many novel features. The UV-visible absorption spectra of many diasteroisomeric pairs were measured in dilute solution, with Beer's law checked to ensure that the spectra were truly those of intramolecular species. Solutions in rigid glasses allow low-temperature spectroscopy while determination of the crystal structures gives details of molecular structure (the rings are generally deformed in these strongly interacting systems) and also shows if there are appreciable intermolecular interactions in the crystals. Approximate molecular orbital calculations aid in the interpretation of the results.

The reader should be warned that what started out as a simple and clearcut means of testing Mulliken's "overlap and orientation" principle has developed complications; Staab, Dohling and Krieger (1991) remark "that a general and strict correlation between CT absorption and ground-state stabilisation by CT interaction (which is still widely taken for granted in organic chemistry) cannot be expected." This will be illustrated towards the end of this chapter.

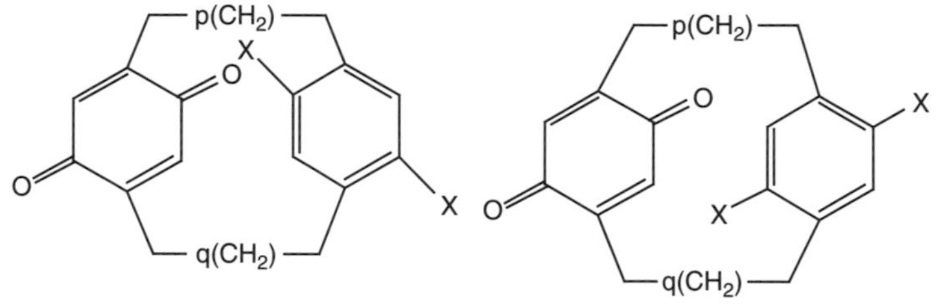

14.2   **14.3**

Fig. 14.1. Schematic diagram of the pseudo-ortho (**14.2**) and pseudo-geminal (**14.3**) diastereo-isomers of [n,m]paracyclophane quinhydrones (X=OH) and related molecules. The methylene linkages between the rings can be equal but are not required to be so; $n = p + 2$ and $m = q + 2$, where p and q have integer values ≥ 0. The para mode of bridging is shown for both rings but meta bridging, and mixed meta-para bridging, are also found, as are other types of ring. **14.1** has X = H, $p = q = 0$; **14.2** and **14.3** have X = OH and $p = q = 0$.

14.2   Molecules of the paracyclophane type

14.2.1   *Molecules derived from [n.n]paracyclophanes*

The experimental results show that para- and metacyclophane systems differ in important respects and we consider them in separate sections, starting with the paracyclophanes. Nomenclature and schematic structures are summarised in Fig. 14.1. Initial emphasis was placed on the quinhydrone systems; analogous crystalline intermolecular systems have been extensively studied (Section 15.7.1). The pseudo-ortho (**14.2**) and pseudogeminal (**14.3**) diastereoisomers were synthesized (Rebafka and Staab, 1973, 1974), followed by other syntheses and studies of physical properties (Staab and Rebafka, 1977; Staab, Herz and Henke, 1977; Staab and Haffner, 1977; Staab and Taglieber, 1977). The absorption spectra of the diastereoisomeric pair **14.2** and **14.3** are shown in Fig. 14.2 and it is clear that the charge transfer band centred at *ca.* 500 nm is much more pronounced for the pseudogeminal than for the pseudo-ortho isomer, and thus the charge-transfer interaction is appreciably stronger when the donor and acceptor moieties are parallel rather than when they are 60° apart. This conclusion is reinforced by a number of other measurements. For example, NMR results (^{1}H NMR, d_6-DMSO solution, coalescence of singlets at $\tau = 4.14$ and 3.98 above 161°C) shows that there is exchange of oxidation states between the two rings with simultaneous proton exchange in **14.3** but not in **14.2** (Rebafka and Staab, 1974). As the crystal structure of **14.3** has not been reported, it is not known whether there is interannular hydrogen bonding in this molecule. The pseudo-ortho configuration of the compound obtained in a synthesis of the paracyclophane diene quinhydrone (rings linked by double rather than single bonds) was inferred from the essential identity of its charge

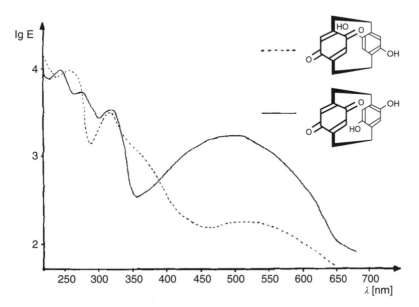

Fig. 14.2. Absorption spectra (dotted **14.2**, full **14.3**) of the diastereoisomeric [2.2]paracyclo-phane quinhydrones (solvent CH_3OH). (Reproduced from Rebafka and Staab, 1974.)

transfer absorption with that of **14.2** (Stobbe, Kirchmeyer, Adiwidjajaj and Meijere, 1986). The orientation dependence of the charge transfer excitations in the [2.2] and [3.3] quinhydrone paracyclophanes has been explained in terms of Hückel molecular orbital theory (Vogler, 1983a,b), and good agreement obtained with experiment.

The enhancement of the charge-transfer band in the pseudogeminal [2.2]para-cyclophane donor–acceptor diastereoisomer (**14.3**) over that in the corresponding pseudo-ortho diastereoisomer (**14.2**) is a quite general phenomenon (see spectral parameters summarized in Table 14.1). The [2.2]paracyclophane quinhydrones have the disadvant-age, in the present context, of very strong interannular interactions, with marked deforma-tions of the rings and distances of $\approx$2.9 Å between their mean planes, compared to the $\approx$3.2 Å found in the crystalline quinhydrones. These complicating features do not appear in the [3.3]paracyclophane quinhydrones (Staab and Herz, 1977b; Staab, Herz, Krieger and Rentea, 1983); in these diastereoisomeric pairs the CT interaction is also strikingly greater in the pseudogeminal than in the pseudo-ortho diastereoisomer.

One of the objectives of the synthetic programmes was the preparation of cyclophanes with very strong donor and acceptor moieties. One such molecule would contain TMPD as donor and TCNQ as acceptor. Paracyclophanes with TCNQ as acceptor and donors of various kinds have been reported (Staab and Knaus, 1979; Staab, Knaus, Henke and Krieger, 1983; Tatemitsu et al., 1978). They show the same orientation effects as described above for the quinhydrones. The importance of the methoxy substituents in enhancing the donor strength is shown by the much stronger CT absorption in the compound with a p-dimethoxybenzene donor group and TCNQ acceptor group than when the donor is a benzene ring (Staab, Knaus et al., 1983). The converse situation where TMPD is donor and there are various acceptors has been achieved in the [2.2] series (Staab, Reimann-Haus et al., 1983) and also in the [3.3]paracyclophane series (Staab, Gabel and Krieger, 1983), where specifically the pseudo-ortho and pseudogeminal diastereoisomers of N,N,N′,N′-tetramethyl[3]-(2,5)p-benzoquinone-[3]paracyclophane-5,8-diamine were prepared and the crystal structure of an analog to the latter compound (in which the p-benzoquinone group was replaced by p-dimethylcyanobenzene) has been determined. One diastereoisomer with the desired TMPD–TCNQ combination has been synthesized – the pseudogeminal N,N,N′,N′-tetramethyl[3]-(2,5)tetracycanoquinodimethane[3]paracyclophane-5,8-diamine   (Staab, Hinz, Knaus and Krieger, 1983). Its absorption spectrum shows a particularly broad CT band stretching over the region from 600–1600 nm. The trans-annular coupling is so strong (because of low I_D and high E_A values) that there is no longer any localization of valence electrons in the two rings and the spectrum does not show features of the TMPD and TCNQ spectra at short wavelengths, such as are found in weaker donor-acceptor systems. There is little dependence of the absorption spectrum on solvent polarity, the C≡N stretching frequency in TCNQ is 2216 cm^{-1} and there is no ESR spectrum; all these facts point to a neutral ground state for the molecule rather than the ionic ground state that might have been anticipated (cf. Section 15.10.2). The TCNQ moiety is also neutral in the corresponding pseudo-ortho diastereoisomer ($\nu_{CN} = 2220$ cm^{-1}).

Another such strong donor-acceptor pair is TTF and TCNQ where considerable pro-gress has been made in the synthesis of the separate systems (TCNQ as before, TTF as below) but the desired combination in the same cyclophane (**14.4**) has not yet been reported. However, [3]tetrathiafulvaleno-[3]paracyclophane (**14.5.1**) and an analogous

Table 14.1. Comparison of spectral parameters for charge-transfer absorption bands of various donor–acceptor cyclophanes in pseudogeminal and pseudo-ortho configurations (note that this distinction does not hold when the donor is a benzene ring). The λ_{max} values are in nm

Donor moiety	Acceptor moiety	Cyclophane	Pseudo-geminal $\lambda_{max}/\varepsilon$	Pseudo-ortho $\lambda_{max}/\varepsilon$	Solvent	Ref
p-Dimethoxy-benzene (DMB)	TCNQ	[2.2]para	695/3225	730/258	CHCl$_3$	SKHK83
DMB	TCNQ	[3.3]para	705/3450	670/117	CHCl$_3$	SK79
Benzene	TCNQ	[3.3]para	530*/795			SKHK83
DMB	p-benzo-quinone (BQ)	[2.2]para	483/1329	438/160 498/142	CHCl$_3$	SHH77
DMB	BQ	[3.3]para	475/3000	500*/100	CHCl$_3$	SDK83
DMB	BQ	[4.4]para	466/360	445/97	CHCl$_3$	SDK83
DMB	BQ	[2]para crown (3)	462/324			SSK83
DMB	BQ	[2]para crown (3) with Na$^+$	478/874			SSK83
TMPD	BQ	[2.2]para	577/1930	595/160	CH$_2$Cl$_2$	SR83
TMPD	BQ	[3.3]para	538/2455	537/76 387/1480$^{\#}$		SGK83
TMPD	TCNQ	[3.3]para	1050/3160			SH83
p-Dihydroxy-benzene (quinol)	BQ	[2.2]para	495/1600	515/170	CH$_3$OH	SR77
quinol	BQ	[3.3]para	462/3210	500*/105	dioxane	SHKR83
p-trimethylsil-oxybenzene	BQ	[2.2]para	444/1710 500*/116	430/205	CHCl$_3$	SHH77
DMB	pyrazine	[2.2]para	440/405	460/103	CF3-COOH	SA81
benzene	BQ	[2.2]para	340/597			CD66
benzene	BQ	[3.3]para	406/407			SIY77
benzene	BQ	[4.4]para	288/1290			CD66
DMB	p-dinitrobenzene	[2.2]para	468/414	475/120	CHCl$_3$	SH77

* shoulder $\quad$ $^{\#}$2nd CT band

References: CD66 – Cram and Day, 1966; SA81 – Staab and Appel, 1981; SDK83 – Staab, Döhling and Krieger, 1981; SGK83 – Staab, Gabel and Krieger, 1983; SH77 – Staab and Haffner, 1977; SH83 – Staab, Hinz, Knaus and Krieger, 1983; SHH77 – Staab, Herz and Henke, 1977; SHKR83 – Staab, Herz, Krieger and Rentea, 1983; SIY77 – Shinmyozu, Inazu and Yoshino, 1977; SK79 – Staab and Knaus, 1979; SKHK83 – Staab, Knaus, Henke and Krieger, 1983; SR77 – Staab and Rebafka, 1977; SR83 – Staab, Riemann-Haus, Ulrich and Krieger, 1983; SSK83 – Staab, Starker and Krieger, 1983.

[4.4] compound have been synthesized (Staab, Ippen *et al.*, 1980) and crystal structure analyses (briefly) reported for both compounds; surprisingly the latter compound was found to be the isomer **14.6** rather than **14.5.2**.

The [2.2]tetrathiafulvalene isomers **14.7** and **14.8** have been synthesized but only the second was obtained in pure form (Ippen, Tao-pen, Starker, Schweitzer and Staab, 1980).

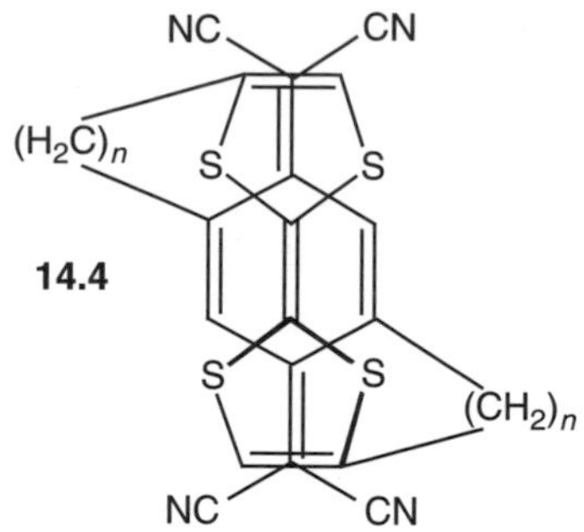

14.4

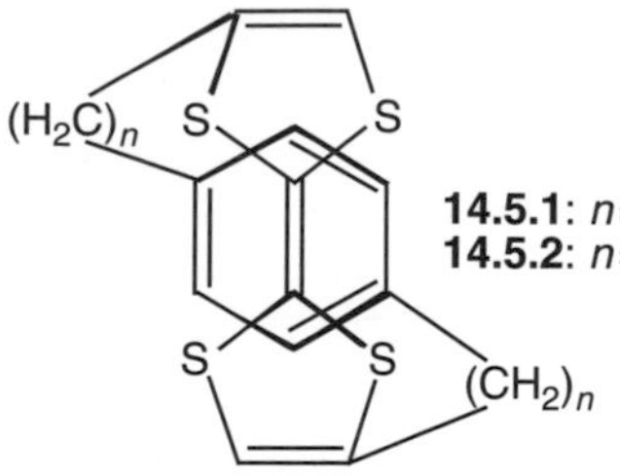

14.5.1: *n*=3
14.5.2: *n*=4

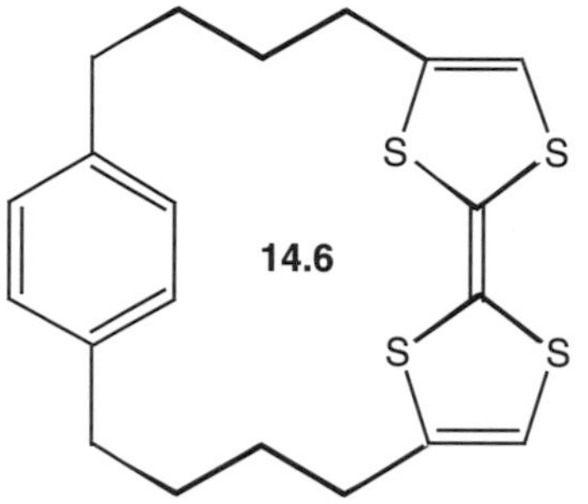

14.6

Its structure has been confirmed crystallographically; the molecule has a step-like *anti* conformation. The [3.3]tetra-thiafulvalenes (**14.9** and **14.10**) have been synthesized but not completely purified. Black needles of a complex of composition [3.3]tetra-thiafulvalene:(TCNQ)$_4$ were obtained and shown to be triclinic but the structure was not reported. Conductivity measurements gave values of about 10^{-2} S/cm along the needle axis at 300K the conductivity at 40K is ten orders of magnitude less than at 400K!

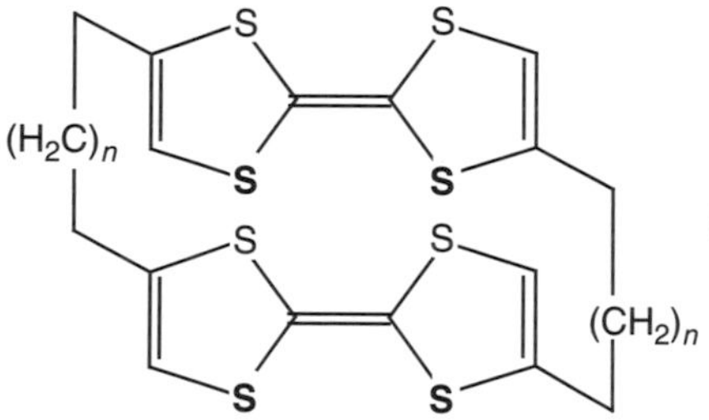

14.7: *n*=0; **14.9:** *n*=1. 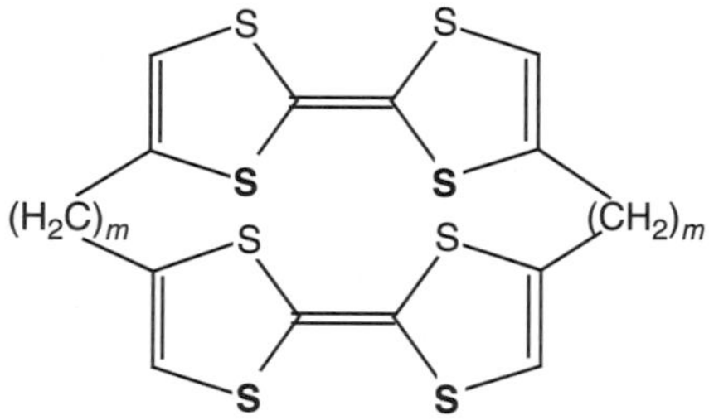   **14.8:** *m*=2; **14.10:** *m*=3.

Sometimes there are complications. For example, the pyrazine moiety acts as an acceptor in the diastereoisomeric 12,15-dimethoxy-4,7-diaza[2.2]paracyclophanes (pseudogeminal and pseudo-ortho) (Staab and Appel, 1981); however, there are only small differences, which are solvent dependent, between the absorption spectra of the two diastereoisomers in the CT region of the spectrum. Thus a straightforward explanation in terms of ring overlap is no longer possible, as already noted in the Introduction.

The generalization that there is greater CT interaction in pseudogeminal than in pseudo-ortho diastereoisomers breaks down for [4.4]paracyclophanes and those with longer methylene bridges (Staab, Döhling and Krieger, 1981; Staab, Starker and Krieger, 1983). Crystal structure analysis of pseudogeminal-6,9,16,19-tetramethoxy[4.4]paracyclophane (Staab, Döhling and Krieger, 1983) and pseudogeminal-7,10,18,21-tetramethoxy-[5.5]paracyclophane (Staab, Starker and Krieger, 1983) shows that the interannular distances are 4.01 and 5.11 Å respectively. There is no distortion of the rings and presumably there is little CT interaction at such large distances.

Further comparison of the relative strengths of the donor–acceptor interaction in the two diastereoisomers can be made through emission spectra and zero-field splitting (ZFS) parameters (Hausser and Wolf, 1976); in these experiments the molecules are held at 1.3K, either in rigid glasses or as their crystals. The diastereoisomers used were those of 4,7-dicyano-12,15-dimethoxy-[2.2]-paracyclophanes (Schweitzer, Hausser et al., 1976) and 12,15-dimethoxy-4,7-diaza[2.2]-paracyclophanes (Staab, Herz and Henke, 1977). Crystal structures have been determined for the first pair of compounds by Irngartinger and Merkert but details do not appear to have been published. Emission spectra were used as follows: the characteristic properties of the excited triplet state of CT compounds originate from the fact that the two triplet excitons have a high probability of being, at a given time, in two different orbitals separated in space, i.e. one in the HOMO of the donor and the other in the LUMO of the acceptor. Consequently the exchange integral is diminished, with the energy of the first excited triplet state being reduced less than that of the corresponding excited singlet state. Thus large spectral overlap occurs between phosphorescence (triplet to ground state transition) and fluorescence (first excited singlet to ground state transition) spectra. Furthermore, the absolute red shifts of the emission spectra in CT-cyclophanes are larger than those in CT molecular compounds because of the stronger interannular interactions in the cyclophanes. The ESR method is to be preferred to emission spectroscopy because the ZFS parameters (the D values) are more sensitive than emission spectra to differences in CT interaction. The larger the separation between the two triplet electrons, the smaller is $|D|$; small $|D|$ values thus correspond to larger CT interactions. The trend in $|D|$ values shown in Table 14.2 is entirely compatible with the absorption spectra and also with the results of Hückel MO calculations (Vogler, Ege and Staab, 1977). However, it is not clear why the $|D|$ values from the crystals are lower than those from the rigid glasses.

The spectroscopic and other results mentioned above are reinforced by the detailed molecular geometries obtained from crystal structure analyses (note that in some instances structures of analogs have been reported rather than those of the immediately-relevant molecules, presumably because of difficulties in obtaining suitable crystals of the molecules of primary interest). Only a limited number of representative structures are discussed here.

The pseudogeminal and pseudo-ortho diastereoisomers of 4,7-dimethoxy-[2](2,5)tetracyanoquinodimethane[2]paracyclophane (Staab, Knaus, Henke and Krieger, 1983) are shown in Figs. 14.3 and 14.4, the actual molecular structures showing a remarkable resemblance to schematic formulae such as **14.1.** The geometries of the two diastereoisomers are very similar (apart from the mutual positioning of the substituents), even in respect to the distortions introduced by the interannular interactions. Thus it is probable that these geometries can be taken as representative of all diastereoisomeric pairs in the [2.2]paracyclophane series.

Table 14.2. Zero Field Splitting (ZFS) parameters measured by Optical Detection of Magnetic Resonance (ODMR); all measurements at 1.3K. n-Octane and PMMA are rigid glasses. The $|D|$ values are given in units of cm^{-1}.

Compound	Component		Paracyclophane		Remarks
	Donor	Acceptor	Pseudogeminal	Pseudoortho	
a. 2,5-dimethylpyrazine		0.177			n-octane
b. 1,4-dimethoxy-2,5- dimethylbenzene	0.116				n-octane
c. 4,7-diaza[2.2]paracyclophane (**Combination a-b**)			0.0967	0.1022	n-octane
d. 1,4-dicyano-2,5- dimethylbenzene		0.1229			PMMA
e. 4,7-dicyano-12,15- dimethoxy[2.2]-paracyclophane (**Combination b-d**)			0.0313 0.0213	0.0642 0.0259	PMMA single crystal

The results of crystal structure analyses of [3.3]paracyclophanes are similar to those in the [2.2] series, and we shall give only one example – pseudogeminal 14,17-dimethoxy[3](2,5)-p-benzoquinone[3]paracyclophane (Staab and Knaus, 1979; Fig. 14.5). There is almost exact overlap of the two rings as in the [2.2] series but less ring distortion. This perhaps explains the somewhat higher extinction coefficients found for [3.3] than for [2.2]cyclophanes (Table 14.1). The methylene bridge is disordered in some of the crystal structures in this group (for example, Fig. 14.5) but not in all of them.

When both rings of the [n·n]paracyclophane are the same and both are para-disubstituted in the same way, then the pseudogeminal diastereoisomer will be centrosymmetric while the pseudo-ortho diastereoisomer will have a $C_2(2)$ axis normal to the mean ring planes (these symmetries can be exact or approximate). If the rings are different then both diastereoisomers will have twofold axes (or even lower symmetry). If the molecules are chiral, there is a possibility of spontaneous resolution if the compound crystallizes in a Sohnke space group (cf. Section 11.2.2.1). This has so far been reported, for paracyclophanes, only for pseudogeminal-N,N,N',N'-tetramethyl[3](2,5)-p-benzoquinone[3]paracyclophane-14,17-diamine (space group $P2_1$, $Z=2$, molecular symmetry C_1-1; Staab, Gabel and Krieger, 1983; BUVRIT). The metacyclophane (see Section 14.4 below) syn-15,18-dimethoxy[3]-p-benzoquinone[3]metacyclophane also crystallizes in space group $P2_1$, $Z=2$; Staab, Herz, Döhling and Krieger, (1980). Correlation of absolute configuration and optical rotation has not been reported for these two compounds but has been effected for a metaparacyclophane (see Section 14.3 below).

A rather detailed comparison of the effects of various factors on donor–acceptor interactions in paracyclophanes has been made possible by the synthesis of a series of [2.2], [3.3] and [4.4] paracyclophanes all containing 1,2,4,5-tetracyanobenzene as acceptor and with a variety of donor moieties (Staab, Wahl and Kay, 1987). As crystal structures have been determined for most of these molecules, the comparisons can be

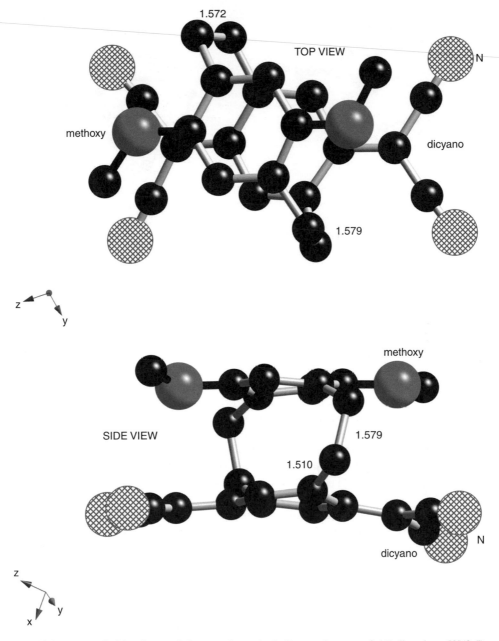

Fig. 14.3. Top and side views of the pseudogeminal diastereoisomer of 4,7-dimethoxy[2](2,5)-tetracyanoquinodimethane[2]paracyclophane (BUZROD). Note the close superpositioning of donor and acceptor moieties, and also their appreciable deformation. Distances in Å. Full molecular dimensions of all crystal structures discussed in this Chapter are given in the original papers. (Data from Staab, Knaus, Henke and Krieger, 1983.)

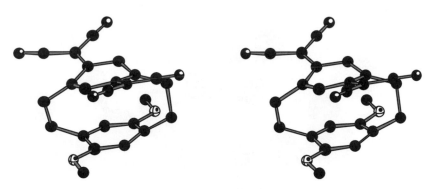

Fig. 14.4. Stereoview view of the pseudo-ortho diastereoisomer of 4,7-dimethoxy[2](2,5)-tetra-cyanoquinodimethane[2]paracyclophane (BICFAU01). The overall molecular structure is very similar to that of the pseudogeminal diastereoisomer in Fig. 14.3. (Diagram produced using data from Staab, Knaus, Henke and Krieger, 1983.)

made for systems of well-defined geometry. The spectroscopic results are summarized in Table 14.3. The crystallographic results (Staab, Krieger, Wahl and Kay, 1987) follow the pattern established previously – considerably more distortion in the [2.2] (FIPLOF, FIPLUL, FIPMAS) than in the [3.3]paracyclophanes (FIPMEW, FIPMIA, FIPMOG), while distortion is negligible in the [4.4]paracyclophanes; almost complete eclipse of the donor and acceptor ring systems in the [2.2]paracyclophanes but with some mutual lateral shift of rings in [3.3]paracyclophanes and rather more variability in [4.4]paracyclophanes; substituents are coplanar with associated rings except for the methoxy groups in the tetramethoxy-substituted [2.2]paracyclophanes where the methyl groups are roughly normal to the ring planes. As Staab, Krieger, Wahl and Kay, (1987) point out, these changes can be followed qualitatively by changes in the colour of the crystals and more quantitatively by changes in λ_{max} and ε of the charge transfer absorption bands (Table 14.3). For example, in the [2.2]paracyclophane series, the benzene donor is weakest while increasing the donor strength by substitution of four methyl groups or two methoxy groups leads to a deepening of the colour of the solutions; however, four methoxy groups cannot be coplanar, thus reducing the mesomeric effect and the donor strength. If the dimethoxy donor is kept constant while the methylene chain length is increased, then the colour weakens from deep violet to dark red to orange as one passes from the [2.2] to the [3.3] and then to the [4.4]paracyclophane.

The paracyclophanes discussed above are intramolecular charge-transfer molecular compounds and, as such, have a donor face and an acceptor face (Janus-like molecules) Thus one can anticipate formation in the crystals of stacks of the following type:

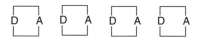

These occur in some crystals but not in others; certainly this type of stacking is not a feature of the crystal structures comparable in importance to the – D A D A D A – arrangement

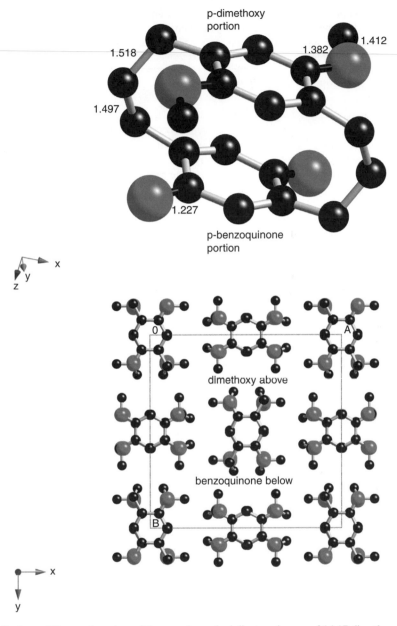

Fig. 14.5. (upper) Perspective view of the pseudogeminal diastereoisomer of 14,17-dimethoxy[3](2,5)-*p*-benzoquinone[3]paracyclophane (CECTIN). The molecule has a twofold axis normal to the mean ring planes. Distances in Å. There is some disorder, not shown, in the methylene bridges. Note the close superpositioning of donor and acceptor moieties, and also some deformation.
(lower) Packing diagram viewed down [001]; tetragonal, 15.891(2), 13.223(2) Å, $I4_1cd$, $Z = 8$. (Data from Staab, Herz, Krieger and Rentzea, 1983.)

Table 14.3. Colours and parameters for charge transfer absorption bands in spectra of [2.2], [3.3] and [4.4]paracyclophanes, with 1,2,4,5-tetracycanobenzene as acceptor moiety and various donors. The data in this Table are taken from pp. 556–557 of Staab, Krieger, Wahl and Kay, (1987). Asterisks indicate shoulders

Donor Moiety	[n.n]	Colour	Charge-transfer band	
			λ_{max}(nm)	ε
1. Benzene	[2.2]	yellow	395*	437
2. p-dimethoxybenzene		deep violet	520	240
3. 1,2,4,5-tetramethylbenzene		red	440	537
4. 1,2,4,5-tetramethoxybenzene		orange red	380*	1318
5. benzene	[3.3]		416	1288
6. p-dimethoxybenzene		dark red	508	347
7. 1,2,4,5-tetramethylbenzene			434	575
8. p-dimethoxybenzene	[4.4]	orange	495	89

found in the mixed stacks of most intermolecular donor–acceptor molecular compounds (see Chapter 15). However, we note that this sort of stacking is found in many of the donor–acceptor complexes discussed in Chapter 3.

One may venture the suggestion that ternary molecular compounds could be formed with stacks having the following arrangement:

$$\boxed{D} \quad \boxed{A} \quad \boxed{D'} \quad \boxed{A} \quad \boxed{D} \quad \boxed{A'} \quad \boxed{D} \quad \boxed{A} \quad \boxed{D'} \quad \boxed{A} \quad \boxed{D}$$

Synthesis of such molecular compounds does not appear to have been attempted – a neat balance of donor and acceptor strengths would appear to be needed for success; however, it should be noted that [3.3]paracyclophane forms a $1:1$ $\pi:\pi^*$ molecular compound with TCNE (Bernstein and Trueblood, 1971; PACTCN10) (see also Section 15.4), so the suggestion may not be entirely fanciful.

It has been argued that cyclophanes containing donor (or alternatively acceptor) moieties in *both* rings would not be likely to form mixed stacks with added acceptor (donor) molecules but that segregated stacks would be favored (Staab, Gabel and Krieger, 1987). This proposal has so far been tested only for pseudogeminal-5,8,14,17-tetrakis(dimethyl-amino)[3.3]paracyclophane, which is formulated as

and which we shall denote for convenience as [TMPD-(CH$_2$)$_3$]$_2$. This material forms a black $1:2$ molecular compound of metallic appearance with TCNQ, the room-temperature conductivity along the "longer crystal axis" being 1.5 S/cm, about 10^6 times as large as that of TMPD···TCNQ (in which there is a mixed stack arrangement of ionized moieties). The crystal structure has not yet been reported.

The compound [TMPD-(CH$_2$)$_3$]$_2$ would be expected to undergo oxidation in four stages, the first giving an analog of Würster's blue cation. Cyclic voltametry shows that oxidation occurs at potentials of −0.242, −0.102 and 0.249 V (two unresolved stages); these values should be compared with potentials of −0.206 and 0.378 V for TMPD and 0.081 and 0.242 V for 2,5-dimethyl-TMPD. Thus oxidation to the radical cation is more facile in [TMPD-(CH$_2$)$_3$]$_2$ than in the other two compounds. Fast electron exchange has been demonstrated in the radical cation according to the following scheme:

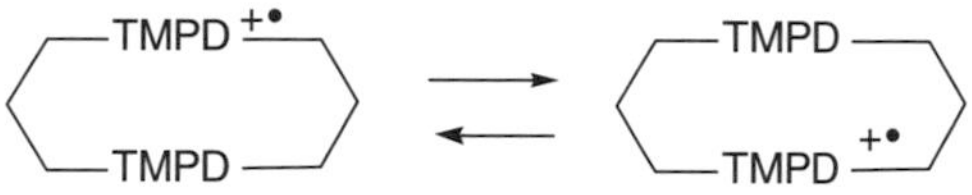

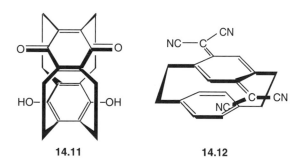

The very rigidly linked [2.2.2.2](1,2,4,5) cyclophane quinhydrone (**14.11**; Staab and Schwendeman, 1978) is strongly deformed, the interannular distance (as judged from the crystal structures of the corresponding cyclophane (Hanson, 1977; CYLOPH) and the tetraquinone (Krieger, 1978) is about 2.69 Å, compared to about 3 Å in **14.3**; the rings are boat-shaped in both reference molecules, with the substituent groups pointing *away* from the transannular region rather than towards it as in the paracyclophanes. Nevertheless the absorption spectrum is very similar to that of **14.3** (for the CT band, $\lambda_m = 491$ nm and $\varepsilon = 1280$, compared to 492 nm and 1600 for **14.3**). Another surprising feature is that there is no hydrogen exchange between donor and acceptor moieties, even on heating to 140 °C; perhaps this difference from the behaviour of **14.3** is due to a different type of deformation in **14.11** which increases the distance between hydroxyl and quinone oxygens beyond the limit for which exchange is possible.

14.11   **14.12**

Crystal structures have been reported (Mizuma, Miki, Kai, Yasuoka and Kasai, 1982) for **14.12** (as its benzene solvate, $P\bar{1}$, $Z = 2$; BICDUM) and for the 14,17-dimethoxy derivative of **14.12** (*Fdd*2, $Z = 16$; pseudo-ortho diastereoisomer; BICFAU). The molecular structures show the distortions familiar from earlier work on cyclophane systems; in particular the six-membered rings have boat forms and the C(CN)$_2$ portions are slightly twisted away from coplanarity with the ring. The packing in both crystals is based on head-to-tail stacking of the molecules (cf. previous paragraph). The solvent molecules in the benzene solvate do not participate in the stacking but are located in sheets about the (100) planes, interleaving double sheets of stacks of cyclophane molecules.

14.2.2 *Systems related to [n·n]paracyclophanes*

Enhancement of the CT absorption was obtained in intramolecular quinhydrones with oligo-oxaparacyclophane structures (**14.13**; Bauer, Briaire and Staab, 1983). The effect was most marked for $n = 3$; the CT absorption in the region of 500 nm was rather weak for the neat molecules but was strongly enhanced when the Na$^+$ crown ether complex was formed, the conclusion being that the complexation had forced the donor-acceptor quinhydrone pair into a more parallel alignment. The crown ether portion of the molecule can be adjusted to be selective for a particular cation, while clathration of the cation is shown by enhanced spectral absorption; thus the principle of a cation-selective ligand with a "built-in" charge-transfer indicator has been demonstrated.

14.13

Note: the –CH$_2$–O–CH$_2$– group is repeated n times (for $n \geq 1$).

Some work has also been done on cyclophanes containing naphthaleno systems, including synthesis of intramolecular quinhydrones based on [2](1,4)naphthaleno[2]-paracyclophane (Herz and Staab, 1977) and [2.2](1,4)naphthalenophane. The *syn* and *anti* isomers **14.14** and **14.15** of the latter type have very similar absorption spectra, both showing broad CT bands between 500 and 700 nm (Staab and Herz, 1977a; not reproduced here). It was inferred that in these molecules direct donor–acceptor interaction through space was less important than the interaction through the strongly coupled [2.2]paracyclophane portions of the molecules. The spectra of **14.16** and **14.17** (Fig. 14.6) show even broader CT bands than **14.14** and **14.15**.

14.14 (*syn* isomer, dec 260 °C)   **14.15** (*anti* isomer)

The crystal structure of **14.16** (space group *Cc*, $Z = 4$) has been reported (Mizuma, Miki, Kai, Tanaka and Kasai, 1982; BIMNOA). One of the six-membered rings in the

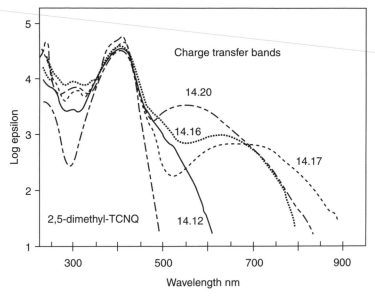

Fig. 14.6. Electronic spectra (in CH_2Cl_2) of the two-layer molecules **14.12**, **14.16** and **14.17**, the three-layer molecule **14.20** and 2,5-dimethyl-TCNQ. (Adapted from Yoshida *et al.*, 1978.)

naphthaleno portion is boat-shaped while the other, which protrudes, is planar. The crystals contain head-to-tail stacks of molecules.

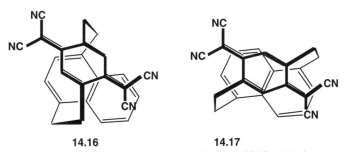

14.16 **14.17**

(Note: some double bonds have been omitted in the TCNQ moieties).

Some work has been done on systems where the two rings have different sizes, one six-membered and the other seven-membered; para systems are discussed here and meta and mixed systems later. The [2]paracyclophane[2](3,7)*p*-tropoquinonophane (**14.18**) shows little evidence of CT absorption, and this is also true of the corresponding [3.3] compound (Kawamata, Fukazawa, Fujise and Ito, 1982a,b). However, there is strong charge transfer when the seven-membered ring is positively charged and so acts as a strong electron acceptor; for example [2.2](1,4)tropyliopara-cyclophane tetrafluoroborate (**14.19**) has a broad CT band centred at about 350 nm, with ε_m about 3100 (Horita, Otsubo, Sakata and Misumi, 1976; O'Connor and Keehn, 1976).

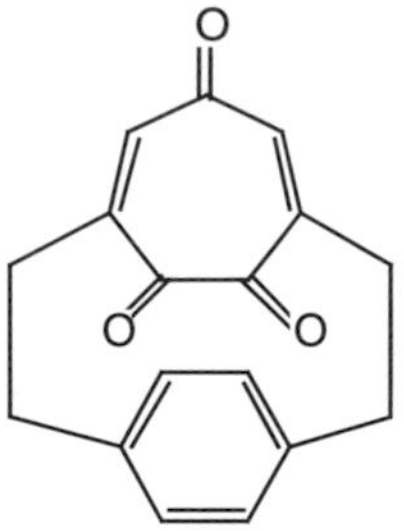

14.18 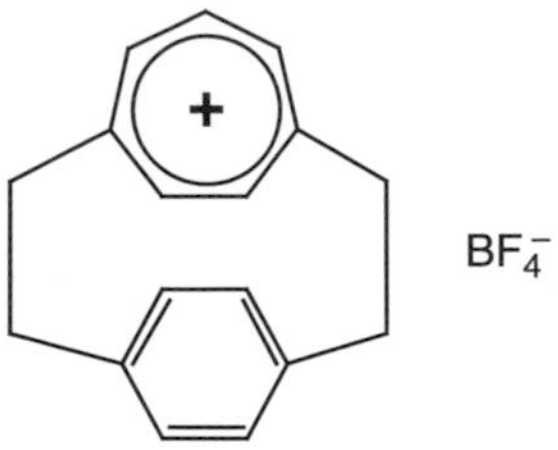   **14.19**

14.2.3 *Multi-layered systems*

Considerable effort has been invested in the synthesis of multilayered systems and, so far, triply and quadruply layered molecules are known. In the triple-layer molecules **14.20–14.25** there are rather similar CT bands (**14.21** and **14.23** have essentially identical spectra), which do not depend on the mutual orientation of donor and acceptor moieties (Staab, Zapf and Gurke, 1977; Machida, Tatemitsu, Sakat and Misumi, 1978; Staab and Zapf, 1978). **14.20** has the absorption maximum of its CT band in the 600 nm region (Fig. 14.6) but replacement of the TCNQ moiety in these molecules by *p*-benzoquinone leads to appreciable bathochromic shifts to the 450–500 nm region. On the basis of the spectroscopic results Machida, Tatemitsu, Sakat and Misumi (1978) inferred that "a sandwiched benzene ring functions as a sort of conductor for intra-molecular donor–acceptor interaction and not as an insulator." Presumably this is the reason why there is a distinct CT absorption for **14.20** but only a barely noticeable shoulder for **14.12** (Fig. 14.6). The generalisation holds also for the triple-layer charged tropylium system **14.24** (Horita, Otsubo, Sakata and Misumi, 1976). Thus it is intriguing to note that the CT interaction in **14.25** is weak (Tatemitsu, Otsubo, Sakata and Misumi, 1975). Unfortunately details of the absorption spectra have not been published for the *p*-dicyano triple-layer analogs (Yoshida, Tatemitsu, Sakata, Misumi, Masuhara and Mataga, 1976) of **14.21–14.23** (where the *p*-benzoquinone group has been replaced by *p*-dicyanobenzene), which appear to be the only other comparable group of molecules to have been synthesized.

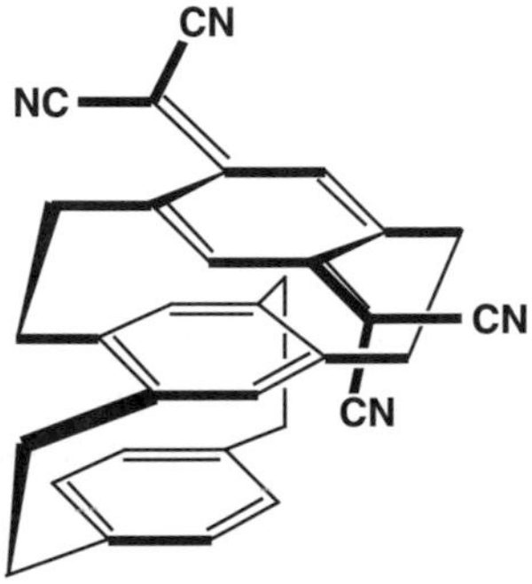

14.20

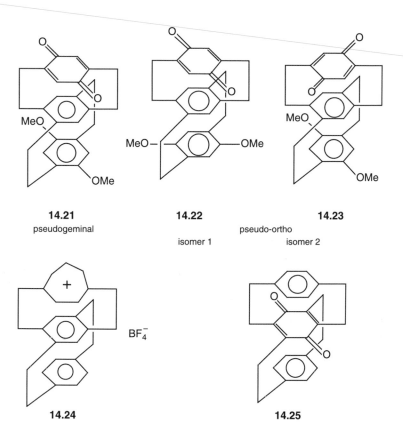

14.21
pseudogeminal

14.22

isomer 1

14.23

pseudo-ortho

isomer 2

BF_4^-

14.24

14.25

There are four-layer systems (Staab and Zapf, 1978) with a *p*-benzoquinone moiety at one end as acceptor and a *p*-dimethoxybenzene moiety at the other end as donor, separated by two benzene rings. These compounds show intense CT absorption bands in the 350–550 nm range (a: $\lambda_m = 447$ nm, $\varepsilon = 2500$; b: $\lambda_m = 450$ nm, $\varepsilon = 2490$); the intramolecular nature of the absorption was checked from the concentration dependence of the spectra. Not only are the intensities of the CT bands higher than those of the comparable triple-layered [2.2]paracyclophane quinhydrones but they also show a distinct bathochromic shift, indicating overall enhancement of the donor strength of the π-electron system interacting with the acceptor *p*-benzoquinone moiety. These molecules were synthesized as a diastereoisomeric pair but there was no evidence from their spectra of any dependence of the absorption on the mutual orientation of donor and acceptor.

Thus the triple- and quadruple-layer systems behave differently in this regard from the simpler [2.2] and [3.3] paracyclophane quinhydrones. Simple Hückel molecular orbital theory breaks down when four-layer cyclophane double-quinhydrones are considered, the states being dominated by important (nearly first order) configuration interaction (Vogler, 1983b).

The crystal structure of **14.25** has been reported (Toyoda, Tatemitsu, Sakata, Kasai and Misumi, 1986; FEFYIY) and also that of the compound in which the *p*-benzoquinone moiety is replaced by bromobenzene (Koizumi, Toyoda, Miki, Kasai and Misumi, 1986; DOHXAZ); the crystals are isomorphous, *Pbcn*, $Z=4$, and the molecules have twofold axes along the O...O (Br...H) vectors. The outer benzene rings are deformed to boats and the inner ring has a twist shape in these triply layered molecules. A similar pattern of distortions for inner and outer rings is found in the centrosymmetric quadruply layered tetramethyl [2.2]cyclophane $C_{40}H_{14}$ (Mizuno *et al.*, 1977; MPCPHT10).

14.3   Molecules of the metaparacyclophane type

The [2.2] metaparacyclophane quinhydrones and derivatives are conveniently treated here because they act as a bridge between the paracyclophanes, which they resemble in rigidity, and the metacyclophanes, to which they are perhaps closer in geometrical structure.   12,15-Dimethoxy[2](2,6)-*p*-benzoquinono[2]paracyclophane (**14.26**; DAL-TOZ) and 13,16-dimethoxy[2](2,5)-*p*-benzoquinono[2]metacyclophane (**14.27**) were first reported by Staab, Jörns and Krieger in 1979 and, in more detail, some years later (Staab, Jörns, Krieger and Rentzea, 1985). The quinhydrone analog of **14.27** has also been reported (Tashiro, Koya and Yamamoto, 1983). **14.26** has a broad CT band from 400–630 nm ($\lambda_m=490$ nm, $\varepsilon=590$, in $CHCl_3$), whereas the CT band of **14.27** is blue-shifted by about 70 nm ($\lambda_m=420$ nm, $\varepsilon=825$, in $CHCl_3$).

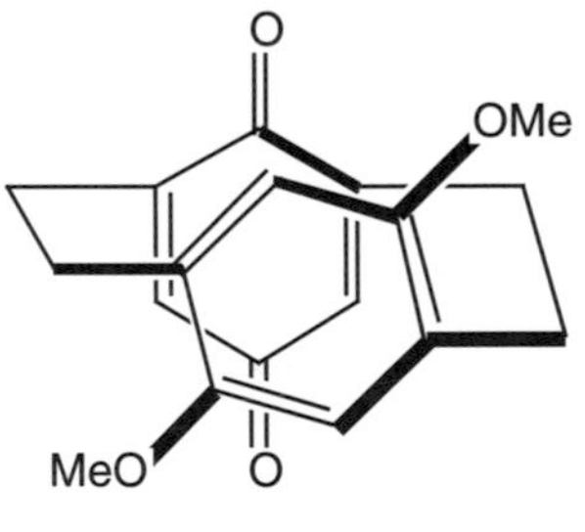

14.26 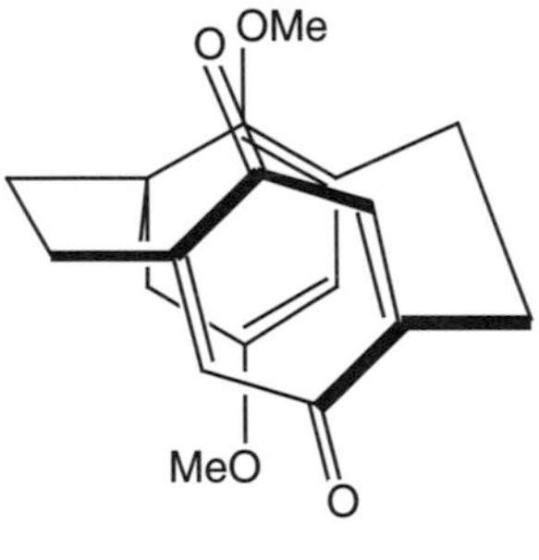   **14.27**

14.26 crystallizes in three polymorphic forms, two racemic and one enantiomorphic ($P2_12_12_1$, $Z=4$) and the stereochemical implications derived from the crystal structure of the latter have been investigated in particularly thorough fashion (Staab, Jörns, Krieger and Rentzea, 1985). The meta-bridged quinone unit shows a much greater deformation from planarity than the para-bridged aromatic moiety (Fig. 14.7). The carbonyl group is located above the aromatic ring in a manner similar to that found in binary CT molecular compounds where the acceptor is a quinone, and also in self-complexes such as naphthoquinones (cf. Section 15.6).

The spontaneous resolution of **14.26** into chiral crystals was exploited by hand-separation of enantiomorphs under the microscope (the Pasteur method) and measurement of optical rotatory dispersion (ORD) and circular dichroism (CD) in $CHCl_3$ solution (Fig. 14.8). In principle, at least, the absolute configurations of the crystals could

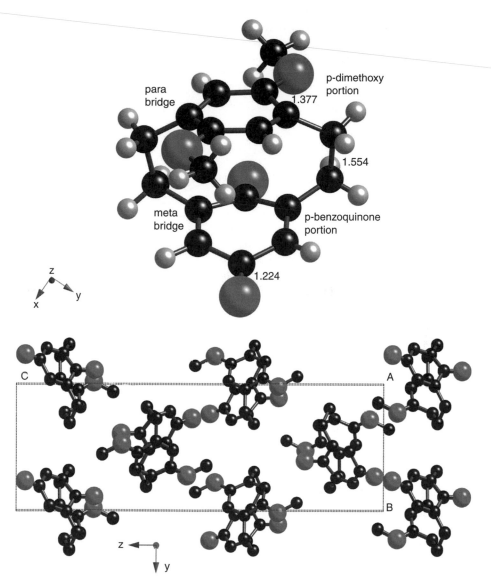

Fig. 14.7. Molecular and crystal structure of **14.26** (DALTOZ). (upper): Perspective view of the molecule. The different C–O distances (Å) in the two portions of the molecule are shown, as well as a typical bridging C–C distance. Bond angle deformation is preferred to alteration of bond lengths. (lower): Packing arrangement shown in projection down [100]; hydrogens omitted for clarity. Orthorhombic; 7.698(1) 8.205(1) 24.025(2) Å, $P2_12_12_1$, $Z = 4$. (Data from Staab, Jörns, Krieger and Rentzea, 1985.)

be determined by the Bijvoet method using anomalous scattering from the oxygen atoms and then related to the signs of the optical rotation; however, this has not yet been done and is a challenging task, as has been demonstrated, for example, by Rabinovich and Hope (1980). Furthermore, "these systems offer, apparently for the first time, the

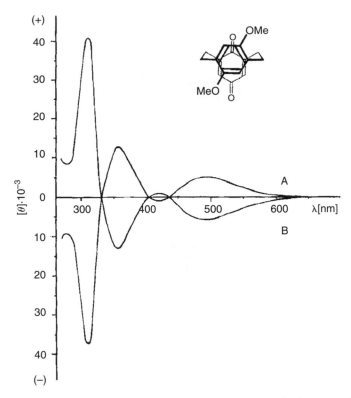

Fig. 14.8. Circular dichroism of two enantiomorphic single crystals (of unknown optical purity) in CHCl$_3$ solution (concentrations 2.8×10^{-6} (A) and 2.1×10^{-6} g/ml (B) respectively). The molar rotations (in deg. at 20°) were [Φ]460 -5215 (A) and $+5811$ (B); [Φ]334 -35164 (A) and $+41720$ (B). (Reproduced from Staab, Jörns, Krieger and Rentzea, 1985.)

opportunity of measuring ORD and CD related to a charge-transfer chromophore with well-defined and rigid donor–acceptor orientations" (Staab, Jörns, Krieger and Rentzea, 1985).

14.26 and **14.27** are, of course, isomers which differ in that the aromatic ring is para bridged in the first of the pair and meta bridged in the second. The crystal structure of **14.27** has not been reported but those of 5,8,12,15-tetra-methoxy[2.2]metaparacyclophane (DALTUE) and the corresponding bis-quinone {[2.2](2,5)-(2,6)-*p*-benzoquinophane (DALVAN) have been determined (Staab, Jörns, Krieger and Rentzea, 1985). All three molecules have similar overall shapes, and it is reasonable to assume that this holds for **14.27** as well.

There have also been a number of investigations of compounds of the [3.3]metapara-cyclophane series and both quinhydrone isomers have been synthesized (as the methoxy derivatives **14.28** and **14.29**) (Staab, Jörns, Krieger and Rentzea, 1985). In the corresponding benzene compounds it was inferred (Staab and Knaus, 1979) from NMR spectra that the rings were not parallel, thus accounting for the rather ill-defined CT bands in the UV-visible spectra. Definitive evidence about molecular structure comes from the crystal structure analysis of **14.28** (Fig. 14.9; DALVER).

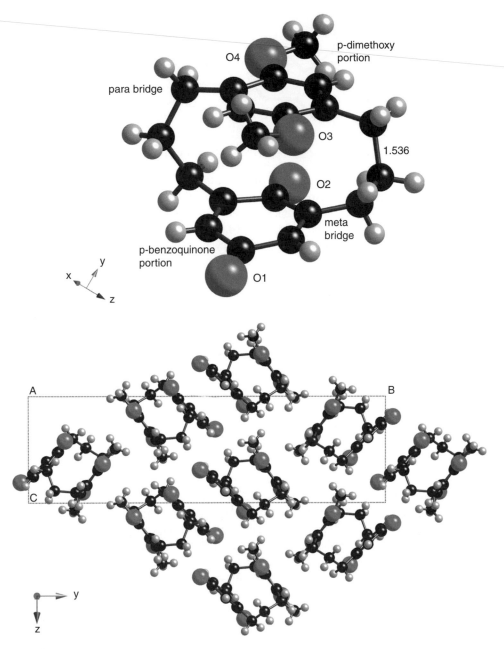

Fig. 14.9. Molecular and crystal structure of **14.28** (DALVER). (upper) Perspective view; distances in Å. The torsion angle O1–O2–O3–04 is 141.5°.

(lower) Pseudo-hexagonal close packing of molecular stacks viewed along [100]; monoclinic, 8.555(1) 25.697(3) 8.370(1) Å, $\beta = 114.87(2)°$, $P2_1/c$, $Z = 4$. (Data from Staab, Jörns, Krieger and Rentzea, 1985.)

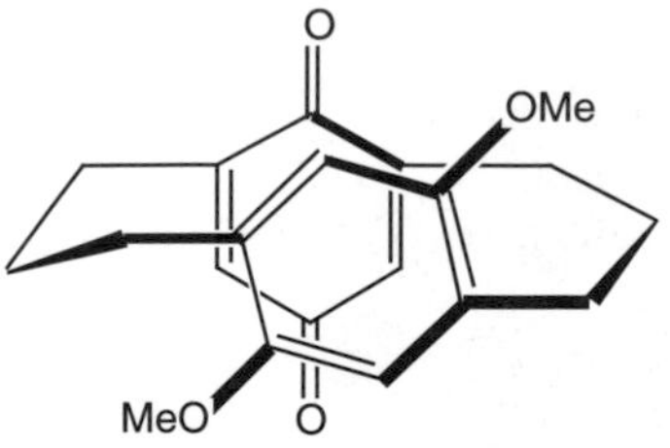

14.28 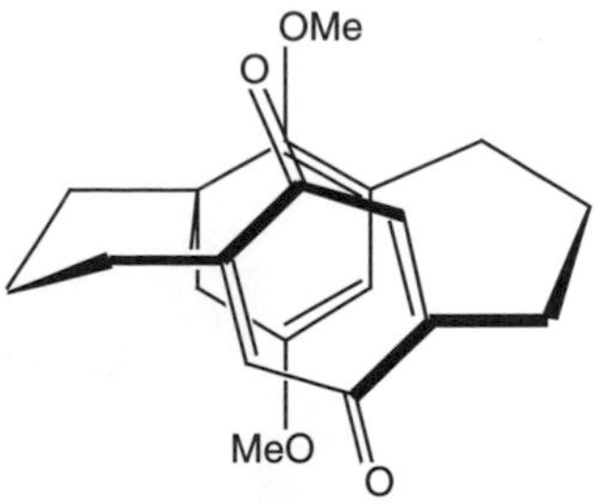   **14.29**

The two rings are inclined at an angle of 12.6°, the individual rings being only mildly deformed; one bridge shows disorder of a methylene group over two sites (occupancy of major site ≈85%) similar to that encountered earlier (see Fig. 14.5). The spectrum of **14.28** has a CT band similar to that of **14.26** ($\lambda_m = 485$ nm, $\varepsilon = 560$, in $CHCl_3$); however, there is also an absorption at shorter wavelength ($\lambda_m = 375$ nm, $\varepsilon = 620$, in $CHCl_3$) which was ascribed to a second CT transition analogous to that found in the electronic spectra of pseudo-ortho [3.3]paracyclophane quinhydrones. The CT band of **14.29** is blue-shifted by about 120 nm ($\lambda_m = 365$ nm, $\varepsilon = 800$, in $CHCl_3$), analogous to the behaviour of the **14.26** and **14.27** diastereoisomers. The rather complicated and distorted geometrical structures of the molecules in this series prevent explanation of the spectra by the simple HMO model that worked fairly well for the [2.2] and [3.3]paracyclophane quinhydrones.

The analogous compounds containing tropoquinono rings have also been reported (Kawamata, Fukazawa, Fujise and Ito, 1982a,b); the spectra are very similar to those of the para compounds (e.g. **14.18**), with ill-defined CT bands.

14.4 Molecules of the metacyclophane type

The *anti*- and *syn*-isomers of a number of [2.2]metacyclophanes, where both donor and acceptor groups are suitably-substituted aromatic moieties, have been investigated (Staab, Schanne, Krieger and Taglieber, 1985); the corresponding quinhydrones are also available (Staab, Reibel and Krieger, 1985). Crystal structures have been reported for representative molecules and there are appreciable geometrical differences between *anti* and *syn* isomers. The structure of *anti*-5,8-dimethoxy-13-nitro[2.2]metacyclophane (Staab, Schanne, Krieger and Taglieber, 1985; DAVHAJ) is shown in Fig. 14.10 and that of *syn*-13,16-dimethoxy-[2](2,6)-p-benzoquinono[2]metacyclophane (Staab, Reibel and Krieger, 1985; DEBZEP) in Fig. 14.11.

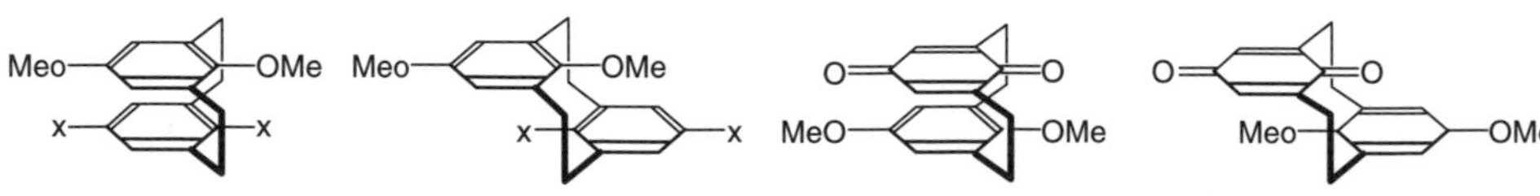

X=COOMe, CN, NO2   SYN   ANTI

The *anti*-isomer shows little overlap of donor and acceptor portions and the tilt between them is relatively small at ≈15°. On the other hand, although the two rings of

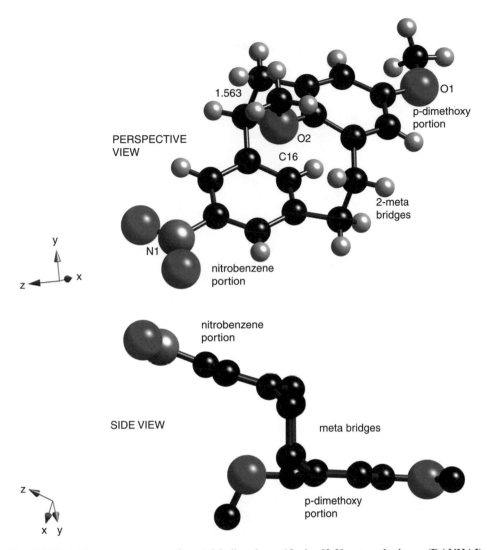

Fig. 14.10. Molecular structure of *anti*-5,8-dimethoxy-13-nitro[2.2]metacyclophane (DAVHAJ). (upper) Perspective view of molecule; the torsion angle O1–O2–C16–N1 is 178.95°; distances in Å. (lower) Side view. A stereoview of the molecule is shown in the original paper. (Data from Staab, Schanne, Krieger and Taglieber, 1985.)

the *syn* molecule appear to be appreciably overlapped in plan view, the side view shows that there is a tilt angle of ≈33° between them and that there are close interannular approaches only on one side of the molecule. Despite these geometrical differences, the spectra of the *anti, syn* pair of diastereoisomeric quinhydrones are remarkably similar (Fig. 14.12; Staab, Reibel and Krieger, 1985) and this holds also for other analogous pairs; investigation of the solvent dependence of the fluorescence from various molecules showed that the long wavelength absorption bands were indeed charge-transfer bands (Staab, Schanne, Krieger and Taglieber, 1985).

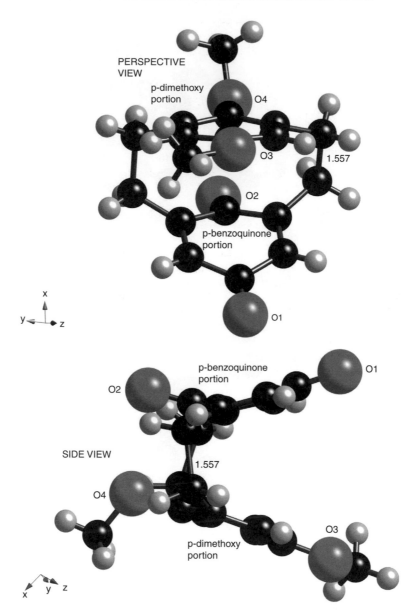

Fig. 14.11. Molecular structure of *syn*-13,16-dimethoxy[2](2,6)-*p*-benzoquinone-[2]metacyclo-phane (DEBZEP). (upper) Perspective view of molecule; the torsion angle O1–O2–O4–O3 is 9.81°; distances in Å. (lower) Side view. A stereo view of the molecule is shown in the original paper. (Data from Staab, Reibel and Krieger, 1985.)

The similarity of the CT absorption bands from *anti* and *syn* donor–acceptor metacyclophanes despite the very different disposition of donor and acceptor moieties has been explained in terms of Hückel molecular orbital theory (Vogler, Schanne and Staab, 1985).

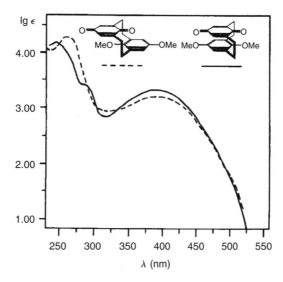

Fig. 14.12. The charge-transfer spectra (in CHCl₃) of the *anti -syn* isomer pair of quinhydrones shown as inserts at the top of the diagram. (Reproduced from Staab, Schanne, Krieger and Taglieber, 1985.)

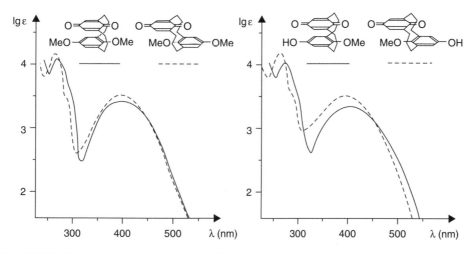

Fig. 14.13. Comparison of the spectra of the *syn* and *anti* diastereoisomers of the two [3.3]metacyclophanes shown as inserts at the top of the figure. Replacement of one methoxy group in each donor moiety by an hydroxyl has little effect on the spectra. (Reproduced from Staab, Herz, Döhling and Krieger, 1980.)

Quinhydrones of the [3.3]metacyclophane series and related molecules have also been studied (Staab, Herz and Döhling, 1979; Staab and Döhling, 1979; Staab, Herz and Döhling, 1980; Staab, Herz, Döhling and Krieger, 1980). Spectra of *anti* and *syn* pairs of isomers show marked resemblances (Fig. 14.13).

The spectrum of *syn*-15,18-dihydroxy[3](2,6)-*p*-benzoquinone[3]meta-cyclophane is very similar to that of pseudogeminal 14,17-dihydroxy[3](2,5)-*p*-benzoquinone-[3]

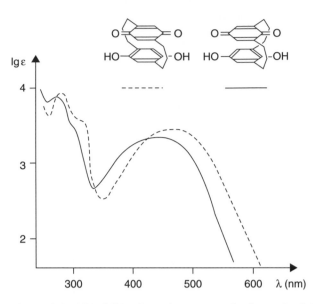

Fig. 14.14. Comparison of the UV-visible absorption spectra (in dioxane) of the pseudogeminal 14,17-dihydroxy[3](2,5)-*p*-benzoquinone[3]paracyclophane (on the left of the insert in the upper portion of the diagram) and of *syn*-15,18-dihydroxy[3](2,6)-*p*-benzoquinone[3]metacyclophane (on the right). (Reproduced from Staab, Herz, Döhling and Krieger, 1980.)

paracyclophane (Fig. 14.14). The overlaps of donor and acceptor portions of these two molecules are not very different, thus accounting for the similarity of their spectra; the structure of *syn*-15,18-dihydroxy[3](2,6)-*p*-benzoquinone[3]metacyclophane is shown in Fig. 3 of Staab, Herz, Döhling and Krieger (1980) while that of pseudogeminal 14,17-dihydroxy-[3](2,5)-*p*-benzoquinone[3]paracyclophane is in Fig. 14.5.

The crystal structures of two *syn* isomers have been reported – of *syn*-15,18-dimethoxy(2,6)-*p*-benzoquinono[3.3]metacyclophane (Staab, Herz, Döhling and Krieger, 1980; MXBQMP) and of *syn*-6,9-dimethoxy-15,18-dinitro-2,11-dithia[3.3]metacyclophane (Staab, Schanne, Krieger and Taglieber, 1985; DAVGUC). As would be expected, the molecules of the [3.3]metacyclophane series are less distorted (but some distortion remains) than those of the [2.2]metacyclophane series.

14.5   Some other systems

We give a few examples of other systems where the principles discussed above are being applied in order to illustrate how one may expect the particular features of cyclophane systems to be exploited in the future. The electron transfer properties of the vertically-stacked porphyrin-quinone(1)-quinone(2) cyclophane, an analog to compounds involved in the primary process of biological photosynthesis, have been studied in order to determine which structural factors favour a consecutive, stepwise electron transfer and which an integrated process relying on electron coupling (Staab, Tercel, Fischer and Krieger, 1994). During the course of the synthesis of compound **14.30**, the crystal structure of **14.31** (which is a substituted pseudo-ortho[3.3]-paracyclophane) was determined (WIHMIJ); the molecular structure resembles that shown in Fig. 14.5.

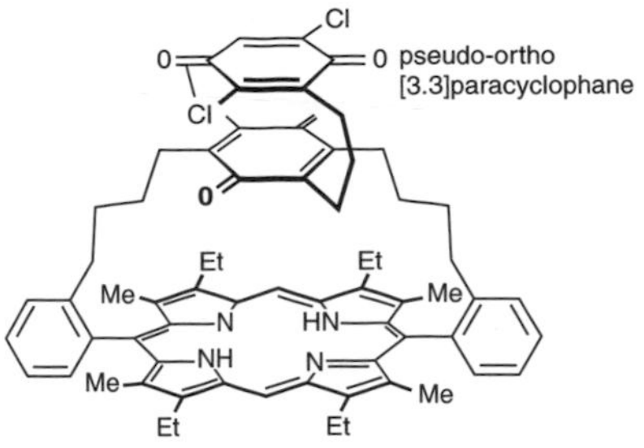

14.30

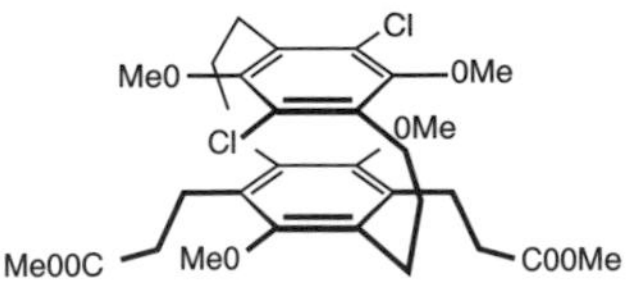

14.31

Another system (Cowan, Sanders, Beddard and Harrison, 1987), involving different dispositions of pyromellitimide (an electron acceptor) and porphyrin rings (the cofacial pair acting as an electron donor), demonstrates that "mere proximity between donor and acceptor is not a sufficient condition for electron transfer. There is also a strong geometrical requirement." Picosecond fluorescence measurements show (by marked quenching of the fluorescence) that there is rapid electron transfer from the excited porphyrin pair to the pyromellitimide electron acceptor in **14.32** whereas there is relatively little fluorescence quenching in **14.33**, and hence slow electron transfer. A rather similar rigid triple-ring molecule with a porphyrin sandwiched between two parallel *p*-benzoquinone units has also been reported (Weiser and Staab, 1984). There are clear analogies to triple-layer CT cyclophanes. Another triple layer system is the crystalline green cationic acceptor–donor–acceptor system prepared by Simonsen *et al.* (1998), with butane-1,4-diyl linkers; there are four PF_6 counter-anions and three MeCN molecules of solvation. The acceptors are bipyridinium moieties and the donor is based on TTF. Crystal structure analysis shows a rather distorted molecule without particularly close interactions. Presumably shorter linker units are needed to enforce geometrical constraints.

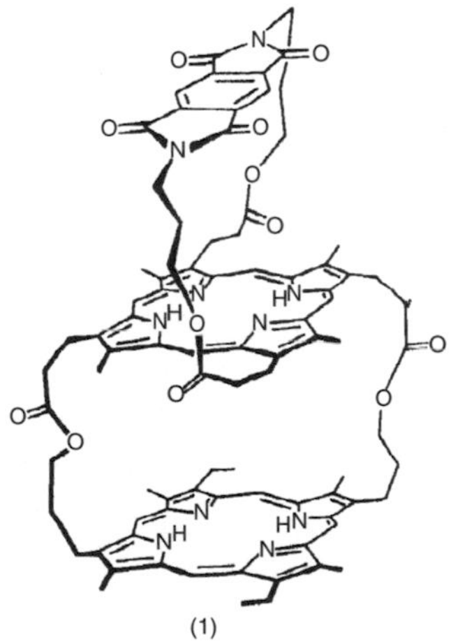

(1)

14.32

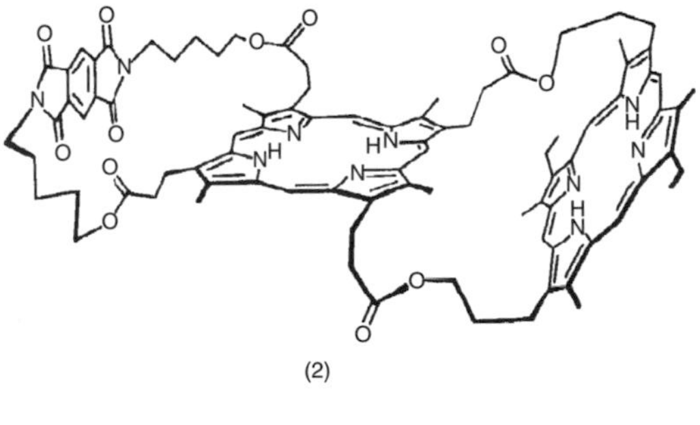

(2)

14.33

14.6 Concluding summary

Thus, in summary, the measurements on the cyclophane intramolecular donor–acceptor compounds have confirmed, for many two-layer molecules, a marked dependence of the degree of charge transfer on the mutual orientation of the donor and acceptor moieties. The charge transfer is greater when the long axes of the moieties are parallel than when they are inclined at an angle of $\approx 60°$. For two-layer molecules the situation in the [2.2]paracyclophanes is complicated by the very strong interaction between the adjacent rings, but the orientation effect is just as marked in the [3.3]paracyclophanes where the distance between the rings is similar to non-bonded distances found in the corresponding crystalline compounds. Charge transfer interaction persists through the rings in three- and four-layer molecules but the orientation effect is lost. The orientation effects are not marked in systems such as the tropoquinophanes and the higher [$n.n$]cyclophanes ($n > 3$), where the CT interaction is rather weak. In the [$n.n$]metacyclophane series the geometrically-different *anti* and *syn* isomers give remarkably similar CT absorption spectra, a result which has been explained by HMO theory.

One may anticipate that a wealth of information remains to be uncovered by spectroscopy at very low temperatures of molecules in rigid glasses, and that these results will eventually be interpreted by *ab initio* calculations based on the detailed geometrical structures obtained from diffraction studies. In addition one should note that cyclophane-type molecules lend themselves to study of a variety of secondary interactions between moieties in defined geometrical situations, and this is likely to be an important growth area in future research.

References

Bauer, H., Briaire, J. and Staab, H. A. (1983). *Angew. Chem. Int. Ed. Engl.*, **22**, 334–335.
Bernstein, J. and Trueblood, K. N. (1971). *Acta Cryst.*, B**27**, 2078–2089.
Cowan, J. A., Sanders, J. K. M., Beddard, G. S. and Harrison, R. J. (1987). *J. Chem. Soc., Chem. Comm.*, pp. 55–58.
Cram, D. J. and Day, A. C. (1966). *J. Org. Chem.*, **31**, 1227–1232.
Hanson, A. W. (1977). *Acta Cryst.*, B**33**, 2003–2007.
Hausser, K. H. and Wolf, H. C. (1976). *Adv. Magn. Reson.*, **8**, 85–121.
Herz, C. P. and Staab, H. A. (1977). *Angew.Chem. Int. Ed. Engl.*, **16**, 394.
Horita, H., Otsubo, T., Sakata, Y. and Misumi, S. (1976). *Tetr. Letts.*, pp. 3899–3902.
Ippen, J., Tao-pen, C., Starker, B., Schweitzer, D. and Staab, H. A. (1980). *Angew. Chem. Int. Ed. Engl.*, **19**, 67–69.
Kawamata, A., Fukazawa, Y., Fujise, Y. and Ito, S. (1982a). *Tetr. Letts.*, **23**, 1083–1086.
Kawamata, A., Fukazawa, Y., Fujise, Y. and Ito, S. (1982b). *Tetr. Letts.*, **23**, 4955–4958.
Koizumi, Y., Toyoda, T., Miki, K., Kasai, N. and Misumi, S. (1986). *Bull. Chem. Soc. Jpn.*, **59**, 239–242.
Krieger, C. (1978). Unpublished.
Machida, H., Tatemitsu, H., Sakata, Y. and Misumi, S. (1978). *Tetr. Letts.*, pp. 915–918.
Mizuma, T., Miki, K., Kai, Y., Tanaka, N. and Kasai, N. (1982). *Bull. Chem. Soc. Jpn.*, **55**, 2026–2028.

Mizuma, T., Miki, K., Kai, Y., Yasuoka, N. K. and Kasai, N. (1982). *Bull. Chem. Soc. Jpn.*, **55**, 979–984.

Mizuno, H., Nishiguchi, K., Toyoda, T., Otsubo, T., Misumi, S. and Morimoto, M. (1977). *Acta Cryst.* **B33**, 329–334.

O'Connor, J. G. and Keehn, P. M. (1976). *J. Am. Chem. Soc.*, **98**, 8446–8450.

Rabinovich, D. and Hope, H. (1980). *Acta Cryst.*, A**36**, 670–678.

Rebafka, W. and Staab, H. A. (1973). *Angew. Chem. Int. Ed. Engl.*, **12**, 776–777.

Rebafka, W. and Staab, H. A. (1974). *Angew. Chem. Int. Ed. Engl.*, **13**, 203–204.

Schroff, L. G., Weerdt, A. J. A.van der, Staalman, D. J. H., Verhoeven, J. W. and de Boer, Th. J. (1973). *Tetr. Letts.*, pp. 1649–1652.

Schroff, L. G., Zsom, R. L. J., Weerdt, A. J. A. van der, Schrier, P. I., Geerts, J. P., Nibbering, N. M. M., Verhoeven, J. W. and de Boer, Th. J. (1976). *Rec. Trav. Chim. Pays-Bas*, **95**, 89–93.

Schwartz, M. H. (1990). *J. Incl. Phenom.*, **9**, 1–35.

Schweitzer, D., Hausser, K. H., Taglieber, V. and Staab, H. A. (1976). *Chem. Phys.*, **14**, 183–187.

Shinmyozu, T., Inazu, T. and Yoshino, T. (1977). *Chem. Letts.*, pp. 1347–1350.

Simonsen, K. B., Thorup, N., Cava, M. P. and Becher, J. (1998). *Chem. Commun.*, pp. 901–902.

Staab, H. A. and Appel, W. (1981). *Liebig's Ann. Chem.*, pp. 1065–1072.

Staab, H. A. and Döhling, A. (1979). *Tetr. Letts.*, pp. 2019–2022.

Staab, H. A. and Haffner, H. (1977). *Chem. Ber.*, **110**, 3358–3365.

Staab, H. A. and Herz, C. P. (1977a). *Angew. Chem. Int. Ed. Engl.*, **16**, 392–394.

Staab, H. A. and Herz, C. P. (1977b). *Angew. Chem. Int. Ed. Engl.*, **16**, 799–801.

Staab, H. A. and Knaus, G. H. (1979). *Tetr. Letts.*, pp. 4261–4264.

Staab, H. A. and Rebafka, W. (1977). *Chem. Ber.*, **110**, 3333–3350.

Staab, H. A. and Schwendemann, V. (1978). *Angew. Chem. Int. Ed. Engl.*, **17**, 756–757.

Staab, H. A. and Taglieber, V. (1977). *Chem. Ber.*, **110**, 3366–3376.

Staab, H. A. and Zapf, U. (1978). *Angew. Chem. Int. Ed. Engl.*, **17**, 757–758.

Staab, H. A., Döhling, A. and Krieger, C. (1981). *Liebig's Ann. Chem.*, pp. 1052–1064.

Staab, H. A., Döhling, A. and Krieger, C. (1991). *Tetr. Letts.*, **32**, 2215–2218.

Staab, H. A., Gabel, G. and Krieger, C. (1983). *Chem. Ber.*, **116**, 2827–2834.

Staab, H. A., Gabel, G. and Krieger, C. (1987). *Chem. Ber.*, **120**, 269–273.

Staab, H. A., Herz, C. P. and Döhling, A. (1979). *Tetr. Letts.*, pp. 791–794.

Staab, H. A., Herz, C. P. and Döhling, A. (1980). *Chem. Ber.*, **113**, 233–240.

Staab, H. A., Herz, C. P. and Henke, H.-E. (1977). *Chem. Ber.*, **110**, 3351–3357.

Staab, H. A., Herz, C. P., Döhling, A. and Krieger, C. (1980). *Chem. Ber.*, **113**, 241–254.

Staab, H. A., Herz, C. P., Krieger, C. and Rentzea, M. (1983). *Chem. Ber.*, **116**, 3813–3830.

Staab, H. A., Hinz, R., Knaus, G. H. and Krieger, C. (1983). *Chem. Ber.*, **116**, 2835–2847.

Staab, H. A., Ippen, J., Tao-pen, C., Krieger, C. and Starker, B. (1980). *Angew. Chem. Int. Ed. Engl.*, **19**, 66–67.

Staab, H. A., Jörns, M. and Krieger, C. (1979). *Tetr. Letts.*, pp. 2513–2516.

Staab, H. A., Jörns, M., Krieger, C. and Rentzea, M. (1985). *Chem. Ber.*, **118**, 796–813.

Staab, H. A., Knaus, G. H., Henke, H. -E. and Krieger, C. (1983). *Chem. Ber.*, **116**, 2785–2807.

Staab, H. A., Krieger, C., Wahl, P. and Kay, K-Y. (1987). *Chem. Ber.*, **120**, 551–558.

Staab, H. A., Reibel, W. R. K. and Krieger, C. (1985). *Chem. Ber.*, **118**, 1230–1253.

Staab, H. A., Reimann-Haus, R., Ulrich, P. and Krieger, C. (1983). *Chem. Ber.*, **116**, 2808–2826.

Staab, H. A., Schanne, L., Krieger, C. and Taglieber, V. (1985). *Chem. Ber.*, **118**, 1204–1229.

Staab, H. A., Starker, B. and Krieger, C. (1983). *Chem. Ber.*, **116**, 3831–3834.

Staab, H. A., Tercel, M., Fischer, R. and Krieger, C. (1994). *Angew. Chem. Int. Ed. Engl.*, **33**, 1463–1466.

Staab, H. A., Wahl, P. and Kay, K-Y. (1987). *Chem. Ber.*, **120**, 541–549.

Staab, H. A., Zapf, U. and Gurke, A. (1977). *Angew. Chem. Int. Ed. Engl.*, **16**, 801–803.

Stöbbe, M., Kirchmeyer, S., Adiwidjaja, G. and Meijere, A. de, (1986). *Angew. Chem. Int. Ed. Engl.*, **25**, 171–173.

Tashiro, M., Koya, K, and Yamato, T. (1983). *J. Am. Chem. Soc.*, **105**, 6650–6653.

Tatemitsu, H., Natsume, B., Yoshida, M., Sakata, Y. and Misumi, S. (1978). *Tetr. Letts.*, pp. 3459–3462.

Tatemitsu, H., Otsubo, T., Sakata, Y. and Misumi, S. (1975). *Tetr. Letts.*, pp. 3059–3062.

Toyoda, T., Tatemitsu, H., Sakata, Y., Kasai, N. and Misumi, S. (1986). *Bull. Chem. Soc. Jpn.*, **59**, 3994–3996.

Vogler, H. (1983a). *Tetr. Letts.*, pp. 2159–2162.

Vogler, H. (1983b). *Z. Naturforschung*, **38B**, 1130–1135.

Vogler, H., Ege, G. and Staab, H. A. (1977). *Mol. Phys.*, **33**, 923–932.

Vogler, H., Schanne, L. and Staab, H. A. (1985). *Chem. Ber.*, **118**, 1254–1260.

Weiser, J. and Staab, H. A. (1984). *Angew. Chem. Int. Ed. Engl.*, **23**, 623–625.

Yoshida, M., Tatemitsu, H., Sakata, Y., Misumi, S., Masuhara, H. and Mataga, N. (1976). *J. Chem. Soc. Chem. Comm.*, pp. 587–588.

Yoshida, M., Tochiaki, H., Tatemitsu, H., Sakata, Y. and Misumi, S. (1978). *Chem. Lett.*, pp. 829–832.

Chapter 15

Crystal chemistry of mixed-stack π–π* molecular compounds

Wie aus zahlreichen Versuchen hervorgeht, vereinigen sich die Nitrokörper der aliphatischen wie aromatischen Reihe mit den verschiedenartigsten organischen Verbindungen zu mehr oder weinigen tieffarbigen Additionsprodukten. Besonders gut charakterisiert sind vor allem die Verbindungen aromatischer Di-und Trinitrokörper mit aromatischen Kohlenwasserstoffen, Amine und Phenolen. Wie ein statistische Ueberschicht der bis heute dargestellten, etwa 700 Molekül-verbindungen der Nitro-körper zeigt, haben diese in den allermeisten Fällen, ganz unabhängig davon, wieviele Nitrogruppen die nitroide Komponente enthält, auch unabhängig von der Zusammensetzung der benzoiden Komponente, die denkbar einfachste Zusammensetzung A_1B_1, indem auf 1 Molekül des Nitrokörper 1 Molekül des Kohlenwasserstoffs bzw. seiner Derivate kommt. Rund 85% der Verbindungen, von dene wenige aufgezählt seien, entsprechen diesen Typus.

Paul Pfeiffer, 1927 (p. 336).

Summary: The room-temperature crystal structures of many mixed stack charge transfer molecular compounds can be grouped into a relatively small number of crystallochemical families. Although the mixed stack arrangement predominates, there are structures which deviate to a greater or lesser extent from such an arrangement, for reasons which are often not clear. We first review these maverick structures and then classify the more common mixed stack arrangements into a number of structural groups. Compounds with quinonoid acceptors sometimes have special structural features because of the mode of interaction of the carbonyl groups with aromatic rings. In the quinhydrone family the (aromatic ring)•••carbonyl interaction is supplemented by hydrogen bonding between carbonyl and hydroxyl oxygens and this leads to considerable structural homogeneity. Many CT compounds where the π–π* interaction is supplemented by hydrogen bonding show special structural features and physical properties different from those with only π–π* interaction. The mixed stack compounds with ionic ground states generally resemble those with neutral ground states in structural terms but have different physical properties. In the isomeric compounds the possibility of having both electron and proton transfer leads to variations on the mixed stack theme.

(Note. The components in the ground states of these molecular compounds are taken to be neutral unless explicitly stated otherwise).

15.1   Introduction

Most π–π* molecular compounds have $1:1$ donor$\cdots$acceptor (D$\cdots$A) ratios and the components crystallize in alternating array in mixed stacks, the stacks being close packed in a quasi-hexagonal arrangement of (approximate) cylinders (Fig. 15.1). We remind the reader of our conventions – donors always come first in the formulation and are linked to the acceptor by$\cdots$, indicating a π–π* charge transfer (CT) interaction; we have not been very strict in our usage of this indicator. Most donor and acceptor molecules are disk-like in shape, i.e. their cross-sectional area is appreciably greater than their thickness. The molecular thickness of aromatic donor and acceptors is about 3.5 Å and thus the periodicity along the stack axis should be 7–8 Å for. . . . DADA. . . . stacks. The interplanar spacing, measured along the normals to the planes of the approximately parallel disks, is often found to be 0.1–0.2 Å less than the sum of the thickness of the components; this is interpreted as evidence for a D···A interaction along the normal to the component planes, or stack axis (see below), which is additional to the ubiquitous van der Waals interactions. There are a number of facts in support of this contention. Firstly, one can compare the transannular atom-to-atom distances, found from crystal structure analyses (Staab *et al.*, 1983) in pseudo-geminal-5,8,14,17-tetramethoxy[3.3]paracyclophane and its quinhydrone analog pseudo-geminal-14,17-dimethoxy-[3](2,5)*p*-benzoquinone-[3]paracyclophane (see Chapter 14 for discussion of these compounds). The molecules have identical conformations but the two rings are 0.13 Å closer in the latter than in the former molecule, in conformity with the existence of charge transfer interactions between the rings in the latter molecule but not in the former. A second line of evidence comes from comparison of Young's modulus values in the direction of the stack axes, which are about ten times larger for crystals of π-molecular compounds than for those of comparable aromatic hydrocarbons (Danno *et al.*, 1967). Thus we choose to emphasize the mixed stacks as the essential and characteristic feature of the packing arrangements in π-molecular compounds. The simple mixed stack description works remarkably well for most of the structures so far reported; however, it tends to break down when donor and acceptor molecules differ appreciably in disk (face) area or when there is additional bonding (e.g. hydrogen bonding or dipolar interactions) between the components. There are some crystals of this type where the structure can equally well be described in terms of layers of donor and acceptors.

The extremes of packing type within the stacks can be described as "overlapped disks" (Fig.15.1(a)) and "slipped disks" (Fig.15.1(b)). However, intermediate situations are also found and there have been a number of essentially similar proposals (Fritchie and Arthur, 1966; Goldberg and Shmueli, 1973b; Visser *et al.*, 1990) for a quantitative description of the arrangement within the stacks; we have adapted the proposal of Visser *et al.* (1990) to mixed stack compounds where the components lack symmetry. Right-handed ortho-normal axial systems **L, M, N** are defined for the donor and acceptor molecules (or convenient portions, such as benzene rings); with **L** along the longest molecular axis,

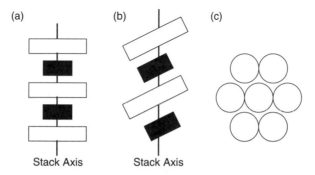

Fig. 15.1. Schematic representation of different types of stacking in crystalline mixed stack π-compounds: (a) overlapped disk stacking, (b) slipped disk stacking and (c) quasi-hexagonal arrangement of stacks.

M perpendicular to **L** and in the best molecular plane, and **N** at right angles to the plane. The offset of the centre of gravity of the donor from the center of gravity of the acceptor is then given in terms of the components along the **L**, **M** and **N** axes of the *acceptor* (an arbitrary choice). A simplified situation for centrosymmetric molecules projected onto their mean planes is shown in Fig. 15.2. Results for a few molecular compounds are summarized in Table 15.1. Unfortunately most authors describe molecular overlap and calculate interplanar distances on a less well-defined basis.

The repetition period of ≈ 7 Å along the stack axis may require amendment in two ways:

(a)   there are a number of crystals (Table 15.3) where the stack periodicity is doubled to ≈ 14 Å and the arrangement along the stack is

$$-------D_1\ A_1\ D_2\ A_2\ D_1\ A_1\ D_2\ A_2-------$$
$$|\ \ \approx 14\ \text{Å}\ \ \ |$$

The subscripts here refer to positional and/or orientational differences between donor (and acceptor) molecules, not to chemical differences. The causes of the ordering

Table 15.1. Examples of different stacking arrangements in mixed stack molecular compounds: I is {phenazine···TCNQ} (Goldberg and Shmueli, 1973c; TCQPEN10); II is {dibenzo-*p*-dioxin···TCNQ} (Goldberg and Shmueli, 1973b; TCQBDX); III is {anthracene···TCNQ} (Williams and Wallwork, 1968; TCQANT). Distances in Å and angles in deg

Parameter	I	II	III
1. Repeat distance along stack axis	8.57	7.04	7.00
2. Angle between stack axis and plane normal	38	11	0
3. Mutual offset along L	1.85	0.51	≈ 0.0
4. Mutual offset along M	1.89	0.48	≈ 0.0
5. Interplanar spacing	3.38	3.46	3.50

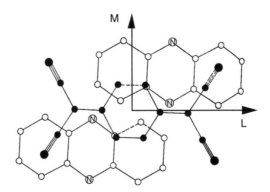

Fig. 15.2. {Phenazine···TCNQ}, showing donor and acceptor molecules in a mixed stack projected onto their common mean plane. The long (L) and short (M) in-plane axes of TCNQ are shown. (Reproduced from Goldberg and Shmueli, 1973b.)

along the stack direction are not always clear; interactions between adjacent stacks are sometimes invoked. Higher degrees of ordering (i.e. periodicities greater than ≈14 Å) do not appear to have been encountered.

(b)   there are a number of crystals where the concept of infinite stacks no longer applies; instead the stack length is limited to one, two or three pairs. These finite stacks can be arranged in different ways.

These ideas will be illustrated in more detail and also extended to π-molecular compounds with $D : A \neq 1$.

15.2   Nonstacked structures containing structural groups of limited size

The smallest structural group of limited size is the donor–acceptor pair, not incorporated in a stack. Such arrangements are found in {benz[a]-anthracene···PMDA} (Foster et al., 1976; BZAPRM10) and {1,10-phenanthroline···TCNQ} (Goldberg and Shmueli, 1977; TCQPAN10). The overall crystal structures can be described in terms of mutually shifted layers of donors and acceptors. Analogous donor–acceptor pairs are found in the crystal structures of tryptamine picrate (TRYPIC) and DL-tryptophan picrate·methanol (TPTPCM) (Gartland et al., 1974). The red colour of these crystals attests to the occurrence of charge transfer interactions within the donor–acceptor pairs; neighbouring pairs are linked by hydrogen bonds and not by π–π* interactions.[1]

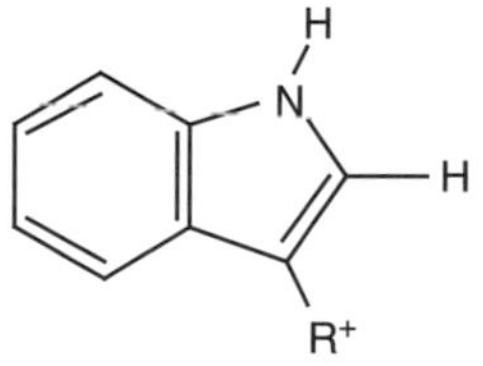

tryptamine: $R^+ = -CH_2CH_2NH_3^+$

tryptophan: $R^+ = -CH_2CH(CO_2H)NH_3^+$

In {(9-ethylcarbazole)$_2$···TCNE} (Lee and Wallwork, 1978; ETCZCE) the structure consists of centrosymmetric DAD units, with a distance of 3.24 Å between donor and acceptor molecules. These sandwiches are arranged in face-centred pleated sheets, with mean plane (010); however, the pleats are alternately parallel to (021) and (0$\bar{2}$1); there being an angle of ≈60° between these planes. Successive sheets are related by a c glide plane perpendicular to [010]. The analogous arrangement found in {(acridine)$_2$···PMDA} (Karl, Binder et al., 1982; BIWVUY) is shown in Fig. 15.3. Centrosymmetric ADA sandwiches are found in the 1 : 2 compound of a Ni(II)etioporphyrin with 2,4,5,7-tetra-nitrofluorenone ($P2_1/n$, $Z = 2$) (Grigg et al., 1978; ETPNFL); and in {2,7-bis(methylthio)-1,6-dithiapyrene-···(tetrahydrobarreleno-TCNQ)$_2$} ($P\bar{1}$, $Z = 1$) (Nakasuji, Sasaki et al., 1988; SANYOV). Effectively-isolated DAD triads are found in (toluene)$_2$···tetra-phenylporphyrinato-M(II) molecular compounds (M = Cr, Mn, Zn; see Chapter 8 and Section 15.9.4). In the 2 : 1 compound of N,N,N',N'-tetramethylbenzidine and chloranil

---

[1] Mixed stacks are found in three tryptophan metabolite–picric acid molecular compounds; their red colour shows that there is charge-transfer interaction (Nagata et al., 1995).

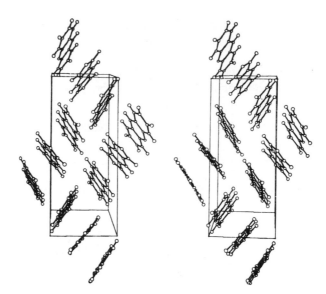

Fig. 15.3. The crystal structure of {(acridine)$_2$···PMDA} at 120K, showing a stereoscopic view down [100] of one layer of the structure ([001] is vertical and [010] runs from left to right). The tilt of the acridine molecules with respect to the PMDA molecules can be seen clearly. (Reproduced from Karl, Binder *et al.*, 1982.)

the two benzene rings of the benzidine moiety have an interplanar (twist) angle of 31° and the chloranil is sandwiched between two parallel rings of successive benzidines, the second ring of each benzidine molecule not participating in the charge transfer interaction (Yakushi, Ikemoto and Kuroda, 1971; TMBCAN).

There are a number of analogous structures where the ground state is ionic and these are discussed in Section 15.10. No overall explanation has yet been put forward for these exceptional structures.

15.3   The crystallochemical families found for 1 : 1 $\pi-\pi^*$ molecular compounds

The crystal structures of *most* of the 1 : 1 $\pi-\pi^*$ molecular compounds can be grouped into a limited number of crystallochemical types; we use number of formula units in the unit cell (Z); space group and stack axis direction as criteria for this classification, restricting ourselves at this point to stack axis periodicities of ≈ 7 Å (however, some exceptions are admitted when the donor or acceptor molecule is thicker than the usual ≈ 3.5 Å). The cell dimensions in directions normal (or approximately so) to the stack axis are determined mainly by the cross-sectional dimensions of the stacks and their mode of arrangement. As noted above, there is generally quasi-hexagonal close packing of the stacks, all stack axes being parallel. A few exceptions have been found to this rule; for example, in {pyrene···*p*-benzoquinone} (Bernstein *et al.*, 1976; PYRBZQ) (tetragonal, space group $P4_1$, Z = 4) "slipped disk" stacking of the usual kind is found but the stacks are arranged in *layers* one stack thick, successive layers being related by the 4_1 axes (Fig. 15.4). No explanation has

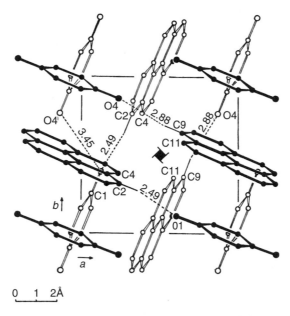

0   1   2Å

Fig. 15.4. Part of the {pyrene···*p*-benzoquinone} structure, viewed down [001]. The stack axes are alternately along the [100] (open circles) and [010] (full circles) axes of the tetragonal unit cell, and the layers are related by the fourfold screw axis along [001]. Some short distances between stacks are shown. In particular, the (C)–H . . . O=C distance of 2.49 Å would now be considered as evidence for a weak hydrogen bond. (Reproduced from Bernstein *et al.*, 1976.)

been put forward for this unusual arrangement, which is chiral (but the absolute config-uration of the crystal used in the analysis was not determined).

A distorted version of this structure type is found in {9-methoxy-5,11-dimethyl-6H-pyrido[4,3-b]carbazole···TCNQ·CH$_3$CN} ($P2_1/c$, $Z = 4$) (Viossat, Dung and Daran, 1988; GEZKIF), where there are mixed stacks with axes approximately along [110] and [1$\bar{1}$0], arranged in successive layers about $z \approx 0$ and 1/2. A similar arrangement is found in 1,3-indandione (tetragonal, $a = 14.361$, $c = 13.631$ Å, $Z = 16$, space group $I4_1/a$) (Bravic *et al.*, 1976; INDDON); which can be considered to be a self-complex, with the two parts of the molecule having donor and acceptor properties respectively. It seems probable that there are both dipole–dipole and π–π* interactions.

Most of the other structures can be classified as shown in Table 15.2; the scheme is based on that developed earlier (Herbstein, 1971; see Table 20) but the numbering of the groups has been changed to match the usual crystallographic hierarchy – triclinic, monoclinic, etc.; primitive, centred unit cells; molecules at special positions, molecules at general positions. Structures have been determined for most but not all of the compounds listed.

The most prolific acceptor and donor components are TCNQ and TTF. There are 1161 hits for TCNQ in the October, 2002 issue of the CSD, and 253 for TTF, both numbers including derivatives and covering binary adducts of all kinds. Table 15.2, which includes only mixed-stack structures, has 48 examples with TCNQ as acceptor and 16 with TTF as donor.

Table 15.2. Classification of $1:1$ $\pi-\pi*$ molecular compounds into structural groups. In each group the donors that form molecular compounds with particular acceptors are listed; in general only the most recent reference is given. Alternative orientations (e.g $P2_1/c$ instead of $P2_1/a$ with alternative stack axis) are not listed separately. Acronyms are listed in Table 14.2. Z is the number of formula units in the unit cell

Acceptor	Donors
Group 1a: Triclinic, $P\bar{1}$, $Z=1$, stack axis [001]	
TCNE	$C_6(CH_3)_6$, (Sahaki *et al.*, 1976; MBZTCE; Maverick *et al.*, 1978; MBZTCE10)
	[3,3]paracyclophane (Bernstein and Trueblood, 1971; PACTCN10); ferrocene (Adman *et al.*, 1967; FERTCE); [2,2]metacyclophane (Cohen-Addad, Renault *et al.*, 1988; GEBREK).
2,5-dimethyl-TCNQ	Octamethylene-TTF (OMTTF) (Chasseau and Leroy, 1981; BESPEU).
fluoranil	TTF (Mayerle *et al.*, 1979; TTFFAN);
	durene (Dahl and Sørensen, 1985);
	N,N-dimethylaniline (disordered over two orientations) (Dahl, 1981b; BAPLEJ).
chloranil	Pd(II)oxinate (Kamenar *et al.*, 1965; CLAQPD);
	bis(8-hydroxyquinoline) (Prout and Wheeler, 1967; HQUCLA);
	perylene (Kozawa and Uchisa, 1983; CAFWAH);
	N,N,N′,N′-tetramethylbenzidine (TMBD) (Yakushi *et al.*, 1973; MBZDCN).
2,5-dibromo-3,6-dichloro-*p*-benzoquinone	Perylene (Kozawa and Uchida, 1979; PERPBQ); acceptor disordered at inversion center.
TCNB	Pd(II)oxinate (Kamenar, Prout and Wright, 1966; PDHQCB);
	TMPD (Ohashi *et al.*, 1967; TPDTCB);
	biphenyl (Pasimeni *et al.*, 1983; BUHSIG);
	acridine (disordered) (Marsh, 1990; KARKAP01).
TCNQ	Naphthalene (Shaanan *et al.*, 1967; TCQNAP);
	chrysene (Munnoch and Wright, 1974; CHRTCQ);
	d^{14}-*p*-terphenyl (Lisensky *et al.*, 1976; TCQDTP10, ND);
	phenazine (Goldberg and Shmueli, 1973b; TCQPEN10);
	Cu(II)oxinate (Williams and Wallwork, 1967; TCQCUH);
	Pt(II)oxinate (Bergamini *et al.*, 1987; FEFCAU);
	bis(1,2-benzoquinonedioximato)-Pd(II) (Keller *et al.*, 1977; BCDPDQ);
	bis(1,2-benzoquinonedioximato)Ni(II) (Keller *et al.*, 1977; ZZZATV);
	bis(propene-3-thione-1-thiolato)Pt(II) (Mayerle, 1977; PRTTCQ);
	dibenzofuran (may be ordered in $P1$) (Wright and Ahmed, 1981);
	TMTSF (Kistenmacher *et al.*, 1982; SEOTCR);
	dibenzotetrathiafulvalene (DBTTF; $\rho=0.47e$) (Kobayashi and Nakayama, 1981; Emge, Wijgul *et al.*, 1982; BALNAD);
	9,9-*trans*-bis-(telluraxanthenyl) (Lobovskaya *et al.*, 1983);
	E-DMDBTTF (Shibaeva and Yarochkina, 1975);
	Octamethylene-TTF (OMTTF) (Chasseau *et al.*, 1982; BESPEU);
	2,2′,5,5′-tetramethoxystilbene (Zobel and Ruban, 1983; TMXSTQ10).
	Tetrakis(methylthio)TTF (Mori, Wu *et al.*, 1987; FIJYEC).
	Bis(ethylenedithio)TTF (Mori and Inkuchi, 1986; FAHLEF). There is also a monoclinic polymorph.

Table 15.2. (*Continued*)

Acceptor	Donors
2,5-difluoro-TCNQ	DibenzoTTF (Emge, Wijgul *et al.*, 1982; BITROL)
PMDA	Anthracene (Robertson and Stezowski, 1978; ANTPML at 153 and 300K);
	phenazine (Bulgarovskaya *et al.*, 1982; Karl, Ketterer and Stezowski, 1982; BECNUS02);
	acridine (disordered) (Binder *et al.*, 1982; BIHBUP10) {previous three examples isomorphous};
	tetracene (Bulgarovskaya *et al.*, 1987a; FILHOK).
C_6F_6	*p*-xylene (Dahl, 1975a; PXYHFB);
	TMPD (Dahl, 1979; MPAHFB);
	$C_6(CH_3)_6$ (at 233K) (Dahl, 1973; MBZFBZ01).
octafluoronaphthalene	Pyrene (Collings *et al.*, 2001; ECUVIH).
bis(cis-1, 2-prefluoromethyl-ethylene-1,2-dithiolato)Ni(II)	Perylene (Schmitt *et al.*, 1969; PERNIT).
2,3,5,6-tetracyano-hydroquinone	Pyrene (Bock, Seitz *et al.*, 1996; TEXPOB10)
p-dinitrobenzene	TTF (Bryce, Secco *et al.*, 1982; BIRDIP).

Group 1b: Triclinic, $P\bar{1}$, $Z = 2$, stack axis [001]

TCNQ	Acenaphthene (Tickle and Prout, 1973c; ACNTCQ);
	5-phenyl-,3-thiaselenole-2-thione (Kaminski *et al.*, 1979; PTSTCQ); dibenzothiophene (Wright and Ahmed, 1981; BAHFEV).
	phenothiazine (Toupet and Karl, 1995; PTZTCQ01).
PMDA	Phenothiazine (Anthonj *et al.*, 1980; PTZBMA).
TCPA	d^8-naphthalene (Wilkerson *et al.*, 1975; DNPCPH at 120K).
TNB	Pyrene (Prout and Tickle, 1973b; PYRTNB);
	acepleiadylene (Hanson, 1966; APANBZ);
	tetrabenznaphthalene (Herbstein *et al.*, 1976);
	dibenzothiophene (Bechtel *et al.*, 1977; DBTTNB);
	TTF (Bryce and Davies, 1987; GASGUC);
	phenanthrene-chromium tricarbonyl (De *et al.*, 1979; CPCTNB);
	12-imino-12H-benzimidazo [2,1-*b*]-[1,3]benzothiazine (Viossat *et al.*, 1995; ZAYQEV).
TCNB	2,3,6,7-tetramethoxythianthrene (Bock, Rauschenbach *et al.*, 1996; RIKYUF)
picric acid	Tetrabenznaphthalene (Herbstein *et al.*, 1976; ZZZAHA).
3,5-dinitrobenzoic acid	Phenothiazine (Fritchie and Trus, 1968; PHTNBA).
BTF	13,14-dithiatricyclo[8,2,1,1^{4,7}]tetradeca-4,6,10,12-tetraene (Kamenar and Prout, 1965; BOXTET).
2,6-dichloro-N-tosyl-1,4-benzoquinonemonoimine	Pyrene (Shvets *et al.*, 1980; PYTQIM).
octafluoronaphthalene	Triphenylene (Collings Roscoe *et al.*, 2001; ECUVON).

Group 2a: Monoclinic, $P2_1/a$, $Z = 2$, stack axis [001]

TCNE	h^{10}-and d^{10}-pyrene (Larsen *et al.*, 1975; PYRTCE);
	[2.2] (9,10)-anthracenophane (Masnovi *et al.*, 1985; DIRKIY).
	Perylene (Ikemoto *et al.*, 1970; PERTCE10)

Table 15.2. (*Continued*)

Acceptor	Donors
TCNB	anthracene (below 200K) (Stezowski, 1980; ANTCYB11); pyrene (Prout, Morley *et al.*, 1973; PYRCBZ at 178 and 300K); $C_6(CH_3)_6$ (Niimura *et al.*, 1968; CYBHMB); *p*-phenylenediamine (PD) (Tsuchiya *et al.*, 1973; PDTCNB); biphenylene (Stezowski *et al.*, 1986; Agostini *et al.*, 1986; DURYUK);, durene (Lefebvre *et al.*, 1989; KARHAM).
HCBD	Perylene (Yamachi *et al.*, 1987).
TCNQ	Perylene (Tickle and Prout, 1973a; PERTCQ); pyrene (Prout, Tickle and Wright, 1973; PYRTCQ); dibenzo-*p*-dioxin (Goldberg and Shmueli, 1973b; TCQBOX); 1,2-di(4-pyridyl)ethylene (Ashwell *et al.*, 1983; BUKXOU) (red, non-conducting crystals; black crystals were also reported); 6,13-diacetyl-5,14-dimethyl-1,4,8,11-tetraazacyclotetradeca-4,6, 11,13-tetraenenickel(II) (Lopex-Morales *et al.*, 1985); 4, 6, 8-trimethylazulene (Hansmann *et al.*, 1997a; ROJYUK) (donor disordered at inversion center). Bis(ethylenedithio) TTF (Mori and Inokuchi, 1987; FAHLEF01); monoclinic polymorph.
Dimethoxy-TCNQ	OMTTF (Chasseau and Hauw, 1980; OMTFNQ).
(tetracyano-2,6-napthoquinodi-methane (TNAP)	2,7-bis(methylthio)-1,6-dithiapyrene (Toyoda *et al.*, 1993; PIGYUK).
PMDA	pyrene (above 160K) (Herbstein and Snyman, 1969; Allen *et al.*, 1989; PYRPMA); carbazole (disordered) (Stezowski, Binder and Karl, 1982; BIWVOS); biphenylene (red, stable above 400K. (Stezowski, Stiegler and Karl, 1986; DURZAR).
1,8-4,5-naphthalene tetracarboxylic dianhyride	Antharacene (Hoier *et al.*, 1993; WABWEB); dibenz[a,h]anthracene (Zacharias, 1993; BZANTC10).
Pyromellitic dithioanhydride	Anthracene (Bulgarovskaya *et al.*, 1974; TPYMAN); acridine (Bulgarovskaya *et al.*, 1976; ACRTMA); dibenzothiophene (Bulgarovskaya *et al.*, 1978; PMTABT); (all three isostructural).
N,N'-dimethyl-pyromellitic di-imide)	Anthracene (Bulgarovskaya *et al.*, 1977; PMEANT).
1,4-dithiintetra carboxylic N,N'-dimethyldiimide	Acridine (Yamaguchi and Ueda, 1984; space group correction by Marsh, 1986; CEJTAM).
p-benzoquinone	TTF (Frankenbach *et al.*, 1991; SIVBAA).
fluoranil	Perylene (Hanson, 1963; PERFAN).
chloranil	Pyrene (Prout and Tickle, 1973c; PYRCLN); 9-methylanthracene (Prout and Tickle, 1973a, MANTCB) (disordered); TTF (Ohashi *et al.*, 1967; Mayerle *et al.*, 1979; TTFCAN).
BTF	copper oxinate (BTF is disordered) (Prout and Powell, 1965; ZZZGDI)

Table 15.2. (*Continued*)

Acceptor	Donors
benzo[1,2-c;4,5-c']-bis[1,2,5]-thiadiazole-4,8-dione	TTF (Gieren *et al.*, 1984; CIYNUT)
1,4,5,8-naphthalenetetrone	Pyrene (Herbstein and Reisner, 1984; CEKBUP)
3,3',5,5'-tetrachloro-diphenoquinone	Anthracene (Starikova *et al.*, 1980; ANPHXN).
octafluoronaphthalene	Naphthalene (Potenza and Mastropaolo, 1975; NPOFNP);
	tolan (Collings, Batsanov, *et al.*, 2001; OCAYIA).
	anthracene (Collings, Roscoe, *et al.*, 2001; ECUTUR).
C_6F_6	dimer of *o*-diethynylbenzene (Bunz and Enkelmann, 1999; JOCRIC);
	Perylene (Boeyens and Herbstein, 1965a; ZZZLJY)
bis(difluoroborondi methylgloximato)Ni(II)	Anthracene (Stephens and Vagg, 1981; BADZOV).

Group 2b: Monoclinic, $P2_1/a$, $Z = 4$, stack axis [001]

TCNB	N,N-dimethylphenylenediamine (Ohashi, 1973; DMPTCN);
	perylene (Bock, Seitz *et al.*, 1996; REHMUM).
PMDA	Phenanthrene (Evans and Robinson, 1977; PENPYM).
TNB	*s*-triaminobenzene (Iwasaki and Saito, 1970; NIBZAM);
	p-iodoaniline (Powell, Huse and Cooke, 1943; IANNOB);
	tricarbonylchromium anisole (Carter *et al.*, 1966; CCATNB);
	3-formylbenzothiophene (Pascard and Pascard-Billy, 1972; TNBFTB);
	DBTTF (Lobovskaya *et al.*, 1983).
picric acid	Anthracene (Herbstein and Kaftory, 1976; ANTPIIC).
TCNQ	dithieno (3,2-b';2',3-d)thiophene ((Zobel and Ruban, 1983; Bertinelli *et al.*, 1984; CAPTOC);
	dithieno [3,4-b : 3',4'd]thiophene (Catellani and Porzio, 1991; VIGTAG);
2,4,7-trinitrofluorenone	$C_6(CH_3)_6$ (Brown, Cheung *et al.*,1974; TNFLMB).
2,4,6-trinitro-anisole	pyrene (disordered over two orientations) (Barnes *et al.*, 1984; CILRAQ).

Group 3a: Monoclinic, $P2_1/a$, $Z = 2$, stack axis [010]

TCNE	Perylene (Ikemoto *et al.*, 1970; PERTCE10);
	DBTTF (Lobovskaya *et al.*, 1983).
TCNB	Hydroquinone (Bock, Seitz *et al.*, 1996; REHNAT)
TCNQ	bis (ethylenedithio)TTF (Mori and Inokuchi, 1987; FAHLEF01) (also triclinic isomer with segregated stack).
$TCNQF_4$	*Trans*-stilbene (Sato *et al.*, 2001; QILZOA).
PMDA	Benzene (Boeyens and Herbstein, 1965a; ZZZKSM);
	naphthalene (orange polymorph) (Bar-Combe *et al.*, 1979; NAPYMA); perylene (Boeyens and Herbstein, 1965b; PERPML);
	trans-stilbene (Kodama and Kumakura, 1974a; PYMAST);
	chrysene (Bulgarovskaya *et al.*, 1987b; FILHIR).

Table 15.2. (*Continued*)

Acceptor	Donors
fluoranil	Pyrene (Bernstein and Regev, 1980; PYRFLR); chrysene (Munnoch and Wright, 1975; CHRFAN).
3,3′,5,5′-tetrahalodi-phenoquinone (halo = Cl,Br)	Pyrene (Shchlegova *et al.*, 1981; Cl–BAZDAH, Br–BAZCUA).

Group 3b: Monoclinic, $P2_1/a$, $Z = 4$, stack axis [010]

TCNQ	(tetramethyl)porphyrinato)Ni(II) (Pace *et al.*, 1982; BEGLUU); dibenzotellurophene (Singh *et al.*, 1984).
TNB	Naphthalene (Herbstein and Kaftory, 1975a; PVVBKP); azulene (Hanson, 1965a; AZUNBZ; Brown and Wallwork, 1965); skatole (133K) (Hanson, 1964; SKINIB); indole (133K)) (Hanson, 1964; INTNIB).
picric acid	Naphthalene (Banerjee and Brown, 1985; PVVBHJ01).
picryl chloride	9-isopropylcarbazole (Cherin and Burack, 1966; ZZZMCI).
picryl bromide	fluoranthene (Herbstein and Kaftory, 1975a; FLABPC).
BTF	Perylene (Boeyens and Herbstein, 1965a; ZZZMHI); Benzene (Boeyens and Herbstein, 1965a; ZZZLDA).

Group 4a: Monoclinic, $C2/m$, $Z = 2$, stack axis [001]

PMDA	naphthalene (ordered yellow form) (Le Bars-Combe *et al.*, 1979; NAPYMA01); 9,10-dibromoanthracene (Bulgarovskaya, Belsky *et al.*, 1987; FILHEN).
TCNE	Naphthalene (Shmueli and Goldberg, 1974; CYENAP01).
TCNB	naphthalene (above 68K) (Kumakura *et al.*, 1967); anthracene (above 200K) (Stezowski, 1980; ANTCYB12); phenanthrene (donor disordered) (Wright *et al.*, 1978; PHTCBZ); α-naphthol (NAPTCC10), β-naphthol (BNATCB10) (both Couldwell and Prout, 1978).
TCNQ	Anthracene (Williams and Wallwork, 1968; TCQANT); benzidine (Yakushi *et al.*, 1974a; BZTCNQ); N-methyl-phenothiazine (Kobayashi, 1973, 1987; TCQNMP10); dithienylethene (Zobel and Ruban, 1983; CAPTAO); *trans*-stilbene (Zobel and Ruban, 1983; STILTQ10); carbazole (donor disordered) (Mitkevich and Sukhodub, 1987). TMPD (Hanson, 1965b; QMEPHE)
C_6F_6	Anthracene (ZZZGMW); pyrene (ZZZGKE) (both Boeyens and Herbstein, 1965a); durene (Dahl, 1975b; DURHFB).
octafluoronaphthalene	Phenanthrene (Collings, Batsanov *et al.*, 2001; ECUVED)

Note: {4,6,8-trimethylazulene···TCNE} is a rare example crystallizing in space group $P2/m$, $Z = 4$, stack axis [010] (Hansmann *et al.*, 1997b).

We cite three examples of structures of the 1 : 1 mixed stack type which crystallize in a space group other than those listed in Table 15.2. {5,8-Dimethoxy-2,11-dithia[3,3]paracyclophane•••TCNE} crystallizes in space group $C2/c$, $Z=4$, with mixed stacks along [010] (Cohen-Addad, Consigny et al., 1988). Both components lie on two fold axes; the donor molecule is chiral and so are individual stacks as successive donors are related by translation; the space group is, of course, centrosymmetric. The interplanar spacings between TCNE and p-dimethoxyphenyl and phenyl rings are respectively 3.15(1) and 3.33(1) Å, indicating, as would be expected, that p-dimethoxyphenyl is a stronger donor than phenyl. {TTF•••m-dinitrobenzene} also crystallizes in space group $C2/c$, $Z=4$, but the mixed stacks lie along [100], with TTF at centres of symmetry and m-dinitrobenzene on two fold axes (Bryce et al., 1988). There is little overlap of components and their planes are mutually inclined at 8°; nevertheless, the black color of the crystals suggests appreciable charge transfer (in the excited state). The low conductivity along the stack axis ($\approx 10^{-9}$ S/cm) and other physical properties show that the ground state is neutral. {Chrysene•••TNB} crystallizes in space group $Pna2_1$, $Z=4$, with both components in general positions; there are mixed stacks along [001] (Zacharias et al., 1991; VIGKIF {3-Methylchrysene•••TNB} (VIGLAY) is isomorphous, with disorder of the donor.

The structures in Table 15.2 all have stack axis periodicities of ≈ 7 Å, except a few examples where donor and/or acceptor thickness is greater than ≈ 3.5 Å. Although {trans-stilbene•••PMDA} (Kodama and Kumakura, 1974a) has a 12.48 Å stack axis periodicity, the arrangement of the components justifies its inclusion in Table 15.2; both molecules lie about centres of symmetry, but there are stepped stacks, with each phenyl group of the donor being overlapped on one side only by an anhydride portion of the PMDA acceptor. However, in about 10% of 1 : 1 donor•••acceptor molecular compounds the stack axis periodicity is ≈ 14 Å, a possibility already noted in Section 15.1. A somewhat random selection of examples is given in Table 15.3. Possibly some of these crystals have order-disorder transitions leading to doubling of the stack periodicity as occurs in {pyrene•••PMDA} (Herbstein and Samson, 1994); but this has hardly been investigated.

15.4   Packing arrangements in n : m π–π* molecular compounds

D : A ratios of 1 : 2 or 2 : 1 are found in a rather small fraction (roughly 5%) of the crystal structures that have been reported. Other ratios (e.g. 3 : 2, 3 : 4) are much less common. The pyrene-picryl chloride system is remarkable in that five molecular compounds with mole ratios of 4 : 1, 3 : 1, 2 : 1, 1 : 1 and 1 : 3 respectively have been reported on the basis of a DSC study (Bando and Matsunaga, 1976) (but see Section 13.4 for conflicting results). Crystal data have been reported only for the 1 : 1 compound (Herbstein and Kaftory, 1975a).

The 1 : 2 (or 2 : 1) compounds are generally found in one or other of two structural groups. In the first group the donor (say) has a much larger cross-sectional area than the acceptor and can thus behave as a bifunctional donor, with two acceptor molecules sandwiched between each pair of donors. This arrangement is found in {copper oxinate[2]•••(TCNB)$_2$} and in {copper oxinate•••(picryl azide)$_2$} (Bailey and Prout, 1965;

---

[2] Copper oxinate is bis (8-hydroxyquinolinato)Cu(II).

Table 15.3. 1 : 1 donor-acceptor molecular compounds with $\approx$14 Å periodicity along the stack axis; see also Section 15.5.1. Only the higher symmetry space group is given when the systematic absences do not unequivocally determine the space group

Molecular compound	Stack axis along	Stack axis periodicity (Å)	Space group	Z	Crystal structure reported?
{naphthalene···picryl chloride} (Herbstein and Kaftory, 1975a);	[001]	13.81	$P\bar{1}$	2	No
{naphthalene···picryl bromide} (Herbstein and Kaftory, 1975a); PVVBHG	[001]	13.94	$P\bar{1}$	2	No
{pyrene···dicyanomethylene-croconate} (Doherty et al., 1982); BEFGIC	[101]	13.8	$P\bar{1}$	2	Yes*
{pyrene···picryl chloride} (polymorph II) (Herbstein and Kaftory, 1975a)	[001]	13.93	$P\bar{1}$	4	No
{pyrene···PMDA} (below 160K) (Herbstein et al., 1994)	[001]	14.5	$P2_1/n$	4	Yes*
{phenanthrene···TNB} (Herbstein and Kaftory, 1975a)	[001]	13.7	$P2_1/c$	4	No
{benzo[c]phenanthrene···DDQ} (Bernstein et al., 1977); BZPCBQ	[001]	13.83	$P2_1/c$	4	Yes
{4-(2-hydroxyethyl)carbazole···DDQ} (Qi et al., 1996); TEJGUK	[001]	14.57	$P2_1/c$	4	Yes
{2,3,7,8-tetramethoxythianthrene···TCNQ} (D'yachenko et al., 1977); Hinrichs and Klar, 1982; MXTTCQ	[001]	13.72	$P2_1/c$	4	Yes
{(pyrene)$_3$···(picryl bromide)$_2$} (Herbstein and Kaftory, 1975b); PYRBPC	[001]	14.61	$P2_1/c$	4	Yes
{guiacol···picric acid} (yellow polymorph) (Herbstein, Kaftory and Regev, 1976); ZZZAHG01	[001]	13.7	$P2_1/c$	8	No
{anthracene···TNB} (at 173K) (Brown et al., 1964); ANCTNB	[001]	13.02	$C2/c$	4	Yes
{trithia[5]heterohelicene···TCNQ} (Konno et al., 1980); THLCTC	[001]	13.84	$C2/c$	4	Yes
{triphenylene···picryl chloride} (polymorph II) (Herbstein and Kaftory, 1975a); PVVBEY01	[001]	14.4	$A2/a$	8	No
{triphenylene···picryl bromide} (polymorph II) (Herbstein and Kaftory, 1975a); PVVBEV	[001]	14.54	$A2/a$	8	No
{benzene···picric acid} (Herbstein, Kaftory and Regev, 1975); ZZZAGV	[001]	14.13	$Pcmm$	8	No
{C$_6$(CH$_3$)$_6$···picryl chloride} (Powell and Huse, 1943)[§]	[100]	14.0	$Amam$	4	No
{resorcinol···p–benzoquinone} (Ito et al., 1970); BZQRES	[100]	14.63	$Pnca$	4	Yes
{5,10–dihydro-5,10-diethyl-phenazinium···TCNQ} (Dietz et al., 1982); BEWBUA	[001]	14.53	$P2_1/c$	4	Yes
{dipyrido-1,3,4,6-tetraazapentalene···TNF} (Groziak et al., 1986); FARMAM	[201]	14.73	$P2_1/n$	4	Yes

Table 15.3. (*Continued*)

Molecular compound	Stack axis along	Stack axis periodicity (Å)	Space group	Z	Crystal structure reported?
{α-2,7-bis(methylthio)-1,6-dithiapyrene•••TCNQ} (Nakasuji *et al.*, 1987); FUDTON01	[100]	15.08	$P2_1/c$	4	Yes

Notes:

* two independent pyrenes at inversion centers; acceptor at general position.

§ diffuse scattering ignored. The $C_6(CH_3)_6$•••picryl bromide and picryl iodide molecular compounds are similar but the diffuse scattering is more complex.

PAZQCU). In these two molecular compounds the stacks can be represented schematically as shown below and thus are essentially 1 : 1 in character; consequently the stack axis periodicity will be about 7 Å, as before. Analogous situations are found in {bis(N-isopropyl-2-oxy-1-naphthylidene-aminato)Cu(II)•••(TCNQ)$_2$} (Matsumoto *et al.*, 1979; IPONTC) and in {9,10-dihydroanthracene•••(TNB)$_2$} (Herbstein *et al.*, 1986; ZZZAGS10). Relative donor and acceptor sizes cannot be the only factor in determining the stability of these 1 : 2 compounds as palladium oxinate forms a 1 : 1 compound with TCNB (Kamenar *et al.*, 1966; PDHQCB).

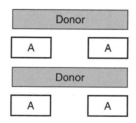

The second group of molecular compounds is characterized by stacks of

-----DAD DAD DAD DAD-----

type (for a 2 : 1 composition). The stack axis periodicity will be about 10.5 Å ($\approx 3 \times 3.5$ Å). A striking example is found in the {(perylene)$_3$•••TCNQ} compound (Fig. 15.5) (Hanson, 1978; TCQPER); the role of the perylene molecule outside the stacks is discussed below (Section 15.5.1). Rather similar stack arrangements (but without the additional interstitial molecules) are found in the isomorphous pair {benzo[c]pyrene•••(TMU)$_2$} (Damiani, Giglio, Liquori and Ripamonti, 1967; TMUBZP10) and {coronene•••(TMU)$_2$} (Damiani, Giglio, Liquori, Puliti and Ripamonti, 1967) and in {stilbene•••(TNB)$_2$} (Bar and Bernstein, 1978; STINBZ), {DBA•••(TNB)$_2$} (Zacharias, 1976), {(HMTSF)$_2$•••TCNQ} (Emge *et al.*, 1982; BOWSUB), {(BTT)$_2$•••TCNQF$_4$} (Sugano *et al.*, 1988; SAJMEV) (BTT is the hexaradialene benzo[1,2-c:3,4-c':5,6-c″]trithiophene), {(HMB)$_2$•••TCNE} (IR study, crystal structure has not been reported; Hall and Devlin, 1967), and {(TTM-TTF)$_2$•••TCNQ} (Mori, Wu *et al.*, 1987; FIJYAY; there is also a 1 : 1 compound FIJYEC).

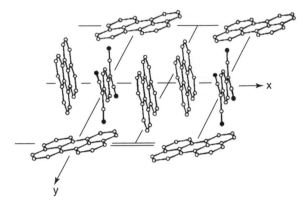

Fig. 15.5. Projection down [100] of the {(perylene)3···TCNQ} crystal structure showing – DADDAD– stacks and interstitial perylene molecules. The crystals are triclinic and the stack axis periodicity is 10.422 Å. (Reproduced from Hanson, 1978.)

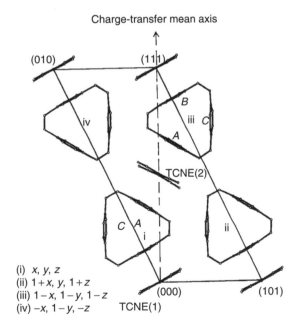

Fig. 15.6. Intermolecular arrangement in {[2.2.2]paracyclophane···TCNE}; the molecules are projected onto the plane defined by the barycentres of the phenyl rings. (Reproduced from Cohen-Addad *et al.*, 1984.)

One method of forcing the formation of ---DADDAD--- type stacks is to use a cyclophane donor where the two potential donor portions have very different donor strengths. Thus in {(5,7-[12]-paracyclophanediyne)$_2$···TCNE}, the TCNE acceptor is sandwiched between benzene rings of two different donor molecules, with an interplanar distance of

3.27 Å, while there is no interaction between triple bonds and TCNE (Harata *et al.*, 1972; PCDTCN). In contrast, mixed stacks of the usual type are found in {1,5-naphthaleno(2)paracyclophane···TCNE}, with the distances between TCNE and benzene and naphthalene planes essentially equal at 3.464 and 3.457 Å (Irngartinger and Goldman, 1978); a similar situation is found in {[2.2.2.2]paracyclophane···TCNE} (Cohen-Addad *et al.*, 1984; COMHOB). An unusual variant on mixed stacking is found in {[2.2.2]paracyclophane···TCNE} (Cohen-Addad *et al.*, 1984; COMHIV), where only two of three benzene rings participate in charge transfer interactions with TCNE (Fig. 15.6).

In the molecular compounds considered up to this point, the donor molecules have all been planar or approximately so. However, in the isomorphous {*trans*-azobenzene···(TNB)$_2$} (ABTNBA) and {N-benzylideneaniline···(TNB)$_2$} (ABTNBB) (Bar and Bernstein, 1981) there are angles of 42 and 48° between the planes of the phenyl rings in the donor molecules and thus there is some disruption of the stacking leading to a tendency to form DAD units. This tendency reaches an extreme in {(N,N,N′,N′-tetramethylbenzidine)$_2$···chloranil} (Yakushi *et al.*, 1971; TMBCAN) where isolated (nonstacked) DAD sandwiches are found (see Section 15.2).

Finally we note that {*trans*-4-methylstilbene···(PMDA)$_2$} has a rather complicated disordered structure (Kodama and Kumakura, 1974b; PYMSTL).

15.5 Some special features of packing arrangements in π–π* molecular compounds

15.5.1 *Crystals where one of the components is also found in interstitial positions*

Perhaps some thirty π-molecular compounds with compositions other than 1 : 1, 1 : 2 or 2 : 1 have been reported (for earlier work see Table 12 of Herbstein, 1971). Some of these compositions require authentication but crystal structure analysis does provide explanations for the unusual compositions of {(pyrene)$_3$···(picryl bromide)$_2$} (Herbstein and Kaftory, 1975b; PYRBPC) and {(perylene)$_3$···TCNQ} (Hanson, 1978; TCQPER). In these two compounds there are respectively 1 : 1 and 2 : 1 donor···acceptor stacks of the usual types, with the additional molecules in interstitial positions where they do not participate in the charge transfer interaction (see Fig. 15.5). There are a number of examples of interstitial donors among the molecular compounds of the flavins and these are noted in Section 15.7.2. The fact that a particular component can play two (or more) different roles in a crystal structure is not new; a classical example already noted is CuSO$_4$·5H$_2$O where four water oxygens and two sulphate oxygens are coordinated octahedrally about Cu and the fifth water molecule is present as solvent of crystallization (Beevers and Lipson, 1934).

It is possible that {(coronene)$_3$···TCNQ} (Truong and Bandrauk, 1977) has a structure similar to that of {(perylene)$_3$···TCNQ}. {5,6-Dihydropyrimidino[5,4-c]carbazole}$_3$···TCNQ·2H$_2$O also comes into this general group. There are mixed....DADDAD... stacks (TCNQs at centers of symmetry) with additional donor molecules in interstitial positions. The two types of donor are mutually hydrogen bonded and also to the water molecules (Dung *et al.*, 1986; DULFAR). {Phenothiazine···(PMDA)$_2$} (Brierley *et al.*, 1982) is another example with 1 : 1 donor···acceptor stacks, with the additional PMDA separating the stacks from one another. The stacking arrangements in the 1:1 stacks are appreciably different in {phenothiazine···PMDA} and in {phenothiazine···(PMDA)$_2$}. {(TMTTF)$_{1.3}$···(TCNQ)$_2$}, which is a segregated stack molecular compound (Chapter 17), is mentioned here because the additional 0.3 molecule of TMTTF is inserted interstitially between the stacks of TMTTF and (TCNQ)$_2$ with its molecular plane parallel to the stack axis (Kistenmacher et al., 1976; SEOTCR01).

The isostructural π-compounds {bis(3,6-dibromocarbazole)···tris(PMDA)} ((Bulgarovskaya et al., 1989; VILFIF) and {bis(N-methyl-3,6-dibromocarbazole)···tris(PMDA)} (Dzyabchenko et al., 1994; WEXKEP) and the (not isostructural) {3,6-dibromocarbazole···bis(PMDA)} (Bulgarovskaya et al., 1989; VILFEB) show interesting resemblances and differences. The crystals are all triclinic and were reported in reduced cells, albeit with unconventional choices of origin. All three have 1:1 mixed donor···acceptor stacks, with, however, different modes of overlap and differing dispositions of the additional PMDA molecules. In the isostructural pair there are sheets of donor···acceptor stacks (stack axis [001]) arranged in the (002) planes, interleaved by sheets of stacks of PMDA molecules, located at crystallographic centres; these PMDA molecules are roughly *coplanar* with the components in the stacks. Thus VILFIF and WEXKEP could be said to have compositions {donor···PMDA}·0.5(PMDA). In VILFEB[3] the additional PMDA molecules are located ·at two independent sets of symmetry centres with markedly different orientations with respect to the mixed stacks, although both have their molecular planes parallel to the stack axes. Thus VILFEB could be said to have composition {(3,6-dibromocarbazole)···PMDA}·[0.5(PMDA1) + 0.5(PMDA2)], where '1' and '2' refer to crystallographically independent PMDAs. This arrangement has similarities to that found in (perylenium)$_2$·PF$_6$·2/3(THF) shown in Fig. 17.6 and is also related to that of {(perylene)$_3$···TCNQ}.

One might guess that triclinic {(fluorene)$_3$···(TNB)$_4$} (Hertel and Bergk, 1936; Herbstein, Kaftory and Regev, 1976; ZZZAGP) has 1:1 stacks with the additional TNB molecules inserted interstitially; however, the crystal structure (Mariezcurrena et al., 1999; ZZZAGP02) shows a remarkable arrangement of mutually-perpendicular 1:1 and 1:2 stacks (Fig. 15.7). The 1:1 stacks, with orientationally-disordered fluorene about inversion centers, are similar to those found in {4-methylchrysene···TNB} (Zacharias et al., 1991; VIGLEC) while the 1:2 stacks resemble those in {1-methylchrysene···(TNB)$_2$} (Zacharias et al., 1991; VIGKOL). Placing this structure description here is clearly somewhat arbitrary.

The structure of the ternary compound {benzidine···TNB}·1/2(C$_6$H$_6$) (Yakushi, Tachikawa et al., 1975; BDTNNB) is somewhat different from those described above. Benzidine and TNB form mixed stacks rather similar to those found in {benzidine···TNB} itself as crystallized from CHCl$_3$ (Tachikawa et al., 1974; BNZTNB); however, there is some alteration in the mutual positioning of the stacks so as to leave channels in which the benzene molecules of solvation are accommodated at centres of symmetry with their molecular planes approximately perpendicular to those of the components in the stacks. The benzene molecules are quite strongly contained within the channels and require application of a vacuum for their removal. Other similar examples are discussed later.

---

[3] VILFEB is incorrectly given a 2:3 composition by the CSD and incorrectly named as 'bis(PMDA) 3,6-dibromocarbazole PMDA solvate'.

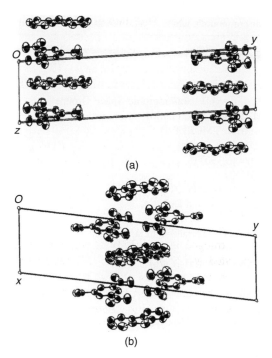

(a)

(b)

Fig. 15.7. Triclinic {(fluorene)$_3$···(TNB)$_4$} ($a = 7.596(7)$, $b = 27.69(2)$, $c = 7.276(11)$ Å, $\alpha = 93.117(9)$, $\beta = 91.114(11)$, $\gamma = 82.374(8)°$; this is the Niggli reduced cell in a non-standard setting)) showing (above) the 1:1 stacking arranged along the [001] axis, and (below) the 1:2 stacking arranged along the [100] axis. Atomic displacement ellipsoids are at an arbitrary level. (Reproduced from Mariezcurrena *et al.*, 1999.)

15.5.2 *Noncentrosymmetric crystals of π-molecular compounds*

The vast majority of π-molecular compounds crystallize in centrosymmetric space groups. Some of the exceptions are listed in Table 15.4, where a separation has been made between the noncentrosymmetrical crystals with enantiomorphic (Sohncke groups) and non-enantiomorphic space groups (see International Tables for X-ray Crystallography, 1965, Vol. I, pp. 41–43, for further discussion). The components in the listed crystals are achiral;[4] thus the chirality for the first group, or lack of a centre for the second group, results from the details of the mixed-stack donor–acceptor arrangement. These crystals may well have interesting physical properties. The example of {1,5-diaminonaphthalene···chloranil} (Tamura and Ogawa, 1977; CANANP) is worthy of note because the component molecules are centrosymmetric; the noncentrosymmetric structure probably results from a compromise between the requirements of hydrogen bonding between different stacks and charge-transfer interactions within stacks. The 1,6-diaminopyrene – bromanil system (Fujinawa *et al.*, 1999) has a number of points of interest. Firstly, the compound is polymorphic and the thermodynamically stable

---

[4] This holds, for some examples, only if the component molecule is planar (a situation not always realized in practice) or there is disorder.

Table 15.4. π-Molecular compounds which crystallize in non-centrosymmetric space groups

Molecular Compound/reference/refcode	Space Group	Z	Crystal Structure Reported ?
I. Enantiomorphic Space Groups			
1. benzo[c]pyrene···(TMU)$_2$ (Damiani, Giglio, Liquori and Ripamonti, 1967); TMUBZP10.	$P1$	2	Yes
2. coronene···(TMU)$_2$ (Damiani, Giglio, Liquori, Puliti and Ripamonti, 1967)	$P1$	2	Yes
3. 7,8-benzoquinoline···TCNQ (Shaanan and Shmueli, 1980); BZQTCQ10.	$P2_1$	2	Yes
4. 1-acetylskatole···TNB (Surcouf and Delettre, 1978); ASKNBZ	$P2_1$	2	Yes
5. TTF···2,7-dintro-9-fluorenone (Soriano-Garcia et al., 1989); KARHOA.	$P2_1$	2	Yes
6. 2-methylchrysene···TNB (Zacharias et al., 1991); VIGKUR.	$P2_1$	2	Yes
7. 5-methylchrysene···TNB (Zacharias et al., 1991); ZEGKIF10.	$P2_12_12_1$	4	Yes
8. carbazole···TNB (Bechtel et al., 1976); CBZTNB	$P2_12_12_1$	4	Yes
9. perylene···TNB (Hertel and Bergk, 1936); ZZZOZO.	$P2_12_12_1$	4	Yes
10. anthracene···BTF (Boeyens and Herbstein, 1965a); ZZZTOS.	$P2_12_12_1$	4	No
11. hydroquinone···naphthaquinone (Thozet and Gaultier, 1977a); NPQHRQ.	$P2_12_12_1$	4	Yes
12. acenaphthene···3,5-dimethylpicric acid (Chantooni and Britton, 1998); PUNYUS.	$P2_12_12_1$	4	Yes
13. pyrene···p–benzoquinone (Bernstein et al., 1976); PYRBZQ.	$P4_1$ (or $P4_3$)	4	Yes
II. Non-enantiomorphic Space Groups			
1. pyrene···TMU (Damiani et al., 1965); MURPYR.	Pc	2	Yes
2. N,N'-dimethyldihydrophenazine···TCNQ (Goldberg and Shmueli, 1973a); TCQMHP.	Cm	2	Yes
3. trans-4-methylstilbene···PMDA (disordered) (Kodama and Kumakura, 1974); PYMSTL.	Cc	2	Yes
4. 1,5-diaminonaphthalene···chloranil (Tamura and Ogawa, 1977); CANANP.	Pn	2	Yes
5. 1,6-diaminopyrene···bromanil (Fujinawa et al., 1999); QADGEH.	Pn	2	Yes
6. phenanthrene···DDQ (Herbstein et al., 1978); PANCYQ.	$Pca2_1$	4	Yes
7. benz[a]anthracene···PMDA (Foster et al., 1976); BZAPRM10.	$Pna2_1$	4	Yes
8. chrysene···TNB (Zacharias et al., 1991); VIGKIF.	$Pna2_1$	4	Yes
9. 3-methylchrysene···TNB (Zacharias et al., 1991); VIGLAY; disordered.	$Pna2_1$	4	Yes

delta-polymorph crystallizes in the noncentrosymmetric space group Pn (QADGEH). This polymorph resembles {1,5-diaminonaphthalene···chloranil}. The other (metastable) polymorph (triclinic $P\bar{1}$, $Z = 4$; QADGEH01) appears to have an intriguing structure but poor crystal quality prevented a definitive determination of its structure. Both polymorphs have interesting physical properties that we shall not discuss.

We have deliberately excluded from consideration here molecular compounds where one or both of the components are themselves chiral; for example, crystals of some flavin compounds (Section 15.7.2) lack centres of symmetry but are not included because the flavin molecules are chiral and also because the charge-transfer interactions are probably weaker than the hydrogen bonding in these crystals.

15.5.3 *Acceptors based on polynitrofluorene*

Four acceptors of this type (2,7-dinitrofluoren-9-one, 2,4,7-trinitrofluoren-9-one, 2,4,5,7-tetranitrofluoren-9-one and 2-(2,4,5,7-tetranitrofluoren-9-ylidenene)-propane-dinitrile) have been found to form donor-acceptor compounds with a variety of aromatic hydrocarbons (Table 15.5). An early use was the formation of molecular compounds with

Table 15.5. Some molecular compounds of aromatic hydrocarbons with acceptors based on substituted polynitrofluorenes. These are all mixed stack structures

Molecular compound	Space group	Reference/refcode
1. 1,12-dimethylbenzo[c]phenanthrene···4-bromo-2,5,7-trinitrofluorenone	Details not given	Ferguson et al., 1969
2. hexahelicene···4-bromo-2,5,7-trinitrofluorenone	Details not given	Ferguson et al., 1969; HELFLU.
3. 2,6-dimethylnaphthalene···2,7-dinitro-9-fluorenone	$P\bar{1}$	Suzuki, Fuji et al., 1992; PARBIT.
4. TTF···2,7-dinitro-9-fluorenone	$P2_1$	Soriano-Garcia et al., 1989; KARHOA.
5. Hexamethylbenzene···2,4,7-trinitrofluorene-9-one	$P2_1/a$	Brown et al., 1974
6. 1-ethylnaphthalene···2,4,5,7-tetranitrofluorene-9-one	$P2_1/c$	Baldwin and Baughman, 1993; LAVFOD.
7. 2-ethylnaphthalene···2,4,5,7-tetranitrofluorene-9-one	$P2_1/c$	Shah and Baughman, 1994; LESZIS.
8. 3,6-dimethylphenanthrene···2,4,5,7-tetranitrofluorene-9-one	$P2_1/n$	Baldwin and Baughman, 1993; LAVFUJ.
9. chlorobenzene···2-(2,4,5,7-tetranitrofluorene-9-ylidene)-propanedinitrile)	$P2_1/c$	Batsanov, Perepichka et al., 2001; TIJTIP.
10. (chlorobenzene)$_2$···2,4,5,7-tetranitrofluorene-9-one	$P2_12_12_1$	Batsanov, Perepichka et al., 2001; TIJTOV.
11. TTF···2-(2,4,5,7-tetranitrofluorene-9-ylidene)-propanedinitrile)	$Pna2_1$	Perepichka, Kuz'mina et al., 1998
12. benzonitrile···2,7-dicyano-9-dicyanomethylene-4,5-dinitrofluorene	$P2_1$	Batsanov and Perepichka, 2003.

overcrowded aromatic hydrocarbons such as hexahelicene (cf. Mackay, Robertson and Sime, 1969). More recent examples have some of the nitro groups replaced by cyano groups (Perepichka, Kuz'mina *et al.*, 1998) or by butylsulfanyl, butylsulfinyl or butyl-sulfoxyl substituents (Perepichka, Popov *et al.*, 2000). Presumably there is a synergistic effect between electron-withdrawing nitro and carbonyl (dicyanoethylene) groups, such as is also found between halogens and carbonyl groups (cf. 1,4-benzoqunone and chloranil).

The two chlorobenzene molecular compounds are unusual in their mode of preparation (see original paper for details) and in the role of chlorobenzene as apparent donor, although it is usually considered a weak acceptor, All these crystal structures are of standard infinite mixed-stack type without aberrant features; overlap of the donor is over the polynitrofluorene ring without the carbonyl (or dicyanoethylene) substituent appearing to play an important role.

15.5.4 Resolution of helicenes by formation of diastereoisomeric charge transfer molecular compounds with enantiomeric acceptors

An acceptor of the type discussed in the previous section has an interesting application. Resolution of the helicenes (Martin, 1974) is an interesting example of the use of charge transfer interactions. Helicenes are nonplanar because of intramolecular overcrowding and hence exist in enantiomeric forms. The first steps towards their resolution were taken in 1956(!) when Newman and Lutz synthesized and resolved TAPA (2-[2,4,5,7-tetranitro-9-fluorenylidene-aminoöxy]propionic acid; Fig. 15.8; see also Newman and Lednicer, 1956). R(–)TAPA forms a red complex *in solution* when mixed with racemic hex-ahelicene and M(–)hexahelicene[5] can be recovered from the solution. These results imply that the P(+)hexahelicene•••R(–)TAPA diastereoisomer is more stable in solution than the M(–)hexahelicene•••R(–)TAPA diastereoisomer. These principles have been adapted to

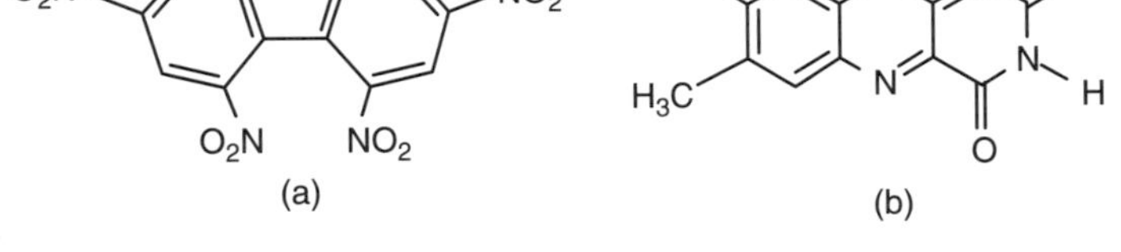

Fig. 15.8. Enantiomeric electron acceptors used in the resolution of helicenes by formation of diastereoisomeric charge transfer molecular compounds (a) TAPA and related compounds: X = methyl, R(–)TAPA; X = ethyl, R(–)TABA; X = isopropyl, R(–)TAIVA; X = butyl, R(–)TAHA.

---

[5] The absolute configurations (e.g. that of TAPA (Kemmer *et al.*, 1976)) were determined after Newman's pioneering studies and have been added here for completeness.

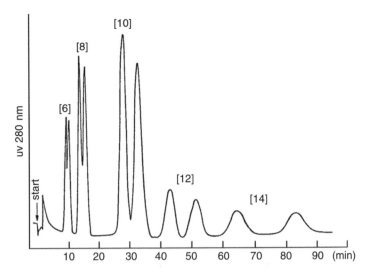

Fig. 15.9. Resolution of a mixture of the racemates of the [6]-, [8]-, [10]-, [12]-and [14]-helicenes. The microsilica column contained 25% R(–)TAPA covalently linked to the silica; the mobile phase was 25% dichloromethane–cyclohexane; U = 0.26 cm/sec. In all instances the more strongly retained enantiomer was the P(+)helicene. Similar results were obtained for the [7]-, [9]-, [11]-and [13]helicenes. [5]-Helicene was resolved only by a multipass technique. (Reproduced from Mikes *et al.*, 1976.)

high performance liquid chromatography with great success (Mikes *et al.*, 1976; Numan *et al.*, 1976), (Fig. 15.9). The HPLC results confirm the deduction from the solution experiments that the P(+)hexahelicene•••R(–)TAPA diastereoisomer is the more stable. Later work has shown that riboflavin, adenosine and adenylic acid coated on microsilica particles can also be used as the stationary phases. The M(–)helicenes are more strongly retained on riboflavin.

Measurements have been made (in tetrachloroethane solution at 248K, using ^{1}H NMR) of the stabilities of the (P)-[7]-thiaheterohelicene•••S(+)-TAPA and (P)-[7]-thiahetero-helicene•••R(–)TAPA diastereoisomeric molecular compounds (Nakagawa *et al.*, 1982); the P•••S diastereoisomer was found to be the more stable by $\Delta\Delta H = -1.03$ kJ/mol and $\Delta\Delta S = 1.40$ J/mol K. A ^{1}H NMR survey of the structures of the two diastereoisomers was said to indicate that the components packed better in the P•••S than in the P•••R diastereoisomer. Analogous crystal-structure comparisons do not appear to have been made.

15.6 Structurally important interactions between polarizable and polar groups

The juxtaposition of polar groups (such as $>C=O$ or $R > C=C < R'$) in acceptors and polarizable groups (such as benzene rings) in donors leads to dipole-induced dipole interaction between them. This interaction, first emphasized in the present context by Prout and Wallwork (1966), can be an important or even dominating influence in determining mutual donor-acceptor arrangement in unary and binary crystals (see also Gaultier *et al.*, (1969)). {Perylene•••fluoranil} (Hanson, 1963; PERFAN) has a typical overlap diagram (Fig. 15.10(a)) which is also found in the molecular compounds of perylene with 2,5-dibromo-3,6-dichloro-*p*-benzoquinone (Kozawa and Uchida, 1979; PERPBQ) and TCNQ (Tickle and Prout, 1973a), and in {(perylene)$_3$···TCNQ} (Hanson, 1978) (Fig. 15.10(c)); and also in the unary crystals of 1,4-naphthoquinone (Gaultier and Hauw, 1965; NAPHQU) and 1,4-anthraquinone (Dzyabchenko and Zavodnik, 1984; COBBIE04) (Fig. 15.10 (d) and (e)). Aromatic ring···carbonyl interactions are also important in forming the DAD triples in {(TMBD)$_2$···chloranil} (Yakushi, Ikemoto and Kuroda, 1971) (Fig. 15.10(f)) and in the mixed stacks of chrysene···fluoranil (Munnoch and Wright, 1975; CHRFAN) and {1,5-diaminonaphthalene···chloranil} (Tamura and Ogawa, 1977; CANANP). Direct overlap of donor and acceptor is also found in solvent-free and in solvated benzidine···TCNQ (see Section 15.7.3.2). One should note the close dimensional correspondence between donor and acceptor molecules in all these molecular compounds; the distance between the centres of the benzene rings in a diphenyl-like portion of perylene or benzidine is closely equal to the O---O distance in the halo-anils or the distance between the centres of the extra-ring double bonds in TCNQ.

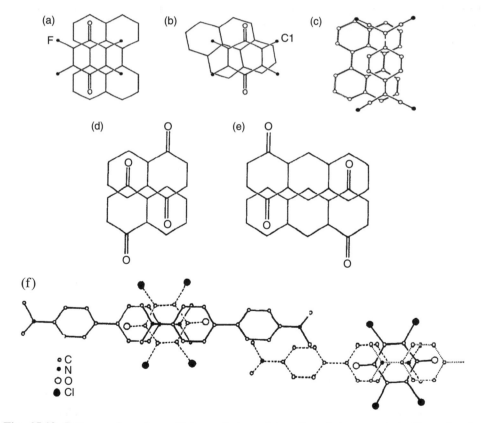

Fig. 15.10. Patterns of overlap which result from interactions between polarizable and polar groups (a) perylene···fluoranil; (b) pyrene···chloranil; (c) perylene···TCNQ in (perylene)$_3$···TCNQ; a somewhat similar overlap diagram is found in perylene···TCNQ; (d) 1,4-naphthaquinone; (e) 1,4-anthraquinone; (f) (TMBD)$_2$···chloranil.

Aromatic ring···polar group overlap is less pronounced but still obvious in the molecular compounds of pyrene with chloranil (Prout and Tickle, 1973c; PYRCLN) (Fig. 15.10(b)), fluoranil (Bernstein and Regev, 1980; PYRFLR), *p*-benzoquinone (Bernstein *et al.*, 1976) and 2,6-dichloro-N-tosyl-1,4-benzoquinonemonoimine (Shvets *et al.*, 1980; PYTQIM), and in {acenaphthene···chloranil} (Tickle and Prout, 1973a) and {acenaphthene···TCNQ} (Tickle and Prout, 1973b; ACNTCQ). We end this section by stressing the obvious – there are many molecular compounds in which interaction between polarizable and polar groups could occur but does not. This is the situation for such molecular compounds as {HMB···chloranil} (Jones and Marsh, 1962; CLAHMB), {TMBD···chloranil} (Yakushi *et al.*, 1973; MBZDCN), {anthracene···3,3′,5,5′-tetrachlorodiphenoquinone} (Starikova *et al.*, 1980; ANPHXN) and {pyrene···3,3′,5,5′-tetrachlorodiphenoquinone} (Shchlegova *et al.*, 1981; BAZDAH). Although there is no direct aromatic ring···carbonyl overlap in the $P2_1/n$ polymorph of 2,3-dichloronaphthazarin (Rubio *et al.* 1985; DCDHNQ01), nevertheless the molecules are arranged in stacks containing symmetry centres that are situated such that the quinoid part of one molecule partially overlaps a benzenoid part of a neighbouring symmetry-related molecule, the interplanar distance being 3.40(3) Å. Consequently this polymorph forms a self-complex in the solid state; spectroscopic data indicate that such close associations also exist in solution. Finally, we note that in the few crystal structures that have been reported for molecular compounds with polynitrofluorenones (TNF and TENF) as acceptors, the acceptor behaves as a polynitroaromatic and aromatic ring···carbonyl interactions do not seem to be important.

15.7   Mixed-stack crystals with both charge transfer and hydrogen bonding interactions

15.7.1   *The quinhydrones as a crystallochemical family*

15.7.1.1   *Crystal structures*

Quinhydrones (Patil *et al.*, 1986) are molecular compounds formed by polyhydroxy-aromatics (as donors) and aromatic quinones (as acceptors); quinhydrone itself is {hydroquinone···*p*-benzoquinone} (hydroquinone is also called quinol or 1,4-dihydroxy-benzene). We define members of the quinhydrone structural family as having crystal structures involving two specific interactions between the components

(i)   hydrogen bonding between hydroxyl and carbonyl groups
(ii) charge transfer interaction between donor and acceptor ring systems.

The crystal structures appear to be determined by an interplay between these factors, whereas the UV–visible spectroscopic properties are a consequence of the charge transfer interactions, which often show themselves structurally by a superposition of the carbonyl group of the acceptor over the (possibly substituted) aromatic ring of the donor. A consequence of this definition is that every molecular compound composed of a polyhydroxy-aromatic and an aromatic quinone is not necessarily a quinhydrone.

Donor : acceptor ratios of 2 : 1, 1 : 1 and 1 : 2 (and, exceptionally, 1 : 3) are known and can be seen to depend on the nature of the components. Much of the systematic preparative work dates back 60–80 years; we quote some of these results (Siegmund, 1908;

Table 15.6. Quinhydrones reported to be formed between polyhydroxy-aromatics and aromatic quinones

Donor	D/A Ratio	Donor	D/A Ratio
(a) With *p*-benzoquinone as acceptor:			
phenol	2 : 1	1,4-dihydroxybenzene(quinol)	1 : 1
p-chlorophenol	2 : 1, 1 : 1	1,2,3-trihydroxybenzene	1 : 3
p-bromophenol	2 : 1, 1 : 1	1-naphthol	2 : 1, 1 : 1
p-nitrophenol	1 : 1	2-naphthol	2 : 1, 1 : 1
1,2-dihydroxybenzene	2 : 1, 1 : 1	*p*-methylaminobenzene	1 : 1
1,3-dihydroxybenzene	1 : 1	2-naphthylamine	2 : 1, 1 : 1

(b) With other acceptors: 1-naphthol···α-naphthoquinone; 1-naphthol···phenanthrenequinone

Meyer, 1909; Kremann *et al.*, 1922) in Table 15.6, which does not give a complete survey of the known compounds. This type of information, taken together with that about crystal structures in Table 15.7 (which is hopefully more comprehensive), shows both what has been learned and what gaps remain. The methods used include crystallization from solutions of the mixed components and determination of binary phase diagrams. Two aromatic amines are included among the donors because their molecular compounds *may* resemble the quinhydrones in structure.

Similarity of intercomponent interactions leads to striking resemblances among many of the crystal structures and justifies treatment of the whole group as a single crystallochemical family. The known crystal structures (Bernstein, Cohen and Leiserowitz (1974) and later work) can be grouped as shown in Table 15.7. In this classification we have emphasized the importance of donor-acceptor stacking and distinguished between 1 : 1 and 1 : 2 (or 2 : 1) compositions of the *stacks*. This means that quinhydrones with overall compositions 1 : 2 (or 2 : 1) are classified according to the D : A ratio in the stack and not in terms of overall composition.

The structural features of Group A are conveniently introduced by reference to the two polymorphs of quinhydrone (α triclinic, $P\bar{1}$, $Z = 1$; β, monoclinic, $P2_1/c$, $Z = 2$). In these crystals a combination of charge-transfer interactions along [100] and hydrogen bonding along [120] gives molecular sheets, one molecule thick, parallel to (001). In the triclinic polymorph (Fig. 15.11(a)) all the molecular sheets are identical, whereas in the monoclinic polymorph (Fig. 15.11(b)) the direction of the hydrogen bonds in successive sheets changes from [120] to [1$\bar{2}$0] in accord with the introduction of the *c* glide plane. The arrangements in the two polymorphs are formally similar to those found in the triclinic and monoclinic polymorphs of *p*-dichlorobenzene (Herbstein, 2001). The volume per formula unit in both quinhydrone polymorphs is less than the sum of the volumes (in their respective crystals) of the two components; the reduction is 9% for the triclinic polymorph and 7% for the monoclinic polymorph, suggesting that the former is the more stable at room temperature and pressure. Analogous reductions are found for almost all the quinhydrones but none is as large as that found for α-quinhydrone; the only exception we have found is {(2,5-dimethylhydroquinone)$_2$···(2,5-dimethyl-*p*-benzoquinone)} (CIKRAP), where there is a volume increase of 1.4%.

The orientations of the stacks alternate in {resorcinol···*p*-benzoquinone} (BZQRES) (space group *Pnca*, $Z = 4$) (Fig. 15.11(c)) and the structure thus resembles that of

Table 15.7. Classification of the quinhydrones according to relationships among their crystal structures

---

Type I: with DA stacks
Group A: (donor : acceptor ratio 1 : 1).
α-quinhydrone (Sakurai, 1968; QUIDON01) and β-quinhydrone (Sakurai, 1965; QUIDON); resorcinol···p-benzoquinone (Ito *et al.*, 1970; BZQRES); hydroquinone···1,4-naphthaquinone (Thozet and Gaultier, 1977a; NPQHRQ); 2-phenyl- (or 2-chlorophenyl)-benzhydroquinone···2-phenyl- (or 2-chlorophenyl)-benzoquinone (Desiraju *et al.*, 1979; PBZQHQ); durohydroquinone···duroquinone [duroquinhydrone] (Patil *et al.*, 1986; Pennington *et al.*, 1986; FADCES),
Group B: (donor : acceptor ratio 1 : 1, donor is a phenol).
p-chloro- (or p-bromo-)phenol···p-benzoquinone (isomorphous) (Shipley and Wallwork, 1967b; BNQCLP, BNQBRP).
Group C: (donor : acceptor ratio 2 : 1; additional donor bridges between DA stacks)
(hydroquinone)$_2$···2,5-dimethylbenzoquinone) (Patil *et al.*, 1984; CISCOW); (hydroquinone)$_2$ ···duroquinone) (Pennington *et al.*, 1986; FADCIW).

Type II: (with --DAD-- or --ADA-- as repeat units in the stacks)
Group D1: (donor : acceptor ratio 2 : 1).
{(phenol)$_2$···p-benzoquinone} [phenoquinone] (Sakurai, 1968; PHENQU); ({p-chloro-(or p-bromo-) phenol}$_2$···p-benzoquinone} (these two molecular compounds are isomorphous) (Shipley and Wallwork, 1967b; BNQDCP, BNQDPB).
Group D2: (donor : acceptor ratio 1 : 2).
{(1,3,5-trihydroxybenzene)···(p-benzoquinone)$_2$} (Sakurai and Tagawa, 1971; PHLBZQ); {1,4-dihydroxynaphthalene···(1,4-naphthoquinone)$_2$} (Artiga *et al.*, 1978b; NPQNHQ); {α-naphthol···(2,3-dichloro-1,4-naphthoquinone)$_2$} (Thozet and Gaultier, 1977b; CNQNPO); {(2,5-dimethyl-1,4-dihydroxybenzene)···(2,5-dimethyl-1,4-benzenoquinone)$_2$} (Patil *et al.*, 1984).

Notes: Cell dimensions have been reported for {(p-cresol)$_2$···p-benzoquinone} and {α-naphthoquinone-quinhydrone} (ZZZOXY) (Anderson, 1937); the structure of {α-naphthol···2-methyl-1,4-naphthoquinone} has been briefly described (Berthelon *et al.*, 1979).

monoclinic β-quinhydrone; polymorphism has not been reported. There is also a remarkable resemblance (Herbstein, 1971, see p. 317) between the structures of {resorcinol···p-benzoquinone} and {anthracene···TNB} (Brown, Wallwork and Wilson, 1964; ANCTNB). There is superpositioning of carbonyl groups over aromatic rings in α- and β-quinhydrones (Fig. 15.11(d)) and this is also found in {p-ClC$_6$H$_4$OH···C$_6$H$_4$O$_2$}, phenoquinone, {resorcinol···C$_6$H$_4$O$_2$} and {(p-ClC$_6$H$_4$OH)$_2$···C$_6$H$_4$O$_2$} (C$_6$H$_4$O$_2$ is p-benzoquinone).

Two structures in this group have noteworthy features. {Hydroquinone···1,4-naphthoquinone} has a structure (Thozet and Gaultier, 1977a; NPQHRQ) based on hydrogen bonding and charge-transfer interactions, with aromatic ring-carbonyl super-positioning. Individual sheets are analogous to those in α-quinhydrone but four criss-crossed sheets are packed in each b-axis period. The crystals are chiral (space group $P2_12_12_1$) although the components are not. The structures of 2-phenylquinhydrone and 2-(p-chloro)-phenylquinhydrone are based on the usual principles of hydrogen bonding and charge transfer interaction but are such that quinol and quinone components cannot

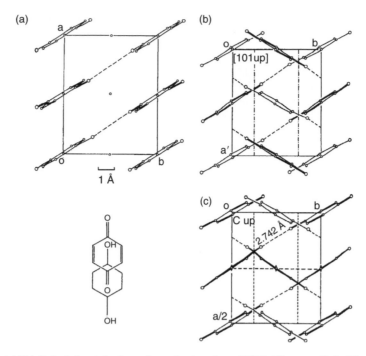

Fig. 15.11. (a) Triclinic (α) quinhydrone in projection down [001]; (b) monoclinic (β) quinhydrone in projection down [101]; (c) {resorcinol···p-benzoquinone} in projection down [001]; the resorcinol molecules are seen edge-on and the p-benzoquinone molecules are seen slightly tilted. The overlap diagram typical of many quinhydrones showing superposition of carbonyl group on aromatic ring is illustrated in the lower left portion of the diagram. Hydrogen bond lengths (d(O...O)) are 2.739 Å in α-quinhydrone and 2.71 Å in the β-polymorph. (Reproduced from Herbstein, 1971.)

be distinguished in the crystal because of disorder (space group $P2_1/c$, $Z=2$); it was suggested that the crystals could be conglomerates of regions with true space group symmetry either $P2_1$ or Pc (Desiraju *et al.*, 1979).

The same structural principles apply in Group B as in Group A, except that the molecules are hydrogen bonded in pairs (Fig. 15.12) rather than ribbons because of the monofunctionality of the phenols.

The structures of {(hydroquinone)$_2$···(2,5-dimethyl-p-benzoquinone)} (CISCOW) and {(hydroquinone)$_2$···(duroquinone)} (FADCIW) form a separate group in which the equimolar DA stacks are preserved with the second hydroquinone molecule bridging between the stacks; the details of the hydrogen bonding by which the bridging is effected differ in the two crystals. We first illustrate for {(hydroquinone)$_2$···2,5-dimethyl-p-benzoquinone)} (Fig. 15.13); there is carbonyl-aromatic ring overlap within the stacks and the stacks are linked together by hydrogen bonding between hydroquinone and benzoquinone to give sheets analogous to those already noted in α- and β-quinhydrones. The interstitial hydroquinones hydrogen bond between hydroquinones of the stacks to give cohesion in the third dimension. On the other hand, the stacks in {(hydroquinone)$_2$···(duroquinone)} are not hydrogen bonded into sheets but the cohesion is given by the bridging hydroquinones

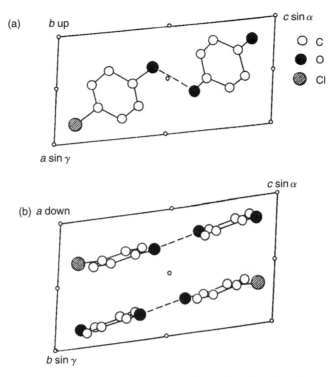

Fig. 15.12. Projections of the crystal structure of {*p*-chlorophenol···*p*-benzoquinone} down (a) [010], (b) [100]. In (b) two hydrogen-bonded donor–acceptor pairs are shown, related by a centre of symmetry; one of these pairs has been omitted in (a) for clarity. (Reproduced from Herbstein, 1971.)

which link four stacks by HQ--HQ and HQ--DQ hydrogen bonds (Fig. 15.14); there is no carbonyl-aromatic ring overlap within the stacks. Thus the first structure is essentially a true quinhydrone in terms of its resemblance to α- and β-quinhydrone while the second is noted here for reasons of chemical rather than structural resemblance.

The 2:1 and 1:2 quinhydrones of Group D have somewhat similar structures and we describe that of phenoquinone ({(phenol)$_2$···*p*-benzoquinone}; Fig. 15.15). The components are hydrogen bonded in groups of three, the minor component being flanked on each side by a molecule of the major component. The stacks have composition

----DAD DAD DAD----

and thus the characteristic periodicity along the stack axis is ≈11 Å; the D---A interplanar distances are somewhat shorter than the D---D distances. Aromatic ring-carbonyl overlap occurs in phenoquinone, there is partial overlap in {(2,5-dimethylhydroquinone)$_2$···(2,5-dimethyl-*p*-benzoquinone)} (CIKRAP) but none in {1,4-dihydroxynaphthalene···(1,4-naphthoquinone)$_2$} (NPQNHQ) nor in {α-naphthol···(2,3-dichloro-1,4-naphthoquinone)$_2$} (CIQNPO). In the latter the α-naphthol molecule is disordered across a centre of symmetry and Cl---O interactions (d = 2.95 Å) are also important.

The {1,3,5-trihydroxybenzene···(*p*-benzoquinone)$_2$} compound (PHLBZQ) has an interesting structure in which there are independent branched zigzag chains of molecules

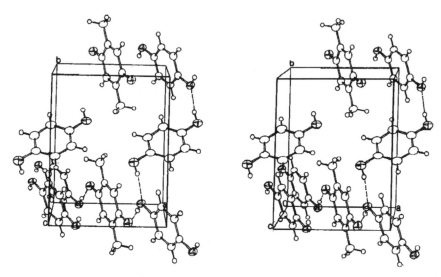

Fig. 15.13. Stereoview of the {(hydroquinone)$_2$···(2,5-dimethyl-p-benzoquinone)} structure, seen approximately down [001]; hydrogen bonds are shown as broken lines; all three component molecules are at independent centres of symmetry, with some symmetry-related molecules omitted to simplify the diagram. Ribbons of hydrogen-bonded hydroquinone···2,5-dimethyl-p-benzoqui-none extend along the [101] direction as shown in the lower ac plane of the cell. The aromatic ring-carbonyl overlap occurs between superimposed ribbons of this type. The ribbons are linked by the "interstitial" hydroquinone molecules shown in the centre of the (100) faces. (Reproduced from Patil, Curtin and Paul, 1984.)

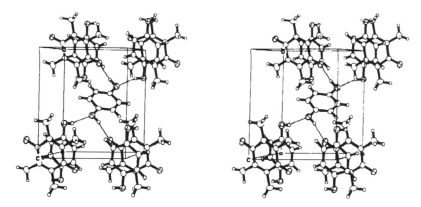

Fig. 15.14. Stereodiagram of the crystal structure of {(hydroquinone)$_2$···(duroquinone)} showing linking of four DA stacks by hydrogen bonding to the interstitial hydroquinone in the center of the cell. (Reproduced from Pennington *et al.*, 1986.)

in each molecular layer (Fig. 15.16). Two of the acceptor molecules bridge between different hydroxyl groups of the donor while the third p-benzoquinone is linked through *one* of its carbonyl groups to the third hydroxyl, while the second is unlinked; this arrangement accounts for the composition. The chains interact only by dispersion forces

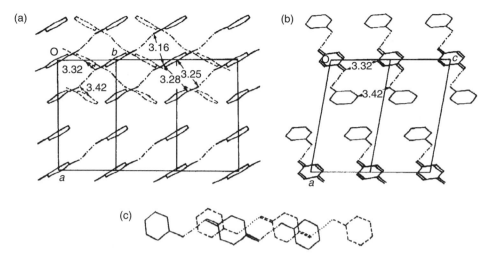

Fig. 15.15. Phenoquinone (a) projection down [001], (b) projection down [010], (c) overlap diagram showing superimposed DAD hydrogen-bonded triples; note the aromatic ring-carbonyl overlap. (Reproduced from Sakurai, 1968.)

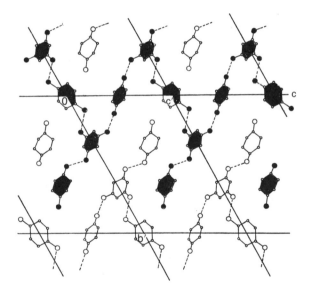

Fig. 15.16. {1,3,5-trihydroxybenzene···(p-benzoquinone)$_2$} projected onto the mean molecular plane. The zigzag hydrogen-bonded ribbons, one of which has been shaded, extend along the axis labelled **c**. There is no hydrogen bonding between neighbouring ribbons in the **b**′ direction. (Reproduced from Sakurai and Tagawa, 1971.)

so that the crystal structure is a two-dimensional analog of the intersecting but non-bonded three-dimensional networks found in the quinol clathrates (Chapter 8) and some other crystals such as α-trimesic acid (Section 9.3) and adamantane-1,3,5,7-tetracarboxylic acid (Ermer, 1988; GEJVEW).

The equimolar molecular compound of 6-hydroxydopamine·HCl and its oxidized *p*-quinonoid form is a quinhydrone (Andersen *et al.*, 1975; DOPAQC). The compound crystallizes in a monoclinic unit cell with $a = 6.815$, $b = 13.108$, $c = 20.365$ Å, $\beta = 95.43°$, space group $P2_1/n$, $Z = 4$. The component cations stack alternately along [100], with an angle of 3.8° between ring planes and interplanar spacings of 3.42 and 3.49 Å, somewhat longer than in most quinhydrones. The two components are respectively colourless and yellow in their neat crystals and the red colour of the molecular compound is credibly ascribed to a charge transfer interaction. There is no carbonyl-aromatic ring overlap.

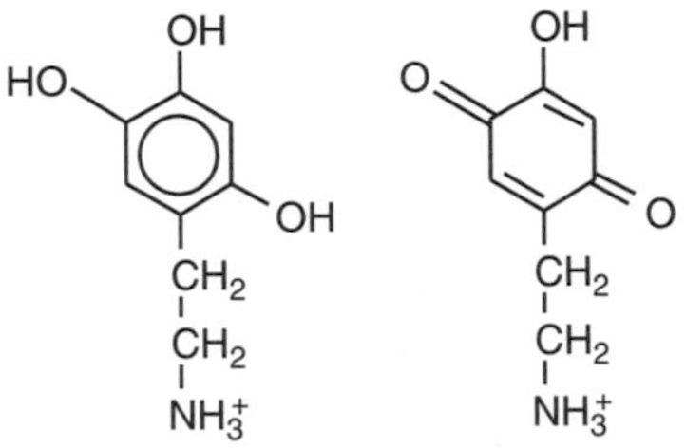

The asymmetric unit consists of the two cations shown and two chlorides as counterions.

Crystal structures have been reported for a number of cyclophane quinhydrones (see Chapter 14 for details and references); for example, a stacked structure was found for pseudogeminal 14,17-dimethoxy[3](2,5)-*p*-benzoquinone, with an

AD AD AD AD

arrangement along the stacks and an average intermolecular separation of 3.42 Å, about 0.23 Å larger than the intramolecular transannular separation (see Fig. 14.5). Hydrogen bonding between moieties in different stacks seems unlikely for steric reasons.

15.7.1.2 *Thermodynamic studies of quinhydrones*

A number of measurements have been made of thermodynamic parameters for various quinhydrones. The enthalpy of formation of a crystalline molecular compound of composition D_mA_n is defined as

$$\Delta H_f(D_mA_n) = H(D_mA_n) - \{mH(D) + nH(A)\},$$

where the enthalpies $H(D_mA_n)$ etc. refer to the crystalline materials at the same (specified) temperature. Analogous equations and conditions apply to free energies $\Delta G_f(D_mA_n)$ and entropies of formation $\Delta S_f(D_mA_n)$. Available results for $\Delta H_f(D_mA_n)$ are summarized in Table 15.8.

We first consider the results for quinhydrone. Three polymorphs of hydroquinone and two of quinhydrone are now known and thus specification of polymorph, with regard to both quinhydrone and hydroquinone, is required; Suzuki and Seki (1953) checked that they used α-hydroquinone but do not report which quinhydrone polymorph was used, while the converse situation applies to the results of Artiga *et al.* (1978a, b). The values of the crystal densities suggest that triclinic quinhydrone is stable with respect to monoclinic at room temperature but the quantitative situation is not known, nor is it known whether

Table 15.8. Values of $\Delta H_f(D_mA_n)$ (kJ/mol) for some quinhydrones as measured by a variety of methods

Molecular Compound	$-\Delta H_f(D_mA_n)$	Method
Quinhydrone	20.2 (303K); 22.6 (297K)	Dissolution (Artiga et al., 1978a; Suzuki and Seki, 1953)
	19.7; 20.6(1) (295K)	vapour pressure (Nitta et al., 1951; Kruif et al., 1981).
	33.9	Combustion (Schreiner, 1925)
	23.2 (297K)	calculated from ΔG, ΔS (Schreiner, 1925)
	22.0 (0K)	Estimated (Suzuki and Seki, 1953)
{hydroquinone···1,4-naphthoquinone}	6.1 (303K) 8.0(2) (320K)	Dissolution (Artiga et al., 1978a, b); vapour pressure (Kruif et al., 1981)
{1,4-dihydroxynaphthalene··· (1,4-naphthoquinone)$_2$}	8.6 (303K) 9.6(4) (320K)	Dissolution (Artiga et al., 1978b); vapour pressure (Kruif et al., 1981)
{1,4-dihydroxynaphthalene··· 1,4-naphthoquinone}	11.6(2) (330K)	vapour pressure (Kruif et al., 1981)

the two polymorphs are enantiotropically or monotropically related. Artiga *et al.*, report that triclinic quinhydrone is thermally stable up to its melting point at 443K, suggesting that the relationship for quinhydrone is monotropic. The precision reported for the various measurements is 1–3%, but intercomparison of the ΔH_f values suggests that systematic errors can occur, apart from the uncertainty about which polymorph was used.

The value of ΔH_f for quinhydrone at 297K is about –21(2) kJ/mol, presumably applying to the more stable α-polymorphs of hydroquinone and quinhydrone. The value of ΔH_f at 0K was estimated Suzuki and Seki (1953) by correcting the 297K value *via* specific heat values measured by Lange (1924) and "put in order" by Schreiner (1925); as the specific heats were measured at rather large temperature intervals, these corrections are somewhat uncertain.

The free energy and entropy of formation of quinhydrone at 297K were measured by an EMF method (Schreiner, 1925) and the following values obtained:

$$\Delta G_f = -15.3\text{kJ/mol} \quad \Delta S_f = -26.3\text{J/mol K}.$$

The corresponding values from vapour pressures (Kruif *et al.*, 1981) [–13.8(1) kJ/mol and –23(5) J/mol K] are in good agreement. One may conclude that the components are more strongly bound in crystalline quinhydrone than in the separate crystals of hydroquinone and *p*-benzoquinone, and that there is a concomitant reduction in the entropy, presumably because of a reduction in the amplitudes of thermal vibration. The negative value of ΔS_f implies that quinhydrone becomes more stable with respect to its separated components as the temperature is lowered, it being assumed that ΔH_f and ΔS_f are only weakly temperature dependent; this is also the situation for anthracene···picric acid (see Chapter 16).

A value of ΔG_f (330K) of –8.6(1) kJ/mol. has been calculated for {1,4-dihydroxy-naphthalene···1,4-naphthoquinone} (crystal structure does not seem to have been reported) from vapour pressure measurements (Kruif *et al.*, 1981) with ΔS_f (330K) calculated as $-9(5)$ J/mol K. The situation is similar to that found for quinhydrone but the molecular compound is less stable with respect to its components than is quinhydrone.

Specific heats have been measured for {1,4-dihydroxynaphthalene$\cdots$(1,4-naphthoquinone)$_2$} and its components from 4–300K and $\Delta S_f(297K)$ calculated as 7.7 J/mol K, leading to a value of ΔG_f (297K) of -10.9 kJ/mol (Artiga et $al.$, 1978b). Vapor pressure measurements (Kruif et $al.$, 1981) give ΔG_f (320K) $= -12.9(10)$ kJ/mol., ΔH_f (320K) $= -9.6(4)$ kJ/mol, and $\Delta S_f(320K) = 10(10)$ J/mol K. The positive value of ΔS_f implies that {1,4-dihydroxynaphthalene$\cdots$(1,4-naphthoquinone)$_2$} becomes less stable with respect to its separated components as the temperature is lowered, contrary to the situation in quinhydrone.

The energies of complexation have been calculated for some of these molecular compounds using atom-atom potentials (Wit et $al.$, 1980); the results were qualitatively correct but not very accurate. This is not surprising as atom-atom potentials hardly seem fitted to the quantitative representation of hydrogen bonding and charge transfer interactions.

15.7.1.3 *Spectroscopy*

Polarized absorption (polymorph not specified) (Nakamoto, 1952) and reflection spectra (monoclinic polymorph) (Anex and Parkhurst, 1963) have been measured from small single crystals of quinhydrone. Nakamoto (1952) interpreted his results in terms of enhanced absorption, due to charge transfer, when the radiation was polarized normal to the ring planes of the two component molecules but Anex and Parkhurst (1963) claimed that the appropriate polarization direction is along the stack axis and not along the ring normals, a result verified by both techniques used by them. They point out that Mulliken's theory in fact suggests that the polarization vector for the charge transfer band should lie along the line joining the centres of the components. However, the occurrence of criss-crossed stacks in monoclinic quinhydrone (Fig. 15.10(b)) forces the polarization to lie along the average normal to the molecular planes, which is the stack axis. Measurements on the triclinic polymorph, where all plane normals are parallel, and inclined to the stack axis, would provide the decisive test. A measurement of this kind made on triclinic {anthracene$\cdots$PMDA} (ANTPML) shows unequivocally that the moment of the principal charge transfer transition lies along the anthracene to PMDA center-to-center vector or [001] axis and not along the plane normal (Merski and Eckhardt, 1981) (cf. Section 17.3).

Polarized single-crystal spectra of {resorcinol$\cdots p$-benzoquinone} have been studied by Amano (quoted by Ito et $al.$, 1970). Intense charge transfer bands were observed with light polarized along [100] while the crystal was almost transparent to light polarized along [010]. The orthorhombic crystal structure requires the charge-transfer moment to be along the stack axis as in monoclinic quinhydrone.

15.7.2 *Molecular compounds of the flavins*

The flavo co-enzymes, which are the non-protein parts of flavoenzymes, are widely involved in oxidation-reduction processes in the cell. The two forms most commonly found in biological systems are flavin mononucleotide (FMN or riboflavin 5$'$-phosphate) and flavin adenine dinucleotide (FAD), whose molecular structures are given in Fig. 15.17.

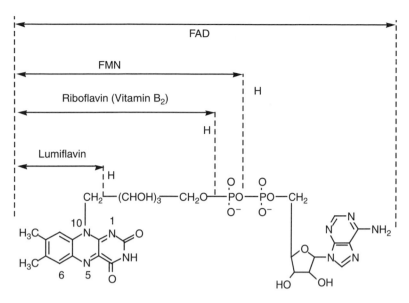

Fig. 15.17. The molecular structures of several flavins. In iso-alloxazine the substituents at C(7), C(8) and N(10) are all hydrogens.

The iso-alloxine or flavin nucleus, which is common to all forms, is the site of electron exchange and is shown in the air-stable, fully oxidized or quinoid form. It is reduced enzymatically by one or two electrons to either the semiquinoid or the hydroquinoid state. Since many of the substrates, including reduced pyridine nucleotide, DPNH, are aromatic or quasi-aromatic, it is possible that the electron interchange reaction occurs by formation of an intermediate, transient, charge transfer complex (Szent-Gyorgi, 1960). In consequence considerable effort has been applied to the preparation of complexes between electron donors and flavins in various oxidation states. Cationic, neutral and anionic species can be obtained depending on the pH of the system (Kosower, 1960; Tollin, 1968). Neutral and charged flavoquinone complexes have been obtained, as well as charged flavosemiquinone complexes.

A fairly complicated series of equilibria has been shown to exist between the fully oxidized and fully reduced forms of flavins (Kuhn and Stroeble, 1937), each intermediate being characterized by "excellent crystallizing power, vivid colours and sharply defined composition." The best results were obtained with riboflavin where the equilibria can be represented as:

flavin → verdoflavin → chloroflavin → rhodoflavin → leucoflavin.

All the crystalline intermediates were paramagnetic (bulk susceptibility measurements), ionic (Na^+ or Cl^- ions were necessary), and supposed to be composed of different mixtures of the various flavin oxidation states.

Most studies to date of complexes of flavins are based on visible and ESR spectra of solutions but some crystalline compounds have been prepared. For convenience, neutral and charged species will be described together. The neutral crystalline compounds are usually orange to orange-red, while the protonated complexes range from green through deep red to black. Charge transfer interaction between donor and flavin acceptor has been suggested for certain compounds, although this may not always have been appropriate.

Neutral crystalline molecular compounds of 2,3- and 2,7-naphthalenediol with ribo-flavin have been prepared (Fleischmann and Tollin, 1965a) with phenol : flavin ratios of 2 : 1, although spectroscopic analysis indicated that the solution compositions were 1 : 1. There was an extra band in the spectra of mineral oil suspensions of powdered crystals, and this has been ascribed to charge transfer. Various indoles, such as 5-methylindole, carbazole and tryptophan, gave neutral 1 : 1 crystalline compounds with lumiflavin but not with riboflavin. Protonated complexes such as lumiflavin-tryptophan-HCl-H_2O have also been obtained. Although the crystalline complexes are more highly colored than the parent flavins, the diffuseness of the additional bands in the spectra makes it difficult to decide whether there is indeed a charge transfer interaction (Pereira and Tollin, 1967). Crystalline complexes have also been prepared with compositions phenol : flavin : HCl, where the phenols include hydroquinone, 4-chloropyrocatechol, resorcinol and 1,2- and 1,4-dihydroxynaphthalene and the flavins riboflavin, lumiflavin and 9-methyl-isoalloxazine (Fleischmann and Tollin, 1965b). The spectra suggest charge-transfer interactions between the components.

Mention must also be made of the paramagnetic semiquinone salts and complexes first prepared by Kuhn and Strobele (1937) and studied later by Fleischmann and Tollin (1965c), who showed that iso-alloxazine derivatives in concentrated hydroiodic acid were reduced to the semiquinone state and that some crystalline flavin hydroiodides could be obtained; ESR measurements showed that the latter had 100% unpaired spins. Riboflavin hydroiodide (no analysis given) gave pink crystalline platelets, while lumiflavin hydro-iodide (composition $FH_2 \cdot 2HI$) was a dark-brown solid which could not be recrystallized. Black crystals (red in thin section) were obtained from saturated solutions of riboflavin in hydroiodic acid to which various phenols (e.g. hydroquinone, 2-naphthol, 2,3-, 1,7-, 2,7-, 1,5- and 1,4-dihydroxynaphthalene) had been added; a typical composition was $FH_3^+ \cdot I^- \cdot 2,3$-dihydroxynaphthalene. There were no indications of charge transfer inter-actions in the spectra (Tollin, 1968).

A number of crystal structures have been determined (Table 15.9) and comparison shows many common features.

1. There is always an extensive array of hydrogen bonds which is probably the primary factor in determining the overall arrangement.

2. Most structures contain infinite stacks of donor and acceptor molecules but two have finite DAAD sequences (Fig. 15.18). The interplanar distances range from ≈ 3.3 to 3.5 Å and there is some correlation between colour of the crystals (a rough measure of charge transfer, in the absence of spectroscopic results) and interplanar distance. However, the nature of the moieties is probably a more important factor here and it seems clear that the protonated flavins are appreciably stronger acceptors than the neutral flavins (in contrast one may note that picric acid molecule and picrate ion have similar acceptor strengths).

3. In four of the compounds (out of eight) there are additional donor molecules in interstitial positions outside the mixed stacks, with the planes of the interstitial molecules approximately normal to those in the mixed stacks. The arrangement is similar to that encountered in a few mixed stack donor–acceptor molecular compounds

Table 15.9. Some structural details for crystalline flavin molecular compounds. Reports of crystal structure determinations after 1975 have not been found.

Complex/reference/refcode	Colour of Crystals	Interplanar Spacings (Å)	Space Group	Z	Stacking Type	Interstitial Donors?
1. Adenine∙∙∙ riboflavin. 3H$_2$O (Fujii *et al.*, 1977); ADRBFT10	dark yellow	≈3.4–3.5	$P2_1$	4	infinite mixed stacks	No
2. (Naphthalene-2,3-diol)∙∙∙lumiflavin (Wells *et al.*, 1974); LUMNPO20	yellow	3.46; 3.48	$P2_1/c$	4	ditto	Yes
3. 5′-Bromo-5′-deoxy-adenosine∙∙∙riboflavin∙3H$_2$O (Voet and Rich, 1971); RIBBAD10	orange-brown	3.45; 3.54	$P2_12_12_1$	4	ditto	No
4. (Naphthalene-2,3-diol)$_2$∙∙∙(10-propyl-isoalloxazine) (Kuo *et al.*, 1974); PALXND10	orange	3.38; 3.46	$A2/a$	8	ditto	Yes
5. (Naphthalene-2,3-diol)2∙∙∙lumiflavin∙3H$_2$O (Fritchie and Johnson, 1975); LUNMPO10	reddish-orange	≈3.3	$P\bar{1}$	2	ditto	Yes
6. (2,6-Diamino-9-ethylpurine)$_2$∙∙∙(lumiflavin)$_2$∙C$_2$H$_5$OH∙H$_2$O (Scarbrough *et al.*, 1976, 1977); LUFAEP	deep red	≈3.4	$P2_1/c$	4	DAAD sequence (cf. Fig. 15.17)	No
7. Quinol∙∙∙lumiflavin chloride (Karlsson, 1972); LUMIHQ	green	quinol/flavin 3.49; flavin/flavin 3.72	$P2_1/c$	8	DAAD sequence (see Fig. 15.17)	No
8. Quinol∙∙∙riboflavin-(HBr)$_2$-(H$_2$O)$_2$ (Bear *et al.*, 1973); BIBHQN10	black	≈3.4 in infinite mixed stack; 3.26 and 3.52 in "paired" stack.	$P2_1$	4	One stack of infinite mixed type; the other has pairs of donor and acceptor moieties.	No
9. (Quinol)$_3$∙∙∙lumiflavinium bromide (Tillberg and Norrestam, 1972); LUMFHQ	blackish	≈3.3	$P2_1/c$	4	Infinite mixed stacks.	Yes
10. (Naphthalene-2,7-diol)$_3$∙∙∙(10-methyl-isoalloxazinium bromide)$_2$∙2H$_2$O (Langhoff and Fritchie, 1970); MAZNDO10	black	Not listed	$P\bar{1}$	1	Charge transfer interaction suggested.	Yes

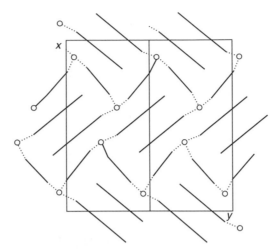

Fig. 15.18. Schematic diagram showing one sheet of the molecular packing in quinol···lumiflavin chloride extended over two unit cells. The long and short full lines represent the lumiflavin and hydroquinone molecules respectively; the dashed lines are hydrogen bonds and the circles Cl⁻ ions. The four-moiety groups, in the sequence hydroquinone···lumiflavin···lumiflavin···hydroquinone, lie approximately in the $(30\bar{1})$ planes. Similar DAAD groupings are found in the structure of (2,6-diamino-9-ethylpurine···lumiflavin)$_2$·C$_2$H$_5$OH.H$_2$O (Scarborough *et al.*, 1977). (Reproduced from Karlsson, 1972.)

(Section 15.5.1). A much larger sample of flavin complexes will have to be examined before drawing conclusions from the high frequency of interstitial donors among the complexes of Table 15.9.

Judging from the colours of the complexes it appears that the flavins act as quasi-acceptors in the first three complexes of Table 15.9 with essentially no charge transfer interaction with the donor; indeed adenine and 5′-bromo-5′-deoxyadenosine are both weak donors. The geometry of the π-systems is rather variable and the flavins appear to belong to that group of acceptors in which the strength of the π-interaction varies little with the mutual lateral position of donor and acceptor molecules. This implies that any part of the flavin nucleus can participate in charge transfer interaction and that the actual overlap that occurs in the crystals depends both on the nature of the partners and the requirements of the hydrogen bonding scheme. This may well be a requirement for a moiety so widely involved in biological and other oxidation-reduction reactions.

The complexes containing riboflavin all crystallize in chiral space groups but this is just a consequence of the chirality of riboflavin itself.

15.7.3 *Other crystals with both charge transfer and hydrogen bonding interactions*

15.7.3.1 *Crystals without solvent of crystallization*

In {1,5-diaminonaphthalene···chloranil} (DAN···CA) (Tamura and Ogawa, 1977; CANANP) there are mixed stacks of DAN and CA molecules along [010] (7.95 Å), with

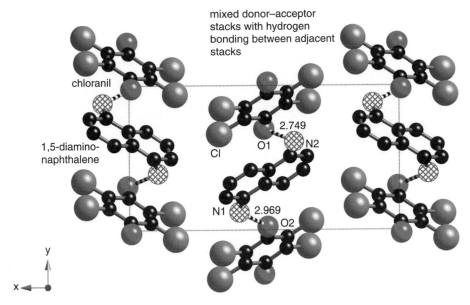

Fig. 15.19. Crystal structure of (DAN···CA) projected down [001]. The N atoms are shown cross-hatched. Distances in Å. The two kinds of hydrogen bonds between amino groups and chloranil oxygens are shown by broken lines. (Data from Tamura and Ogawa, 1977.)

hydrogen bonding between carbonyl and amino groups of molecules in different stacks separated by [001]. Thus sheets of molecules, linked by both charge transfer interactions (there is carbonyl group–aromatic ring overlap) and hydrogen bonds, are formed in the (100) planes. These sheets are then stacked along [100] (Fig. 15.19). The component molecules are, of course, centrosymmetric and so is their arrangement within the sheets, at least to a very good approximation. However, the space group is Pn, which is non-centrosymmetric (see Table 15.4). It seems that most efficient packing of the sheets is obtained when they are mutually offset along [010].

One may perhaps anticipate that both charge transfer interactions and hydrogen bonds will be found in {durendiamine···chloranil} (Matsunaga, 1964) and in the two poly-morphs of {1,6-diaminopyrene···chloranil} (Matsunaga, 1966) but crystal structures have not been reported.

Hydrogen bonding interactions have been used in an attempt to convert the herring-bone arrangements of charge-transfer crystals into more planar arrangements, hopefully increasing the interaction between donor and acceptor moieties by reducing the inter-planar distance and thus affording observable changes in moiety geometries (Bock, Seitz *et al.*, 1996). Specifically the herring-bone structure of {perylene···1,2–4,5-tetracyanobenzene} ($P2_1/c$, $Z = 4$; REHMUM) was compared with that of {pyrene···2,3,5,6-tetracyanohydroquinone} ($P\bar{1}$, $Z = 1$; TEXPOB10). Indeed, both planarization and (minor) shortening were achieved (Fig. 15.20), but there was no effect on moiety geometries. Perhaps more interesting is the way in which the formation of CN...HO hydrogen bonds gives a structure composed of pyrene and hydrogen-bonded tetracyano-hydroqunone layers without removing the integrity of the donor–acceptor stacks.

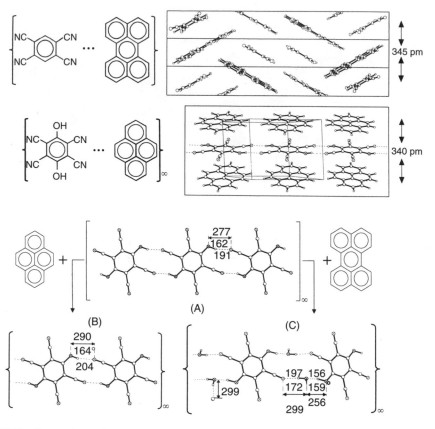

Fig. 15.20. Comparison of the crystal structures of {perylene···1,2–4,5-tetracyanobenzene} ($P2_1/c$, $Z=4$) and {pyrene···2,3,5,6-tetracyanohydroquinone} ($P\bar{1}$, $Z = 1$). Both are viewed normal to the stack axes. Note the central layer of hydrogen-bonded tetracyanohydroquinone moieties. In the lower part of the figure the hydrogen-bonding patterns found with the 2,3,5,6-tetracyano-hydroquinone molecular compounds with pyrene (left) and perylene + 2H$_2$O are shown. (Reproduced from Bock, Seitz *et al.*, 1996.)

When perylene is co-crystallized with 2,3,5,6-tetracyanohydroquinone, two molecules of water are incorporated in the crystals (TEXPUH10) and the dimerized hydrogen-bonding pattern found in {pyrene···2,3,5,6-tetracyanohydroquinone} is replaced by an interrupted pattern (Fig. 15.20 above). The donor–acceptor stacks remain.

15.7.3.2 *Crystals with solvent of crystallization*

For some donor–acceptor combinations, the combined occurrence of charge transfer interactions and hydrogen bonding allows formation of a more open arrangement of molecules, with interstices or channels which can be occupied by solvent. This occurs in the TCNQ compounds of benzidine (BD), toluidine (TL) and diaminobenzidine, where solvent-free and solvated types are found (Ohmasa *et al.*, 1971). The BD/TCNQ system is the most thoroughly studied and three types of crystal have been obtained.

I – solvent-free BD···TCNQ (crystal structure reported (Yakushi *et al.*, 1974a; BZTCNQ);

II – with aliphatic guests such as acetone, acetonitrile, CH_2Cl_2, CH_2ClCH_2Cl and CH_2BrCH_2Br. These crystals are all isostructural and the structure of BD··· $TCNQ·1.8CH_2Cl_2$ has been determined in detail (Ikemoto, Chikaishi *et al.*, 1972; BDTCQC10);

III – with aromatic guests such as benzene and substituted benzenes ($X = CH_3$, Cl, Br, NO_2, CN). The crystal structure of BD···$TCNQ·C_6H_6$ has been determined (Yakushi, Ikemoto and Kuroda, 1974b; BDTCNB10).

The crystal structures of {BD···$TCNQ·1.8CH_2Cl_2$} and {BD···$TCNQ·C_6H_6$} are based on similar arrangements of hydrogen-bonded BD and TCNQ molecules in layers, as illustrated for {BD···$TCNQ·1.8CH_2Cl_2$} in Fig. 15.21. The stacking of these layers is rather different in the two types of inclusion complex. In the CH_2Cl_2 complex the layers are directly superimposed so that channel and stack axes coincide, while in the benzene complex a BD···TCNQ pair forms the repeat unit in the stack and channel and stack axes are mutually inclined at an angle of $\approx 30°$.

The enthalpy of formation of crystalline {BD···$TCNQ·1.8CH_2Cl_2$} has been measured (Ohmasa, Kinoshita and Akamatsu, 1971), the reaction being:

$$BD···TCNQ(s) + 1.8CH_2Cl_2(g) \rightarrow BD··· TCNQ·1.8CH_2Cl_2(s) + 83.6\,kJ.$$

The reaction is thus exothermic and the inclusion complex is stable with respect to its components. The composition with 1.8 molecules of CH_2Cl_2 is the most stable, and the 10%

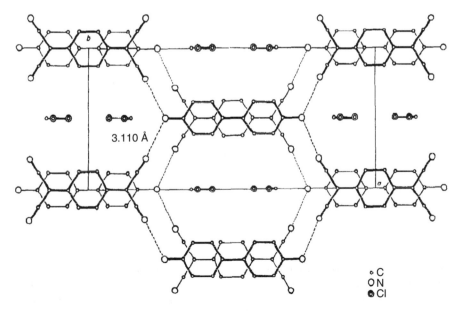

Fig. 15.21. Projection of the {BD···$TCNQ·1.8CH_2Cl_2$} structure onto (001). The arrangement of BD and TCNQ molecules in a *single* layer of the benzene complex is very similar. (Reproduced from Ikemoto *et al.*, 1972.)

Table 15.10. Conductivities of benzidine/TCNQ inclusion complexes

Type	Formula	Conductivity (ohm cm)$^{-1}$	Activation energy (eV)
I	{BD···TCNQ}	10^{-9}	0.54
II	BD···TCNQ·1.8CH$_2$Cl$_2$	10^{-4}	0.11
III	BD···TCNQ·C$_6$H$_6$	10^{-6}	0.28

of vacancies on solvent sites must contribute some entropy stabilization as well. All solvent sites are occupied in the benzene complex. It seems unlikely, however, that (in both types of complex) the solvent molecules are as ordered as depicted in the structure diagrams.

Very different electrical conductivities and activation energies are found for the three types of crystal (Takahashi *et al.*, 1976), the conductivities of the inclusion complexes being much larger than those of the solvent-free crystals (Table 15.10).

The conductivity of {BD···TCNQ·xCH$_2$Cl$_2$} is strongly dependent on the concentration of solvent vacancies; in the most stable crystals ($x = 1.8$) there is extrinsic semiconduction due to the solvent vacancies. It is the presence of these vacancies that probably accounts for the larger conductivity and smaller activation energy of Type II as compared to Type III crystals.

Polarized reflectance spectra of single crystals of {BD···TCNQ} and {BD···TCNQ·1.8CH$_2$Cl$_2$} have been measured at 295 and 30K at ambient pressure and also at 295K, 56.7 kbar for {BD···TCNQ}. The degree of charge transfer at 295K, 1 bar was estimated to be ≈ 0.3 and increased by $\approx 30\%$ on cooling or application of pressure; however, the crystals retained their neutral ground states (Yakushi *et al.*, 1985).

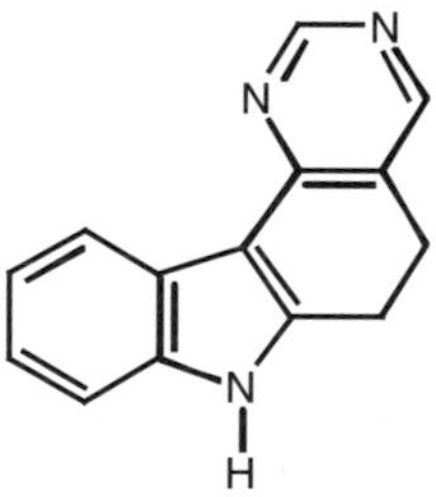

DPC

Monoclinic {[bis(dihydro-5,6-pyrimidino[5,4-*c*]carbazole)-···TCNQ]·-dihydro-5,6-pyrimidino[5,4-c]carbazole dihydrate} ($P2_1/c$, Z = 2; {[DPC·TCNQ·DPC]·DPC·2H$_2$O} for short; Dung *et al.*, 1986; DULFAR) has mixed...DAD...stacks (shown within the square brackets) along [100] (11.04 Å) linked by DPC and water molecules. Only the first-mentioned DPC molecules are involved in donor–acceptor interactions. The mixed stacks retain their integrity despite the considerable perturbation of the structure.

15.8 Mixed-stack crystals with both delocalized and localized charge transfer interactions

By analogy with the structures discussed in Section 15.7, one may anticipate mixed-stack structures in which the hydrogen bonding between stacks is replaced by localized charge transfer interactions, acting in addition to the delocalized charge transfer interactions

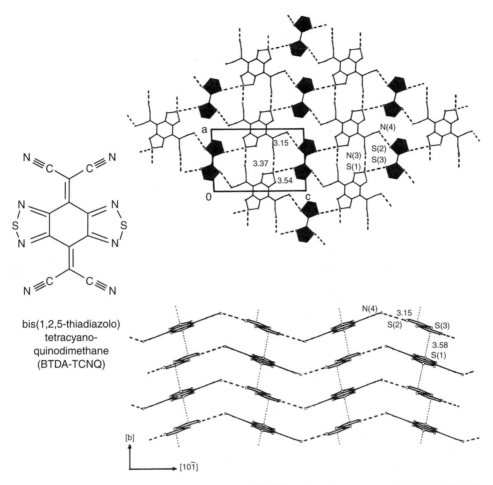

Fig. 15.22. Crystal structure of {TTF•••(BDTA-TCNQ)}, where the acceptor is bis(1,2,5-thiadia-zolo)tetracyanoquinodimethane. (upper) "Sheet-like" network in the *ac* plane showing the secondary N•••S interactions between TTF and acceptor molecules and also between the acceptor molecules themselves; the TTF molecules are shaded.
(lower) Arrangement of the corrugated sheets along the [010] (stack) axis, with π-π* interactions in the [010] direction. (Adapted from Susuki, Kabuto *et al.*, 1987.)

within the stacks. This is a striking feature of the structural arrangements in the Bechgaard salts, important as the first organic superconductors (Williams *et al.*, 1985) but outside the scope of this book. So far only one example is known among the mixed stack structures – {TTF•••(BTDA-TCNQ)}, where preparation of the acceptor (and of a number of its molecular compounds (Yamashita *et al.*, 1985)), determination of the crystal structure of the acceptor (Kabuto *et al.*, 1986; FARSOG[6]) and that of the TTF molecular compound (Suzuki, Kabuto *et al.*, 1987; FUVYEA), have been described. The neat acceptor (formula in Fig. 15.22) forms a two-dimensional sheet-like network with strong

[6] For reinterpretation in space group C2/m see FARSOG01 (Suzuki, Fujii *et al.*, 1992).

S...N≡C– interactions. The equimolar molecular compound is monoclinic ($a = 9.660(1)$, $b = 7.231(1)$, $c = 14.628(2)$ Å, $\beta = 91.312(1)°$, $Z = 2$, space group $P2_1/n$) and has a neutral ground state. The interactions among the components are such that a three-dimensional network is formed.

15.9 Donors and acceptors with special chemical features

We use the term 'quasi-acceptor' to describe the second component in binary systems where a donor and this component form mixed stack crystals but other evidence (generally spectroscopic) for a charge transfer interaction between the components is lacking.

15.9.1 *Fluorinated aromatics as quasi-acceptors*

Considerable effort has been devoted to the study of systems of aromatic hydrocarbons and polyfluorinated aromatics (Fenby, 1972; Swinton, 1974; Table 15.11); the spectroscopy and thermodynamics of solutions have been studied, phase diagrams have been reported and a number of crystal structures determined (Table 15.12; we include some other quasi-acceptors in this Table for convenience). Resurgence of interest in this area has been prompted by (expressed in currently fashionable terms) recognition of the arene-perfluoroarene interaction as an important supramolecular synthon (Dai, Nguyan *et al.*, 1999) with potential applications for solid-state chemistry, crystal engineering, molecular electronics, liquid crystals etc. Congruently melting 1:1 compounds have been found in the phase diagrams of C_6F_6 with benzene, toluene, *p*-xylene and mesitylene (Duncan and Swinton, 1966) and congruently melting 1:1 compounds in the phase diagrams of benzene and naphthalene with perfluoronaphthalene, and of benzene with perfluorobiphenyl, while the {biphenyl–C_6F_6} compound melts incongruently (McLaughlin and Messer, 1966). Equimolar molecular compounds have been found in most binary phase diagrams but not other compositions. No evidence of $\pi-\pi^*$ charge transfer interaction has been

Table 15.11. Melting points of equimolar molecular compounds formed from various donors and hexafluorobenzene or octafluoronaphthalene

Donor/hexafluorobenzene	M.pt. (K)	Donor/ octafluoronaphthalene	M.pt. (K)
Benzene (Patrick and Prosser, 1960)	297	Anthracene (Collings, Roscoe *et al.*, 2001)	447
1-methylnaphthalene (Griffith *et al.*, 1983)	373	Phenanthrene (Collings, Roscoe *et al.*, 2001)	445
2-methylnaphthalene (Jackson and Morecombe, 1986)	335.7	Pyrene (Collings, Roscoe *et al.*, 2001)	522
2-ethylnaphthalene (Dahl, 1988)	306.7	Triphenylene (Collings, Roscoe *et al.*, 2001)	478
Tetralin (Dahl, 1988)	266.7		
Quinoline (Dahl, 1988)	318.1		
iso-quinoline (Dahl, 1988)	285.4		

found in solutions of aromatic hydrocarbons and fluoroaromatics but there may be n–π*
charge transfer when nitrogen-containing heteroaromatics are used as donors (Beaumont
and Davis, 1967); there is some crystallographic evidence for specific interactions
between nitrogens of heteroaromatic donors and quasi-acceptors.

Mixed stacks of donor and quasi-acceptor molecules have been found in almost all the
crystal structures so far reported and thus these structures resemble those of π-molecular
compounds insofar as packing arrangements are concerned (Dahl, 1988). However, the
absence of charge-transfer bands in the solution spectra, the colourless nature of the
crystals, and the fact that lattice energy minimizations based on atom–atom potentials do
not reproduce the observed crystal structures very well (Dahl, 1990) all indicate that
directional forces of another kind must be invoked to account for the retained occurrence
of mixed stacks despite replacement of true acceptors by quasi-acceptors. One possibility
is localized dipole–dipole interaction between C-H and C-F bonds, expressed in terms of
quadrupole interactions. The fact that both benzene and hexafluorobenzene have large
molecular quadrupole moments (MQM), of similar magnitude but of opposite sign[7], led to
the suggestion (Williams, 1993) that the stacking motif is dictated by electrostatic
quadrupole-quadrupole interactions (see also Clyburne et al., 2001). The negative values
for aromatic hydrocarbons indicate that the negative charge is clustered in the center and
above the molecular plane, while that for perfluoroaromatics is located at the molecular
perimeter and therefore there is a depletion of electron density at the center of the ring
(Fig. 15.23). Illustrations, apparently based on Williams (1993), have been given by
Dunitz (1996) and Collings, Roscoe et al. (2001).

Many of the crystals show low temperature phase transitions; for example, the structures of four 1 : 1 benzene–hexafluorobenzene phases have been determined (Table 15.13).
The structural progression has been described as follows (Williams et al., 1992) "In
phase I the molecules are rotating about the column axis, and the system behaves like
parallel cylindrical rods. On lowering the temperature the molecules first tilt with respect to
the column axis, leading to a monoclinic phase, followed by a distortion to a triclinic phase
due to the freezing out of the rotations of the heavier C_6F_6 molecules. At this stage the
unfavorable interaction between the columns is compensated by the rotational freedom of

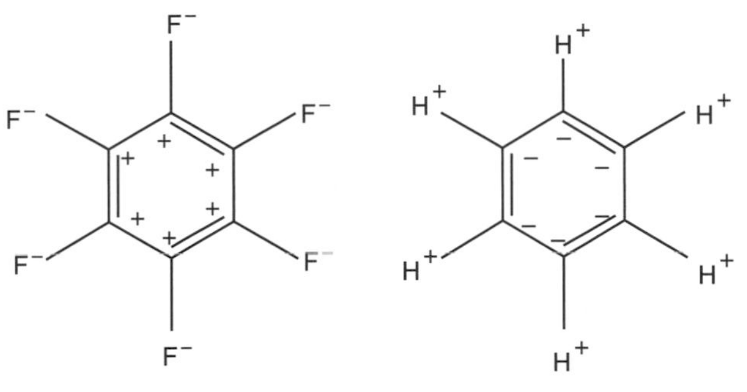

Fig. 15.23. Schematic charge distributions in hexafluorobenzene and benzene.

[7] MQM of benzene −29.0, naphthalene −45, ferrocene −30 and hexafluorobenzene +31.7, all × 10^{-40} C m^2.

Table 15.12. Summary of information available about mixed-stack crystal structures of molecular compounds formed between aromatic hydrocarbons and quasi-acceptors

Aromatic hydrocarbon/reference/refcode	Quasi-acceptor	Space group	Z	Remarks
1. Anthracene (Boeyens and Herbstein, 1965a); ZZZGMW	C_6F_6	$C2/m$	2	Only cell dimensions and space groups (at 300K) reported for these three compounds.
2. Pyrene (Boeyens and Herbstein, 1965a); ZZZGKE	C_6F_6	$C2/m$	2	
3. Perylene (Boeyens and Herbstein, 1965a); ZZZLJY	C_6F_6	$P2_1/a$	2	
4. p-Xylene (Dahl, 1975a); PXYHFB	C_6F_6	$P\bar{1}$	1	
5. Mesitylene (Dahl, 1971a); MTYHFB	C_6F_6	$Pnma$	4	
6. Durene (Dahl, 1975a); DURHFB	C_6F_6	$C2/m$	2	triclinic below 223K
7. $C_6(CH_3)_6$; MBZFBZ	C_6F_6	$R\bar{3}m, P\bar{1}$	3	at 298K (Dahl, 1971b);
			1	at 233K (Dahl, 1973)
8. N,N-dimethylaniline (Dahl, 1981b); BAPLEJ	C_6F_6	$P\bar{1}$	1	
9. N,N-dimethyl-p-toluidine (Dahl, 1981a); METOFB	C_6F_6	$I2/m$	2	
10. TMPD (Dahl, 1979); MPAHFB	C_6F_6	$P\bar{1}$	1	
11. $Trans$-stilbene (Batsanov, Howard et al., 2001); TIJTUB	C_6F_6	$P2_1/c$	2	Colourless plates
12. Perdeuterobenzene (Overell and Pawley, 1982); BICVUE	C_6F_6	$R\bar{3}m$, or $R3m$	1	at 299K; phase transitions at lower temperatures (see Table 15.13)
13. $Cr(\eta^6\text{-}C_6H_6)_2$ (Aspley et al., 1999); FIBGUS	C_6F_6	$P\bar{1}$	1	Red-pink blocks; there are also yellow plates of rather similar structure not described in detail.
14. o-diethynylbenzene dimer (Bunz and Enkelmann, 1999); JOCRIC	C_6F_6	$P2_1/a$	4	At 165K.
15. hexamethylmelamine (Aroney et al., 1987)	C_6F_6	$C2/m$	2	Poor, disordered crystals
16. Naphthalene (Potenza and Mastropaolo, 1975); NFOFNP	octafluoronaphthalene	$P2_1/c$	2	
17. anthracene (Collings, Roscoe et al., 2001); ECUTUR	octafluoronaphthalene	$P2_1/a$	2	At 120K

# Compound (reference); Refcode	Second component	Space group	Z	Notes
18. phenanthrene (Collings, Roscoe et al., 2001); ECUVED	octafluoronaphthalene	$P2_1/a$	2	Below 250K
19. pyrene (Collings, Roscoe et al., 2001); ECUVIH	octafluoronaphthalene	$P\bar{1}$	1	At 120K
20. triphenylene (Collings, Roscoe et al., 2001); ECUVON	octafluoronaphthalene	$P\bar{1}$	2	At 120K
21. TTF (Batsanov, Collings et al., 2001); TIJVAJ	octafluoronaphthalene	$P2_1/c$	2	At 120K
22. Diphenylacetylene (Collings, Batsanov et al., 2001; at 100K); OCAYIA	octafluoronaphthalene	$P2_1/a$	2	
23. 2,6-dimethylnaphthalene (Birtle and Naae, 1980); BPHFPC	octafluoronaphthalene	$P\bar{1}$		
24. 1,8-diaminonaphthalene (Batsanov, Collings, Howard and Marder, 2001); EDAWAH	octafluoronaphthalene	$P2_1$	2	Pseudo-isostructural with naphthalene·octafluoronaphthalene
25. ferrocene (Clyburne et al., 2001); YEBQOL	1.5(Octafluoronaphthalene)	$P\bar{1}$	2	At 223K.
26. ferrocene (Burdenuic et al., 1997)	Perfluorophenanthrene	$P\bar{1}$	2	
27. bis(decamethylferrocene) (Beck et al., 1998); SIBQOJ	Perfluorophenanthrene	$P2_1/c$	4	
28. bis(N-(2,6-dimethylphenyl)N'-(1,3,4,5,6,7,8-heptafluoro-naphth-2-yl)sulfurdiimide (Lork et al., 2001); NEMHOC	Octafluoronaphthalene	$P\bar{1}$	1	At 173K.
29. biphenyl (Lin and Naase, 1978); ZZZBRD	2,3,4,5,6-pentafluorobiphenyl	$C*/c$		
30. Biphenyl (Naae, 1979); BPPFBP	decafluorobiphenyl	$C2/c$	4	
31. 4-bromobiphenyl (Birtle and Naase, 1980); BPHFBP	decafluorobiphenyl	$P2_1/c$		
32. 4-methylbiphenyl (Birtle and Naase, 1980); BPHFPA	decafluorobiphenyl	$P2_1/c$		
33. Naphthalene (Foss et al., 1984); CEKYUM	decafluorobiphenyl	$C2/c$	4	
34. triphenylene (Weck et al., 1999); CUKXIP	Perfluorotriphenylene	$C2/c$	8	
35. Trans-stilbene (Bruce et al., 1987); SERQAH	trans-decafluoro-azobenzene	$P\bar{1}$	1	Orange crystals, m.pt. 428–430K.
36. Benzene (Hazell, 1978); DCLPYR	decachloropyrene	$P2_1/c$	4	
37. Diphenylbutadiyne (Coates et al., 1997). M.pt. 360K;	Decafluorodiphenyl butadiyne. M.pt. 387K.	$P\bar{1}$	1	M.pt. 425K.

Table 15.12. (*Continued*)

Aromatic hydrocarbon/reference/refcode	Quasi-acceptor	Space group	Z	Remarks
38. Benzalazine (Vangala, Nangia and Lynch, 2002); EGAWEO	Bis(pentafluoro-phenylmethylidene) hydrazone	$P\bar{1}$	2	
39. Pyrene (Damiani, De Santis *et al.*, 1965)	TMU	Pc	2	
40. Benzo[c]pyrene (Damiani, Giglio, Liquori and Ripamonti, 1967)	(TMU)$_2$	$P1$	1	
41. Coronene (Damiani, Giglio, Liquori, Puliti and Ripamonti, 1967)	(TMU)$_2$	$P1$	1	
42. Acridine (Yamaguchi and Ueda, 1984; Marsh, 1986); CEJTAM	1,4-dithiintetra-carboxylic-N,N′-dimethyldiimide	$P2_1/n$	2	
self complex	2,3,4,5,6-penta- fluorobiphenyl (Brock *et al.*, 1978); PFBIPH	$C222_1$	4	

Notes:

(1) TMU is 1,3,5,7-tetramethyluric acid.

(2) There is no evidence for charge transfer interaction in compound #43.

(3) Beck *et al.* (1999) note the preparation of {(ferrocene)$_4$·octafluoronaphthalene} but a structure was not reported.

Table 15.13. Crystal data (Å, deg.) for the four C_6H_6–C_6F_6 phases (angle values are given only if different from $90°$)

Phase	$T(K)$	a/α	b/β	c/γ	V/formula unit (Å^3)	Z	Space group
I (Overell and Pawley, 1982)	279	11.952	11.952	7.238/120	299	3	$R\bar{3}m$
II (Williams et al., 1992)	260	6.631	12.330/99.67	7.302	295	2	$I2/m$
III (Williams et al., 1992)	215	6.380/93.99	12.338/96.37	7.395/91.85	284	2	$P\bar{1}$
IV (Williams et al., 1992)	30	9.516	7.429/95.60	7.537	265	2	$P2_1/a$

the C_6H_6 molecules. Reduction of their rotational energy leads to a realignment of the columns and the unusual transition from a triclinic to a monoclinic phase with decreasing temperature . . . The columnar nature of the structure[s] is a result of the strong attraction between the quadrupole moments of opposite phase of the benzene and hexafluorobenzene molecules." Shorter range electrostatic interactions determine the details of the packing. The structure determinations are noteworthy because they were carried out by combined use of X-ray (synchrotron) and neutron diffraction on *polycrystalline* samples.

The electron affinity of hexafluorobenzene has been measured as 0.86(3) eV (Wentworth *et al.*, 1987). The enthalpy of formation of crystalline $\{C_6H_6 \cdots C_6F_6\}$ has been measured by a differential scanning calorimetric method as $+1.0(3)$ kJ/mol (Brennan *et al.*, 1974). This implies that the entropy of formation must be positive and greater than 3.3 J/mol K (at 300K). In contrast to this situation, most π-molecular compounds are enthalpy-stabilized (see Section 16.5). The enthalpy of formation of $\{p$-xylene$\cdots C_6F_6\}$ has been reported to be -0.08 ± 0.20 kJ/mol (Ott *et al.*, 1976).

The {triphenylene–perfluorotriphenylene} molecular compound, first prepared by Smith and Massey (1969), has a melting point of 524K, compared with 472K for triphenylene and 382K for perfluorotriphenylene (structure by Hursthouse, Smith and Massey, 1977; $Fdd2$, $Z = 8$; PPTRPH). The DSC trace of the molecular complex shows first order transitions at 294 and 377K. The crystal structure reported (but the phase was not specified) is particularly interesting (Fig. 15.24; Weck *et al.*, 1999; 7.390 20.987 16.998 Å, $\beta = 95.26°$, $C2/c$, $Z = 8$; CUKXIP)) and has many features typical of the mixed-stack structures. There are mixed stacks of the two components, seen edge-on in the upper part of Fig. 15.24. This also shows the appreciable distortions from planarity of both components; Weck *et al.* summarize these as 33° and 16° tilts in the skeletons of perfluorotriphenylene and triphenylene in the cocrystal, to be compared with 40° and 2° tilts in the neat crystals. In the lower part of the figure one notes the almost ideal superpositioning of the component molecules, an unusual situation in crystal packing arrangements. The centroid to centroid distances for the central rings is 3.698 Å, while the shortest intermolecular C . . . C distance is 3.369 Å.

Most of the structures listed in Table 15.12 are of the mixed stack type (as illustrated in Figs. 15.24 and 15.25), some having particular features needing comment. The deca-fluorobiphenyl ($C_{12}F_{10}$) molecule is not planar; the dihedral angles between the two rings are 59.5° in neat crystalline $C_{12}F_{10}$ (Gleason and Britton, 1976; DECFDP01) (and 64.4° in

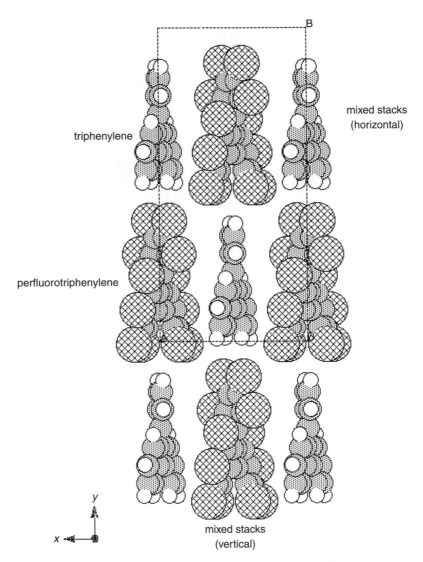

Fig. 15.24. The triphenylene ... perfluorotriphenylene crystal structure viewed (above) edge-on to the stacks (fluorines are large cross-hatched circles and hydrogens are open circles) and (below) normal to the mean molecular planes, showing the close-packed arrangement of the mixed stacks; perfluorotriphenylene has been removed from the central mixed stack. (Data from Weck *et al.*, 1999.)

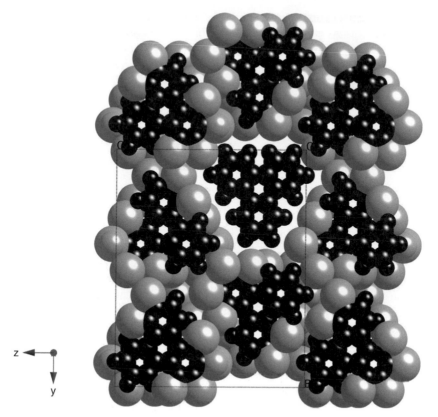

Fig. 15.24. (*Continued*)

the gas phase (Bastiansen *et al.*, 1989)), 55.3° in the naphthalene compound, 50.8° in the
biphenyl compound and 52.9° in the analogous molecule 2,3,4,5,6-pentafluorobiphenyl
(C_6H_5–C_6F_5) (PFBIPH). In {$C_{12}H_{10}$···$C_{12}F_{10}$} (BPPFBP) the biphenyl molecule is also
nonplanar (dihedral angle 36.6°); however, mixed stacks are formed with an angle of 7.1°
between juxtaposed halves of each molecule. A similar mixed stack arrangement is found
in 2,3,4,5,6-pentafluorobiphenyl with superpositioning of C_6H_5 and C_6F_5 portions of the
molecule; this compound must therefore be classed as a self-complex. Stacks are not
formed in {naphthalene-decafluorobiphenyl}, where naphthalene is planar and deca-
fluorobiphenyl nonplanar; instead the naphthalene molecule is sandwiched between C_6F_5
halves of neighboring $C_{12}F_{10}$ molecules, with an interplanar angle of 6°. The naphthalene
in the complex is phosphorescent, in contrast to its lack of phosphorescence in its neat
crystals. The mixed-stack {naphthalene···octafluoronaphthalene} molecular compound
has been studied by Raman spectroscopy at ambient pressure down to 10K, and at ambient
temperature in the pressure range 1–80 kbar and found to be stable (i.e. no phase transitions or chemical reactions) under these conditions (Desgreniers *et al.*, 1985). The unusual
composition of {ferrocene·1.5(octafluoronaphthalene)} (Clyburne *et al.*, 2001) is

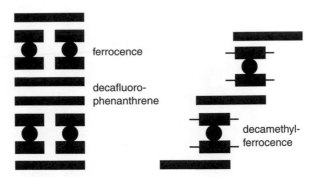

Fig. 15.25. Comparison of the packing arrangements found in {ferrocene···deca fluorophenanthrene} (on the left: a $2+2$ supersandwich) and {decamethylferrocene···deca fluorophenanthrene} (SIBQOJ; Beck *et al.*, 1998) (on the right: mixed stacks). (Adapted from Burdeniuc *et al.*, 1997.)

explained by noting that there are $1:1$ mixed stacks of ferrocene and octafluoro-naphthalene separated by "octafluoronaphthalene of crystallization", the molecules being located at centers of symmetry. The {ferrocene···decafluorophenanthrene} structure (Burdeniuc *et al.*, 1997; no refcode) is unusual in that there is a $2+2$ supersandwich with the ferrocenes located between the wingtip arene rings of the decafluorophenanthrene (Fig. 15.25).

The {naphthalene···(p-iodotetrafluorobenzene)$_2$} ($P\bar{1}$, $Z = 1$) (Law and Prasad, 1982) molecular compound possibly belongs in this group. The crystal structure has not been reported but a Raman study shows that the crystals are ordered, with the components linked mainly by van der Waals interactions; no low-lying charge transfer band was found. Our inclusion of {benzene–decachloropyrene} in Table 15.12 is questionable; although mixed stacks are formed the material could be an inclusion complex of an unusual type. A spectroscopic study is needed to establish whether the orange color of {*trans*-stilbene···*trans*-decafluoroazobenzene} is due to charge transfer interaction.

15.9.2 *1,3,5,7-Tetramethyluric acid (TMU) as quasi-acceptor*

TMU forms crystalline molecular compounds with a number of aromatic hydrocarbons and some crystal structures have been reported (Table 15.10). These are all mixed stack structures based on ---DQDQ--- and ---QDQ QDQ--- sequences (where Q represents the quasi-acceptor TMU). There is considerable disorder and low-temperature studies would appear to be necessary for understanding the interaction between the components. No charge transfer bands are found in the solution or solid state spectra.

15.9.3 *Acceptor is a metal coordination complex*

Examples of metal chelate complexes which act as electron acceptors have been briefly noted in Chapter 13 and this material will now be expanded. The 1,2-ethylenebis (dithiolene)-M system (see Section 13.3.3, Table 13.4 for formula) forms many different salts and also molecular compounds with ionic and neutral ground states. For example, equimolar perylene and pyrene molecular compounds have been prepared with Ni dithiete ($X = CF_3$) as acceptor and found to have mixed stack structures with neutral ground states (Schmitt *et al.*, 1969; PERNIT). The crystals are semiconductors with room-temperature resistivities of $\approx 10^5$ ohm cm; the activation energy for conduction in the perylene molecular compound is anisotropic with minimum activation energy along the stack axis.

$\alpha,\beta,\gamma,\partial$-Tetraphenylporphyrinato-Zn(II) forms an equimolar neutral ground state molecular compound with Ni dithiete (from spectroscopy, crystal structure not known) but the analogous Co(II) porphyrin links to the Ni dithiete by a bridging Co–S bond to give a covalent molecule rather than a molecular compound (Shkolnik and Geiger, 1966). Ni dithiete is the anion in some ion-radical salts to be discussed later (Section 15.10.2).

Bis(difluoroboronbenzimidazole)Ni(II) ($Ni(dmgBF_2)_2$; formula in Table 13.4) forms equimolar molecular compounds with the donors anthracene (Stephens and Vagg, 1981; BADZOV) and benzimidazole (Stephens and Vagg, 1980; FBGLNJ). The acceptor exists in the solid state and in solution as a weakly bonded dimer, with π–π* interaction between the two halves of the dimer. It also occurs as a dimer in the mixed stack molecular compound with benzimidazole. However, in the anthracene molecular compound it is the monomer which acts as acceptor. Analogous examples of the effect of molecular compound formation on donor or acceptor structure have been found in the σ-dimerization of TCNQ in some ion-radical salts (Table 15.12 below). In both anthracene and benzimidazole molecular compounds donor and acceptor are mutually located so as to avoid interference with the protruding fluorine atom of the BF_2 groups. The metal atoms do not play a special role in any of these molecular compounds.

In contrast to these nonconducting molecular compounds, $\{(perylene)_2 \cdots MS_4C_4(CN)_4\}$ (M = Ni, Cu, Pd) compounds have been found to be fairly good conductors (Alcacer and Maki, 1974); for example, the Pd molecular compound has a room temperature resisitivity of ≈ 50 ohm cm. Structures are not known but there are presumably stacks of $(perylene)_2^+$ moieties with the anions between the stacks (cf. Section 17.3).

Molecular compounds are known with a 2 : 1 ratio of toluene to tetraphenylporphyrinato-M(II) [M = Mn (Kirner *et al.*, 1977); Cr (Scheidt and Reed, 1978); Zn (Scheidt *et al.*, 1978)]. The Mn and Zn compounds are triclinic and isomorphous ($P\bar{1}$, $Z = 1$), while that with Cr is monoclinic ($P2_1/c$, $Z = 2$). All three molecular compounds contain DAD sandwiches, with toluene and porphyrin planes approximately parallel and ≈ 3.5 Å apart; the DAD triples are not stacked but are effectively isolated from one another (cf. Section 15.2). The bis(toluene)porphyrin arrangement is thought to be indicative of donor-acceptor interaction. We have discussed this group of molecular compounds in Chapter 8.

The bis(arene)Fe(II) dication has been shown to act as an electron acceptor in the formation of mixed stack charge transfer crystals with the electron donor arenes ferrocene and durene, PF_6^- acting as counterion. The structures of the carmine tetragonal crystals of $[Cp_2Fe, (durene)_2Fe^{2+}, (PF_6^-)_2]$ (VIPJUZ) and of the red-orange triclinic crystals of $[durene, (hexamethylbenzene)_2Fe^{2+}, (PF_6^-)_2]\cdot$acetone (VIPKAG) have been determined. There is spectroscopic evidence for the effectiveness of other arenes such as mesitylene, pentamethylbenzene, 1,4-dimethylnaphthalene and 9-methylanthracene (Lehmann and Kochi, 1991).

15.9.4 *Donor is a metal coordination complex*

15.9.4.1 *Metal chelates as donors*

The structures of some seven metal oxinate$\cdots$acceptor molecular compounds have been determined (for references see Table 15.2), as well as those of a number of other metal chelate$\cdots$acceptor molecular compounds. The formulae of the metal chelates are shown in Fig. 15.26. These are all equimolar (or effectively equimolar) mixed-stack, neutral ground state molecular compounds, diamagnetic (except where the metal atom is paramagnetic) and with high resistivities. The interaction between metal atom and acceptor is weak and plays little, if any, role in determining the crystal structures. This is shown dramatically by the resemblances between the component arrangements in {Pd(II) oxinate$\cdots$chloranil} (Kamenar, Prout and Wright, 1965; CLAQPD) and in the metal-free {(8-hydroxy-quinolinol)$_2$$\cdots$chloranil}.(Prout and Wheeler, 1967; HQUCLA). However, there is at least one exception (Matsumoto *et al.*, 1979) to this generalization.

Three members of this group show special features. Using bis(N-alkyl-2-oxynaphthylidene-aminato)Cu(II) and Ni(II) chelates as donors, a series of 1:2 molecular compounds was prepared (Matsumoto *et al.*, 1979) with TCNQ and chloranil as acceptors, and crystal structures were determined for Cu(L-*i*-pr)$_2$ itself and for {Cu(L-*i*-pr)$_2$$\cdots$(TCNQ)$_2$}, (IPONTC) where L-*i* pr is the isopropyl substituted ligand. In Cu(L-*i*-pr)$_2$ itself the coordination at Cu is distorted tetrahedral, with a dihedral angle of 39° between the two intersecting Cu(NO) planes; this distortion from the expected square planar arrangement was ascribed to steric hindrance between the two bulky *iso*-propyl groups. In {Cu(L-*i*-pr)$_2$$\cdots$(TCNQ)$_2$} there are two antiparallel sets of mixed stacks, with "naphthalene" portions of the chelate and TCNQ molecules in alternating array. The geometry of the metal chelate has changed to a centrosymmetric stepped structure with square planar coordination about Cu. It was suggested (Matsumoto *et al.*, 1979) that the overall arrangement represents a compromise in which the favoured coordination about Cu, as indicated by the arrangement taken up in the neat compound, is distorted minimally in order to achieve the most favourable stacking of "naphthalene" and TCNQ moieties. This is one of the most striking examples known of the effect of molecular compound formation on donor (or acceptor) geometry.

In {[Cu(salphen)$_2$]$_2$$\cdots$TCNQ} (Cassoux and Gleizes, 1980; PSALTQ) there appear, at first sight, to be segregated stacks of metal chelate and TCNQ molecules. However, the conductivity of powders is 10^{-8} S/cm and the moieties (especially TCNQ, as judged from bond lengths) appear to be neutral. The paradox is resolved by reference to the stereo-diagram of the packing (Fig. 15.27). This shows that the two halves of the centrosymmetric TCNQ molecules interact with benzene rings of salicylaldiaminato groups of two different donors to form two separate ---DADA--- stacks, where A = 1/2(TCNQ). The second salicylaldiaminato groups of each of these donors form a quasi-segregated stack, presumably without charge transfer interaction.

In {Ni(gh)$_2$$\cdots$TCNQ} (Megnamisi-Belome and Endres, 1982; BEXZIN) there is an alternating arrangement of neutral moieties (as judged from the TCNQ bond lengths) but there is so little overlap between them that one can hardly refer to "mixed stacks." Instead the black colour of the crystals is ascribed to Ni–NC interactions (d(Ni $\ldots$ N) = 3.357(5) Å) mediated through the conjugated system of the TCNQ molecules.

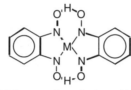

bis(8-hydroxyquinolato) M(II)
metal chelates (M = Cu, Ni, Pd)

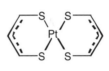

bis(N-alkyl-2-oxynaphthylidene-
aminato M(II) with M = Cu, Ni and
R = CH$_3$, C$_2$H$_5$, i-C$_3$H$_7$, n-C$_3$H$_7$,
t-C$_4$H$_9$.

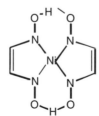

bis(1,2-benzoquinondioximato)M(II)
with M(II) = Ni, Pd. Abbreviated as
M(BCD)$_2$

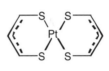

bis(propene-3-thione-
1-thiolate)Pt(II)

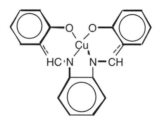

N,N'-(1,2-phenylene)bis(salicylaldiaminato) Cu(II)
Abbreviated as Cu(salphen).

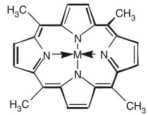

bis(ethanediol dioximato)Ni(II)
Ni(gh)$_2$

tetramethylporphyrinato Ni(II)

Fig. 15.26. Formulae of metal chelate compounds which act as donors in various molecular compounds.

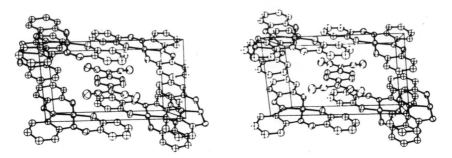

Fig. 15.27. Stereodiagram of [Cu(salphen)$_2$]$_2$···TCNQ structure viewed down [010]. The [100] axis is horizontal. (Reproduced from Cassoux and Gleizes, 1980.)

15.9.4.2 *Metal porphyrins as donors*

Some one hundred molecular compounds of various porphyrins with a variety of electron acceptors have been reported, most by Treibs (1929) and some by Hill *et al.* (1967). The molecular compounds are in the main coloured powders but some single crystals were obtained. The crystal structure of {(tetramethylporphyrinato)Ni(II)···TCNQ} has been determined (Pace *et al.*, 1982; BEGLUU); it has a mixed stack structure with a neutral ground state, is diamagnetic and has a needle axis conductivity of $<10^{-5}$ S/cm. Influence of the metal atom on structure and properties appears to be minimal. The {Zn tetraphenylporphyrin···Ni thiete} compound has been discussed previously and we have also noted that tetraphenylporphyrinatoM(II) (M = Cr, Mn, Zn) appears to behave as an acceptor in its toluene molecular compounds (Chapter 8 and Section 15.9.3).

15.9.4.3 *Other metal compounds as donors*

The structures of {ferrocene···TCNE} (Adman *et al.*, 1967; FERTCE) and {tricarbonyl-chromium-anisole···TNB} (Carter *et al.*, 1966; CCATNB) have been reported. Both have mixed stack structures with neutral ground states and the donor molecules behave structurally as typical aromatic moieties. Indeed the two very different faces of the tricarbonyl-chromium-anisole molecule interact in much the same way with the TNB acceptor.

Metallocenes and analogous molecules also form many molecular compounds with ionic ground states and these are discussed later.

15.9.5 *Donors based on phenazine*

Donors based on phenazine provide a wide variety of interrelated systems in which both electron transfers and proton transfers among the various moieties are possible (Soos *et al.*, 1977, 1978; Keller and Soos, 1986). The phenazine-based moieties appear in molecular compounds and salts as neutral molecules, neutral radicals and cation radicals; there are now sufficient experimental results to allow comparison with the expected structures summarized in Fig. 15.28.

(i)   Phenazine is a stable, diamagnetic molecule with 14 π electrons; and has been found to be planar and aromatic in a number of structures, starting from phenazine itself (two polymorphs PHENAZ04 (Wozniak *et al.*, 1991) and PHENAZ11 (Jankowski and Gdaniec 2002) and including the neutral ground state mixed-stack π-compound phenazine$\cdots$TCNQ (P$\cdots$TCNQ) (Goldberg and Shmueli, 1973c; TCQPEN10). Experiment and expectation coincide.

(ii)   5,10-Dihydrophenazine is expected to be a stable, diamagnetic molecule with 12 π electrons, folded about the N–N axis. In {5,10-dimethylphenazine$\cdots$TCNQ} (M$_2$P$\cdots$TCNQ) (Goldberg and Shmueli, 1973a; TCQMHP) the M$_2$P moiety is folded about the N–N axis with a dihedral angle of 165°. This fits the expectation from Fig. 15.28, but the authors inferred a "markedly ionic (dative) state" so the matter remains to be resolved. It has been emphasized (Soos *et al.*, 1977, 1978) that phenazine$\cdots$TCNQ compounds must be close to the neutral-ionic boundary for mixed stack charge transfer compounds.

(iii) {N-methylphenazinium$\cdots$TCNQ} has segregated stacks and a high conductivity (Fritchie, 1966; MPHCQM); the 12-π electron cation is diamagnetic, planar and aromatic, thus agreeing with Fig. 15.28. Butler *et al.*, (1975) have suggested that a more accurate formulation would be NMP$^{\gamma+}$ TCNQ$^{\gamma-}$, where $\gamma = 0.94$.

(iv)   The cations in {bis(N-methylphenazinium)$\cdots$(σ-(TCNQ)$_2$)} (Morosin, Plastas *et al.*, 1978; EPZTCD) are slightly bowed with the least squares planes through the two sets of six C and two N atoms forming an angle of 2.1°. This deviation from planarity seems hardly large enough to be in contradiction to Fig. 15.28.

(v)   The structures of a number of salts containing (variously) substituted N, N'-dihydrophenaziniumyl cation radicals (HMP$^{+\bullet}$) (13 π-electrons) have been determined. According to Fig. 15.28, this cation would be expected to be folded about the N$\ldots$N axis. The cation radical is N-hydro-N'-methylphenaziniumyl (HMP)$^{+\bullet}$ in two of these salts (Morosin, 1978). In (HMP)$^{+\bullet}$ClO$_4^-$ (HMPZPC10) the cation radical has a small twist ($\approx 2°$) forming a flattened propeller. In the mixed-stack ionic ground state {HMP$\cdots$TCNQ} molecular compound (MPHCQM12) the cation radical is folded about the N–N axis with a dihedral angle of 174°.[8] Somewhat similar folding is found in 5,10-dihydro-5,10-dimethylphenazinium triiodide where the cation has a dihedral angle of 165.5°; the ESR spectrum shows that this compound is paramagnetic as expected (Keller *et al.*, 1978; HDMPZI; $P2_1/n$, $Z = 4$). However, in 5,10-dihydro-5,10-dimethyl-phenazinium$^{+\bullet}\cdots$TCNQF$_4^{-\bullet}$ (Soos *et al.*, 1981; BEXZOT) both components are planar (space group $P\bar{1}$, $Z = 1$); here the cation has a structure contradicting expectation.

(vi)   The cation in 5,10-dihydro-5,10-diethylphenazinium triiodide is centrosymmetric and planar (Keller, Moroni *et al.*, 1978; HETPZI; $P2_1/n$, $Z = 2$). while that in {5,10-dihydro-5,10-diethylphenazinium$\cdots$TCNQ} (Dietz *et al.*, 1982; BEWBUA) is somewhat twisted. Coordinates are not available for {5,10-dihydro-5,10-diethyl-phenazinium$\cdots$TCNQ}·phenazine (Dietz *et al.*, 1982; BEWCAH).

(vii) In {5,10-dihydro-2,3,5,7,8,10-hexamethyl-phenaziniumyl$\cdots$TCNE} the M$_6$P$^{+\bullet}$ cation radical is also planar ($P2_1/n$, $Z = 2$); an ionic ground state for this mixed stack molecular compound was inferred from the occurrence of an ESR spectrum and from the length of the C-C bond in the TCNE moiety (Flandrois *et al.*, 1983;

---

[8] Morosin (1978) has corrected his earlier report (1976) that the latter compound was a mixed–stack form of N-methylphenazinium$\cdots$TCNQ.

CEGYUI). According to the expectations of Fig. 15.28 these cations should be folded.

The evidence from the geometrical structures of the phenazine (phenazinium) moieties in general favours the predictions as the deviations from planarity are small; one difficulty is that one does not know to what extent packing effects can affect the structures, nor does one know what 'dihedral angles' are to be expected. Physical evidence, as from ESR spectra, is helpful where it is available.

$\{M_2P\cdots TCNQ\}$ appears to undergo a reversible neutral to ionic transition on heating to 390K; the high-temperature ionic form appears to be stabilized by spin contributions to

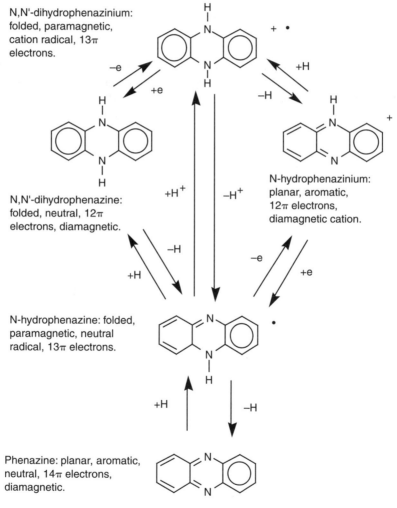

N,N'-dihydrophenazinium: folded, paramagnetic, cation radical, 13π electrons.

N,N'-dihydrophenazine: folded, neutral, 12π electrons, diamagnetic.

N-hydrophenazinium: planar, aromatic, 12π electrons, diamagnetic cation.

N-hydrophenazine: folded, paramagnetic, neutral radical, 13π electrons.

Phenazine: planar, aromatic, neutral, 14π electrons, diamagnetic.

Fig. 15.28. Relationships among different donor species based on phenazine. Hydrogen is used as the substituent for purposes of illustration but can be replaced by alkyl groups. It has been convenient to draw the formulae in terms of particular canonical forms but the actual structures are resonance hybrids. (Adapted from Soos *et al.*, 1978.)

the entropy, accidental degeneracy of the energies of the neutral and ionic forms being inferred. This is an unusual situation as ionic states are generally energy-stabilized at low temperatures and/or high pressures (Chapter 17 and Section 15.10.1).

The crystal structures of many of the ionic compounds with phenazine-based cations are discussed in Section 15.10.

15.10   Mixed-stack donor–acceptor molecular compounds with ionized ground states

15.10.1   *Mixed-stack closed shell charge transfer salts*

Many salts have been investigated where the components are closed-shell ions – the cations are based on pyridinium, quinolinium or pyrrylium while the anions (shown below) are derived from polycyanopropene acids.

The crystal structures of 2,4,6-triphenylpyrrylium–1,1,3,3-tetracyanopropenide (Tamamura, Yamani *et al.*, 1974; PYLTCP), bis(quinolinium)–2-dicyanomethylene-1,1,3,3-tetracyanopropanediide (Sakanoue *et al.*, 1971; QUPRCN10). and N,N′-dimethyl-4,4′-bipyridinium–2-dicyanomethylene-1,1,3,3-tetracyanopropanediide   (Nakamura *et al.*, 1981; BELTER) have been determined. There are $- D^+ A^{2-} D^+ A^- -$ stacks in the first of these salts but the relative positioning of cation and anion appears to depend more on ionic than on charge transfer interactions. In the second salt there are $- D^+ A^{2-} D^+ D^+ A^{2-} D^+ -$ stacks with charge transfer interactions between the moieties in the $D^+ A^{2-} D^+$ sandwiches. Occurrence of charge transfer interactions along the stack axes is confirmed for both salts by single-crystal spectra using polarized radiation. The third salt has mixed stacks of alternating D^{2+} and A^{2-} ions.

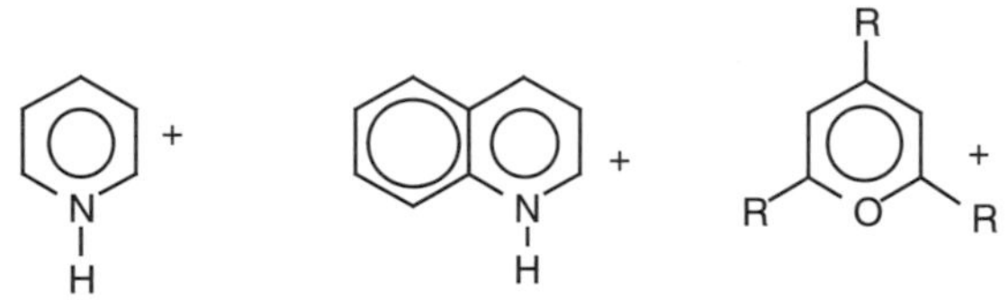

pyridinium   quinolinium   pyrrolinium

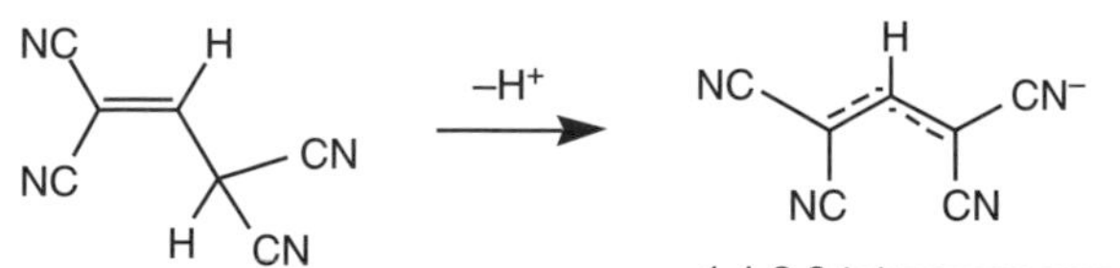

1,1,3,3-tetracyanopropenide

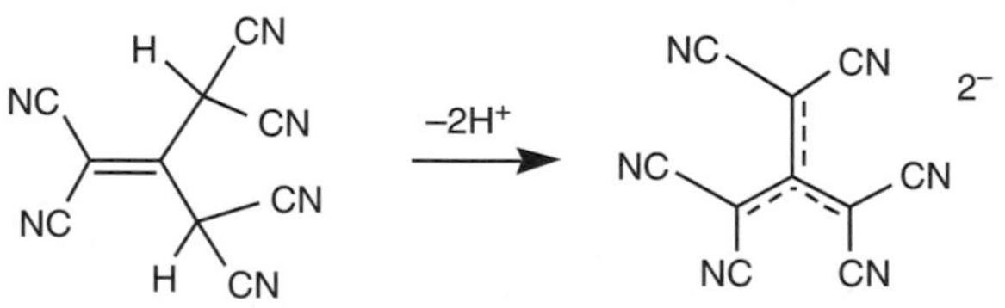

2-dicyanomethylene-1,1,3,3-tetracyanopropenide

15.10.2 Ion-radical salts

The crystal structures of some twenty mixed stack ion radical salts have been determined (Table 15.14). In *overall* structural arrangement there is little to distinguish these ionic materials from their neutral fellows; indeed neutral-to-ionic phase transformations are found in some on cooling or application of pressure (see Chapter 13). The ionic nature of the phase is generally demonstrated by magnetic measurements, spectroscopy (ESR, UV–visible, resonance Raman or infrared) or from bond lengths. The three latter types of measurement allow one to infer the degree of charge transfer (Z) in the molecular compounds; in some compounds the degree of charge transfer is inferred to be nonintegral, in contradiction to earlier theoretical predictions (see Chapter 13). However, it has been shown, using valence-bond methods, that intermediate values of Z are possible (Soos *et al.*, 1979) and that there is a useful limiting relationship between the charge on the moieties and the magnetic energy gap ΔE_m. In a mixed stack compound with fully ionic ground state ($Z = 1$) the material would be a regular Heisenberg antiferromagnet, with $\Delta E_m = 0$. However, static magnetic susceptibility ($\chi(T)$) measurements show that this is not so for molecular compounds such as {TMPD$\cdots$TCNQ} and {TMPD$\cdots$CA} where the following temperature dependence is found in the range $200 < T < 350$K:

$$T\chi(T) \propto \exp(-\Delta E_m/kT).$$

The magnetic gap ΔE_m is found to be ≈ 0.1 eV, which is almost as large as the charge transfer integral $t = \; <D^+ A^-|H|$ D A $>$. Soos *et al.* (1979) have shown that in a regular charge transfer solid $\Delta E_m > 0$ when $Z < 0.68 \pm 0.01$. For example, ΔE_m is measured as ≈ 0.07 eV in {TMPD$\cdots$TCNQ}, which corresponds to $Z \approx 0.60 - 0.65$, in good agreement with values obtained by other methods. The larger values of $\Delta E_m \approx 0.13$ eV in TMPD$\cdots$CA and PD$\cdots$CA imply somewhat smaller values of Z. When $Z > 0.68$ then ΔE_m is predicted to be zero and the solid becomes paramagnetic.

The {TMPD$\cdots$CA} structure is unusual in that there is complete eclipsing of donor and acceptor ions in the quasi-hexagonally close-packed stacks; such eclipsing is found elsewhere only in the (neutral ground state) molecular compound between 2,4,6-tris(dimethylamino)-*s*-triazine and TNB (Williams and Wallwork, 1966). Both X-ray diffraction and ESR measurements show that {TMPD$\cdots$CA} undergoes a phase change on cooling below ≈ 250K but the low-temperature structure has not been reported.

Another unusual feature is σ-dimerization[9] of TCNQ in the formation of molecular compounds with N-ethylphenazinium and bis(dipyridyl)Pt(II) (nos. 15–17 in Table 15.14). The two parts of the (TCNQ–TCNQ)$^{2-}$ ion are joined by a covalent bond of length 1.63 Å, ≈ 0.09 Å longer than the standard C–C single bond length; the arrangement at the linked carbons is approximately tetrahedral. Each half of the (TCNQ–TCNQ)$^{2-}$ ion, represented as vertical lines in the diagram, participates in charge transfer interactions in separate stacks while the central portion of the dianion only links the stacks.

---

[9] σ-Dimerization is used here to indicate covalent bond formation between two moieties; π-dimerization is used to describe pairing of moieties by HOMO–LUMO interaction without formation of a covalent bond.

Table 15.14. Some mixed-stack ion radical salts for which crystal structures have been reported. $(tfd)_2$ is sometimes called dithiete. Phenazine moieties are discussed earlier and, apart from one example, have been omitted

Ion radical salt/Reference/refcode	Z	Remarks
1. Tropylium···Ni(tfd)$_2$ (Wing and Schlupp, 1970); TRFSNI	≈1	disorder of cations; Curie–Weiss magnetic behavior, i.e. two independent spins per $D^{+\bullet} A^{-\bullet}$ pair.
2. Phenoxazinium···Ni(tfd)$_2$ (Singhabandu et al., 1975); FMENPX	≈1	
3. TTF···Pt(tfd)$_2$ (Kasper and Interrante, 1976); FMEPTF	≈1	
4. TMPD···chloranil (Boer and Vos, 1968); TMABCA	≈0.60	
5. TTF···fluoranil (Torrance, Vasquez et al., 1981); TTFFAN		Ionic form is stable at low temperatures or above 9 kBar at 300K
6. TTF···chloranil (Torrance, Vasquez et al., 1981); TTFCAN		
7. p-Phenylenediamine···chloranil (Hughes and Soos, 1968)	≈0.5	From ΔE_m; crystal structure known only in outline
8. Decamethylferrocenium$^{+\bullet}$···DDQH$^-$ (Gebert et al., 1982); MEFEQU10	≈1	Both ions disordered
9. TMPD···TCNQ (Hanson, 1965b); QMEPHE	≈0.60	
10. TMBTP···TCNQ (Darocha et al., 1979); MBPTCR	≈1	
11. Decamethylferrocenium···TCNQ (Miller et al., 1987); MCFECT01	≈1	
12. DBTSeF···TCNQ (Emge, Bryden et al., 1982); BOWSUB	≈0.47	
13. DBTTF···2,5-TCNQF$_2$ (Emge, Wijgul et al., 1982); BITROL	≈0.6 − 0.7	
14. OMTTF···DBTCNQ (Akhtar et al., 1985)	≈1	

σ-bonded TCNQ dimers

15. (N-ethylphenazinium$^{+\bullet}$)2···-(TCNQ–TCNQ)$^{2-}$ (Morosin et al., 1978) EPZTCD		
16. [Bis(dipyridyl)Pt(II)]$^{2+}$···-(TCNQ–TCNQ)$^{2-}$ (Dong et al., 1977) TCQDPT;		structure not stacked
17. [Bis(2,9-dimethyl-1,10-phenanthroline)Cu(I)]···(TCNQ–TCNQ)$^{2-}$ (Hoffmann et al., 1983; CABKEV)		structure not stacked

The ESR spectrum of $(NEP^+)_2 \cdots (TCNQ–TCNQ)^{2-}$ has been studied (Harms, Keller et al., 1981); in addition to a number of features that were not clearly understood, there are thermally activated triplet spin exciton (TSE) lines (line width ≈ 0.5 mT) which show an orientation-dependent fine structure splitting that was identified with an $S = 1$ excitation of the dianion, and was assigned to a transition to an excited state in which the long σ-bond was broken. The TSE is quasi-immobilized on the dianion, in contrast to TSEs in other TCNQ salts, which show extreme line narrowing because of the fast diffusional or hopping nature of the paramagnetic excitation along the TCNQ anion stack. In the $\{Pt(2,2'-dipy)_2 \cdots (TCNQ-TCNQ)_2\}$ salt there is a phase transition at $87\,°C$, accompanied by a color change and an enormous increase in paramagnetism (corresponding to two unpaired electrons per formula unit) which is ascribed to breaking of the long σ-bond.

There are a number of molecular compounds where the components have ionic ground states and the stacks are mixed but where a simple alternation of moieties is not found. There are five examples of a $-D^+D^+A^-A^-D^+D^+A^-A^-$ arrangement, which can be further subdivided in terms of the interactions between the moieties. The first example is {phenothiazine$\cdots$Ni(tfd)$_2$}, which is a cation radical, anion radical salt (Geiger and Maki, 1971; Singhabhandhu et al., 1975; FMENPZ). The cation is nonplanar, with a dihedral angle of $172°$ and there is strong cation–anion interaction with an interplanar distance of 3.36 Å; the anion–anion distance is 3.83 Å and the cation–cation distance varies from 3.4–3.9 Å because of cation folding. Thus the structure is based on individual cation . . . anion pairs, largely isolated from other such pairs, and could be represented schematically as

$$-D^+ \; D^+A^- \; A^-D^+ \; D^+A^- \; A^--.$$

The magnetic susceptibility is consistent with a spin-paired singlet ground state and a thermally populated triplet excited state; a detailed ESR study has not yet been reported. {Phenoxazine$\cdots$Ni(tfd)$_2$}, which could be expected to be isostructural, has in fact a different stacking $(-D^+A^- \; D^+A^- \; D^+A^-)$ and very different physical properties (Singhabhandhu et al., 1975; FMENPX).

Phenothiazine

M = Ni, X = CF$_3$ Bis(cis-1,2-perfluoro-methylethylene-1,2-dithiolato)Ni(II), abbreviated as Ni(tfd)$_2$.

M = Pt, X = CN Bis(dicyanoethylene-1,2-dithiolato)Pt(II).

The three compounds of the second subset {5-(1-butylphenazinium)$\cdots$tetrafluoro-TCNQ} (NBP$\cdots$TCNQF$_4$) (Metzger et al., 1982; BISWAB10); {NBP$\cdots$TCNQ} (Gundel et al., 1983; BISWUU10); {N,N'-dimethylbenzimidazolinium$\cdots$TCNQ} (Chasseau et al., 1972; MBZTCQ) are all characterized by strong anion . . . anion interactions. Although {NBP$\cdots$TCNQF$_4$} and {NBP$\cdots$TCNQ} are not isomorphous (space groups $P2_1/c$ and $P\bar{1}$, respectively) the structures of their stacks are very similar. The anion radicals interact strongly (interplanar spacings ≈ 3.15 Å) and both have ring-external bond (R-EB) overlap while the other interactions are much weaker. ESR studies show that there are "quasi-immobile" Frenkel excitons localized on the pairs of adjacent anion radicals. A similar cation-anion radical arrangement is found in {N,N′-dimethylbenzimidazolinium⋯ TCNQ}, where the interplanar spacing between adjacent anion radicals is remarkably short at 3.07 Å, implying strong coupling. Physical properties have not been measured apart from conductivity which is very low at 10^{-10} S/cm. These compounds with strongly bound TCNQ π-dimers are also noted in Section 17.4.2.

The briefly-described {tetrakis(methylthio)TTF⋯TCNQ}[10] has a mixed stack –DDAADDAA– arrangement (Mori,Wu *et al.*, 1987; FIJYEC) with d(D . . . D) = 3.48 Å, d(A . . . A) = 3.41 Å, d(D . . . A) = 3.58 Å; thus π-dimers do not appear to be formed. A similar stacking arrangement is found (Iwasaki, Hironaka, Yamazaki and Kobayashi, 1992) in {TTF⋯4,8-bis(dicyanomethylene)−4,8-dihydrobenzo-[1,2-*b* : 4,5-*b*′]dithiophene} with d(D . . . D) = 3.62 Å, *d*(A . . . A) = 3.49 Å, d(D . . . A) = 3.36 Å, indicating that the strongest interaction is between donor and acceptor units. The bond lengths in TTF suggest that the moieties are present as ions.

There is a six-molecule periodicity along the stack axis in {4,4′,5,5′-tetraethyl-tetrathiofulvalene⋯(TCNQ)$_2$} {(TETTF)⋯(TCNQ)$_2$} (Galigné *et al.*, 1977; ETFTCQ). The space group here is *C2/c* and the stack axis is [001] (=22.61 Å). The arrangement can be represented schematically as:

∘ TCNQ   TETTF   TCNQ ∘ TCNQ   TETTF   TCNQ ∘ TCNQ   TETTF   TCNQ ∘
$\bar{1}$ C_2 $\bar{1}$ C_2 $\bar{1}$ C_2 $\bar{1}$

where $\bar{1}$ represents a crystallographic centre of symmetry and C_2 a crystallographic two fold axis. The mixed stack arrangement explains the low conductivity. It was inferred, from the bond lengths in TCNQ, that 0.4 *e* had been transferred from donor to acceptor. The interplanar distance between TCNQ moieties is 3.36 Å and the angle between TETTF and TCNQ planes is 9.3°. Thus a possible description is of weak TCNQ π-dimers separated by TETTF moieties, with the π-dimers having laterally displaced R/R overlap rather than the more usual R/EB overlap.

Mixed stacks of neutral (TNF) and charged (TCVPDM$^-$) moieties are found in the crystal structure of {tetramethylammonium *p*-tricyanovinylphenyl-dicyanomethide⋯2,4,7-trinitrofluorenone} {(CH$_3$)$_4$N$^+$·TCVPDM$^-$·TNF} (Sandman *et al.*, 1980; TCVPDA), with the tetramethylammonium acting as counterion. The charge transfer interaction is

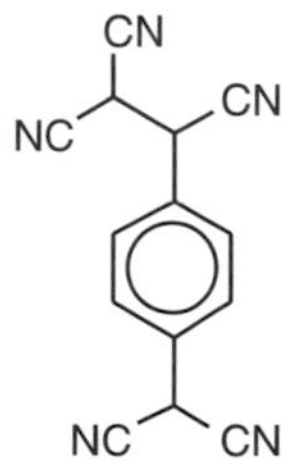

TCVPDM$^-$ 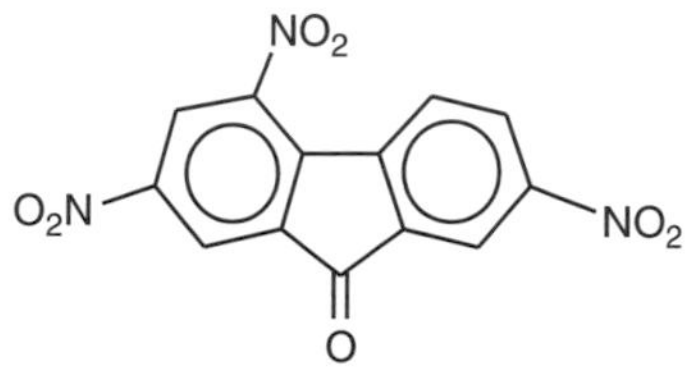   TNF

[10] {bis(tetrakis(methylthio)TTF)⋯TCNQ}has also been studied(FIJYAY).

somewhat limited as there is an angle of 16° between the planes of the two moieties in the stacks; it was considered that TCVPDM⁻ behaved as a closed-shell mono-anion electron donor and TNF as a neutral closed-shell electron acceptor (which is its usual role in π-molecular compounds).

15.11 Isomeric (polymorphic) molecular compounds

There are a number of examples of binary molecular compounds of the same overall chemical formula occurring in different crystal structures. This is the usual definition of polymorphism, which does not take into account that the nature of the chemical entity in two (or more) polymorphs can vary, without change of chemical formula, from "hardly different" to "very different" (Herbstein, 2001). The charge transfer molecular compounds provide many interesting examples of this range of possibilities, leading to 'isomerism' of the molecular compounds, where the state of the components may differ appreciably in the two polymorphs (or isomers), usually as a result of electron or proton transfer. This is evidence that different types of interaction between the components predominate as the crystals are formed under different (but not always well defined) conditions. We call these "isomeric (polymorphic) molecular compounds", with 'polymorphic' usually being dropped for brevity. We restrict ourselves in this chapter to molecular compounds of the charge-transfer type, where three types of isomerism are known:

> Type 1: If donor and acceptor can interact in different ways without appreciable change in their individual (chemical) structures, then different crystal structures can ensue, depending on which of the different types of intercomponent interaction (e.g. $\pi-\pi^*$; $n-\pi^*$; hydrogen bonding) predominates in a particular isomeric molecular compound.
> Type 2: Here the components differ chemically in the different isomers; among the examples are those where the components are neutral in the ground state of one isomeric molecular compound but ionic in the ground state of the other.
> Type 3: Here the isomeric molecular compounds are distinguished by the occurrence of proton transfer from acceptor to donor in one isomer but not in the other, where electron transfer (usually virtual electron transfer, with the ground state neutral) takes place.

A variety of isomeric molecular compounds is possible in principle in Type 1 but only pairs of isomers occur in Types 2 and 3.

It is convenient to include here, as an extension of the discussion of Type 3 isomeric molecular compounds, a group of molecular compounds where both charge transfer and proton transfer occur; these CPT molecular compounds (Section 15.11.4) are not isomers.

15.11.1   Type 1 – isomerism due to different types of interaction without change of moiety structure

Two clearcut examples are currrently known: firstly, in {(9,10-diazaphenanthrene)^{2}···TCNE} there is $n-\pi^*$ interaction in the triclinic polymorph and $\pi-\pi^*$ in the monoclinic polymorph (Shmueli and Mayorzik, 1980) (see also Section 12.4). Secondly, decamethylferrocene and TCNQ form two 1:1 salts (Miller, Zhang *et al.*, 1987); the monoclinic salt (MCFETO01) is dark green, metamagnetic and has mixed...DADA... stacks, while the triclinic salt is purple, paramagnetic and has discrete stacks of DAAD dimeric units (MCFETO02; cf. Section 17.4.2). Formation of the mixed stack structure is kinetically controlled and that of the dimeric stacks thermodynamically.

Table 15.15. Cell dimensions at 120K for the high (red; HT) and low-temperature (black; LT) phases of biphenylene···PMDA. The stack axis is [001] for both phases

Phase	a(Å)	b(Å)	c(Å)	β(deg.)	Volume/ formula unit	Calculated density $g\,cm^{-3}$
Red (HT); DURZAR	9.280(1)	11.869(2)	7.293(1)	98.68(2)	794.0 Å^3	1.548
Black (LT); DURZAR01	13.368(1)	5.809(1)	10.443(1)	102.30(1)	792.3	1.552

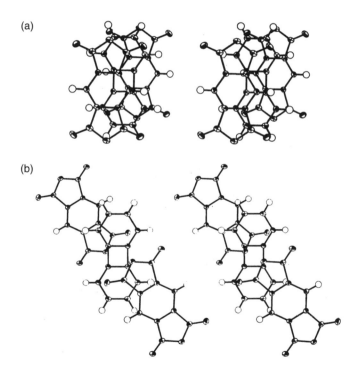

(a)

(b)

Fig. 15.29. ORTEP stereodiagrams of (a) donor–acceptor pair in red {biphenylenePMDA}; (b) donor–acceptor pair in black {biphenylene···PMDA}. In both diagrams projection is onto the biphenylene plane. (Reproduced from Stezowski, Stigler and Karl, 1986.)

A more subtle example, essentially polymorphic in nature, is provided by {biphenylene···PMDA}, which has a first-order phase transition (with hysteresis) at ≈ 400K (Stezowski, Stigler and Karl, 1986). Cell dimensions (at 120K) for both phases are given in Table 15.15.

The red phase has almost complete overlap of donor and acceptor molecules (Fig. 15.29(a)), with an interplanar angle of 4.1° and appears to be a $\pi-\pi^*$ donor–acceptor molecular compound of the standard type; there is typical overlapped disk packing. The black phase has donor and acceptor displaced along their long molecular axes, with an angle of 9.9° between molecular planes; this is typical slipped disk packing. Carbonyl groups of PMDA overlap benzene rings of biphenylene and it seems that both components behave as though composed of two virtually noninteracting halves. Thus one isomer has standard $\pi-\pi^*$ interaction whereas the second has more localized interaction between polar carbonyl and polarizable benzene ring. Typical slipped disk packing is also shown, for example, by {trans-stilbene···PMDA} (Koduma and Kumakura, 1974a; PYMAST).

15.11.2 Type 2 – isomerism due to electron transfer

There are two structural groups to be considered. In the first group mixed stack π-compounds undergo neutral $\Leftrightarrow$ ionic polymorphic transitions at temperatures and pressures depending on the components involved, the mixed stack structure remaining largely unchanged through the transitions (see Section 13.2.1). The best studied example is {TTF···chloranil} which has a mixed stack structure at room temperature and pressure ($P2_1/n$, $Z = 2$) (Mayerle et al., 1979; TTFCAN) and undergoes a neutral $\Leftrightarrow$ ionic polymorphic transition at 84K and atmospheric pressure (Batail et al., 1981). The conductivity of {TTF···chloranil} at 300K is 8×10^{-4} S/cm, in conformity with its mixed stack structure (Torrance, Mayerle et al., 1979). An extended discussion is given in Section 16.9.

The second group has pairs of isomers, one of which has a mixed stack structure and the other a segregated stack structure. There is a growing number of examples. In the TMTSF/TCNQ system the red, semiconducting, and the black, conducting phases (1 : 1 compositions) were crystallized under rather similar conditions. The black form (Bechgaard, Kistenmacher, Bloch and Cowan, 1977; SEOTCR) has a segregated stack structure with charge transfer of ≈ 0.6 e and volume/formula unit of 272.0 Å^3, while the red form has a mixed stack structure with a charge transfer of ≈ 0.2 e and volume/formula unit of 282.5 Å^3 (Kistenmacher, Emge et al., 1982; SEOTCR01). The red form is obtained by recrystallization of either black or red forms from hot CH_3CN (Bechgaard et al., 1977). The {bis(ethylenedithio)tetrathiafulvalene/TCNQ} {BEDT-TTF/TCNQ} system provides another example. One isomer has a mixed stack structure (space group $P2_1/n$, $Z = 2$; volume/formula unit of 610.9(5) Å^3), a resistivity of 10^6 ohm-cm and a sharp IR spectrum, charge transfer $\approx$ 0.2–0.3 e from TCNQ bond lengths (Mori and Inokuchi, 1987; FAHLEF01); the other has segregated stacks (space group $P\bar{1}$, $Z = 1$; volume/formula unit of 598.5(2) Å^3), high conductivity, a broadened IR spectrum and a metal $\Leftrightarrow$ insulator transition at around room temperature (Mori and Inokuchi, 1986; FAHLEF). Similar polymorphism has been reported in {2,7-bis(methylthio)-1,6-dithiapyrene/TCNQ} {MTDTPY/TCNQ}; Nakasuji et al., 1987). The black mixed-stack crystals (monoclinic; FUDTON01) have a volume per formula unit of 608.0(2) $Å^3$ while that for the segregated stack structure (triclinic; FUDTON) is 596.4(5) $Å^3$. In all these examples the segregated stack phase has a lower volume/formula unit than the mixed stack phase, suggesting that the former has the lower enthalpy at 300K; it remains to be seen whether this is a general feature confirmed for more examples and by direct measurement of the enthalpy differences. In {1,2-di(4-pyridyl)ethylene)···TCNQ} there is a red, semi-conducting form with a typical mixed stack, neutral ground state structure (BUKXOU) and a black, conducting form, the structure of which was not reported (Ashwell *et al.*, 1983).

Matsunaga (1978) has reported that {7-methylbenzo[*a*]phenazinium····-TCNQ} is dark green when crystallized from ethanol and violet when crystallized from CH_3CN. Recrystallization of the green form from CH_3CN gave the violet form. Examination of the electronic, vibrational and ESR spectra suggested that the violet form is {$NMBP^+$···$TCNQ^{-•}$} and that the green form is non-ionic with a neutral ground state i.e. {$NMBP^•$···TCNQ}. There was no N-H stretching vibration in the region of 3000–3500 cm^{-1} in the spectrum of either sample, thus eliminating the possibility that reduction of the diamagnetic $NMBP^+$ cation to the $HNMBP^{+•}$ cation radical had occurred during preparation and that this was the cause of the apparent dimorphism (cf. Section 15.8.5). Thermodynamic results do not appear to be available for any of these systems.

There are a number of actual or potential isomeric molecular compounds of the mixed stack/segregated stack type among the various TTF/haloanil binary systems. For example, {TTF···fluoranil} (Mayerle *et al.*, 1979; Torrance, Mayerle *et al.*, 1979; TTFFAN) has a structure similar to that of {TTF···chloranil} although triclinic ($P\bar{1}, Z = 1$). {TTF···fluoranil} also appears as a micro-crystalline phase with $\sigma_{RT} = 10$ S/cm and, presumably, a segregated stack structure in analogy with that of {TMTTF···bromanil} ($\sigma_{RT} = 1$ S/cm) (Mayerle and Torrance, 1981; TMFBRQ10).

15.11.3   *Type 3 – isomerism due to proton transfer or to π–π* electron transfer*

When the components of a binary system are acids and bases in both the Lowry-Brønsted (proton transfer from acid to base) and Lewis (electron transfer from base to acid) senses, then both types of interaction can occur. For the donor-acceptor combinations considered here the proton transfer will be actual for the ground state moieties but the electron transfer will be virtual, occurring only in the excited state. The comparative strengths of the Lowry–Brønsted and Lewis acid–base interactions will determine which structure occurs at a particular temperature, as the two interactions are temperature-dependent in the crystalline state, with the salt-like structure favoured at lower temperatures. Thus, taking an amine–phenol pair as an example, the salt-like structure {RNH_3^+ – $^-OR'$} will be favoured at lower temperatures but this may transform to a neutral π-molecular compound form {[RNH_2]···[HOR']} on heating. The transition temperature can be anywhere between a very low temperature and the melting point, depending on the details of the system. In crystallographic terms the two isomers are polymorphs, the relationship being enantiotropic in most known examples but monotropic in some.

Certain organic moieties function effectively as electron acceptors in both neutral and anionic states; for example, the picric acid–picrate anion pair (Matsunaga and Saito, 1972; Saito and Matsunaga, 1972; Saito and Matsunaga, 1973a, b). Thus *both* charge and proton transfer can take place in the same crystalline species in many binary systems: these are the CPT complexes discussed in the next section.

Early suggestions that 'isomeric complexes' could occur were elaborated in a series of experimental studies by Hertel and coworkers in the 1930s. Later Briegleb and coworkers used UV–visible and IR spectroscopy to identify the salt-like and molecular-compound forms of various isomeric complexes. This work has been summarized by Herbstein (1971). Many of the more recent developments are due to Matsunaga and his colleagues.

It does not seem possible to give a clearcut relation between the type of molecular compound formed and the pK_a values of acid and base (for summaries see Kortum *et al.*, 1961 and Perrin (1965)). This is because of the competition between proton transfer and charge transfer and the temperature-dependence of these processes; also many pK_a values are not known. In illustration we note that anilinium picrate and pyridinium picrates are salts (pK_a of aniline, pyridine and picric acid are 4.51, 5.23 and 0.42 respectively). There are cations and anions in the crystal structure of pyridinium picrate; the original structure determination (Talukdar and Chaudhuri, 1976; PYRPIC) is wrong and has been corrected (Botoshansky *et al.*, 1994; PYRPIC02). The picrates of *o*-chloroaniline ($pK_a = 1.96$), *m*-nitroaniline ($pK_a = 2.46$), 3-nitro-4-methyl-aniline ($pK_a = 2.96$) and N,N-dimethylamino-*p*-benzaldehyde ($pK_a = 1.62$) are yellow crystalline salts which give red solutions in benzene and red melts. Thus proton transfer occurs in the solid state but charge transfer interactions predominate in (nonpolar) solutions and in the melt (Saito and Matsunaga, 1971).

Charge transfer molecular compounds are formed at room temperature when the bases involved are very weak, as in the picric acid compounds of carbazole, skatole and indole (Briegleb and Delle, 1960) and these are also found with bases such as *o*-nitroaniline ($pK_a = -0.62$) and *p*-nitroaniline ($pK_a = 0.99$); similarly *p*-dimethylaminobenzaldehyde ($pK_a = 1.62$) and *o*-nitroaniline give molecular compounds with 2,4,6-trinitrobenzoic acid

Table 15.16. Polymorphic transitions in crystals of some isomeric compounds. All transitions are from a salt-like structure at lower temperatures to a charge-transfer molecular compound stable above the transition temperature. pK_a values are given in brackets

Components	$t_{tr}(°C)$	ΔH_{tr} (kJ/mol)
A. Picric acid (0.42) as acceptor:		
1. 2,5-dichloroaniline (1.57) (Matsunaga *et al.*, 1974)	74	17.6
2. *o*-bromoaniline (2.55) (Komorowski *et al.*, 1976)	102	24.7
3. *o*-iodoaniline (2.60) (Matsunaga *et al.*, 1975)	100–107	28.0
4. 1-chloro-2-naphthylamine (Matsunaga *et al.*, 1974)	128–138	33.5
5. 1-bromo-2-naphthylamine (Matsunaga *et al.*, 1975)	118–125	34.3
6. 1,6-dibromo-2-naphthylamine (Matsunaga *et al.*, 1975)	99–108	14.2
B. 2,4,6-trinitrobenzoic acid (0.65) as acceptor:		
1. *o*-chloroaniline (2.65) (Matsunaga *et al.*, 1974)	133	9.2
C. 2,6-dinitrophenol(3.57) as acceptor:		
1. 4-bromo-1-naphthylamine (3.21) (Hertel and Frank, 1934)	monotropic	
2. 4-chloro-1-naphthylamine (Matsunaga *et al.*, 1974)	76	

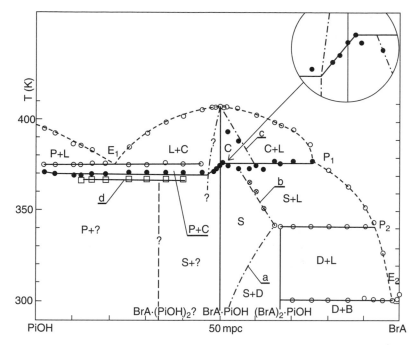

Fig. 15.30. Phase diagram of the *o*-bromoaniline(BrA)–picric acid (PiOH or P) system. L = liquid; C = solid solution of charge transfer molecular compound type; S = solid solution of salt type; D = (BrA)$_2$···PiOH; insert – the region of the solid state transition in expanded scale. (Reproduced from Komorowski *et al.*, 1976.)

(pK$_a$ = 0.65) (Matsunaga and Osawa, 1974)). Results of studies of polymorphic transitions in isomeric compounds are summarized in Table 15.16.

A complete study of a particular binary system that includes isomeric complexes does not appear to have been made, but some partial studies, extending beyond identification by spectroscopic techniques, have been reported. For example, the phase diagram of the *o*-bromoaniline–picric acid system has been determined in some detail (Fig. 15.30) (Komorowski *et al.*, 1976). Equimolar {*o*-bromoaniline···picric acid} is an isomeric complex with a salt ⇔ molecular compound transition temperature at 102 °C while the 2 : 1 compound is of the CPT type discussed in the next section. The appearance of solid solutions in both polymorphs of the equimolar material is quite unexpected and should be checked.

Thermodynamic studies have been made of 4-bromo-1-naphthylamine···2,6-dinitrophenol (Hertel and Schneider, 1931), where the yellow (stable; m.pt. 91.5°; measured density 1.654 g cm^{-3}) and red (metastable; m.pt. 84.5°; measured density 1.56 g cm^{-3}) phases are monotropically related (Hertel and Frank, 1934). The enthalpies of formation of the yellow and red phases were determined calorimetrically (temperature not specified but presumably ≈300K) to be −13.4 and 0.4 kJ/mol respectively. The yellow phase has thus both lower enthalpy and higher density than the red phase, as would be expected from the stronger (Coulombic) binding between the components. The crystal structure of the (metastable) red phase has been determined at 298K (space group $P2_1/a$, $Z=4$, stack axis [010]); the familiar mixed stacks were found and there is no hydrogen bonding between the components (Carsten-Oeser *et al.*, 1968; PABRAN).

15.11.4 *Isomerism stabilized by both charge (π–π*) and proton transfer (CPT compounds)*

In these molecular compounds the charge and proton transfer can be carried out in a number of different ways and currently three types can be identified:

(a) CPT-A compounds: these are 1 : 1 (electron) donor–acceptor molecular compounds, where a proton is transferred from electron acceptor to donor, thus giving rise to the possibility of cation–anion and charge transfer interaction occurring in the same binary adduct; examples are tryptophan picrate (Matsunaga, 1973) and seretonin picrate (Thewalt and Bugg, 1972); see also Section 15.2. The red colours of both molecular compounds are due to charge transfer interactions between the cation donors and the picrate anions, acting as acceptors. This has been demonstrated directly for seretonin picrate by determination of the crystal structure of the monohydrate ($P2_1/c$, $Z=4$) that contains mixed stacks along [010] (cf. Section 15.2).

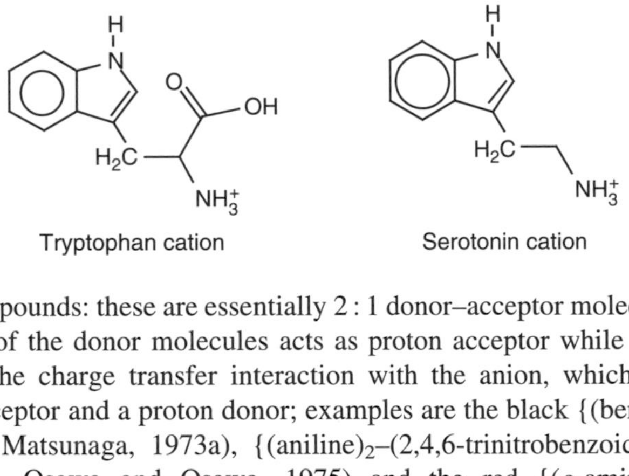

Tryptophan cation   Serotonin cation

(b) CPT-B compounds: these are essentially 2 : 1 donor–acceptor molecular compounds, where one of the donor molecules acts as proton acceptor while the second participates in the charge transfer interaction with the anion, which acts both as an electron acceptor and a proton donor; examples are the black {(benzidine)₂ picrate} (Saito and Matsunaga, 1973a), {(aniline)₂–(2,4,6-trinitrobenzoic acid} (TNBA)) (Matsunaga, Osawa and Osawa, 1975) and the red {(*o*-aminobenzoic acid)₂ picrate} (Matsunaga and Usui, 1980). More complicated compositions have been encountered in other systems (e.g. {(2,4,5-trimethylaniline)₃(3,3′,5,5′-tetranitrobiphenyl-4,4′-diol (TNBP))₂} (Matsunaga and Osawa, 1974) and the dark orange {(N,N-diethylaniline)₅ (TNBP)₃} (Saito and Matsunaga, 1973b; Lloyd and Sudborough, 1899); spectroscopic studies show that these are CPT compounds but crystal structure analyses are needed to show how the charge and proton transfers are distributed among the moieties.

One can distinguish a subgroup in which these functions are combined in a single molecule but localized in different parts thereof; here one half of a monoprotonated (electron) donor acts as proton acceptor and the other half participates in charge-transfer interaction with the anion acceptor; examples are the picrates of *o*-dianisidine and of tetramethylbenzidine (Saito and Matsunaga, 1973a) and the 3 : 2

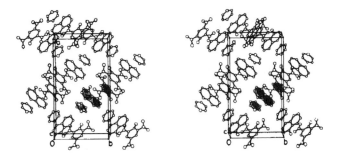

Fig. 15.31. ORTEP stereoview of pyridinium-1-naphthylamine–picrate crystal structure. The asymmetric unit has been marked. Only one orientation of the disordered 1-naphthylamine molecule is shown. (Reproduced from Bernstein *et al.*, 1980.)

      compound of benzidine (pK$_a$ = 4.95) with 2,6-dinitrophenol (pK$_a$ = 4.09) (Saito and Matsunaga, 1974).

(c)   CPT-C compounds : here the functions are localized in different molecules and thus a ternary composition can be expected. Köfler's (1940) ternary complex {pyridine–1-naphthylamine–picric acid} {PY-NA-PC} was shown, by spectroscopy, to be a CPT compound (Matsunaga and Saito, 1972). Crystal structure analysis (Bernstein *et al.*, 1980; PYNPCR) showed that it was in fact {pyridinium–1-naphthylamine–picrate}, in which there was a herringbone arrangement of centrosymmetric plane-to-plane stacks of composition PY$^+$ NA PC$^-$ ($\bar{1}$) PC$^-$ NA PY$^+$ (Fig. 15.31). This arrangement has similarities to that found in {(acridine)$_2$ ... PMDA} (Fig. 15.3). Thus the proton transfer is from picric acid to pyridine while the principal charge transfer is from 1-naphthylamine to picrate anion. The 1-naphthylamine molecule was disordered over (at least) two orientations.

15.12 Self-complexes

In a molecule that has regions with electron-donor properties and other regions with electron-acceptor properties, intramolecular and/or intermolecular charge transfer can occur. If nitrogen and/or oxygen is present, then there can be n–π* interaction. Alternatively, with appropriate molecular conformation, π–π* interactions could occur. If these interactions are intramolecular, then charge-transfer bands would be expected in both solution and solid state spectra. If the charge transfer interactions are intermolecular then one would expect a charge-transfer band in the spectrum of the solid but not (or to a far lesser extent) in solution, and a pairwise or stacked arrangement in the solid, with propinquity of donor and acceptor regions of different molecules. Many examples have been studied and all the possibilities outlined above have been encountered (Table 15.17). Classification can be conveniently made in terms of acceptor type and it will be noted that virtually all the well-known acceptor types are represented. The molecular arrangement in the red polymorph of 2-(4'-methoxyphenyl)-1,4-benzoquinone is shown in Fig.15.32 (Desiraju *et al.*, 1977; PANQUO). Early and more recent work has been reviewed (Bleidelis, Shvets and Freimanis, 1976; Chitkina and Belskii, 2002).

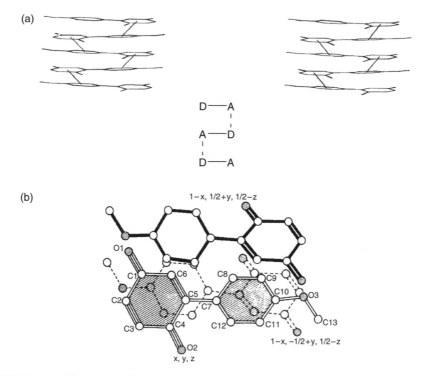

Fig. 15.32. π–π* Interactions in the red polymorph of 2-(4'-methoxyphenyl)-1,4-benzoquinone. (a) stereopair diagram of the alternating donor–acceptor interaction in a molecular stack with the schematic arrangement shown in the center; (b) Details of pairwise overlap. The molecules are projected onto the plane of the central benzoquinone ring (line shaded), with the molecule above darkened and that below shown with broken lines. (Reproduced from Desiraju *et al.*, 1977.)

A few structures that do not fit easily into the framework of Table 15.17 will be discussed separately. The crystal structure of 1-(2-indol-3-ylethyl)-3-carbamido-pyridinium chloride monohydrate has been determined (tetragonal, space group $P4_1$) (Herriott *et al.*, 1974; INYECP); it serves as an intramolecular model of the nicotinamide adenine dinucleotide – tryptophan charge transfer compound. In the crystal the molecules are in the fully extended *transoid* conformation, with the two planar portions (the donor and acceptor portions) of a particular molecule essentially mutually perpendicular. An indole ring of one molecule is stacked between nicotinamide rings of two other molecules, and conversely (Fig. 15.33; p. 1061). Thus there are two sets of mutually perpendicular donor–acceptor stacks in the crystal. Within a stack the donor and acceptor molecules are approximately parallel, with an interplanar distance of ≈3.7 Å.

Table 15.17. Self-complexes classified according to acceptor type

(a) Polynitrobenzene acceptor portions

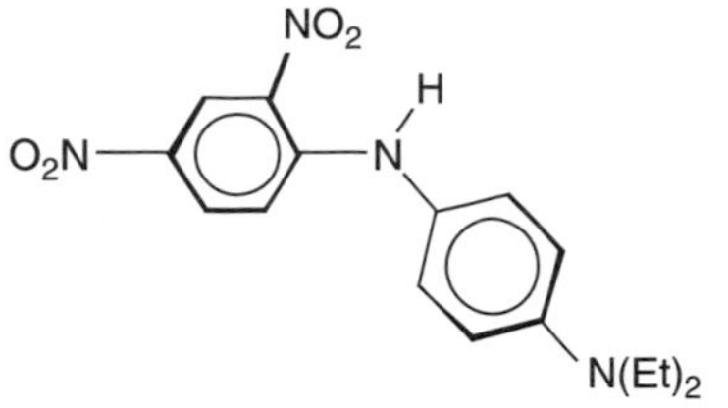

1. Donor portion of one molecule overlaps with acceptor of adjacent molecule to form a stack (Shvets *et al.*, 1974; NETPHN10).

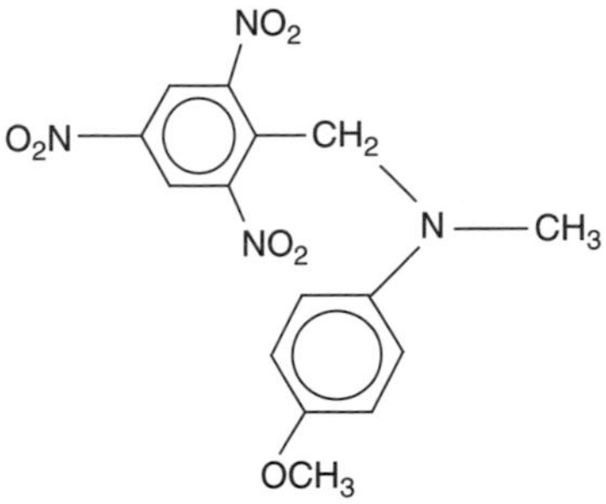

2. Donor and acceptor portions are approximately perpendicular, with hinge at tertiary nitrogen. ---DA DA DA--- stacks are formed (Shvets *et al.*, 1975a).

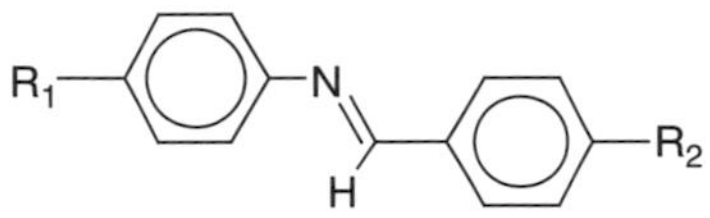

3. R1 = N(CH₃)₂; R₂ = NO₂. Dark brown crystals (Nakai *et al.*, 1976; two polymorphs MABZNA (*P*1, *Z* = 4) and NBZMAA (*P*2₁/*c*, *Z* = 4)).

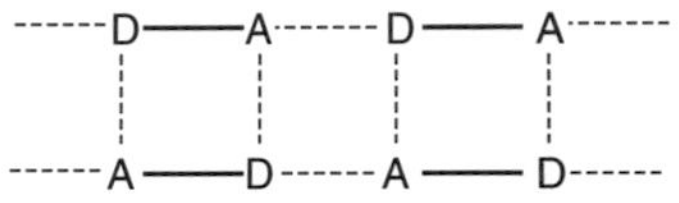

4. As for #3 but with R_1 = NO₂; R2 = N(CH₃)₂. Orange crystals. Stacking as for #3 (Shvets *et al.*, 1974).

(b) *p*-Benzoquinone acceptor portions

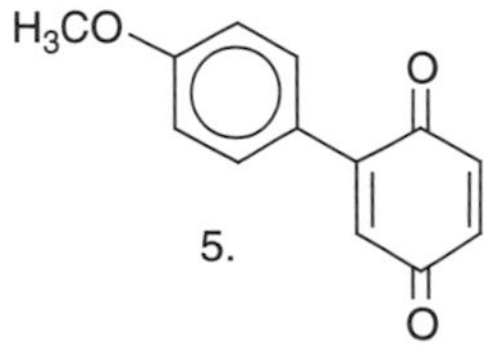

5. Red crystals; stacks with overlap of donor and acceptor rings, with π-π* interaction between the rings. There is no overlap of carbonyl groups and aromatic rings (Desiraju, Paul and Curtin, 1977; PANQUO).

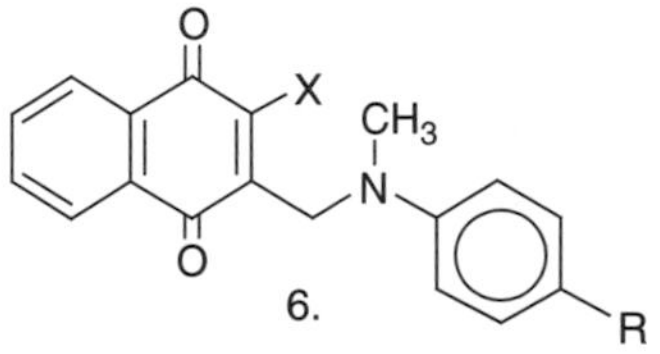

X = CH3, R = H.
6. Adjacent molecules are stacked antiparallel about two fold screw axes, thus allowing D---A interactions along the stack (Prout and Castellano, 1970).

7. As for #6 but with X = Cl, R = COOC₂H₅; orange crystals. Stacking as in #6 (Shvets *et al.*, 1975b).

8. As for #6 but with X = Cl, R = CH₃; orange crystals. Stacking as in #6, but with additional Cl...O interaction, d(Cl...O) = 3.29 Å (Shvets *et al.*, 1979; MAMNAQ).

Table 15.17. *(Continued)*

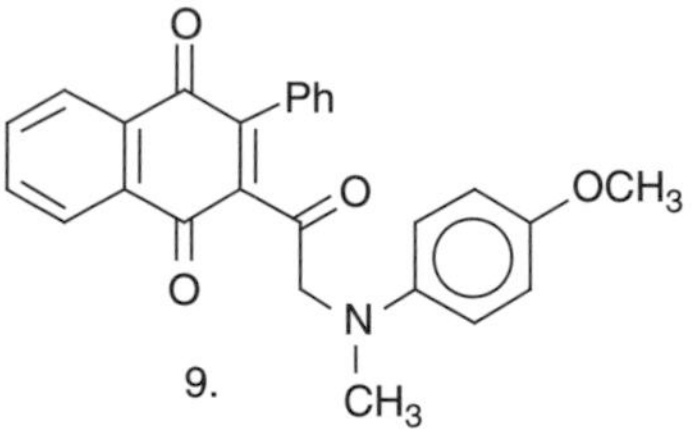

9. Black crystals with metallic lustre. Molecules are grouped in isolated donor-acceptor pairs (Shvets *et al.*, 1978).

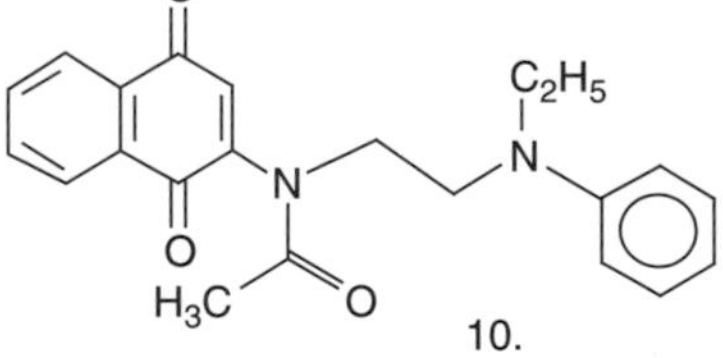

10. Molecule folded back on itself to allow intramolecular donor–acceptor interaction (Shvets *et al.*, 1975c; AEPCNQ10).

11. 2,3-Dichloronaphthazarin (DCDHNQ01) – see Section 15.6.

(c) TCNE type acceptor (for review see Chetkina and Bel'skii, 2002).

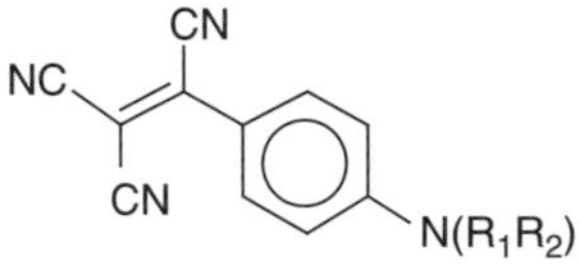

12. $R_1 = R_2 = CH_3$. Violet crystals. Molecules are stacked antiparallel with donor–acceptor interactions between adjacent molecules (Chetkina *et al.*, 1976; TCYVMA).

13. $R_1 = Ph$, $R_2 = H$. Dark-violet crystals. Partial overlap of donor and acceptor portions of the molecules (Popova *et al.*, 1978).

(d) TCNQ type acceptor

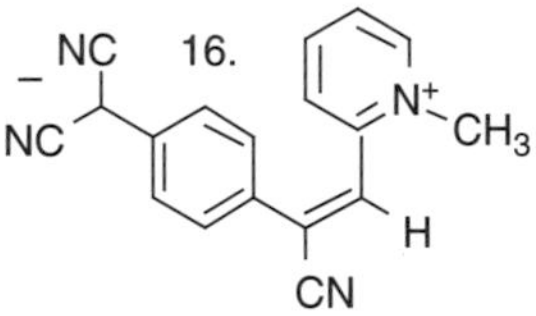

14. R = n-propyl; dark green crystals with metallic lustre. Pairwise interactions between TCNQ portions (Povot'eva *et al.*, 1981; BAFGOE).
15. R = CH_3; black crystals. Pairwise interactions between donor and acceptor portions of molecule (Povot'eva *et al.*, 1981).

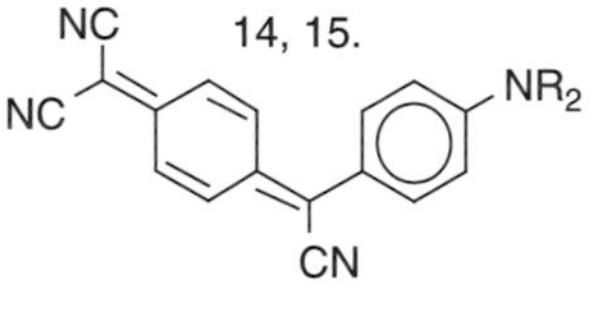

16. Picolyl-tricyanoquinodimethan. Zwitterionic structure; diamagnetic. Angle of 30° between planes of benzene and pyridinium rings. The pyridinium nitrogen stacks above the dicyanomethide portion of the next molecule along the b axis (Popova, Chetkina and Bespalov, 1981).

(e) Hexafluorobenzene type acceptor
17. Pentafluorobiphenyl, see Section 15.9.1.
18. Two polymorphic forms, both of which have stacking arrangements similar to those shown for #3 and 4 earlier in this Table. (Lindeman *et al.*, 1981; BANGOM).

Table 15.17. (*Continued*)

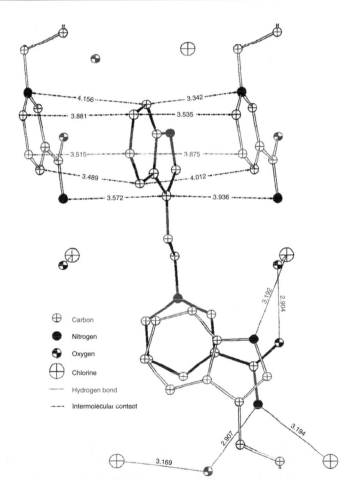

(f) Various
19. 1,3-Indanedione (Bravic *et al.*, 1976; INDDON). There are stacks of antiparallel molecules, thus allowing maximum interaction between donor and acceptor portions of the molecules and also maximum dipolar interaction. The crystals are tetragonal and the arrangement of the stacks resembles that found in {pyrene•••*p*-benzoquinone} (see Fig. 15.4).

Fig. 15.33. Projection of part of crystal structure of 1-(2-indol-3-ylethyl)-3-carbamido-pyridinium chloride monohydrate onto (100); two mutually-perpendicular donor–acceptor stacks are shown. (Reproduced from Herriott *et al.*, 1974.)

The π-orbital overlap between adjacent donor and acceptor portions appears nearly optimal, and the permanent dipoles are strongly coupled. It was suggested (Herriott *et al.*, 1974) that unsymmetrical $\pi-\pi^*$ and dipole–dipole interactions could lead to a specific three-dimensional mutual orientation, provided that one of the faces of the molecule is blocked, and that such a mechanism could be important in enzyme-coenzyme interactions.

The molecule of 2-(2-pyridylmethyldithio)benzoic acid has a hinged conformation in the solid, with pyridyl and phenyl rings approximately parallel and arranged in stacks along [001] ($=7.37$ Å) (Karle *et al.*, 1969; PYSBAC). Presumably the pyridyl ring acts as electron donor and the carboxyl-substituted phenyl ring as acceptor, and there are both intramolecular and intermolecular charge transfer interactions. The hinged conformation with parallel donor and acceptor portions is a feature that appears in many of the compounds discussed in this section. The compound 3,3'-diacetyl-5,5'-diethoxy-carbonylglaucyrone gives black crystals from benzene, whose structure shows a step-wise arrangement of conjugated molecules connected by self charge transfer interactions (Baker *et al.*, 1980; ETGLAU). The right hand side of the molecule in the diagram is considered to behave as the electron–donor portion and the left hand side as the electron–acceptor portion. These two portions are superposed in alternating fashion in the crystal.

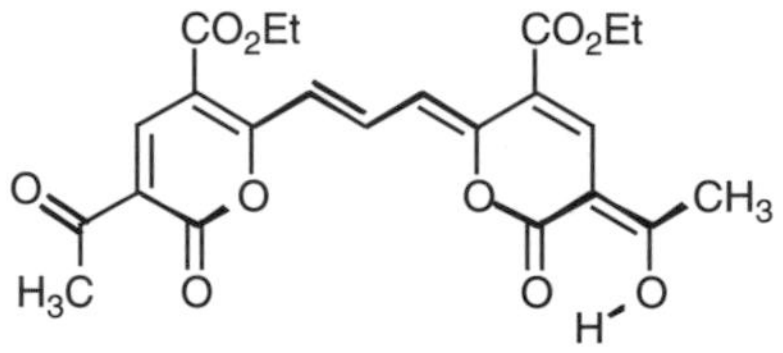

3,3'-diacetyl-5,5'-diethoxycarbonylglaucyrone

15.13   Conclusions

15.13.1   *Structural variety in $\pi-\pi^*$ molecular compounds*

One sees, from the survey above, that there is considerable structural variety in this family of molecular compounds. The simple picture of parallel mixed stacks of donors and acceptors in alternating array, although applicable to the majority of $\pi-\pi^*$ molecular compounds, and also to the quasi-acceptor molecular compounds, needs considerable emendation and extension. Stacks of limited size are found, stack axes are not necessarily parallel, and components with suitable functional groups can give rise to important lateral interactions. In particular, hydrogen bonding between like and/or unlike components, added to the primary stacking structural synthon, can provide possibilities for crystal engineering similar to those described in Chapter 14.

Although the most usual donor : acceptor ratio is 1 : 1, other compositions are found. These generally maintain mixed stacks with the component in excess accommodated in various ways, often as 'solvent of crystallization' filling space but not playing any structural role.

15.13.2 *How should the packing arrangements in π–π* molecular compounds be described?*

Despite the considerable structural variety in this family of molecular compounds, the essential feature appearing in virtually all structures is a plane-to-plane interaction between the two components. For most of the molecular compounds considered here, this is a donor–acceptor interaction leading to color changes on formation, and physical properties that stem from the anisotropic arrangement. However, the same feature appears also in the structures containing quasi-acceptors rather than true acceptors; as there is no sensible *structural* difference between these two groups, they can be treated together. A natural consequence is to describe the structures in terms of mixed stacks containing donors and acceptors (or quasi-acceptors) in alternating array. Some authors have, however, preferred a mixed-layer to a mixed-stack description. The matter could be settled if one could compare, for a particular structure, the one-dimensional interaction energy within mixed stacks to the two-dimensional interaction within layers but this information is not available. Description is a matter of choice, and we use three examples, from the extremes and the center, to illustrate the dilemmas. The interaction of a benzene ring (a polarizable donor group) with a polar acceptor group (e.g. a carbonyl group) often leads to a characteristic superposition (overlap diagram) within a mixed stack. An example is (monoclinic) {perylene•••fluoranil} (Hanson, 1963; PERFAN; Section 15.6 and Fig. 15.10), which has a typical herring-bone structure also shown by (monoclinic) {perylene•••1,2–4,5-tetracyanobenzene} (Fig. 15.20; Bock, Seitz *et al.*, 1996; REHMUM). How could one visualize the growth of such a crystal in formal (not necessarily physical) terms? Perhaps as preformed quasi-cylindrical stacks, with a given mutual donor-acceptor disposition, that then aggregate into a quasi close-packed arrangement of the quasi-cylindrical stacks. The mutual orientation of the stacks is determined by secondary cross-stack interactions; in the particular example of {perylene ... fluoranil} these could be weak C–H ... O=C hydrogen bonds.

At the other extreme one has a hydrogen-bonded layer such as that found in {hexamethylbenzene•••1,3,5-tricyanobenzene} (LAGNAI) (Fig. 15.34; Reddy, Goud *et al.*, 1993; Desiraju and Steiner, 1999; see p. 327 *et seq.*); each layer contains only a single component. Here the layers would be formed first and then stacked one over the other to give optimal plane-to-plane overlap of hexamethylbenzene donor and tricyanobenzene acceptor. {Pyrene•••1,2–4,5-tetracyanohydroquinone} (Fig. 15.20; Bock, Seitz *et al.*, 1996; TEXPOB10) provides a similar example, again with the layers each containing a single component and the –OH ... NC–C hydrogen bonding in the tetracyanohydroquinone layer providing the dominant structural interaction.

What happens in the middle? One example is {dibenz[*a,c*]anthracene•••-1,3,5-trinitrobenzene} ({DBA•••TNB}; Carrell and Glusker, 1997; RULLUF). Here the layers, which are somewhat corrugated, contain both components, with molecules of the smaller TNB acceptors surrounded by molecules of the larger DBA donors (Fig. 15.35). Carrell and Glusker provide a well-balanced description of the structure as having "... *layers* containing both DBA and TNB molecules, interconnected within a layer by C–H ... O interactions. Layers *stack* on one another so that DBA molecules are sandwiched between TNB molecules and vice versa. The average distance between molecules in these sandwiches is 3.23 Å" (italics added). Are the (relatively) many weak hydrogen bonds more

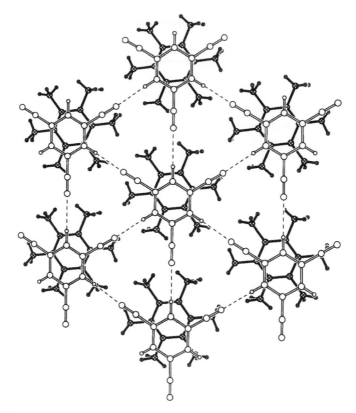

Fig. 15.34. The crystal structure of {hexamethylbenzene···1,3,5-tricyanobenzene} (15.207 8.839 14.460 Å, 110.4$\bar{8}$, $Z = 4$, $C2/c$; the HMB molecules are at inversion centers and the 1,3,5-tricyano-benzene molecules on two fold axes). The overlapped layers of hexamethylbenzene (darkened, below) and the hexagonal 1,3,5-tricyanobenzene network mediated by weak C–H...N hydrogen bonds (above) are shown. The weak C–H...NC–C hydrogen bonds (d(C...N) = 3.47, 3.52 Å) are indicated by dashed lines. The layers are parallel to (110). (Reproduced from Desiraju and Steiner, 1999.)

important in determining the overall structure than the plane-to-plane π interaction? At present there does not seem to be an answer to this question. However, stacking appears to be a feature occurring in all the structures of this family whereas layers can only be identified when there are appreciable lateral interactions. Thus, in our view, 'stacking' is the primary feature with 'layering' a possibly important secondary attribute.

15.13.3 *Structural consequences of π–π* interactions*

It is widely recognized that the donor–acceptor interactions in π–π* molecular com-pounds are too weak to cause changes in bond lengths that are measurable at current levels of precision. This applies even to measurements made at very low temperatures (e.g. {pyrene···PMDA} at 19K (Herbstein, Marsh and Samson, 1994; PYRPMA04).

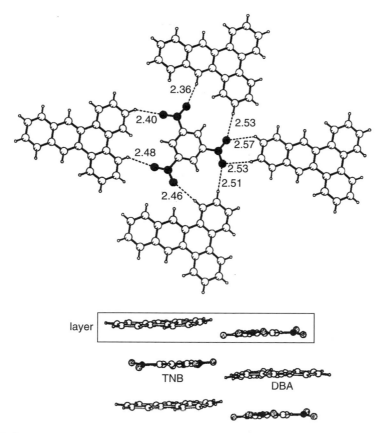

Fig. 15.35. Structural features of the triclinic DBA···TNB crystals ($a = 7.277(2)$, $b = 11.237(6)$, $c = 13.902(5)$ Å, $\alpha = 104.13(4)$, $\beta = 96.04(3)$, $\gamma = 95.15(2)°$, space group , Z = 2. (above) One layer of the structure showing TNB surrounded by DBA molecules and linked by CH···O interactions. (below) Stacking of layers showing that DBA and TNB are not entirely coplananr. (Reproduced from Desiraju and Steiner, 1999.)

Charge density studies (Chap. 17.7), comparing neat components with such components in the molecular compounds, may provide evidence of interaction but this remains a task for the future. Deviations from planarity in flexible molecules are more easily accessible. A clearly discernible effect has been found in {9,10-dihydroanthracene···TNB}, where the DHA molecule is folded (dihedral angle 146°) in its neat crystals (DITBOX) but planar in the molecular compound (ZZZAGS10) (Herbstein, Kapon and Reisner, 1986). Appreciable differences in the shapes of components in their neat crystals and in their molecular compounds has been found in {triphenylene–perfluorotriphenylene} (Weck *et al.*, 1999; CUKXIP; Fig. 15.24) and, to a lesser extent, for (nonplanar) benzo[c] phenanthrene in its neat crystals (Lakshman *et al.*, 2000; BZPHAN01) and its DDQ molecular compound (Bernstein, Herbstein and Regev, 1977; BZPCPQ).

References

Adman, E., Rosenblum, M., Sullivan, S. and Margulis, T. N. (1967). *J. Am. Chem. Soc.*, **89**, 4540–4542.

Agostini, G., Corvaja, C., Giacometti, G., Pasimeni, L., Clement, D. A. and Bandoli, G. (1986). *Mol. Cryst. Liq. Cryst.*, **141**, 165–178.

Akhtar, S., Tanaka, J., Nakasuji, K. and Murata, I. (1985). *Bull. Chem. Soc. Jpn.*, **58**, 2279–2284.

Alcacer, L. and Maki, A. H. (1974). *J. Phys. Chem.*, **78**, 215–217.

Allen, C. C., Boeyens, J. C. A. and Levendis, D. C. (1989). *S. Afr. J. Chem.*, **42**, 38–42.

Andersen, A. M., Mostad, A. and Rømming, C. (1975). *Acta Chem. Scand.*, B**29**, 45–50.

Anderson, J. S. (1937). *Nature, Lond.*, **140**, 583–584.

Anex, B. G. and Parkhurst, L. J. (1963). *J. Am. Chem. Soc.*, **85**, 3301–3302.

Anthonj, R., Karl, N., Robertson, B. E. and Stezowski, J. J. (1980). *J. Chem. Phys.*, **72**, 1244–1255.

Aroney, M. J., Hambley, T. W., Patsilades, E., Piernes, R. K., Chan, M.-K. and Gonda, I. (1987). *J. Chem. Soc., Perkin Trans.* 2, pp. 1747–1752.

Artiga, A., Gaultier, J., Haget, Y. and Chanh, N. B. (1978a). *J. Chim. Phys.*, **75**, 379–383.

Artiga, A., Gaultier, J., Hauw, C. and Chanh, N. B. (1978b). *Acta Cryst.*, B**34**, 1212–1218.

Ashwell, G. J., Kennedy, D. A. and Nowell, I. W. (1983). *Acta Cryst.*, C**39**, 733–734.

Aspley, C. J., Boxwell, C., Buil, M. L., Higgitt, C. L., Long, C. and Perutz, R. N. (1999). *Chem. Commun.*, 1027–1028.

Bailey, A. S. and Prout, C. K. (1965). *J. Chem. Soc.*, pp. 4867–4881.

Baker, S. R., Begley, M. J. and Crombie, L. (1980). *J. Chem. Soc. Chem. Commun.*, pp. 390–392.

Baldwin, S. L. and Baughman, M. C. (1993). *Acta Cryst.*, C**49**, 1840–1844.

Bando, M. C. and Matsunaga, Y. (1976). *Bull. Chem. Soc. Jpn.*, **49**, 3345–3346.

Banerjee, A. and Brown, C. J. (1985). *Acta Cryst.*, C**41**, 82–84.

Bar, I. and Bernstein, J. (1978). *Acta Cryst.*, B**34**, 3438–3441.

Bar, I. and Bernstein, J. (1981). *Acta Cryst.*, B**37**, 569–575.

Barnes, J. C., Chudek, J. A., Foster, R., Jarrett, F., Mackie, F., Paton, F. and Twiselton, D. R. (1984). *Tetrahedron*, **40**, 1595–1601.

Bars-Combe, M. le, Chion, B. and Lajzerowicz-Bonneteau, Janine. (1979). *Acta Cryst.*, B**35**, 913–920.

Bastiansen, O., Gunderson, S. and Samdal, S. (1989). *Acta Chem. Scand.*, **43**, 6–10.

Batail, P., La Placa, S. J., Mayerle, J. J. and Torrance, J. B. (1981). *J. Am. Chem. Soc.*, **103**, 951–953.

Batsanov, A. S. and Perepichka, I. F. (2003). *Acta Cryst.*, E**59**, o1318–o1320.

Batsanov, A. S., Collings, J. C., Howard, J. A. K. and Marder, T. B. (2001). *Acta Cryst.*, C**57**, 1306–1307.

Batsanov, A. S., Collings, J. C., Howard, J. A. K., Marder, T. B. and Perepichka, I. F. (2001). *Acta Cryst.*, E**57**, o950–o952.

Batsanov, A. S., Howard, J. A. K., Marder, T. B. and Robins, E. G. (2001). *Acta Cryst.*, C**57**, 1303–1305.

Batsanov, A. S., Perepichka, I. F., Bryce, M. R. and Howard, J. A. K., (2001). *Acta Cryst.*, C**57**, 1299–1302.

Bear, C. A., Waters, J. M. and Waters, T. N. (1973). *J. Chem. Soc. Perkin Trans.*2, pp. 1266–1271.

Beaumont, T. G. and Davis, K. M. C. (1967). *J. Chem. Soc. (B)*, pp. 1131–1134.

Bechgaard, K., Kistenmacher, T. J., Bloch, A. N. and Cowan, D. O. (1977). *Acta Cryst.*, B**33**, 417–422.

Bechtel, F., Chasseau, D. and Gaultier, J. (1976). *Cryst. Struct. Commun.*, **5**, 297–300.

Bechtel, F., Chasseau, D., Gaultier, J. and Hauw, C. (1977). *Cryst. Struct. Commun.*, **6**, 699–702.

Beck, C. M., Burdeniuc, J., Crabtree, R. H., Rheingold, A. L. and Yap, G. P. A. (1998). *Inorg. Chim. Acta*, **270**, 559–562.

Beevers, C. A. and Lipson, H. (1934). *Proc. Roy. Soc. Lond.*, A**146**, 570–582.

Bergamini, P., Bertolasi, V., Ferretti, V. and Sostero, S. (1987). *Inorg. Chim. Acta*, **126**, 151–155.

Bernstein, J. and Regev, H. (1980). *Cryst. Struct. Commun.*, **9**, 581–586.

Bernstein, J. and Trueblood, K. N. (1971). *Acta Cryst.*, B**27**, 2078–2089.

Bernstein, J., Cohen, M. D. and Leiserowitz, L. (1974). "The Structural Chemistry of the Quinones", in *The Chemistry of the Quinonoid Compounds*, S. Patai, Editor, J. Wiley and Sons, New York.

Bernstein, J., Herbstein, F. H. and Regev, H. (1977). *Acta Cryst.*, B**33**, 1716–1724.

Bernstein, J., Regev, H. and Herbstein, F. H. (1980). *Acta Cryst.*, B**36**, 1170–1175.

Bernstein, J., Regev, H., Herbstein, F. H., Main, P., Rizvi, S. H., Sasvari, K. and Turcsanyi, B. (1976). *Proc. Roy. Soc. Lond.*, A**347**, 419–434.

Bertholon, G., Perrin, R., Lamartine, R., Thozet, A., Perrin, M., Caillet, J. and Claverie, P. (1979). *Mol. Cryst. Liq. Cryst.*, **52**, 589–596.

Bertinelli, F., Costa Bizzarri, F., Della Casa, C., Marchesini, A., Pelizzi, G., Zamboni, R. and Taliani, C. (1984). *Mol. Cryst. Liq. Cryst.*, **109**, 289–302.

Binder, W., Karl, N. and Stezowski, J. J. (1982). *Acta Cryst.*, B**38**, 2915–2916.

Birtle, S. L. and Naae, D. G. (1980). *ACA*, Ser., 2, **7**, 10.

Bleidelis, J., Shvets, A. E. and Freimanis, J. (1976). *Zh. Strukt. Khim.*, **17**, 1096–1110.

Bock, H., Rauschenbach, A., Nather, C., Kleine, M. and Bats, J. W. (1996). *Phosphorus, sulfur, silicon, related elements*, **115**, 51–83.

Bock. H., Seitz, W., Sievert, M., Kleine, M. and Bats, J. W. (1996). *Liebigs Ann.*, 1929–1940.

Boer, J. L. de, and Vos, A. (1968). *Acta Cryst.*, **24**, 720–725.

Boeyens, J. C. A. and Herbstein, F. H. (1965a). *J. Phys. Chem.*, **69**, 2153–2159.

Boeyens, J. C. A. and Herbstein, F. H. (1965b). *J. Phys. Chem.*, **69**, 2160–2176.

Botoshansky, M., Herbstein, F. H. and Kapon, M. (1994). *Acta Cryst.*, B**50**, 191–200.

Bravic, G., Bechtel, F., Gaultier, J. and Hauw, C. (1976). *Cryst. Struct. Commun.*, **5**, 1–4.

Brennan, J. S., Brown, N. M. D. and Swinton, F. L (1974). *J. Chem. Soc. Farad. Trans. I*, **70**, 1965–1970.

Briegleb, G. and Delle, H. (1960). *Z. Elektrochem.*, **64**, 347–355.

Brierley, C., Barton, R., Robertson, B. E. and Karl, N. (1982). ACA Summer Meeting, La Jolla, Abstract PB-7.

Brock, C. P., Naae, D. G., Goodland, N. and Hamor, T. A. (1978). *Acta Cryst.*, B**34**, 3691–3696.

Brown, D. S. and Wallwork, S. C. (1965). *Acta Cryst.*, **19**, 149.

Brown, D., Wallwork, S. C. and Wilson, A. (1964). *Acta Cryst.*, **17**, 168–176.

Brown, J. N., Cheung, L. D., Trefonas, L. M. and Majeste, R. J. (1974). *J. Cryst. Mol. Struct.*, **4**, 361–374.

Bruce, M. I., Snow, M. R. and Tiekink, E. R. (1987) *Acta Cryst.* C**43**, 1640–1641.

Bryce, M. R. and Davies, S. R. (1987). *Synth. Metals*, **20**, 373–374.

Bryce, M. R., Davies, S. R., Hursthouse, M. R. and Motevalli, M. (1988). *J. Chem. Soc., Perkin Trans. II*, pp. 1713–1716.

Bryce, M. R., Secco, A. S., Trotter, J. and Weiler, L. (1982). *Can. J. Chem.*, **60**, 2057–2061.

Bulgarovskaya, I. V. and Zvonkova, Z. V. (1976). *Sov. Phys. Cryst.*, **21**, 335–337; *Kristallografiya*, **21**, 597–599.

Bulgarovskaya, I. V., Smelyanskaya, E. M., Federov, Yu. G. and Zvonkova, Z. V. (1977). *Sov. Phys. Cryst.*, **22**, 104–106; *Kristallografiya*, **22**, 184–187.

Bulgarovskaya, I. V., Smelyanskoya, E. M., Federov, Yu. G. and Zvonkova, Z. V. (1974). *Sov. Phys. Cryst.*, **19**, 157–160; *Kristallografiya*, **19**, 260–265.

Bulgarovskaya, I. V., Belsky, V. K. and Vozzhennikov, V. M. (1987). *Acta Cryst.*, C**43**, 768–770.

Bulgarovskaya, I. V., Vozzhennikov, V. M., Krasavin, V. P. and Kotov, B. V. (1982). *Cryst. Struct. Commun.*, **11**, 501–504.

Bulgarovskaya, I. V., Zavodnik, V. E. and Vozzhennikhov, V. M. (1987a). *Acta Cryst.*, **C43**, 764–766.

Bulgarovskaya, I. V., Zavodnik, V. E. and Vozzhennikov, V. M. (1987b). *Acta Cryst.*, **C43**, 766–768.

Bulgarovskaya, I. V., Zavodnik, V. E., Bel'skii, V. K. and Vozzhennikov, V. M. (1989). *Sov. Phys. Cryst.*, **34**, 203–207; *Kristallografiya*, **34**, 345–352.

Bulgarovskaya, I. V., Zvonkova, Z. V. and Kolninov, O. V. (1978). *Sov. Phys. Cryst.*, **23**, 665–669; *Kristallografiya*, **23**, 1175–1182.

Bunz, U. H. F. and Enkelmann, V. (1999). *Chem. Eur. J.*, **5**, 263–266.

Burdeniuc, J., Crabtree, R. H., Rheingold, A. L. and Yap, G. P. A. (1997). *Bull. Soc. Chim. France*, **134**, 955–958.

Butler, M. A., Wudl, F. and Soos, Z. G. (1975). *Phys. Rev.*, **B12**, 4708–4719.

Carrell, H. L. and Glusker, J. P. (1997). *Struct. Chem.*, **8**, 141–147.

Carstensen-Oeser, E., Göttlicher, S. and Habermehl, G. (1968). *Chem. Ber.*, **101**, 1648–1655.

Carter, O. L., McPhail, A. T. and Sim, G. A. (1966). *J. Chem. Soc. (A)*, pp. 822–838.

Cassoux, P. and Gleizes, A. (1980). *Inorg. Chem.*, **19**, 665–672.

Catellani, M. and Porzio, W. (1991). *Acta Cryst.*, **C47**, 596–599.

Chantooni, M. K., Jr. and Britton, D. (1998). *J. Chem. Cryst.*, **28**, 329–333.

Chasseau, D. and Hauw, C. (1980). *Acta Cryst.*, **B36**, 3131–3133.

Chasseau, D. and Leroy, F. (1981). *Acta Cryst.*, **B37**, 454–456.

Chasseau, D., Gaultier, J. and Hauw, C. (1972). *Compt. rend. Acad. Sci., Paris, Ser. C*, **274**, 1434–1437.

Chasseau, D., Gaultier, J., Fabre, J. M. and Giral, L. (1982). *Acta Cryst.*, **B38**, 1632–1635.

Cherin, P. and Burack, M. (1966). *J. Phys. Chem.*, **70**, 1470–1472.

Chetkina, L. A. and Bel'skii, V. K. (2002). *Cryst. Reps.*, **47**, 581–602.

Chetkina, L. A., Popova, E. G., Kotov, B. V., Ginzburg, S. L. and Smelyanskaya, E. M. (1976). *Zh. Strukt. Khim.*, **17**, 1060–1066.

Clyburne, J. A. C., Hamilton, T. and Jenkins, H. A. (2001). *Cryst. Engin.*, **4**, 1–9.

Coates, G. W., Dunn, A. R., Henling, L. M., Dougherty, D. A. and Grubbs, R. H. (1997). *Angew. Chem. Int. Ed. Engl.*, **36**, 248–251.

Cohen-Addad, C., Consigny, M., D'Assenza, G. and Baret, P. (1988). *Acta Cryst.*, **C44**, 1924–1926.

Cohen-Addad, C., Lebars, M., Renault, A. and Baret, P. (1984). *Acta Cryst.*, **C40**, 1927–1931.

Cohen-Addad, C., Renault, A., Communandeur, G. and Baret, P. (1988). *Acta Cryst.*, **C44**, 914–916.

Collings, J. C., Batsanov, A. S., Howard, J. A. K. and Marder, T. B. (2001). *Acta Cryst.*, **C57**, 870–872.

Collings, J. C., Roscoe, K. P., Thomas, R. Ll., Batsanaov, A. S., Stimson, L. M., Howard, J. A. K. and Marder, T. B. (2001). *New J. Chem.*, **25**, 1410–1417.

Couldwell, M. C. and Prout, C. K. (1978). *J. Chem. Soc. Perkin II*, pp. 160–164.

Dahl, T. (1971a). *Acta Chem. Scand.*, **25**, 1031–1039.

Dahl, T. (1971b). *Acta Chem. Scand.*, **26**, 1569–1575.

Dahl, T. (1973). *Acta Chem. Scand.*, **27**, 995–1003.

Dahl, T. (1975a). *Acta Chem. Scand.*, **A29**, 170–174.

Dahl, T. (1975b). *Acta Chem. Scand.*, **A29**, 699–705.

Dahl, T. (1979). *Acta Chem. Scand.*, **A33**, 665–669.

Dahl, T. (1981a). *Acta Cryst.*, **B37**, 98–101.

Dahl, T. (1981b). *Acta Chem. Scand.*, **A35**, 701–705.

Dahl, T. (1988). *Acta Chem. Scand.*, **A42**, 1–7.

Dahl, T. (1990). *Acta Cryst.*, **B46**, 283–288.

Dahl, T. and Sørensen, B. (1985). *Acta Chem. Scand.*, **B39**, 423–428.

Dai, C., Nguyen, P., Marder, T. B., Scott, A. I., Clegg, W. and Viney, C. (1999). *Chem. Commun.* 2493–2494.

Damiani, A., De Santis, P., Giglio, E., Liquori, A. M. and Ripamonti, A. (1965). *Acta Cryst.*, **19**, 340–348.

Damiani, A., Giglio, E., Liquori, A. M. and Ripamonti, A. (1967). *Acta Cryst.*, **23**, 675–681.

Damiani, A., Giglio, E., Liquori, A. M., Puliti, R. and Ripamonti, A. (1967). *J. Mol. Biol.*, **23**, 113–115.

Danno, T., Kajiwara, T. and Inokuchi, H. (1967). *Bull. Chem. Soc. Jpn.*, **40**, 2793–2795.

Darocha, B. F., Titus, D. D. and Sandman, D. J. (1979). *Acta Cryst.*, B35, 2445–2448.

De, R. L., Seyerl, J. von, Zsolnai, L. and Huttner, G. (1979). *J. Organometall. Chem.*, **175**, 185–191.

Desgreniers, S., Kourouklis, G. A., Jayaraman, A., Kaplan, M. L. and Schmitt, P. H. (1985). *J. Chem. Phys.*, **83**, 480–485.

Desiraju, G. D., Paul, I. C. and Curtin, D.Y. (1977). *J. Am. Chem. Soc.*, **99**, 1594–1601.

Desiraju, G. R., Curtin, D. Y. and Paul, I. C. (1979). *Mol. Cryst. Liq. Cryst.*, **52**, 259–266.

Desiraju, G. R. and Steiner, T. (1999). The Weak Hydrogen Bond in Structural Chemistry and Biology, IUCr Monographs on Crystallography, Oxford University Press.

Dietz, K., Endres, H., Keller, H. J., Moroni, W. and Wehe, D. (1982). *Z. Naturforsch.*, **37b**, 437–442.

Doherty, R. M., Stewart, J. M., Mighell, A. D., Hubbard, C. R. and Patiadi, A. J. (1982). *Acta Cryst.*, B38, 859–863.

Dong, V., Endres, H., Keller, H. J., Moroni, W. and Nöthe, D. (1977). *Acta Cryst.*, B33, 2428–2431.

Duncan, W. A. and Swinton, F. L. (1966). *Trans. Farad. Soc.*, **62**, 1082–1089.

Dung, N.-H., Viossat, B., Lancelot, J.-C. and Robba, M. (1986). *Acta Cryst.*, C42, 843–847.

Dunitz, J. D. (1996). *Persp. Supramol. Chem.*, **2**, 1–30.

D'yachenko, O. A., Atovmyan, L. O., Kovalev, A. A. and Soboleva, S. V. (1977). *J. Struct. Chem.*, **18**, 713–719; *Zh. Strukt. Khim.*, **18**, 898–907.

Dzyabchenko, A. V., Bulgarovskaya, I. V., Zavodnik, V. E. and Stash, A. I. (1994). *Kristallografiya*, **39**, 434–438.

Dzyabchenko, A. V. and Zavodnik, V. E. (1984). *Zh. Strukt. Khim.*, **25**, 177–179.

Emge, T. J., Bryden, W. A., Cowan, D. O. and Kistenmacher, T. J. (1982). *Mol. Cryst. Liq. Cryst.*, **90**, 173–184.

Emge, T. J., Wijgul, F. M., Chappell, J. S., Bloch, A. N., Ferraris, J. P., Cowan, D. O. and Kistenmacher, T. J. (1982). *Mol. Cryst. Liq. Cryst.*, **87**, 137–161.

Ermer, O. (1988). *J. Am. Chem. Soc.*, **110**, 3747–3754.

Evans, D. L. and Robinson, W. T. (1977). *Acta Cryst.*, B33, 2891–2893.

Fenby, D. V. (1972). *Rev. Pure Appl. Chem.*, **22**, 55–65.

Ferguson, G., Mackay, I. R., Pollard, D. R. and Robertson, J. M. (1969). *Acta Cryst.*, A25, S132.

Flandrois, S., Ludolf, K., Keller, H. J., Nöthe, D., Bondeson, S. R., Soos, Z. G. and Wehe, D. (1983). *Mol. Cryst. Liq. Cryst.*, **95**, 149–164.

Fleischmann, D. E. and Tollin, G. (1965a). *Proc. Nat Acad. Sci.*, **53**, 38–46.

Fleischmann, D. E. and Tollin, G. (1965b). *Proc. Nat Acad. Sci.*, **53**, 237–242.

Fleischmann, D. E. and Tollin, G. (1965c). *Biochim. Biophys. Acta*, **94**, 248–257.

Foss, L. I., Syed, A., Stevens, E. D. and Klein, C. L. (1984). *Acta Cryst.*, C40, 272–274.

Foster, R., Iball, J., Scrimgeour, S. N. and Williams, B. C. (1976). *J. Chem. Soc. Perkin 2*, pp. 682–685.

Frankenbach, G. M., Beno, M. A. and Williams, J. M. (1991). *Acta Cryst.*, C47, 762–764.

Fritchie, C. J., Jr. (1966). *Acta Cryst.*, **20**, 892–898.

Fritchie, C.J., Jr. and Arthur, P., Jr. (1966). *Acta Cryst.*, **21**, 139–145.

Fritchie, C. J., Jr. and Johnston, R. M. (1975). *Acta Cryst.*, B31, 454–461.

Fritchie, C. J., Jr. and Trus, B. L. (1968). *J. Chem. Soc. Chem. Commun.*, pp. 833–834.

Fujii, S., Kawasaki, K., Sato, A., Fujiwara, T. and Tomita, K.-I. (1977). *Arch. Biochem. Biophys.*, **181**, 363–370.

Fujinawa, T., Goto, H., Naito, T., Inabe, T., Akutagawa, T. and Nakamura, T. (1999). *Bull. Chem. Soc. Jpn.*, **72**, 21–26.

Galigne, T. L., Fabre, J. M. and Giral, L. (1977). *Acta Cryst.*, B**33**, 3827–3831.

Gartland, G. L., Freeman, G. R. and Bugg, C. E. (1974). *Acta Cryst.*, B**30**, 1841–1849.

Gaultier, J., Hauw, C. and Breton-Lacombe, M. (1969). *Acta Cryst.*, B**25**, 231–237.

Gaultier, J. and Hauw, C. (1965). *Acta Cryst.*, **18**, 179–183.

Gebert, E., Reis, A. H., Jr., Miller, J. S., Rommelman, H. and Epstein, A. J. (1982). *J. Am. Chem. Soc.*, **104**, 4403–4410.

Geiger, W. E., Jr. and Maki, A. H. (1971). *J. Phys. Chem.*, **75**, 2387–2394.

Gieren, A., Lamm, V., Hübner, T., Rabben, M., Neidlein, R. and Droste, D. (1984). *Chem. Ber.*, **117**, 1940–1953.

Gleason, W. B. and Britton, D. (1976). *Cryst. Struct. Commun.*, **5**, 483–488.

Goldberg, I. and Shmueli, U. (1973a). *Acta Cryst.*, B**29**, 421–431.

Goldberg, I. and Shmueli, U. (1973b). *Acta Cryst.*, B**29**, 432–440.

Goldberg, I. and Shmueli, U. (1973c). *Acta Cryst.*, B**29**, 440–448.

Goldberg, I. and Shmueli, U. (1977). *Acta Cryst.*, B**33**, 2189–2197.

Griffith, G., Jackson, P. R., Kenyon-Blair, E. and Morcom, K. W. (1983). *J. Chem. Thermodynam.*, **15**, 1001–1002.

Grigg, R., Trocha-Grimshaw, J. and King, T. J. (1978). *J. Chem. Soc. Chem. Commun.*, pp. 571–572.

Groziak, M. P., Wilson, S. R., Clauson, G. R. and Leonard, N. J. (1986). *J. Am. Chem. Soc.*, **108**, 8002–8006.

Gundel, D., Sixl, H., Metzger, R. M., Heimer, N. E., Harms, R. H., Keller, H. J., Nöthe, D. and Wehe, D. (1983). *J. Chem. Phys.*, **79**, 3678–3688.

Hall, B. and Devlin, J. P. (1967). *J. Phys. Chem.*, **71**, 465–466.

Hansmann, C., Foro, S., Lindner, H. J. and Fuess, H. (1997a). *Z. Kristallogr.*, **212**, 79–80.

Hansmann, C., Foro, S., Lindner, H. J. and Fuess, H. (1997b). *Z. Kristallogr.*, **212**, 81–82.

Hanson, A. W. (1963). *Acta Cryst.*, **16**, 1147–1151.

Hanson, A. W. (1964). *Acta Cryst.*, **17**, 559–568.

Hanson, A. W. (1965a). *Acta Cryst.*, **19**, 19–26.

Hanson, A. W. (1965b). *Acta Cryst.*, **19**, 610–613.

Hanson, A. W. (1966). *Acta Cryst.*, **20**, 97–102.

Hanson, A. W. (1978). *Acta Cryst.*, B**34**, 2339–2341.

Harata, K., Aono, T., Sakabe, N. and Tanaka, J. (1972). *Acta Cryst.*, A**28**, S14.

Harms, R. H., Keller, H. J., Nöthe, D., Werner, M., Gundl, D., Sixl, H., Soos, Z. G. and Metzger, R. M. (1981). *Mol. Cryst. Liq. Cryst.*, **65**, 179–196.

Hazell, A. C. (1978). *Acta Cryst.*, B**34**, 466–471.

Herbstein, F. H. (1971). "Crystalline π-Molecular Compounds: Chemistry, Spectroscopy, and Crystallography", in *Perspectives in Structural Chemistry*, edited by J. D. Dunitz and J. A. Ibers, Wiley, New York, Vol. IV, pp. 166–395.

Herbstein, F. H. (2001). "Varieties of polymorphism." Advanced Methods in Strucutre Analysis, pp. 114–154. Eds. R. Kuzel and J. Hasak, CSCA. http://www.xray/ecm/book/.

Herbstein, F. H. and Kaftory, M. (1975a). *Acta Cryst.*, B**31**, 60–67.

Herbstein, F. H. and Kaftory, M. (1975b). *Acta Cryst.*, B**31**, 68–75.

Herbstein, F. H. and Kaftory, M. (1976). *Acta Cryst.*, B**32**, 387–396.

Herbstein, F. H. and Reisner, G. M. (1984). *Acta Cryst.*, C**40**, 202–204.

Herbstein, F. H. and Samson, S. (1994). *Acta Cryst.*, B**50**, 182–191.

Herbstein, F. H. and Snyman, J. A. (1969). *Phil. Trans. Roy. Soc. Lond.*, A**264**, 635–666.

Herbstein, F. H., Kaftory, M. and Regev, H. (1976). *J. Appl. Cryst.*, **9**, 361–364.

Herbstein, F. H., Kapon, M. and Reisner, G. M. (1986). *Acta Cryst.*, B**42**, 181–187.

Herbstein, F. H., Kapon, M., Rzonzew, G. and Rabinovich, D. (1978). *Acta Cryst.*, B**34**, 476–481.

Herbstein, F. H., Marsh, R. E. and Samson, S. (1994). *Acta Cryst.*, B**50**, 174–181.

Herriott, J. R., Camerman, A. and Deranleau, D. A. (1974). *J. Am. Chem. Soc.*, **96**, 1585–1589.

Hertel, E. and Bergk, H. W. (1936). *Z. Phys. Chem.*, B**33**, 319–333.

Hertel, E. and Frank, H. (1934). *Z. Physik. Chem.*, B**27**, 460–466.

Hertel, E. and Schneider, K. (1931). *Z. Physik. Chem.*, B**13**, 387–399.

Hill, H. A. O., MacFarlane, A. J. and Williams, R. J. P. (1967). *J. Chem. Soc. Chem. Commun.* p. 605.

Hinrichs, W. and Klar, G. (1982). *J. Chem. Res. (S)*, pp. 336–337.

Hoffman, S. K., Corvan, P. J., Singh, P, Sethulekshmi, C. N., Metzger, R. M. and Hatfield, W. E. (1983). *J. Am. Chem. Soc.*, **105**, 4608–4617.

Hoier, H., Zacharias, D. E., Carrell, H. and Glusker, J. P. (1993). *Acta Cryst.*, C**49**, 523–526.

Hughes, R. C. and Soos, Z. G. (1968). *J. Chem. Phys.*, **48**, 1066–1076.

Hursthouse, M. B., Smith, V. B. and Massey, A. G. (1977). *J. Fluor. Chem.*, **10**, 145–156.

Ikemoto, I., Chikaishi, K., Yakushi, K. and Kuroda, H. (1972). *Acta Cryst.*, B**28**, 3502–3506.

Ikemoto, I., Yakushi, K. and Kuroda, H. (1970). *Acta Cryst.*, B**26**, 800–806.

Irngartinger, H. and Goldmann, A. (1978). *Z. Kristallogr.*, **149**, 97.

Ito, T., Minobe, M. and Sakurai, T. (1970). *Acta Cryst.*, B**26**, 1145–1151.

Iwasaki, F. and Saito, Y. (1970). *Acta Cryst.*, B**26**, 251–260.

Iwasaki, F., Hironaka, S., Yamazaki, N. and Kobayashi, K. (1992). *Bull. Chem. Soc. Jpn.*, **65**, 2180–2186.

Jackson, P. R. and Morcom, K. W. (1986). *J. Chem. Thermodynam.*, **18**, 75–80.

Jankowski, W. and Gdaniec, M. (2002). *Acta Cryst.*, C**58**, o181–o182.

Jones, N. B. and Marsh, R. E. (1962). *Acta Cryst.*, **15**, 809–810.

Kabuto, C., Suzuki, T., Yamashita, Y. and Mukai, T. (1986). *Chem. Letts.*, pp. 1433–1436.

Kamenar, B. and Prout, C. K. (1965). *J. Chem. Soc.*, pp. 4838–4851.

Kamenar, B., Prout, C. K. and Wright, J. D. (1965). *J. Chem. Soc.*, pp. 4851–4867.

Kamenar, B., Prout, C. K. and Wright, J. D. (1966). *J. Chem. Soc.*, (A) pp. 661–664.

Kaminskii, V. F., Shibaeva, R. P., Aldoshina, M. Z., Lyubovskaya, R. N. and Khidekel, M. L. (1979). *J. Struct. Chem.*, **20**, 130–133; *Zh. Strukt. Khim.*, **20**, 157–160.

Karl, N., Binder, W., Kollet, P. and Stezowski, J. J. (1982). *Acta Cryst.*, B**38**, 2919–2921.

Karl, N., Ketterer, W. and Stezowski, J. J. (1982). *Acta Cryst.*, B**38**, 2917–2919.

Karle, J., Karle, I. L. and Mitchell, D. (1969). *Acta Cryst.*, B**25**, 866–871.

Karlsson, R. (1972). *Acta Cryst.*, B**28**, 2358–2364.

Kasper, J. S. and Interrante, L. V. (1976). *Acta Cryst.*, B**32**, 2914–2916.

Keller, H. J. and Soos, Z. G. (1985). *Top. Curr. Chem.*, **127**, 169–216.

Keller, H. J., Leichert, I., Megnamisi-Belombe, M., Nothe, D. and Weiss, J. (1977). *Z. anorg. allgem. Chem.*, **429**, 231–236.

Keller, H. J., Moroni, W., Nöthe, D., Scherz, M. and Weiss, J. (1978). *Z. Naturforsch.*, **33b**, 838–842.

Kemmer, T., Sheldrick, W. S. and Brockmann, H. (1976). *Angew. Chem. Int. Ed. Engl.*, **15**, 115.

Kirner, J. F., Reed, C. A. and Scheidt, W. R. (1977). *J. Am. Chem. Soc.*, **99**, 1093–1101.

Kistenmacher, T. J., Emge, T. J., Bloch, A. N. and Cowan, D. O. (1982). *Acta Cryst.*, B**38**, 1193–1199.

Kistenmacher, T. J., Phillips, T. E., Cowan, D. O., Ferraris, J. P., Bloch, A. N. and Poehler, T. O. (1976). *Acta Cryst.*, B**32**, 539–547.

Kobayashi, H. and Nakayama, J. (1981). *Bull. Chem. Soc. Jpn.*, **54**, 2408–2411.

Kobayashi, H. (1973). *Bull. Chem. Soc. Jpn.*, **46**, 2945–2949.

Kobayashi, H. (1987). *Synth. Metals*, **19**, 475–480.

Kodama, T. and Kumakura, S. (1974a). *Bull. Chem. Soc. Jpn.*, **47**, 1081–1084.

Kodama, T. and Kumakura, S. (1974b). *Bull. Chem. Soc. Jpn.*, **47**, 2146–2151.

Kofler, A. (1940). *Z. Elektrochem.*, **50**, 200–207.

Komorowski, L., Krajewska, A. and Pignon, K. (1976). *Mol. Cryst. Liq. Cryst.*, **36**, 337–348.

Konno, M., Saito, Y., Yamada, K. and Kawazura, H. (1980). *Acta Cryst.*, **B36**, 1680–1683.

Kortüm, G., Vogel, W. and Andrussow, K. (1961). "Dissociation Constants of Organic Acids in Aqueous Solutions." Butterworths, London.

Kosower, E.M. (1966). In "Flavins and Flavoproteins", E. C. Slater, editor, Elsevier, Amsterdam, pp. 1–14.

Kozawa, K. and Uchida, T. (1979). *Bull. Chem. Soc. Jpn.*, **52**, 1555–1558.

Kozawa, K. and Uchida, T. (1983). *Acta Cryst.*, **C39**, 1233–1235.

Kremann, R., Sutter, S., Sitte, F., Strzelba, H. and Dobotzky, A. (1922). *Monatsh.*, **43**, 269–313.

Kruif, C. G. de, Smit, E. J. and Govers, H. A. J. (1981). *J. Chem. Phys.*, **74**, 5838–5841.

Kuhn, R. and Strobele, R. (1937). *Ber. Deut. Chem. Ges.*, **73**, 753–760.

Kumakura, S., Iwasaki, F. and Saito, Y. (1967). *Bull. Chem. Soc. Jpn.*, **40**, 1826–1833.

Kuo, M. C., Dunn, J. B. R. and Fritchie, C. J., Jr. (1974). *Acta Cryst.*, **B30**, 1766–1771.

Lakshman, M. H., Kole, P. L., Chaturvedi, S., Saugier, J. H., Yeh, H. J. C., Glusker, J. P., Carrell, H. L., Katz, A. K., Afshar, C. E., Dashwood, W-M, Kenniston, G. and Baird, W. M. (2000). *J. Am. Chem. Soc.*, **122**, 12629–12636.

Lange, F. (1924). *Z. Physik. Chem.*, **110**, 343–362.

Langhoff, C. A. and Fritchie, C. J., Jr. (1970). *J. Chem. Soc. Chem. Commun.*, pp. 20–21.

Larsen, F. K., Little, R. G. and Coppens, P. (1975). *Acta Cryst.*, **B31**, 430–440.

Law, K. S. and Prasad, P. N. (1982). *J. Chem. Phys.*, **77**, 1107–1113.

Lee, D. and Wallwork, S. C. (1978). *Acta Cryst.*, **B34**, 3604–3608.

Le Bars-Combe, M., Chion, B. and Lajzerowicz-Bonneteau, J. (1979). *Acta Cryst.*, **B35**, 913–920.

Lefebvre, J., Miniewicz, A. and Kowal, A. (1989). *Acta Cryst.*, **C45**, 1372–1376.

Lehmann, R. E. and Kochi, J. K. (1991). *J. Am. Chem. Soc.*, **113**, 501–512.

Lin, T. and Naae, D. G. (1978). *Tetr. Letts.*, pp. 1653–1656.

Lindeman, S. V., Andrianov, V. G., Kravcheni, S. G., Potapov, V. M., Potekhin, K. A. and Struchkov, Yu. T. (1981). *J. Struct. Chem. USSR*, **22**, 578–585; *Zh. Strukt. Khim.*, **23**, 123–131.

Lisensky, G. C., Johnson, C. K. and Levy, H. A. (1976). *Acta Cryst.*, **B32**, 2188–2197.

Lloyd, L. L. and Sudborough, J. J. (1899). *J. Chem. Soc.*, **75**, 580.

Lobovskaya, R. M., Shibaeva, R. P. and Ermenk, O. N. (1983). *Sov. Phys. Cryst.*, **28**, 161–163.

Lopez-Morales, M. E., Soriano-Garcia, M., Gomez-Lara, J. and Toscano, R. A. (1985). *Mol. Cryst. Liq. Cryst.*, **125**, 421–427.

Lork, E., Mews, R., Shakirov, M. M., Watson, P. G. and Zibarev, A. V. (2001). *Eur. J. Inorg. Chem.*, pp. 2123–2134.

Mackay, I. R., Robertson, J. M. and Sime, J. G. (1969). *J. Chem. Soc., Chem. Commun.*, pp. 1470–1471.

Mariezcurrena, R. A., Russi, S., Mombrú, A. W., Suescun, L., Pardo, H., Tombesi, O. L. and Frontera, M. A. (1999). *Acta Cryst.*, **C55**, 1170–1173.

Marsh, R. E. (1986). *Acta Cryst.*, **B42**, 193–198.

Marsh, R. E. (1990). *Acta Cryst.*, **C46**, 1356–1357.

Martin, R. H. (1974). *Angew. Chem. Int. Ed. Engl.*, **13**, 649–660.

Masnovi, J. M., Kochi, J. K., Hilinski, E. F. and Rentzepis, P. M. (1985). *J. Phys. Chem.*, **89**, 5387–5395.

Matsumoto, N., Nonaka, Y., Kida, S., Kawano and Ueda, I. (1979). *Inorg. Chim. Acta*, **37**, 27–36.

Matsunaga, Y. (1964). *J. Chem. Phys.*, **41**, 1609–1613..

Matsunaga, Y. (1966). *Nature, Lond.*, **211**, 183–184.

Matsunaga, Y. (1973). *Bull. Chem. Soc. Jpn.*, **46**, 998–999.

Matsunaga, Y. (1978). *Bull. Chem. Soc. Jpn.*, **51**, 3071–3072.

Matsunaga, Y. and Osawa, R. (1974). *Bull. Chem. Soc. Jpn.*, **47**, 1589–1592.

Matsunaga, Y. and Saito, G. (1972). *Bull. Chem. Soc. Jpn.*, **45**, 963–964.

Matsunaga, Y. and Usui, R. (1980). *Bull. Chem. Soc. Jpn.*, **53**, 3085–3088.

Matsunaga, Y., Osawa, E. and Osawa, R. (1975). *Bull. Chem. Soc. Jpn.*, **48**, 37–40.

Matsunaga, Y., Saito, G. and Sakai, N. (1974). *Bull. Chem. Soc. Jpn.*, **47**, 2873–2874.

Maverick, E., Trueblood, K. N. and Bekoe, D. A. (1978). *Acta Cryst.*, **B34**, 2777–2781.

Mayerle, J. J. (1977). *Inorg. Chem.*, **16**, 916–919.

Mayerle, J. J. and Torrance, J. B. (1981). *Acta Cryst.*, **B37**, 2030–2034.

Mayerle, J. J., Torrance, J. B. and Crowley, J. I. (1979). *Acta Cryst.*, **B35**, 2988–2995.

McLaughlin, E. and Messer, C. E. (1966). *J. Chem. Soc.* (A), pp. 1106–1110.

Megnamisi–Belombe, M. and Endres, H. (1982). *Acta Cryst.*, **B38**, 1826–1828.

Merski, J. and Eckhardt, C. J. (1981). *J. Chem. Phys.*, **75**, 3705–3718.

Metzger, R. M., Heimer, N. E., Gundel, D., Sixl, H., Harms, R. H., Keller, H. J., Nöthe, D. and Wehe, D. (1982). *J. Chem. Phys.*, **77**, 6203–6214.

Meyer, K. H. (1909). *Chem. Ber.*, **42**, 1149–1153.

Mikes, F., Boshart, G. and Gil-Av, E. (1976). *J. Chromatogr.*, **122**, 205–221.

Miller, J. S., Zhang, J. H., Reiff, W. M., Dixon, D. A., Preston, L. D., Reis, A. H., Jr., Gebert, E., Extine, M., Troup, J., Epstein, A. J. and Ward, M. D. (1987). *J. Phys. Chem.*, **91**, 4344–4360.

Mitkevich, V. V. and Sukhodub, L. F. (1987). *Sov. Phys. Cryst.*, **31**, 483–484; *Kristallografiya*, **31**, 815–817.

Mori, T. and Inokuchi, H. (1986). *Solid State Commun.*, **59**, 355–359.

Mori, T and Inokuchi, H. (1987). *Bull. Chem. Soc. Jpn.*, **60**, 402–404.

Mori, T., Wu, P., Imaeda, K., Enoki, T., Inokuchi, H. and Saito, G. (1987). *Synth. Metals*, **19**, 545–550.

Morosin, B. (1976). *Acta Cryst.*, **B32**, 1176–1179.

Morosin, B. (1978). *Acta Cryst.*, **B34**, 1905–1909.

Morosin, B., Plastas, H. J., Coleman, L. B. and Stewart, J. M. (1978). *Acta Cryst.*, **B34**, 540–543.

Munnoch, P. J. and Wright, J. D. (1974). *J. Chem. Soc. Perkin II*, pp. 1397–1400.

Munnoch, P. J. and Wright, J. D. (1975). *J. Chem. Soc. Perkin II*, pp. 1071–1074.

Naae, D. G. (1979). *Acta Cryst.*, **B35**, 2765–2768.

Nagata, H., In, Y., Doi, M. and Ishida, T. (1995). *Acta Cryst.*, **B51**, 1051–1068.

Nakagawa, H., Tanaka, H., Yamada, K-I. and Kawazura, H. (1982). *J. Phys. Chem.*, **86**, 2311–2314.

Nakai, H., Shiro, M., Ezumi, K., Sakata, S. and Kubota, T. (1976). *Acta Cryst.*, **B32**, 1827–1833.

Nakamoto, K. (1952). *J. Am. Chem. Soc.*, **74**, 1739–1742.

Nakamura, K., Kai, Y., Yasuoka, N. and Kasai, N. (1981). *Bull. Chem. Soc. Jpn.*, **54**, 3300–3303.

Nakasuji, K., Sasaki, M., Kotani, T., Murata, I., Enoki, T., Imaeda, K., Inokuchi, H., Kawamoto, A. and Tanaka, J. (1987). *J. Am. Chem. Soc.*, **109**, 6970–6975.

Nakasuji, K., Sasaki, M., Murata, I., Kawamoto, A. and Tanaka, J. (1988). *Bull. Chem. Soc. Jpn.*, **61**, 4461–4463.

Newman, M. S. and Lednicer, D. (1956). *J. Am. Chem. Soc.*, **78**, 4765–4770.

Newman, M. S. and Lutz, W. B. (1956). *J. Am. Chem. Soc.*, **78**, 2469–2473.

Niimura, N., Ohashi, Y. and Saito, Y. (1968). *Bull. Chem. Soc. Jpn.*, **41**, 1815–1820.

Nitta, I., Seki, S., Chihara, H. and Suzuki, K. (1951 *Sci. Pap. Osaka Univ.* No. 29.).

Numan, H., Helder, R. and Wynberg, H. (1976). *Rec. Trav. Chim. Pays-bas*, **95**, 211–212.

Ohashi, Y. (1973). *Acta Cryst.*, **B29**, 2863–2871.

Ohashi, Y., Iwasaki, H. and Saito, Y. (1967). *Bull. Chem. Soc. Jpn.*, **40**, 1789–1796.

Ohmasa, M., Kinoshita, M. and Akamatu, H. (1971). *Bull. Chem. Soc. Jpn.*, **44**, 391–395.

Ott, J. B., Goates, J. R and Cardon, D. L (1976). *J. Chem. Thermodynam.*, **8**, 505–512.

Overell, J. S. W. and Pawley, G. S. (1982). *Acta Cryst.*, **B38**, 1966–1972.

Pace, L. J., Ulman, A. and Ibers, J. A. (1982). *Inorg. Chem.*, **21**, 199–207.

Pascard, R. and Pascard-Billy, C. (1972). *Acta Cryst.*, B**28**, 1926–1935.

Pasimeni, L., Guella, G., Corvaja, C., Clemente, D.A. and Vincentini, M. (1983). *Mol. Cryst. Liq. Cryst.*, **91**, 25–38.

Patil, A. O., Curtin, D. Y. and Paul, I. C. (1984). *J. Am. Chem. Soc.*, **106**, 4010–4015.

Patil, A. O., Curtin, D. Y. and Paul, I. C. (1986). *J. Chem. Soc. Perkin Trans.2*, pp. 1687–1692.

Patil, A. O., Pennington, W. T., Desiraju, G. R., Curtin, D. Y. and Paul, I. C. (1986). *Mol. Cryst. Liq. Cryst.*, **134**, 279–304.

Patil, A. O., Wilson, S. R., Curtin, D. Y. and Paul, I. C. (1984). *J. Chem. Soc. Perkin Trans. 2*, pp. 1107–1110.

Patrick, C. R. and Prosser, G. S. (1960). *Nature, Lond.*, **187**, 1021.

Pennington, W. T., Patil, A. O., Curtin, D. Y. and Paul, I. C. (1986), *J. Chem. Soc. Perkin Trans.2*, pp. 1693–1700.

Pereira, J. F. and Tollin, G. (1967). *Biochim. Biophys. Acta*, **143**, 79–87.

Perepichka, I. F., Kuz'mina, L. G., Perepichka, D. F., Bryce, M. R., Goldenberg, L. M., Popov, A. F. and Howard, J. A. K. (1998). *J. Org. Chem.*, **63**, 6484–6493.

Perepichka, I. F., Popov, A. F., Orekhova, T. V., Bryce, M. R., Andrievskii, A. M., Batsanov, A. S. and Howard, J. A. K. (2000). *J. Org. Chem.*, **65**, 3053–3063.

Perrin, D. (1965). "Dissociation Constants of Organic Acids in Aqueous Solutions." Butterworths, London.

Popova, E. G., Chetkina, L. A. and Bespalov, B. P. (1981). *J. Struct. Chem. USSR*, **22**, 586–589; *Zh. Strukt. Khim.*, **22**, 132–136.

Popova, E. G., Chetkina, L. A. and Kotov, B. V. (1978). *J. Struct. Chem. USSR*, **19**, 914–919; *Zh. Strukt. Khim.*, **19**, 1071–1079.

Potenza, J. and Mastropaolo, D. (1975). *Acta Cryst.*, B**31**, 2527–2529.

Povot'eva, Z. P., Chetkina, L. A. and Bespalov, B. P. (1981). *J. Struct. Chem. USSR*, **22**, 386–390; *Zh. Strukt. Khim.*, **22**, 94–99.

Powell, H. M. and Huse, G. (1943). *J. Chem. Soc.*, pp. 435–437.

Powell, H. M., Huse, G. and Cooke, P. W. (1943). *J. Chem. Soc.*, pp. 153–157.

Prout, C. K. and Castellano, E. E. (1970). *J. Chem. Soc.* (A), pp. 2775–2778

Prout, C. K. and Powell, H. M. (1965). *J. Chem. Soc.*, pp. 4882–4887.

Prout, C. K. and Tickle, I. J. (1973a). *J. Chem. Soc. Perkin II*, pp. 731–734.

Prout, C. K. and Tickle, I. J. (1973b). *J. Chem. Soc. Perkin II*, pp. 734–737.

Prout, C. K. and Tickle, I. J. (1973c). *J. Chem. Soc. Perkin II*, pp. 1212–1215.

Prout, C. K. and Wallwork, S. C. (1966). *Acta Cryst.*, **21**, 449–450.

Prout, C. K. and Wheeler, A. G. (1967). *J. Chem. Soc.* (A), pp. 469–475.

Prout, C. K., Morley, T., Tickle, I. J. and Wright, J. D. (1973). *J. Chem. Soc. Perkin II*, pp. 523–527.

Prout, C. K., Tickle, I. J. and Wright, J. D. (1973). *J. Chem. Soc. Perkin II*, pp. 528–530.

Qi, F., Yu, W.-T., Hong, L., Lin, X.-Y. and Yu, Z.-P. (1996). *Acta Cryst.*, C**52**, 2274–2277.

Reddy, D. S., Goud, B. S., Panneerselvam, K. and Desiraju, G. R. (1993). *J. Chem. Soc., Chem. Commun.*, pp. 663–664.

Robertson, B. E. and Stezowski, J. J. (1978). *Acta Cryst.*, B**34**, 3005–3011.

Rubio, P., Florencio, F., Garcia-Blanco, S. and Rodriguez, J. G. (1985). *Acta Cryst.*, C**41**, 1797–1799.

Sahaki, M., Yamada, H., Yoshioka, H. and Nakatsu, K. (1976). *Acta Cryst.*, B**32**, 662–664.

Saito, G. and Matsunaga, Y. (1971). *Bull. Chem. Soc. Jpn.*, **44**, 3328–3335.

Saito, G. and Matsunaga, Y. (1972). *Bull. Chem. Soc. Jpn.*, **45**, 2214–2215.

Saito, G. and Matsunaga, Y. (1973a). *Bull. Chem. Soc. Jpn.*, **46**, 714–718.

Saito, G. and Matsunaga, Y. (1973b). *Bull. Chem. Soc. Jpn.*, **46**, 1609–1616.

Saito, G. and Matsunaga, Y. (1974). *Bull. Chem. Soc. Jpn.*, **47**, 1020–1021.

Sakanoue, S., Yasuoka, N., Kasai, N. and Kakudo, M. (1971). *Bull. Chem. Soc. Jpn.*, **44**, 1–8.

Sakurai, T. (1965). *Acta Cryst.*, **19**, 320–330.

Sakurai, T. (1968). *Acta Cryst.*, B**24**, 403–412.

Sakurai, T. and Tagawa, H. (1971). *Acta Cryst.*, B**27**, 1453–1459.

Sandman, D. J., Grammatica, S. J., Holmes, T. J. and Richter, A. F. (1980). *Mol. Cryst. Liq. Cryst.*, **59**, 241–252.

Sato, A., Okada, M., Saito, K. and Sorai, M. (2001). *Acta Cryst.*, C**57**, 564–565.

Scarbrough, F. E., Shieh, H.-S. and Voet, D. (1976). *Proc. Nat. Acad. Sci.*, **73**, 3807–3811.

Scarbrough, F. E., Shieh, H.-S. and Voet, D. (1977). *Acta Cryst.*, B**33**, 2512–2523.

Scheidt, W. R. and Reed, C. A. (1978). *Inorg. Chem.*, **17**, 710–714.

Scheidt, W. R., Kastner, M. E. and Hatano, K. (1978). *Inorg. Chem.*, **17**, 706–710.

Schmitt, R. D., Wing, R. M. and Maki, A. H. (1969). *J. Am. Chem. Soc.*, **91**, 4394–4401.

Schreiner, E. (1925). *Z. physik. Chem.*, **117**, 57–87.

Shaanan, B. and Shmueli, U. (1980). *Acta Cryst.*, B**36**, 2076–2082.

Shaanan, B., Shmueli, U. and Rabinovich, D. (1976). *Acta Cryst.*, B**32**, 2574–2580.

Shah, M. C. and Baughman, M. C. (1994). *Acta Cryst.*, C**50**, 1114–1117.

Shchlegova, T. M., Starikova, Z. A., Trunov, V. K., Lantratova, O. B. and Pokrovskaya, I. E. (1981). *J. Struct. Chem.*, **22**, 553–557; *Zh. Strukt. Khim.*, **22**, 93–97.

Shibaeva, R. P. and Yarochkina, O. Y. (1975). *Sov. Phys. Dokl.*, **20**, 304–305.

Shipley, G. G. and Wallwork, S. C. (1967a). *Acta Cryst.*, **22**, 585–592.

Shipley, G. G. and Wallwork, S. C. (1967b). *Acta Cryst.*, **22**, 593–601.

Shkolnik, G. M. and Geiger, W. E., Jr. (1966). *Inorg. Chem.*, **5**, 1020–1025.

Shmucli, U. and Goldberg, I. (1974). *Acta Cryst.*, B**30**, 573–578.

Shmueli, U. and Mayorzik, H. (1980). Abstract 1-A-37, ECM-6, Barcelona.

Shvets, A. E., Bleidelis, Ya. Ya. and Freimanis, Ya. F. (1974). *J. Struct. Chem. USSR*, **15**, 430–433; *Zh. Strukt. Khim.*, **15**, 504–508.

Shvets, A. E., Bleidelis, Ya. Ya. and Freimanis, Ya. F. (1975a). *Zh. Strukt. Khim.*, **16**, 98–103.

Shvets, A. E., Bleidelis, Ya. Ya. and Freimanis, Ya. F. (1975b). *J. Struct. Chem. USSR*, **16**, 386–390; *Zh. Strukt. Khim.*, **16**, 604–606.

Shvets, A. E., Bleidelis, Ya. Ya. and Freimanis, Ya. F. (1975c). *J. Struct. Chem. USSR*, **16**, 592–596; *Zh. Strukt. Khim.*, **16**, 640–644.

Shvets, A. E., Bleidelis, Ya. Ya. and Freimanis, Ya. F. (1975d). *Zh. Strukt. Khim.*, **16**, 415–419.

Shvets, A. E., Bleidelis, Ya. Ya., Freimanis, Ya. F. and Bundule, M. F. (1978). *J. Struct. Chem. USSR*, **18**, 84–87; *Zh. Strukt. Khim.*, **18**, 107–111.

Shvets, A. E., Bleidelis, J. J., Markava, E. J., Freimanis, J. F. and Kanepae, D. V. (1980). *J. Struct. Chem.*, **21**, 559–563; *Zh. Strukt. Khim.*, **21**, 190–195.

Shvets, A. E., Malmanis, A. J., Freimanis, Ya. F., Bleidelis, Ya. Ya. and Dregeris, J. F. (1979). *J. Struct. Chem. USSR*, **20**, 414–419; *Zh. Strukt. Khim.*, **20**, 491–497.

Siegmund, W. (1908). *Monatsh.*, **29**, 1089–1109.

Singh, N. B., McWhinnie, W. R., Ziolo, R. F. and Jones, C. H. W. (1984). *J. Chem. Soc., Dalton Trans.*, pp. 1267–1273.

Singhabandhu, A., Robinson, P. D., Fang, J. H. and Geiger, W. E., Jr. (1975). *Inorg. Chem.*, **14**, 318–321.

Smith, V. B. and Massey, A. G. (1969). *Tetrahedron*, **25**, 5495–5501.

Soos, Z. G., Keller, H. J., Ludolf, K., Queckbörner, J., Wehe, D. and Flandrois, S. (1981). *J. Chem. Phys.*, **74**, 5287–5294.

Soos, Z. G., Keller, H. J., Moroni, W. and Nothe, D. (1977). *J. Am. Chem. Soc.*, **99**, 5040–5044.

Soos, Z. G., Keller, H. J., Moroni, W. and Nothe, D, (1978). *Ann. N. Y. Acad. Sci.*, **313**, 442–458.

Soos, Z. G., Mazumdar, S. and Cheung, T. P. P. (1979). *Mol. Cryst. Liq. Cryst.*, **52**, 92–102.

Soriano-Garcia, M., Toscano, R. A., Robles Martinez, J. G., Salméron, U. A. and Lezama, R. R. (1989). *Acta Cryst.*, **C45**, 2442–2444.

Staab, H. A., Herz, C. P., Krieger, C. and Rentea, M. (1983). *Chem. Ber.*, **116**, 3813–3830.

Starikova, Z. A., Shchegoleva, T. M., Trunov, V. K., Lantratova, O. B. and Pokrovskaya, J. E. (1980). *J. Struct. Chem.*, **21**, 181–184; *Zh. Strukt. Khim.*, **21**, 73–76.

Stephens, F. S. and Vagg, R. S. (1980). *Inorg. Chim. Acta*, **43**, 77–82.

Stephens, F. S. and Vagg, R. S. (1981). *Inorg. Chim. Acta*, **51**, 163–167.

Stezowski, J. J. (1980). *J. Chem. Phys.*, **73**, 538–547.

Stezowski, J. J., Binder, W. and Karl, N. (1982). *Acta Cryst.*, B**38**, 2912–2914.

Stezowski, J. J., Stigler, R.-D. and Karl, N. (1986). *J. Chem. Phys.*, **84**, 5162–5170.

Sugano, T., Hashida, T., Kobayashi, A., Kobayashi, H. and Kinoshita, M. (1988). *Bull. Chem. Soc. Jpn.*, **61**, 2303–2308.

Surcouf, E. and Delettré, J. (1978). *Acta Cryst.*, B**34**, 2173–2176.

Suzuki, K. and Seki, S. (1953). *Bull. Chem. Soc. Jpn.*, **26**, 372–380.

Suzuki, T., Fujii, H., Yamashita, Y., Kabuto, C., Tanaka, S., Harasawa, M., Mukai, T. and Miyashi, T. (1992). *J. Am. Chem. Soc.*, **114**, 3034–3043.

Suzuki, T., Kabuto, C., Yamashita, Y., and Mukai, T. (1987). *Bull. Chem. Soc. Jpn.*, **60**, 2111–2115.

Swinton, F. L. (1974). "Interactions in binary systems containing aromatic fluorocarbons", Molecular Complexes, edited by R. Foster, London, Elek, Vol. II, pp. 63–106.

Szent-Gyorgi, A. (1960). "An Introduction to a Submolecular Biology", New York: Academic Press, p. 197.

Tachikawa, N., Yakushi, K. and Kuroda, H. (1974). *Acta Cryst.*, B**30**, 2770–2772.

Takahashi, N., Yakushi, K., Ishii, K. and Kuroda, H. (1976). *Bull. Chem. Soc. Jpn.*, **49**, 182–187.

Talukdar, A. N. and Chaudhuri, B. (1976). *Acta Cryst.*, B**32**, 803–808.

Tamamura, T., Yamane, T., Yasuoka, N. and Kasai, N. (1974). *Bull. Chem. Soc. Jpn.*, **47**, 832–837.

Tamura, H. and Ogawa, K. (1977). *Cryst. Struct. Commun.*, **6**, 517–520.

Thewalt, U. and Bugg, C. E. (1972). *Acta Cryst.*, B**28**, 82–92.

Thozet, A. and Gaultier, J. (1977a). *Acta Cryst.*, B**33**, 1052–1057.

Thozet, A. and Gaultier, J. (1977b). *Acta Cryst.*, B**33**, 1058–1063.

Tickle, I. J. and Prout, C. K. (1973a). *J. Chem. Soc. Perkin II*, pp. 720–723.

Tickle, I. J. and Prout, C. K. (1973b). *J. Chem. Soc., Perkin II*, pp. 724–727.

Tickle, I. J. and Prout, C. K. (1973c). *J. Chem. Soc. Perkin II*, pp. 727–731.

Tillberg, O. and Norrestam, R. (1972). *Acta Cryst.*, B**28**, 890–898.

Tollin, G. (1968). In "Molecular Associations in Biology," *Proceedings of an International Symposium*, B. Pullman, editor, Academic Press, New York, pp. 393–409.

Torrance, J. B., Mayerle, J. J., Lee, V. J. and Bechgaard, K. (1979). *J. Am. Chem. Soc.*, **100**, 4747–4748.

Torrance, J. B., Vazquez, J. E., Mayerle, J. J. and Lee, V. Y. (1981). *Phys. Rev. Letts.*, **46**, 253–257.

Toupet, L. and Karl, N. (1995). *Acta Cryst.*, C**51**, 249–251.

Toyoda, J., Oda, A., Murata, I., Kawamoto, A., Tanaka, J. and Nakasuji, K. (1993). *Bull. Chem. Soc. Jpn.*, **66**, 2115–2117.

Treibs, A. (1929). *Ann.*, **476**, 1–60.

Truong, K. D. and Bandrauk, A. D. (1977). *Can. J. Chem.*, **55**, 3712–3716.

Tsuchiya, H., Marumo, F. and Saito, Y. (1973). *Acta Cryst.*, B**29**, 659–666.

Vangala, V. R., Nangia, A. and Lynch, V. M. (2002). *Chem. Commun.*, pp. 1304–1305.

Viossat, B., Dung, N.-G. and Daran, J. C. (1988). *Acta Cryst.*, C**44**, 1797–1800.

Viossat, B., Tomas, A., Dung, N.-G., Mettey, Y. and Viervond, J. -M. (1995). *Acta Cryst.*, C**51**, 1896–1898.

Visser, R. J. J., Bouwmeester, H. J. M., Boer, J. L. de. and Vos, A. (1990). *Acta Cryst.*, **C46**, 852–856.

Voet, D. and Rich, A. (1971). *Proc. Nat. Acad. Sci.*, **68**, 1151–1156.

Weck, M., Dunn, A. R., Matsumoto, K., Coates, G. W., Lobkovsky, E. and Grubbs, R. H. (1999). *Angew. Chem. Int. Ed. Engl.*, **38**, 2741–2745.

Wells, J. L., Trus, B. L., Johnston, R. M., Marsh, R. E. and Fritchie, C. J., Jr. (1974). *Acta Cryst.*, **B30**, 1127–1134.

Wentworth, W. E., Limero, T. and Chen, E. C. M. (1987). *J. Phys. Chem.*, **91**, 241–245.

Wilkerson, A. K., Chodak, J. B. and Strouse, C. E. (1975). *J. Am. Chem. Soc.*, **97**, 3000–3004.

Williams, J. H. (1993). *Acc. Chem. Res.*, **26**, 593–598.

Williams, J. H., Cockcroft, J. K. and Finch, A. N. (1992). *Angew. Chem. Int. Ed. Engl.*, **31**, 1655–1657.

Williams, J. M., Beno, M. A., Wang, H. H., Leung, P. C. W., Emge, T. J., Geiser, U. and Carlson, K. D. (1985). *Acc. Chem. Res.*, **18**, 261–267.

Williams, R. M. and Wallwork, S. C. (1966). *Acta Cryst.*, **21**, 406–412.

Williams, R. M. and Wallwork, S. C. (1967). *Acta Cryst.*, **23**, 448–455.

Williams, R. M. and Wallwork, S. C. (1968). *Acta Cryst.*, **B24**, 168–174.

Wing, R. M. and Schlupp, R. L. (1970). *Inorg. Chem.*, **9**, 471–475.

Wit, H. G. M. de, Klauw, C. H. M. van der, Derissen, J. L., Govers, H. A. J. and Chanh, N. B. (1980). *Acta Cryst.*, **A36**, 490–492.

Wozniak, K., Kariuki, B. and Jones, W. (1991). *Acta Cryst.*, **C47**, 1113–1114.

Wright, J. D. and Ahmed, Z. A. (1981). *Acta Cryst.*, **B37**, 1848–1852.

Wright, J. D., Yakushi, K. and Kuroda, H. (1978). *Acta Cryst.*, **B34**, 1934–1938.

Yakushi, K., Ikemoto, I. and Kuroda, H. (1971). *Acta Cryst.*, **B27**, 1710–1718.

Yakushi, K., Ikemoto, I. and Kuroda, H. (1973). *Acta Cryst.*, **B29**, 2640–2641.

Yakushi, K., Ikemoto, I. and Kuroda, H. (1974a). *Acta Cryst.*, **B30**, 835–837.

Yakushi, K., Ikemoto, I. and Kuroda, H. (1974b). *Acta Cryst.*, **B30**, 1738–1742.

Yakushi, K., Tachikawa, N., Ikemoto, I. and Kuroda, H. (1975). *Acta Cryst.*, **B31**, 738–742.

Yakushi, K., Uesaka, T. and Kuroda, H. (1985). *Mol. Cryst. Liq. Cryst.*, **125**, 355–363.

Yamachi, H., Saito, G., Sugano, T., Katayama, C. and Tanaka, T. (1987). *Synth. Metals*, **19**, 533–536.

Yamaguchi, Y. and Ueda, I. (1984). *Acta Cryst.*, **C40**, 113–115.

Yamashita, Y., Suzuki, T., Mukai, T. and Saito, G. (1985). *J. Chem. Soc. Chem. Commun.*, pp. 1044–1045.

Zacharias, D. E. (1976). Abstracts of ACA Winter Meeting, Clemson, S. Carolina, p. 12.

Zacharias, D. E. (1993). *Acta Cryst.*, **C49**, 1082–1087.

Zacharias, D. E., Prout, K., Myers, C. B. and Glusker, J. P. (1991). *Acta Cryst.*, **B47**, 97–107.

Zobel, D. and Ruban, G. (1983). *Acta Cryst.*, **B39**, 638–645.

Chapter 16

Crystal (structural) physics of mixed stack
π–π* molecular compounds

Be not curious in unnecessary matters for
more things are shewed to you than men
understand.

Ecclesiaticus 3:23

Summary: Thermodynamic measurements for a limited sample of crystalline mixed-stack
1 : 1 π-molecular compounds show that most are enthalpy-stabilized, some entropy-stabilized,
and a few both enthalpy and entropy-stabilized. Correlation of these thermodynamic results
with crystal structures remains a task for the future. Combination of optical spectroscopic methods at
very low temperatures and electron spin resonance measurements have provided proof of Mulliken's
theory also for the solid state. The room temperature crystal structures of 1 : 1 π-molecular compounds are not necessarily representative of the entire range of pressure–temperature behavior of
these materials. There are often hints of disorder (usually of the donor) in the room temperature
structures, and these have been correlated for a few systems with disorder-to-order transitions
(thermodynamically second-order, following Ehrenfest) that occur on cooling; these have been
studied by a combination of calorimetric, diffraction and resonance techniques. Despite overall
similarities, each system surveyed has its own individual characteristics. A number of 1 : 1
π-molecular compounds have been shown to transform to quasi-plastic phases on heating. A small
number of mixed stack 1 : 1 π-molecular compounds with neutral ground states have been shown to
transform to ionic structures on cooling or application of pressure.

16.1   Introduction

The mixed-stack model of crystalline π-molecular compounds described in the preceding chapters requires amplification in many respects for it to provide a full picture of current knowledge. In this chapter we shall concentrate on some aspects of the structural physics of mixed-stack $1:1$ π-molecular crystals; treatment is limited to the $1:1$ composition and infinite mixed stack arrangement because information is not available for other compositions and types of structure. We begin by considering the thermodynamic parameters reported for ambient temperature and pressure; this will give some overall feeling for the strengths of the interactions leading to the formation of the molecular compounds, and, in particular, whether these are enthalpy or entropy stabilized (or both) with respect to the individual components. In the next section we discuss low temperature spectroscopic studies that give information about the degree of charge transfer in the excited state and hence provide a test of the applicability of Mulliken's theory to the solid state. The types of solid state transformation that occur on cooling are then discussed in some detail. Most of the transformations studied so far are of the second order, but there are also some first-order transformations ('order' defined below). This leads us to consideration of the nature of the structural disorder found in many crystalline π-molecular compounds at room temperature. Then two types of solid state transformation pertinent to only a limited number of π-molecular compounds are discussed – for some, to a quasi-plastic phase at high temperatures, and for some others to an ionic ground state on cooling or application of pressure.

One aspect of the crystal physics of mixed-stack π-molecular compounds has been entirely omitted – their electronic properties. This is not because of lack of importance but because it would move us too far from our essentially structural theme.

16.2 Thermodynamic parameters

We note in Appendix 1 (section 1) that a crystalline binary adduct may be enthalpy and entropy stabilized with respect to its crystalline components (situation (i)), or entropy stabilized (situation (ii)), or enthalpy stabilized (situation (iii)). Values of ΔH_c and ΔS_c for three groups of π-molecular compounds are plotted in Fig. 16.1. These values refer nominally to 298K and were determined by the electrochemical method (Appendix 1, Section 2.1.1), where ΔH_c and ΔS_c (assumed constant over the range 278–318K) are derived from the measured temperature dependence of ΔG_c. The precision of the thermo-dynamic parameters is not high. Although the errors of the free energy values range from 0.01 to 0.18 kJ/mol, those of the derived ΔH_c and ΔS_c values range from $\approx$5% up to $\approx$40%. Fig. 16.1 shows that all three permitted quadrants of ΔH_c – ΔS_c space are populated, but not in equal measure. For the available sample, which is not necessarily randomly selected from the total population, most of the π-molecular compounds are enthalpy stabilized, fewer are entropy stabilized and even fewer are both enthalpy and entropy stabilised. Abdel-Rehiem et al. (1975) pointed out that there is an approximately linear relationship between ΔH_c and ΔS_c for compound formation between a series of monosubstituted naphthalenes and picric acid; this is illustrated for a wider range

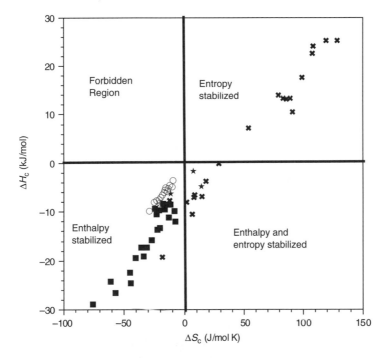

Fig. 16.1. Plot of ΔH_c against ΔS_c for 60 π-molecular compounds of the following groups: (a) crosses – Picric acid compounds of various aromatic hydrocarbons [19 values] (Shahidi and Farrell, 1980), denoted by diagonal crosses; (b) filled squares – Picric acid compounds of substituted naphthalenes [21 values] (Abdel-Reheim et al., 1975); (c) circles – Styphnic acid (1,3-dihydroxy-2,4,6-trinitrobenzene) compounds of substituted naphthalenes [17 values] (Shahidi, Farrell and Westwood, 1980); (d) five-pointed stars – aromatic hydrocarbons with TNB (3 values). The detailed numerical values and identities of the compounds are given in the cited references.

of compounds in Fig. 16.1. The values for the picric acid compounds of substituted naphthalenes and of aromatic hydrocarbons all fall on the same straight line; the first of these
groups is enthalpy stabilized all the latter either enthalpy and entropy stabilized or entropy
stabilized. The line for the styphnic acid compounds (of much the same group of substituted
naphthalenes) is parallel but displaced to higher ΔH_c values; these are all enthalpy stabilized.

The linear relationship between ΔH_c and ΔS_c follows from the form of the free energy
equation recast as

$$\Delta H_c = T\Delta S_c + \Delta G_c$$

A linear plot of ΔH_c against ΔS_c requires that the slope should be T ($= 298\text{K}$ for the present set of data) and that ΔG_c should be constant; using all the points in Fig. 16.1, we obtain

$$\Delta H_c \text{ (kJ/mol)} = 250 \text{ (K) } \Delta S_c \text{ (J/mol K)} - 6.13 \text{ (kJ/mol)}.$$

The derived slope ($R^2 = 0.917$) is reasonably close to the required value. The distribution of $-\Delta G_c$ is shown in Fig. 16.2, where the different groups of molecular compounds have not been distinguished. If this is done then one finds that the mean for the
picric acid compounds of substituted naphthalenes is $-6.54(2.4)$ kJ/mol), $-1.1(0.7)$ kJ/
mol for the styphnic acid compounds of substituted naphthalenes and $-10.5(2.7)$ kJ/mol
for the picric acid compounds of aromatic hydrocarbons. The picric acid compounds of
substituted naphthalenes and of aromatic hydrocarbons do not differ significantly in terms
of free energy values but are very different when enthalpies and entropies are compared;
the least stable are the styphnic acid compounds of substituted naphthalenes.

The deviations of individual points from the overall linear relationship must have a
physical explanation, as must the location of particular types of compound in different
quadrants of the diagram; for example (with the exceptions of indene and acenaphthylene)
all the {aromatic hydrocarbon···picric acid} compounds have large positive entropies of
formation. Presumably the principal distinctions will have to be made among the compounds lying in the different quadrants, in terms of the different types of stabilization
(enthalpy and/or entropy) that apply. Unfortunately there is not much overlap between the
sample whose crystal structures have been determined and that for which thermodynamic

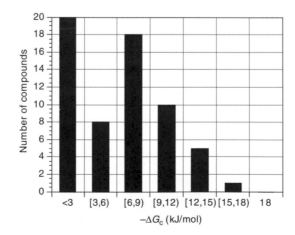

Fig. 16.2. Distribution of $-\Delta G_c$ values for the whole sample. The values along the abscissa are for
the limits of the bins, in kJ/mol.

parameters have been measured and thus attempts to find such explanations now would appear to be premature.

The linear relationship found between ΔH and ΔS in a wide range of experimental solution studies is called 'enthalpy–entropy compensation'. There are some thousands of references under this heading in *Chemical Abstracts*, starting from 1925 (Liu and Guo, 2001a). Briefly stated, "Entropy–enthalpy (SH) compensation occurs when a small change in ΔG is caused by large, and nearly compensatory, changes in ΔH and ΔS. It is considered a ubiquitous property of reactions in water" (Beasely, Doyle *et al.*, 2002). Nevertheless, "Enthalpy–entropy compensation remains a mystery in chemistry and biophysics" (Liu and Guo, 2001b). Dunitz (1995) has used a simple thermodynamic argument applied to hydrogen bonding of a water molecule to a very large molecule in solution and concluded that enthalpy–entropy compensation is a general property of weak intermolecular interactions and not limited to reactions in water. The relationship does not appear to have been widely studied in solid state interactions; the pioneering contribution of Abdel-Rehiem *et al.*, (1975) was not set in the context of enthalpy–entropy compensation. We present Figs. 16.1 and 16.2 as empirical findings without attempting to assess their significance.

In addition to the information summarized in Fig. 16.1, individual thermodynamic parameters have been measured for a number of molecular compounds and it is sometimes possible to combine ΔG_c and ΔH_c values from different sources to give ΔS_c; for example ΔG_c has been measured electrochemically for {naphthalene···TNB} and {anthracene··· TNB} (and also for other arene···TNB and TNT compounds) (Hammick and Hutchinson, 1955) while ΔH_c has been measured by a dissolution method (Suzuki and Seki, 1955). Some of the available values are listed in Table 16.1.

Table 16.1. Thermodynamic parameters (at 298K) for some mixed stack π–π* molecular compounds. ΔG_c and ΔH_c in kJ/mol and ΔS_c in J/mol K. The errors (bracketed) have been taken from the original publications

Molecular compound					
Donor	Acceptor	ΔG_c	ΔH_c	ΔS_c	Reference
1. Naphthalene	picric acid	−8.60(4)	−8.03(15)	−1.9(4)	SF80
2. [^{2}H$_8$]-naphthalene	picric acid	−8.85(2)	−6.44(15)	−8.1(5)	SF80
3. Anthracene	picric acid	−2.05	−9.25	−24.3	H71
4. Naphthalene	TNB	−8.66	−4.69	13.4	H71
5. Anthracene	TNB	−3.52	−1.43	7.1	H71
6. Benzene	TNB		−2.93		H71
7. Phenanthrene	picric acid	−8.59(7)	−0.1(1.89)	28.5(6.1)	SF80
8. 0.65 (Phenanthrene)/ 0.35 (anthracene)	picric acid	−6.99			
9. Anthracene	TNT	−0.59			HH55
10. Naphthalene	TNT	−5.98			HH55
11. Naphthalene	styphnic acid	−2.66(2)	−6.10(45)	−11.5(1.4)	SF78
12. [^{2}H$_8$]-naphthalene	styphnic acid	−2.74(2)	−6.60(35)	−13.0(1.1)	SF78

References:
H71 – Herbstein (1971), Tables 10 and 11 where references to earlier work are given; HH55 – Hammick and Hutchison, 1955; SF78 – Shahidi and Farrell, 1978; SF80 – Shahidi and Farrell, 1980.

Some individual ΔH_c values are of special interest. Thus a combustion method has been used to measure enthalpies of formation at 298K for the 1:1 TCNQ π-molecular compounds of naphthalene, anthracene and TMPD, giving values of -30.0 ± 7.6, -30.3 ± 5.3 and -82.1 ± 4.3 kJ/mol (Metzger and Arafat, 1983). The first two values are at the lower end of the range of values included in Fig. 16.1, and indicate appreciable enthalpy stabilization, although the ground state is neutral for both compounds. The much more negative value for {TMPD···TCNQ} presumably indicates that an even greater degree of enthalpy stabilization is achieved when the ground state is ionic (see Section 15.10).

16.3 Spectroscopic measurements on the excited state[1]

The originally colorless solutions of the individual donor and acceptor compounds give, on mixing, a vividly colored solution; similarly, highly colored crystals of the molecular compound appear on crystallization from a mixture of the colorless components. In quantitative terms, a new broad, structureless absorption band appears in the UV – visible region of the spectrum which is otherwise, to a good approximation, made up of the sum of the spectra of the individual components. The Mulliken theory of the spectra and stabilization of charge transfer compounds postulates an essentially neutral ground state and an essentially ionic (or dative) first excited state. Light absorption in the charge transfer band leads to electron transfer from the HOMO of the donor to the LUMO of the acceptor (for references see Section 13.1). A possible arrangement of energy levels is shown schematically in Fig. 16.3.

The various possible situations according to the relative energies of CT and locally excited states (based on Iwata *et al.*, 1967) are summarized as follows (for brevity the energy of a state is given, for example, by $^1(D^+A^-)$):

Situation I	(a)	$^1(DA) < {}^1(D^+A^-) \ll {}^3(DA*), {}^3(D*A)$
	(b)	$^1(DA) < {}^3(D^+A^-) \ll {}^3(DA*), {}^3(D*A)$
Situation II	(a)	$^1(DA) < {}^3(DA*) < {}^3(D^+A^-) \approx {}^1(D^+A^-)$
	(b)	$^1(DA) < {}^3(D*A) < {}^3(D^+A^-) \approx {}^1(D^+A^-)$
Situation III	(a)	$^1(DA) < {}^3(D^+A^-) \approx {}^3(DA*)$
	(b)	$^1(DA) < {}^3(D^+A^-) \approx {}^3(D*A)$
Situation IV		$^1(D^+A^-) \approx {}^3(D^+A^-) < {}^1(DA)$

In the first three of these 'Situations' the no-bond singlet is the ground state of the molecular compound (molecular compounds of this type are discussed in Section 15.3); the first excited state is shown as a charge transfer singlet $^1(D^+A^-)$ in I(a) and a charge transfer triplet $^3(D^+A^-)$ in I(b), which are the usual situations. Situation II(b) with a locally excited donor triplet state is also found, but the locally excited triplet state of the acceptor is usually at much higher energies. III(a) and (b) are limiting cases of I(b). In Situation IV the ionic state is the ground state, and the triplet state may be thermally accessible (molecular compounds of this type are discussed in Section 15.10).

There are very few experimental demonstrations of the nature of the excited state. The degree of charge transfer in the $^1(D^+A^-)$ state (Z_s) has been measured for hexamethylbenzene···TCPA in solution (Czekalla and Meyer, 1961); values of $\Delta\mu$, the change in dipole moment following the creation of the excited charge transfer state, of

---

[1] This section draws heavily on the reviews of Ponte Goncalves (1980) and Krzystek and von Schutz (1993).

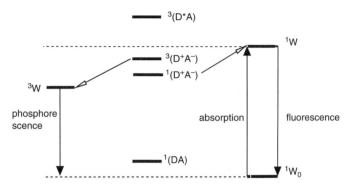

Fig. 16.3. Schematic diagram of a possible arrangement of the energy levels of a charge transfer compound. The zero-order (unmixed) charge transfer states are shown in the center of the diagram, and the effects of mixing on the left (^{3}W is the lowest triplet CT state) and right (^{1}W is the lowest singlet CT state), together with the transitions involved in phosphorescence, absorption and fluorescence. The arrangement of energy levels corresponds to Case II(b) below. (Adapted from Iwata *et al.*, 1967.)

≈ 10 D were obtained, corresponding to $Z_s \approx 1$. The first demonstration of the occurrence of charge transfer triplet states followed the finding that the phosphorescence spectra of several $1:1$ aromatic hydrocarbon-acceptor combinations in glassy media at 77K were considerably red-shifted with respect to the phosphorescence spectra of the individual components (the aromatic donors were benzene, toluene, mesitylene, durene, hexa-methylbenzene, phenanthrene and triphenylene; the acceptors were 1,2,4,5-tetracyanobenzene (TCNB), phthalic anhydride and pyromellitic dianhydride (PMDA)).

Analogous measurements for singlet excited states of crystalline CT compounds (discussed in a general context by Ponte Goncalves (1980)) have been made, only for {anthracene$\cdots$PMDA}, but there are many investigations involving triplet states. The triclinic crystal structure of {anthracene$\cdots$PMDA} (deep red crystals, $P\bar{1}$, $Z=1$, at T = 153K, Robertson and Stezowski, 1978; Table 15.2; ANTPML) consists of sheets of stacks (Fig. 16.4). The molecules of the two components are at crystallographic centers of symmetry and related by translations; the interplanar distance is ≈ 3.33 Å and the normal to the plane of the central ring of anthracene makes an angle of 20.7° with [001], which is the stack axis. All donor molecules are mutually parallel, as are all acceptor molecules. Donor and acceptor molecules are almost parallel, with an angle of 3.0° between their planes, and of 4.2° between their long axes. As there is no evidence for a phase change between 300 and 2K, we may assume that the spectroscopic studies, carried out at ≈ 2K (Haarer, 1977) are of this structure.

The emission (fluorescence) and absorption spectra associated with the lowest CT transition of crystalline {anthracene$\cdots$PMDA} at 2K are shown in Fig. 16.5 (Haarer, 1977). There is a phonon continuum, and sharp lines that correspond to vibrations of the donor and acceptor moieties. The zero-phonon electronic transition centred at 18320 cm^{-1} (5459 Å) (identified as such from the mirror image relation of the vibronic structure in emission and absorption) was shown to be strongly polarized along the [001] stack axis. A zero-phonon line is intense when the coupling between electronic transition and lattice vibrations is small ($S \lesssim 1$; S is the average number of phonons coupled to the electronic transition); when the coupling strength increases ($S \approx 10$ is strong electron–phonon coupling) the intensity is redistributed into a broad phonon wing

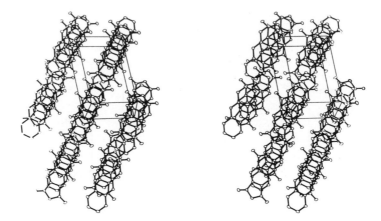

Fig. 16.4. Stereodiagram of the molecular arrangement in {anthracene···PMDA}, viewed normal to the moiety planes, i.e. approximately along **c***; the *a* axis points to the right and the *b* axis points upwards, both being in the plane of the page. The reduced cell can be obtained from that in the diagram by the transformation (001/100/111). (Reproduced from Robertson and Stezowski, 1978.)

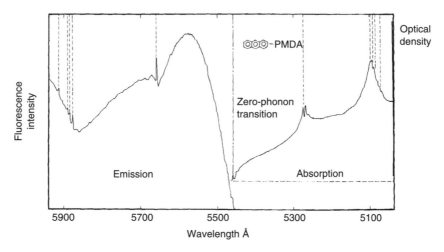

Fig. 16.5. Emission and absorption spectra of crystalline {anthracene···PMDA} at 2K. The sharp lines correspond to anthracene and PMDA vibrations as determined by Raman spectroscopy. (Reproduced from Haarer, 1977.)

and the zero-phonon transition cannot be detected under normal experimental conditions. The intensity of the zero-phonon line from {anthracene···PMDA} is strongly temperature dependent and the line is not detectable above ≈20K (Haarer, 1974). S was estimated from the temperature dependence as 6.0 ± 0.3 and is thus in the intermediate regime.

The electric field dependence of the frequency of the zero-phonon line (the Stark effect) has been measured and gives an approximate value for the change in dipole moment which characterizes the creation of the ionic charge transfer state as $\Delta\mu = 11 \pm 1$ D (Haarer, 1975; Haarer *et al.*, 1975). Thus $Z_s \approx 1$, and Mulliken's theory receives experimental confirmation also for the solid state.

The broadened (and hence structureless) nature of the charge transfer band of π-molecular compounds in solution is ascribed to the variety of geometrical approaches, in terms of both intermolecular distances and orientations, that occur between the donor and acceptor moieties. Variations of ≈ 0.01 Å can lead to energy fluctuations of ≈ 100 cm^{-1} (Czekalla et al., 1957). However, the broadening due to such geometrical inhomogeneities is small or negligible in crystals. Here Haarer (1975) has suggested that the additional Coulombic attraction that occurs in the polar excited state leads to a contraction of the interplanar distance between adjacent donor and acceptor molecules by ≈ 0.1 Å; this gives rise to the observed broadening of the CT band, and also to the observed Stokes shift between absorption and emission spectra.

The nature of the excited triplet state in single crystals of donor–acceptor molecular compounds has been extensively investigated by electron paramagnetic resonance, which leads to determination of Z_t, the fraction of charge transferred in the triplet state. This brief account of the theory closely follows Ponte Goncalves (1980). The dipole–dipole interaction between two unpaired spins in a molecular triplet state is given by

$$H_{ss} = \mathbf{S} \cdot \mathbf{T} \cdot \mathbf{S} = D(S_z^2 - 1/3\mathbf{S}^2) + E(S_x^2 - S_y^2).$$

For the planar molecules considered here we choose x, y, z (the principal axes of the interaction tensor T) with z normal to the molecular plane, and x and y determined by the in-plane symmetry (if any) of the molecule (these axes correspond to the $\mathbf{L}$, $\mathbf{M}$, $\mathbf{N}$ axes defined in Section 15.1). The dipole-dipole interaction results in the splitting (even in the absence of an external field) of the three spin sublevels of the triplet state into levels at $(-2D/3)$ and $(D/3 \pm E)$, where D and E are the zero-field splitting parameters. Their values are given by

$$D = (3g^2\beta^2/4) \langle {}^3\psi \mid (r_{12}^2 - 3z_{12}^2) r_{12}^{-5} \mid {}^3\psi \rangle$$
$$E = (3g^2\beta^2/4) \langle {}^3\psi \mid (y_{12}^2 - x_{12}^2) r_{12}^{-5} \mid {}^3\psi \rangle$$

D and E can be measured by optical detection of magnetic resonance (ODMR), which involves monitoring the changes induced in phosphoresence intensity by application of resonant microwaves to induce transitions between the zero-field spin sublevels. Because in-plane delocalization of spin is generally larger than that normal to the molecular plane, the first term in the equation for D is usually much larger than the second, leading to fairly facile estimation of D. E, however, is given by the difference between two terms of similar magnitude and hence is more difficult to calculate.

We now turn to estimation of Z_t. Assuming Situation III(b) above to apply, we write the wave function for the triplet state in terms of configuration mixing between the donor triplet state and the charge transfer triplet state:

$${}^3\psi = (1 - Z_t)^{1/2} \, {}^3\psi(D^*A) + (Z_t)^{1/2} \, {}^3\psi(D^+A^-)$$

The value of D is then, assuming a common magnetic axis system for donor and compound, given by

$$D = (1 - Z_t) D_d + Z_t D_{ct} + [Z_t (1 - Z_t)]^{1/2} D_{d,ct}$$

where D_d refers to the individual donor molecule, D_{ct} refers to the fully charge transferred state, and $D_{d,ct}$ is a hybrid contribution (zero if the two-center approximation applies). Thus, if it is found that $D \approx D_d$, then one may conclude that $Z_t \approx 0$ and that the triplet state is localized on the donor molecule. This situation occurs in the triplet state of crystals of highly

Table 16.2. Values of D and E measured for anthracene in various environments compared with those found in vapour-grown crystals of {anthracene···PMDA}

Host	phenazine	diphenyl	anthracene-d_{10}	{anthracene···PMDA}
D (cm^{-1})	0.07055	0.07156	0.06945	0.0695(3)
E (cm^{-1})	-0.00791	-0.00844	-0.00836	$-0.0079(3)$

purified {anthracene···PMDA} (Haarer and Karl, 1973) grown by sublimation; these crystals show (at 1.6K) an EPR spectrum on irradiation and a phosphorescence spectrum (crystals grown by the Bridgman method show only very weak phosphorescence and EPR signals). Comparison of values of D and E for anthracene in various environments (Table 16.2) shows that the triplet state in {anthracene···PMDA} is localized on the donor molecule.

More sophisticated methods must be used for determination of Z_t when this is not small.

EPR methods have given valuable information about D and E parameters, triplet states immobilized at low temperatures, kinetics of spin state migration and hyperfine interactions. We shall consider below particularly those measurements relevant to the study of phase transitions in charge transfer molecular compounds.

16.4   Crystals with disorder $\Rightarrow$ order transformations on cooling – modern treatments of second order phase transitions

16.4.1   *General introduction*

Room temperature x-ray diffraction photographs of crystalline π-charge transfer molecular compounds are generally (but not always) characterized by a strong reduction in reflection intensity with increasing scattering angle; this has been interpreted (see Herbstein (1971) for summary) to mean that donor and/or acceptor molecules are dynamically or statically disordered. This in turn suggests that disorder $\Rightarrow$ order transformations may occur on cooling, as been demonstrated experimentally by diffraction and other physical techniques for a number of systems; these transitions are (generally) second order (as defined below) and can be related to the classical treatments of Bragg and Williams (1935) and Landau (1937). We plan to integrate experiment and theory (at its simpler levels) for the systems that have been studied, first introducing concepts encountered in modern discussions. The various experimental techniques are complementary while the various systems maintain a similar overall behavior with individual quirks.

Specific heat–temperature curves provide a concise qualitative and quantitative picture of thermal behaviour showing whether phase transformations occur and allowing some deductions about their nature; they also provide values of the thermodynamic functions over the temperature range of measurement (see Appendix 1). The relatively few C_p–T measurements reported for π-molecular compounds show a diversity of effects. The yellow polymorph of {naphthalene··· PMDA} shows no transition over the range 0–300K (Boerio-Goates and Westrum, 1979; Appendix 1); {pyrene···PMDA} shows a second-order transition at $\approx$155K (Boerio-Goates and Westrum, 1980b; Dunn *et al.*, 1978); {naphthalene··· TCNB} shows a broad second-order transition over the range 40–75K (Boerio-Goates and Westrum, 1979); {naphthalene···TCNE} (Boerio-Goates and Westrum, 1980a) shows a combination of a broad second-order transition over the range 150–210K and a sharp transition over the range 155–163K, with the added complication that the sharp peak is split; these detailed and intricate results for {naphthalene···TCNE} should be useful in guiding diffraction studies of this substance beyond the room temperature results currently available.

Although our approach is quite standard, it is perhaps desirable to clarify certain concepts and set out our treatment in a general context. The term 'order' is an abbreviation for 'long range order' and has its origins in studies of metallic alloys; a general account is given by Ziman (1964). Here we assume that A and B in a binary (metallic) crystal AB are completely segregated onto two different sublattices in a state of perfect long range order at $T = 0$K (A, B are different types of atom). A classic example is β-brass (ideally CuZn) with the B2 CsCl structure, in which Cu atoms have only Zn nearest neighbors and conversely. As the temperature increases, some B atoms are found on the A sublattice (i.e. in the wrong positions) and the degree of long range order is reduced, due to the interplay of enthalpy (favoring order) and entropy (favoring disorder) contributions to the overall Gibbs free energy. The degree of long range order Q is defined as the fraction of atoms in their right positions minus the fraction of atoms in their wrong positions; here $Q = |1 - 2p|$, where p is the probability of finding an A atom on the A sublattice. When $p = 1$, then $Q = 1$ and the long range order is complete; when $p = 1/2$, then $Q = 0$ and the long range order disappears. The temperature of disappearance is T_c, the critical temperature. Long range order gives rise to superlattice reflections in the diffraction pattern, with intensities proportional to Q. Although it is not taken into account in the simple theory, short range order can exist above T_c, i.e. the probability of finding a B atom as nearest neighbor to an A atom is greater than statistical; short range order gives rise to diffuse scattering in the diffraction pattern, generally centered at the superlattice reciprocal lattice points.

Although we have used *chemical* ordering as the basis for our description above, the concepts are general and can refer, for example, to ordering of spins in magnetic systems, or to ordering of orientations in mixed stack π-molecular compounds, which is the context relevant here[2]. We shall find that, in the systems discussed below, the donor molecule can take up two *orientations* on its sublattice sites; at 0K these orientations are arranged in completely ordered fashion but wrong orientations begin to appear as the temperature is increased until the long range order (together with the accompanying superlattice reflections) disappears at T_c. Thus the transitions (on heating from 0K) to be discussed are from orientational order to orientational disorder.

16.4.2 *The Ehrenfest order of a phase transition*

The order of a phase transition was first defined by Ehrenfest (1933): in an nth order phase transition the nth and higher derivatives of G (Gibbs free energy) with respect to T and P show discontinuities. Examples relevant here are:

- First order – discontinuities in V $(= (\partial G/\partial P)_T)$, S $(= (\partial G/\partial T)_P)$, H, U, etc.
- Second order – discontinuities in α $(= (1/V)(\partial V/\partial T)_p = \{(1/V)[\partial/\partial T(\partial G/\partial P)_T]\}_P)$ and C_P $(= [T(\partial S/\partial T)_P] = [-T(\partial^2 G/\partial T^2)_P]$.

---

[2] Chemical ordering does not occur in these structures except perhaps of mixed donors or acceptors in the ternary systems noted in Section 13.5.

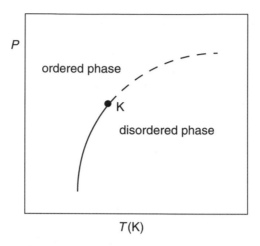

Fig. 16.6. The line representing the transition between two polymorphic phases of a one-component system, drawn in the P–T plane. The continuous line represents a first order transition and the broken line a second order transition. The point K is the tricritical point. (Adapted from Fig. 66 of Lifshitz and Pitaevskii (1980).)

Higher order phase transitions can be defined similarly but it is not clear whether they exist as other than mathematical concepts.

We give a brief definition of first order transitions for completeness. The two phases (A and B) have different structures and the transition shows hysteresis (i.e. superheating and supercooling are possible). Because of the structural difference, the transition mechanism is generally of the nucleation and growth type and one would expect a single crystal of phase A to transform into a polycrystal of phase B on heating, and conversely on cooling; twinning often occurs. In another nomenclature, these are called discontinuous transitions.

In a second order transition the two phases have structures that merge into one another and the transition does not show hysteresis. Because of the structural similarity, the transition mechanism is generally from a single crystal of the A phase to a single crystal of the B phase, with preservation of component orientations; these are often called continuous transitions. Transitions between phases of *similar* structure can be either first or second order depending on external conditions (here T and P); this is illustrated by a diagram adapted from Lifshitz and Pitaevskii (1980, Chapter 10, Fig. 66, p. 493) (Fig. 16.6).

16.4.3 *Landau theory of phase transitions*

Phase transitions are often described in terms of a theoretical approach originally developed by Landau (1937) and, at a deeper level that takes fluctuations into account, in terms of renormalization groups (Wilson, 1971). We shall restrict ourselves to Landau theory, which is a mean field theory and has been detailed in a number of texts (Lifshitz and Pitaevskii, 1980; Salje, 1990; Stanley, 1971; Stokes and Hatch, 1988; Tolédano and Tolédano, 1987) and many articles. Our brief description is primarily based on Salje and

Stokes and Hatch, and we consider only those aspects relevant to the present chapter. Landau theory is generally considered to apply only to second order (continuous) changes but its extension to discontinuous (first order) changes has been discussed (see Stokes and Hatch (1988), pp. 2–3 for references).

The excess Gibbs free energy $G(T)$ is the difference in Gibbs free energy between high symmetry and low symmetry phases of the crystalline compound under consideration. Assuming that the excess enthalpies (H) and entropies (S) are not temperature dependent, $G(T)$ can be written as

$$G(T) = 1/2A\ (T - T_c)Q^2 + 1/4BQ^4 + 1/6CQ^6 \qquad (16.1)$$

where the parameters A, B, C are not temperature dependent; T_c is the temperature of the phase transition (the critical temperature) and Q is the order parameter defined above in Section 16.4.1. $G(T)$ as given above is called a 2-4-6 potential; when $B = 0$ (the 2-6 potential), the transition is *tricritical*, representing the intermediate stage between continuous $(B > 0)$ and discontinuous $(B < 0)$ transitions (see Fig. 16.6). The condition for thermodynamic equilibrium is determined by the minima of the potentials, i.e. for $\partial G/\partial Q = 0$. The solutions for second order (2-4) and tricritical (2-6) potentials are

$$\text{2-4: } Q^2 = A/B\ (T_c - T)$$

and

$$\text{2-4-6: } Q^4 = A/C\ (T_c - T),\ T < T_c.$$

In Landau's original work the implicit assumption was made that the polynomial expansion of the potential was only valid in the vicinity of T_c, and hence for small values of Q, but Salje (1990) has proposed that "the polynomial expansion of G is a good approximation over an extended temperature interval and that the approximation also holds for larger values of the order parameter." We consider in the next section how the results summarised here can be used to determine whether there is justification for applying Salje's extension of Landau theory to binary π molecular compounds.

16.4.4 *The critical exponents*

In a more general approach, the behaviour of the system is described in terms of critical exponents λ, the values of which depend on the physical property under investigation and the nature of the transition (Berry, Rice and Ross, 1980, p. 863–864; Lifshitz and Pitaevskii, 1980; Stanley, 1971; Tolédano and Tolédano, 1987). Very general relationships among the various critical exponents can be derived that do not depend on the physical system.

A physically measurable quantity such as spontaneous strain (defined below), intensity of a superlattice reflection (for these two quantities the symbol β is generally used for the exponent) or the excess specific heat ΔC_p (here α is generally used for the exponent) can be given as

$$f(-\varepsilon) = A(-\varepsilon)^\lambda \{1 + B(-\varepsilon)^x + \cdots\} \quad (x > 0),\ \text{where } \varepsilon = (T - T_c)/Tc.$$

The correction terms drop out on taking the limit in order to obtain the critical exponent, which is defined as

$$\lambda = \lim_{\varepsilon \to 0} [\ln f(-\varepsilon)/\ln(-\varepsilon)].$$

λ is expressed as a function of temperature in the form $f(1-T/T_c)$, which is given (to a first approximation) as

$$f(1 - T/T_c) = A(1 - T/T_c)^\lambda \tag{16.2}$$

or, recast in log-log form, as

$$\log[f(1 - T/T_c)] = \log A + \lambda \log[(1 - T/T_c)]. \tag{16.3}$$

The critical exponents have generally been obtained from measurements made very close to T_c (perhaps within 0.5–1°) because then the effect of fluctuations can be investigated. Ignoring the correction terms, one has $f(-\varepsilon) = A(-\varepsilon)^\lambda$. We have used this power law form to fit various order parameters and obtain accompanying values for the exponents. The Salje extension implies that data over the whole available temperature range can be used; thus equation (16.3) should give a linear plot with slope λ. We shall find that this holds well for {pyrene···PMDA} (PYRPMA), but that there are problems, discussed below, with {naphthalene···TCNB} (NAPTCB) and {anthracene···TCNB} (ANTCYB); however, these are hardly *critical* exponents in the sense of the theory, particularly because experimental values sufficiently close to T_c are usually lacking. We consider them convenient for description but their physical significance has still to be established.

16.4.5　*The permitted symmetries of a low symmetry phase derived from a particular high symmetry phase*

The permissible space groups of the low temperature phase derived from a particular high temperature phase after a second order phase transition were first obtained by Landau and Lifshitz in 1937–1939 (see Lifshitz and Pitaevski, 1980, XIV §145) and, more extensively, by Lyubarski (1960) and Koci′nski (1983). The latter authors have treated some tetragonal and hexagonal space groups in detail. However, space groups of mixed stack π-molecular compounds are usually monoclinic and, to the best of our knowledge, those relevant to second order phase transitions have been treated only by Bernstein (1967). We quote a limited set of Bernstein's results, assuming that the high temperature phase has space group $C2/m$ or $P2_1/a$, that the moieties are located at crystallographic centers, and that the transition occurs with preservation of axial *directions*.

I. High temperature phase has space group $C2/m$. The conditions and possible space groups for the low temperature phase are:
1. Unit cell remains centered and cell volume remains unchanged
   (a) Cm, (b) $C2$, (c) $C\bar{1}$.
2. Unit cell becomes primitive but cell volume remains unchanged
   (a) $P2/m$, (b) $P2_1/a$, (c) $P2/a$, (d) $P2_1/m$.
   Examples of I.2(b) are {naphthalene···TCNB}, {anthracene···TCNB}, {pyrene···C_6F_6}, and (perhaps) {anthracene···TCNQ}.
3. Unit cell remains C-centered, cell volume is doubled and lattice becomes triclinic (P). The axial directions are not preserved in the reduced cell.
4. Unit cell becomes body centered monoclinic, c axis and cell volume are doubled. The axial directions are not preserved in the reduced cell.
   (a) $I2/m$, (b) $I2/c$.
   Example of I.4(b) is {anthracene···C_6F_6}.
5. Unit cell remains C-centered monoclinic, c axis and cell volume are doubled.
   (a) $C2/m$, (b) $C2/c$.

II. High temperature phase has space group $P2_1/a$. The conditions and possible space groups for the low temperature phase are:

1. Unit cell remains primitive monoclinic, but center of symmetry disappears
    (a) $P2_1$, (b) Pa.
    The C-polymorph of naphthazarin (of course, not a π-molecular compound) belongs to II.1(b) (Herbstein *et al.*, 1985), as does (TTF-CA} (§16.9.2).

2. Unit cell becomes primitive triclinic, without change of volume
    (a) $P\bar{1}$
    A possible example of II.2(a) is {naphthalene···TCNE}.

3. Unit cell remains primitive monoclinic, *c* axis and cell volume are doubled.
    (a) $P2_1/a$, (b) $P2_1/n$.
    Example of II.3(b) is {pyrene···PMDA}.

4. Unit cell becomes pseudo-centered monoclinic, but the actual space group is $P\bar{1}$. The axial directions are not preserved in the reduced cell.
    (a) pseudo-*C*-centered (b) pseudo-*I*-centered.

The piecemeal contributions described above, important as they were, have now been consolidated into the tables of Stokes and Hatch (1988). The parent (high symmetry) polymorphic phase with space group G_0 transforms, with a physical generalized distortion $\Delta\rho(r)$, to a low symmetry phase with space group G, where G is a subgroup of G_0. Stokes and Hatch (1988) have derived the G space groups for all the 230 G_0 space groups. The G_0 space groups of interest here are $C2/m$ and $P2_1/c$ and we excerpt in Table 16.3 the relevant information from the Stokes and Hatch tables; this is discussed together with phase transitions of the various compounds.

The label 'Spec(ies)' in Table 16.3 refers to the ferroic species of the transition. For the three examples of interest here, this label is 'nf' nonferroic. A 'nonferroic' or 'co-elastic' crystal has "elastic and strain anomalies... correlated with the structural phase transition" (Salje, 1990).

The Bernstein and Stokes and Hatch prescriptions can be reconciled when one remembers that in Bernstein the transition occurs with preservation of axial *directions*

Table 16.3. Information from the Stokes and Hatch (1988) tables for the G_0 space groups $C2/m$ and $P2_1/c$. The definitions of the various column headings are given by Stokes and Hatch. The columns relevant in the present context have been emphasized

Space group G_0	Parent Irrep.	Image	Lan	Lif	Subgroup G	Spec.	Dir	Size	Basis	Origin
C2/m	Y_2^+	A2a	0	0	**14** $P2_1/c$	nf	$P1^{**}$	2	(001); ($0\bar{1}0$); (100)	(000)
	A_2^+	A2a	0	0	**15** $C2/c$	nf	$P1^{**}$	2	(100); (010); (002)	(000)
P2₁/c	Y_2^+	A2a	0	0	**14** $P2_1/c$	nf	$P1^{**}$	2	(001); (010); ($\bar{2}0\bar{1}$)	(000)

while the space group can change (as we have done here for {pyrene···PMDA}, whereas Stokes and Hatch maintain space group and allow axial directions to change.

16.4.6 *Temperature dependence of the order parameter*

We now define "spontaneous strain" and show how it is measured, basing ourselves extensively on Salje (1990; see Chapter 4). In a ferroelastic (not relevant here) or co-elastic transition the shape of the crystal is changed and this creates a macroscopic spontaneous strain. Ignoring microstructures (justified by the absence of transition-induced twinning in the crystals considered here), the macroscopic spontaneous strain can be replaced by the structural spontaneous strain, usually called the spontaneous strain, which is measured as the volume average of the deformation of the unit cell. Originally introduced for ferroelastic systems by Aizu (1970), this definition has been expanded to include all structural phase transitions that lead to variation of the shape of the crystal-lographic unit cell, especially for co-elastic systems. In order to measure the spontaneous strain the lattice parameters of the high-symmetry phase have to be extrapolated into the temperature regime of the low-symmetry phase, i.e. the high symmetry phase is the reference phase. This extrapolation represents that part of the thermal expansion that is not related to the structural phase transition and therefore does not contribute to the excess spontaneous strain. The numerical values of the spontaneous strain are now defined by the strain tensor that relates the low-symmetry unit cell to the high-symmetry unit cell when extrapolated to the same temperature. The spontaneous strain in the (low-symmetry) ordered phase is conveniently given in terms of the Vogt coefficients e_j, calculable as a function of temperature from the measured cell dimensions of the ordered phase (not sub-scripted) and the extrapolated cell dimensions of the disordered phase (subscripted) using

$$
\begin{aligned}
e_1 &= a/a_0 - 1, \\
e_2 &= b/b_0 - 1, \\
e_3 &= (c \sin \beta^*/c_0 \sin \beta_0^*) - 1, \\
e_5 &= (a \cos \beta^*/a_0 \sin \beta_0^*) - (c \cos \beta^*/c_0 \sin \beta_0^*).
\end{aligned}
\tag{16.4}
$$

We give the equations only for monoclinic crystals as the three present examples are all monoclinic; Salje (1990) gives the equations for all the crystal families. The (scalar) spontaneous strain in the ordered phase is $e_s = (\Sigma e_j^2)^{1/2}$.

The long range order parameter Q refers to the crystal structure averaged over many unit cells; it can be determined from measurements on the intensities of superlattice reflections, from the spontaneous strain and from the ESR spectrum. Spontaneous strain in the ordered phase, and intensities of the superlattice reflections are both proportional to Q^2.

There is also a (structural) short range order parameter referring to the crystal structure averaged over a few unit cells, which will be only briefly considered in what follows; it can be determined from measurements on the intensities of the diffuse scattering of x-rays (or neutrons). The diffraction aspects of these parameters are clearly discussed by Warren (1969; see Chapter 12). The physical interpretation (i.e. at the molecular level) of the measured values is often controversial. The disordered state above T_c is usually described in terms of one or other of two extremes, statistical 'static' disorder where the potential energy diagram for a molecule in the field of its neighbors is a double-well potential with the barrier height appreciably higher than kT, or as 'dynamic' disorder where the potential energy diagram is a flat-bottomed well (definitions adapted from Lefebvre et al., 1989). One is able to build up a fairly satisfactory picture of the ideally ordered structures below the transition temperature but it is much more difficult to define the nature of the high temperature, disordered structure, and particularly to decide whether the disorder is dynamic or static. Some progress has been made by use of rigid-group refinement of diffraction data and solid state NMR methods, and by model calculations. These results are described in the final parts of this section.

One important concept, mentioned here only in passing, is that of the 'antiphase domain' (Warren, 1990; pp. 216–227). These are well-ordered regions of the crystal separated by 'change step' boundaries. As the name suggests, neighbouring domains are mutually out-of-phase. The concept has been applied mostly to metallic alloys (e.g. Cu_3Au) and hardly to organic systems.

16.4.7 *Pressure dependence of critical temperature for ordering*

Applying Fig. 16.6 to PYRPMA, the point K has the coordinates $P = 1$ atm. (≈ 1 bar), $T_c \approx 165K$, the line separating ordered and disordered phases has a positive slope and, following Fig. 16.6, one would expect increase of pressure to favour ordering. The pressure dependence of T_c. can be calculated from Ehrenfest's second order analog to the first order Clapeyron equation given by Pippard's (1964), equations (8.19) and (9.1.) Ehrenfest's equation can be written as

$$dT_c/dP = T_c[(\partial V/\partial T)_2 - (\partial V/\partial T)_1]/[(C_P)_2 - (C_P)_1] \qquad (16.5)$$

where the subscripts 1 and 2 refer to the state of the system just below and just above T_c. A value of $[(\partial V/\partial T)_2 - (\partial V/\partial T)_1]$ is obtained from Fig. 16.10 and a value of $[(C_P)_2 - (C_P)_1,]$ from Fig. 16.7; using these values we calculate that $dT_c/dP \approx +17$ K/kbar, if the specific heat values given by Boerio-Goates and Westrum (1980) are used, and about half this amount using those of Dunn et al. (1978). Thus application of a pressure of a few kbar should produce the ordered phase of PYRPMA at room temperature. Anticipating later discussion we note that $[(\partial V/\partial T)_2 - (\partial V/\partial T)_1]$ is negative for PYRPMA and ANTCYB and positive for NAPTCB. As $[(C_P)_2 - (C_P)_1]$ is negative, application of pressure should increase T_c for PYRPMA and ANTCYB, and reduce it for NAPTCB. The only experimental test is for ANTCYB (Ecolivet et al., 1988; see §16.5.3(d)).

16.5   Thermodynamic, structural and kinetic investigations of various systems showing second order transitions on cooling[3]

16.5.1   *The crystal structure of {Pyrene···PMDA} (PYRPMA) and evidence for an order ⇔ disorder phase transition at ≈160K*

(a) Introduction.   The melting point diagram (Herbstein and Snyman, 1969; HS69) shows three molecular compounds, the equimolar compound being by far the most stable of the three. The 160K phase transition in PYRPMA was first demonstrated by x-ray

---

[3] This treatment is largely based on Herbstein (1996).

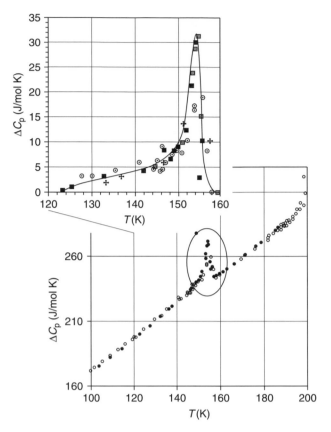

Fig. 16.7. Part of the $C_p - T$ curve for {pyrene···PMDA}, showing the broad transition centred about ≈155K which corresponds to the second-order transition studied by x-ray diffraction. The filled circles show the results from Boerio-Goates and Westrum (1980) and the open circles those from Dunn *et al.*, (1978). The specific heat is a smooth function of temperature outside the range shown. The enlarged inset shows ΔC_P after subtraction of background. The values of C_{p1}, and C_{p2}, the specific heats just below and above T_c, are 272 and 245 J/mol K. The curves are guides to the eye. (Adapted from HS93.)

diffraction (HS69) but many of its parameters were established by two independent sets of calorimetric measurements (Dunn, Rahman and Staveley, 1978; Boerio-Goates and Westrum, 1980; Fig. 16.7), These showed a small λ-type anomaly at ≈155K, with the enthalpy and entropy of the transition calculated as 222 J/mol and 1.34 J/molK by Boerio-Goates and Westrum. There is another transition at 353K about which very little is known.

In a much more detailed study (Herbstein, Marsh and Samson, 1993; HMS93; Herbstein and Samson, 1993; HS93) the earlier structural results were confirmed, the structure of PYRPMA at 19K was determined and cell dimensions measured as a function of temperature, as well as intensities of some superlattice reflections. No hysteresis of intensities or cell dimensions was encountered. The experimental measurements were analysed using Ehrenfest's criteria for determining the order of a transition and then, more quantitatively, in terms of Landau theory, using the temperature dependence of spontaneous strain and superlattice intensities to determine the nature of the transition. The excellent agreement obtained for these two parameters between theory and experiment was then checked against the excess specific heat, but here agreement was incomplete. Detailed studies by resonance techniques and measurements of elastic constants as functions of temperature are lacking for PYRPMA.

(b) Crystal structure of PYRPMA in ordered and disordered phases. The earlier x-ray studies on PYRPMA (HS69) showed that there were mixed stacks (stack axis [001]) of alternating pyrene and PMDA molecules; in the 295K structure, both moieties were located at crystallographic centres (space group $P2_1/a$, $Z=2$). The c axis was found to have doubled on cooling to ≈ 110K and the space group changed to $P2_1/n$. There were now two pairs of pyrenes at independent centres, with the four PMDA molecules in the cell at general positions, but only slightly displaced from their 295K positions. The transition was single crystal to single crystal, with conservation of axial directions. The crystal structure of PYRPMA at 295K has been redetermined by Allen, Boeyens and Levendis (1989) and described in terms of a model in which the pyrenes are statically disordered over three orientations.

(c) Experimental study of the order–disorder transition. The most striking feature of the cell dimension–T curves (Fig. 16.9) is the *expansion* of b on cooling from 300 to ≈ 165K, followed by the more usual contraction on further cooling. $b - T$ and β–T curve have a cusps at ≈ 165K; a–T and c–T curves show only changes of slope in this region. As $V = abc \sin \beta$, the V–T curve must also show a cusp at T_c. The cell dimensions given in Fig. 16.9 for the disordered phase were extrapolated ("by eye") into the region of stability of the ordered phase for use in the calculation of the spontaneous strain.

Superlattice reflections, corresponding to a doubling of the c axis (i.e. those with l odd in the ordered structure), appear below 165K (Fig. 16.11). A smooth increase of relative superlattice intensities $(I_T/I_0)_{hkl}$ from zero at T_c to unity at $T = 0$K, all $(I_T/I_0)_{hkl}$ showing the same behavior, is characteristic of the occurrence of a second order phase transformation (Cowley, 1980). The behavior of the superlattice intensities on cooling is quite different from the usual temperature dependence of Bragg reflections; these may increase or decrease as the crystal is cooled, in accord with small changes in atomic positions, but uniform behavior is not found (the different behavior of fundamental and superlattice intensities is illustrated below for ANTCYB (Fig. 16.25)).

(d) The Ehrenfest order of the transition. The following items of evidence taken together indicate that the transition is of the second order as defined by Ehrenfest. The V–T curve (Figs. 16.9 and 16.10) is continuous but has a change in slope at $T_c = 160$K; all three thermal expansion coefficients (not shown) have discontinuities at ≈ 165K; there are no superheating or supercooling effects (i.e. no hysteresis); the specific heat C_P has a peak (Fig. 16.7) but not a discontinuity. The detailed behaviour of the cell dimensions (Fig. 16.9) suggests that the real physical situation is more complicated than that implied by the formal definition.

(e) Application of Landau theory. Analysis following Stokes and Hatch (1988; see p. 1–10) gives many possible subgroups G when the high-symmetry space group (G_0)

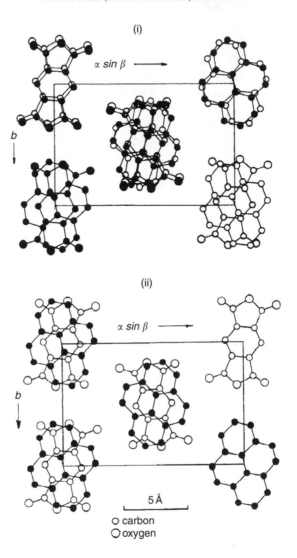

Fig. 16.8. Crystal structure of {pyrene···PMDA}, showing the projections down [001] in (i) the low-temperature ordered structure (ii) the room-temperature disordered structure. In (i) the atoms of the PMDA molecules near c/4 are represented by shaded circles and those near 3c/4 by open circles; the length of [001] is 14.4 Å (at 19K). The length of [001] is 7.3 Å (at 300K). For clarity two molecules at each corner of the projection have been omitted in (i) while in (ii) one pyrene and one PMDA molecule have been omitted; hydrogens omitted throughout for clarity. (Reproduced from Herbstein and Snyman, 1969.)

is $P2_1/c$ (Table 16.3). The description used here has maintained analogous cell axes for both phases with consequent differences in the space groups, whereas Stokes and Hatch follow the converse path. On our basis one of the possibilities for the space group of the low temperature ordered phase is $P2_1/n$ with one axis doubled as reported by HS69 and HS93. The physically irreducible representations for this transition is Y_2^+ and

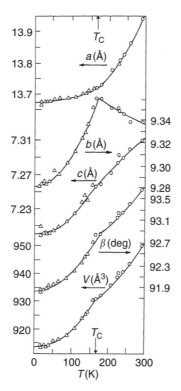

Fig. 16.9. $a(T)$ plotted against $T(K)$; there is a change of slope in the region of Tc ($\approx 165K$) but no cusp. $b(T)$ plotted against $T(K)$; there is a cusp at T_c. $c(T)$ and $c(T)/2$ plotted against $T(K)$. The ordinate values below T_c must be doubled. There are indications of a change of slope at T_c (167K). $\beta(T)$ plotted against $T(K)$. There is a cusp at T_c. $V(T)$ (or $V(T)/2$) plotted against $T(K)$. There is a cusp at T_c (160K). (Reproduced from HS93.)

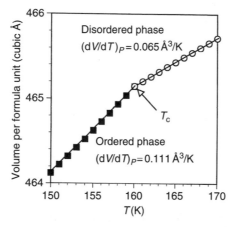

Fig. 16.10. The variation in (volume/formula unit) of the ordered and disordered phases in the vicinity of T_c (=160K), as calculated from quadratic expressions fitted to the cell dimension curves. (Adapted from HS93.)

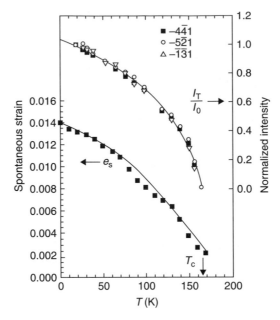

Fig. 16.11. Upper graph – variation of the normalized intensities (I_T/I_0) of superlattice reflections as a function of temperature (I_0 was estimated for each reflection from a smooth extrapolation of the *I*–*T* curve to $T = 0$K). Lower graph – the spontaneous strain in the ordered phase of PYRPMA plotted against temperature. The curves shown are guides to the eye. (Reproduced from HS93.)

a continuous change is permitted by both Landau and Lifshitz conditions and also under renormalization group analysis. The relationship between the space groups was discussed in HS69.

The usual method of deriving the critical exponents depends on fitting a power law equation of the type given above with experimental values measured as close as possible to the critical temperature. As the measurements for PYRPMA covered a wide temperature range but were sparse close to T_c, the Salje extension of classical Landau theory to large values of the order parameter was used. The (scalar) spontaneous strain $e_s = (\Sigma e_j^2)^{1/2}$ in the ordered phase calculated from the Vogt coefficients e_j as a function of temperature is proportional to Q^2, as is the superlattice intensity. Both curves have similar shapes (Fig. 16.11). If the phase transition is tricritical (the intermediate stage between continuous and discontinuous transitions), then it follows from Landau theory that the fourth power of the order parameter is linearly proportional to T; this is indeed found in the plot of $(e_s)^2$ against T (Fig. 16.12). A further test of the tricritical nature of the transition can be made using the intensities of the superlattice reflections. The squares of the normalized intensities against T give a linear plot (Fig. 16.12).

Finally dependence of the excess specific heat on temperature (Fig. 16.7 (insert)) was tested for compatibility with a tricritical transition by plotting log (ΔC_P) against log ($T_c - T$), where $T_c = 155$K. The log–log plot (Fig. 16.13) shows that ΔC_P as a function of $(1 - T/T_c)$ is not well represented by the power law equation $\Delta C_P = A[1 - (T/T_c)]^\alpha$. A forced linear fit gives $\alpha = -0.77$, whereas $\alpha = -0.5$ for a tricritical transition (Salje (1990, p. 120)).

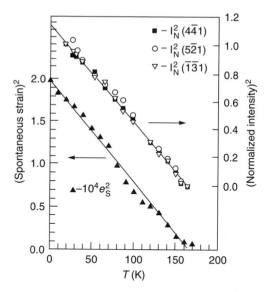

Fig. 16.12. Upper graph – the square of the normalised intensity $(I_T/I_0)^2$ plotted against temperature for three superlattice reflections of the ordered phase of PYRPMA. The origin has been moved up for clarity. Lower graph – the square of the spontaneous strain $(e_s)^2$ in the ordered phase of PYRPMA plotted against temperature. (Reproduced from HS93.)

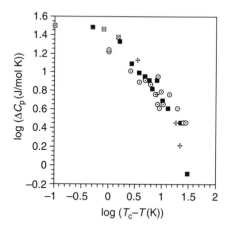

Fig. 16.13. Log–log plot of the excess heat capacity (ΔC_p in Fig. 16.7) against deviation from critical temperature in the vicinity of the phase transition. (Reproduced from HS93.)

A problem with this calculation is that the two independent sets of C_P measurements agree well over the range 0–300K *except* in the region of the peak. A standard calculation of the entropy of transition gives 2.87 J/molK (as there are two orientations, but for only one of the components, a specific heat anomaly of $1/2R \ln 2$ would be predicted); this is about twice the measured entropy of transition of 1.34 J/molK.

The theory of an almost tricritical phase transition with a simple one-component order parameter indicates that the power-law exponents β, obtained from the temperature dependence of the spontaneous strain and the intensities of the superlattice reflections, should be 0.25 and α, obtained from (ΔC_P), should be -0.5 (Salje, 1990). Good agreement with experiment was found for the first of these predictions but not for the second. There is a physical difference in that the diffraction measurements cover the whole temperature range below T_c while (ΔC_p) is restricted to between T_c–35 and T_c.

(f) Crystal structures of disordered and ordered phases. One way of describing the process in which the long range order of the ordered phase is decreased is by comparing ordered and disordered unit cells through their cell dimensions and molecular arrangements. Perhaps a natural approach is to note that the stack axis ([001]) doubles at T_c and to infer that one is dealing with a Peierls phenomenon,[4] due to some change in the π-electron HOMO–LUMO interaction between donor pyrene and acceptor PMDA molecules. However, this would appear to be ruled out by the fact that the [001] axial length changes smoothly with temperature. The most striking change in cell dimension behaviour at T_c is that in [010], where (abnormal) expansion on cooling (of the disordered phase) changes to (normal) contraction (of the ordered phase). This could imply that the interaction between adjacent stacks in the direction of [010] is normal from 0–165K, and then becomes (in some unspecified way) abnormal when the long range order along the stack axis disappears. Expansion on cooling is unusual but not unprecedented; however, neither NAPTCB nor ANTCYB behave in this way. More definite, but still tentative, indications were inferred from an analysis of the molecular arrangement at 19K (Herbstein, 1996, which should be consulted for details).

There is as yet no consensus about the nature of the driving force for ordering. Wide-line NMR studies have been made of {pyrene···PMDA} (Fyfe, 1974a, b). Line width and $(\Delta H)^2$ begin to fall at $\approx$195K, about 30° above T_c but there is no sign of any change in line width or $(\Delta H)^2$ at T_c itself nor at 353K where there is a first-order transition to a phase of unknown structure. The measured value of $(\Delta H)^2$ is 4.9 G^2 at 77K, which is about 20% smaller than the calculated value for a rigid lattice based on the crystal structure, suggesting that there is still appreciable reorientational motion below T_c. A similar lack of correlation between wide-line NMR results and studies of solid-state transitions appears also for other systems discussed here. The explanation (Darmon and Brot, 1967) is that the proportion of molecules jumping from one orientation to another at any given instant is very small and the contribution of these molecules to the thermodynamic functions is essentially negligible. The values of $(\Delta H)^2$ at high temperatures (2.0–2.4 G^2 at 300–420K) are explicable (but still not entirely) only if reorientation of the pyrene molecules takes place by both small angle ($\approx$20°) and large angle ($\approx$160°) in-plane jumps. Measurements of spin-lattice relaxation times (Fyfe *et al.*, 1976) give an activation energy of $\approx$57 kJ/mol for the large angle process; E_a could not be measured for the small angle jumps.

---

[4] The Peierls phenomenon (Peierls, 1954, see Chapter 5) refers to the doubling of the periodicity of an ... ABABAB ... stack at low temperatures because the 'dimerized' arrangement ... AB AB AB ... has lower energy.

16.5.2   The crystal structure of {Naphthalene···TCNB} and evidence for an order ⇔ disorder phase transition around 72K

(a) Introduction.   The order-disorder transitions in the π-molecular compounds {naph-thalene···TCNB} (NAPTCB) and {anthracene···TCNB} (ANTCYB) are rather similar (Lefebvre, Odou, Muller, Mierzejewski and Luty, 1989 (LOMML89); Ripmeester, (1995)). We first discuss NAPTCB and then ANTCYB; comparisons are also made with the (somewhat different) behavior of PYRPMA. The order–disorder behavior in NAPTCB and ANTCYB is similar in the sense that the same space group changes occur, but T_c values are appreciably different at ≈76 and ≈212K (Fig. 16.14).

(b) Crystallographic background and x-ray diffraction measurements of temperature dependence of degree of long range order.   Cell dimensions have been measured by XRD at 294, 95 and 65K (LOMML89) and by neutron diffraction over the range 300–15K

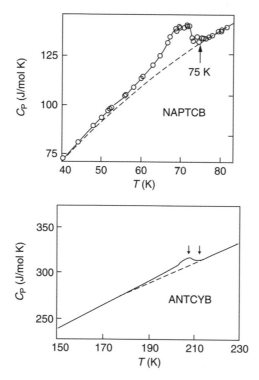

Fig. 16.14. (Upper panel) the calorimetrically measured specific heat of NAPTCB in the region of the order–disorder transition. The C_P–T curve is smooth over the rest of the temperature range 0–300K. (Lower panel) the specific heat of ANTCYB in the region of the order–disorder transition, measured by differential scanning calorimetry (DSC). Ecolivet *et al.* (1988) suggest that there may be a double peak (shown by arrows) in the DSC curve, corresponding to two events, but this remains to be proved. (Reproduced from Ecolivet, Lemée, Delugeard, Girard, Bertault, Collet and Mierzejewski, 1988.)

(Czarniecka *et al.*, 1985) (Fig. 16.15). There are systematic differences between XRD and ND values and closer temperature intervals are required before definite inferences can be drawn about the order of the transition; the indications are that there is no discontinuity in cell volume at T_c, in accordance with a second order character for the transition. Calculation of the temperature dependence of the spontaneous strain requires a better set of cell dimensions.

The NAPTCB and ANTCYB molecular compounds have closely related structures but are not isomorphous. An early room temperature structure analysis on a twinned crystal of NAPTCB (Kumakura, Iwasaki and Saito, 1967) showed that the components were arranged in the usual donor–acceptor mixed stacks, with stack axes along [001] (Fig. 16.16). These results were confirmed (and extended) in a later single crystal study of the structures at 300, 95 and 65K (LOMML89). The same formal description applies to NAPTCB and ANTCYB although the arrangement of the molecules, and their relationship to the cell axes, is different for the two compounds; one structural difference is that in NAPTCB the molecules lie very nearly in (102) planes, but in ANTCYB in ($\bar{1}$02) planes (Fig. 16.17).

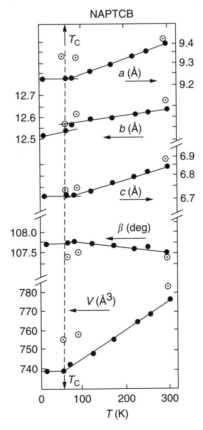

Fig. 16.15. NAPTCB – cell dimensions as a function of temperature, measured by XRD (centered circles, LOMML89) and ND (solid circles, Czarniecka *et al.*, 1985).

At 300K the naphthalene molecules are disordered in equal measure over two orientations separated by 36°, while the TCNB molecules are completely ordered. On cooling, the space group changes from $C2/m$ in the disordered phase to $P2_1/a$ in the ordered phase, without appreciable change in unit cell dimensions. In the $C2/m$, $Z=2$ structure, the naphthalene centres are at Wyckoff positions (a) (000, 1/2 1/2 0) and TCNB centres at Wyckoff positions (c) (00 1/2, 1/2 1/2 1/2); the site symmetry of (a) and (c) positions is $2/m$. In the $P2_1/a$, $Z=2$ structure, the naphthalene centres are at Wyckoff positions (a) (000, 1/2 1/2 0) and the TCNB centres at Wyckoff positions (b) (001/2, 1/2 1/2 1/2); the site symmetry of (a) and (b) positions is $\bar{1}$. The donor molecules take up one (averaged) orientation above T_c and two below; the TCNBs do not change. Above T_c the two stacks

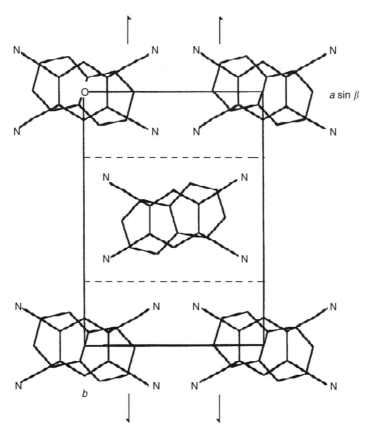

Fig. 16.16. The ordered structure of NAPTCB, space group $P2_1/a$ (hydrogens omitted for clarity). In the diagram the centers of symmetry at cell corners, centers of cell edges and at cell center have been omitted. The site symmetry is for both components. The naphthalenes are centred at Wyckoff positions (a) [000; 1/2,1/2,0] and the TCNB's at Wyckoff positions (b) [00,1/2; 1/2,1/2,1/2]. The molecules lie nearly in ($10\bar{2}$) planes. The disordered structure (space group $C2/m$) has essentially the same cell dimensions, but the units at cell corners and center are now identical (i.e. related by translations). The naphthalene molecule is disordered over the orientation shown and its mirror image, with equal occupancies. The site symmetry is $2/m$ for both components, the two fold axis running along [010] and the mirror plane lying in (010). (Adapted from Kumakura *et al.*, 1967.)

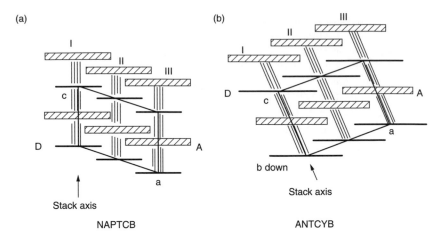

Fig. 16.17. Comparison of the stack arrangements in NAPTCB (at 300K: 9.420 12.684 7.31Å, 107.4°, $Z = 2$, $C2/m$) and ANTCYB (at 300K: 9.526 12.780 7.440 Å, 92.36°, $Z = 2$, $C2/m$), with the donor molecules (respectively naphthalene and anthracene) labelled as D and denoted by thick lines, while the acceptor molecules (TCNB) are labelled A and denoted by hatched rectangles. Stacks I and III are translationally equivalent, while stack II is shifted by $b/2$ behind the plane of the page and is related to I by the a-glide. (Adapted from LOMML89.)

(stack axis [001]) in the unit cell are related by the C-centering operation while below T_c the two stacks are related by the a-glide. The structure can be described as a close packed arrangement of stacks of elliptical cross section, with the packing of the stacks shown in Fig. 16.16. Within a particular stack, all donor molecules are translationally equivalent, as are all acceptor molecules; this applies to both disordered and ordered structures. In the disordered structure the arrangement of components along a stack is

$$\ldots\ldots\langle D\rangle \ A \ \langle D\rangle \ A \ \langle D\rangle \ A \ \langle D\rangle \ldots$$

where $\langle D\rangle$ represents the averaged orientation of the donor molecules in a stack. In the fully ordered structure the arrangement of components along stack I (Fig. 16.16) is

$$\ldots\ldots D_1 \ A \ D_1 \ A \ D_1 \ A \ D_1 \ldots \text{ and that in stack II is} \ldots D_2 \ A \ D_2 \ A \ D_2 \ A \ D_2 \ldots$$

where D_1 and D_2 are the two orientations of the donor molecules, related by the a-glide; the TCNBs have the same orientations in both stacks but this is not required by the crystal symmetry. Thus the disordering process, viewed from "within stack I", is to produce sections of

$$\ldots D_1 \ A \ D_2 \ A \ D_1 \ A \ D_1 \ldots$$

and analogously for stack II.

Analysis following Stokes and Hatch (1988; see p. 1.8–9) gives many possibilities for a contiuous transition from the high symmetry (G_0) space group $C2/m$; one of these is to $P2_1/a$, without change of cell dimensions or shift of origin, the physically irreducible

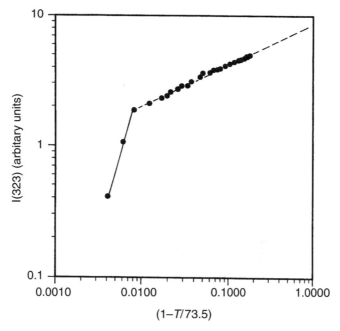

Fig. 16.18. $I(323)$ against $(1 - T/73.5)$ for NAPTCB, both on a logarithmic scale; the intensity measurements (in arbitary units) extend over the range 60–75K. Apart from the two points closest to T_c, the measurements are well fitted by $I(323) \propto (1 - T/73.5)^{0.334}$.

representation being Y_2^+ and a continuous change being permitted by both Landau and Lifshitz conditions and also under renormalization group analysis. The relationship between the space groups is discussed above.

The critical exponent can be determined from the temperature dependence of the intensities of the superlattice reflections; the necessary measurements (by LOMML89) are available only for reflection (323). We have noted above (equation (16.2)) that

$$Q^2 \propto I_{\text{superlattice}} = A(1-T/T_c)^{2\beta}$$

The log/log plot (Fig. 16.18) shows that the exponent $2\beta = 0.334$ except in the region closest to T_c, where a higher value could be appropriate. $\beta = 0.33$ is generally considered to indicate an order–disorder transition so that there is a considerable discrepancy here, perhaps due to the fact that I(323) was measured over a range of only 13.5K (compared to ranges of $\approx$145K for the analogous measurements for PYRPMA and ANTCYB). The values of the order parameter Q at 65K from various sources can be compared as follows: an extrapolated value of $I(323) = 9$ at 0K is obtained from Fig. 16.18, with $I(323) = 4.3$ at 65K (from Fig. 6 of LOMML89). Thus $Q(65K) = (4.3/9)^{1/2} = 0.7$, which is not in good agreement with the NMR value (essentially unity) or that of LOMML89 from XRD ($Q(65K) = 0.90$). The diffraction measurements of superlattice intensities at different temperatures for ANTCYB are more extensive than those for NAPTCB and we discuss extraction of order parameters and critical exponents below.

(c) Other physical measurements.   Specific heat measurements (Boerio-Goates, Westrum and Fyfe, 1978) (Fig. 16.14(a)) show that there is a gradual transition over the range 40–75K ($\Delta H_{trans} = 192 \pm 21$ J/mol, $\Delta S_{trans} = 3 \pm 0.3$ J/mol K); ΔH_{trans} is not very different from the value found for the transition in PYRPMA (222 J/mol) but the ΔS_{trans} values differ by a factor of ≈ 2 because of the difference in T_c values. On the basis of the structural results given above, we note that ordering of the molecules below T_c would give a specific heat anomaly of $1/2R \ln 2$, because there are two orientations, but only for one of the components. This prediction (2.87 J/molK) is in excellent agreement with the measured entropy of transition. Unfortunately the same argument applied to PYRPMA resulted in a discrepancy of 100%! The anomalous regions in the C_P–T curves of NAPTCB and ANTCYB have strikingly similar shapes that differ from the classical λ shape found in PYRPMA. The significance of these differences is not clear.

The occurrence of a phase transition was confirmed by Raman spectroscopy which placed T_c at 69K and 62K for the $C_{10}H_8$ and $C_{10}D_8$ molecular compounds respectively (Bernstein, Dalal, Murphy, Reddoch, Sunder and Williams, 1978).

The donor molecule (labelled D in Fig. 16.17) at 000 (for NAPTCB) or 100 (for ANTCYB) impinges on the TCNB molecule at 1,0,1/2 (labelled A in Fig. 16.17); the H...N interactions between these two components are responsible for the ordering. This is illustrated for NAPTCB at 65K in Fig. 16.19.

There are three N...H–(C) interactions at the $\Phi = 0°$ equilibrium position while there are only two such interactions if the naphthalene is rotated in its own plane to $\Phi = 36°$. At both these positions the N...H nonbonded distances are close to the sum of the van der Waals radii. If the naphthalene molecule took up an intermediate orientation at $\Phi = 18°$, the closest N...H nonbonded distances would be 0.2–0.3 Å *less* than the sum of the van der Waals distances. This suggests that the transition is driven by intermolecular inter-actions ("packing effects"), in accordance with the lowering of T_c when hydrogen is replaced by deuterium (Bernstein, Dalal, Murphy, Reddoch, Sunder and Williams, 1978; Dalal, Ripmeester *et al.*, 1978). These qualitative considerations can be rendered quant-itative by theoretical calculations and the results of measurements by various resonance techniques.

A double well potential energy curve for naphthalene in the field of the surrounding TCNB molecules was calculated by Shmueli and Goldberg (1973) with barrier height about 12kJ/mol and minima separated by 36°, in accordance with the 300K crystal structure; naphthalene to TCNB interactions were found to be the most important.

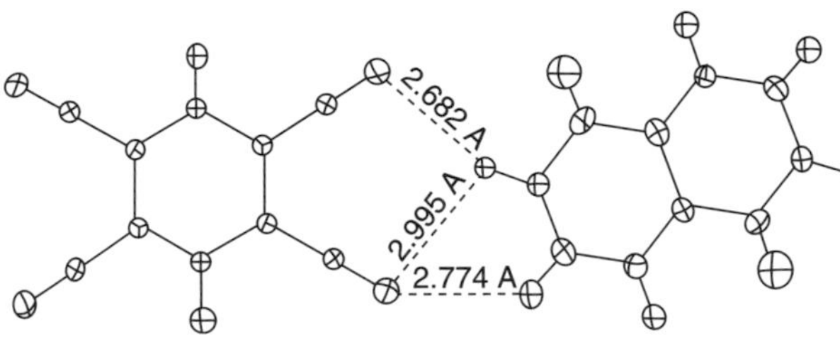

Fig. 16.19. Two adjacent molecules in the $(10\bar{2})$ plane of {naphthalene···TCNB} (almost the plane of the molecules), showing the closer N...H distances in the ordered crystal structure. (Reproduced from Lefebvre *et al.*, 1989.)

A converse, quasi-experimental, analysis was carried out by LOMML89. They calculated structure factors for the disordered structure at 294 and 95K in terms of an orientational probability function for the disordered molecule. For NAPTCB the best fit between observed and calculated structure factors was obtained when the orientational probability function had two maxima at $\pm16°$ (at 294K) or $\pm18.5°$ (at 95K). An angle of $\approx36°$ between the energy minima was measured by ^2H NMR spectroscopy (Ripmeester, 1982). Thus these three approaches agree in describing the disordered structure as statically disordered, with the two orientations separated by a barrier height of ≈12kJ/mol.

Earlier wide line (Fyfe, 1974a, b) and pulsed NMR measurements (Fyfe, Harold-Smith and Ripmeester, 1976) have since been extended to lower temperatures, and supplemented by ESR measurements (Ripmeester, Reddoch and Dalal, 1981). We consider all these results together. Pulsed NMR measurements of relaxation times (Fig. 16.20; $T1$, the spin–lattice relaxation time, and $T_{1\rho}$ the spin–lattice relaxation time in the rotating frame) were used to obtain activation energies of 9.4 (E_a) and 42.6 kJ/mol (E_b) for small and large angle reorientations respectively, to be compared with calculated values of 8 and 46 kJ/mol (Fyfe, Harold-Smith and Ripmeester, 1976); only $T_{1\rho}$ changes noticeably at T_c.

The spin–lattice relaxation times (Fig. 16.20) were also used to estimate ΔE (the energy difference between sites 1 and 2) as a function of temperature (Ripmeester, Ratcliffe et al.,

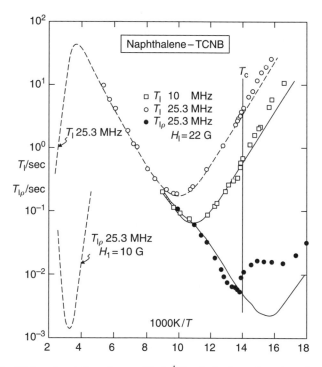

Fig. 16.20. Static (T_1) and rotating frame ($T_1\rho$) ^1H spin-lattice relaxation times measured for NAPTCB. (Spin–lattice relaxation times vs. 1000/T.) Dashed lines represent measurements from Fyfe, Harold-Smith and Ripmeester, 1976. (Reproduced from Ripmeester, Dalal and Reddoch, 1981.)

1995). These values increase from 1.5 kJ/mol to 7.15 kJ/mol over the range 71.5–63K, and can be fitted to $\Delta E = -6.72T + 49.54$ ($R^2 = 0.99$) (Fig. 16.20). Extrapolation gives $T_c = 73.7$K ($\Delta E = 0$), in good agreement with the value obtained from temperature of disappearance of the 323 superlattice reflection (Fig. 16.18; 73.5K; LOMML89). The anomalous increase of the specific heat occurs (on cooling) at 75K (Fig. 16.14(a)). *Linear* extrapolation, which seems inherently unlikely, would give $\Delta E = 49.5$ kJ/mol at $T = 0$K. The ratio of the occupancies of the two sites 1 and 2 is given by $p_2/p_1 = \exp(\Delta E/RT)$ and can be calculated from the ΔE–T dependence. At 71.4K, $p_1 = 7.2\%$, falling to 0.2% at 68.6K and is essentially zero (i.e. complete ordering) at lower temperatures. LOMML89 measured $p_1 = 4.8\%$ from their 65K crystal structure analysis, using the heights of residual (difference) electron density peaks in the alternative orientation; possibly their temperature was slightly underestimated. A diagram of potential energy as a function of orientation of the naphthalene molecule in the unit cell is shown in Fig. 16.22 for the ordered and disordered phases.

The correlation times (average time between instantaneous 36° jumps) have been measured by NMR (Ripmeester, Dalal and Reddoch, 1981), ESR (Grupp, Wolf and Schmid, 1982) and incoherent quasi-elastic neutron scattering (QNS) (Czarniecka *et al.*, 1985). The values obtained by the different methods for the disordered structure all fall on the same straight line in a plot of log τ against $1000/T$ (Fig. 16.23). However, a distinct break occurs at T_c. The QNS measurements give an average residence time at room temperature of 8 ps. Density of states curves for the components and the molecular compound have been obtained by incoherent inelastic neutron scattering.

The 15K ESR measurements of Erdle and Möhwald (1979) showed that the triplet state was largely localized on naphthalene ($Z_t \approx 0.25 \pm 0.05$) and that the normal to

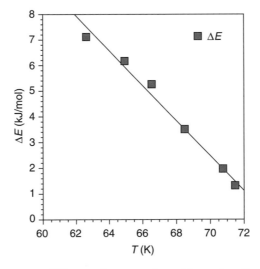

Fig. 16.21. ΔE (the energy difference between favorable and unfavorable orientations of naphthalene (sites 1 and 2) in the ordered phase) as a function of temperature for NAPTCB; these values were redrawn from those given by Ripmeester, Ratcliffe *et al.* (1995).

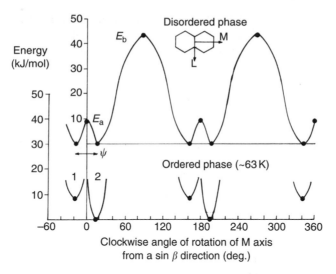

Fig. 16.22. Potential for the reorientation of the naphthalene molecule in the disordered and ordered phases of NAPTCB; E_a, E_b and ΔE are defined in the text. ψ is 36°. In the disordered phase, a particular naphthalene molecule is located in the isoenergetic (and hence equally populated) minima about 0° (and 180°) and separated by ψ. In the ordered phase these two minima no longer have equal energies (energy difference as function of temperature shown in Fig. 16.21) and one is favored and the other disfavored. (Adapted from Ripmeester, Ratcliffe *et al.* (1995).)

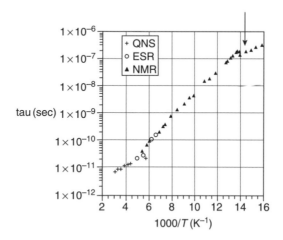

Fig. 16.23. Correlation time as a function of temperature for NAPTCB. NMR measurements from Ripmeester, Dalal and Reddoch, 1981; ESR from Grupp, Wolf and Schmid, 1982; QNS from Czarniecka *et al.*, 1985. The latter authors note ... "we conclude that the correlation times determined by NMR and QNS methods corroborate each other, and we are inclined not to consider the difference in slopes of the NMR and QNS results as real.." Arrow marks T_c. (Adapted from Czarniecka *et al.*, 1985.)

the plane of the triplet state naphthalene molecules was tilted by $10°$ with respect to [001]. However, this tilt angle was less than $2°$ in the 65K structure. This indicates that formation of the excited state leads to a geometrical change, a phenomenon for which there is considerable evidence from other sources (Ponte Goncalves, 1977).

(d) The phase transition in summary. A rather complete picture emerges when the results from the various experimental methods are concatenated. The transition can be inferred to be second order (in agreement with LOMML89) from the form of the specific heat curve (Fig. 16.14(a)) and from the absence of a discontinuity in cell volume at T_c (=73.5K) (Fig. 16.15). At very low temperatures the naphthalene and TCNB molecules are completely ordered in space group $P2_1/a$, each moiety being located at a crystallographic centre of symmetry. The long axes of the two naphthalenes are $36°$ apart and their mean molecular planes are mutually tilted by $2°$. The TCNB molecules are so located that they are all parallel, although this is not required by the space group. The temperature dependence of the intensity of a superlattice reflection indicates that the long range order begins to fall below its 0K value of unity as the crystal is heated. There is some disagreement about the temperature at which this becomes appreciable, one interpretation of the XRD measurements suggesting fairly low temperatures, while another suggests $\approx 65K$, while NMR relaxation times indicate that this process occurs even closer to the critical temperature of $\approx 73K$. The average energy difference between a naphthalene molecule favorably and unfavorably oriented at an inversion centre is $\approx 7kJ/mol$ at 63K, falling to zero at T_c and rising to some tens of kJ/mol at 0K. There is a second-order 'order to disorder' transition at $\approx 73K$, accompanied by an anomalous additional specific heat corresponding to a measured entropy of transition of 3 ± 0.3 J/mol K, which is very close to the expected value of $1/2R \ln 2$. As the temperature approaches T_c from below, the probability of finding an unfavorably oriented naphthalene molecule increases. Unfavorably oriented molecules lead to a decrease in the energy difference between sites with correctly and unfavorably oriented molecules, and this cooperative effect leads to complete loss of long range order as T_c is reached. The driving force for disordering on heating is the small energy difference between correctly and unfavourably oriented molecules in a particular lattice site, which decreases as the temperature approaches T_c from below.

Above T_c the space group changes to $C2/m$, without change in cell dimensions. The site symmetry of the sites occupied by naphthalene and TCNB molecules increases to $2/m$. This is achieved for naphthalene by statistical occupation of a particular site by the molecule in two orientations separated by $36°$, while the orientation of the TCNB molecule (which hardly changes from that in the ordered structure) is now fixed by the $2/m$ symmetry of its site. Above T_c the potential barrier between the two naphthalene orientations, which are of equal energy, is 8.0 kJ/mol. Nothing seems to be known about possible existence of short range order above T_c – in other words, the diffuse scattering in the diffraction pattern requires investigation.

One anomaly awaits explanation – the pressure dependence of T_c. One can make an approximate calculation using the results given in Figs. 16.14(a) and 16.15 ('approximate' because of the nature of the experimental evidence) similar to that carried out for PYRPMA and this gives $dT_c/dP = -31$ K/kbar, contrary to the expectations from Fig. 16.6. No explanation has been offered.

16.5.3 The crystal structure of {Anthracene···TCNB} and evidence for an order ⇔ disorder phase transition at ≈213K

(a) Introduction. The molecular compound {anthracene···TCNB} (ANTCYB) has been extensively studied. The specific heat has been measured by DSC (Fig. 16.14(b); Ecolivet, Bertault, Mierzejewski and Collet, 1987) and the anomalous region is remarkably similar in shape to that found for NAPTCB and shown in Fig. 16.14(a); $\Delta H_{trans} = 150$ J/mol and $\Delta S_{trans} = 0.7$ J/mol K. The enthalpy of transition is similar to that found for NAPTCB (192 ± 21 J/mol) but the entropy of transition is smaller by a factor of 4, again (as in PYRPMA) a reflection of the higher transition temperature. A Brillouin scattering study of the elastic anomalies at the phase transition confirms that $T_c = 212 \pm 0.5$K. This study also allowed determination, in the ordered phase, of the order parameter relaxation time as $[5 \times 10^{-11}/(T_c - T)]$ s K^{-1} (Ecolivet and Mierzejewski, 1990).

(b) Crystallographic background. The crystal structure at 300K was first determined by Tsuchiya, Marumo and Saito (1972) and structures above *and* below T_c first by Stezowski (1980; crystal structures at 297, 234, 226, 202, 170, 138 and 119K), and then by LOMML89 (crystal structures at 294, 225 and 65K). There is a comprehensive series of cell dimension – temperature measurements for ANTCYB that show very weak cusps at T_c but no discontinuities (Fig. 16.24). These results show that the spontaneous strain in the ordered phase is almost entirely in the [100] direction. The disordered structure is isostructural with that of NAPTCB but the two ordered structures differ in the gross sense (as noted above) that the naphthalene molecules are in (102) planes but the anthracene molecules in ($\bar{1}$02) planes, and in a subtle sense in that the TCNBs in ANTCYB change their orientations slightly (by ≈2°) on ordering but not in NAPTCB. Stezowski used his intensity measurements to calculate molecular libration amplitudes using the rigid body model. Electron density plots at the different temperatures were also calculated, from which the angles between various axes of symmetry-related molecules were obtained; these angles are in good agreement with values deduced from ESR measurements by Möhwald, Erdle and Thaer (1978). However, this approach seems to ignore the essential order-disorder features of the system, as pointed out by Park and Reddoch (1981); in particular, the electron density plots and the (inferred) libration amplitudes will contain contributions from both thermal vibrations and disorder.

The LOMML89 and Stezowski measurements of the superlattice ($h + k$ odd) reflection (140) and the fundamental reflection (240) are in excellent agreement; the different modes of temperature dependence of these different reflection types are shown in Fig, 16.25. A check on whether the superlattice reflections all show the same dependence of F_{obs} on T, as required by theory, shows a considerable spread and more closely spaced measurements like those made for (140) appear to be required.

(c) ESR measurements of the order parameter as a function of temperature. The temperature dependence of the order parameter has also been determined by ESR and can be compared with the diffraction measurements. Measurements on the ESR spectra of the triplet excitons produced by optical irradiation of ANTCYB in its charge transfer band give the temperature dependence of the long range order parameter. Below T_c, for a

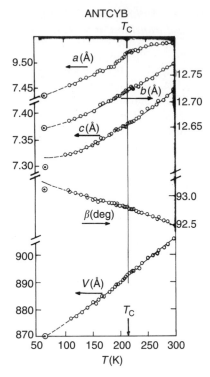

Fig. 16.24. ANTCYB – cell dimensions as a function of temperature, including 65K values (centered circles), are from LOMML89. The cell dimensions given by Stezowski (1980) at 297, 234, 226, 202, 170, 138 and 119K are in satisfactory agreement (to within ≈0.01 Å, 0.2°) with those of LOMML89. The curves are guides to the eye.

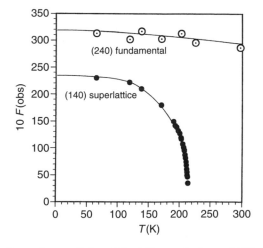

Fig. 16.25. ANTCYB: Dependence of F_{obs} of the 140 superlattice and 240 fundamental reflections on temperature. The measured values are from LOMML89 and Stezowski (1980). The curves are guides to the eye.

general orientation, the ESR spectrum for each of the $\Delta m_s = \pm 1$ transitions consists of two lines, corresponding to the fine structure tensors of the two sublattices (which are related, as noted earlier, by the a glide plane). As the temperature is raised, these two lines approach and broaden, coalescing near T_c into a single line that becomes narrower at higher temperatures. The separation between the two lines gives the long range order parameter $S = S_0(1 - 2p)$, where S_0 would be the separation for a perfectly ordered crystal, and p is the fraction of sites occupied by unfavorably oriented molecules. This experiment was first done for ANTCYB by Möhwald, Erdle and Thaer (1978), who identified their order parameter with the angle between the long axes of the anthracene molecules. Park and Reddoch (1981) pointed out that this approach suffers from the same deficiencies as that of Stezowski.

We quote from Park and Reddoch "An exciton moving rapidly along a chain [stack in our nomenclature] will then sample this disorder [of the orientations of the anthracene molecules along the stack]. In the fast limit the position of its resonance line will be the average of the lines for the two orientations, weighted by the relative populations within a given chain. Such a fast-limit resonance can thus shift with temperature, but need not broaden. If this fast moving exciton now jumps at a slow rate to a neighboring chain which is predominantly of opposite orientation, line broadening may be expected... One way to treat this problem is to consider a four-site exchange model with two rate constants and four sites, consisting of the two orientations in each of the two sublattices... The separation [of the lines] S is... a statistical average weighted by the number of correctly [x_A] and incorrectly [x_B] oriented anthracenes in the chain [$(x_A + x_B) = 1$]. Thus

$$S = S_0 \, (x_A - x_B) = S_0(1 - 2x_B)$$

where S_0 would be the separation for a perfectly ordered crystal." Park and Reddoch derived x_B (our p) from their experimental measurements as a function of $1/T$ and their values have been replotted as Q ($=S/S_0$) against T, to which we have added the XRD values of Q derived from reflection 140 (see Fig. 16.26, which should be compared with Fig. 16.25). There is excellent agreement between Q(ESR) and Q(XRD) (from 140) showing that both have the same functional dependence on T/T_c.

We next plot $\log(I_T/I_0)$ against $\log(1 - T/T_c)$ (a form that allows direct comparison with the NAPTCB analog in Fig. 16.18) to test whether the power law equation (16.2) is applicable; Fig. 16.27 shows that a linear relation is not obtained. Park and Reddoch reported $\beta = 0.34$ for the range 198–208.3K and made a comprehensive comparison of their experimental results with various theoretical treatments and came to the conclusion that the renormalization group calculation gave best agreement with their value for β. This is not surprising as there is evidence for the occurrence of fluctuations in the temperature region for which their value was derived.

Park and Reddoch determined the rate constant for interchain hopping as $\omega_{12} = [1.6 \times 10^7 + 1 \times 10^{11} \exp(-E_a/kT)] \text{ s}^{-1}$, where $E_a = 1051 \text{ cm}^{-1}$ ($= 12.6$ kJ/mol). The second term was tentatively interpreted as due to scattering of triplet excitons by anthracene molecules undergoing large amplitude vibrations. E_a is the barrier between A and B orientations. The corresponding activation energy in NAPTCB was given as 9.4 kJ/mol.

(d) The phase transition in summary. The weight of the evidence suggests that the driving force for the phase transition comes from intermolecular packing interactions, and

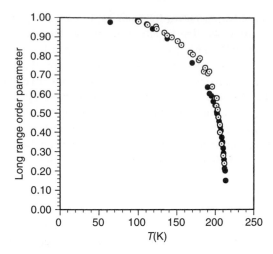

Fig. 16.26. ANTCYB – long range order parameter Q as a function of temperature. The open circles show values derived from the splitting of ESR lines and are replotted from Fig. 1 of Park Reddoch (1981); the dots show XRD values for $Q(140)$ and are transferred from Fig. 16.25.

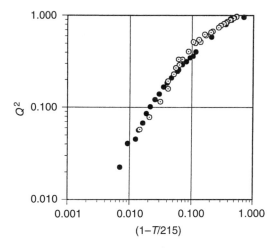

Fig. 16.27. ANTCYB – ESR and XRD values of Q^2 plotted against $(1-T/T_c)$ ($T_c = 215K$); the scales are logarithmic. The data are replotted from Fig. 16.26.

not from orientation dependence of charge transfer interactions. One important indication comes from the lowering of T_c in ANTCYB with increasing deuteration of anthracene (h_{10} – 214.5 (± 0.5) K; β-d_4 – 202.5; α-d_4 – 201.5; d_{10} – 198.5), that has been ascribed (Dalal, Haley, Northcott, Park, Reddoch, Ripmeester, Williams and Charlton, 1980) to reduction in intermolecular repulsions on deuteration ($d(C-D) < d(C-H)$). A similar effect found for NAPTCB has already been noted. There are resemblances between the NAPTCB and ANTCYB situations but details differ. At the two equilibrium positions separated by an in-plane rotation angle of 16° (Φ) the *average* N . . . H distance is close to the sum of the van der Waals radii (in NAPTCB $\Phi = 36°$, the *average* N . . . H distance again being close to the sum of the van der Waals radii). If the anthracene molecule took up an intermediate orientation at $\Phi = 8°$, the closest N . . . .H non-bonded distances would be 0.18 Å less than the sum of the van der Waals distances (0.28 Å in NAPTCB). The difference of 0.1 Å could lead one to expect dynamic disorder in ANTCYB instead of the static disorder found in NAPTCB. The potential energy curves calculated for an anthracene molecule librating in the field of its nearest neighbours have been found to show a broad single-well minimum rather than the double potential well noted above for NAPTCB. The orientational probability function determined by LOMML89 shows a single maximum at $\Phi = 0°$. The differences between the two structures have led LOMML89 to suggest that the phase transition at ≈ 213K has elements of both a 'displacive' and a 'disorder to order' transition. In this sense 'displacive' means that the molecules (here only anthracenes) undergo a continuous change in orientation as the temperature is lowered below T_c (this is similar to the points of view expressed by Möhwald and Stezowski) while 'disorder to order' means that the fraction of molecules in the correct orientation for a particular sub-lattice increases as T falls below T_c. The matter is not yet settled.

A semiquantitative analysis of the pressure dependence of T_c can be carried out using equation (16.4). From Figs. 16.14(b) and 16.24, we calculate $dT_c/dP = 80$ K/kbar; a measured value of 32 K/kbar has been reported (Ecolivet, Lemée, Delugeard, Girard, Bertault, Collet and Mierzejewski, 1988). The agreement seems reasonable considering the difficulties of interpreting the specific heat curve (Fig. 16.14(b)) and of inferring values of $\partial V/\partial T$ from Fig. 16.24, where the slopes must be measured close to T_c and not averaged over the whole 0–300K temperature interval, which would give zero slope difference.

We shall not discuss the many preliminary theoretical calculations made on ANTCYB but give two leading references (Brose, Luty and Eckhardt, 1990; Kuchta, Luty and Etters, 1990).

16.5.4 *Other examples of second order transitions*

There is clear evidence for two orientations for anthracene in {anthracene···TCNQ} at 300K (Williams and Wallwork, 1968). These crystals are isomorphous with those of the disordered structure of {naphthalene···TCNB}, with [001] stack axis in both examples. It seems reasonable to infer that {anthracene···TCNQ} will show a disorder-to-order transition on cooling, perhaps also to space group $P2_1/a$ without appreciable change of cell dimensions.

Heat capacity measurements on {pyrene···TCNB} showed an approximately symmetrical anomaly in the range 220–250K; the details were difficult to reproduce even with samples of the same thermal history and there was facile supercooling of the 290K crystals down to at least 150K (Clayton *et al.*, 1976). Thermal hysteresis of $\approx 40°$ has been reported and shattering of the crystals, on cooling through a transition at 183 ± 3K. The crystal structures at 290 and 178K were found to be similar (Prout, Morley *et al.*, 1973) and it is possible that supercooled crystals were used for the 178K diffraction measurements. The structure analysis showed the pyrene molecules to be azimuthally disordered and the TCNB molecules ordered; the available NMR measurements (Fyfe, 1974a, b) are not detailed enough for comparisons to be made with the x-ray diffraction and C_p results. Further study is needed before conclusions can be drawn. There are also indications of pyrene disorder in {pyrene···TCNQ} (Prout, Tickle and Wright, 1973).

Low temperature phase transformations without change of crystal system have been demonstrated in {anthracene···C_6F_6} and {pyrene···C_6F_6} (see Bernstein (1967)) – these are probably disorder to order transitions. Measurements of reorientational motion of both components have been made in {pyrene···C_6F_6}, using ^{19}F in addition to proton resonance. The two molecules behave rather similarly. A decrease in the second moment of the ^{19}F line above 130K was attributed to onset of reorientation about the hexad axes of hexafluorobenzene, which would not be detectable by x-ray diffraction or C_p measurements; such reorientation has been detected by NMR in {benzene···C_6F_6} (Gilson and McDowell, 1966).

The crystal structures of two polymorphs (yellow and orange) of {naphthalene··· PMDA}, which are presumably monotropically related, have been reported (LeBars-Combe et al., 1979). The crystal structure shows that the yellow polymorph is ordered at 293K; specific heat and Raman scattering measurements (Macfarlane and Ushioda, 1977) show that there are no transitions in the range 2–300K. Donor and acceptor molecules have their long axes parallel over the whole temperature range; such an eclipsed mutual orientation is unusual. In the orange polymorph, for which only crystallographic results are available, two orientations (separated by 42°) are found for the naphthalene molecules at a particular site, with unequal populations of 72 and 28%. Although one might expect a disorder to order transition on cooling (perhaps analogous to that in {pyrene···PMDA), none has been found in the range 300–148K.

16.6   Crystals with first order phase transformations on cooling

16.6.1   {Cycl[3.2.2]azine···TNB}

The first order transformation in {cycl[3.2.2]azine···TNB} (orange needles elongated along [001]) is monoclinic to monoclinic (Table 16.4), with clear discontinuities in the cell dimensions at T = 143 ± 3K (Fig. 16.28) (Hanson, 1978). The transition was

Table 16.4. {Cycl[3.3.2]azine···TNB} – unit cell dimensions (Å, deg.) at 293 and 91K. Describing the 91K structure in terms of a non-standard B-centered cell shows that the a and c dimensions have approximately doubled as a result of the transformation

	B 293K	A 91K	
		centered	primitive
a	14.37(1)	28.474(15)	15.528(8)
b	15.99(1)	15.636(8)	15.636(8)
c	6.682(4)	13.102(6)	13.102(6)
β	92.17(5)	91.39(4)	113.56(5)
V(Å^3)	1534.3	5831.5	2915.9
Z	4	16	8
Space group	$P2_1/n$	$B2_1/d$	$P2_1/c$

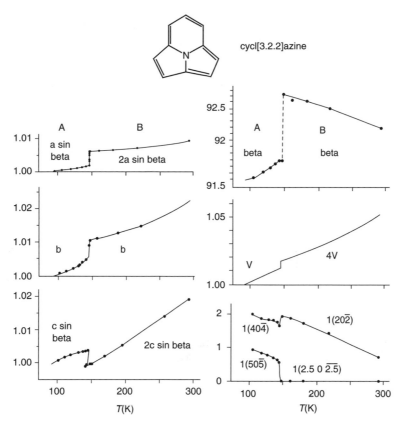

cycl[3.2.2]azine

Fig. 16.28 {Cycl[3.2.2]azine···TNB} – variation of some crystal properties with temperature. B refers to the $P2_1/n$ disordered structure and A to the $B2_1/d$ ordered structure given in Table 16.4. Cell lengths and volume are expressed as multiples of the values at 91K; the temperature dependence of one fundamental reflection and one "superlattice" reflection is illustrated. (Adapted from Hanson, 1978.)

reversible but there was a small hysteresis between the transition temperature on cooling (140K) and heating (146K); the transition was single crystal to single crystal.

The 293K structure has mixed stacks of alternating cycl[3.2.2]azine and TNB molecules along [001]; the former take up three (or more) orientations in disordered fashion, and calculations suggest that there is no steric reason why the cycl[3.2.2]azine molecule should not rotate (presumably in hindered fashion) in its own plane. In the ordered structure the stack axis periodicity has doubled and the two TNB molecules differ in orientation by 12°; the two cycl[3.2.2]azine molecules also take up (to a good approximation) two orientations. The two TNB molecules change their orientations slightly as the ordered form is heated, and become indistinguishable above 143K; at the same time the constraints on the orientations of the cycl[3.2.2]azine molecules are relaxed and these molecules rotate, at random, into their high temperature orientations. Some of these rotations are through angles as large as 134°. There are resemblances to the {pyrene···PMDA} system.

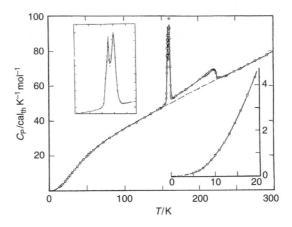

Fig. 16.29. Measured heat capacity for {naphthalene···TCNE}; the dashed line shows the lattice heat capacity. The splitting of the peak at ≈155K is shown at top left, and the (very) low-temperature specific heat at lower right. (Reproduced from Boerio-Goates and Westrum, 1980a)

16.6.2 *Other examples*

There are transformations from monoclinic room temperature structures to triclinic low temperature structures in {TMPD···chloranil} (Boer and Vos, 1968), {naphthalene···TCNE} (Bernstein, 1967) and {TDT[5]···TNB} (Williams and Wallwork, 1966). These are first order transformations but may also have disorder–order features; low temperature crystal structures are not known. [14]N NQR measurements show that there are two independent N atoms in {naphthalene···TCNE} at 77K (Onda *et al.*, 1973), which is compatible with the triclinic symmetry found by x-ray diffraction at a somewhat higher temperature. However, NMR measurements (Fyfe, 1974b) do not give any indication of the complicated phase behavior revealed by C_p measurements (Fig. 16.29; Boerio-Goates and Westrum, 1980a).

16.7 Physical nature of the disordered phase

Disorder of one or other of the components is a not-unusual feature of mixed stack π-molecular compounds. A key question is whether such disorder is static or dynamic. A condition of static disorder at low temperature may well change to one of dynamic disorder at higher temperature. It is perhaps worth emphasizing that it is much more difficult to describe a disordered than an ordered state.

The experimental methods available can be divided into three groups:

(a)   Those sensitive to the presence of disorder, but not to reorientational motions. The principal methods are x-ray and neutron diffraction (as employed for crystal structure analysis by use of Bragg reflections).

(b)   Those sensitive to reorientational motions but not to the presence of disorder. NMR line width and spin–lattice relaxation time measurements come into this category.

---

[5]   2,4,6-tris(dimethylamino)-1,3,5-triazine

(c) Those sensitive both to disorder and to reorientational motion. Raman scattering and the behavior of triplet excitons are suitable methods. Elastic and inelastic scattering of neutrons are also powerful techniques in the study of phase transformations (Axe, 1971) but have hardly been applied as yet to the study of π-molecular compounds.

The most direct method of demonstrating the occurrence of disorder is *via* x-ray (or neutron) crystal structure analysis, particularly if the molecular compound has been studied at only one temperature, the most common current situation. Under these circumstances the disorder is revealed by the atoms of one or other of the components having abnormally large displacement factors. The room temperature structure of {naphthalene···TCNE} ($P2_1/a$, $Z = 2$, both components at centers of symmetry) provides a convenient example (Williams and Wallwork, 1966), which has been much discussed. The atomic displacement factors are much larger for naphthalene than for TCNE (Fig. 16.30) and the difference increases for atoms at the periphery of the molecule, suggesting large librational motions, or static disorder.

These results can be shown in another way through electron density and difference electron density syntheses; the electron density synthesis is essentially independent of any postulated model. A comparison is made for the disordered naphthalene and ordered TCNE molecules in Fig. 16.31.

Two models can be postulated at this stage:

(a) large in-plane librations of naphthalene about the normal to the molecular plane, corresponding to the dynamic disorder defined above, (Fig. 16.32(a)); or

(b) static disorder of naphthalene in two orientations (Fig. 16.32(b)).

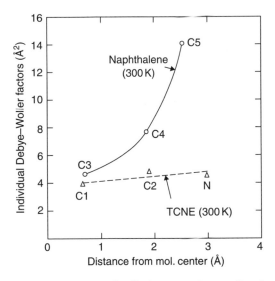

Fig. 16.30. The equivalent isotropic atomic displacement factors for the individual atoms in {naphthalene···TCNE} at room temperature, plotted against the distance of the atom from the molecular centre. Separate curves are shown for the two molecules (naphthalene – C3 to C5; TCNE – C1 to N). (Reproduced from Herbstein and Snyman, 1969.)

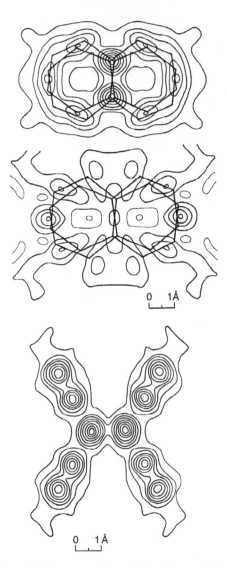

Fig. 16.31. Electron density and difference syntheses in the planes of (a) naphthalene and (b) TCNE molecules in the crystals of {naphthalene•••TCNE} at room temperature. The contours of electron density are at intervals of 1 e$\AA^{-3}$ and start at 0 e$\AA^{-3}$, while the contours of difference density are at intervals of 0.5 e$\AA^{-3}$ (thick lines zero and positive contours, thin lines negative contors). Two naphthalenes, mutually rotated by $\approx 12°$, give a reasonable fit to the electron density and difference density contours. (Reproduced from Herbstein and Snyman, 1969.)

A distinction can be made between the two models by calculating the potential energy (PE) curves for libration of the naphthalene molecule in the {naphthalene•••TCNE} structure (Shmueli and Goldberg, 1974). The complete PE calculations show minima at two orientations separated by 90° in-plane rotation (the N atoms of TCNE lie nearly at the

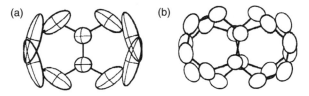

Fig. 16.32. {Naphthalene···TCNE} – ORTEP diagrams illustrating (a) dynamic disorder (b) static disorder. (Reproduced from Shmueli and Goldberg, 1974.)

corners of a square). The first orientation has a double well, indicating model (b). The second orientation is ≈8 kJ/mol higher in energy and is not populated in {naphthalene··· TCNE}. However, in {[3,3]paracyclophane···TCNE} these two positions are occupied in 3:1 ratio (Bernstein and Trueblood, 1971) and in {(d_8-pyrene)···TCNE} at 105K in 93:7 ratio (Larsen *et al.*, 1975), corresponding to at least 2.3 kJ/mol energy difference between the potential minima. Supporting evidence for static disorder in {naphthalene···TCNE} came from the results of a constrained refinement of the x-ray diffraction data, in which the molecules were treated as rigid bodies of standard dimensions whose translational and librational motions were refined. The constrained refinement gave an *R*-factor of 9.0%, compared to 12.8% for a conventional refinement, in which positions and anisotropic displacement factors of the individual atoms were refined. The physical reason for favoring static disorder is that N. . . . H contacts are more repulsive for model (a) than (b), a situation similar to that illustrated above for {naphthalene··· TCNB} (Fig. 16.16).

Two points should be noted about the calculations of librational potential energy. Firstly changes in charge transfer energy as a function of libration angle have been neglected; quantum mechanical calculations (Kuroda, Amano *et al.*, 1967) indicate that the charge transfer interaction energy between naphthalene and TCNE increases by only 0.4 kJ/mol when naphthalene is rotated in-plane by 12° from the model (a) structure. Secondly, as Shmueli and Goldberg (1974) point out, the possible relaxation of the surroundings of a molecule during its libration was not taken into account in their PE calculations. Such correlation of molecular motions was included in later calculations by Allen, Boeyens and Levendis (1989).

The disorder discussed above is a thermodynamic disorder characteristic of the equilibrium state of the crystal and thus temperature dependent. However, there are also some examples of nonequilibrium disorder, presumably frozen-in during growth of the crystals, and thus dependent on their history and not alterable by changes in temperature. One could expect two orientations, widely separated in azimuth, for one (or both) of the components, possibly unequally populated and perhaps not interconvertible by in-plane rotation. This type of disorder has been postulated for {azulene···TNB} (structure determined at 300K (Brown and Wallwork, 1965) and 178K (Hanson, 1965)) and for {indole··· TNB} at 133K. However, the evidence does not seem to be conclusive – in {azulene···TNB} there are indications of a phase change at lower temperatures, while in {indole···TNB} the existence of the disorder depends on a distinction made between C and NH groups. A more established example is provide by depends on a distinction made between C and NH groups. A more established example is provide by {benzo[c] phenanthrene···DDQ} where there is disorder of the Cl and CN groups in the acceptor (Bernstein *et al.*, 1977).

Orientational disorder must certainly occur in the quasi-plastic phases of charge transfer molecular compounds (Section 16.8) and substitutional disorder in the solid solutions discussed earlier (Section 13.4) but detailed studies have not yet been reported in either of these areas.

16.8 Transformations to quasi-plastic phase(s) on heating

The occurrence of disorder to order transformation on cooling in many charge transfer molecular compounds has been noted above. Conversely, it has been found that many charge transfer molecular compounds transform on heating to a quasi-plastic phase stable at temperatures close to their melting points (Inabe *et al.*, 1981; see also Section 13.4). The systems investigated and the results obtained are summarized in Fig. 16.33(a). Three donors (pyrene, fluoranthene and phenanthrene) have been studied in combination with

(a)

D \ A	DNF	DNC	DNP	DNT	TNB	TNC	TNP	TNT	NPA	PMDA	NBF	DNBF	BTF
Py	st	st	st	st	st	st	st	st	st	st	st	st	st
Fl			st		st	st	st	st	st	st	mst	st	st
Ph					st		st	mst	mst			st	

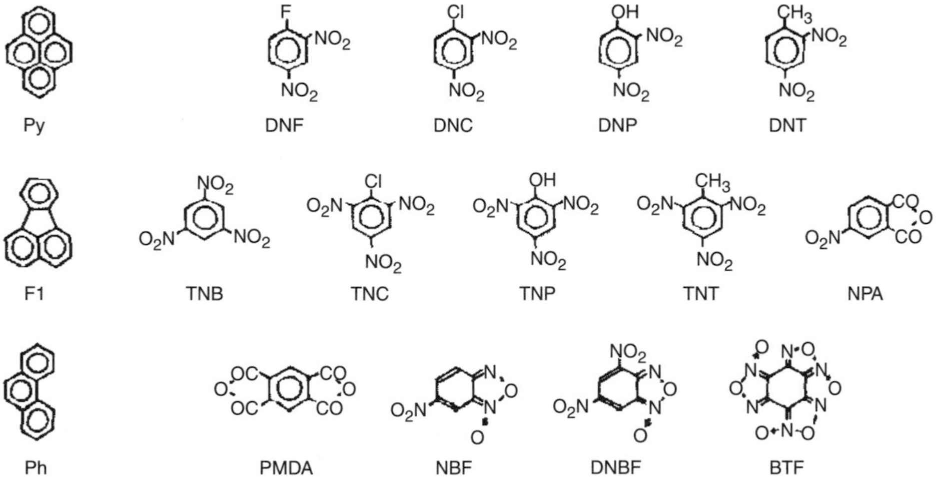

Fig. 16.33. (a) The π-molecular compounds that give isomorphous quasi-plastic phases on heating, and the structural formulae of the compounds; 'st' refers to stable and 'mst' to metastable molecular compounds. (b) The structure proposed for the quasi-plastic phases. The molecules located at 000 and 2/3,1/3, 1/3 are different from those at 00,1/2; 2/3, 1/3, 5/6 and 1/3, 2/3, 1/6. (Reproduced from Inabe *et al.*, 1981.)

(b)

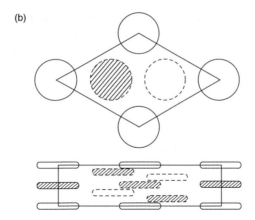

Fig. 16.33. (*Continued*)

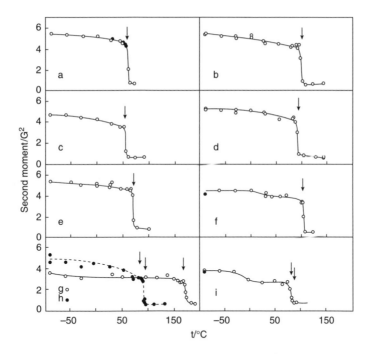

Fig. 16.34. Second moments of the proton resonance for the fluoranthene compounds with (a) DNP; (b) TNB; (c) TNC; (d) TNP; (e) TNT; (f) NPA; (g) PMDA; (h) DNBF and (i) BTF. The vertical arrows indicate the transition temperatures to the quasi-plastic phases. (Reproduced from Inabe *et al.*, 1981.)

13 different acceptors; of the 39 possible combinations, 28 transform to quasi-plastic phases characterized by very simple Debye–Scherrer x-ray diffraction patterns, of the type illustrated in Fig.13.9(b) (P_γ and I_∂). The crystal structure proposed for the isomorphous quasi-plastic phases is shown in Fig. 16.33(b); the donor–acceptor stacks continue to exist

but both donor and acceptor molecules are supposed to be azimuthally disordered so that the differences between the various donors, on the one hand, and the various acceptors, on the other, are eliminated by the rotational disorder. Such rotational disorder is feasible because of the disk-like shapes of the donor and acceptor molecules. The proton NMR spectra (Fig. 16.34) indicate that the disorder is dynamic rather than static; the second moments at lower temperatures ($\approx 4\,G^2$) are consistent with those crystal structures that are known, while the much reduced values of $\approx 1\,G^2$ above the temperatures of transition to the quasi-plastic phases are consistent with rotation of the donor molecules in their own planes (there are too few protons in the acceptors for conclusions to be drawn about their state of motion).

Another indication of the quasi-plastic nature of the high-temperature phases comes from the values measured for the entropies of transformation to these phases, which range up to 57 J/K mol and are comparable with the entropies of fusion of many of the compounds (46–67 J/K mol).

Similar results have been obtained in a high-temperature infrared study of {pyrene•••4-nitrophthalic anhydride} (Swamy et al., 1983).

16.9 Transformation of the ground state from neutral ⇒ ionic on cooling and/or application of pressure (NI transitions)

16.9.1 Introduction

We have already noted in Chapter 13, following McConnell, Soos and Torrance and their coworkers, that crystalline $1:1$ mixed stack π-molecular compounds are quite sharply divided into those with nominally neutral and those with nominally ionic ground states. However, some of the nominally neutral group are close to the neutral–ionic border (Fig. 13.2) and ten of the compounds in Table 13.1 (those with values of P_c appended) have been shown to transform reversibly to an ionic state on the application of pressure (pressure-induced NIT or PINIT; Torrance, Vasquez et al., 1981), as could well be expected on the basis of McConnell's original ideas.[6] The transformation pressure was determined as the onset of a distinct colour change from yellow or green (neutral) to red or brown (ionic). This phenomenon of a neutral ⇔ ionic transformation in the solid state has received considerable attention. Originally {TTF•••chloranil} was the only example of a transition induced on cooling (temperature-induced NIT or TINIT) but others have since been added, most having 1:1 ratios and being composed of TTF and chloranil derivatives or analogs. Transition temperatures are mostly below 100K, while transition pressures are a few kbar; the ionic states achieved by cooling or pressure may well differ in detail. Spectroscopic techniques are usually employed to show that a transition does occur and permit inferences about its nature. Diffraction studies appear mandatory for a proper understanding of the system but are still quite rare. The transition can also be photo-induced (Tanimura and Koshihara, 2001) but this will not be discussed here. Experiment shows that both first-order and second-order NITs are found. Most

[6] Girlando, Painelli et al. (1993) state that investigations carried out on members of the original series have not confirmed the occurrence of transitions. Discontinuous NI transitions have been found only for TTF-CA, TTF – fluoranil and tetramethylbenzidine – TCNQ.

information is available for {TTF···chloranil} (summarized by Le Cointe, Lemeé-Cailleau, Cailleau and Toudic, 1996; see also Bernstein (2002), pp. 195–197); this is one of the few organic crystals to have been studied over a range of temperatures and pressures and by a variety of experimental techniques.

We shall find that there are two primary parameters describing (perhaps determining) the occurrence of a transition and its nature. The first is the charge transfer from neutral to ionic state, leading to a difference in ionicity between the two states. The N state has an ionicity of about 0.3 (quasi-neutral could be a preferable descriptor) and that in the I state is larger. The second parameter is the degree of dimerization – donor and acceptor molecules are equally spaced along the stack axis in the N state but unequally spaced in the I state. As the difference between the two distances is 0.2 Å or less, dimerization is an unfortunate term but too well-established to be dislodged; some *schematic* diagrams should be viewed with caution. The ionicity is best determined spectroscopically and the degree of dimerization from comparison of N and I crystal structures.

16.9.2   *{TTF···chloranil}*

The first study of the molecular compound (needles crystallized from acetonitrile) showed a mixed stack, neutral ground state crystal structure at 300K ($a = 7.411$, $b = 7.621$, $c = 14.571$ Å, $\beta = 99.20°$, $P2_1/n$, $Z = 2$; stack axis [100]; degree of charge transfer $\lesssim 20\%$; Mayerle et al., 1979; TTFCAN). In parallel studies, powder patterns (temperature range 10–300K at 1 bar) were indexed in terms of the same unit cell throughout; no additional reflections appeared on cooling to 4K nor was there evidence for a triclinic distortion. There were also cell dimension measurements, from single crystals at 300K, over the pressure range 0–20 kbar. Distinct changes of slope were seen at 84K and 1 bar, which was identified as the 1 bar neutral ⇔ ionic transformation temperature, and at 11 kbar and 300K, identified as the 300K neutral ⇔ ionic transformation pressure.[7]

This work was followed by a detailed neutron diffraction study using co-sublimed crystals (Le Cointe, 1994[8]; Le Cointe, Lemeé-Cailleau, Cailleau, Toudic, Toupet et al., 1995; TTFCAN); other techniques (C_P–T, [35]Cl NQR, IR and Raman spectroscopy) were also used. The cell dimensions at various temperatures and pressures (up to 5 kbar) are shown in Fig. 16.35 (Le Cointe, 1994); at atmospheric pressure there are abrupt changes in [010] and [001] around 81K, and to a lesser extent in β and V, while [100] is continuous.

The P–T phase diagram derived from these (and other) measurements (Lemée-Cailleau, Le Cointe et al., 1997) is shown in Fig. 16.36.

The space group below 84K is determined by the behaviour of the 030 reflection which is systematically absent above 84K.[9] The intensity of this reflection increases

---

[7] Some of the earlier results must be treated with caution. We give a few examples. Batail et al. (1981) give a curve of cell volume against T with a cusp at 84K, i.e. $\Delta V = 0$ (their Fig. 1). This suggests that the NI transition is second order, contrary to all later views. Unfortunately the later neutron diffraction measurements are not complete enough to allow calculation of V–T curves. Kagoshima et al. (1985) give a (heating) curve (their Fig. 2) of I ($31\bar{1}$) against T with an abrupt fall to zero at 78.5 (5)K; it is not clear how this can be reconciled with the known structures.

[8] I am grateful to Dr Marylise Buron-Le Cointe for a copy of her doctoral thesis (University of Rennes I).

[9] Le Cointe, Lemeé-Cailleau, Cailleau, Toudic, Toupet et al., 1995; (see III B#2, Ionic phase; p. 3378) remark that "only the (070) superstructure reflection was clearly extracted from the background"; however, it is the temperature dependence at 1 bar of I(030) that is shown in Fig. 2(a) of this reference. The temperature and pressure dependence of I(030) is shown in Fig. 3 of Le Cointe (1994), from which our Fig. 16.37 has been adapted.

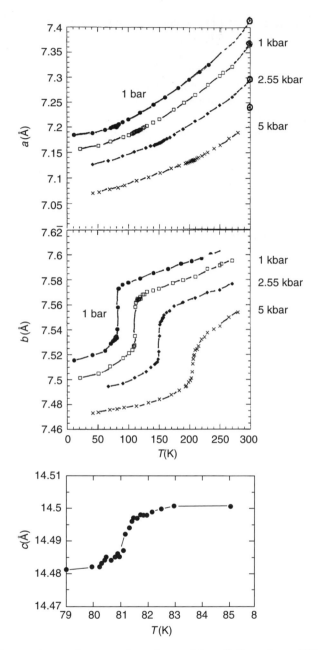

Fig. 16.35. The temperature and pressure dependence of the cell dimensions of TTF-CA (values for [001] were given only for the temperature range shown; β is 99.1° at 300K and 98.6° at 50K (both at 1 bar). From Le Cointe (1994).

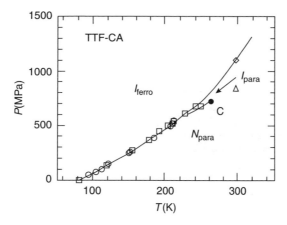

Fig. 16.36. TTF-CA pressure-temperature phase diagram – circles ND, squares NQR, diamonds vibrational spectroscopy. The subscripts para and ferro are abbreviations for paraelectric and ferroelectric. C is the estimated critical point. Lines are guides to the eye. (Reproduced from Lemée-Cailleau, Le Cointe *et al.*, 1997.)

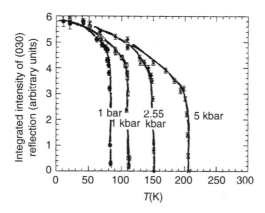

Fig, 16.37. Neutron diffraction intensity of TTF-CA (030) reflection as a function of temperature and pressure. (Adapted from Le Cointe (1994).)

monotonically with falling temperature; similar behaviour is found over the range 0 to 5 kbar (Fig. 16.37). Thus the transition involves a change of space group from $P2_1/n$, $Z = 2$ (N phase) to Pn, $Z = 2$ (I phase). In the N phase the D and A molecules are located at centres of symmetry but this requirement is relaxed in the I phase. Confirmation of the loss of centrosymmetry comes from the ^{35}Cl NQR spectra – two independent resonances in the N phase and four in the I phase (Gallier *et al.*, 1993; Gourdji *et al.*, 1991; Le Cointe, Gallier *et al.*, 1995).

   The shape of the curves of $I(030)$ against T (Fig. 16.37) requires some comment. These start out as though the transition was second order (cf. Fig. 16.25) and then fall abruptly to

Table 16.5. Calorimetric measurements on TTF-CA

Reference	ΔH J/mol	ΔS J/mol K
Kawamura *et al.* (1997)	504(5)	6.12(6)
Lemée-Cailleau *et al.* (1997)	461(32)	5.49(9)

Note:

ΔS measured on ≈ 0.2 mg samples was given as ≈ 4 J/mol K (Wolfe, 1982).

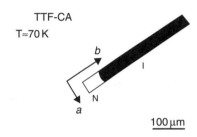

Fig. 16.38. TTF-CA – coexistence of the N and I phases at 70K observed by optical microscopy with unpolarized light. (Reproduced from Buron-Le Cointe *et al.* (2003; Fig. 6).)

zero, as one would expect for a first order transition. The overall shape, which appears to be qualitatively the same up to at least 5 kbar, is reminiscent of that found for long range order in Cu_3Au (see Fig. 12.4 of Warren (1969)).

There are a number of measurements of specific heat for TTF-CA; Kawamura *et al.* (1997) found a single peak while Lemée-Cailleau, Le Cointe *et al.* (1997) found a split peak in the same temperature region; nevertheless, the transition enthalpy and entropy values are in reasonably good agreement (Table 16.5) and the thermodynamic parameters are in the expected range for a first order transition.

The Clapeyron equation ($dP/dT = \Delta H/T_c\Delta V$, where dP/dT is the slope at T_c) provides a test of various interrelated quantities. A precise value of dP/dT ($= 4.17$ Mpa/K) is obtained from Fig. 16.36 and T_c is precise at 81K. However, neither ΔH nor ΔV is that precisely defined. We have estimated ΔV from the cell dimensions above and below T_c (Fig. 16.35) and obtain $\Delta V = 7.2$ Å^3 (Lemée-Cailleau, Le Cointe *et al.* (1997) give 4–5 Å^3). Using a mean value for ΔH (Table 16.5) we obtain $(dP/dT)_{calc} = 1.37$ Mpa/K, some 40% of the phase diagram value. Lemée-Cailleau, Le Cointe *et al.* (1997) give a phase diagram value of 3.1 Mpa/K (3.6 Mpa/K from Fig. 8 of Le Cointe (1994)) and a derived value of 3.2 ± 0.4 Mpa/K. The source of the discrepancies is not known.

Buron-Le Cointe, Lemée-Cailleau *et al.* (2003) show a micrograph of a TTF-CA needle crystal in which the interface ((010) plane) between the N and I phases can be clearly seen in Fig. 16.38. This is an example of Mnyukh's (2001; see pp. 121–143 and Figs. 2.34 and 2.43) category of epitaxial growth when the two phases resemble one another. The hysteresis and related phenomena can all be described in terms of Mnyukh's treatment of first-order enantiotropic phase transitions (see his Fig. 2.30) and have no direct connection with the neutral-to-ionic nature of the phase transformation.

The details of the crystal structures of the N and I phases of TTF-CA (Le Cointe, Lemeé-Cailleau, Cailleau, Toudic, Toupet et al., 1995) require some comment. The N phase structure was determined by neutron diffraction at 300 and 90K; we consider only the 90K results (1223 independent reflections, $R = 3.4\%$, 118 parameters, goodness of fit 1.40, all nonhydrogen atoms refined anisotropically). Both component molecules have D_{2h}–mmm symmetry within the precision of the measurements; bond lengths and angles have standard values. The I phase structure was determined by neutron diffraction at 40K (1636 independent reflections, $R = 5.0\%$, 105 parameters, goodness of fit 1.48, all atoms refined isotropically). There are large differences between chemically equivalent bonds; for example, the (intraring) C=C bond lengths in TTF are given as 1.323(1) and 1.364(1) Å, with $\Delta/\sigma = 29$. Thus either there is a remarkable (and unprecedented) polarization effect in the I structure or the standard uncertainties have been underestimated by a factor of about 10;[10] the second alternative seems more likely. Le Cointe, Lemeé-Cailleau, Cailleau, Toudic, Toupet et al. (1995) give considerable attention to differences in some intermolecular distances between N and I structures. They also assign NQR frequencies to specific Cl atoms in the I phase on the basis of C–Cl bond lengths (d(C–Cl) = 1.724, 1.705,1.711,1.714 Å, all with s.u.s of 0.001 A). These differences are less impressive if the true standard uncertainties are ≈ 0.01 Å.

The main difference between N and I structures lies in the separation of the TTF and CA molecules along the [100] stack axis; these (center to center, not plane to plane) distances are equal in the N structure at 3.61 Å (at 90K, $= a/2$) but unequal at 3.50 and 3.69 Å in the I structure at 40K. This alternation is called 'dimerization'. Batail et al. (1981) pointed out that there was C–H...O hydrogen bonding along the [001] direction, at a time when such bonding was controversial. However, it seems disputable that the transformation is due to changes in C–H...O hydrogen bonding because hydrogenated and deuterated molecular compounds transform in essentially the same way (Ayache and Torrance, 1983), On the other hand, Oison et al. (2001) contend that charge transfer produces a strengthening of these hydrogen bonds. Lack of precision prevents the use of component dimensions to estimate the degree of charge transfer (ionicity) in the I phase, and this must be done by spectroscopic techniques (see below).

The temperature (Girlando et al., 1983) and pressure (Tokura et al., 1986; Girlando et al., 1986) induced neutral to ionic phase transitions in {TTF•••chloranil} have been studied by IR spectroscopy (Fig. 16.39). The differences between the spectra of the neutral and ionic phases are shown by comparing the upper spectrum with those in the centre and lower positions. The resemblances between temperature-induced and pressure-induced ionic phases is shown by a comparison of centre and lower parts – the low-temperature and high-pressure ionic phases are, at least, very similar in nature.

The Le Cointe 300K ND measurements (Fig. 16.35, up to 5 kbar) indicate a first order transition but one that deviates somewhat from classic expectations. These measurements show a linear dependence of a and b cell dimensions on pressure in both phases; the values

---

[10] Le Cointe et al. (1995) remark "...the standard deviations have particularly small values at 40K, which can be explained as an effect of the isotropic refinement." It does not seem reasonable that a structure refined isotropically (even at lower temperature) will be more precise than one refined anisotropically, conditions of measurement being essentially the same.

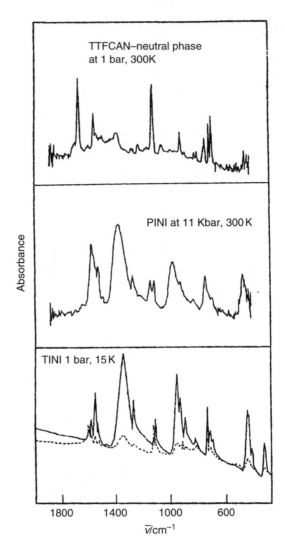

Fig. 16.39. IR spectra of {TTF···chloranil}. (top), the neutral phase at room temperature and pressure; (center) the ionic phase at 300K and 11 kbar; (bottom) the ionic phase at 15K, 1 bar (Girlando *et al.*, 1986).

The spectra shown in the centre and lower panels were obtained with polarized IR – full line, electric vector parallel to stack axis; dashed line, perpendicular to it). The resemblance between the central and lower spectra implies that the same ionic phase is obtained by cooling to 15K at 1 bar and by compression to 11 kbar at 300K. (Reproduced from Girlando *et al.* (1986).)

of da/dP are -0.041 and -0.025 Å/kbar in the N and I phases, and the corresponding values for the b dimension are -0.013 and -0.012 respectively. At 1 bar the (020) reflection shows definite hysteresis between heating and cooling curves (Fig. 16.40; this is also found by other techniques such as ^{35}Cl NQR and specific heat measurements), in

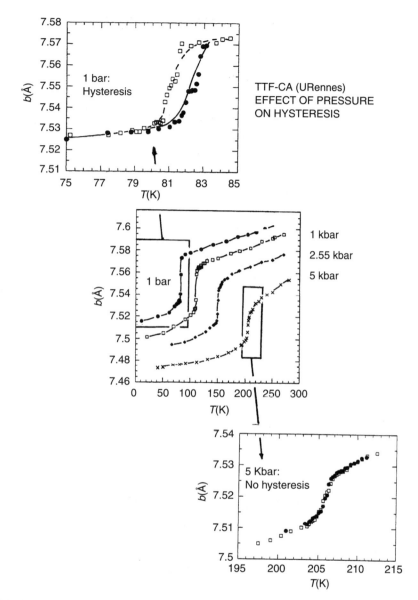

Fig. 16.40. The effect of pressure on the TINI in TTF-CA. The central panel is taken from Fig. 16.35, while the upper and lower panels are from the Le Cointe (1994) thesis (Figs. 7 and 16). The filled circles are for the heating regimen and the open circles for cooling.

contrast to what is found at 5 kbar (Fig. 16 from Le Cointe (1994)), where there is no hysteresis. This confirms the first order nature of the transition at 1 bar and suggests that the transition at 5 kbar is second order. If so, the tricritical point in the phase diagram (Fig. 16.36) must be shifted.

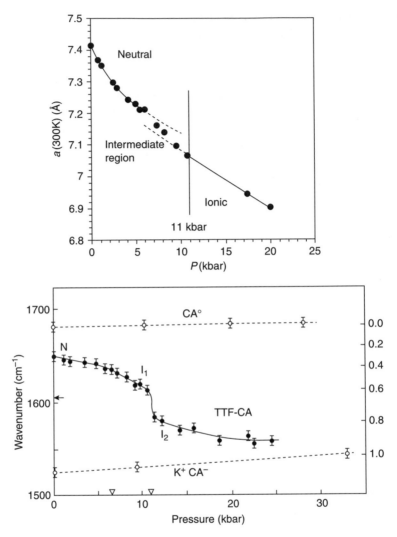

Fig, 16.41. *Upper panel*: Diffraction study of PINI transition in TTF-CA at 300K; the diagram has been redrawn from Metzger and Torrance (1985) who quote results ("to be submitted") of King *et al*. The King results do not appear to have been submitted.

Lower panel: frequencies of the C=O stretch band at 290K as a function of hydrostatic pressure; the pressure dependences for neutral chloranil and for K^+CA^- were used for calibration purposes. Approximate values deduced for the ionicity are shown by the right-hand ordinate (Tokura *et al*., 1986). (Adapted from Metzger and Torrance (1985) and Tokura *et al*., (1986).)

The PINI transition at 300K, 11 kbar is shown by diffraction (the variation of the [100] cell dimension with pressure; upper panel of Fig. 16.40) and by IR spectroscopy (lower panel of Fig. 16.41; Tokura *et al*., 1986); the transition is essentially first order but neither diffraction nor spectroscopy shows an ideally sharp transition.

Some care is needed in the preparation of {TTF···chloranil} as it crystallizes from solution as plates or needles, and, by co-sublimation, as parallelepipeds of up to 20 mm^3 in volume (Kawamura *et al.*, 1997). It is the structure of the co-sublimed form (also obtainable from solution) that is discussed above. Two other phases of {TTF···chloranil}, named as I and II, have been reported; I and II both appeared as black needles, with phase II shown to have a 1:1 composition, and this was presumed to hold also for phase I (Matsuzaki *et al.*, 1983). The analogous black "snowflakes" (Girlando *et al.*, 1983) appear, from comparison of 300K IR spectra, to be essentially II, perhaps with a small contamination of I. Kawamura *et al.* (1997) noted that their specific heat measurements required correction for the presence of more than one phase. It was inferred (Girlando *et al.*, 1986) that phase I was a mixed valence ionic solid but detail is lacking. Phase II is a fully ionic solid containing $(TTF^+)_2$ and $(CA^-)_2$ self dimers, present either in distorted stacks or as discrete units (e.g. as in TTF.Br) (Girlando *et al.*, 1986; Matsuzaki *et al.*, 1983); IR and Raman spectra show that there is no phase transition between 300 and 15K (Matsuzaki *et al.*, 1983). Electronic spectra for {TTF···bromanil} suggest that phases I and II are also found in this system. No crystallographic information is available for these phases.

16.9.3 *{DMTTF···chloranil}*

A TINI transition at 65 ± 5K in {2,6-dimethyltetrathiafulvalene···chloranil} was first proposed by Aoki *et al.* (1993) on the basis of studies of polarized visible reflection and IR spectra; a "most striking feature" was the appearance of a coexisting neutral-ionic phase at lower temperatures. Definitive structure determinations at 300, 75 and 45K by x-ray diffraction (Collet *et al.*, 2001) clarified earlier confusions (Nogami *et al.*, 1995) about the low temperature structure. Cell dimensions are given in Table 16.6; there is excellent agreement among the three independent 300K measurements and this applies also to the two crystal structure determinations. Nogami *et al.* have reported unusual behaviour of the cell dimensions at low temperatures (their Fig. 2, reproduced here as Fig. 16.42); the Collet values at 75 and 40K agree well with those Nogami, thus suggesting that the full set of Nogami cell dimensions can be relied upon to give valuable information about the nature of the transition.

Information about the nature of the transition is also obtained from spectroscopy. The optical conductivity spectra for the $C = O$ stretch mode are shown in Fig. 16.43(left), while the wave numbers of the peaks of the spectra at different temperatures are shown in Fig. 16.43(right). The transition is not abrupt but is spread over some 50K, suggesting that it is second order. Similar conclusions can be drawn from the temperature dependence of the IR spectra of DMTTF-CA (Fig. 16.44) and from the cell dimension–T curves (Fig. 16.41),

The low temperature phase has a unit cell doubled along [001] with noncentrosymmetric space group $P1$; thus there are two crystallographically independent molecules of each component in the unit cell. Detailed crystal structures have been reported by Collet *et al.* at 300, 75 and 40K. We compare the 75K (2395 independent reflections, $R = 2.82\%$, 125 parameters, goodness of fit 1.04, all nonhydrogen atoms refined anisotropically) and 40K (4693 independent reflections,[11] $R = 3.44\%$, 485 parameters,

---

[11] 2777 with $F_0^2 > 2\sigma(F_0^2)$; 2362 of the 4693 were superstructure reflections, 765 with $F_0^2 > 2\sigma(F_0^2)$. In the 40K determination the coordinates of one atom should have been fixed and the Friedel parameter should have been included in the refinement. The low internal R value for Friedel opposites (2.16%) suggests the possibility of polysynthetic twinning.

Table 16.6. Cell dimensions (reduced cell; Å, °, Å^3) for DMTTF-CA from various sources and at different temperatures. The space group is $P\bar{1}$, $Z = 1$ at all temperatures except for 40K, where it is $P1$, $Z = 2$

T(K)	a	b	c	α	β	γ	V	Reference
300	7.272	7.666	8.512	95.91	103.89	91.89	457.4	Aoki et al. (1993)
300	7.285	7.678	8.521	95.90	103.89	91.91	459.4	Horiuchi et al. (2001); UDEQUP01
300	7.269	7.673	8.514	95.87	103.91	91.87	457.2	Collet et al. (2001)
75	7.121	7.586	8.476	95.87	104.07	90.92	441.4	Collet et al. (2001).
75	7.118	7.581	8.464	96.00	104.02	90.83		Nogami et al. (1995)*
40#	7.099	7.563	16.937	95.77	104.21	91.02	876.2	Collet et al. (2001).
40	7.090	7.556	8.450 X2	95.83	104.17	90.98		Nogami et al. (1995)*

* Values read off Fig.2 of Nogami et al.; volumes deliberately omitted.
\# Reduced cell 7.100 7.564 16.680 Å, 83.78 79.84 88.99°.

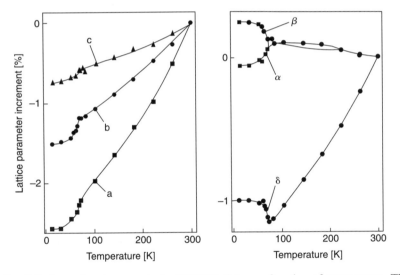

Fig. 16.42. Cell dimension decrements for DMTTF–CA as a function of temperature. The space group is $P\bar{1}$ above ≈ 65K and $P1$ below. (Reproduced from Nogami et al., 1995.)

goodness of fit 0.93, all nonhydrogen atoms refined anisotropically) structures. At 75K there is a ... DMTTF..CA ... stack along [100], the components being separated by 3.56 Å ($= a/2$). At 40K there are two crystallographically independent ... DMTTF–CA ... stacks, with the pairs of components separated by 3.49 and 3.61 Å in one stack, and 3.46 and 3.64 Å in the second. Collet et al. describe these as neutral and ionic stacks respectively but critical scrutiny suggests that the component dimensions are not precise enough for such an identification. Although it was reported (Nogami et al., 1995) that structures had been determined at 293, 109, 48 and 29K, details do not appear to have been published.

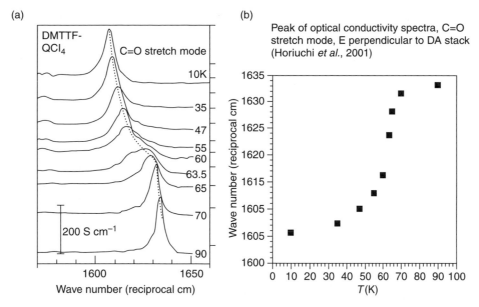

Fig. 16.43. (left) The optical conductivity spectra of DMTTF-CA (E perpendicular to DA stack; C=O stretch mode) over a range of temperatures. The dashed line is a guide to the eye. (right) Wave numbers of spectral peaks (from the dashed line in the left diagram) plotted against temperature. These values can be compared with the values given for CA^0 ($1685\,cm^{-1}$) and CA^{-1} (in K CA) ($1520\,cm^{-1}$) in Fig. 16.40 (lower panel). (Reproduced from Horiuchi *et al.*, 2001.)

Aoki *et al.* (1995) report that the N phase transforms into a fully ionized I phase above 12 kbar at 300K. Thus there may be a difference in pressure and temperature induced transitions in this system.

The NI transition in TTF-CA is essentially first order in nature while that in DMTTF–CA appears to be essentially second order.

16.9.4   *Other examples*

(a) TTF – 2,5-dichloro-p-benzoquinone.   This molecular compound (TTF–2,5-Cl$_2$BQ) crystallizes as dark green triclinic needles (7.935 7.216 6.844 Å, 106.94 97.58 93.66°, $Z = 1$, space group $P\bar{1}$) (Girlando, Painelli *et al.*, 1993). There are resemblances to the TTF-CA crystal structure. Raman and polarized IR spectra allowed the evaluation of many microscopic parameters involved in the contrasting behaviour of TTF-CA and TTF–2,5-Cl$_2$BQ. However, we shall not dwell on these as study over a wide range of temperatures and pressures is lacking. A putative lower-symmetry structure was studied theoretically by Katan and Koenig (1999).

(b) 2-chloro-5-methyl-p-phenylenediamine–2,5-dimethyldicyanoquinone-diimine.   This molecular compound (abbreviated as ClMePD–DMeDCNQI) shows unique features not yet encountered in other molecular compounds that show neutral-to-ionic transitions. IR

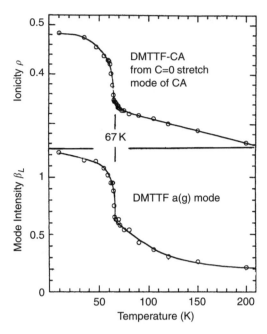

Fig. 16.44. Upper curve – ionicity estimated from frequency of C=O stretch mode versus temperature. Note that the ionicity extrapolates to 0.5 at 0K. Lower curve–mode intensity from DMTTF mode. (Adapted from Horiuchi *et al.* (2001).)

and visible spectra of powdered crystals over the range 50–350K led to the conclusion that a continuous change in molecular ionicity over a range of 200K is accompanied by dynamic distortions of the stacks at low temperatures (Aoki and Nakayama, 1997). The crystals (from dichloromethane) are triclinic needles, 7.463 7.504 7.191 Å, 91.23 112.19 96.91°, $Z = 1$ (presumably at 300K). The space group was given as $P1$, although "to obtain suitable values of thermal factors in the analysis, 2- and 5-substitutional sites [in ClMePD] were assumed to be occupied by Cl atoms and methyl groups with equal probability." The ClMePD and DMeDCNQI molecules formed mixed-stack columns in a direction parallel to the *b* axes. This description suggests that the space group could be $P\bar{1}$; however, no structural details have been published so this remains a speculation. These spectroscopic results were confirmed by a Raman and polarized IR study over the range 80–400K (Masino *et al.*, 2001). Spectroscopic studies at various pressures have been carried out by Masino *et al.*, (2003). A low temperature diffraction study seems essential.

(c) 3,3',5,5'-Tetramethylbenzidine–TCNQ.   A TINI transition was found in this mixed-stack molecular compound at around 205K. At 300K the golden-yellow cosublimed crystals are monoclinic, 6.722 21.873 8.108 Å, 100.19°, $Z = 2$, space group $P2_1/n$ (Iwasa *et al.*, 1990; details of the structure not reported). Polarized visible and IR spectra show a first-order transition on cooling at 200K and on heating at 228K; thus there is appreciable hysteresis. The ionicity is 0.59 in the high temperature N phase and

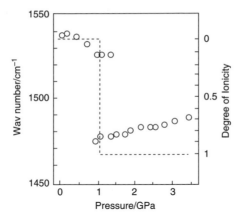

Fig. 16.45. Pressure dependence of the C=N stretching IR peak frequency at 300K (left hand scale) of JIXWES and ionicity (right hand scale). Note the hint of phase coexistence around 1 Gpa. (Reproduced from Aumüller *et al.* (1991).)

0.69 in the low temperature I phase; we retain the N/I nomenclature despite its limitations here. Details of the spectra suggest that there are two types of TCNQ molecule in the I phase. A micrograph at 180K shows that the I phase has a striated appearance. "Narrow diagonal I domains appear in (**a**, **c**) planes whose number increases as the temperature decreases. Such coexistence is observed over several [tens of] degrees Kelvin. However, when large parts of the crystal are transformed,.. the transition is always accompanied by a sharp breaking of the crystal along the **a** and **c** axes." (Buron-Le Cointe *et al.*, 2003). It is difficult to obtain diffraction-quality crystals for further studies.

There are two PINI transitions at 300K, one at 6 and the other at 20 kbar. (Iwasa *et al.*,1993). The 6 kbar transition is to an equal mixture of "ionic" and "neutral" molecules, while there are only ionic entities above 20 kbar.

(d) TTF–N,N'-dicyano-2,5-dimethyl-1,4-benzoquinone-di-imine.

N,N'-dicyano-2,5-dimethyl-1,4-benzoquinone-di-imine

The 300K crystal structure of this mixed-stack neutral molecular compound was reported by Aumüller *et al.* (1991) (6.172 7.831 17.998 Å, 101.41°, $Z=2$, $P2_1/c$; JIXWES). The high-pressure FT-IR and electronic spectra of powdered samples (300K, up to 3.5 GPa;

Fig. 16.45) were measured, and also the electronic conductivity along the needle axis of a single crystal.

There is a sharp PINI transition at 1.0 GPa from an essentially neutral state to an almost completely ionic state. The electrical conductivity increases by five orders of magnitude above about 1.5 GPa. One could anticipate a first order transition to an ionic state on cooling but this has not been investigated. JIXWES appears to resemble TTF-CA more closely than any of the other examples discussed above.

16.9.5   *Concluding summary*

What started off as a rather limited enterprise, exemplified only by TTF-CA, has blossomed into a multifacetted study with a growing number of subjects, each differing from TTF-CA in one or more aspects. The experimental work summarized above has been accompanied by many theoretical studies, mostly employing the twin concepts of 'ionicity' and 'dimerization' and neglected here for reasons of space.

References

Abdel–Rehiem, A. G., Farrell, P. G. and Westwood, J. V. (1975). *J. Chem. Soc. Faraday Trans., 1*, **71**, 1762–1771.

Aizu, K. (1970). *J. Phys. Soc. Jpn.*, **28**, 706–716.

Allen, C. C., Boeyens, J. C. A. and Levendis, D. C. (1989). *S. Afr. J. Chem.*, **42**, 38–42.

Aoki, S. and Nakayama, T, (1997). *Phys. Rev.*, **56**, R2893–R2896.

Aoki, S., Nakayama, T, and Miura, A. (1993). *Phys. Rev.*, B**48**, 626–629.

Aumüller, A., Erk, P., Hünig, S., Hädicke, E., Peters, K. and Schnering, H. G. von (1991). *Chem. Ber.*, **124**, 2001–2004.

Axe, J. D. (1971). *Trans. Amer. Cryst. Assoc.*, **7**, 89–106.

Ayache, C. and Torrance, J. B. (1983). *Solid State Commun.*, **47**, 789–793.

Basaki, S. and Matsuzaki, S. (1995). *Synth. Met.*, **70**, 1239–1240.

Batail, P., LaPlaca, S. J., Mayerle, J. J. and Torrance, J. B. (1981). *J. Am. Chem. Soc.*, **103**, 951–953.

Beasley, J. R., Doyle, D. F., Chen, L., Cohen, D. S., Fine, B. R. and Pielak, G. (2002). *Proteins: Structure, Function and Genetics*, **49**, 398–402.

Berry, R. S., Rice, S. A. and Ross, J. (1980). Physical Chemistry. John Wiley and Sons, New York etc.

Bernstein, H. J., Dalal, N. S., Murphy, W. F., Reddoch, A. H., Sunder, S. and Williams, D. F. (1978). *Chem. Phys. Lett.*, **57**, 159–162.

Bernstein, J. (2002). Polymorphism in molecular crystals. Clarendon Press, Oxford. pp. xiv +410.

Bernstein, J. and Trueblood, K. N. (1971). *Acta Cryst.*, B**27**, 2078–2089.

Bernstein, J., Regev, H. and Herbstein, F. H. (1977). *Acta Cryst.*, B**33**, 1716–1724.

Bernstein, T. (1967). M.Sc. Thesis, Department of Chemistry, Technion – Israel Institute of Technology, Haifa, Israel.

Boer, J. L. de and Vos, A. (1968). *Acta Cryst.*, B **24**, 720–725.

Boerio–Goates, J. and Westrum, E. F., Jr. (1979). *Mol. Cryst. Liq. Cryst.*, **50**, 259–268.

Boerio–Goates, J. and Westrum, E. F., Jr. (1980a). *Mol. Cryst. Liq. Cryst.*, **60**, 237–248.

Boerio–Goates, J. and Westrum, E. F., Jr. (1980b). *Mol. Cryst. Liq. Cryst.*, **60**, 249–266.

Boerio–Goates, J., Westrum, E. F., Jr. and Fyfe, C.A. (1978). *Mol. Cryst. Liq. Cryst.*, **48**, 209–218.

Bragg, W. L. and Williams, E. J. (1935). *Proc. Roy. Soc., London.*, A**151**, 540–546.

Brose, K.–H., Luty, T. and Eckhardt, C. J. (1990). *J. Chem. Phys.*, **93**, 2016–2031.

Brown, D. S. and Wallwork, S. C. (1965). *Acta Cryst.*, **19**, 149.

Buron-Le Cointe, M., Lemée-Cailleau, M. H., Cailleau, H., Toudic, B., Moréac, A., Moussa, F., Ayache, C and Karl, N, (2003). *Phys. Rev.*, B**68**, 064103-1–064103-7.

Clayton, P. R., Worswick, R. D. and Staveley, L. A. K. (1976). *Mol. Cryst. Liq. Cryst.*, **36**, 153–163.

Collet, E., Buron-Le Cointe, M., Lemée-Cailleau, M. H., Cailleau, H., Toupet, L., Meven, M., Mattauch, S., Heger, G, and Karl, N. (2001). *Phys. Rev.*, B**63**, 054105-1–054105-12.

Cowley, R. A. (1980). *Adv. Phys.*, **29**, 1–110.

Czarniecka, K., Janik, J. M., Janik, J. A., Krawcyzk, J., Natkaniec, I., Wasicki, J., Kowal, R., Pigon, K. and Otnes, K. (1985). *J. Chem. Phys.*, **85**, 7289–7293.

Czekalla, J. and Meyer, K. O. (1961). *Z. Physik. Chem.* (NF), **27**, 185–198.

Czekalla, J., Briegleb, G., Herre, W. and Glier, R. (1957). *Z. Elektrochem.*, **61**, 537–546.

Dalal, N. S., Ripmeester, J. A., Reddoch, A. H. and Williams, D. F. (1978). *Mol. Cryst. Liq. Cryst. Letts.*, **49**, 55–59.

Dalal, N., Haley, L. V., Northcott, D. J., Park, J. M., Reddoch, A. H., Ripmeester, J. A., Williams, D. F. and Charlton, J. L. (1980). *J. Chem. Phys.*, **73**, 2515–2517.

Darmon, I. and Brot, C. (1967). *Mol. Cryst.*, **2**, 301–321.

Dunitz, J. D. (1995). *Chem. Biol.*, **2**, 709–712.

Dunn, A. G., Rahman, A. and Staveley, L. A. K. (1978). *J. Chem. Thermodynam.*, **10**, 787–796.

Ecolivet, C. and Mierzejewski, A. (1990). *Phys. Rev.*, B**42**, 8471–8481.

Ecolivet, C., Bertault, M., Mierzejewski, A. and Collet, A. (1987). In *Dynamics of Molecular Crystals*, edited by J. Lacombe, pp. 187–192. Amsterdam: Elsevier.

Ecolivet, C., Lemeé, M. H., Delugeard, Y., Girard, A., Bertault, M., Collet, A. and Mierzejewski, A. (1988). *Mater. Sci.*, **14**, 55–58.

Ehrenfest, P. (1933). *Proc. Acad. Sci. Amsterdam*, **36**, 153–157.

Erdle, E. and Möhwald, H. (1979). *Chem. Phys.*, **36**, 283–290.

Fyfe, C. A. (1974a). *J. Chem. Soc. Farad. Trans. 2*, **70**, 1633–1641.

Fyfe, C. A. (1974b). *J. Chem. Soc. Farad. Trans. 2*, **70**, 1642–1649.

Fyfe, C. A., Harold–Smith, D. and Ripmeester, J. (1976). *J. Chem. Soc. Farad. Trans. 2*, **72**, 2269–2282.

Gallier, J., Toudic, B., Délugeard, Y., Cailleau, H., Gourdji, M., Péneau, A. and Guibé, L. (1993). *Phys. Rev.*, B**47**, 11688–11695.

Gilson, D. F. and McDowell, C. A. (1966). *Canad. J. Chem.*, **44**, 945–952.

Girlando, A., Marzola, F., Pecile, C. and Torrance, J. B. (1983). *J. Chem. Phys.*, **79**, 1075–1085.

Girlando, A., Painelli, A., Pecile, C., Calestani, G., Rizzoli, C., and Metzger, R. M. (1993). *J. Chem. Phys.*, **98**, 7692–7698.

Girlando, A., Pecile, C., Briliante, A. and Syassen, K. (1986). *Solid State Commun.*, **57**, 891–896.

Gourdji, M., Guibé, L. Péneau, A., Gallier, J., Toudic, B. and Cailleau, H., (1991). *Z. Naturforsch.*, **47a**, 257–260.

Grupp, A., Wolf, H. C. and Schmid, D. (1982). *Chem. Phys. Lett.*, **85**, 330–334.

Haarer, D. and Karl, N. (1973). *Chem. Phys. Lett.*, **21**, 49–53.

Haarer, D. (1974). *Chem. Phys. Lett.*, **27**, 91–95.

Haarer, D. (1975). *Chem. Phys. Lett.*, **31**, 192–194.

Haarer, D. (1977). *J. Chem. Phys.*, **67**, 4076–4085.

Haarer, D., Philpott, M. R. and Morawitz, H. (1975). *J. Chem. Phys.*, **63**, 5238–5245.

Hammick, D. Ll. and Hutchinson, H. P. (1955). *J. Chem. Soc.*, pp. 89–91.

Hanson, A. W. (1964). *Acta Cryst.*, **17**, 559–568.

Hanson, A. W. (1965). *Acta Cryst.*, **19**, 19–26.

Hanson, A. W. (1978). *Acta Cryst.*, B**34**, 2195–2200.

Herbstein, F. H. (1971). "Crystalline π-molecular compounds" in *Perspectives in Structural Chemistry*, Wiley: London and New York, Vol. IV, pp. 166–395.

Herbstein, F. H. (1996). *Cryst. Revs.*, **5**, 181–226.

Herbstein, F. H. and Samson, S. (1994). *Acta Cryst.*, B**50**, 182–191.

Herbstein, F. H. and Snyman, J. A. (1969). *Phil. Trans. Roy. Soc. Lond.* A**264**, 635–666.

Herbstein, F. H., Kapon, M., Reisner, G. M., Lehmann, M. S., Kress, R. B., Wilson, R. B., Shau, W.–I., Duesler, E. N., Paul, I. C. and Curtin, D. Y. (1985). *Proc. Roy. Soc. Lond.*, A**399**, 295–319.

Herbstein, F. H., Marsh, R. E. and Samson, S. (1994). *Acta Cryst.*, B**50**, 174–181.

Horiuchi, S., Okimoto, Y., Kumai, R. and Tokura, Y. (2001). *J. Am. Chem. Soc.*, **123**, 665–670.

Inabe, T., Matsunaga, Y. and Nanba, M. (1981). *Bull. Chem. Soc. Jpn.*, **54**, 2557–2564.

Iwasa, Y., Koda, T., Tokura, Y., Kobayashi, A., Iwasawa, N. and Saito, G. (1990). *Phys. Rev.*, B**42**, 2374–2377.

Iwasa, Y., Watasnabe, N., Koda, T. and Saito, G. (1993). *Phys. Rev.*, B**47**, 2920–2377.

Iwata, S., Tanaka, J. and Nagakura, S. (1967). *J. Chem. Phys.*, **47**, 2203–2923.

Kagoshima, S., Kanai, Y., Tani, M., Tokura, Y. and Koda, T. (1985). *Mol. Cryst. Liq. Cryst.*, **120**, 9–15.

Kanai, H., Tani, M., Kagoshima, S., Tokura, Y. and Koda, T. (1984). *Synth. Metals*, **10**, 157–160.

Katan, C. and Koenig, C. (1999). *J. Phys. Cond. Matter*, **11**, 4163–4177.

Kawamura, T., Miyazaki, Y. and Sorai, M. (1997). *Chem. Phys. Lett.*, **273**, 435–438.

Koci′nski, J. (1983). Theory of Symmetry Changes at Continuous Phase Transitions, Elsevier, Amsterdam etc. and PWN–Polish Scientific Publications, Warsaw.

Krzystek, J. and Schutz, J. von (1993). *Adv. Chem. Phys.*, **86**, 167–329.

Kuchta, B., Luty, T. and Etters, R. D. (1990). *J. Chem. Phys.*, **93**, 5935–5939.

Kumakura, S., Iwasaki, F. and Saito, Y. (1967). *Bull. Chem. Soc. Jpn.*, **40**, 1826–1833.

Kuroda, H., Amano, T., Ikemoto, I. and Akamatu, H. (1967). *J. Amer. Chem. Soc.*, **89**, 6056–6063.

Landau, L. D. and Lifshitz, E. M. (1980). Statistical Physics, being Volume 5 of Course of Theoretical Physics, Third Edition, revised and enlarged by E. M. Lifshitz and L. P. Pitaevskii, translated by J. B. Sykes and M. J. Kearsley, Pergamon Press, Oxford etc., Chapter XIV.

Larsen, F. K., Little, R G. and Coppens, P. (1975). *Acta Cryst.*, B**31**, 430–440.

Le Bars–Combe, M., Chion, B. and Lajzerowicz–Bonneteau, J. (1979). *Acta Cryst.*, B**35**, 913–920.

Le Cointe, M. (1994). "La Transition neuter-ionique dans le TTF – *p*-chloranile: Aspects structuraux." University of Rennes I, France.

Le Cointe, M., Lemée-Cailleau, M. H., Cailleau, H., and Toudic, B. (1996). *J. Mol. Struct.*, **374**, 147–153.

Le Cointe, M., Gallier, J., Cailleau, H., Gourdji, M., Péneau, A., and Guibé, L. (1995). *Sol. State Commun.*, **94**, 455–459.

Le Cointe, M., Lemée-Cailleau, M. H., Cailleau, H., Toudic, B., Toupet, L., Heger, G., Moussa, F., Schweiss, P., Kraft, K. H. and Karl, N. (1995). *Phys. Rev.*, B**51**, 3374–3386.

Lefebvre, J., Odou, G., Muller, M., Mierzejewski, A. and Luty, T. (1989). *Acta Cryst.*, B**45**, 323–336.

Lemée-Cailleau, M. H., Le Cointe, M., Cailleau, H., Moussa, F., Roos, J., Brinkmann, D., Toudic, B., Ayache, C and Karl, N, (1997). *Phys. Rev. Lett.*, **79**, 1690–1693.

Liu, L. and Guo, Q-X. (2001a). *Chem. Rev.*, **101**, 673–695.

Liu, L. and Guo, Q-X. (2001b). *Chin. J. Chem.*, **19**, 670–674.

Lyubarski, G. Ya. (1960). The Application of Group Theory in Physics, Pergamon Press, Oxford.

MacFarlane, R. M. and Ushioda, S. (1977). *J. Chem. Phys.*, **67**, 3214–3220.

Masino, M., Girlando, A., Farina, L. and Brillante, A. (2001). *Phys. Chem. Chem. Phys.*, **3**, 1904–1910.

Masino, M., Farina, L., Brillante, A. and Girlando, A. (2003). *Synth. Mets.*, **133–134**, 629–631.

Matsuzaki, S., Moriyama, T., Onomichi, M. and Toyoda, T. (1983). *Bull. Chem. Soc. Jpn.*, **56**, 369–374.

Mayerle, J. J., Torrance, J. B. and Crowley, J. I. (1979). *Acta Cryst.*, B**35**, 2988–2995.

Metzger, R. M. and Arafat, E. S. (1983). *J. Chem. Phys.*, **78**, 2696–2705.

Metzger, R. M. and Torrance, J. B. (1985). *J. Am. Chem. Soc.*, **107**, 117–121.

Mohwald, H., Erdle, E. and Thaer, A. (1978). *Chem. Phys.*, **27**, 79–87.

Mnyukh, Y. (2001). Fundamentals of solid-state phase transitions, ferromagnetism and ferroelectricity. First Books.

Nogami, Y., Taoda, M., Oshima, K., Aoki, S., Nakayama, T. and Miura, A. (1995). *Synth. Mets.*, **70**, 1219–1220.

Oison, V., Katan, C. and Koenig, C. (2001). *J. Phys. Chem.*, A**105**, 4300–4307.

Onda, S., Ikeda, R. Nakemura, D. and Kubo, M. (1973). *Bull. Chem. Soc. Jpn.*, **46**, 2878–2879.

Park, J. M. and Reddoch, A. H. (1981). *J. Chem. Phys.*, **74**, 1519–1525.

Peierls, R. E. (1954). Quantum Theory of Solids. Oxford: Clarendon Press.

Pippard, A. B. (1964). The Elements of Classical Thermodynamics. Cambridge University Press, London and New York.

Ponte Goncalves, A. M. (1980). *Prog. Solid State Chem.*, **13**, 1–88.

Ponte Goncalves, A. M. (1977). *Chem. Phys.*, **19**, 397–405.

Prout, C. K., Morley, T., Tickle, I. J. and Wright, J. D. (1973). *J. Chem. Soc. Perkin II*, pp. 523–528.

Prout, C. K., Tickle, I. J. and Wright, J. D. (1973). *J. Chem. Soc. Perkin II*, pp. 528–530.

Ripmeester, J. A. (1982). *J. Chem. Phys.*, **77**, 1069–1070.

Ripmeester, J. A., Ratcliffe, C. I., Enright, G. and Brouwer, E. (1995). *Acta Cryst.*, B**51**, 513–522.

Ripmeester, J. A., Reddoch, A. H. and Dalal, N. S. (1981). *J. Chem. Phys.*, **74**, 1526–1533.

Robertson, B. E. and Stezowski, J. J. (1978). *Acta Cryst.*, B**34**, 3005–3011.

Salje, E. K. H. (1990). Phase transitions in ferroelastic and co-elastic crystals. Cambridge University Press: London, New York.

Shahidi, F. and Farrell, P. G. (1978). *J. Chem. Soc. Chem. Commun.*, pp. 455–45X.

Shahidi, F. and Farrell, P. G. (1980). *J. Chem. Research (S)*, pp. 214–215.

Shahidi, F., Farrell, P. G. and Westwood, J. V. (1980). *J. Chem. Research (S)*, p. 357.

Shmueli, U. and Goldberg, I. (1973). *Acta Cryst.*, B**29**, 2466–2471.

Shmueli, U. and Goldberg, I. (1974). *Acta Cryst.*, B**30**, 573–578.

Stanley, H. E. (1971). Introduction to Phase Transitions and Critical Phenomena, Oxford University Press, New York, Oxford.

Stezowski, J. J. (1980). *J. Chem. Phys.*, **73**, 538–547.

Stokes, H. T. and Hatch, D. M. (1988). Isotropy subgroups of the 230 crystallographic space groups, World Scientific, Singapore, New Jersey, London, Hong Kong.

Suzuki, K. and Scki, S. (1955). *Bull. Chem. Soc. Jpn.*, **28**, 417–421.

Swamy, H. R., Ganguly, S. and Rao, C. N. R. (1963). *Spectrochim. Acta*, **39**A, 23–28.

Tanimura, K. and Koshihara, S. (2001). *Phase Transitions*, **74**, 21–34.

Tokura, Y., Okamoto, H., Koda, T., Mitani, T. and Saito, G. (1986). *Solid State Commun.*, **57**, 607–610.

Toledano, J.-C. and Toledano, P. (1987). The Landau Theory of Phase Transitions. World Scientific, Singapore.

Torrance, J. B., Vazquez, J. E., Mayerle, J. J. and Lee, V. Y. (1981). *Phys. Rev. Lett.*, **46**, 253–257.

Tsuchiya, H., Marumo, F. and Saito, Y. (1972). *Acta Cryst.*, B**24**, 1935–1941.

Warren, B. E. (1990). X-ray Diffraction. Dover–Mineola, N. Y.

Williams, R. M. and Wallwork, S. C. (1966). *Acta Cryst.*, **21**, 406– 412.

Williams, R. M. and Wallwork, S. C. (1968). *Acta Cryst.*, B**24**, 168–174.

Wilson, K. G. (1971). *Phys. Rev.*, B**4**, 3184–3205.

Wolfe, C. R. (1982). *Mol. Cryst. Liq. Cryst.*, **85**, 337–343.

Ziman, J. M. (1964). Principles of the Theory of Solids, Chapter 10, Magnetism, University Press, Cambridge.

Chapter 17

Segregated stack π-molecular complexes

7,7,8,8-Tetracyanoquinodimethane (TCNQ) is a strong π-acid which forms stable crystalline radical salts of type $M^+TCNQ^{-\bullet}$ and a new class of complex salts represented by $M^+TCNQ^{-\bullet}$ TCNQ which contain formally neutral TCNQ. The complex anion-radical salts have the highest electrical conductivities known for organic compounds, exhibiting volume electrical resistivities as low as 0.01 ohm cm at room temperature. Both the conductivity and electron paramagnetic absorption are anisotropic as determined by measurements along major crystal axes.

Melby, Harder, Hertler, Mahler, Benson and Mochel, 1962

The term 'organic metal' is a misnomer because these solids are neither ductile nor malleable; they are frail organic crystals or relatively brittle polymers. The only properties reminiscent of metals are high reflectivity and relatively low room temperature resistivity, which *decreases* with decreasing temperature over a certain temperature range.

F. Wudl, 1984

Summary: The unexpected phenomenon of high electrical conductivity in some organic crystals appears to be associated with stacked arrangements of radical ions of electron donors and/or acceptors. A variety of chemical types has been developed, principally based on the parent donor tetrathiafulvalene (TTF) and the parent acceptor tetracyanoquinodimethane (TCNQ). TTF cation radical salts and TCNQ anion radical salts have stacked structures with marked anisotropy of arrangement and physical properties and hence are termed 'quasi-one dimensional'. However, lateral interactions between stacks lead to a measure of two-dimensional character in some of these salts, and this aspect is taking on increasing importance with the development of some of the newer types of donor and acceptor. The degree of stacking ranges from π-dimeric pairs to stacks of infinite length; the longer stacks often contain dimeric or tetradic subgroupings, particularly in the TCNQ radical anion salts. In the cation radical-anion radical salts, of which the most famous example is {[TTF][TCNQ]}, there are segregated stacks of cation and anion radicals. {[TTF][TCNQ]} itself has monad stacks in its averaged structure, but there are also examples of diad stacks. The phase transitions in {[TTF][TCNQ]} below 54K are of an unusual type, the drastic drop in stack axis conductivity on cooling below 54K not being accompanied by appreciable changes in average moiety arrangement. Studies of the very weak diffuse scattering above 54K and the weak satellite reflections below 54K lead to a model in which the high resistivity below 54K is accounted for in terms of pinned charge density waves (CDW), which become mobile above the phase transitions. In the temperature region from 54K up to 300K the conductivity can be explained semiquantitatively by a combination of CDW and single phonon scattering.

17.1   Introduction

Many, many studies have been stimulated by the discovery (Ferraris *et al.*, 1973; Coleman *et al.*, 1973) of a large, temperature-dependent electrical conductivity in {[TTF][TCNQ]} (nomenclature is discussed below) and the association of this striking physical property with the (then-novel) segregated (and *not* mixed-stack) arrangement of donor and acceptor moieties in the crystal. TTF (tetrathiafulvalene) was indexed in *Chemical Abstracts* under "$\Delta^{2,2'}$-Bi-1,3-dithiole" until 1972 and thereafter as "1,3-Dithiole, 2-(1,3-dithiol-2-ylidene)''; registry number 31366–25–3) and TCNQ (7,7,8,8-tetracyanoquinodimethane;

first under "2,5-Cyclohexadiene-$\Delta^{1,a:4,a'}$-dimalonitrile" and then as "Propanedinitrile, 2,2'-(2,5-cyclohexadiene-1,4-diylidene)bis"; registry number 1518-16-7); the two-component complex has registry number 51159-15-0. Considerable effort has been invested in chemical modification of TTF and TCNQ and in the study of analogous systems in the hope of finding materials with new or better physical properties. The many new materials synthesized can be classified as cation-radical salts (i.e. with closed shell anions), as anion-radical salts (i.e. with closed shell cations) and as cation-radical anion-radical salts. Crystal structures have been determined for many of these new materials and their physical properties (especially electronic properties such as conductivity, thermo-electric power and magnetic susceptibility) have been measured over a wide range of temperatures and (sometimes) pressures.

The donors and acceptors that form cation and anion radicals and give segregated stack crystal structures can also participate as neutral or ionic moieties in mixed-stack crystals, and also sometimes present crystal structures of new, albeit related, types. The behaviour of individual moieties in a binary system depends on the properties and interactions of *both* components. As noted in Chapter 1, the structure determining influences in segregated-stack binary molecular *complexes* are A...A and B...B and not A...B – hence 'complexes' and not 'compounds'. A different symbolism is also needed. Thus we emphasize the segregation by replacing the inappropriate {TTF···TCNQ} by {[TTF][TCNQ]}, in more general terms {[donor][acceptor]}; in some instances only one of the components is stacked and only it is placed in the square brackets. Problems arise with known structures that do not fit into the simple mixed stack/segregated stack framework and with unknown structures: then we replace 'complexes' and/or 'compounds' by the noncommittal 'adduct' and use a {donor–acceptor} symbolism.

The sequence of events leading to the discovery of high conductivity in {[TTF][TCNQ]} seems to have been as follows. The synthesis of TCNQ and its ability to form highly-conducting organic salts (conductivities of $\approx 10^2$ S/cm) had been reported in 1962 (Melby *et al.*, 1962); rapid progress was made during the next decade in determining and relating crystal structures (Shibaeva and Atovmyan, 1972) and physical properties of relevant materials. Dibenzo-TTF (then called 2,2'-bis-1,3-benzdithiolene) had been prepared in 1926 (Hurtley and Smiles, 1926). TTF itself was reported by a number of groups towards the end of the 1960s (Prinzbach *et al.*, 1965; Wudl *et al.*, 1970). TTF$^{+\bullet}$Cl$^-$ was prepared by Wudl and coworkers in 1970, its conductivity at 25 °C being reported as ≈ 0.2 S/cm in 1972 (Wudl *et al.*, 1972). The crucial step of combining TTF and TCNQ was taken very soon thereafter, essentially simultaneously, by research groups at Monsanto, Johns Hopkins and the University of Pennsylvania (who published their results (Miles *et al.*, 1972; Ferraris *et al.*, 1973; Coleman *et al.*, 1973) and at SUNY-Buffalo (who did not). To quote from Wudl's lively account: (Wudl, 1984)

"Having established that TTF salts are highly conducting, it did not take long before the marriage of the young donor with the old acceptor took place. The exact location of the ceremony is somewhat obscure. It is clear it was a fertile coupling as judged from the number of papers it generated between 1973 and the present. The sudden interest (1973–1975) in this solid was caused by two, almost simultaneous, events: the discovery of metal-like conductivity between room temperature and 56K and the report of 'superconductivity' just above 56K. Of these only the former survived scrutiny."

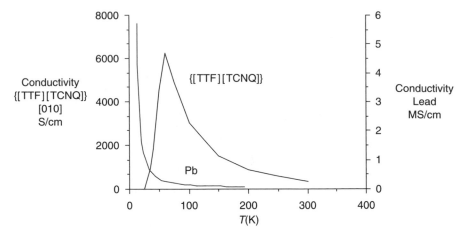

Fig. 17.1. Schematic representations of the conductivities of Pb (Onnes and Tuyn, 1929) (right-hand scale; the superconductivity of Pb below 7.2K has been omitted from the figure) and {[TTF][TCNQ]} (along the stack axis; left-hand scale) as functions of absolute temperature; note that Pb is about 1000 times more conductive than {[TTF][TCNQ]}·Pb has been chosen for comparison because its Debye temperature is not very different from that of {[TTF][TCNQ]}, which is considered to have metallic behaviour down to ≈60K because its σ–T curve follows the same course as that of Pb and other metals.

While much interesting physics and chemistry has been developed in this area since 1973, it is perhaps fair to state that no system has yet been found with properties more striking than those of the classical {[TTF][TCNQ]} molecular complex, although the discovery of superconductivity in the Bechgaard salts $(TMTSF)_2X$ (Friedel and Jerome, 1982) may require some modification of this judgment.

Our primary concern in this chapter is with the *structures* of segregated stack molecular complexes and related ion radical salts and the properties that derive from these structures rather than with the phenomenon of high conductivity as such, which has been the main interest of most investigators in this field (Howard, 1988). Thus we do not discuss many important high conductivity materials, such as the essentially inorganic materials reviewed elsewhere (Miller and Epstein, 1976). However, we have considered the TTF and TCNQ salts with closed shell counterions (and analogs) to come within our boundaries, both because of their intrinsic structural interest in the context of stacked arrangements and because of their relevance to {[TTF][TCNQ]} and related molecular complexes; similar latitude has been extended to some other ion radical salts.

A rough picture of the changing interest in TCNQ, TTF and {[TTF][TCNQ]} can be obtained from Fig. 17.2, where the total number of entries under each of these three headings (in practice using the Registry Numbers) in the quinquennial indexes of Chemical Abstracts has been plotted against the end-year of the quinquennium. The interest in TCNQ and TTF appears to have peaked in the early 1990's, as other donors (many based on TCNQ and TTF) and other types of ion radical salt have taken over part of the stage. {[TTF][TCNQ]} seems to have disappeared from the area of active research but surely it is not yet completely understood.

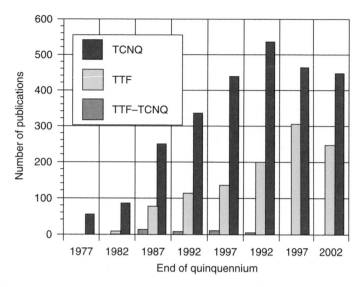

Fig. 17.2. The numbers of references to TCNQ, TTF and {[TTF][TCNQ]} in Chemical Abstracts for 5-year periods are plotted against the end-year of the quinquennium. The numbers of references have been obtained from the Registry Numbers for the three materials using Scifinder Scholar[TM] but no attempt has been made to check their relevance or eliminate duplication.

17.2   Chemistry of donors and acceptors that participate in segregated stacks

17.2.1   *Introduction*

The donors and acceptors of interest here are oxidized or reduced by multistage electron transfers without the formation or rupture of electron pair single bonds. Such reactions have been summarized in terms of three equations by Deuchert and Hünig (1978), whose treatment we follow closely. Their structural principle is stated as follows:
"Reversible redox reactions with transfer of two electrons in two separate steps are to be expected with compounds in which the end groups X and Y of the reduced form

1. have free electron pairs or π-systems available, and
2. are connected by vinylene groups ($n = 0, 1, 2 \ldots$)."

The various stages are represented by the reduced (RED), radical ('semiquinone' or SEM) and oxidized (OX) forms. The systems A and B differ only by two charges; otherwise they are iso-π-electronic. The SEM radical appears as a cation in A and an anion in B. System C, where the SEM stage is a neutral radical, is less relevant in the present context. All three redox systems are capable of many variations. Deuchert and Hünig (1978) define two particular types of redox system which contain most of the examples of donors and acceptors to be considered below:

Two-stage Würster-type redox systems: two-stage redox systems are said to be of the Würster type if their end groups are located outside a cyclic π-system that has aromatic character in the reduced form.

	E_1		E_2	
	$-e$		$-e$	
RED^a	$\leftrightarrow$	SEM^{a+1}	$\leftrightarrow$	OX^{a+2}
	$+e$		$+e$	

Equation A

	$-e$		$-e$	
$X-(-CH=CH-)_n-X$	$\leftrightarrow$	$\{X-(-CH=CH-)_n-X^{\circ+}$	$\leftrightarrow$	$^+X-(-CH=CH-)_n-X^+$
		$\leftrightarrow\ ^{\circ+}X-(-CH=CH-)_n-X\}$		
	$+e$		$+e$	

Equation B

	$-e$		$-e$	
$^-Y-(-CH=CH-)_n-Y^-$	$\leftrightarrow$	$\{^-Y-(-CH=CH-)_n-Y^{\bullet}$	$\leftrightarrow$	$Y-(-CH=CH-)_n-Y$
		$\leftrightarrow\ ^{\bullet}Y-(-CH=CH-)_n-Y^-\}$		
	$+e$		$+e$	

Equation C

	$-e$		$-e$	
$X-(-CH=CH-)_n-Y^-$	$\leftrightarrow$	$\{X-(-CH=CH-)_n-Y^{\bullet}$	$\leftrightarrow$	$^+X-(-CH=CH-)_n-Y$
		$\leftrightarrow\ ^{+\bullet}X-(-CH=CH-)_n-Y^-\}$		
	$+e$		$+e$	

Notes:

1. (electron loss (oxidation) occurs for the forward directions of the arrows, and conversely).

2. SEM is shown as a resonance hybrid.

Two-stage Weitz-type redox systems: two-stage redox systems are said to be of the Weitz type if their end groups form part of a cyclic π-system that has aromatic character in the oxidized form.

In the discussion that follows the chemical formulae of the donors and acceptors are given together with partial reference to the appropriate equations and values of n.

17.2.2   *Donors*

By the early 1970s it had become clear that the following donors had a high tendency to appear as cation radicals in appropriate systems (ionization potentials are given in Table 13.5):

TMPD	N,N,N',N'-tetramethylphenylenediamine
TTF	tetrathiafulvalene
TTT	tetrathiotetracene.

Some effort has also been invested in the synthesis of related systems, especially those based on pyranylpyrans. An extensive review, especially of newer examples, is given by Yamashita and Tomura (1998).

TMPD forms stacks in some of its cation-radical salts (see Section 17.3.4) but mixed rather than segregated stacks in most molecular compounds with acceptors. It has a Würster type redox system.

Equation A:

reduced form   oxidised form

The major synthetic efforts applied to TTF, and to many other donors, have been in two directions:

(i) replacement of S by its congeners Se and Te; for example, in the TTF and TTT systems the Se and Te congeners have been synthesized. Three important trends to be expected upon replacement of S or Se by Te have been noted (McCullough *et al.*, 1987). We quote:

> The more diffuse p and d orbitals centred on tellurium should give larger conduction bandwidths due to increased interstack interactions and result in materials with reduced electron scattering and enhanced metallic electrical conductivity. In addition this increase in orbital spatial extension ought to increase the interchain interactions giving rise to more two- or three-dimensional character. This extended dimensionality should help suppress the various instabilities which often lead to insulating ground states in quasi-one-dimensional organic conductors. Finally, the greater polarizability of tellurium should reduce the on-site Coulombic repulsion and help support doubly-charged species. Unless the molecular component can support doubly-charged species, only a correlated type of conductivity is possible.

(ii) preparation of substituted TTFs and congeners; in particular, alkylthio substituents could alter the dimensionality of the crystals by increasing interstack interactions.

The relevant complexes are shown below, together with notes and references; synthetic details are in the original papers. Most synthetic studies have been accompanied by the preparation of donor – TCNQ molecular complexes of various compositions and measurements of electrical conductivities on compacted powders or single crystals; a wide range of conductivities is often found (a list of over 400 crystalline conducting organic quasi-one-dimensional molecular complexes has been compiled by Howard (1988)). Other acceptors have sometimes also been used. Structural information for correlation with physical properties is slowly being accumulated.

(i) 1,2,4,5-Tetrakis(dimethylamino)benzene is a powerful donor of the Würster type (Eqn. A, $n = 0$) (Elbl *et al.*, 1986). 1,6-Diaminopyrene – TCNQ has a conductivity of 2 S/cm (Scott *et al.*, 1965); its structure is not known. 2,7-Bis(dimethylamino)-tetrahydropyrene – TCNQ has a conductivity of 0.4 S/cm and has been estimated from the CN stretching frequency to have a charge transfer of 0.57; its structure is not known. A number of related molecular complexes of low conductivity have been prepared (Ueda *et al.*, 1983).

Fully aromatic 2,7-bis(dimethylamino)- and 1,3,6,8-tetrakis-(dimethylamino)-pyrene and partially-reduced 2,7-bis(dimethylamino)pyrene have been shown to be good electron donors (Sakata *et al.*, 1984).

(ii) Dithiapyrenes (Tilak, 1951) (Eqn. A, $n = 2$) and dithiaperylenes (Eqn. A, $n = 3$). These are Weitz type systems.

1,6-dithiapyrene (DTPY; $R_1 = R_2 = H$).
When $R_1 = SCH_3$, $R_2 = H$ the complex is
2,7-bis(methylthio)-1,6-dithiapyrene and
when $R_1 = H$, $R_2 = SCH_3$ it is
3,8-bis(methylthio)-1,6-dithiapyrene.

3,10-dithiaperylene (DTPR)

{[DTPY][TCNQ]} has a segregated stack crystal structure (see Table 17.10) (Thorup et al., 1985; DAKTIS); the structures of some salts of 2,11-diphenyl-DTPR have been reported (Nakasuji et al., 1986) and also of some molecular complexes of substituted DTPYs (see Table 17.10). 3,4:3',4'-Bibenzo[b]thiophene (BBT), the thiophene analog of DTPR (i.e. with the S atoms in five-membered rings) has been prepared (Wudl et al., 1978) and is isoelectronic in its neutral state with perylene. Structures of molecular complexes have not been reported.

4,5:9,10-bis(ethanediylthio)-
1,6-dithiapyrene (Nakasuji et al., 1986)

2,3:7,8-bis(ethanediylthio)-1,6-dithiapyrene
(ETDTPY) (Nakasuji et al., 1986)

The crystal structures of the TCNQ (FUDTON) and chloranil (FUDTUT) adducts of 3,8-bis(methylthio)-1,6-dithiapyrene have been reported (Nakasuji *et al.*, 1987).

(iii) Benzotrichalcophenes (Cowan *et al.*, 1982):

X = S (BTT) (Hart and Sasaoka, 1978), Se (BTS), Te (BTTe)

BTT forms intensely coloured, moisture-stable crystalline 1 : 1 charge transfer adducts with TCNE, DDQ, TCNQ and chloranil. Structures and physical properties do not appear to have been reported.

(iv) Systems based on TTF and congeners (Eqn. A, $n = 1$; these are Weitz type systems); syntheses have been extensively reviewed (Narita and Pitman, 1976; Krief, 1986; Schukat *et al.*, 1987); Nielsen *et al.*, 2000; Yamada and Sugimoto, 2004). The basic structure (below left) has been substituted in many ways, some of which are illustrated and others referenced (Fanghängel *et al.*, 1983).

TTF : Ch = S, R = H;
TSF (or TSeF) : Ch = Se, R = H
TTeF : Ch = Te, R = H (Narita and Pitman, 1976)
Tetrakis(alkylthio)TTF (Mizuno *et al.*, 1978): Ch = S, R = SCH₃;
TMTTF : Ch = S, R = CH₃;
TMTSF : Ch = Se, Te, R = CH₃ (Iwasawa *et al.*, 1987)

HMTSF (Berg *et al.*, 1976) : Ch = Se, R = CH₂[1];
HMTTeF (Saito *et al.*, 1983) : Ch = Te, R = CH₂
DTTSF (Elbl *et al.*, 1986): Ch = Se, R = S

Bis(2,5-dimethylpyrrolo[3,4-d]tetrathiafulvalene: R = H (BP-TTF); Ph (BPP-TTF) (Chen *et al.*, 1988)

[1] 2, 2′-bis(2,4-diselenabicyclo[3.3.0]octylidene).

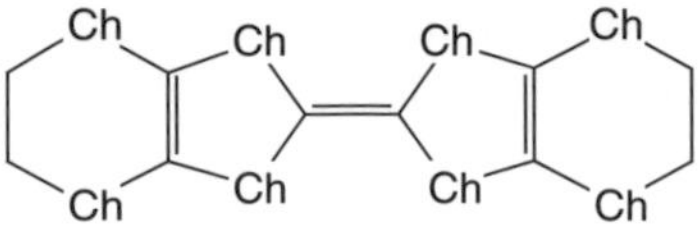

Bis(ethylenedithiolo)TTF (BEDT-TTF)
Ch = S (Saito et al., 1982); Ch = Se
(Lee et al., 1983).

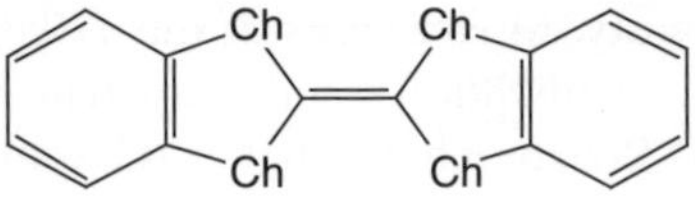

DBTTF : Ch = S (Spencer et al., 1977)
(Unsymmetrical versions with only one
benzene ring have been prepared, and also
symmetrically (E) and unsymmetrically (Z)
substituted dimethyl complexes) (Shibaeva
and Yarochkina, 1975; Tanaka et al., 1983;
Nakano et al., 1989)
DBTSF : Ch = Se

(iii) Systems based on pyran derivatives:

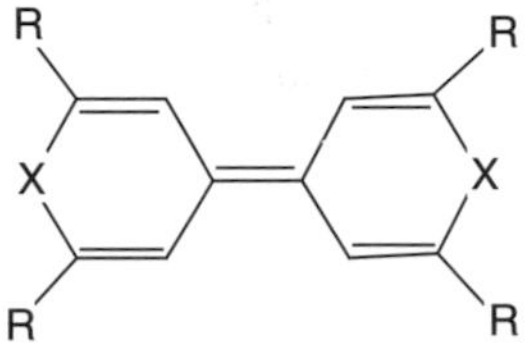

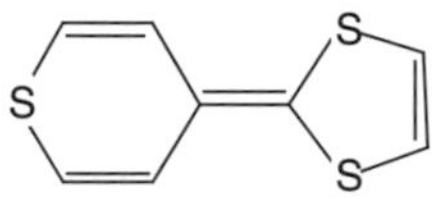

X = O, S, Se, Te; (Alizon et al., 1976; Detty et al., 1983); X = S, R = H or CH_3 (Sandman,
Epstein et al., 1977). See also Sandman, Fisher et al., 1977.
4,4'-Bithiopyranylidene (BTP, X = S, R = H) is isoelectronic with TTF.

Note: when X = O and R = $C_6H_4C_{12}H_{25}$, then charge transfer salts (counterions BF_4, ClO_4, TCNQ) with
appreciable mesophase (liquid crystalline) temperature ranges are found (Saeva et al., 1982).

(iv) Systems based on p-quinobis(1,3-dithiole):

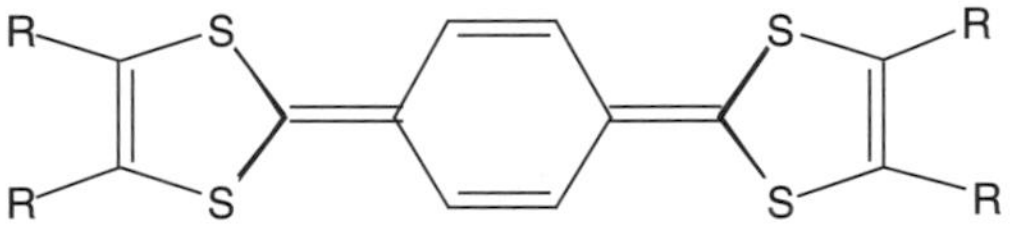

R = CH_3, TMCHDT (Bis(4,5-dimethyl-2H-1,3-dithiolylidene-2)-1,4-cyclohexa-2,5-diene)
(Fabre et al., 1978).

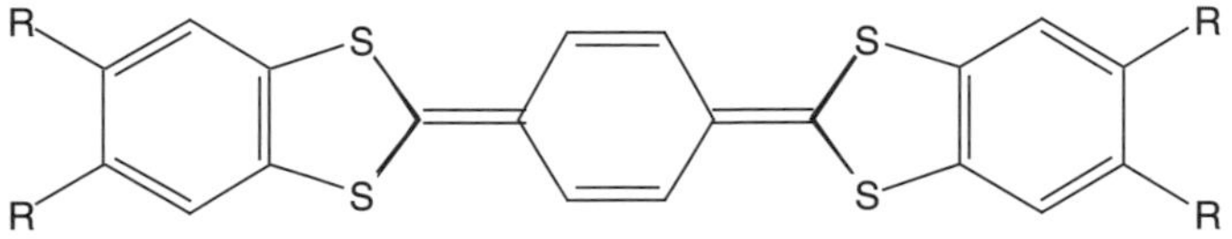

R = H (Sato et al., 1978), H, CH_3 (Ueno et al., 1978). Also analogues based on the
anthracenediylidene system (Bryce et al., 1990). Dithiadiazafulvalenes have been shown to
be strong electron acceptors (Tormos et al., 1995).

(v) Systems based on tetrathiotetracene:

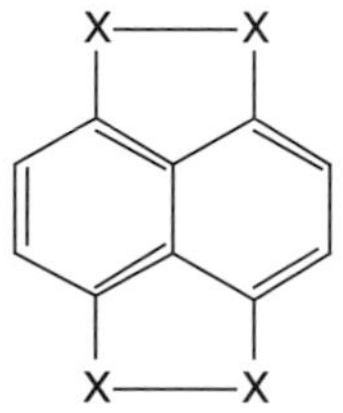

X = S TTN (dehydrotetrathianaphthazarin).
Se TSN (naphtho[1,8-cd:4,5-c′d′]bis(1,2-diselenole)

TTN was shown electrochemically to be a much poorer donor than TTF; however, {[TTN][TCNQ]} crystallizes as black needles with a single crystal conductivity of 40 S/cm between 200–300K, which is higher than that of {[TTT][TCNQ]} (1 S/cm) and the same as the microwave conductivity of {[TTT][(TCNQ)$_2$]} (Wudl *et al.*, 1976). Crystal structures are discussed later.

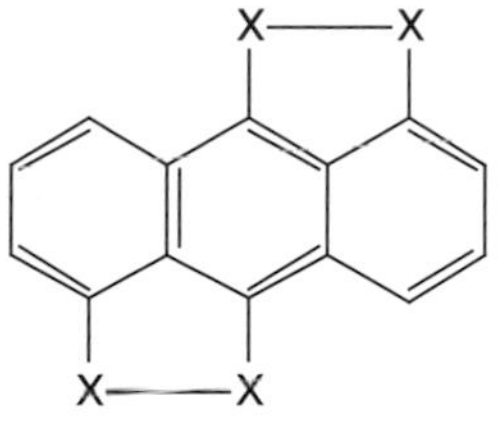

Equation A, $n = 2,3$.
Se TSA (tetraseleno-anthracene;
anthra[9,1-cd:10,5-c′d′]bis(1,2-diselenole))
(Endres *et al.*, 1982)

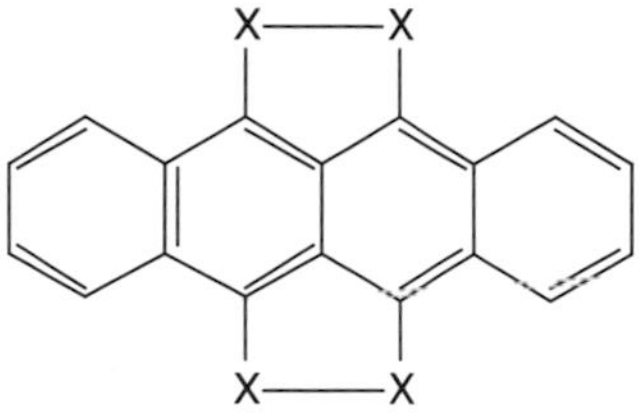

X = S TTT (naphthacene[5,6-cd:
11,12-c′d′]bis(1,2-dithiole))
Se TSeT (Khidekel and Zhilyaeve, 1981)
Te TTeT (Sandman *et al.*, 1982)

(vi) Systems with both N and S atoms:

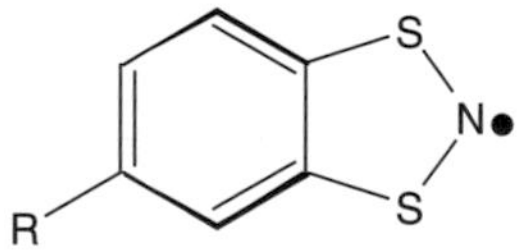

R = H, CH$_3$ Benzo-1,3,2-dithiazol-2-yl and derivatives (Wölmerhauser *et al.*, 1984).

17.2.3 *Acceptors*

The synthesis of TCNQ (Acker and Hertler, 1962) was accompanied by a thorough investigation of its ion radical salts (Melby *et al.*, 1962) (the abstract of this paper is quoted at the head of this chapter); there are examples (Yamaguchi *et al.*, 1989) of alternative syntheses. A number of substituted TCNQs have been synthesized as well as analogs of various kinds. A new family of segregated stack crystals is based on TTF with tetrahalo-*p*-benzoquinones as acceptors (Torrance *et al.*, 1979). Bis(dithiolene)metal complexes behave as acceptors with many cations (Interrante *et al.*, 1975); the crystal structures are rather different from those of TCNQ molecular complexes and the materials have interesting physical properties. The various acceptors are compiled below; structures and properties are discussed at appropriate places later in this chapter.

(i) TCNQ and related complexes:
   Equation B:

reduced form

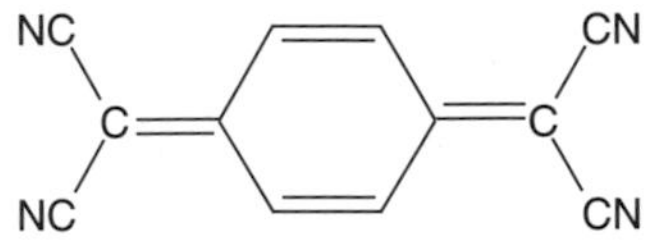

oxidized form

7,7,8,8-tetracyanoquinodimethane (TCNQ) (Equation B, n = 3; these are Würster type systems). Among the derivatives synthesized are monofluoro-TCNQ (Ferraris and Saito, 1979), 2,5-difluoro-TCNQ (Saito and Ferraris, 1979), tetrafluoro-TCNQ (TCNQF$_4$) (Wheland and Martin, 1975), and methyl-, ethyl- and 2,5-dimethyl-TCNQ (Andersen and Jorgensen, 1979).

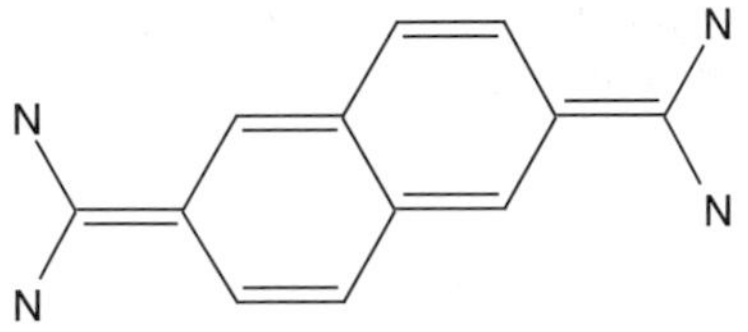

11,11,12,12-tetracyano-naphthoquinodimethane (TNAP) (Diekman *et al.*, 1963)
(Equation B, *n* = 4)

Some three-dimensionally modified TCNQ derivatives such as dihydro- and tetrahydrobarreleno-TCNQ, monobenzo- and dibenzobarrelleno-TCNQ have been

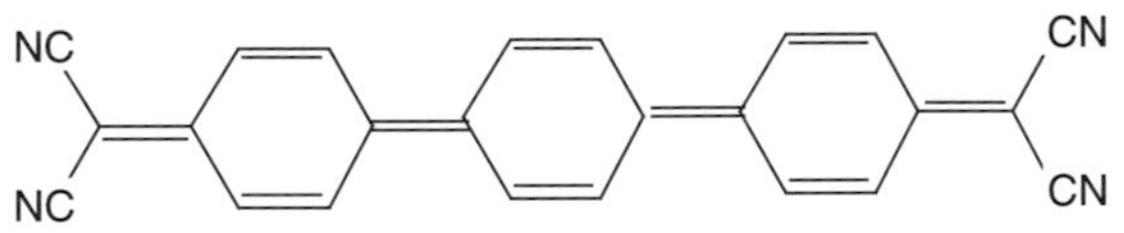

Tetracyanodiphenoquinodimethane (Aharon-Shalom *et al.*, 1979) (Equation B, *n* = 7)

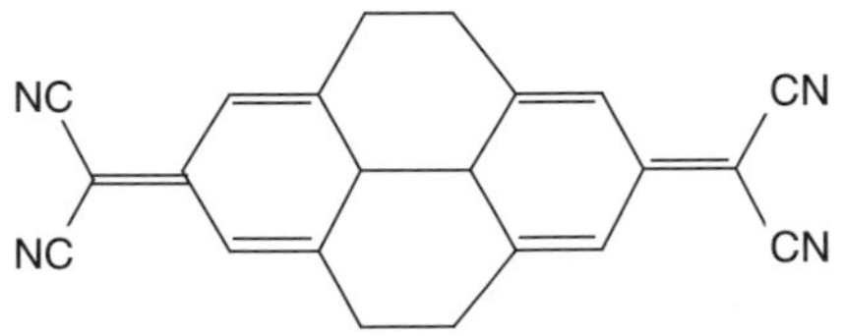

13,13,14,14-Tetracyano-4,5,9,10-tetrahydropyrenequinodimethane (TCNTP) (Aharon-Shalom *et al.*, 1979; Maxfield *et al.*, 1979) (Equation B, *n* = 5); see also Suguira *et al.*, 2000.

prepared and shown to form molecular complexes of varying degrees of ionicity with TTF derivatives (Nakasuji *et al.*, 1986).

11,11,12,12,13,13,14,14-octacyano-1,4:5,8-anthradiquinotetramethane{OCNAQ;1,4:5,8-tetrakis-(dicyanomethylene)anthracene} has been prepared (Mitsuhasi *et al.*, 1988) and shown to form molecular complexes with TTF, pyrene, phenothiazine, TMTTF and TTT. Structures (Inabe *et al.*, 1988) are discussed later. The acceptor is somewhat nonplanar because of steric hindrance between adjacent $C(CN)_2$ groups.

(ii) Other polycyano complexes:

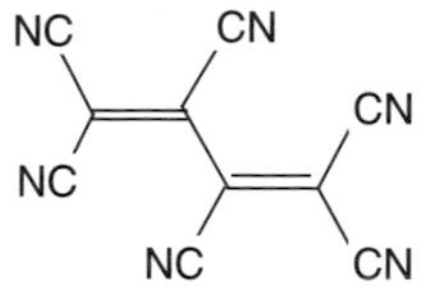

Hexacyanobutadiene (HCBD), also called tetracyanomuconitrile (Webster, 1964).

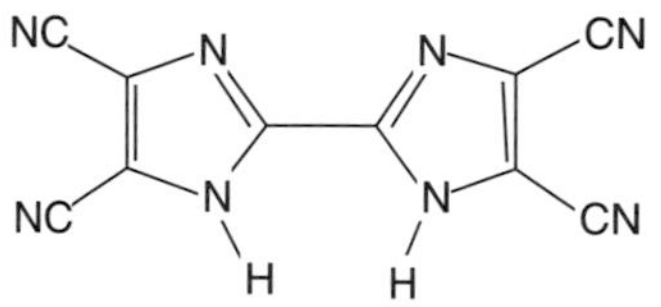

Tetracyanobi-imidazole (Rasmussen *et al.*, 1982).

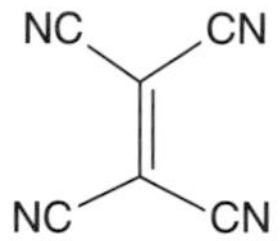

Tetracyanoethylene (TCNE) (Equation B, $n = 1$)

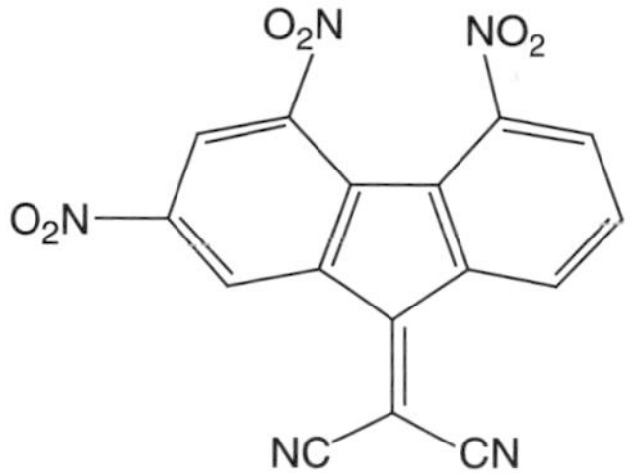

2,4,5-Trinitro-9-(dicyanomethylene)fluorene (Dupuis and Neel, 1969)

(iii) S-heteroquinoid acceptors:

The tetracyanothieno[3,2-b]thiophene system, where the tetracyano groups bracket a heteroquinoid nucleus, has been used as a basis for acceptors (Yui *et al.*, 1989). The corresponding dicyano complexes (2,5-bis(cyanoimino)-2,5-dihydrothieno[3,2-b]thiophenes (DCNTTs)) are also good acceptors in charge transfer complexes and, especially, in highly conductive salts (Aumüller *et al.*, 1988).

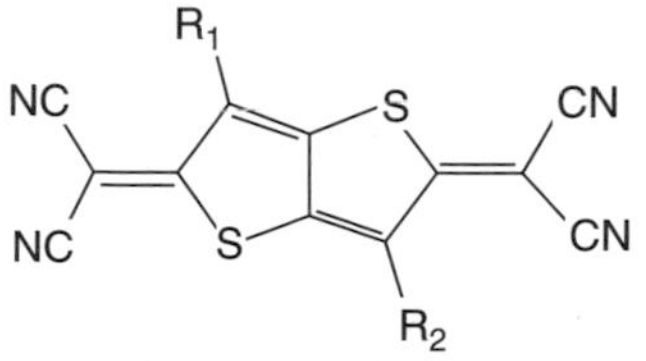

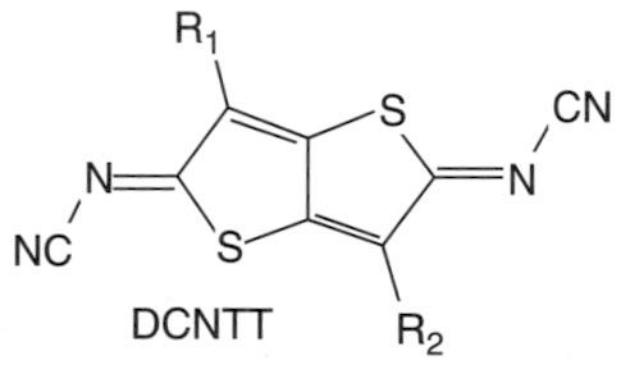

R$_1$ and R$_2$ are various combinations of H, CH$_3$, Cl, Br.

(iv) Other acceptors containing S and N atoms:

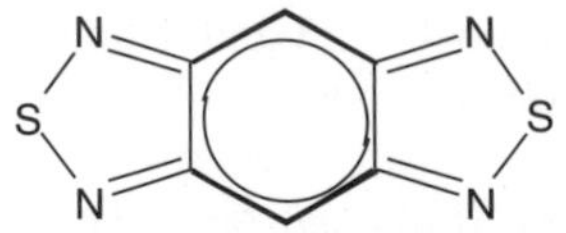

BDTA

Bis([1,2,5]thiadiazolo)[3,4-b;3′,4′-e]pyrazine (BDTA; the formula is drawn to indicate that the molecule is best represented as a resonance hybrid) is a 14π electron heterocycle (Yamashita *et al.*, 1988), with a higher electron affinity than that of BDTA-TCNQ below. Its crystal structure has been determined but formation of molecular complexes and radical ion salts has not been reported in detail.

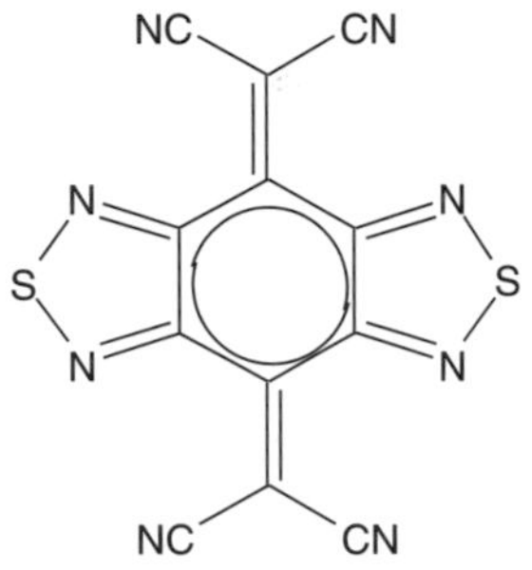

BDTA-TCNQ 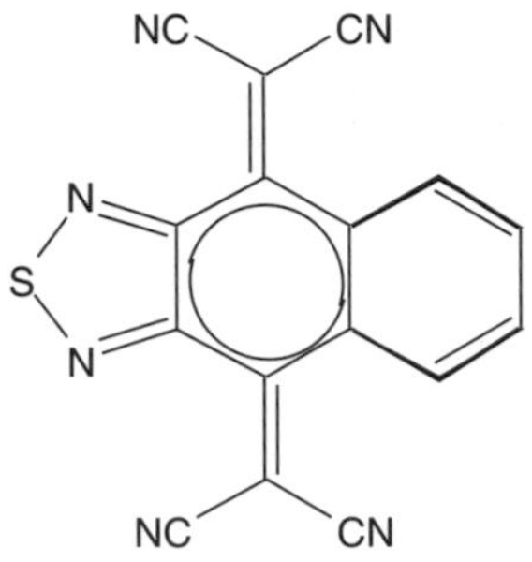   TDA-TCNNQ

Two related acceptors are (bis[1,2,5]thiadiazolo)TCNQ (BDTA-TCNQ) (Se and S, Se analogs have also been reported (Suzuki *et al.*, 1987)) and [1,2,5]thiadiazolo)-tetracyanonaphtho-quinodimethane (TDA-TCNNQ). The electron donors TTF, TMTTF and TMTSF gave very low conductivity ($\approx 10^{-9}$ S/cm) adducts with BDTA-TCNQ and these are probably mixed stack compounds; however, the TTN molecular compound had a conductivity of ≈ 1 S/cm and could have segregated stacks (Yamashita *et al.*, 1985). Structures have not been reported.

(v) Tetrahalo-*p*-benzoquinones and analogs:

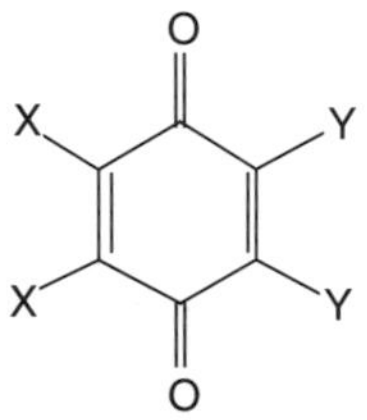

X = Y F, Cl, Br, I
X = CN, Y = Cl 2,3-Dichloro-5,6-dicyano-*p*-benzoquinone (DDQ)

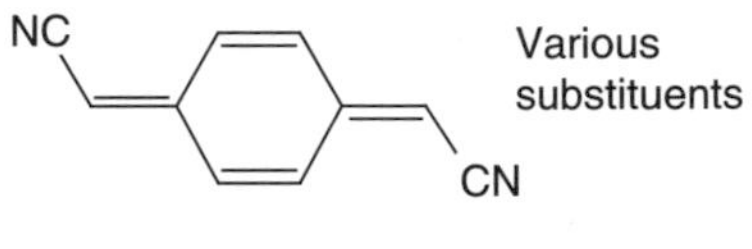

N,N′-dicyanoquinone diimines (Aumüller and Hünig, 1986a,b)

(vi) Aromatic anhydrides:

1,4,5,8-naphthalenetetracarboxylic anhydride (NDTA) forms conducting salts with various cations. The salt $(5,6$-dihydro4a,6a-phenanthrolinium$)_2$(NDTA)$_5$ has a stacked structure (Heywang *et al.*, 1989; VARKEE, $P\bar{1}$, $Z = 1$; Born and Heywang, 1991; VARKEE10; see Section 17.4.5).

(vii) 1,2-Ethylenebis(dithiolene)metal complexes (see Section 17.5.2):

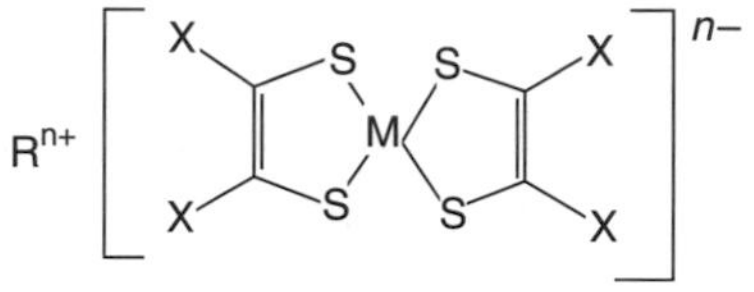

$n = 0, 1, 2.$
X = H, CH$_3$, CN (abbreviated as mnt which is maleonitrile dithiolate or cis-2,3-dimercapto-2-butenedinitrile), CF$_3$ (abbreviated as tfd), C$_6$H$_5$ etc.
M = Ni, Pd, Pt, Fe, Co, Cu, Au etc.

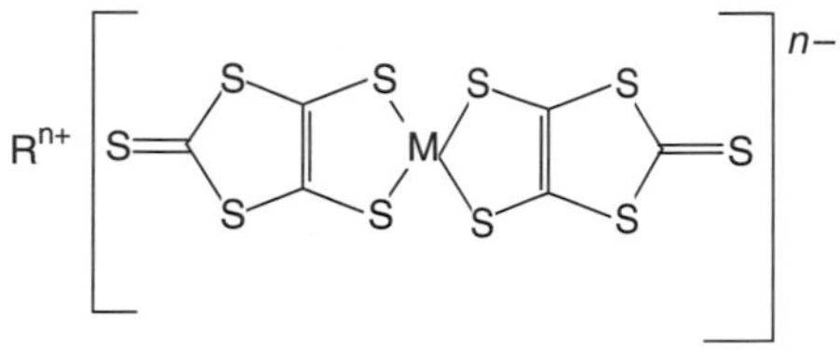

$n = 0.5, 1, 2.$
Abbreviated as dmit.
H$_2$(dmit) = 4,5-dimercapto-1,3-dithiole-2-thione.

17.2.4 Preparation of crystals

A feature of most synthetic papers is that some charge transfer radical ion salts crystallize without difficulty, generally as needles of intense colour (deep green, blue or black) showing metallic reflectivity; on the other hand, some products are obtained only as microcrystalline powders. The reason for such behaviour is generally not understood. Standard methods are used for growth of crystals or recrystallization of powders – slow cooling of saturated solutions, slow evaporation, interdiffusion of components in U-tubes (Kaplan, 1976), simultaneous sublimation of components on to cold surfaces (Andersen, Engler and Bechgaard, 1978). However, many beautiful crystals show disappointing diffraction patterns, the apparent external perfection being accompanied by appreciable disorder at the unit cell level.

Electrocrystallization (Chiang *et al.*, 1971; Ristagro and Shine, 1971; Rosseinsky and Kathirgamanathan, 1982) is a technique of increasing popularity for growing crystals of highly conductive materials. It was first used for the perchlorates of pyrene, perylene and azulene and later extended to {[perylene][Ni dithiolate]} (Alcacer and Maki, 1974) and various salts of TCNQ and analogs (Kathirgamanathan *et al.*, 1979, 1980, 1982). Both potentiostatic and galvanostatic arrangements have been used (Fig. 17.3), and it has been shown that the crystals growing on the anode function directly as electrodes (Enkelmann, Morra *et al.*, 1982). The nature of the solvent can also be important (Anzai *et al.*, 1982).

The highly conductive materials considered here are usually very anisotropic, with the metallic conductivity generally restricted to the stack axis. Thus chemical impurities and lattice imperfections would be expected to strongly influence the one-dimensional electron transport, and more so than in three-dimensional metals. This anticipation is reinforced by the many reports that the conductivity of {[TTF][TCNQ]} is sample dependent (Fig. 17.46). The matter is somewhat controversial. One group (McGhie *et al.*, 1974, 1978) claims that extreme precautions (gradient sublimation under argon, use of quartz

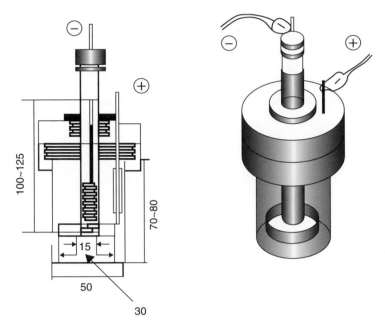

Fig. 17.3. Example of a crystal growth cell used for growing highly conductive crystals in a nitrogen atmosphere. Concentrations of $\approx 10^{-4}$ M were used, with voltages in the range 1–5 V and currents in the range 1–10 mA. Dimensions in mm. (Reproduced from Anzai *et al.*, 1982.)

apparatus) are essential while another (Gemmer *et al.*, 1975) suggests that the simple standard techniques of recrystallization and sublimation are adequate to yield TTF and TCNQ of excellent chemical purity (impurities at the ppm level as judged by the use of high performance liquid chromotography (HPLC)). Gemmer *et al.* (1975) conclude that the variations in the conductivity of different {[TTF][TCNQ]} samples are due to lattice defects. McGhie *et al.* (1974, 1978) emphasize that large well formed crystals of {[TTF][TCNQ]} are obtained only when high purity components are used and that the physical perfection of the crystals depends both on the chemical purity of the components and the growth conditions.

17.3   Structures of cation-radical salts

17.3.1   *Introduction*

The structures of a fairly large number of cation-radical salts have been determined in the last few years and many of these are important in the context of high conductivity (Shchegolev and Yagubskii, 1982; Shibaeva, 1982)), or even superconducting (Williams, Beno *et al.*, 1985) materials. Space limits us to considering here only the cation-radical salts derived from polycyclic aromatic hydrocarbons because of their fundamental position in the crystal chemistry of charge transfer molecular complexes, and the salts of TTF, which illustrate some of the principles found also in other cation radical salts, and where the cation participates in the much-studied {[TTF][TCNQ]}.

17.3.2   *Cations are polycyclic aromatic hydrocarbons*

We shall treat these as one structural group with subdivisions; however, there are already enough structural types to suggest that the true picture may well be more complicated. Three structural subdivisions can be discerned – those in which all the radical cations are in stacks, those in which some are stacked and others not, and a non-stacked group.

17.3.2.1   *Stacked radical cations*

{(Naphthalene)$_2^{+\bullet}$ PF$_6^-$} forms dark red-violet conducting ($\sigma = 0.12$ S/cm) crystals (grown by electrocrystallization; tetragonal, $a = 11.56$, $c = 6.40$ Å, space group $P4_2/n$, $Z = 2$) that are stable at low temperature but decompose on warming to room temperature (Fritz *et al.*, 1978; NAPHFP10); the AsF$_6$ salt (ZZZBTG) is isomorphous. The structure consists of segregated stacks of (C$_{10}$H$_8$)2$^{+\bullet}$ cations and PF$_6^-$ anions along [001]. The

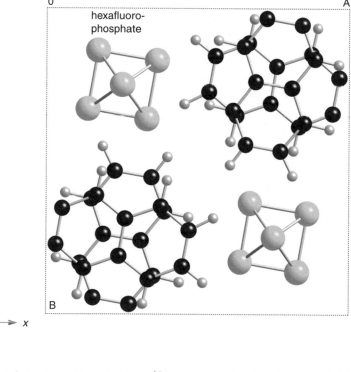

Fig. 17.4. Crystal structure of {(naphthalene)$_2^{+\bullet}$ PF$_6^-$} at 223K projected down [001]. The diagram shows the mutual arrangement of nearly coplanar cations and anions. The 1-9 and 2-3 bonds in the naphthalene cation are significantly shorter than those in neutral naphthalene at 90K (Brock and Dunitz, 1982) (1.408(6) as against 1.423(2) Å and 1.386(6) as against 1.415(2) Å). (Data from Fritz *et al.*, 1978.)

cations lie in (002) planes and are separated by the remarkably small interplanar distance of 3.20 Å, successive cations along the 4_2 axis of a stack being mutually rotated by 90°; presumably there is a random arrangement of neutral molecules and cations in each stack and this accounts for the conductivity. The packing arrangement (Fig. 17.4) is similar to those of bis(diphenylglyoximato)Ni(II) iodide and Ni(phthalocyanine) iodide (Marks, 1978).

The preparation of some of the above and related salts [e.g. (triphenylene)$_2$X, X = PF$_6$ or AsF$_6$; (perylene)$_2$BF$_4$] has been reported (Kröhnke *et al.*, 1980). Conductivities ranged from 10^{-3} to 10^{-1} S/cm. The salts (fluoranthene)$_2^{+\bullet}$ X$^-$ (X = PF$_6$ (FANTHP10), AsF$_6$ (BOSJUO), SbF$_6$ (BUNCAO)) are isomorphous (a = 6.50, b = 12.49, c = 14.75 Å, β = 104.0°, $P2_1/c$, Z = 2 for hexafluoroarsenate at 120K) and, judging from the curves of cell dimensions *versus* T, all show second (or perhaps higher) order phase transitions at $\approx$200K (Enkelmann, Morra *et al.*, 1982). The small differences between high and low temperature structures derive from small rotations of cations and anions; adjacent stacked fluoranthenium cation radicals are separated by 3.22 and 3.28 Å. The dimensions found for the fluoranthenium cation radical are not precise enough for comparison with those of the neutral molecule. The salts are quasi-metallic above $\approx$200K and semiconducting below. In both (naphthalene)$_2^{+\bullet}$ X$^-$ and (fluoranthene)$_2^{+\bullet}$ X$^-$ the cation planes are exactly or nearly normal to the stack axes.

Three isomorphous bis(perylenium) salts [(pe)$_2$(X)$_x$·(Y)$_y$, with X=PF$_6$ (CUW-BAX01), AsF$_6$ (PITDOL), x = 0.8 − 1.5; Y = CH$_2$Cl$_2$, C$_4$H$_8$O, y = 0.5, 0.8] have been studied (Keller, Nöthe *et al.*, 1980); the black needles were grown by electrocrystallization. These have structures (orthorhombic, a = 4.285, b = 12.915, c = 14.033 Å, space group *Pnmn*, Z = 1) different in many respects from that of (naphthalene)$_2^{+\bullet}$ PF$_6^-$. The translationally equivalent perylenium cation radicals are stacked along [100] with an interplanar spacing of 3.40 Å and an inclination of 38° to (100) and are mutually shifted along their long axes. The channels between the cation stacks are filled with an ill-defined melange of anions and solvent molecules. The stack axis DC conductivities are in the range 10^2–10^3 S/cm at room temperature, follow a metallic regime down to $\approx$200K and then diminish rapidly. Keller, Nöthe *et al.* (1980) suggest that the rather high conductivities are due to enhancement of the mixed valence state of the cation radical stacks by non-stoichiometry arising from replacement of some anions by neutral solvent molecules. Other perylene charge transfer salts are discussed in Section 17.5.2.

17.3.2.2 *Structures in which not all cations are stacked*

There is a second conducting form of (pe)$_2$(PF$_6$)·2/3(THF) ($\sigma \approx 10^2$ S/cm; CUWBAX) that has infinite stacks of perylene moieties, with a slight tendency to the formation of tetrads (Endres *et al.*, 1985). However, only *half* the perylene moieties are in stacks, while the others flank the stacks on all four sides, the planes of the flanking perylenes being virtually perpendicular to the planes of those in the stacks (Fig. 17.5). Endres *et al.* (1985) note that "intuition would suggest that the charges are distributed in the stacks but there are some physical arguments against this idea." Thus we shall not venture an explanation for the conductivity.

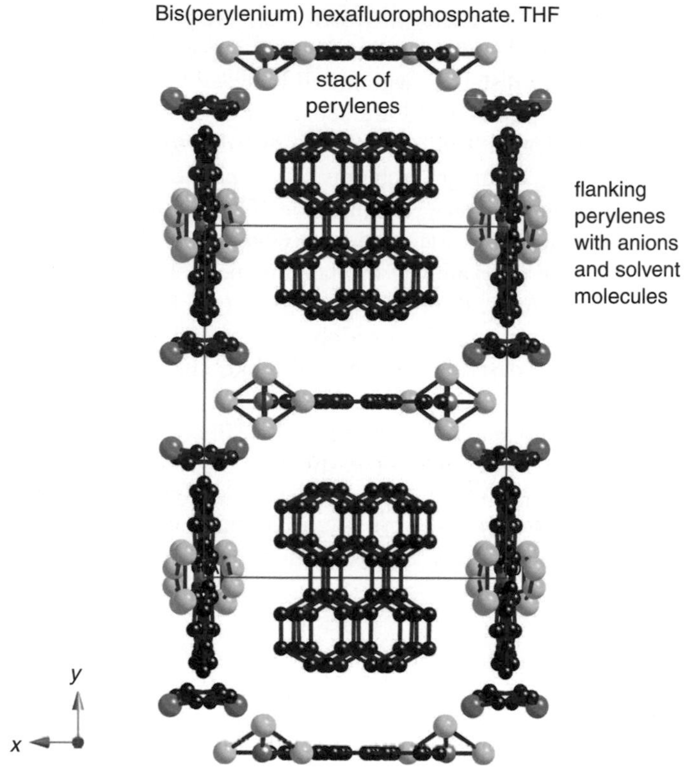

Bis(perylenium) hexafluorophosphate. THF

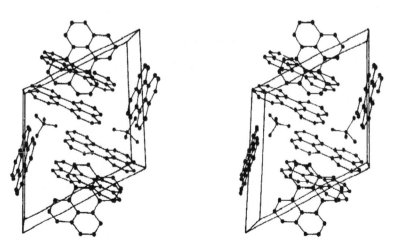

stack of
perylenes

flanking
perylenes
with anions
and solvent
molecules

y

x

Fig. 17.5. Crystal structure of $(pe)_2(PF_6) \cdot 2/3(THF)$ viewed down [001] ($a = 13.076$, $b = 14.159$, $c = 13.796$ Å, $\beta = 110.87°$, space group $P2/m$, $Z = 3$). (Data from Burggraf *et al.*, 1995.)

Fig. 17.6. Stereoview of the $(pe)_3ClO_4$ structure viewed along [001], showing the stacked but jogged tetrad and the flanking perylene moieties. The ClO_4 anions are ordered. (Reproduced from Endres *et al.*, 1985.)

(Perylene)$_3$ClO$_4$ is a semiconductor in which there are two offset pairs (π-dimers) of perylene moieties per triclinic unit cell, thus forming jogged stacks (Endres *et al.*, 1985; CUWBEB); the interplanar distances within and between the π-dimers are not sensibly different at 3.37(5) Å. Adjacent stacks are separated by flanking perylenes (Fig. 17.6), very much as in (pe)$_2$(PF$_6$)·2/3(THF). It was not possible to infer moiety charges from bond lengths because of disorder problems. In {(perylene)$_4$[Co(mnt)$_2$]$_3$} there are three perylenes stacked as trimers, a trinuclear [Co(mnt)$_2$]$_3$ moiety and a flanking perylene approximately normal to the perylenes (Gama *et al.*, 1993; see Section 17.5.2). Thus there are a number of examples of flanking perylenes and analogs have been encountered in other structures; (TTF)$_2$-[Ni(S$_4$C$_4$H$_4$)] (see Section 17.3.7) is a possibly relevant example for comparison. The overlap of the perylene pairs or triads, whether stacked or not, has been described as 'graphite-like' but 'ring-over-bond' is more appropriate.

The structure of (quaterphenyl)$_{12}$(quaterphenyl)$_4$(SbF$_6$)$_{10}$ has been briefly reported (Enkelmann, Göckelmann *et al.*, 1985). The structure consists of stacks of quaterphenyl cations (the first group) separated by layers of anions, in which the second group of cations is incorporated. Thus there appear to be resemblances to (pe)$_2$(PF$_6$)·2/3(THF) but a detailed comparison is not possible.

17.3.2.3 *Structures in which the cations are not stacked*

(Pe)$_6$ClO$_4$ has three independent π-dimers arranged about centers of symmetry of the triclinic unit cell; the overlap diagrams are very similar and the interplanar spacings are 3.38(2), 3.44(2) and 3.46(3) Å (Fig. 17.7). This material is a semiconductor ($\sigma \approx 10^{-2}$ S/cm (Endres *et al.*, 1985; CUWBIF).

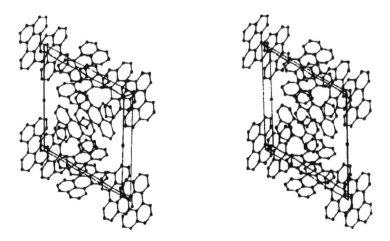

Fig. 17.7. (Pe)$_6$ClO$_4$, stereoview along [100], showing the herring-bone arrangement of π-dimers. The disordered anions (at symmetry centres) have been omitted for clarity. (Reproduced from Endres *et al.*, 1985.)

17.3.2.4   *Various*

The crystal structure of $(\text{pyrene})_2^{+\bullet}(\text{ClO}_4)^-$ has not been reported. In (phenothiazine)$_2^{+\bullet}\text{SbCl}_6^-$ (BUFZEH) the cation radicals are isolated in the crystals and do not interact (Uchida *et al.*, 1983).

17.3.2.5   *Conclusions*

The structures of the aromatic hydrocarbon cation-radical salts show some interesting resemblances to the cation-radical salts of TTF discussed below. There are non-stacked and stacked π-dimers and infinite monad and diad stacks. Formation of a cation-radical anion-radical salt analogous to {[TTF][TCNQ]}, in which the cation radical is an aromatic hydrocarbon and the anion radical is a suitable open shell acceptor and the stacks are segregated, has so far been achieved only for perylene–M(mnt)$_2$ derivatives (see Section 17.5.2).

17.3.3   *TTF and related compounds as cations*

TTF forms salts with many closed shell anions. Quite complex relations were found early on (Scott, La Placa *et al.*, 1977) for the TTF halides (Fig. 17.8) and the situation becomes even more complicated when TTF salts of polyatomic anions are considered. Our classification is based on the overall structural arrangement, and we make a major distinction between nonstacked and stacked arrangements of the TTF moieties.

Type A: Non-stacked arrangements of TTF moieties.

Group 1: Salts of type $(\text{TTF}^{2+})(\text{X}^-)_2$ are isostructural for X = Cl (TTFDCL), Br (TTFDBR). These are tetragonal with $a = 13.56$, $c = 10.10$ Å (for X = Cl), space group

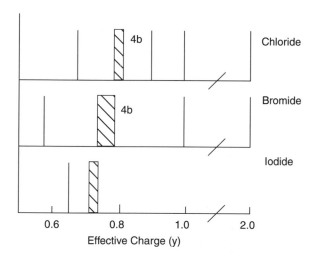

Fig. 17.8. Phases observed in the TTF-halide systems, indicated by vertical lines. The abscissa y is the halogen content defining the effective or average charge per cation site, thus the compositions are given as TTF·X$_y$. The rectangles show the composition ranges found for some salts. (Adapted from Scott *et al.*, 1977.)

$I4_1/acd$, $Z = 8$ (Scott, La Placa *et al.*, 1977). $(TTF^{2+})(ClO_4^-)_2$ is monoclinic (Ashton *et al.*, 1999, who give references to other (TTF^{2+}) structure determinations). In both instances, the cations are nonplanar with the two halves of the moiety mutually rotated by $\approx 60°$ about the central C–C bond. Ashton *et al.* comment "We believe, in all probability, that the conformation of (TTF^{2+}) is determined by the multiple interactions that determine its supramolecular order."

TTF$(OCN)_2$ has been reported as a microcrystalline insulating powder (Kathirgamanathan and Rosseinsky, 1980).

Group 2: Salts of type TTF·X, where X = Cl (TTFMCL), Br (TTFMBR). The isomorphous halide salts are orthorhombic, with $a = 11.073$, $b = 11.218$, $c = 13.95$ Å (for X = Cl), space group *Pbca*, $Z = 8$ (Scott *et al.*, 1977). There are centrosymmetric pairs of eclipsed cations with interplanar spacing 3.34 Å (π-dimers), and ordered anions (Fig. 17.9). Their importance in the present context is to suggest that adjacent neutral TTF molecules and $TTF^{+\bullet}$ cation radicals can be present in a stack without seriously perturbing the stack.

Other examples are TTF·I$_3$, (Teitelbaum *et al.*, 1980; TTFIOD), TTF$_3$·Sn(CH$_3$)$_2$Cl$_3$ (Matsubayashi *et al.*, 1980) and TTF·DDQ (Mayerle and Torrance, 1981b). The TTF·I$_3$ structure has pairs of cations $(TTF^{+\bullet})_2$ surrounded by noninteracting triiodide ions, while TTF$_3$·Sn(CH$_3$)$_2$Cl$_3$ has similar $(TTF^{+\bullet})_2$ dimers in a matrix of centrosymmetric μ_2 chlorine-bridged dimeric [Cl$_2$(CH$_3$)$_2$Sn(Cl)$_2$Sn(CH$_3$)$_2$Cl$_2$] moieties. TTF·DDQ (dark red triclinic crystals, $a = 10.272(4)$, $b = 12.195(5)$, $c = 6.609(2)$ Å, $\alpha = 77.51(2)$, $\beta = 81.93(2)$, $\gamma = 87.30(4)°$, $P\bar{1}$, $Z = 2$; Fig. 17.10) has eclipsed TTF π-dimers (interplanar spacing ≈ 3.4 Å) and also stacks of DDQ π-dimers with very short interplanar spacings of 2.97 Å within the π-dimers and 3.56 Å between them. The axes of the two kinds of stacks

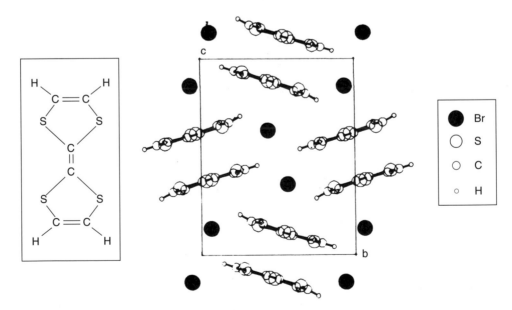

Fig. 17.9. Bounded a axis projection of the orthorhombic crystal structure of TTF·Br; TTF moieties viewed edge-on. The chloride is isostructural. (Reproduced from Scott *et al.*, 1977.)

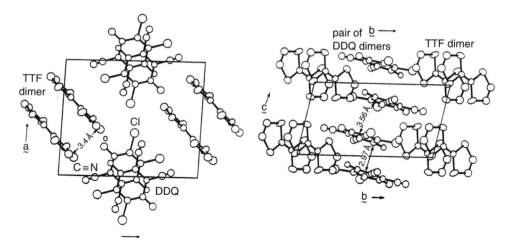

Fig. 17.10. Two projections of the crystal structure of TTF·DDQ. The interplanar spacing between the DDQ moieties is 2.97 Å. (Reproduced from Mayerle and Torrance, 1981b.)

are considerably inclined to one another. In these salts the moieties are fully ionized and the conductivities are correspondingly low ($\sigma \approx 10^{-8}$ S/cm). TTF.DDQ is probably better described as a DDQ anion-radical salt, with the DDQ moieties arranged in jogged tetrad stacks and the $(TTF^+)_2$ π-dimers acting as (possibly spin paired) cations.

TTF·ClO$_4$ (ZZZBWA10) has a rather complicated structure ($a = 16.762(1)$, $b = 20.906(2)$, $c = 12.538(1)$ Å, $Z = 16$, space group $Pbca$) (Yakushi et al., 1980) in which there are pairs of essentially eclipsed cations (π-dimers with interplanar spacing of 3.41 Å) arranged in tetrads. The structure is not stacked.

TTF(OSO$_2$CH$_3$), TTF(SCN)$_{1.4}$ and TTF(HSO$_4$)$_{1.2}$ have also been reported (Kathirgamanathan and Rosseinsky, 1980). but structures are not known.

π-Dimers are characteristic of many structures of this group but $(TTF)_3^{2+}$ triads arranged between $(SnCl_6)^{2-}$ anions are found in $(TTF)_3^{2+}(SnCl_6)^{2-}$ (Fig. 17.11) (Kondo et al., 1984; CELREQ); $(TTF)_3^{2+}(PtCl_6)^{2-}$ (DERKAM) is isomorphous. Salts with $[SnCl_4R_2]^{2-}$ anions (R = methyl, ethyl) are isostructural (Matsubayashi et al., 1985; ethyl is DER-JUF). As the dimensions of the two independent TTF moieties of the triad do not differ significantly, delocalization of the charge across the triad was inferred. Powder conductivity of 2.4×10^{-3} S/cm was ascribed to in-layer transport through sulfur–sulfur contacts.

Type B: Stacked arrangements.

Group 3: (stacks with the long axes of the TTF moieties mutually perpendicular): This group comprises the salts TTF·X$_y$, where X = Cl, NO$_3$, SCN (disordered anions), Br, I (ordered anions). The values of y vary from 0.55 to ≈ 0.7 and depend on the nature and degree of ordering of the anion. The overall gross structural pattern common to all these salts is rather simple but there are many complications of detail. The basic structural pattern has stacks of eclipsed TTF cation moieties (with average charge $Z < 1$) arranged as shown in Fig. 17.12 (broken-line cell in the centre of the diagram). The alignment of TTF moieties in adjacent stacks is mutually perpendicular. The channels between the stacks of Fig. 17.12 contain the anions; as Johnson and

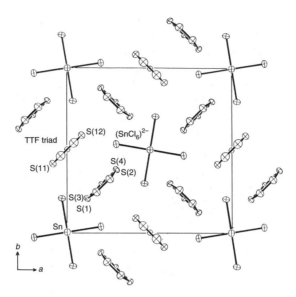

Fig. 17.11. The non-stacked structure of $(TTF)_3SnCl_6$ showing the layers of $(TTF)_3^{2+}$ triads interspersed among the anions. The crystals are tetragonal, 11.801, 11.861 Å, $P4/mbm$, $Z=2$. The interplanar spacing in the triad is 3.49 Å. (Reproduced from Kondo *et al.*, 1984.)

Watson (1976) originally pointed out, anions in these channels with ionic radii greater than 1.8 Å will distort the arrangement of the TTF stacks. The prototype structure is $TTF \cdot Cl_{0.7}$ (MTTFCL), which has a tetragonal unit cell with $a = 11.12$, $c = 3.595$ Å, $Z = 2$, space group $P4_2/mnm$ (we use the highest symmetry space group among the possibilities, unless there is good reason to the contrary). The salts $TTF \cdot Br_{0.59}$ (ZZZBGJ), $TTF \cdot I_{0.69}$ (Scott *et al.*, 1977) and $TTF \cdot (SCN)_{0.57 \pm 0.03}$ (Kobayashi and Kobayashi, 1977) have essentially the same structure, while there is a small monoclinic distortion in $TTF \cdot (NO_3)_{0.55}$. Two different modes of distortion come into play when the halide content increases slightly. In $TTF \cdot Cl_{0.7}$ the unit cell distorts to orthorhombic (see Fig. 17.12) while in $TTF \cdot Br_{0.74-0.77}$ (LaPlaca *et al.*, 1975) and $TTF \cdot I_{0.70-0.72}$ (ZZZBGD) the TTF and halide subcells become incommensurable (Table 17.1). The discussion of the mutual interaction of TTF and I sublattices in $TTF \cdot I_{0.70}$, as inferred by Johnson and Watson (1976) from the details of the very complicated x-ray diffraction patterns, is remarkable for its thoroughness but too complex for summary here. Measurements of the peak shapes of the S 2p peaks in the x-ray photoelectron spectra (Ikemoto, Yamada *et al.*, 1980) indicate that the fraction of cations in $TTF \cdot Br_{0.7}$ is 0.71(5) but 0.52(5) in $TTF \cdot I_{0.7}$, suggesting that some larger polyiodide anions may be present in the latter; both salts contain TTF cations and neutral molecules but in different amounts.

The system $TTF \cdot Cl_y$ ($0.67 \leq y \leq 0.70$) has been investigated in some detail over a range of temperature (Williams, Lowe Ma *et al.*, 1980). All crystals in this composition range show the same temperature dependence of conductivity; however, x-ray diffraction (oscillation) photographs from crystals at the two extremes of composition show slightly different diffuse layer lines. These have their origins in the disordered Cl^-

arrangements; the TTF and Cl^- sublattices are commensurable for $y = 0.67$ (ZZZBGG) and incommensurable for $y = 0.70$ (indeed the composition range was inferred from the detailed spacings of the diffuse layers). There is a phase change at $\approx 250K$ for $TTF \cdot Cl_{0.67}$ (cf. the conductivity behaviour in Fig. 17.13) and the diffuse lines sharpen and resolve into individual Bragg reflections indicating ordering of the Cl^- sublattice. The ordered monoclinic phase has composition $(TTF)_3(Cl)_2$ and commensurable TTF and Cl sublattices. Orientation relations between the monoclinic and tetragonal phases can be

Table 17.1. The TTF and halide sub-cells for $TTF \cdot Br_{0.76}$ and $TTF \cdot I_{0.72}$. These phases, shown as 4b in Fig. 17.9, have small ranges of composition and the values of c and β vary with composition

Subcell parameter	$TTF \cdot Br_{0.76}$ MTTFBR		$TTF \cdot I_{0.72}$ MTTFID	
	TTF	Br	TTF	I
a (Å)	15.617	17.368	15.998	8.19
b	15.627	15.623	16.114	16.11
c	3.573	4.538	3.558	4.871
β(deg)	91.23	116.01	90.96	102.82

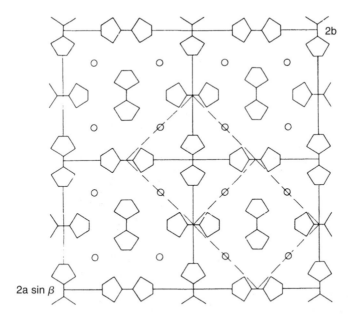

Fig. 17.12. Projection down [001] of the structures of all the $TTF \cdot Xy$ ($y < 1$) phases, with the exception of $TTF \cdot Cl_{0.92}$. The broken-line cell in the center of the diagram is for the disordered tetragonal salts of Group 4_a such as $TTF \cdot Cl_{0.68}$ ($a = 11.12$ Å). The full broken-line rectangle shows the orthorhombic cell of $TTF \cdot Cl_{0.77}$ ($a = 10.77$, $b = 22.10$ Å), while the TTF sublattices of $TTF \cdot Br_{0.76}$ and $TTF \cdot I_{0.72}$ are shown in the top left quarter of the overall diagram. Minor distortions are not shown in the diagram. For all the cells $c \approx 3.6$ Å. (Adapted from Scott *et al.*, 1977.)

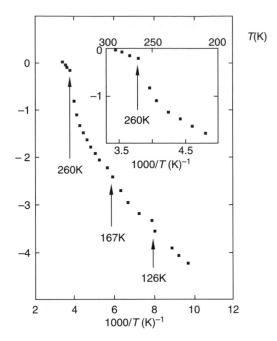

Fig. 17.13. Temperature dependence of the conductivity of TTF·$Cl_{0.67}$. The ordinate (σ/σ_{RT}) is on a logarithmic scale, and the approximate temperatures of the breaks have been inserted into the diagram. (Adapted from Williams *et al.*, 1980.)

summarized as follows:

$$a_{mono} = b_{mono} = \sqrt{2}a_{tetr}; \quad c_{mono} \approx 3c_{tetr}; \quad \beta_{mono} \approx 92° \text{ (temperature dependent)}.$$

Because the four $\langle 110 \rangle$ directions of the tetragonal cell are equivalent, four equally probable orientations will be obtained for the monoclinic cell. Thus the low temperature phase is a quintuple twin, consisting of the domains of the four monoclinic orientations and of unchanged tetragonal phase (twinning often accompanies first-order phase changes). Overlap of diffraction patterns prevented determination of the details of the low temperature structure. When TTF·$Cl_{0.70}$ is cooled, it forms two types of low temperature phase with compositions $(TTF)_3(Cl)_2$ and $(TTF)_7(Cl)_5$ respectively; the TTF and Cl sublattices are commensurable in both.

The ordered structure of $(TSeF)_3(ClO_4)_2$ is a monoclinic distortion of the tetragonal subhalide structure, in which the (TSeF) moieties are arranged in triads with their long axes approximately mutually perpendicular (Shibaeva, 1983; CENBIG).
Group 4: (stacks with the long axes of the TTF moieties mutually parallel): This group comprises the salts $(TTF)_3(BF_4)_2$ (Legros *et al.*, 1983; CELVIJ) and $(TTF)_3(I_3^-)_2$ (Teitelbaum *et al.*, 1980). The first of these (Fig. 17.14) has slightly jogged stacks of TTF moieties and orientationally disordered anions; bond lengths are said to differentiate between TTF cations and neutral molecules, and XPS measurements of the shape of sulphur 2p peak indicate that 2/3rds of the TTF moieties are cations (Ikemoto *et al.*, 1980). The powder conductivity is 2×10^{-5} S/cm. Only a gross structure has been reported for $(TTF)_3(I_3^-)_2$ (Johnson *et al.*, 1975).

TTF·HgCl$_3$ has a remarkable structure that combines in one crystal features encountered above in separate structures (Kistenmacher, Rossi *et al.*, 1980; FTFHGC10). There are two types of layer (Fig. 17.15); in the layer about $y = 0$ there is an inorganic polymer of composition HgCl$_3^-$ and a stack of TTF cation radicals with an interplanar spacing of $\approx$3.6 Å, while in that about $y = 1/2$ there are isolated (Hg$_2$Cl$_6$)$^{2-}$ ions and (TTF$^+$)$_2$ π-dimers, with an interplanar spacing of 3.43 Å. The resonance Raman spectrum shows a splitting of the TTF ν_3 mode, indicating the presence of two types of TTF cation; the difference between their charges was estimated to be $<0.2e$.

The subhalide phases show high conductivities at room temperature ($\approx$1–5 $\times$ 10^2 S/cm) while both (TTF)$_3$(I$_3$)$_2$ and TTF·I$_3$ have conductivities of $\approx$10^{-3} S/cm (Warmack *et al.*, 1975; Teitelbaum *et al.*, 1980), (TTF)·(N$_3$)$_{0.74}$ and TTF·(CH$_3$COO)$_{0.70}$ have been reported as fairly highly conductive phases ($\sigma \approx 1$ S/cm) but their structures are not known.

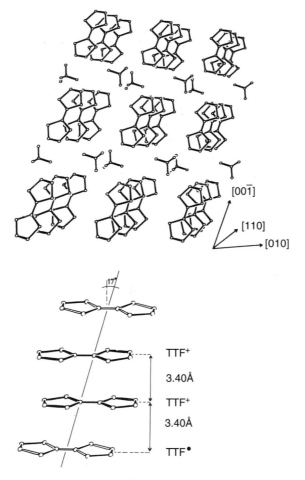

Fig. 17.14. Crystal structure of triclinic $P\bar{1}$, ($Z = 1$) (TTF)$_3$(BF$_4$)$_2$ (a) perspective view of the structure showing the stacks of TTF moieties (hydrogen atoms omitted) with their long axes parallel. The disordered anion is shown in its major (30%) orientation. (b) Side view of stacking along [110] of TTF moieties (tilted by 15° with respect to the page normal). (Reproduced from Legros *et al.*, 1983.)

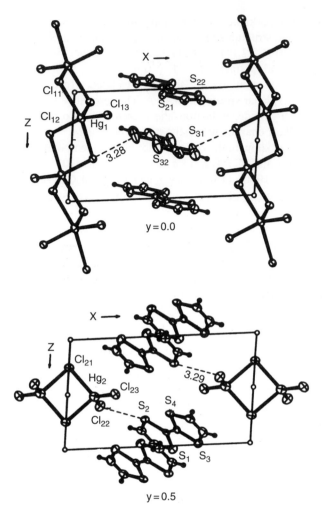

Fig. 17.15. (a) The structure of TTF·HgCl$_3$ ($P\bar{1}$, Z = 1; 12.661, 15.969, 7.416 Å, 98.69, 95.73, 120.01°, cell not reduced). The section about $y=0$ shows the polymeric HgCl$_3$ species and the TTF stacks, which are ring-over-bond, while the section about $y=0.5$ shows the Hg$_2$Cl$_6$ anions and the eclipsed TTF π-dimers. (Reproduced from Kistenmacher *et al.*, 1978.)

Bis(tetrathiafulvalene)-tetracyanomuconitrile[2] ((TTF)$_2$-TCM) crystallizes in long blue needles with a conductivity of 10^3 S/cm at 298K (Wudl and Southwick, 1974), similar to that of {[TTF] [TCNQ]}. The structure has not been reported.

The free energies and entropies of formation of the various TTF iodides have been measured by determining the voltages of suitable solid state electrochemical cells over the temperature range 20–40 °C (Euler *et al.*, 1982). The reaction is:

$$\text{TTF(s)} + (x/2)\text{I}_2(\text{s}) = \text{TTF·I}_x(\text{s}); \quad x = 0.71, 2, 3 \text{ for the phases discussed earlier.}$$

---

[2] Also known as 'hexacyanobutadiene'.

Table 17.2. Thermodynamic formation quantities at 298K for the TTF iodides. "Uncertainties" are given as $\Delta G° \pm 0.4$ kJ/mol, $\Delta S° \pm 1.2$ J/mol K, $\Delta H° \pm 0.8$ kJ/mol. Data from Euler et $al.$, 1982

Phase	$\Delta G°$ (kJ/mol)	$\Delta S°$ (J/mol)	$-298\Delta S°$ (kJ/mol)	$\Delta H°$ (kJ/mol)
TTF·I$_{0.71}$	-7.9	-20.9	6.2	-14.2
(TTF)$_3$(I$_3$)$_2$	-16.3	-74.9	22.3	-38.5
TTF·I$_3$	-21.7	-106.6	31.8	-53.5

The results are summarized in Table 17.2.

The experimental results show that the TTF iodides are enthalpy-stabilized but entropy-destabilized (see Appendix I); the two stacked phases (TTF·I$_{0.71}$ and (TTF)$_3$(I$_3$)$_2$) are less stable than the non-stacked TTF·I$_3$ phase. The enthalpy of formation of {[TTF] [TCNQ]} has been measured as -37.4kJ/mol (see Section 16.7), which is not very different from the values for the TTF iodides. In mixed stack π-molecular compounds enthalpy of formation values range from -29 to 25 kJ/mol (see Section 16.5). The negative entropies of formation of the TTF iodides are presumably due to reduced thermal vibration consequent on the stronger ionic interactions in the iodides as compared to those between the neutral moieties in TTF and I$_2$.

17.3.4 TMPD salts containing π-dimerized cation radicals

The TMPD moiety is the prototype Würster type redox system (Section 17.2.1). Discrete (TMPD$^{+\bullet}$)$_2$ π-dimers are found in TMPD·ClO$_4$ (Würster's blue perchlorate), where the π-dimers are stacked, and in (TMPD)$_2$ {bis(maleonitriledithiolato)Ni(II)} [(TMPD)$_2$ {Ni(mnt)$_2$}], where the π-dimers are not stacked. There is a first-order phase change in TMPD·ClO$_4$ at 186K (Fig.17.16); above this temperature the crystals are orthorhombic (TMPDPC) and the cations are arranged in regular stacks with EB/EB overlap and an interplanar spacing of 3.55 Å (Boer and Vos, 1972a), similar to the arrangement found in TMPD iodide (Boer et $al.$, 1968; TMDABI). Below 186K the cations are arranged in stacks, but with distances of 3.10 and 3.67 Å between planes of successive cations (Boer and Vos, 1972b; TMPDPC01). π-Dimerization has occurred, with coupling of the unpaired electrons of the pairs of closer cations. The variation of paramagnetic susceptibility with temperature (Fig. 17.17) can be explained in terms of a theory due to Soos (1965) in which the stacks are treated as linear alternating Heisenberg antiferromagnets. The agreement of theory and measurements in the low-temperature region can be improved by allowing the exchange integral J to increase as the temperature is reduced; evidence for such an increase is provided by the increased intensity of the CT band at $\approx$12 kK on cooling (Fig. 17.16(c)). The experimental singlet-triplet separation (Thomas, Keller et $al.$, 1963) is 246 ± 20 cm^{-1}.

In (TMPD)$_2$[Ni(mnt)$_2$] (space group $P\bar{1}$, $Z = 1$) (Hove et $al.$, 1972; TMDNMI) the cations are arranged as π-dimers with an interplanar spacing of 3.25 Å; the distance between neighbouring ions of two different π-dimers is 6.57 Å. The interaction between π-dimers is small because of the large distance between them but is not zero, as is shown by the absence of hyperfine splitting in the ESR spectrum. The (TMPD)$_2$ moieties form a strongly alternating chain, that could be called a 'jogged stack.'

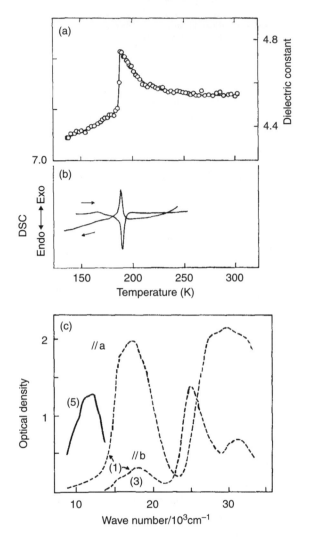

Fig. 17.16. (a) Temperature dependence of the dielectric constant of TMPD·ClO$_4$. An approximate scale of the dielectric constant is shown on the right hand side of the upper part of the diagram. (b) The reversible differential scanning calorimetry (DSC) heating and cooling curves are shown in the lower part of the diagram. (c) Temperature dependence of the crystal spectra of TMPD·ClO$_4$ at 285K (1) and 83K (3). Spectra at intermediate temperatures are not much different. The different absorption spectra obtained when the electric vector of the plane polarized light is parallel to **a** and **b** axes respectively are shown. (Reproduced from Ishii *et al.*, 1976.)

TMPD dimensions have been determined in crystals of different types. Neutral TMPD has a benzenoid structure with a slightly pyramidal arrangement of bonds about N (sum of bond angles at N, $\Sigma = 352°$) (Ikemoto *et al.*, 1979). However, in TMPD$^{+\bullet}$ the quinoid form is the major contributor to the resonance hybrid, with a planar disposition of bonds about N ($\Sigma = 359.3°$). PPP calculations of bond lengths in TMPD and its singly- and doubly-charged cations show excellent agreement with experiment, the degree of quinoid

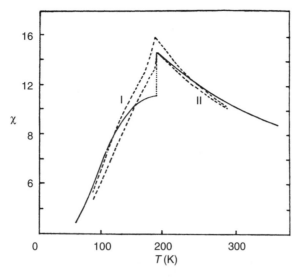

Fig. 17.17. The dashed lines show two independent measurements of the paramagnetic susceptibility (in units of 10^{-4} e.m.u./mol) of TMPD·ClO$_4$ as a function of temperature, while the full line shows calculated values due to Soos (1965). (Reproduced from Boer and Vos, 1972b.)

character increasing with increasing charge on the cation (Kracht, quoted in Boer and Vos (1972a), p. 835).

17.4   Structure of TCNQ anion-radical salts

17.4.1   *Mutual arrangements of approximately plane-parallel TCNQ moieties*

We shall devote most of our attention to TCNQ anion-radical salts, which currently constitute the major part of the larger family of anion-radical salts; hopefully it will be possible to generalize many of the broad features of structures and properties of the TCNQ salts to other members of the family, although the details are bound to differ. The relatively few salts of other anion radicals (such as the haloanils and Ni(dmit)$_2$) are discussed later; the stacked arrangements there emphasize that stacking is not a prerogative of TCNQ alone. Future studies are likely to furnish many other examples.

The closed shell counter ions of TCNQ anion-radical salts include alkali-metal and many organic cations. TCNQ moieties not interacting with other TCNQ moieties have been found only in two examples of composition A$_2$(TCNQ)$_3$ and will be discussed separately below. Otherwise the invariant feature of TCNQ anion-radical salt structures seems to be the appearance of face-to-face arrangements of pairs or larger numbers of TCNQ moieties. Such paired (or larger) aggregates form stacks of limited or unlimited length. Thus the TCNQ salts generally have quasi-one-dimensional structures, and this is reflected in many of their physical properties. However, there are a growing number of examples where interstack interactions are also important and this quasi-two dimensional group is likely to be more important in the future, especially in systems other than TCNQ.

We distinguish between groups with stacks of limited or unlimited length. In both groups we classify the various structures firstly on the basis of the average charge per TCNQ moiety. This parameter depends on the cation charge (which is unequivocally defined for closed shell cations), and on the composition (i.e. the cation/TCNQ ratio) for stacks of limited length. Here a monad unit will contain one TCNQ moiety, a diad unit two, and so on. Further distinctions can be made on the basis of the type of TCNQ overlap within the diad, etc. It is often convenient to describe stacks of limited length as isolated (to a greater or lesser extent) diads, triads, tetrads, etc.

When the stacks are of unlimited length the stack periodicity (i.e. the number of TCNQ moieties in a crystallographic repeat along the stack axis) must also be considered. A monad stack has one TCNQ per period, a diad stack two TCNQs per period, and so on; triad, tetrad, pentad and heptad stacks have also been reported and it would seem that an upper limit to the periodicity depends only on cation shape, size and charge. The composition and cation charge are chemical invariants for a particular compound while the stack periodicity is a physical feature which depends (at least in principle) on external conditions such as temperature and pressure. The 1:2 composition with singly-charged closed-shell cations can be taken as an example. A stack axis periodicity of ≈ 4 Å implies monad stacks, with translationally equivalent $TCNQ^{-1/2}$ ions, presumably better described as a (dynamically or statically) disordered array of $TCNQ^0$ and $TCNQ^{\bullet -}$ moieties. If there are diad stacks (stack periodicity now ≈ 7–8 Å) then the $TCNQ^{-1/2}$ moieties are no longer crystallographically equivalent, but it is not always possible to distinguish them as $TCNQ^0$ and $TCNQ^{\bullet -}$. Tetrad stacks (stack periodicity now ≈ 14 Å) are also found for the 1:2 composition; there may be pairs of π-dimers. Again it may not be possible to differentiate between $TCNQ^0$ and $TCNQ^{\bullet -}$. The relations between physical properties and temperature have been explored for many TCNQ systems, but only a few parallel structural studies have been made.

Some examples of the cations used in preparing TCNQ anion radical salts are shown below:

$NH R_3^+$ R = methyl, ethyl, etc.

$PHPh_3^+$ triphenyphosphonium

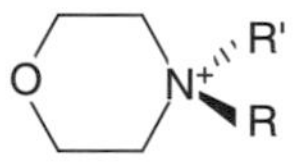

substituted morpholiniums

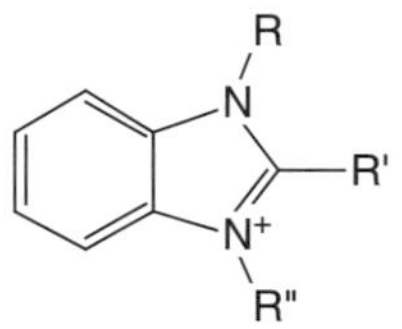

substituted benzamidazoliums

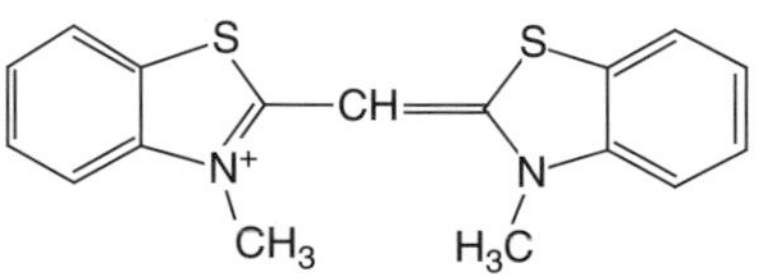

3,3'-dimethylcyanine
and related compounds

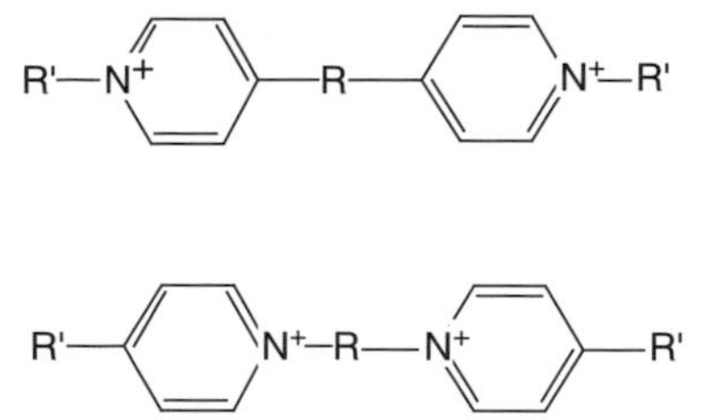

R'—N$^+$⟨⟩—R—⟨⟩—N$^+$—R'

R = $(CH_2)_3$ or $CH_2C_6H_4CH_2$
are examples. Many cations of
these types have been used by
the Nottingham group in a long
series of systematic studies.

R'—⟨⟩—N$^+$—R—N$^+$—⟨⟩—R'

$[Ph_3P=N=PPh_3]^+$

The mutual arrangement of approximately parallel TCNQ moieties can be succinctly described in terms of the overlap parameters of adjacent moieties, in much the same way as has been done for mixed stacks (Section 15.1); we describe the situation for stacks of unlimited length but the same principles apply when the stacks are of limited length. Consider a pair of TCNQ moieties; idealized types of overlap are defined below in terms of displacements with respect to the orthogonal molecular axes L, M and N:

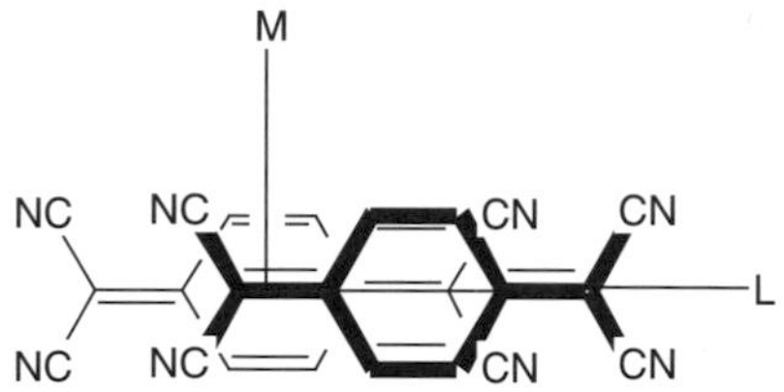

The axial system is centred at the centre of the lower molecule, with N upwards out of the page. Ring-external bond overlap (R/EB) is illustrated, with $\Delta L = 2.0$, $\Delta M = 0$, $\Delta N = 3.3$ Å (the interplanar distance). In ring–ring (R/R) overlap $\Delta L = \Delta M = 0$, $\Delta N = 3.3$ Å, i.e. the rings are eclipsed; in displaced R/R overlap $\Delta L = 0$, $\Delta M = 1.4$, $\Delta N = 3.3$ Å. The absolute values of the signs of the displacements are arbitrary in the usual situation of centrosymmetric crystals; however, the correct relative values must be used in the description of the stacks. This is exemplified for the two arrangements shown below, both with R/EB overlap.

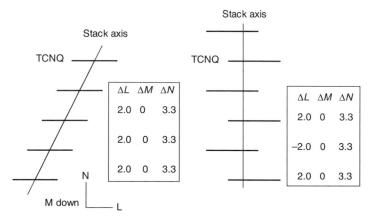

The TCNQ molecules are represented by the thick horizontal lines and the values of ΔL, etc. are given with respect to each successive molecule as one progresses upwards along the stack axes, starting from the bottom molecule in each part of the diagram. In the example on the left the stack is monad and linear, while a jogged or zigzag diad stack is shown on the right. Jogged diad and tetrad stacks are fairly common. Unfortunately, most authors only report the overall type of overlap and the interplanar spacing, i.e. ΔN but not ΔL or ΔM. One should also note that the TCNQ moieties are usually slightly bow-shaped, with the $=C(CN)_2$ groups both displaced (in the same direction) out of the plane of the central quinoid ring, and there is also sometimes a slight twisting of these groups. The structural significance of these distortions is not clear at present.

We shall find it useful to show the moiety arrangement in a stack or elsewhere in the following abbreviated fashion, illustrated for a triad:

$$
\begin{array}{ccc}
\bar{1} & & \bar{1} \\
\text{B'---------A----------B} & \quad & \text{B'---------A'----------B} \\
3.15 \quad 3.45 & & \\
\text{R/EB} \quad \text{R/EB} & &
\end{array}
$$

Here A (at a crystallographic centre) and B (at a general position) are crystallographically independent TCNQ moieties, with A' and B' their congeners related as shown by the crystallographic centres of symmetry ($\bar{1}$) and translations. The type of overlap is shown as well as the interplanar spacing ($\mathring{A}$); the three moieties of the triad are shown as linked. Obviously much more information is conveyed by suitable stereodiagrams or even projections of a crystal structure.

The crystal structure found for a particular TCNQ salt will be that for which the free energy is lowest (ignoring metastable polymorphs). This free energy will be composed of contributions from three sources:

1. the free energy of the anion arrangement, which will depend on the type of stacking of TCNQ moieties and the degree of overlap in the stacks;
2. the free energy of the cation arrangement; and
3. the free energy derived from cation–anion interactions.

In organic crystals the differences between the free energies of different polymorphs are small – up to a few kJ/mol; the TCNQ salts are expected to behave similarly. Thus the crystal structure (which determines such properties as lattice specific heat and thermal expansion) will depend on all three contributions, and it is unlikely that any one will be decisive. However, physical properties such as electrical conductivity, thermoelectric power and magnetic susceptibility depend to a good approximation only on the anion arrangement.

17.4.2 *Structures with stacks of limited length*

17.4.2.1 *Salts in which each TCNQ moiety bears an average charge of $-e$*

(i) Isolated diad structures: here π-bonded pairs of TCNQ moieties are found, isolated to a greater or lesser extent from other such pairs. We use the term "π–dimer" (well entrenched in the literature) to distinguish these pairs from the covalently-bonded σ-dimer anions described earlier (Section 15.9.2). The isolated π-dimers found in {1,2,3-trimethyl-benzimidazolium TCNQ} (Fig. 17.18) (Chasseau *et al.*, 1972; MBITCQ) provide an illustrative if extreme example, the TCNQ π-dimers being enclosed in channels between stacks of cations. A similar example of isolated π-dimers comes from {(2,4,4,6,8,8-hexamethyl-3,4,7,8-tetrahydroanthracene-1,5-dione)·TCNQ·CH$_3$CN} (Brook and Koch, 1997; PULWAU), although here the overlap is displaced R/R despite an interplanar distance of 3.14 Å. Other examples are {3,3'-diethylthiacyanine TCNQ} (Saakyan *et al.*, 1972; ECYTCN10), where there is R/EB overlap in π-dimers with interplanar spacing 3.23 Å and{[Fe$_2$(η^5-C$_5$Me$_5$)$_2$(μ-SEt)$_2$(CO)$_2$]$^+$}$_2$·(TCNQ$_2$)2$^-$} (Büchner *et al.*, 1997; NEHSEY) where there are isolated units of composition D$^+$·(A$_2$)$^{2-}$ D$^+$. and the TCNQ dimers have R/EB overlap and interplanar spacing of 3.26 Å. Weaker π-dimers (displaced R/EB overlap, interplanar spacing 3.40 Å) are found in S-methylthiouronium TCNQ (COFRIY10) and its isomorphous Se analog (Abashev *et al.*, 1987; COFROE10).

Isolated π-dimers bounded on both sides by cations are found in the paramagnetic decamethylferrrocenium TCNQ salt formulated as {[Fe(C$_5$Me$_5$)$_2$$^{+•}$]}$_2$–(TCNQ)$_2^{2-}$ (Reis *et al.*, 1979; Miller, Zhang *et al.*, 1987); MCFETC, in {N,N'-dimethylbenzimidazolinium TCNQ} (Chasseau *et al.*, 1972; MBZTCQ) and also in {[(MeCp)$_5$V$_5$S$_6$](TCNQ)$_2$} (Bollinger *et al.*, 1986; no refcode). The first of these contains isolated units of composition [Fe(C$_5$Me$_5$)$_2$](TCNQ)(TCNQ)[Fe(C$_5$Me$_5$)$_2$]. (Fig. 17.19); bond lengths suggest that the TCNQ moieties have charge -1. There is strong antiferromagnetic coupling within the π-dimer, with b_{2g}–b_{2g} interactions giving a filled a_u bonding orbital. There is

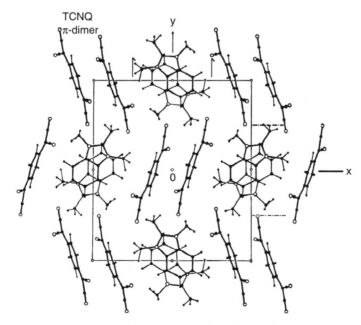

Fig. 17.18. {1,2,3-Trimethylbenzimidazolium TCNQ} (MBITCQ), projection down [100]. The TCNQ anions are associated in pairs across centers of symmetry, with R/EB overlap and interplanar spacings of 3.12 Å; these pairs are the π-dimers. (Reproduced from Chasseau *et al.*, 1972.)

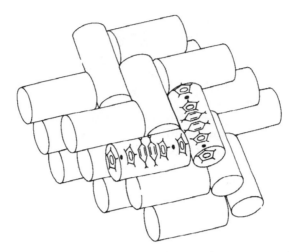

Fig. 17.19. Crystal structure of $\{[Fe(C_5Me_5)_2^{+\bullet}]\}_2(TCNQ)_2^{2-}$ represented schematically; each cylinder depicts an A:B:B:A formula unit. (Reproduced from Reis *et al.*, 1979.)

slightly displaced R/R overlap of the moieties of the π-dimer, with an interplanar spacing of 3.15 Å. Mössbauer studies show that the cations contain Fe(III). The third salt appears to have R/R overlap (described as "cofacial pairs") with an interplanar spacing of 3.37(1) Å.

In the TCNQ salt of tris[(di-μ-chloro)(hexamethylbenzene)niobium] of composition $\{[Nb_3(\mu\text{-}Cl)_6(C_6Me_6)_3]^{2+}(TCNQ)^{2-}\}$ there are zigzag chains of alternating trimer cations and TCNQ π-dimers (R/EB overlap, interplanar spacing 3.10 Å). The odd electron is assigned to the cation cluster, while the π-dimer has strong coupling of the TCNQ moieties, with two electrons in an a_g bonding orbital (Goldberg, Spivak *et al.*, 1977; HMBNBQ).

The ferrocenium salt $\{[Fe(C_5H_5)_2]_2[(TCNQF_4)]_3\}$ contains isolated TCNQF$_4$ π-dimers with displaced R/R overlap and 3.16 Å interplanar spacing enclosed in a matrix of ferrocenium cations and neutral TCNQF$_4$ molecules (Miller *et al.*, 1987; FETBEL). The presence of the latter as "TCNQF$_4$ of crystallization" is unusual but not unprecedented (cf. Section 15.5.1).

Thus, in summary, a variety of situations is encountered – more examples are known of R/EB overlap than R/R overlap but the latter group are present; interplanar spacings range from 3.1 Å upwards.

17.4.2.2 *Average charge of $-2/3e$ on TCNQ moieties*

There are a number of structures containing isolated triads. In $\{[(5\text{-propyl-phenazinium})^+]_2 (TCNQ)_3\}$ $\{(NPP)_2 (TCNQ)_3\}$ (Harms *et al.*, 1982; BIXHUL) the mixed stacks contain an alternating arrangement of pairs of NPP moieties and triads of TCNQ (Fig. 17.20). There is strong interaction within the TCNQ triads and weak interaction between the NPP moieties. Charge balance requires that two of the TCNQ's are anion radicals and one is neutral; however, the distribution of charge over the triad is not known.

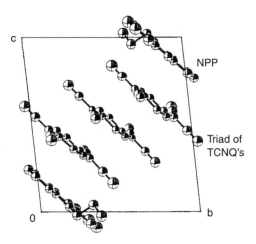

Fig. 17.20. Projection of the crystal structure of {(5-propylphenazinium)$_2$(TCNQ)$_3$} down the [100] axis of the triclinic cell, showing the arrangement of a pair of NPP moieties and a triad of TCNQ moieties in a mixed stack. The TCNQ triad is centrosymmetric and there is R/EB overlap. The precision of the analysis is too low to allow assignment of charges to the TCNQ moieties through bond lengths. The NPP pair is also centrosymmetric. (Reproduced from Harms *et al.*, 1982.)

Parenthetically we note that {DBTTF$^{+•}$ I$_3^-$} (Shibaeva, Rozenberg, Aldoshina *et al.*, 1979) (Fig.17.21), which should be included among the cation radical salts, can be considered, insofar as the stacks are concerned, as an anti-structure of (NPP)$_2$ (TCNQ)$_3$. There are mixed stacks, consisting of sequences of three DBTTF$^{+•}$ cation radicals and then two I$_3^-$ anions; additional, non-interacting I$_3^-$ anions are located between the stacks. The conductivity is fairly low [σ_{293} (powder) $\approx 2 \times 10^{-2}$ S/cm] and not at all anisotropic. There is a very similar . . . D A A A D D A A A D D . . . arrangement of moieties in (4,4′-bithiopyranylidene)$_2$(TCNQ)$_3$} {(BTP)$_2$(TCNQ$_3$)} (Sandman *et al.*, 1980; TCNBTP). Bond lengths in the TCNQ moieties suggest that the charge distribution is A$^{-•}$ AA$^{-•}$, the mutual arrangement being analogous to that in Cs$_2$TCNQ$_3$ (see below).

4,4′-bithiopyranylidene (BTP)

Perhaps the most unusual of the structures of nominal composition A$_2$TCNQ$_3$ is {(trimethylammonium)$_2^{2+}$ (TCNQ)$_3^{2-}$} (Fig. 17.22(a)), where there are discrete pairs of strongly-interacting TCNQ$^{-•}$ anions (R/R overlap with an interplanar spacing of 3.26 Å; π-dimers) as well as single molecules of neutral TCNQ (Kobayashi, Danno and Saito, 1973; TATCNQ). A given π-dimer interacts neither with other π-dimers nor with neutral TCNQ. The only other example known to us of an analogous structure is {(quinuclidinium)$_2$(TCNQ)$_3$}, where there is a very similar sheet arrangement of

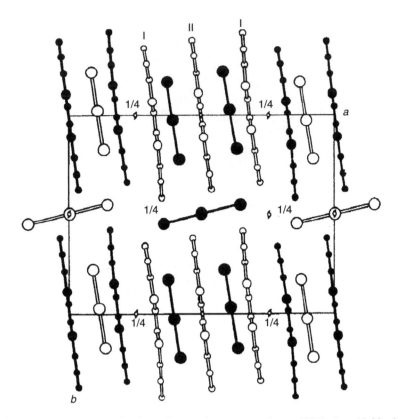

Fig. 17.21. {DBTTF I_3^-}: projection of crystal structure down [001] ($a = 19.89$, $b = 14.55$, $c = 9.29$Å, $\beta = 90.03(3)°$, $Z = 6$, space group $B2_1/m$). The three DBTTF moieties are labelled I (symmetry m), II (symmetry $2/m$). I and II are seen edge-on; they have an eclipsed overlap and the interplanar spacing is ≈3.5–3.6 Å. The moieties of stacks separated by **c** (normal to the page) are differentiated by open and filled circles. (Reproduced from Shibaeva *et al.*, 1979.)

TCNQ π-dimers and $TCNQ^0$ moieties (Bandrauk, Truong *et al.*, 1985) (Fig.17.22(b)). The type of overlap and the interplanar spacing differ from what has been found in most other π-dimers. The IR spectrum of {$(QND)_2(TCNQ)_3$} shows the presence of $TCNQ^{-\bullet}$ and $TCNQ^0$; despite their different structures, the IR spectra of $(QND)_2(TCNQ)_3$ and $Cs_2(TCNQ)_3$ (see Section 17.4.6 below for discussion of this structure) are very similar (Truong *et al.*, 1985).

Here we note a resemblance between the structures of {$[(TMA)_2]^{2+} [(TCNQ)_3]^{2-}$} and {$[TTF]^+.[\{Pt(dmit)_2\}_3]^-$} (Bousseau *et al.*, 1986; DOPMUG) despite the difference in formal charges on the anions. The latter contains $Pt(dmit)_2$ monomers and $[Pt(dmit)_2]_2$ dimers, in which the moieties are linked by a Pt–Pt bond of length 2.935 Å. These dimers resemble analogous dimers (d(Pt–Pt) = 2.748 Å) found in the neutral $PtS_4C_4H_4$ complex (Browall *et al.*, 1978; PTETOL10); dimerization of acceptors has previously been noted in bis(difluoroboron-benzimidazole)Ni(II) (Section 15.9.3). The TTF moieties are located

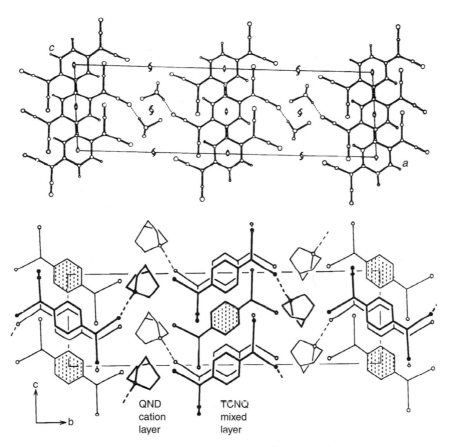

Fig. 17.22. (a), above): Projection on (010) of the $\{(TMA)_2^{2+}(TCNQ)_3^{2-}\}$ structure (26.662, 9.767, 7.725 Å, 93.98°, $C2/m$, $Z=2$) showing the segregation of cations and anions in sheets parallel to (100). There are sheets of TCNQ moieties in both structures. In $\{(TMA)_2^{2+}(TCNQ)_3^{2-}\}$ there are pairs of eclipsed (i.e. R/R overlap) TCNQ moieties at 0,1/3,1/2 and 0,2/3,1/2, with an interplanar distance of 3.26 Å. The bond lengths indicate that these are anions; they constitute the π-dimers. The TCNQ centred at the origin appears to be a neutral molecule and does not make close contacts to other TCNQ moieties. The dashed lines indicate cation...anion hydrogen bonds. (Reproduced from Kobayashi *et al.*, 1973.) (b), below): Projection on (100) of the $\{(QND)_2^{2+}(TCNQ)_3^{2-}\}$ structure (9.906 28.562 7.791 Å, 92.75°, $Z=2$, $P2_1/n$). The TCNQ moieties are in sheets normal to [010], separated by sheets of QND cations. There are no stacks in the TCNQ sheets. The monomers at the centres of symmetry (shaded rings) are essentially neutral while the π-dimers (with R/R overlap and 3.26 Å interplanar spacing), about the centers, have charge −1. (Reproduced from Bandrauk, Ishii *et al.*, 1985.)

interstitially between the stacks (Fig. 17.23); the charge distribution is uncertain. The compound has a room temperature conductivity of ≈20 S/cm, which Bousseau *et al.* (1986) describe as "surprisingly high" in view of the structure; lateral interactions between stacks may be an important factor.

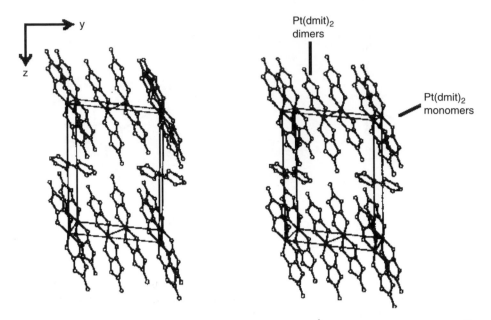

Fig. 17.23. Stereodiagram of the unit cell (6.314 11.660 16.810 Å, 84.50 89.60 81.90°, $Z = 1, P\bar{1}$) of TTF.[Pt(dmit)$_2$]$_3$ viewed in a direction close to [100]. This diagram should be rotated by 90° clockwise for comparison with Fig. 17.22. (Reproduced from Bousseau *et al.*, 1986.)

17.4.2.3 *Average charge of –4/7e on TCNQ moieties*

One of the most striking examples of a stack of limited length is found in $\{\{N(n\text{-}$butyl)pyridinium$\}_4^{4+}$ (TCNQ)$_7^{4-}\}$ where there are isolated heptads of TCNQ moieties (Fig. 17.24) (Konno and Saito, 1981; PYTCNB10).

This is the largest isolated grouping of TCNQ moieties so far reported. The authors suggest, on the basis of bond lengths, that TCNQ A and C carry more negative charge than B and D. The n-butyl groups of the two crystallographically-independent cations are extended, and disordered, in different ways. This may be important in the first-order insulator-to-insulator (I → I) phase transformation that occurs above room temperature (Murakami and Yoshimura, 1980). The high temperature structure has not yet been reported.

17.4.2.4 *Average charge of −0.5e on TCNQ moieties*

A π-tetrad of TCNQs, bounded at each end by planar cations, has been found in the triclinic polymorph of {3,3'-diethylthiacarbocyanine (TCNQ)$_2$} (Kaminski *et al.*, 1973; DTCTCQ11) and in {1-methyl-3,3-dimethyl-2-(*p*-N,N-methyl-*β*-chloroethylstyryl)-indole-(TCNQ)$_2$}(Shibaeva, Rozenberg and Atovmyan, 1973; MSTYTQ10). There are some structures in which π-dimers are arranged in stacks, but with virtually no overlap between successive π-dimers. These are discussed below.

Two independent sets of isolated π-dimers of TCNQF$_4$ (interplanar spacing 3.18 Å, displaced R/R overlap) are found in the complex salt {[(10-methyl-5,10-dihydro-phenarsazine-10-oxide)$_2$H]$^+$[(TCNQF$_4$)$_2$]$^-$(10-methyl-5,10-dihydrophenarsazine-10-oxide) (acetonitrile)}{[(C$_{13}$H$_{12}$AsNO)$_2$H]$^+$[(C$_{12}$N$_4$F$_4$)$_2$]$^-$(C$_{13}$H$_{12}$-AsNO)·CH$_3$CN} (Dietz, Endres, Keller and Moroni, 1984; CELYUN). The structure is ionic and contains slightly folded

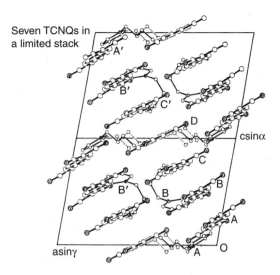

Fig. 17.24. Projection of the structure of $(NBP_y^+)_4(TCNQ)_7^{4-}$ along **b**. Disordered butyl groups of $NBPy^+$ are shown by dotted lines. The stack of seven TCNQ moieties has the following arrangement, running diagonally down from top left hand corner. (Reproduced from Konno and Saito, 1981.)

$$[A \quad B \quad C \quad \overset{\bar{1}}{D} \quad C' \quad B' \quad A']$$
$$3.22 \quad 3.33 \quad 3.36$$
$$R/EB \quad R/EB \quad R/EB$$

$[(C_{13}H_{12}AsNO)_2H]^+$ cations and π-dimeric $[(C_{12}N_4F_4)_2]^-$ anions. The spaces between the large ions are filled with neutral $(C_{13}H_{12}AsNO)_2$ dimers and a molecule of acetonitrile.

$$C_{13}H_{12}AsNO$$

The cation consists of two $C_{13}H_{12}AsNO$ moieties *syn* linked by a very strong $O\ldots H^+\ldots O$ hydrogen bond ($d(O\ldots O) = 2.39(2)$ Å), while the neutral dimer has the two $C_{13}H_{12}AsNO$ moieties *anti*-overlapped, which would add a dipole–dipole component to any π–π^* interaction.

17.4.3 *TCNQ anion-radical salts in which the cations are metals*

The so-called 'simple' salts $M^+(TCNQ)^{-\bullet}$ ($M = Li$, Na, K, Rb, NH_4) are discussed here while the so-called 'complex' $(M^+)_2(TCNQ)_3^{2-}$ salts are discussed later with other triad stacked salts in Section 17.4.6. Although free energies of formation have been determined for the Li, Na, K, Ba, Ag, Cu, Ni and Pb TCNQ salts (Aronson and Mitelman, 1981) and electronic absorption spectra (3100–24000 Å between 10–300K) have been studied for the K, Cs and Ba salts of TCNQ (Michaud *et al.*, 1979), the crystallographic information

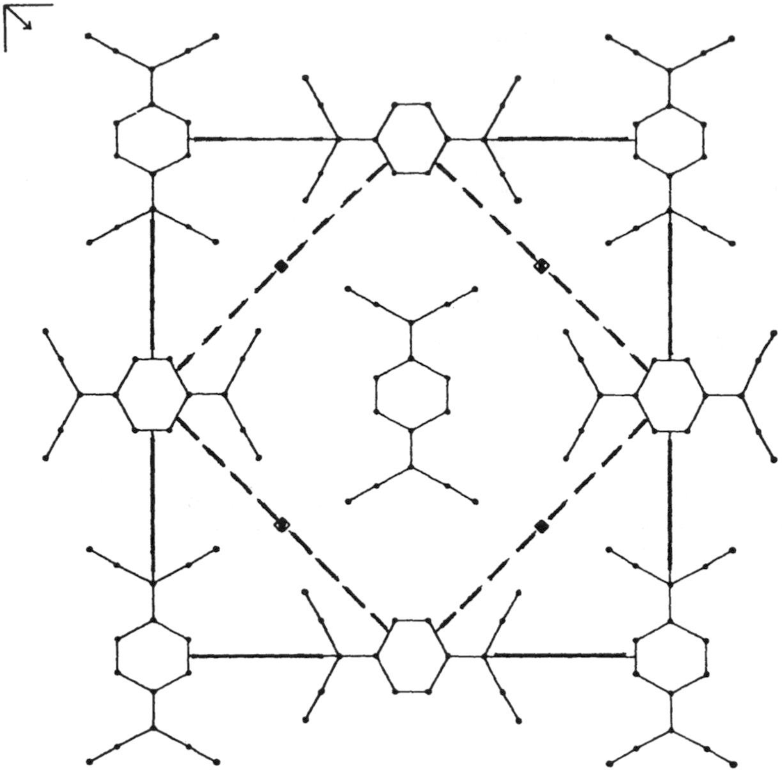

Fig. 17.25. Projection of the crystal structures of two polymorphs of Rb(TCNQ). The unit cell of phase I (projection on (100); $a = 7.19$ Å at 113K) is outlined by broken lines and that of phase III (projection on (001); $c = 3.86$ Å at 298K) by full lines. The Rb ions are denoted by small squares. This diagram should be compared with Fig. 17.12. (Adapted from Bodegom, de Boer and Vos, 1977.)

(Table 17.3) is more limited. Two groups can be distinguished – the first group is anti-structural to TTF·Cl_y, with cations and anions interchanged. This group contains all except one of the known structures. There are two sets of TCNQ stacks, the stack axes being parallel while the long axes of the TCNQs are mutually perpendicular (Fig. 17.25). Salts of different cations differ in their coordination and in the details of TCNQ stacking. In Na(TCNQ) the sodium ions are enclosed by a nearly octahedral arrangement of six nitrogens from different TCNQ moieties, while the larger potassium ions of K(TCNQ) (in both low and high temperature polymorphs) are at the centre of a distorted cube of nitrogens from surrounding TCNQs. The crystal structure analyses were complicated by problems of twinning and disorder (in particular in the phases of NH_4(TCNQ)). Most of the salts show phase transitions on cooling, which are generally associated with changes from monad to diad stacking; there are accompanying changes in magnetic (Fig. 17.28) (Kommandeur, 1980) and other properties. In K(TCNQ) the structural changes have been correlated with the temperature-dependent magnetic susceptibility using the Heisenberg model. However, more work is needed to complete the correlations between physical properties and crystal structures, particularly for Rb(TCNQ). The triclinic phase II of

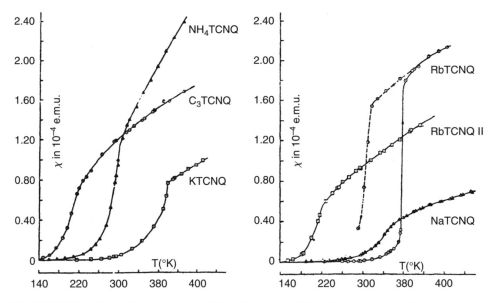

Fig. 17.26. Variation with temperature of the spin susceptibility (shown on the ordinate in units of 10^{-4} emu) for a number of crystalline M(TCNQ) salts. A Curie-type 'impurity' contribution has been subtracted from each curve. (Adapted from Kommandeur, 1980.)

Rb(TCNQ) constitutes, at present, the second structural group; here all the TCNQ stacks are translationally equivalent and hence parallel. Rb(TCNQ)-I has a room-temperature conductivity of 10^{-5} S/cm while that of Rb(TCNQ)-II is 10^{-2} S/cm.

17.4.4   Stacked structures with $-e$ average charge on the TCNQ moieties

17.4.4.1   Monad stacks

The blue-black crystals of (oxamide oximato) (oxamide oxime)Pt(II)TCNQ {[Pt(oaoH$_2$)-(oaoH$_2$)]$^+$(TCNQ)}, where 'oaoH$_2$' is oxamide oxime $HON = C(NH_2) - C(NH_2) = NOH$ ($P\bar{1}, Z = 1, c = 3.797$ Å) (Endres, 1982 BIWWAF) contain segregated stacks of cations and anions along [001], with interplanar spacings of 3.41 and 3.20 Å respectively and an angle of 36° between cation and anion planes. The TCNQs show R/EB overlap. There is hydrogen bonding between the stacks. The powder conductivity is low at $\approx 10^{-3}$ S/cm, which is compatible with integral charges on the anions.

$$
\left[\begin{array}{c} H_2N \\ \\ H_2N \end{array} \begin{array}{c} OH \quad O \\ | \quad\quad \| \\ N \quad N \\ \diagdown Pt \diagup \\ N \quad N \\ \| \quad\quad | \\ O \quad HO \end{array} \begin{array}{c} NH_2 \\ \\ NH_2 \end{array} \right]^{2+}
$$

N-methylphenazinium TCNQ attracted early attention because of its high conductivity (1.5×10^2 S/cm) along the needle axis (Melby, 1965). The crystal structure contains stacks

Table 17.3. Crystal data (Å, deg., Å^3) for the 1:1 alkali-metal salts of TCNQ. The stacking axis is in bold

Salt	a/α	b/β	c/γ	Vol.	Z	Space group	Interplanar spacings
TTF·Cl$_{0.77}$	10.77	22.10	**3.56**	847.3	4		
Group I: Antistructures to TTF·Cl$_{0.77}$							
Na TCNQ at 296K	**6.99**	23.71	12.47	2044	8	$C\bar{1}$	3.21/3.49
(Konno and Saito,	90.1	98.6	90.8				Displaced R/R
1974); NATCNQ							
Na TCNQ at 353K	**3.51**	11.87	12.47	514	2	$P2_1/n$	3.39
(Konno and Saito,		98.2					Displaced R/R
1975); NATCNQ01							
K TCNQ at 298K	**7.08**	17.77	17.86	2240	8	$P2_1/n$	3.24/3.57
(Konno, Ishii and Saito,		95.0					Displaced R/R
1977)*; KTCYQM01							
K TCNQ at 413	**3.59**	12.68	12.61	570	2	$P2_1/c$	3.48
(Kobayashi *et al.*,		96.4					Displaced R/R
1973); KTCYQM02							
Rb TCNQ-I at 113K	**7.19**	12.35	13.08	1147	4	$P2_1/n$	3.16 (R/R) 3.48
(Hoekstra *et al.*,		98.9					(R/EB)
1972); RBTCNQ							
Rb TCNQ-III at 298K #	17.65	17.65	**3.86**	1203	4	$P4/n$	3.33(R/EB)
(Bodegom *et al.*,							
1977); RBTCNQ02							
NH$_4$ TCNQ-I at 293K§	18.19	17.75	**7.19**	2321	8	orthorhombic	no structure
NH$_4$ TCNQ at 311K§	18.23	17.80	**3.61**	1171	4	orthorhombic	no structure
NH$_4$ TCNQ-II at 298K;	12.50	12.50	**3.82**	5972	2	$P4/mbm$	(R/EB)
AMTCNQ§						average structure	
Group II							
Rb TCNQ-II# (Shirotani	9.91	7.20	**3.39**	275	1	$P\bar{1}$	3.43(R/EB)
and Kobayashi, 1973);	92.7	86.2	97.7				
RBTCNQ01							

Notes:

* A very similar structure at 298K, differing only in that $a = 3.54$ Å, has been reported (Richard *et al.*, 1978; KTCYQM). The interplanar spacing is 3.44 Å.

\# Region of temperature stability not stated.

§ NH$_4$ TCNQ polymorphs studied by Kobayashi (1978).

of TCNQ$^{-•}$ ions along [100] ($=3.868$ Å), the interplanar spacing being 3.26(1) Å. As there is one formula unit in space group $P\bar{1}$, the planar but noncentrosymmmetric cations must be disordered over two orientations. The high conductivity is difficult to reconcile with the apparently integral charge on the anions. Thus it seems probable that the crystal structure reported by Fritchie (1966; MPHCQM) is only an average structure and disorder must be taken into account in any explanation of the physical properties (Kobayashi, 1975). This does not yet seem to have been worked out in detail.

17.4.4.2 *Isolated diad stacking*

$\{[Pt(NH_3)_4]^{2+}\ (TCNQ)_2^{2-}\}$ contains centrosymmetric π-dimers with an interplanar spacing of 3.14 Å and displaced R/R overlap ($\Delta L = 0$, $\Delta M = 1.1$ Å) (Endres *et al.*, 1978a; TCQMPT). Limited precision prevented assignment of charges to TCNQ moieties *via* bond lengths. The ESR spectra show one exchange-narrowed line of about 5G width without fine structure; the paramagnetism increases with temperature. If it is assumed that a thermally activated triplet state becomes occupied, then the activation (or exchange) energy is ≈ 0.05 eV ($400\,\mathrm{cm}^{-1}$); this is similar to the singlet-triplet separation of $246(20)\ \mathrm{cm}^{-1}$ in TMPD.ClO$_4$ (Section 17.3.4).

{Dibenzenechromium TCNQ} appears to have a similar structure with isolated diads (displaced R/R overlap, interplanar spacing 3.28 Å) stacked along [001] (Shibaeva, Shvets and Atovmyan, 1971; BCRLQM).

17.4.4.3 *Diad stacks*

Four examples have been reported. In {4-oxo-6-iodoquinolinium TCNQ} there are stacks along the *a* axis (7.16 Å), with TCNQ interplanar spacings of 3.09 and 3.32 Å (Keller, Steiger and Werner, 1981; BARTIX). In {ditoluenechromium TCNQ} (Shibaeva, Atovmyan and Rozenberg, 1975; DTCRCQ10) there are diad stacks along [100] (7.00 Å), with incipient π-dimerization in the following sense. There is R/R overlap of two centrosymmetrically related TCNQs, the interplanar spacing being 3.33 Å; such pairs are themselves related by centres of symmetry, the interplanar spacing here being 3.42 Å with some sideways displacement from an eclipsed arrangement. The conductivity is low (4×10^{-2} S/cm), in agreement with integral charges on the cations and anions. There is more definite π-dimerization in {morpholinium TCNQ} (Sundaresan and Wallwork, 1972e; MORTCD), where the almost eclipsed TCNQs of the centrosymmetric π-dimer (R/R overlap) are separated by 3.28 Å, while adjacent π-dimers in a stack hardly overlap across centres of symmetry, with an interplanar spacing of 3.61 Å. Each morpholinium cation forms two pairs of weak bifurcated N . . . N hydrogen bonds (d(N . . . N) ranges from 2.92 to 3.11 Å) to four TCNQ anions. There is a phase change at 424K, with $\Delta H = 2.1(1)$ kJ/mol; ESR studies suggest there is only a minor structural change and essentially no change in the spin state of the anions (Bailey and Chesnut, 1969). Rather similar π-dimerization is found in $\{[Fe(C_5H_5)_2][(TCNQF_4)]\}$ (FETBEL) where TCNQF$_4$s of the π-dimer (displaced R/R overlap) are separated by 3.22 Å, while adjacent π-dimers in a stack overlap somewhat less (but also in displaced R/R fashion) with an interplanar spacing of 3.68 Å.

17.4.4.4 *Tetrad stacks*

These have been briefly reported in {chlorpromazine TCNQ} (Metzger *et al.*, 1974). More detail is available for $\{(1,1'\text{-ethylene-}2,2'\text{-bipyridylium})^{2+}(TCNQ^-)_2\}$ (Sundaresan and Wallwork, 1972d; EBPTCQ), where the TCNQ stack contains two crystallographically-independent π-dimers well separated from one another (Fig. 17.27).

In {diethylmorpholinium (2,5-dichloro-TCNQ)} (DEM·TCNQCl$_2$) (Bryce and Howard, 1982; BIPJAL10; $P2_1/n$, $Z = 8$) there is a sequence of R/EB (3.38 Å) and displaced R/R (3.28 Å) overlaps, the latter being across a crystallographic centre of symmetry. Thus π-dimers can hardly be identified.

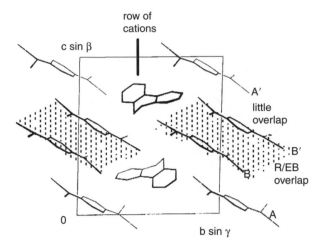

Fig. 17.27. The moiety arrangement in $(1,1'\text{-ethylene-}2,2'\text{-bipyridylium})^{2+}(TCNQ^-)_2$ viewed along [100]. The stack axis is [001] (14.625 Å) and the two anions denoted as A and B are crystallographically independent. The interplanar distances A'A and B'B (shaded pair) are respectively 3.22 and 3.26 Å and the overlap is R/EB. The A and B anions hardly overlap, the nominal interplanar spacing being 3.59 Å. The dihedral angle between the two pyridinium rings of the cation is $\approx 22°$. (Reproduced from Sundaresan and Wallwork, 1972d.)

17.4.5 Stacked structures with −0.8e average charge on the TCNQ moieties

The cation is 1,2-di(*N*-ethyl-4-pyridinium)ethane (DEPA) in the only TCNQ example, $\{(DEPA^{2+})_2(TCNQ)_5^{4-}\}$,(Ashwell, Eley, Wallwork, Willis, Welch and Woodward, 1977; EPETCR). The crystals are triclinic, $P\bar{1}$, $Z = 1$. There are pentad stacks along [001] (16.24 Å). Although the interplanar spacings between adjacent TCNQs are essentially the same (3.24(5) Å), there are different kinds of overlap, as shown below:

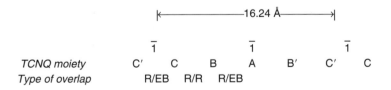

The stack can thus be considered as made up of a π-dimer (CC′) and a centrosymmetric triad (BAB′). The CSD formulation as $\{(DEPA^{2+})_2 (TCNQ)(TCNQ^-)_4\}$ does not seem to be justified, The conductivities along the three crystallographic axes are 6.7×10^{-4}, 5.6×10^{-2} and 4×10^{-2} S/cm respectively. As the bond lengths in the various TCNQ moieties were not significantly different (but precision was not high), it was inferred that the four electrons are delocalized over the five TCNQ moieties. The ESR spectra were complex and not completely interpretable. The modest conductivity along the stack axis presumably is a consequence of the R/R overlap separating the π-dimer and the triad in the stack.

1,4,5,8-Naphthalenetetracarboxylic anhydride and diimide analogs have been shown to form semiconducting anion radical salts with various cations (Heywang *et al.*, 1989).

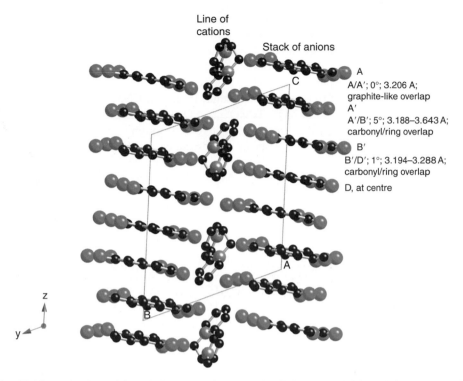

Fig. 17.28. Projection of {[(5,6-dihydro-4a,6a-phenanthrolinium)₂][(1,4,5,8-naphthalenetetracarb-oxylic anhydride)₅]} structure down [100]. The reduced cell (used in the structure analysis) is 8.935 12.785 15.995 Å, 112.38 92.03 93.68°. $Z = 1$)). The molecules are denoted A, B . . . following Born and Heywang (1991), and interplanar spacings, angles between successive pairs of molecular planes and modes of overlap are shown in the diagram. (Data from Born and Heywang, 1991.)

Pentad stacking is found in the centrosymmetric triclinic crystals of {[(5,6-dihydro-4a,6a-phenanthrolinium)₂][(1,4,5,8-naphthalenetetracarboxylic   anhydride)₅]}   (Born   and Heywang, 1991; VARKEE10) (Fig. 17.28). Conductivity measurements suggest a phase change at ≈150K; only the 300K structure has been determined. The carbonyl bond lengths vary between 1.202 and 1.210 Å, with a mean of 1.206 Å. This suggests delocalization of charges. The carbonyl bond length in neat 1,4,5,8-naphthalene-tetracarboxylic anhydride is 1.186 Å (Born and Heywang, 1990; KENDEM).

The importance of this structure lies in the demonstration that radical anions other than TCNQ and some others can form stacks. It is not clear whether attempts were made to prepare cation radical anion radical salts. Does a crystal chemistry analogous to that of TCNQ exist for 1,4,5,8-naphthalenetetracarboxylic anhydride, and perhaps other acceptors?

17.4.6   *Stacked structures with −2/3e average charge on the TCNQ moieties*

Nine examples are discussed, six as {(M⁺)₂(TCNQ)₃²⁻} and three as {M²⁺(TCNQ)₃²⁻}. The arrangements of the TCNQ moieties in the triads are summarized in Table 17.4; these triads can be isolated (three examples) or stacked (six examples). The formation of triads is characteristic of a Peierls distortion of a uniform stack when $2n$ excess electrons are shared by $3n$ moieties (Su and Schrieffer, 1981; Gammell and Krumhansl, 1983). IR spectroscopy has potential usefulness for demonstrating the presence of $TCNQ^0$ and $TCNQ^-$; two distinct absorptions, at $1453\,cm^{-1}$ ($TCNQ^0$) and $1386\,cm^{-1}$ ($TCNQ^-$), have been found for {(morpholinium)$_2$(TCNQ)$_3$} (Sundaresan and Wallwork, 1972a; MORTCQ) and $Cs_2(TCNQ)_3$ (Fritchie and Arthur, 1966; CSSCNM10). The more isolated triads can be described as ($TCNQ^-$ $TCNQ^0$ $TCNQ^-$) but the interplanar distances in Rb_2TCNQ_3 and Cs_2TCNQ_3 *suggest* the alternative description of π-dimers separated by neutral TCNQs in the pseudomonad stacks of these salts:

$$\ldots(TCNQ^-\ TCNQ^-)\quad TCNQ^0\quad (TCNQ^-\ TCNQ^-)\ldots$$

Perhaps there is a gradual transition from one arrangement to the other. The π-dimer – $TCNQ^0$ – π-dimer description is compatible with the zero-splitting parameters in Cs_2TCNQ_3 ($D = 280.4(6)$, $E = 45.3(1.4)$ Mc/sec), which indicate that the two electrons of the TCNQ triad are present as a tightly bound or Frenkel exciton, probably located on the triad itself (Chesnut and Arthur, 1962). Very similar values have been obtained for the zero-splitting parameters in {(morpholinium)$_2$TCNQ$_3$} (Bailey and Chesnut, 1969).

Considerable attention has been paid to Cs_2TCNQ_3. This salt is a semiconductor with $\sigma \approx 10^{-3}$ S/cm at room temperature, only slightly higher than that of K(TCNQ). There are two CT transitions, one at $\approx 3900\,cm^{-1}$ between the neutral molecule and anions of a triad, and one at $\approx 10000\,cm^{-1}$ between anions of different triads. The electronic charge along a stack is described as a "pinned charge density wave," presumably induced by the Coulomb potential of the cations. The IR and Raman spectra in the region of the intramolecular modes have been reinterpreted, taking interactions between electronic and molecular vibrations into account (Painelli *et al.*, 1986). On this basis a charge of $\approx 0.1e$ has been assigned to the formally neutral TCNQ moiety at centres of symmetry and of $\approx 0.9e$ to the anions in general positions.

The crystal structure of {3-(*p*-methoxyphenyl-1,2-dithiolylium)$_2$(TCNQ)$_3$}, which has not been included in Table 17.4, has been briefly (Amzil *et al.*, 1986) and more extensively reported (Mathieu, 1986; DUXRET). The interplanar distance within the triad is 3.17 Å (R/EB overlap, $\Delta L = 2.03$, $\Delta M = 0.03$ Å) and that between triads is 3.43 Å (displaced R/R overlap, $\Delta L = 0.51$, $\Delta M = 1.69$ Å). Bond lengths suggest that the charge is equally distributed over all three TCNQ moieties.

3-(*p*-methoxyphenyl-1,2-dithiolylium)

Triads are also found in the 2:3 charge-transfer salt of tetraethylammonium and 4,8-bis(dicyanomethylene)-4,8-dihydrobenzo[1,2-*b*:4,5-*b*′]dithiophene ($P2_1/n$, $Z = 2$) (Yasui *et al.*, 1992; PANVOX). The salt shows the relatively high conductivity of 0.46 S/cm. One of the anions (A) is at a centre of symmetry and the other (B) at a general position. Bond lengths suggest that A is neutral and B singly charged. The A...B

Table 17.4. Arrangement of TCNQ moieties in $M_2(TCNQ)_3$ anion radical salts. The structures were determined at room temperature unless stated otherwise; the interplanar distances (Å) are not strictly comparable because the various authors do not use a standard method of calculation

Compound	Stacking arrangement	Remarks
1. (N-methylphenazinium)$_2$ (TCNQ)$_3$ (Sanz and Daly, 1975); MPZTCQ	$\bar{1}$ $\bar{1}$ **B'** **A** (3.15)**B**(3.26)**B'** **A'** R/EB	No B/B' overlap, thus isolated triads; A, B charges not assignable.
2. [1,4-di(N-quinolinium-methyl)benzene](TCNQ)$_3$ (Ashwell, Eley, Wallwork, Willis, Peachy and Wilkos, 1977); AMBTCQ	$\bar{1}$ $\bar{1}$ **B'** **A**(3.12)**B** **B'** **A'** R/EB	No B/B' overlap, thus isolated triads; A, B charges not assignable.
3. [Bis(triphenyl-phosphoranediyl) (TCNQ)3·2CH$_3$CN (Halfpenny, 1985); COVMIJ	$\bar{1}$ **B'** **A**(3.30)**B** R/EB	Triads widely separated; A assigned as TCNQ^0, B as TCNQ$^-$.
4. 1,1'-bi-cobaltocene-[Co^{III}Co^{III}](TCNQ)$_3$ (Lau et al., 1982); BEHGEA	$\bar{1}$ $\bar{1}$ **B'** (3.15)**A**(3.15)**B**(3.45) **B'** R/EB mixed	Triad stacks; A, B charges not assignable.
5. [Fe$_2$(η-C$_5$H$_5$)$_2$(μ-CO)$_2$ {μ-Ph$_2$PN-(Et)PPh$_2$}] (TCNQ)$_3$·CH$_3$CN (Bell et al., 1991); KIRDUK	**B'** (3.24)**A** (3.24)**B**(3.39)**B'** R/EB displaced R/R	Triad stacks; A, B charges not assignable.
6. Bis(2,2'-bipyridyl)-Pt(II) (TCNQ)$_3$ (Endres et al., 1978b); PYPTCR	$\bar{1}$ $\bar{1}$ **B'** (3.23)**A** (3.23)**B**(3.33)**B'** R/EB displaced R/R	Triad stacks; A, B charges not assignable.
7. Rb$_2$(TCNQ)$_3$ at 294K (Wal and Bodegom, 1978); RBTCNR	$\bar{1}$ $\bar{1}$ **B'** **A** (3.28)**B**(3.12)**B'** R/EB displaced R/R	Triads arranged in pseudo-monad stacks.
8. Rb$_2$(TCNQ)$_3$ at 113K (Wal and Bodegom, 1979); RBTCNR01	$\bar{1}$ $\bar{1}$ **B'** **A** (3.23)**B**(3.07)**B'** R/EB displaced R/R	Triads arranged in pseudo-monad stacks.
9. Cs$_2$(TCNQ)$_3$ at 294K (Fritchie and Arthur, 1966); CSSCNM10	$\bar{1}$ $\bar{1}$ **B'** **A** (3.27)**B**(3.12)**B'** R/EB displaced R/R	Triads arranged in pseudo-monad stacks.
10. (Morpholinium)$_2$ (TCNQ)$_3$; (Sundaresan and Wallwork, 1972d); MORTCQ	$\bar{1}$ $\bar{1}$ **B'** **A** (3.25)**B**(3.24)**B'** R/EB displaced R/R	Triads arranged in pseudo-monad stacks.

Notes:

Rb$_2$(TCNQ)$_3$ and Cs$_2$(TCNQ)$_3$ are isomorphous; in # 7, 8, 9, 10 A = TCNQ^0 and B = TCNQ$^-$, from bond lengths and IR spectra.

separation in a stack is 3.47 Å, with R/EB overlap, and the B . . . B separation 3.39 Å, with displaced R/R overlap. The interpretation of these intrastack separations is ambiguous because of the presence of the bulky S atoms in the anions, and of side-by-side linkages between stacks by short S . . . S contacts of 3.570(3) and 3.529(3) Å.

(Triethylenediammonium)$_2$(TCNQ)$_3$ [(TEDA)$_2$(TCNQ)$_3$] (also not in Table 17.4) requires special mention. The IR spectrum shows only one absorption, at $1416 \, cm^{-1}$, leading to the inference that this triad is delocalized. The crystal structure (Bandrauk, Ishii et al., 1985) has irregular stacks of TCNQ moieties arranged in centrosymmetric triads with displaced R/R overlap within a triad (interplanar spacing 3.22 Å) and a more accentuated R/R overlap (3.45 Å) between triads). The bond lengths suggest that the central TCNQ bears a charge of $0.8e$ and that the flanking TCNQs each have a charge of $0.6e$; this is quite different from the approximately e, 0, e distributions of some of the M$_2$TCNQ$_3$ structures listed in Table 17.4. The difference was ascribed to the much weaker interaction between cations and anions in (TEDA)$_2$(TCNQ)$_3$ than in, say, Cs$_2$TCNQ$_3$.

17.4.7   Stacked structures with $-0.5e$ average charge on the TCNQ moieties

Two major groups can be distinguished. In the first group TCNQ diads (π-dimers) or tetrads are arranged in stacks but with virtually no overlap between successive diads or tetrads. We call these isolated diad or tetrad stacks. In the second group the diads or tetrads, although distinguishable to a greater or lesser extent, quite clearly overlap and interact.

17.4.7.1   Isolated diad or isolated tetrad stacking

The following salts have been reported with isolated-diad stacking at room temperature (for #1–10 only cations are listed, the anions all being (TCNQ)$_2^-$).

1. 3,3'-diethylthiacarbocyanine (monoclinic polymorph) (Kaminskii et al., 1973a); ECYTCN10
2. tetraphenylphosphonium (Goldstein et al., 1968); PPHTCQ
3. 5,5'-dimethylphosphonium (Ashwell, Allen et al., 1982); BINREV
4. tetraethylammonium (Shibaeva, Kaminskii and Simonov, 1980)
5. N-(n-propyl)pyridinium (Konno and Saito, 1981); PYTCNA10
6. N-methyl-N-ethylmorpholinium (at 113K) (Bosch and Bodegom, 1987); MEMTCQ10
7. 1,2,4-trimethylpyridinium (at 300K; Rizkallah et al., 1983; BOHNOB. At 173K Graja et al., 1983)
8. 4,4'-diethylmorpholinium (Morssink and Bodegom, 1981); DEMTCN
9. 1-methyl-1,4-dithianium (Bryce et al., 1988); JACPUY
10. N^2-methylcinnalinium (Daoben et al., 1985); CUMHAT10
11. N^1-methylcinnalinium (Daoben et al., 1985); CUMHEXT10
12. (1,1'-tetramethylene-dipyridinium)$^{2+}$ (TCNQ)$_4^{2-}$ (cation is also called 'dipyridylbutene' DBP) (Waclawek et al., 1983); PYBUTQ10
13. (N,N,N,N',N',N'-hexamethylhexa-methylenediammonium)$^{2+}$ (TCNQ)$_4^{2-}$ (Flandrois et al., 1979); HMACQM

The arrangement of the anions is remarkably similar in all these salts – the π-dimers are centrosymmetric (although not necessarily located at a crystallographic centre of symmetry e.g. entries #4 and 6), and have interplanar spacings lying between 3.17 and 3.23 Å, with R/EB overlap between the two moieties. In some of the examples it was not possible to distinguish between the two TCNQ moieties of the π-dimer on the basis of bond lengths, while in others the two TCNQ moieties were crystallographically equivalent; the usual inference was that there was real electron delocalization. The conductivities were found to be intermediate, i.e. $\approx 10^{-3}$ S/cm. At room temperature #5 has two independent π-dimers in the unit cell, with an angle of 17.5° between their planes. The bond lengths in the TCNQ moieties are identical so both are described as $TCNQ^{-1/2}$. There is a first-order insulator-metal transition at 388K; the high-temperature phase has a uniform zigzag stacking of TCNQ moieties. #6 has been studied over a wide range of temperatures and is discussed in more detail in Section 17.4.8. $(DEM)(TCNQ)_2$ (#8) requires special mention because there are two kinds of TCNQ arrangement in sheets parallel to (010), separated by disordered cations (Fig. 17.29(upper)); there are both isolated π-dimers (A'A) and isolated tetrads (BB"B"B), the angle between the normals to the planes of A and B being 57.4(5)°. Such a marked difference between the arrangements of two sets of TCNQ moieties in the same crystal is most unusual. The so-called AA' 'π-dimer' is rather weakly bound, with the interplanar spacing of 3.41 Å being typical of R/R overlap; the planes of these TCNQ moieties are approximately parallel to (001). The B moieties have their planes approximately parallel to $(11\bar{1})$. The 'isolated tetrad' is probably better described as consisting of two poorly-overlapping π-dimers, which have the internal spacing of 3.14 Å typical of strongly bound moieties with R/EB overlap. The types of overlap are shown in Fig. 17.29(lower) and are described by the displacements given in the caption to this figure. Various morpholinium derivatives have the formula shown below; thiomorpholinium has O replaced by S.

(thio)morpholinium formulae

$$\left[\begin{array}{c} O \quad N \begin{array}{c} R_1 \\ {}'\!'R_2 \end{array} \end{array} \right]^{+}$$
$R_1 = R_2 =$ methyl: DMM , 4,4'-dimethylmorpholinium;
$R_1 =$ methyl, $R_2 =$ ethyl: MEM;
$R_1 = R_2 =$ ethyl: DEM;
$R_1 =$ methyl, $R_2 =$ H: HMM.

An arrangement similar to that in $\{DEM(TCNQ)_2\}$ is found in N-methylmorpholinium $(TCNQ)_2$ $\{HMM(TCNQ)_2\}$ (space group $P4_1$, $Z = 4$); the successive TCNQ stacks along [001], interleaved by cations, are equivalent but mutually rotated by 90° (Visser et al., 1990; DESGIR11). Indeed $\{HMM(TCNQ)_2\}$ and $\{pyrene \cdots p\text{-benzoquinone}\}$ (Section 15.2) have structural resemblances. There is another remarkable structure (Endres, 1984; CURTIS) with a resemblance to $(DEM)(TCNQ)_2$ in the sense of having non-parallel stacks. This is bis[(oxamide oximato)(oxamide oxime)Pt(II)] [tris(tetracyanoquinodimethane)diide] or $\{[Pt(oaoH)(oaoH_2)]^{+}\}_2(TCNQ)_3^{2-}$ (for formula see Section 17.4.4(a)). The triclinic crystals (space group $P\bar{1}$ $Z = 1$) have two sets of mutually-perpendicular stacks. One is a diad stack containing crystallographically-equivalent TCNQ moieties with average charge of $-0.5e$ arranged in zigzag fashion with alternating interplanar spacings of 3.08 and 3.27 Å. The second stack is mixed and contains a

$$\ldots\ldots D \ D \ A^{-} \ D \ D \ A^{-} \ D \ D \ldots\ldots$$

arrangement. The DC conductivity is 12 S/cm at 300K. This is the only structure known (to me) in which there are both segregated and mixed stacks.

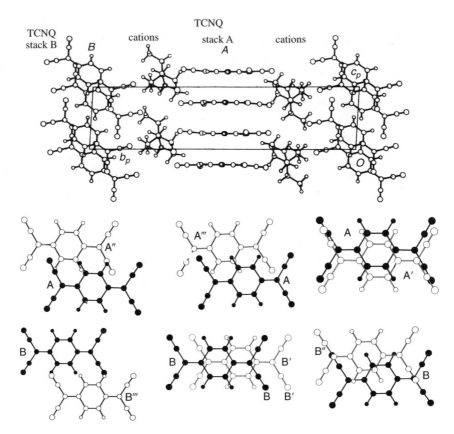

Fig. 17.29. (upper) The crystal structure of (DEM)(TCNQ)$_2$ projected down [100]. The A and B stacks are inclined at an angle of 57.4(5)°; (lower) the types of overlap. The A″ and A‴ moieties are diagonally displaced above A, as are B″ and B‴ with respect to B; the BB′ overlap is that of a standard π-dimer. The mutual displacements of various pairs of moieties are given below:

Pair of moieties	$\Delta L(\text{Å})$	$\Delta M(\text{Å})$	$\Delta N(\text{Å})$	Type of overlap
AA′	−0.14	−0.76	3.41	Shifted R/R
B′B 'π-dimer'	−2.02	0.07	3.14	R/EB
BB″	1.17	2.08	3.36	Diagonal shift between R/R and R/EB

(Adapted from Morssink and Bodegom, 1981)

(ii) Isolated-tetrad stacking: the known structures are listed in Table 17.5; two groups can be distinguished. In the first of these (Type I) the tetrads can be separated into two π-dimers, with displaced R/R overlap between them, while in the second group (Type II) the four TCNQ moieties of a tetrad are separated by essentially identical interplanar spacings and have R/EB overlap.

The salts [1,3-di(*N*-pyridinium)propane](TCNQ)$_4$ (Ashwell, Bartlett *et al.*, 1977; PYPTCQ) and [*N, N*′-diethyl-4,4′-bipyridylium](TCNQ)$_4$ (Ashwell, Eley *et al.*, 1975;

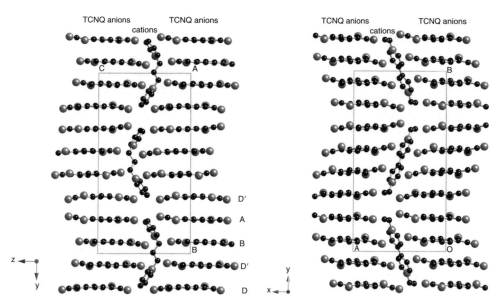

Fig. 17.30. Left: Crystal structure of [1,3-di(*N*-pyridinium)propane](TCNQ)$_4$ projected down [100]. The crystals are monoclinic, space group *P*2$_1$; this is a rare example of a TCNQ salt crystallizing in a Sohncke space group. Note the eight molecules of TCNQ in a stack period, and the sinusoidal profile of the stack, conforming to the conformation of the cation. Right: analogous projection (down [001]) for [*N*, *N'*-diethyl-4,4'-bipyridylium](TCNQ)$_4$; the crystals are monoclinic, space group *P*2$_1$/*c*. (Data from Ashwell, Bartlett *et al.*, 1977 and Ashwell, Eley *et al.*, 1975.)

EPMTCQ) require special mention because they have the largest (room temperature) stack periodicities (both 25.3 Å) yet encountered. These are octad stacks with two tetrads in the 25.3 Å period, each tetrad consisting of two π-dimers. The sinusoidal profiles of the stacks (Fig. 17.30) suggest that there is a strong mutual interaction between the detailed arrangements of the cation and anion stacks, but it is not clear which of these has the predominating influence. The conductivities along the stack axes have intermediate values, consistent with the jogs in the stacks.

17.4.7.2 *Fully stacked arrangements*

(i) Monad or pseudomonad stacking: There are a number of examples in this category (Table 17.6); we note that a monad stack will have a periodicity of $\approx$3.6 Å along the stack axis, while in a pseudomonad stack the periodicity is $\approx$7.2 Å but the overlaps are crystallographically equivalent.

(ii) Diad or pseudo-diad stacking: The examples reported are summarized in Table 17.7. Presumably the distinction between this group and those classified earlier as isolated diads is gradual rather than sharp. π-Dimers can be identified in most of the present group and the entries in Table 17.7 have been arranged so as to proceed from salts with greater separation (isolation) between π-dimers to those in which this separation tends towards that in the π-dimer itself.

Table 17.5. Salts with isolated-tetrad stacking (all measurements at room temperature, unless stated otherwise)

Salt	Stacking arrangement	Remarks
TYPE I:		
1. [1-methyl-3,3-dimethyl-2-(p-N-methyl-N-β-chloro-ethylstyryl)indole]$^+$(TCNQ)$_2^-$ (Shibaeva, Rozenberg and Atovmyan, 1973)	**A**(3.12)**B**(3.45)**B′ A′** R/EB displ R/R	
2. [1,2-di(N-methyl-4-pyridinium)ethane]$^{2+}$-(TCNQ)$_4^{2-}$ (Ashwell, Eley et al., 1977); MPETCQ	**A**(3.20)**B**(3.39)**B′** R/EB displ R/R	
3. [1,3-di(N-pyridinium)propane]$^{2+}$-(TCNQ)$_4^{2-}$ (Ashwell, Bartlett et al., 1977); PYPTCQ	**A**(3.20)**B**(3.25)**C** (3.18) **D** R/EB displR/EB R/R	Two such tetrads per period along stack axis.
4. [N, N′diethyl-4,4′-bipyridylium)]$^{2+}$-(TCNQ)$_4^{2-}$ (Ashwell, Eley et al., 1975); EPMTCQ	**B**(3.19)**A**(3.16)**A′** R/EB R/EB	Two such tetrads per period along stack axis.
TYPE II:		
1. [3,3′-diethylthio-carbocyanine]$^+$ (TCNQ)$_2^-$ (Kaminskii et al., 1973b)	**A**(3.28)**B**(3.28) **B′** R/EB R/EB	
2. [N, N′-dibenzyl- 4,4′-bipyridylium)]$^{2+}$- (TCNQ)$_4^{2-}$ (Sundaresan and Wallwork, 1972c); BPYTCQ	**B**(3.16) **A**(3.24) **A′ B′** R/EB R/EB	The tetrads are separated by BB′ overlap of displaced EB/EB type and interplanar spacing of 3.62Å.
3. [1,2-Di(N-ethyl-4-pyridinium)-ethylene]$^{2+}$(TCNQ)$_4^{2-}$ (Ashwell, Eley, Fleming et al., 1976); EPETCQ	**B**(3.32) **A**(3.15) **A′**displ R/EB R/EB	The tetrads do not overlap directly.
4. [1,1′-bis(p-cyano-phenyl-4,4′-bipyridylium)]$^{2+}$ (TCNQ)$_4^{2-}$ (Ashwell, Cross et al., 1983); CAVJIS	**B**(3.36) **A**(3.28) **A′** displ R/EB R/R	No overlap between adjacent tetrads. A is neutral (0.3e) and B is anion (0.7e) from bond lengths.
5. [Pd(NCCH$_3$)$_4$]$^{2+}$(TCNQ)$_2$-2CH$_3$CN (Goldberg, Eisenberg et al., 1976); ICPTCQ	**B**(3.29)**A**(3.32) **A′ B′** displR/EB R/EB 7.4° dihedral angle	Solvate molecules poorly defined. No overlap between adjacent tetrads.
6. [1,4-di(diphenyl-methyl-phosphonium)-benzene]$^{2+}$(TCNQ)$_4^{2-}$ (Batail et al., 1985); DOZPAJ	**A**(3.25)**B**(3.33) **B′ A′** R/EB R/EB	Overlap of tetrads not described.

Table 17.6 TCNQ arrangements in $M(TCNQ)_2$ ion radical salts with monad or pseudomonad stacks. $D(\text{Å})$ is the interplanar spacing, and σ (S/cm) is the conductivity, both along the stack direction

Salt	TCNQ stack arrangement	$D(\text{Å})$	σ (S/cm)	Remarks
1. N-methylphthalazinium (TCNQ)$_2$ at 130K (Yan and Coppens, 1987); FOVZUL	A......A A R/EB	3.178(2)	120	§1 and §2 isomorphous, C2/c, Z = 4. Monad stack with TCNQs translationally equivalent.
2. Quinolinium (TCNQ)$_2$ (Kobayashi, Marumo and Saito, 1971); QUTCNQ	$\bar{1}$ $\bar{1}$ A......A' A R/EB	3.22(1)	100	§1 has [010] = 3.77 Å and §2 has [010] = 3.838 Å.
3. Ditoluene-chromium (TCNQ)$_2$ at 153K (Shibaeva, Atovmyan and Ponomarev, t 1975).	$\bar{1}$ $\bar{1}$ $\bar{1}$ A......A' A R/EB	3.16(1)	2	TCNQs at independent centres of symmetry.
4. Acridinium (TCNQ)$_2$ (Kobayashi, 1974); ARDTCQ	$\bar{1}$ $\bar{1}$ A......A' A R/EB	3.246		Possible 2nd order transition at $\approx$180K.
5. 5-t-butyl-3-methylthio-1,2-dithiolyl-(TCNQ)$_2$ (Amzil et al., 1986);	A......A'......A R/EB R/EB	3.24, 3.24	4	Zigzag pseudomonad stack.
6. [η^5-Cyclo-pentadicnyl-η^6-hexamethylbenzene-Fe(II)] (TCNQ)$_2$ (Lequon et al., 1985);	A......A'......A R/EB R/EB	3.34, 3.34	4	Zigzag pseudomonad stack.
7. (Diphenylphosphonium-cyclohexadiene)$^{2+}$ (TCNQ)$_4^{2-}$ (Batail et al., 1985); DOZPEN	A......A'......A......A' R/EB R/EB R/EB	3.20, 3.27	$\gg 1$	Zigzag pseudomonad stack. Phase change at 205K.

(iii) Tetrad or pseudotetrad stacking: this group (Table 17.8) follows on gradually from the structures summarized in Table 17.7.

The structures and physical properties of the {(thio)morpholinium (TCNQ)$_2$} complexes have been reviewed (Visser, de Boer and Vos, 1993) and the abstract bears quotation in full:

Crystal structures of the title compounds are compared. Classes I, II and II' with 2, 4 and 8 TCNQ moieties per translation period, respectively, are distinguished. For class I a subclassification is made according to: the number of inequivalent stacks (1 or 2); cation disorder [dynamic (d) or static (s)]; and chain directions [parallel (p) or crossed (c)]. Crystals of classes II and II' appear to be (1, d, p). Disorder of the cations is a frequent phenomenon. Generally, changes in the ordering of the cations play an important role in the phase transitions. Magnetic susceptibility curves χ(T) turn out to be different for the various (sub)classes. Within each (sub)class the electrical conductivity decreases with increasing calculated band gap. A quantitative interpretation of the electrical transport properties is considered impossible because of the interaction between charge carriers and the dynamic

Table 17.7. Salts with diad or pseudo-diad stackings. All structures determined at room temperature unless stated otherwise. Interplanar spacings are in Å

Salt	Stacking arrangement				Remarks
1. Trimethyleneferro-cenium (TCNQ)$_2$ (Willi *et al.*, 1980)	**A**...(3.14)...**A'**	(3.54)	**A**...**A'**		Formally a cation-radical anion- radical salt.
2. (3,3'-dimethylthia-cyanine)(TCNQ)$_2$ (Shibaeva *et al.*, 1974)	**A**...(3.25)...**B** displR/EB	(3.38) **A'**....**B'** displR/R			Zigzag stack; A, B crystallographically independent; A, A' and B, B' related by 2$_1$ axis.
3. (*N*-ethyl-*o*-phenanthrolinium) (TCNQ)$_2$ (Chasseau *et al.*, 1976); EOPTCO	**A''**...(3.17)...**A** R/EB	(3.40) displR/EB	**A'**		Cations disordered across centre of symmetry.
4. (4,4'-dimethyl-morpholinium) (TCNQ)$_2$ (DMM)(TCNQ)$_2$ (Kamminga and Bodegom, 1981); DMMTCN	**A''**...(3.25)...**A** displR/R	(3.29) displR/EB	**A'**		Cations disordered across mirror plane. Little overlap between π-dimers.
5. (1-*N*-methylcinnalinium) (TCNQ)$_2$ (Daoben *et al.*, 1985); CUMHEX10	**A''**...(3.27)...**A** R/EB	(3.24) R/EB	**A'**		

Notes:

1. *N,N'*-dimethylthiomorpholinium (TCNQ)$_2$ (DMTM)(TCNQ)$_2$; Visser, de Boer and Vos, 1990b; DAW-VOM02) is isomorphous with DMM(TCNQ)$_2$.
2. Diad stacks are also found in *N*-methyl-*N*-ethylthiomorpholinium (TCNQ)$_2$ (DESFOW03) and *N*-methyl-*N*-butylthiomorpholinium (TCNQ)$_2$(VEJPIJ) (Visser, Bouwmeester, de Boer and Vos, 1990a).
3. Diad stacks are also found in *N*-methyl-*N*-propylmorpholinium (TCNQ)$_2$ (DESFUC02) and *N*-methyl-*N*-butylmorpholinium (TCNQ)$_2$ (VEJFEV) (Visser, de Boer and Vos, 1990a).
4. Diad stacks are also found in *N*-butylthiomorpholinium (TCNQ)$_2$ (DESFUC02) and *N*-methyl-*N*-butylmorpholinium (TCNQ)$_2$ (SEMZUF) (Visser, Smaalen, de Boer and Vos, 1990).

lattice as a whole. Large unpredictable variations in crystal structure are observed for chemically small modifications of the cations. Therefore, crystals with a priori desired properties cannot be designed in a systematic way.

The descriptions and conclusions can be carried over without much change to the wider variety of cations considered in the various parts of this chapter.

17.4.8 *Stacked structures with −0.4e average charge on the TCNQ moieties*

Two examples, of formula $A^{2+}(TCNQ)_5^{2-}$, are known; the cations are respectively [1,2-di(*N*-benzyl-4-pyridinium)ethylene] (Ashwell, Eley, Harper *et al.*, 1977; BPETCQ) and

Table 17.8. $M^+(TCNQ)_2^-$ and $M^{2+}(TCNQ)_4^{2-}$ salts with tetrad or pseudotetrad stacking. All structures determined at room temperature unless stated otherwise

Salt	Stacking arrangement	Remarks
1. Methyltriphenyl-phosphonium (TCNQ)₂ (at 326K) (Konno and Saito, 1973; MPPTCQ01)	$\bar{1}$ $\qquad$ $\bar{1}$ **B′** (3.32) **A′** (3.26) **A** (3.32) **B** (3.55) **B′** R/EB $\quad$ R/EB $\quad$ displR/R	First order transformation at 316K; $\Delta H \approx 2\,kJ/mol$.
2. Methyltriphenyl-phosphonium (TCNQ)₂ (at 298K) (Konno and Saito, 1973; MPPTCQ02); McPhail *et al.*, 1971; MPPTCQ)	$\bar{1}$ $\qquad$ $\bar{1}$ **B′** (3.20) **A′** (3.20) **A** (3.20) **B** (3.58) **B′** R/EB $\quad$ R/EB $\quad$ displR/R	$\sigma \approx 2 \times 10^{-5}\,S/cm$. Isomorphous with arsonium analogue (MPATCQ)
3. [N-(n-propyl)phosphonium (TCNQ)₂ (Sundaresan and Wallwork, 1972b); PQTCNQ	$\bar{1}$ $\qquad$ $\bar{1}$ **B′** (3.24) **A′** (3.28) **A** (3.24) **B** (3.43) **B′** R/EB $\quad$ R/EB $\quad$ displR/R	
4. 4-Ethylmorpholinium (TCNQ)₂ (Bodegom and Boer, 1981); HEMTCN	$\bar{1}$ $\qquad$ $\bar{1}$ **B′** (3.24) **A′** (3.42) **A** (3.24) **B** (3.31) **B′** R/EB $\quad$ R/EB $\quad$ displR/R	
5. [N-ethyl-2-methyl-thiazolinium (TCNQ)₂ (Shibaeva and Ponomarev, 1975);	$\bar{1}$ $\qquad$ $\bar{1}$ **B′** (3.30) **A′** (3.30) **A** (3.30) **B** (3.30) **B′** R/EB $\quad$ R/EB $\quad$ displR/EB	
6. Triethylammonium (TCNQ)₂ (Filhol, Zeyen *et al.*, 1980); TCQETA02; ND at 40K.	$\bar{1}$ $\qquad$ $\bar{1}$ **B′** (3.24) **A′** (3.30) **A** (3.24) **B** (3.32) **B′** R/EB $\quad$ R/EB $\quad$ R/EB	See text for detailed discussion (#17.4.9).
7. [1-Methyl-3-ethylbenzimid-azolinium](TCNQ)₂·CH₃CN (Chasseau *et al.*, 1973a); TCQMIM	$\bar{1}$ $\qquad$ $\bar{1}$ **B′** (3.28) **A′** (3.32) **A** (3.28) **B** (3.28) **B′** R/EB $\quad$ R/EB $\quad$ R/EB	§7 TCQMIM and §8 TCQMEJ are isostructural by comparison of reduced cells.
8. [N,2-Dimethyl-N′-ethylbenzimidazolinium] (TCNQ)₂·CH₃CN (Chasseau *et al.*, 1973b); TCQMEJ	$\bar{1}$ $\qquad$ $\bar{1}$ **B′** (3.21) **A′** (3.27) **A** (3.21) **B** (3.38) **B′** R/EB $\quad$ R/EB $\quad$ R/EB	Zigzag pseudomonad stack.
9. [8-Hydroxyethoxy-5-chloro-1-methyl-quinolinium] (TCNQ)₂·CH₃CN (Bunzel *et al.*, 1984); CINKEP	$\bar{1}$ $\qquad$ $\bar{1}$ **B′** (3.18) **A′** (3.18) **A** (3.18) **B** (3.41) **B′** R/EB $\quad$ R/EB $\quad$ displR/R	
10. (C₂₄H₂₃N₂O)⁺·(TCNQ)₂⁻ Malatesta *et al.*, 1995 ZAGCEP	$\bar{1}$ $\qquad$ $\bar{1}$ **A′** (3.20) **B** (3.28) **B″** (3.20) **A″** R/EB $\quad$ R/EB $\quad$ R/EB	

Table 17.5. (*Continued*)

Salt	Stacking Arrangement		Remarks
11. [1,4-Di(*N*-pyridinium methyl)benzene]$^{2+}$ (TCNQ)$_4^{2-}$ (Ashwell, Wallwork *et al.*, 1975); PYMBTQ	$\bar{1}$ **B′** (3.21) **A′** (3.31) R/EB	$\bar{1}$ **A** (3.21) **B** (3.48) **B′** displR/R displR/EB	A is anion, B is neutral, from bond lengths.
12. (Diethyl-morpholinium) (TCNQBr$_2$)$_2$ (Bryce and Howard, 1982); BIPJAL10	$\bar{1}$ **B′** (3.29) **A′** (3.42) R/EB	$\bar{1}$ **A** (3.32) **B** (3.55) **B′** R/EB displR/R	Twisted; isomorphous with dichloro analogue.

Note: 1. Tetrad stacks are also found in *N*-ethyl-*N*-butylmorpholinium (TCNQ)$_2$ (DESFUC02) and *N*-ethyl-*N*-butylthiomorpholinium (TCNQ)$_2$ (VEJFEV) (Visser, Bouwmeester, de Boer and Vos, 1990b).

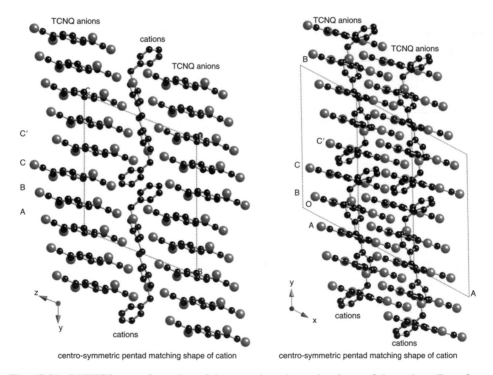

centro-symmetric pentad matching shape of cation   centro-symmetric pentad matching shape of cation

Fig. 17.31. BPETCQ – conformation of the pentad stacks to the shape of the cation. (Data from Ashwell, Eley, Harper *et al.*, 1977.)

[1,2-di(*N*-benzyl-4-pyridinium)-ethane] (Ashwell, Eley, Drew *et al.*, 1978; BPYETC). Comparison of the reduced cells[3] (BPETCQ: 8.043 14.558 16.103 Å, 110.17 95.04 101.75°; BPYETC: 8.109 14.607 16.165 Å, 109.04 95.19 101.92°; both $P\bar{1}$, Z = 2) shows that these two salts are isomorphous and thus can be discussed together. Within the

---

[3] the cells in the original reports were not reduced.

pentads there is R/EB overlap between adjacent moieties, with the same direction of staggering and interplanar spacings of 3.23(2) Å; between pentads there is R/EB overlap but with the direction of staggering reversed and the interplanar spacing now 3.40 Å (Fig. 17.31). The pentads of TCNQ conform to the disposition of the cation, as has been found in other examples (Fig. 17.30).

17.4.9 *Systems studied over a wide range of temperatures*

The structural results summarized and classified in the previous sections generally refer to measurements made at room temperature. Physical measurements have often been made over a wide range of temperatures. However, very few systems have been studied both structurally and in terms of physical properties over a wide range of temperatures. The few results available illustrate the complexities of individual systems and how important it is to base interpretations of properties on detailed structural studies of *well defined* chemical systems. Two obvious but often neglected points are emphasized first and we then discuss four systems that have been studied extensively.

(a) ordered systems are easier to describe and understand than disordered systems. Thus, where possible, we first discuss the most ordered state of the system and then extrapolate towards the disordered situation.

(b) the state of the crystal at any particular temperature is determined by interactions *between* cation and anion partial structures[4] as well as by interactions within these partial structures. These are coupled systems; while the three types of interaction can be described separately, it is not easy to ascribe causes – for example, is ordering of the cations or an electronic transition in the anion partial structure responsible for a particular phase transition?

(i) Triethylammonium (TCNQ)$_2$ {[(TEA)(TCNQ)$_2$]} The crystal structure has been determined at 345, 295, 234, 173, 110 (all by XRD) and 40K (neutron diffraction; TCQETA02) and the results, including earlier work, have been summarized by Filhol and Thomas (1984), whom we follow quite closely. The crystals are triclinic, $P\bar{1}$, with $Z = 2$ and a TEA cation and two TCNQs, A and B, in the asymmetric unit; the mean charge per TCNQ is $-0.5e$. The planar TCNQ moieties are stacked face to face with the repeat sequence BA($\bar{1}$)AB leading to three independent spacing distances (d_{AA}, d_{BB}, d_{AB}) and three independent overlapping modes (see Table 17.8). The packing may thus be described as a set of stacks of tetrads parallel to **c** and spaced in the **a** direction by chains of disordered TEA cations; the overlap within and between the tetrads is R/EB, but there is a jog between one tetrad and the next. There are no abrupt changes in cell dimensions as a function of temperature (Fig. 17.32(a)), but there are discontinuities in the principal coefficients of expansion. Thus the transition that occurs around 200K appears to be one of higher order, without change of space group. The structure at 40K is shown in

---

[4] We avoid the term 'sublattice' because 'lattice' is a precisely defined mathematical concept; we avoid the term 'substructure' because of its use in studies of crystal texture and mosaicity.

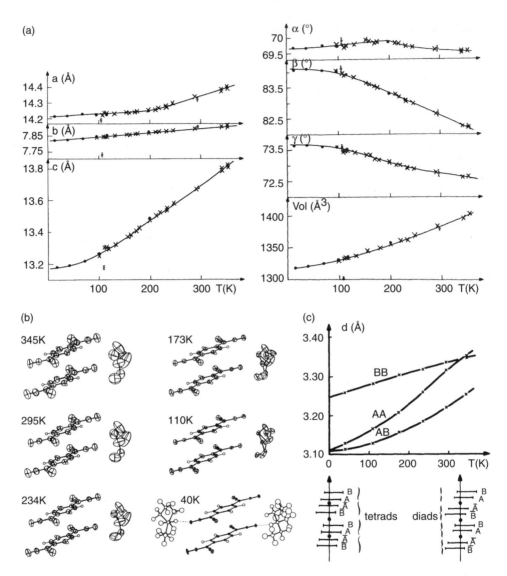

Fig. 17.32. (a) Cell parameters of (TEA)(TCNQ)$_2$ as a function of temperature; results from a number of sources; the lines are guides to the eye. (b) ORTEP diagrams (50% probabilities) of the asymmetric unit at 300K viewed in projection on the *ac* plane. The two fold disorder of the TEA cation has been partially resolved at 173 and 110K and fully at 40K, where the two orientations are shown separately to left and right. (c) Temperature dependence of the interplanar spacing between adjacent TCNQ molecules in the stacks. The transition from tetrads to diads is shown. (Reproduced from Filhol and Thomas (1984).)

Fig. 17.32(b). The details of the overlap mode, but not its general form, change with temperature.

At 0K (by extrapolation) the tetrad stacks have equally spaced TCNQ moieties, with an interplanar spacing of 3.11 Å; these tetrads are separated by a BB′ jog with a spacing of

3.25 Å (the tetrads are represented schematically by the sequence

$$...\text{B}'\;\text{A}'(\bar{1})\;\text{A}\;\text{B}\;(\bar{1})\;\text{B}'...,$$

as shown in Entry #6 in Table 17.8)). The tetrads separate into π-dimers as the temperature rises, with an internal (AB) interplanar spacing of 3.22 Å at 300K, and separated by an intra-diad AA' overlap of 3.29 Å and a between-diad (BB') spacing of 3.31 Å. The mutual ΔM displacement of adjacent A moieties increases from ≈ 0.25 Å at 0K to 0.45 Å at 300K (Fig. 17.32 (c)). This has led Farges (1985a) to describe the change in structure with temperature as a transverse shift of π-dimers within a jogged tetrad stack. It is not clear whether the TCNQ moieties are ordered into anions and neutral molecules at low temperatures, as suggested by Farges (1985b), or disordered (with respect to charge) over the whole temperature range, as preferred by Filhol and Thomas (1984). There is two fold orientational disorder of the TEA cations, which is reported to be static below ≈ 200K and dynamic above this temperature; the corresponding species (TEA' and TEA″) have different structural roles. Two half-populated hydrogen bonds link molecules TCNQ(A) to TEA' and TCNQ(B) to TEA″ respectively. Thus there is coupling between cation and anion partial structures but the cations do not couple adjacent TCNQ stacks.

Extensive measurements (Farges, 1980) have been made of electrical conductivity along three perpendicular directions: axis **1** is parallel to the TCNQ stack axis (i.e. along **c**), **2** is in the TCNQ layers perpendicular to the stacks and **3** is perpendicular to the alternate layers of TEA and TCNQ ions. At room temperature $\sigma_1 = 7.4$ S/cm, $\sigma_1/\sigma_2 = 164$ and $\sigma_1/\sigma_3 = 2850$; thus there is a remarkable anisotropy at 300K which decreases with decreasing temperature. This can be explained in terms of the structural changes described above.

(ii) {*N*-ethyl-*N*-methylmorpholinium (TCNQ)$_2$} [(MEM)(TCNQ)$_2$]. Three phases have been found from studies of structure and physical properties over a wide range of temperatures:

	19 K		340 K	
IA	$\Leftrightarrow$	**IB**	$\Leftrightarrow$	**II**
	2nd order		1st order	
	large change in		σ increases by 10^3	
	magnetic properties			

It has also been suggested (Kobayashi, 1982) that there is a second order transition at 315K; this has not been included as the structural change appears to be very small. The structure of phase IA at 6K has been determined by neutron diffraction; the *c*-axis is doubled with respect to IB (Visser, Oostra *et al.*, 1983) and there are small changes in the stacking arrangement. However, this requires clarification as the doubling of *c* was not confirmed in an XRD study at 10K (Figgis *et al.*, 2001); radiation damage to the crystal used or insufficient measurement sensitivity were invoked as possible reasons for the discrepancy. The crystal structure of phase IB at 113K (Bosch and Bodegom, 1977; MEMTCQ10) is shown in Fig. 17.33 (see Section 17.4.7.1, #6).

There are AB π-dimers (R/EB overlap) in tetrad stacks, with little overlap between successive π-dimers. The values of ΔL, ΔM and ΔN (at 113K) change only by ≈0.05 Å at 323K, slightly decreasing the AB overlap and increasing the AB′ overlap.

	ΔL	ΔM	ΔN	Overlap type
AB	1.97	0.12	3.15	R/EB
BB′	2.57	2.21	3.27	displR/EB

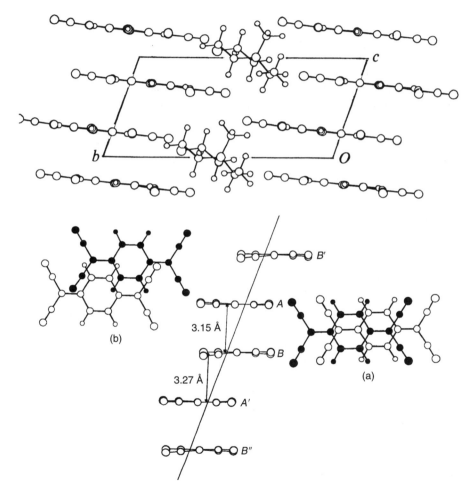

Fig. 17.33. [(MEM)(TCNQ)$_2$] at 113K (Phase IB). The triclinic (not reduced) cell is 7.824 15.426 6.896 Å 113.59 73.27 112.71°, $Z=1$. As the crystals are piezoelectric, the space group is $P1$. (a) Projection of the structure on the plane perpendicular to [100]. The TCNQ stacks are separated by lines of cations. (b) TCNQ molecules seen along the average direction of the longest axis. The AB (3,15 A, on the right) and B′A (3.27 Å, on the left) overlap diagrams are shown. (Adapted from Bosch and Bodegom, 1977.)

The principal change between 19 and 340K is an increasing disorder of the MEM cation. To a first approximation there are two preferred orientations with occupancies x and $(1 - x)$; x decreases from 1.0 at 113K through 0.84 at 294K to 0.63 at 323K.

Above the first order transition at 340K, an approximately monad stack is found with the following overlap shifts (in Å) (Bodegom and Bosch, 1981):

	ΔL	ΔM	ΔN	Overlap type
AB	−1.96	0.10	3.28	R/EB
AB′	−2.02	0.40	3.30	R/EB

The transformation occurs with change of space group from $P1$ (in phases IA and IB) to $P\bar{1}$ in phase II; there are small and abrupt changes in the axial lengths and angles and a small volume contraction ($\Delta a \approx -0.1$ Å, $\Delta b \approx +0.4$ Å, $\Delta c \approx -0.4$ Å, $\Delta\alpha \approx -6°$, $\Delta\beta \approx +5.5°$, $\Delta\gamma \approx +1°$; $\Delta V \approx -0.3\%$); a and b do not change their mutual orientation but the tilt of c to these axes does change. Thus phases IB and II differ in two ways:

(i) the cation in II is disordered over two orientations (i.e. $x = 1/2$) whereas in phase IB x decreases from 1.0 at 113K to ≈ 0.6 just below 340K;

(ii) the change in TCNQ stacking pattern from offset π-dimers in IB (right up to 340K) to essentially monad stacks in II.

Similar changes in TCNQ stacking patterns occur in the first order phase changes in Na(TCNQ) at 345 K (Konno and Saito, 1974, 1975), K(TCNQ) at 395K (Konno, Ishii and Saito, 1977) and {methyltriphenyl-phosphonium (TCNQ)$_2$}at 326K (Konno and Saito, 1973; MPPTCQ02). However, the changes in conductivity which occur at T_c in these salts are much smaller than the thousand-fold increase found in {(MEM)(TCNQ)$_2$}, for which a detailed explanation has still to be given.

(iii) N, N-dimethylmorpholinium (TCNQ)$_2$ {(DMM)(TCNQ)$_2$}

This salt provides a (so-far) rare opportunity to study phase changes in two polymorphs. The first crystals obtained from acetonitrile were monoclinic but only one such batch has so far been prepared (Kamminga and Bodegom, 1981; DMMTCN); nevertheless, they have been studied extensively and show phase transformations on cooling (Middeldorp et al., 1985). Later crystallizations gave only triclinic crystals. The smaller volume per formula unit at ≈ 100K for the triclinic phase (1288 Å^3) than for the monoclinic phase (1299 Å^3) suggests that the former is more stable than the latter; presumably the two phases are monotropically related.

The 300K monoclinic phase changes to another monoclinic phase at ≈ 120K, but with 4 formula units in a $P2_1/c$ cell and there is a further change to another monoclinic phase at ≈ 95K with 8 formula units in a $P2_1/c$ cell.

At 300K there are stacks of π-dimers with R/R overlap and an interplanar spacing of 3.25 Å; adjacent π-dimers overlap only to a small extent. The DMM cation, in chair conformation, is disordered across the mirror plane. The structure has also been determined at 95K and the principal change is an ordering of the DMM cations, which is considered to be the major driving force for the transition. Minor changes in TCNQ positions lead to reduction of the π-dimer spacing to 3.19 Å.

Table 17.9. Cell dimensions (Å, °) for the various phases of the two polymorphs of {(DMM) (TCNQ)$_2$}

Phase/refcode/ T(K)	a/α	b/β	c/γ	Z	Volume of formula unit	Space group
Monoclinic phases						
DMMTCN; 300 RT phase	7.792	26.814 58.44	7.594	2	1352	$P2_1/m$
DMMTCN02; 120 beta phase	7.730	26.500 120.90	14.780	4	1299	$P2_1/c$
DMMTCN01; 95 alpha phase	13.280	26.497 92.35	14.782)	8	1299	$P2_1/c$
Triclinic phases						
DMMTCN04; 294	16.487 103.24	12.909 99.60	6.783 102.95	2	1333	$P\bar{1}$
DMMTCN03; 95	16.452 103.18	12.838 98.79	6.783 103.52	2	1288	superspace group $P^P\bar{1}$.

Unit cells of the three phases of monoclinic {DMM (TCNQ)$_2$] viewed down [010]

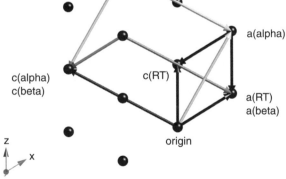

Fig. 17.34. Relations between the unit cells of the three phases of monoclinic {DMM·(TCNQ)$_2$} viewed down their common [010]. Small dimensional differences have been ignored. The following vectorial relationships hold between the axes of the three monoclinic cells (axes without subscripts refer to 300K cell): (Adapted from data given for DMMTCN and companion structures.)

$$\beta \text{ and RT:} \quad \mathbf{a}\beta = \mathbf{a}; \qquad \mathbf{b}\beta = \mathbf{b}; \quad \mathbf{c}\beta = 2[\mathbf{c} - \mathbf{a}].$$
$$\alpha \text{ and RT:} \quad \mathbf{a}\alpha = \mathbf{a} + \mathbf{c}; \qquad \mathbf{b}\alpha = \mathbf{b}; \quad \mathbf{c}\alpha = 2[\mathbf{c} - \mathbf{a}].$$
$$\alpha \text{ and } \beta: \quad \mathbf{a}\alpha = 2\mathbf{a}\beta - \mathbf{c}\beta; \quad \mathbf{b}\alpha = \mathbf{b}\beta; \quad \mathbf{c}\alpha = \mathbf{c}\beta.$$

Crystallization from acetonitrile also gave triclinic crystals (cell dimensions in Table 17.9). In the 300K structure the TCNQ moieties are arranged in diad stacks along [001] ($\approx$6.6 Å) with R/EB overlap while the randomly (and dynamically) disordered DMM cations are located between the anion stacks (Visser *et al.*, 1994; DMMTCN04). The marked differences in the dimensions of the two crystallograpically independent TCNQ moieties have been ascribed to a difference of 0.7e between the charges on the moieties, i.e. one is neutral and the other an anion. At 207K satellite reflections appear in the diffraction patterns, showing incipient formation of an incommensurately modulated phase whose structure has been determined at 99K. The driving force for modulation is the need to accommodate the ordered DMM cations within the confines of the TCNQ 'partial structure' (see footnote on p. 1203). The overall structure is considered to be of an intergrowth type whose elastic energy is minimized by incommensurate modulation (Steurer *et al.* (1987)).

(iv) Trimethylammonium TCNQ Iodine (TMA-TCNQ-I). Perhaps the most complicated TCNQ structure yet studied is that of the ternary salt $[(CH_3)_3NH^+]_3(3TCNQ)^{2-}(I_3^-)$, which is a representative of a wider family of isostructural ternary anion radical salts of general formula $[R(CH_3)_2NH^+]_3(3TCNQ)^{2-}(I_3^-)$ (Dupuis *et al.*, 1978). A comprehensive study (Gallois *et al.*, 1985) is remarkable for its integration of x-ray and neutron diffraction techniques with the first use of high voltage (2000 kV, $\lambda = 0.0054$ Å) electron microdiffraction in the study of TCNQ salts. There are three phase transitions, at $\approx$150, 95 and 65K, associated with different modulations of the lattice. The structural relationships are too complex for detailed discussion here and the reader is referred to Gallois *et al.* (1985), who also give references to earlier work (MATCQI to MATCQI09).

17.4.10   *Conclusions drawn from a survey of structural results for TCNQ anion-radical salts*

Crystal structures have been reported for more than one hundred TCNQ anion radical salts, mostly at room temperature. The fundamental feature common to all these structures is plane-to-plane stacking of TCNQ moieties in segregated stacks of limited or unlimited length. We focus our attention in this section on the π-dimers, whose arrangement in the crystals ranges from isolated to stacked, and distinguish between π-dimers carrying single and double charges. The overlap of the two TCNQ moieties can, to a good approximation, be classified as either R/R (eclipsed) or R/EB (slipped), with deviations from the ideal overlaps being ascribed to packing effects rather than to electronic interactions within the dimers. The available results are summarized in Table 17.10. Stacks with mean charges of 4/7 and 2/3 per TCNQ moiety have also been found but there are not enough examples for conclusions to be drawn.

R/R overlap seems to be about twice as frequent as R/EB when the π-dimer is doubly charged, but only R/EB overlap is found when the π-dimer is singly charged. No trends are discernible in the interplanar spacings that lie in the range 3.1–3.3 Å, apart from a few outliers. The interplanar distance is thus about 0.5–0.3 Å *less* than that found in neutral TCNQ. From this we infer that there is an attractive interaction operating between anions in stacks additional to that operating between neutral molecules. Furthermore, this additional attraction does not appear to depend (at least to a first approximation) on the

Table 17.10. TCNQ π-dimers classified according to overlap type and mean charge. References have been given in previous tables. The interplanar spacings (D) are in Å
Group I. Average charge of $-e$ on each TCNQ moiety (doubly charged π-dimer).

R/EB Overlap	D	R/R Overlap	D
1. (1,2,3-trimethyl-benzimidazolinium) $(TCNQ)_2$	3.12	1. $(TMA)_2(TCNQ)_3$	3.26
2. $[Nb_3(m\text{-}Cl)_6(C_6Me_6)_3]$ $(TCNQ)_2$	3.10	1a. (quinuclidinium)$_2$ $(TCNQ)_3$	3.26
3. (1,1'-ethylene-2,2'-bipyridylium) $(TCNQ)_2$	3.22 3.26	2. $[Fe(C_5Me_5)_2]_2$ $(TCNQ)_2$	3.15
4. N,N'-dimethyl-benzimidazolinium) (TCNQ)	3.07	3. Na(TCNQ) at 300K	3.22 3.20
		4. K(TCNQ) at 300K	3.24
		5. Rb(TCNQ) at 113K	3.16
		6. $[Pt(NH_3)_4]^{2+}(TCNQ)^{2-}$	3.14
		7. {Morpholinium (TCNQ)}	3.28
		8. $[(MeCp)_5V_5S_6]$ $(TCNQ)_2$	3.37

Group II. Average charge on each TCNQ moiety $-0.5e$ (singly charged dimers). All these examples have R/EB overlap and have been listed as follows:
Section 17.4.7, Group I (i) 11 examples, interplanar spacings in the range 3.15–3.24 Å.
Table 17.5, Type I: 4 examples, interplanar spacings within the π-dimer in the range 3.12–3.20 Å.
Table 17.7: 3 examples, interplanar spacings within the π-dimer in the range 3.14–3.25 Å.

mean charge of the TCNQ moiety. The Coulomb repulsion energy between two planar singly-charged anions 3.3 Å apart would be about 400 kJ/mol. Theoretical approaches to the preferred overlap modes of various donors and acceptors are discussed later (Section 17.7).

TCNQ salts with different donors show a wide variety of physical properties despite the essential structural resemblance that results from the stacked arrangement of the TCNQ moieties. A broad classification into four types (Fig. 17.35) has been proposed (Torrance, 1978) on the basis of stoichiometry and physical properties. Classes I and II are 'simple' TCNQ salts with cation : TCNQ ratio 1 : 1 but differing in conductivity by a factor of $\approx 10^6$. Class II salts are also discussed in more detail below under the heading of 'Cation-radical Anion-radical salts.' Classes III and IV are 'complex' TCNQ salts with cation:TCNQ ratio 1 : r (where r is the ratio of two small integers and $r > 1$)), the two classes differing in conductivity by a factor of 10^2–10^6, the value depending on the temperature at which the comparison is made.

Taking the overall segregated stack structure as given, Torrance (1978) considered that there were two factors primarily responsible for determining details of structure and two others for determining physical properties. The first of the structure-determining factors is the cation–anion interaction, which we consider to be of predominant importance in the salts containing closed-shell cations. An example of the way in which the stacking of

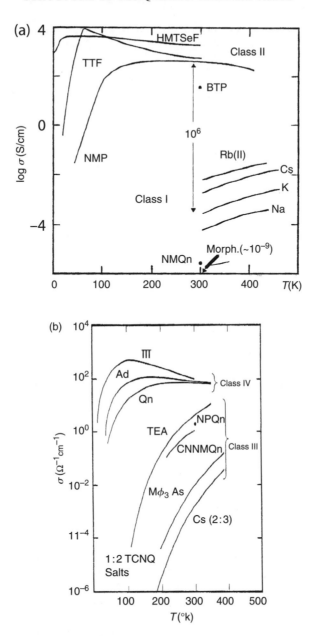

Fig. 17.35. (a) The temperature dependence of the conductivity for a number of simple TCNQ salts (1 : 1), showing how they fall into two distinct classes I and II; (b) the temperature dependence of the conductivity for a number of complex TCNQ salts (1 : r, where r is the ratio of two small numbers and $r > 1$), showing how they fall into two distinct classes III and IV. Some acronyms are: BTP $\Delta 4,4'$-bithiopyranium; NMP N-methylphenazinium; NMQn N-methyl-quinolinium; Ad acridinium; TEA triethylammonium; CNNMQ$_n$ 4-cyano-N-methylquinolinium, TTT tetrathiotetracene. (Reproduced from Torrance, 1978.)

π-dimers depends on the nature of the cation is shown rather vividly in Figs. 17.30 and 17.31 where long cations are threaded through the channels between TCNQ stacks, that conform themselves to the overall cation shape. Cations can play an important role in the overall structure in two other ways – firstly, there is often hydrogen bonding between cation and nitrogens of the TCNQ anion, and, secondly, the degree of disorder of the cations affects physical properties. Torrance's second structure-determining factor is the bandwidth, $4t$, associated with overlap of π-orbitals of adjacent TCNQ moieties in the stacks. A variety of types of overlap is found, and hence $4t$ is structure sensitive. Torrance suggests that a reasonable average value for $4t$ is 0.5 ev (≈ 50 kJ/mol).

The other two factors relate more to physical properties and particularly to the conductivity. The first of these is the effective Coulomb repulsive energy $U = U_0 - V_1$, where U_0 is the Coulomb repulsion energy between two electrons on the same molecule (here the TCNQ anion) and V_1 is the Coulomb repulsion energy between two electrons on neighbouring molecules. Although U_0 and V_1 may both depend on the nature of molecule and environment, their difference appears to be reasonably constant and Torrance has suggested that $U \approx 1$ ev (≈ 100 kJ/mol). The key parameter affecting conductivity is Z, the degree of charge transfer or the average number of unpaired electrons per TCNQ moiety (see Section 14.3.6). The conductivity is very low in those simple salts where $Z = 1$ because the energy required to activate conduction is the effective Coulomb repulsion energy U; this is the situation in the alkali metal TCNQ salts. However, as we shall see below, $Z = 0.59$ in {[TTF][TCNQ]}, and an electron can travel from anion radical to adjacent neutral molecule in the same stack (the pair then exchanging roles) without an activation energy, thus accounting, in principle, for the high conductivity. The details are, of course, more complicated (see Section 17.9).

17.5   Other anion-radical salts

17.5.1   *Alkali-metal chloranil salts*

Physical properties of these salts were first extensively investigated around 1970 (Ishii *et al.*, 1976). There appear to be two groups (chloranil is abbreviated as CA). $Li^+CA^{-\bullet}$ and $Na^+CA^{-\bullet}$ show a strong intermolecular CT band in the near IR, whereas $K^+CA^{-\bullet}$ and $Rb^+CA^{-\bullet}$ each show a weak CT band and a band corresponding to monomeric radical anion absorption. Only the crystallography of $K^+CA^{-\bullet}$ has been studied (Konno, Kobayashi, Marumo and Saito, 1973). There are several polymorphs, two of which have rather similar cell dimensions and are unusual in that they crystallize in the Sohncke space groups $P2_12_12_1$ (α) and $P2_12_12$ (β) although the components are centrosymmetric. The structure of the α-polymorph has been studied (but only roughly because of poor quality crystals, $R = 0.13$ for 146 reflections); there are stacks of chloranils with interspersed K^+ ions. There is little overlap between successive anions in a stack, thus accounting in broad terms for the room temperature electronic spectra. Dielectric constant and spectroscopic studies over a range of temperature show a phase transition at ≈ 210K (presumably first order because of the appreciable hysteresis). The striking increase in the intensity of the CT band on cooling has been interpreted in terms of π-dimerization (pairing) of the anions in the low temperature phase, and a similar explanation (augmentation of electronic polarization consequent on pairing) has been proposed for the increase in dielectric constant on cooling; however, the structure of the low temperature phase has not been reported. There are resemblances to the behaviour of TMPD·ClO$_4$.

17.5.2   M(dmit)$_2$ and M(mnt)$_2$ as anion radicals in various guises

The increasing number of structures available for the M(dmit)$_2$ and M(mnt)$_2$ (M = Co, Ni, Pd, Pt, Au; see Section 17.2.3 for formulae of (dmit)) and (mnt)[5]) anion radical salts (and also one example for a 1,4,5,8-naphthalenetetracarboxylate salt) suggest that their stacking arrangements are analogous to those found in TCNQ anion radical salts. This encourages one to believe that TCNQ is only the first of a number of acceptors that behave similarly. However, we emphasize that each acceptor is likely to show its own special features in addition to fitting into a general pattern. There are also some examples of cation radical anion radical salts containing M(mnt)$_2$, and it is convenient, breaking our usual pattern, to consider all these examples together even though some clearly belong in the next section.

The anion radical salts of M(dmit)$_2$ There are diad stacks along the [010] axis of triclinic (Et$_4$N)$_{0.5}$[Ni(dmit)$_2$][6] ($P\bar{1}$, $Z = 2$), with interplanar distances of 3.44 and 3.76 Å; there are also many interstack S...S distances somewhat less than the van der Waals distance of $\approx$3.6 Å (Kato, Mori *et al.*, 1984; CEVFIS). The cations are disordered and located between the stacks. The anions bear charges of $-0.5e$ and the material is a semiconductor with a maximum conductivity of $\approx 4 \times 10^{-2}$ S/cm along [010]. In (AsPh$_4$)[Ni(dmit)$_2$]$_4$ (Valade, Legros, Cassoux and Kubel, 1986; DIBNAD20) there are tetrad stacks with displaced-eclipsed overlap within the tetrads and metal-over-ring overlap between them, and also lateral S...S approaches (closest is 3.55 Å) that suggest weak interactions. The conductivity along the stack axis is $\approx$15 S/cm at room temperature. The crystals show appreciable diffuse scattering so fluctuations from this average structure are large.

A more complicated arrangement of anions in stacks is found in (n-Bu$_4$N)$_2$ [Ni(dmit)$_2$]$_7$.2CH$_3$CN (Valade, Legros, Bousseau *et al.*, 1985; DEWKOF) (crystal structure at 118K; triclinic, $a = 13.425(2)$, $b = 22.791(3)$, $c = 24.183(4)$ Å, $\alpha = 108.49(1)$, $\beta = 103.02(1)$, $\gamma = 89.82(1)°$, V $= 6818$ Å^3, $Z = 2$, space group $P\bar{1}$). The stacks contain alternating tetrads and triads of Ni(dmit)$_2$ moieties (Fig. 17.36) and are located in thick layers parallel to (001), which are separated by sheets of cations and solvent molecules. There are also some interstack S...S distances of less than 3.6 Å. Relations between physical properties and the rather complicated crystal structure are discussed by Valade *et al.* (1985).

The structures of [NBu$_4$]$_2$[Ni(dmit)$_2$] and of [Nbu$_4$][Ni(dmit)$_2$] have been studied (Lindqvist, Sjölin *et al.*, 1979; DBNTNI10); Lindqvist, Andersen *et al.*, 1982; ITTNBU01)

---

[5] H$_2$(dmit) is 4,5-dimercapto-1,3-dithiole-2-thione and (mnt) is maleonitriledithiolate.
[6] tetraethylammonium bis(bis(trithionedithiolato))Ni(II).

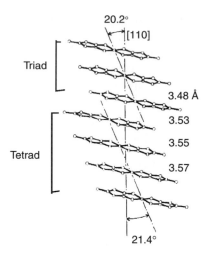

Fig. 17.36. Part of the crystal structure of $(Bu_4N)_2$ [Ni(dmit)$_2$]$_7$·2CH$_3$CN showing one of the two crystallographically independent stacks of planar Ni(dmit)$_2$ moieties along [110]. Both triads and tetrads are centrosymmetric and have their stack axes equally inclined to [110]; there is little variation in the interplanar spacings. (Reproduced from Valade et al., 1985.)

but do not appear to have any features noteworthy in the present context, presumably because the [Ni(dmit)$_2$] moieties do not have fractional charges. The structure of (Et$_4$N)[Ni(dmit)$_2$] is highly one-dimensional (Kramer et al., 1987; FEMNIU10).

The cation radical anion radical salts of perylene with M(mnt)$_2$ The cation radicals are perylenium or modified perylenium. Thus perylene is the first aromatic hydrocarbon to function as a cation radical in cation radical anion radical salts. Crystallographic data are summarized in Table 17.11.

The {(per)$_2$[M(mnt)$_2$]} salts (per = perylene, M = Pd, Pt, Au) are closely isomorphous and have segregated monad stack structures (Fig. 17.37). Unfortunately, coordinates are not available for any of these and so detailed comparisons are not possible. PAJWUS was solved as an averaged structure, so again some detail is lacking. The perylenes in a stack are separated by 3.36 Å and have ring-bond overlap similar to that found in other perylene stacks of this group. Mössbauer spectroscopy (295, 80 and 15K) of {(per)$_2$-[Fe(mnt)$_2$]} gave spectra very similar to those obtained for alkylammonium salts of [Fe(mnt)$_2$]} and it was inferred that there were actually ([Fe(mnt)$_2$]$_2$) dimers in all these materials.

{(per)[Co(mnt)$_2$]} has been prepared but the structure is not known; the dichloromethane solvate (Gama, Henriques, Bonfait, Almeida et al., 1992; SUCCOI) has segregated stacks with perylenes in general positions; however, the two independent interplanar distance are essentially the same at 3.27 Å; these are pseudo-monad stacks. The anions are arranged in polymeric chains along [100].

Table 17.11. The cation radical anion radical salts of perylene with $[M(mnt)_2]$. Measurements at 295K unless stated otherwise

Metal	a/α	b/β	c/γ	V	Z	Space group
Group of three						
Pd	16.469	4.1891	26.640	3846	2	$P2_1/n$
		95.07				
Pt (Alcacer et al., 1980)	16.612	4.1891	26.583	1846	2	$P2_1/n$
PRLNPT		94.54				
Au	16.602	4.194	26.546	1841	2	$P2_1/n$
		94.58				
Group of four						
Ni	17.44	4.176	25.18	1833	2	$P2_1/n$
		91.57				
Cu (Alcacer, 1985;	17.6	4.17	25.5	1856	2	$P2_1/n$
Gama, Almeida et al., 1991)						
		91.4				
Co (PAJXAZ)	17.75	8.22	25.88	3778	4	$P2_1/n$
		92.0				
Fe (Gama, Henriques,	50.571	8.212	17.726	7354	8	$C2/c$
Bonfait, Pereira et al., 1992;		92.43				(averaged
PAJWUS)						as $P2_1/n$)
$\{(pet)_3[Ni(mnt)_2]_2\}$ (3)	10.2972	11.5037	13.3297	1543.7	1	$P1$
SOHMAD	78.320	87.096	87.785			
$\{(per)_4[Co(mnt)_2]_3\}$ (4)	12.093	20.912	16.633	4194.5	2	$P2_1/n$
HAKJEI		94.290				
$\{(per)[Co(mnt)_2]\}\cdot$	6.551	11.732	16.481	1251	2	$P\bar{1}$
CH_2Cl_2 (5)	92.08	95.30	94.62			
SUCCOI						

Notes:

1. The Fe compound (PAJWUS) has been indexed in the standard space group $C2/c$; interchange of a and c axes (space group $A2/a$) makes the comparison of cell dimensions for the first four compounds more obvious.
2. the 'group of three' and the 'group of four' are isostructural, not isomorphous; however, taken together they constitute the α phase. There is also a poorly-defined semi-conducting β phase
3. pet = perilo[1,12-b,c,d]thiophene
4. per = perylene
5. there is a modulated structure with wave vector $0.22a^*$, $-0.13b^*$, $-0.36c^*$. This has not been taken into account in the structure solution which thus refers to an averaged structure.

Some physical properties of these compounds have been measured (Henriques, Alcåcer et al., 1984; Henriques, Almeida et al., 1987).

Perylene also forms a 4:3 complex (Gama, Henriques, Almeida et al., 1993; HAKJEI) with $[Co(mnt)_2]$ that has perylene trimers, perylene of crystallization (almost perpendicular to the perylene trimers) and trimeric $[Co(mnt)_2]_3$ moieties in a complicated

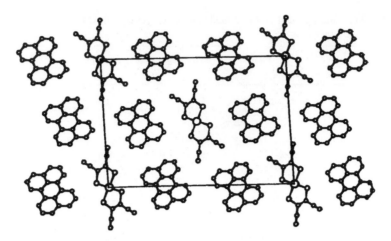

Fig. 17.37. Projection down [010] for the {(per)₂[M(mnt)₂] (per = perylene, M = Pd, Pt, Au) structures. This diagram provides an approximate representation of the M = Ni, Cu, Co and Fe structures but does not take into account possible ordering along the stack axis or in lateral directions. (Reproduced from Gama, Almeida *et al.*, 1991.)

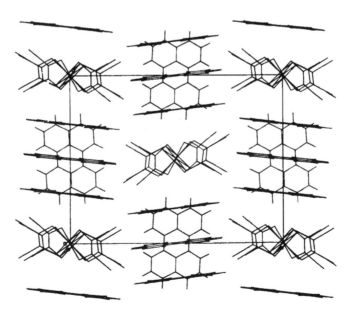

Fig. 17.38. Projection down [100] for the {(per)₄[Co(mnt)₂]₃} structure. The perylene trimer is seen edge-on and the perylene of crystallization plane-on. The centrosymmetric {[Co(mnt)₂]₃} trimer is at the corners of the unit cell. (Reproduced from Gama *et al.*, 1993.)

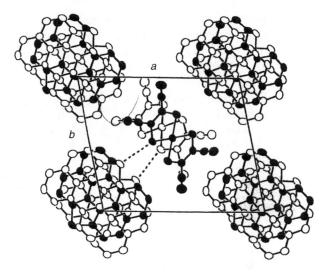

Fig. 17.39. Projection of the SOHMAD structure down [100]. The dashed lines show short S . . . S distances between the two components. (Reproduced from Morgado *et al.*, 1997.)

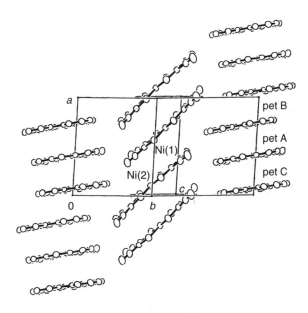

Fig. 17.40. SOHMAD: Side view of the [Ni(mnt)$_2$]$_2$ stacks locked between the pet stacks. The pet trimers and pseudo-monad [Ni(mnt)$_2$]$_2$ stacks are clearly visible. (Reproduced from Morgado *et al.*, 1997.)

arrangement (Fig. 17.38) that has structural features encountered previously. There is one perylene in a general position, one (half) at a center of symmetry, the three molecules constituting the trimer, while the (half) perylene of crystallization is at an independent center of symmetry. Within experimental uncertainty (≈ 0.01 Å), all perylene moieties are planar and have identical bond lengths.

Perilo[1,12-b,c,d]thiophene (pet) is the perylene derivative in which two hydrogens of one bay region are replaced by a sulfur atom. This forms a triclinic non-centro-symmetric 3:2 complex with [Ni(mnt)$_2$] in which there are jogged segregated stacks of (pet) trimers and [Ni(mnt)$_2$] dimers (Fig. 17.39; Morgado *et al.*, 1997; SOHMAD).

The interplanar spacings within a pet trimer are 3.37 (petA–petB) and 3.23 Å (petA–petC) and between them (petC*–petB) 3.47 Å; the overlap modes also differ (Fig. 17.40).

17.6   Structures of cation-radical anion-radical salts

17.6.1   *General survey*

The criterion for inclusion in this group is that both the cation radicals and the anion radicals should form (separate) segregated stacks. Salts of this kind, typified by {[TTF][TCNQ], appear to be fairly rare, although many different chemical types have been observed. There is some, but not much, crystallographic similarity among the various salts (Table 17.12). Our classification, as elsewhere, is based on structural rather than chemical resemblances and we distinguish between salts with different cation: anion ratios and between monad and diad stack periodicities. We start with 1:1 cation: anion ratios, where both monad and diad stacks are found.

17.6.2   *Cation : anion ratio 1 : 1; monad stacks*

This group of monad-stack structures shows considerable overall uniformity in the arrangement of the moieties within the stacks, with the radical cations and radical anions both being located face-to-face in separate segregated homologous stacks. Overlap of unsubstituted TTF and TCNQ moieties is invariably ring-double bond (R/DB) and ring-external bond (R/EB) respectively; however, substitution can lead to changes in the type of overlap. The values of D_C and D_A differ and hence (because both stacks have the same periodicity) the angles of inclination of the moiety planes to the stack axis also differ. The resemblances and differences among the various structures arise from the relationships between the stacks. We first consider the group of triclinic structures. These structures all crystallize in space group $P\bar{1}$ with one formula unit in the unit cell; thus both moieties are located at crystallographic centres of symmetry but there are otherwise no symmetry restrictions on their arrangement. In our descriptions we give priority to the concept of approximately close-packed stacks but modification of this approach is necessary when there are strong interstack interactions. The stacks have elliptical cross-sections because of the elongated shapes of the

Table 17.12. Crystal data (Å, deg., Å³) for cation-radical anion-radical salts with monad-stack structures and 1 : 1 compositions. The triclinic cells have been reduced so that interaxial angles are all either acute (Type I) or obtuse (Type II). Stack axis is in bold. Standard uncertainties of cell parameters are given in the original publications. D_C and D_A are interplanar spacings (Å) in cation-radical and anion-radical stacks respectively

Salt	a/α	b/β	c/γ	V	Remarks
(i) Triclinic (all examples have space group P1̄; Z = 1)					
1. {[TMTSeF][TCNQ]} (Bechgaard et al., 1977); SEOTCR	**3.88** 79.18	7.645 87.78	18.85 85.37	547	Black conducting polymorph. Metal to insulator transition at 59K. D_C = 3.60; D_A = 3.27.(see Fig. 17.36.)
2. {[TTF][TNAP]} (Berger et al., 1975); TTFNAP	**3.76** 85.54	8.100 83.98	16.568 88.00	500	Two-dimensional analysis. Poor quality crystals.
3. {[DEDMTSeF][TCNQ]}* (Andersen et al., 1982);	**3.93** 91.57	7.598 88.22	19.76 94.45	588	Crystals twinned (R = 0.15); D_C = 3.59, D_A = 3.35.
4. {[BMDT-TTF][TCNQ]} (Kobayashi et al., 1986); FERCAG	**3.872** 81.45	7.362 89.87	18.643 87.60	525	Metal to insulator transition at 45K.
5. {[DBTTF][2,5-dichloro-TCNQ]} (Soling, Rindorf and Thorup, 1981); BABCIQ	**3.76** 91.13	7.867 93.28	20.09 91.46	593	At 295 and 115K; no phase change. D_C = 3.51; D_A = 3.41.
6. {[TMTSeF] [2,5-dimethyl-TCNQ]} (Andersen et al., 1979); SEFTCQ	**3.94** 97.3	8.085 98.12	18.96 91.37	592	Metal to insulator transition at 42K (Pouget et al., 1980; Pouget, 1981). D_C = 3.64; D_A = 3.31.
7. {[TTF][2,5-diethyl-TCNQ]} (Schultz, Stucky, Blessing and Coppens, 1976); TFETCQ	**3.859** 83.75	10.094 85.91	13.654 83.00	524	Metal to insulator transition at 111K. D_C = 3.60; D_A = 3.26.

* not reduced

Table 17.12. (*Continued*)

Salt	a/α	b/β	c/γ	V	Remarks
8. {[β-MTDTPY]@ [TCNQ]} (Nakasuji, Sasaki et al., 1987); FUDTON	**4.370** 92.80	8.286 90.12	16.957 103.48	596	
9. {[1,6-dithiapyrene] [TCNQ]} (Thorup et al., 1985); DAKTIS	**3.83** 89.77	8.106 83.96	15.61 84.77	480	$D_C = 3.39$; $D_A = 3.27$.
10. {[1,6-pyrenediamine] [TCNQ]} at 118 K (Inabe, Okinawa et al., 1993). WEMHEB	**3.851** 82.76	7.923 88.60	16.363 85.93	494	$D_C = 3.23$; $D_A = 3.19$. #9 and 10 are isostructural
11. {[β-MTDTPY]@ [chloranil]} (Nakasuji, Sasaki et al., 1987); FUDTUT	**3.797** 85.24	9.994 89.32	14.585 80.60	536	
12. {[DBTTF][Ni(dmit)$_2$]} (Kato, Kobayashi et al., 1985); CUWCIG	**3.830** 73.74	12.256 89.84	14.29 89.97	644	Crystals twinned, data collected as monoclinic.

Salt	a	b/β	c	V	S. G.	Z	Remarks
(ii) Monoclinic 1. {[TTF][TCNQ]} (Kistenmacher et al., 1974; TTFTCQ01) (for neutron study of fully deuterated salt at 10^{-1} MPa and 460 MPa see Filhol, Bravic et al., 1981; TTFTCQ05 and 06).	12.30	**3.82** 104.4	18.47	840	$P2_1/c$	2	$D_C = 3.47$; $D_A = 3.17$. Three transitions at 38, 48 and 54K.

Compound (reference; refcode)	a	b	β	c	V	Sp. gr.	Z	Notes
2. {[DSeTDTF#][TCNQ]} (Etemad et al., 1975)	12.41	**3.85**	104.3	18.49	855	$P2_1/c$	2	$T_c = 64$ K.
3. {[TSeF][TCNQ]} (Etemad et al., 1975; Corfield and LaPlaca, 1996) ZUGRUO	12.505	**3.872**	104.13	18.504	869	$P2_1/c$	2	$T_c = 40$ K; crystals 1, 2 and 3 are isomorphous.
4. {[TMTF][TCNQ]} (Phillips et al., 1977); THOTCQ	18.82	**3.85**	103.7	15.08	1062	$P2/c$	2	$D_C = 3.53$; $D_A = 3.27$.
5. {[TTTF][TCNQ]} (Chasseau, Gaultier, Hauw, Fabre et al., 1978); MYFLTC	40.88	**3.82**	100.9	12.28	1883	$C2$	4	$D_C = 3.58$; $D_A = 3.20$; d(S...N) = 3.11, 3.21. Sohncke space group.
6. {[TMTF][bromanil]} (Mayerle and Torrance, 1981a); TMFBRQ10	31.32	**3.95**	121.2	19.30	2042	$C2/c$	4	$D_C = 3.60$; $D_A = 3.37$.
7. {[TTMTTF][HCBD]} (Katayama, Honda et al., 1985); DATNIV10	30.16	**4.04**	117.5	23.41	2535	$C2/c$	4	$D_C = 3.63$; TTMTTF planar.
8. {[TMTTF][HCBD]} (Katayama, Honda et al., 1985); DATNER10	12.80	21.62	108.1	**4.02**	1058	$C2/m$	2	Components ionic. $D_C = 3.61$; $D_A = 3.43$.
9. {[TMTSeF][Ni(dmit)$_2$]} (Kobayashi et al., 1985); DALPUB	15.23	**34.6**	105.7	25.46	$P2_1/n$	18		Transverse sinusoidal modulation of **b**, fundamental period 3.84 Å.
10. {[HMTSeF][TCNQ]} (Phillips et al., 1976); SEOTCQ	22.00	12.57	90.29	**3.89**	1076	$C2/m$	2	Disorder along [100]. $D_C = 3.6(1)$; $D_A = 3.2(1)$.

Table 17.12. (*Continued*)

Salt	a	b/β	c	V	S. G.	Z	Remarks
(iii) Orthorhombic							
1. {[HMTTF][TCNQ]} (Silverman and LaPlaca, 1978; Chasseau *et al.*,1978); HMTFCQ	12.47	**3.91**	21.60	1053	*Pmna*	2	$D_C = 3.57$; $D_A = 3.25$.
2. {[HMTSeF][TCNQF₄]} (Torrance *et al.*, 1980)	12.61	**4.07**	21.40	1098	*Pmna*	2	$D_C = 3.62$; $D_A = 3.27$.

@ 2,7-bis(methylithio)-1,6-dithiapyrene. There is an isomeric mixed stack (chloranil) compound (Table 15.3)
# *cis/trans*-diselenadithiafulvalene.

moieties. In the triclinic structures an anion radical stack (for example) will have four cation radical stacks and two anion radical stacks as nearest neighbours, and conversely for the cation radical stacks. The stacks in the group of four are identical as they are related by translations, and this also holds for the group of two. The tilts of the moieties to the stack axes are independent in the two types of stack; for most of the present structures the tilts have opposite signs. In {[TMTSeF][2,5-dimethyl-TCNQ]} the long axes of the two moieties are roughly perpendicular (Fig. 17.41) while in {[DBTTF][TCNQCl$_2$]} (Soling et al., 1981) they are approximately parallel. In {[DBTTF][TCNQCl$_2$]} the interstack approach distances are larger than the sums of the van der Waals distances and both structures are described as 'one-dimensional'. In some structures there are interstack distances (usually chalcogen...N) shorter than the sum of the van der Waals distances (S...N ≈ 3.35 Å; Se...N ≈ 3.45 Å) leading to 'two-dimensional character'. In {[DAP][TCNQ]} (DAP is 1,6-pyrenediamine; WEMHEB) there is hydrogen bonding between amino groups of the cation-radical and cyano groups of the anion-radical. In isostructural {[1,6-dithiapyrene][TCNQ]} (DAKTIS) a S...NC approach presumably fills the same function. Both TCNQ bond lengths and IR spectroscopy indicate that WEMHEB is a simple salt rather than a cation-radical anion-radical salt.

Similar descriptions can be used for the monoclinic and orthorhombic structures, care being necessary to take the symmetries of the various space groups into account. The monoclinic structures can be divided into two subgroups, the larger having the stack axis along monoclinic b, and the smaller having c as stack axis. The only two examples in the orthorhombic group are isomorphous. The structures of {[TTF][TCNQ]} (Fig. 17.48) and some related salts will be described in more detail later.

Resemblances in structure do not mean that there are necessarily resemblances in electrical conductivity and related physical properties, and in their dependence on temperature and pressure. A partial classification, based on conductivities, of cation radical-anion radical salts (where the cation radicals are TChF (Ch = chalcogen) and substituted TChF's and the anion radicals are TCNQ and substituted TCNQ's) has been given by Bechgaard and Andersen (1980); we use their classification and note how the salts chosen for illustration (ambient pressure behaviour only) fit into Table 17.9 (crystal type and number given in brackets):

Class I. Semiconductors. The presence of strong acceptors causes full charge transfer, leading to Mott insulators with $\sigma_{RT} \ll 1$ S/cm. An example is {[HMTTF][TCNQF$_4$]} (ortho-#2)·{[DAP][TCNQ]} (tri-#10) fits into this category from a structural viewpoint but the transport properties are discordant.

Class II. Intermediate semiconductors. Metal–insulator transition occurs at relatively high temperatures. Examples are {[TTF][DETCNQ]} (tri-#8), {[TTF][TNAP]} (tri-#3), {[DBTTF][TCNQCl$_2$]} (tri-#5).

Class III. Metals/insulators, with transitions occurring at relatively low temperatures; $\sigma_{RT} \approx 200$–1500 S/cm, $T_C < 100$K. Examples are {[TTF][TCNQ]} (mono-#1), {[TMTSeF][DMTCNQ]} (tri-#7), {[HMTTF][TCNQ]} (ortho-#1), {[TMTSeF][TCNQ]} (mono-#2).

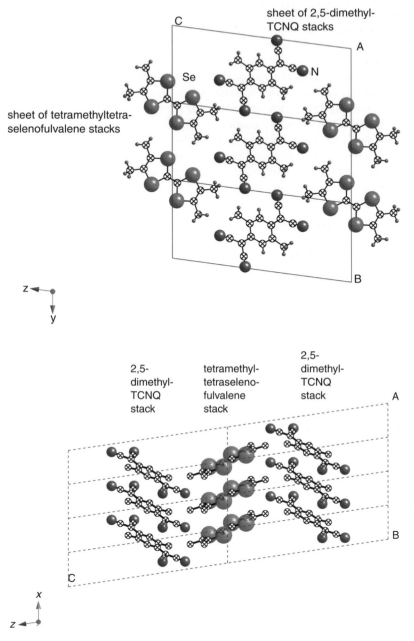

Fig. 17.41. Projections down [100] (upper) and [010] (lower) of the structure of {[TMTSeF]-[2,5-dimethyl-TCNQ]} (SEFTCQ). (Data from Andersen, Bechgaard *et al.*, 1978.)

Class IV. Metals/semimetals, still highly conducting below T_C. One example is {[HMTSeF][TCNQ]} (mono-#7) and {[DEDMTSeF][TCNQ]} (tri-#4) appears to be another.

The structures of {[HMTSeF][TCNQ]}, {[HMTTF][TCNQ]} and {[HMTTF]-[TCNQF$_4$]} require some comment. All three compounds have very similar cell dimensions (Table 17.10) and {[HMTTF][TCNQ]} and {[HMTTF][TCNQF$_4$]} are isomorphous, while {[HMTSeF][TCNQ]}, with a different space group, is not isomorphous with the other two. It has a parallel arrangement of stacks when viewed along [100] while the two isomorphous structures have a herring-bone arrangement. The room temperature conductivities of {[HMTSeF][TCNQ]}, and {[HMTTF][TCNQ]} are about 10^6 times as large as that of {[HMTTF][TCNQF$_4$]}. The conductivity difference is ascribed (Torrance *et al.*, 1980) to partial charge transfer in {[HMTSeF][TCNQ]}, ($Z \approx 0.72$) and also in {[HMTTF][TCNQ]} leading to high conductivity, while the much lower conductivity of {[HMTTF][TCNQF$_4$]} is a consequence of complete ionization in the ground state due to the greater electronegativity of fluorine compared to hydrogen. {[HMTSeF][TCNQ]} is classified as 'two-dimensional' because of the short interstack Se...N distance of 3.10(3) Å.

In some structures lateral interactions between stacks appear to be more important than those within stacks. Thus in {[TTF][[Ni(dmit)$_2$]$_2$]} (Bousseau *et al.*, 1986) the monoclinic crystals ($a = 46.22$, $b = 3.73$, $c = 22.86$ Å, $\beta = 119.3°$, $C2/c$, $Z = 4$) have segregated stacks of TTF (at centres of symmetry) and Ni(dmit)$_2$ moieties (at general positions) leading to an arrangement of alternating sheets of the moieties in the (100) planes (Fig. 17.42). The interplanar spacings within the stacks are much larger than usual (3.65 and 3.55 Å respectively) while there are sixteen crystallographically independent S...S approaches, ranging from 3.68–3.38 Å, which are less than the standard van der Waals diameter of sulphur (taken as 3.70 Å on the basis of the intermolecular contacts in orthorhombic sulphur (Abrahams, 1955). These results, taken together, suggest that the crystal structure is determined by a quasi-three dimensional network of intermolecular S...S interactions. However, no conclusions could be drawn about the charges on the donor and acceptor moieties. The compound has metallic conductivity down to 4K, with $\sigma_{RT} \approx 300$ S/cm and σ_4K ≈ 105 S/cm. The isomorphous Pd compound has $\sigma_{RT} \approx 750$ S/cm but shows a metal-to-semiconductor transition at ≈ 220K.

17.6.3 *Cation : anion ratio 1 : 1; diad stacks*

Four examples are discussed. The first of these, {[TTF] [2,5-TCNQF$_2$]} (Emge, Wijgul *et al.*, 1981; BERYOM) is triclinic ($a = $**7.082**, $b = 8.761$, $c = 14.447$ Å, $\alpha = 87.94$, $\beta = 81.74$, $\gamma = 84.54°$, $P\bar{1}$, $Z = 2$; the crystals were twinned and measurements were made on one individual of the twin). There are stacks along [001] with $D_C = 3.34$ and 3.60 Å, and $D_A = 3.23$ and 3.55 Å, showing clear dimerization within both types of stack. There is nearly eclipsed overlap within the dimerized pairs; moiety dimensions suggest full charge transfer.

The second example, {[DBTTF][TCNQF$_4$]} (Emge, Bryde *et al.*, 1982; BOMGIT) is also triclinic ($a = $**7.533**, $b = 10.094$, $c = 13.703$ Å, $\alpha = 63.47$, $\beta = 77.11$, $\gamma = 74.02°$, $P\bar{1}$,

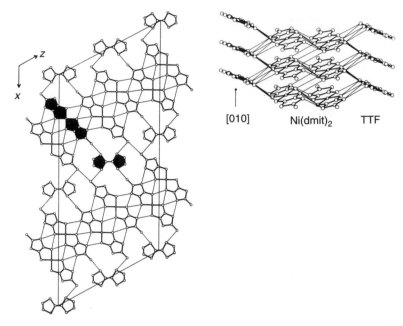

[010]   Ni(dmit)$_2$ TTF

Fig. 17.42. (a) Projection of the crystal structure of {[TTF][[Ni(dmit)$_2$]$_2$]} onto (010), showing S . . . S distances less than 3.70 Å as thin lines and the sheets of TTF stacks (shaded); (b) view in the plane of the TTF moieties showing the stacking arrangement and the lateral interactions. (Reproduced from Bousseau *et al.*, 1986.)

$Z = 2$). There are stacks along [001] with $D_C = 3.35$ Å (eclipsed overlap) and 3.68 Å (R/DB overlap), and $D_A = 3.18$ Å (displaced R/EB overlap) and 3.54 Å (displaced R/R overlap), showing even clearer dimerization within both types of stack. Moiety dimensions suggest full charge transfer, providing an explanation for the insulating qualities of the salt.

The third example, {[BEDT.TTF][TCNQ]}, is polymorphic, one structure having mixed stacks (see Section 15.11.2) and the other segregated stacks (Mori and Inokuchi, 1986; FAHLEF). The latter crystals are triclinic ($a = 6.650$, $b = 7.817$, $c = 23.915$ Å, $\alpha = 89.18$, $\beta = 85.67$, $\gamma = 74.90°$, $P\bar{1}$, $Z = 2$) and have the unusual feature that there are stacking motifs with approximately-perpendicular stacking axes. The TCNQ moieties are stacked along [001] in a diad arrangement with interplanar spacings of 3.34 Å (displaced R/EB overlap) and 3.24 Å (displaced R/R overlap); this stack is one-dimensional in character. The cations are stacked along [010] with interplanar spacings of 3.63 Å (R/DB overlap) and 3.87 Å (sideways displaced R/EB overlap). However, lateral interactions appear to be more important than those within the stacks. Thus it seems that the crystal has a structure that is a compromise between the conflicting requirements of the one-dimensional interactions characteristic of TCNQ and the two-dimensional interactions characteristic of BEDT·TTF.

Dibromo-DCNTT

The fourth example is {[TTF][dibromo-DCNTT]} ($a = 13.82$, $b = 19.19$, $c = 6.97$ Å, *Pccn*, $Z = 4$), which has segregated stacks of planar donors and acceptors, with equidistant interplanar separations of 3.48 Å in both stacks. The stack axes are parallel to [001], the TTF moieties are almost exactly eclipsed while the unsymmetrical acceptors alternate in orientation along the stack. The single crystal conductivity (presumably along the stack axis) is 2×10^{-2} S/cm (Gunther *et al.*, 1990; SETWOD).

17.6.4   *Cation : anion ratio 2 : 1 and 1 : 2; monad stacks*

We treat these two cation : anion ratios together because, in a formal sense, many of the structures are related as structure and anti-structure (i.e. roles of cation and anion are interchanged). Relatively few examples are known. {[(Perylene)$_2$][Pt(mnt)$_2$]}, which could have been discussed in Section 17.5.2, is monoclinic with space group $P2_1/c$, $Z = 2$ (stack axis [010] $= 4.194(1)$ Å) (Alcacer, Novais *et al.*, 1980; PRLNTP; Alcacer, 1985); thus the perylenes are at general positions and the Pt(mnt)$_2$ moieties at centers of symmetry. Six perylene stacks surround a Pt(mnt)$_2$ stack. The (perylene)$^{+\bullet}$ moieties overlap in graphite-like fashion with an interplanar spacing of 3.32 Å; bond lengths were not reported. The Pt(mnt)$_2$ moieties have a metal-over-ring overlap, separated by 3.65 Å. The room temperature conductivity is ≈ 50 S/cm, rising to 1250 S/cm at ≈ 20K, where there is a metal-to-insulator transition to a very high resistivity material. The positive thermopower (≈ 28 mV/K) shows that the predominant charge carriers are holes on the perylene stacks, the electrons on the [Pt(mnt)$_2$]$^{2-}$ moieties being localized. (TTF)$_2$···HCBD, the crystal structure of which has not been reported, possibly has an analogous structure; $\sigma_{RT} \approx 10^3$ S/cm.

There are two analogous anti-structures. {[(TMTTF)$_{1.3}$][(TCNQ)$_2$]} is monoclinic, space group $P2_1/n$, with TMTTF moieties at centres of symmetry and TCNQs at general positions (Kistenmacher *et al.*, 1976; MDTTCQ); the additional TMTTFs are located in channels between the stacks and do not appear to play any part in determining the electronic properties of the crystals. The material is a semiconductor with $\sigma_{RT} \approx 10$ S/cm. The TCNQ stacks have R/EB overlap with an interplanar spacing of 3.24 Å and the TMTTF stacks have R/DB overlap with an interplanar spacing of 3.59 Å. The second example is {[TTT])···[(TCNQ)$_2$]}; the stacking was reported as monad ($c = 3.754$ Å, space group $P\bar{1}$, $Z = 1$), with TTT cations at centers of symmetry and TCNQ moieties at general positions but translationally equivalent. The TCNQ stacks have R/EB overlap with an interplanar spacing of 3.18 Å and the TTT stacks have graphite-like overlap with an interplanar spacing of 3.52 Å. The material shows metallic behaviour down to 90K, below which the conductivity decreases rapidly; $\sigma^{RT} \approx 10$ S/cm. Studies of its physical properties appear to be complicated by presence of solvent (perhaps 1/2 molecule of nitrobenzene per formula unit) and poor crystal quality (Shchegolev and Yagubskii, 1982).

17.6.5   Cation : anion ratio 2 : 1 and 1 : 2; diad stacks

The crystals of {[(TTF)$_2$][OCNAQ^7]} are triclinic, space group $P\bar{1}$, $Z = 1$.(Inabe et al., 1988; SAJNIA). The interplanar spacings in the cation stack are 3.42 Å, with weak dimerization and little overlap between dimers while there is R/EB overlap (spacing 3.30 Å) between halves of the anions. There is metallic conductivity of ≈ 10 S/cm down to ≈ 43K and then a marked reduction in conductivity. The crystals of {[(TTT)$_2$] [OCNAQ]}·DMF are also triclinic, space group $P\bar{1}$, $Z = 1$. (Inabe et al., 1988; GAX-CUD10). The TTT moieties are weakly dimerized in stacks along [001] (interplanar spacings 3.26 and 3.38 Å) while there is limited overlap of the OCNAQ moieties, whose mean planes are steeply inclined to those of the cations. There is metallic conductivity down to ≈ 250K below which the material is a semiconductor; $\sigma_{RT} \approx 60$ S/cm. {[(TMTTF)$_2$][HCBD]}, although monoclinic, has a rather similar structure with the anions in channels between the stacked (and weakly dimerized) cations (Katayama, 1985); the anion planes are parallel to the axis of the cation stacks. This material is a semiconductor with $\sigma_{RT} \approx 3 \times 10^{-4}$ S/cm. A similar structure is found for {[(TMTSeF)$_2$][(2,5-TCNQBr$_2$)]} (Stokes et al., 1982; BEVHUF), where there is considerable disorder of the anions located between the TMTSeF diad (but nearly monad) stacks, whose mean interplanar spacing is given as ≈ 3.6 Å. The nitrile stretching frequency corresponds to unit charge on the anions and hence to an average half-unit charge on the cation moieties. The DC conductivity at room temperature is ≈ 100 S/cm and is ascribed to phonon-assisted hopping.

{[TMPD][(TCNQ)$_2$]} has a DA$_2$ anti-structure compared to the above group of D$_2$A structures; the crystals are triclinic (space group $P\bar{1}$, $Z = 1$; stack axis $c = 6.488$ Å) (Hanson, 1968; TCQPDA) with the TCNQ moieties forming a zigzag pseudomonad stack while the TMPD's are inclined at a large angle to the TCNQ planes (Fig. 17.43(a)). Measurements of conductivity and other properties as a function of temperature on powder samples confirm that only the TCNQ stacks contribute to the transport properties (Somoano et al., 1975). There is evidence for structural transitions at 200 and 70K but parallel crystallographic studies have not been reported. {[E$_2$P][(TCNQ)$_2$]} (E$_2$P is 5,10-dihydro-5,10-diethylphenazine (Section 15.9.5)) provides an analogous example. The crystals are triclinic (stack axis $a = 6.843$ Å) with diad stacks; there are weakly overlapping π-dimers (R/EB overlap, interplanar spacing 3.25 Å) separated by 3.54 Å (Fig. 17.43(b)) (Dietz et al., 1981). Thus, in structural terms, both {[TMPD][(TCNQ)$_2$]} and {[E$_2$P][(TCNQ)$_2$]} could be better described as anion radical salts, as could some of the other examples included for convenience in this section. Because of differences in the TCNQ stacking patterns, differences are to be anticipated in appropriate physical properties.

We conclude this section by noting another structure in which lateral interactions appear to determine the overall crystal structure; this should be compared with the situation in

---

[7] OCNAQ is 11, 11, 12, 12, 13, 13, 14, 14-octacyano-1,4,5,8-anthradiquinotetramethane.

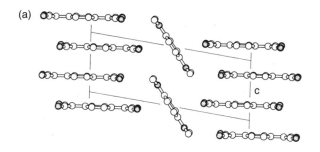

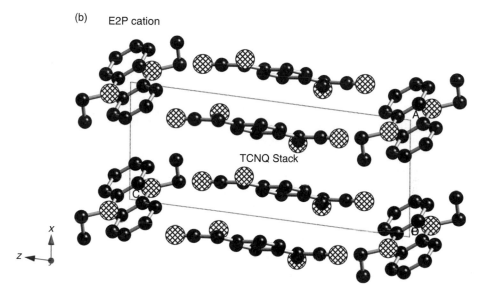

Fig. 17.43. (a) {[TMPD][(TCNQ)$_2$]} viewed normal to the stack axis, showing the zigzag pseudomonad stackof TCNQ anions seen edge-on, and the cations steeply inclined to the stack. (Reproduced from Hanson, 1968.) (b) {[E$_2$P][(TCNQ)$_2$]} viewed normal to the stack axis, showing the stacked TCNQ π-dimers (within the confines of the unit cell) and the cations steeply inclined to the stack. (Data from Dietz *et al.*, 1981.)

the stacked structures of the M(dmit)$_2$ salts discussed in Section 17.5.2. In {[HMTTeF)$_2$][(Pt(dmit)$_2$)$_2$]}[8] (triclinic, $P\bar{1}$, $Z = 2$, $a = 15.47$, $b = 13.53$, $c = 10.59$ Å, $\alpha = 92.85$, $\beta = 102.50$, $\gamma = 75.91°$) the HMTTeF moieties form nonstacked tetrads, which are linked diagonally to other tetrads by Te . . . Te contacts (Kobayashi, Sasaki *et al.*, 1986; FIFRUH). These form a framework enclosing two Pt(dmit)$_2$ moieties (Fig. 17.44) and the salt has strong *structural* resemblances to the channel inclusion complexes. The temperature dependence of the conductivity (≈ 20 S/cm at room temperature) suggests that the material is a semiconductor. Thus we have travelled a long way from our

---

[8] This notation is hardly applicable here. Also, there is a monoclinic crystal FIFRUH01, the structure of which has not been reported.

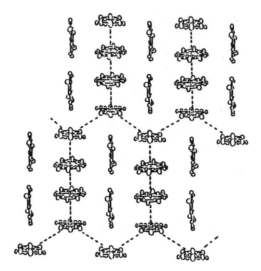

Fig. 17.44. Partial view of the crystal structure of (HMTTeF)$_2$···Pt(dmit)$_2$ showing the tetrad of HMTTeF moieties in the centre of the cell and individual Pt(dmit)$_2$ moieties on the flanks; the planes of the two components are approximately mutually perpendicular and the view direction is approximately along the long axis of the Pt(dmit)$_2$ moiety. The shorter Te...Te distances range from 3.33 to ≈3.9. (Reproduced from Kobayashi *et al.*, 1986.)

starting point – structures based on segregated stacks of donors and acceptors, arranged with stack axes parallel.

17.7 Electron density studies of some segregated stack complexes

The first attempt at investigating the interaction between the components of an ion radical salt through the medium of electron density studies was for {[TMSF]$_2$AsF$_6$} at 300K (Wudl, Nalawajek *et al.*, 1983). This was at an early stage in the development of these techniques and it is perhaps no surprise that the criticism was made that the conclusions drawn were not supported by the experimental measurements (Dunitz, 1985). The only other study of this kind is by Espinosa *et al.* (1997) on the {[bis(thiodimethylene)-TTF)][TCNQ]} complex ({[BTDMTTF][TCNQ]}) using measurements made at 130K and the more recently-developed computational techniques. Crystal data are in Table 17.9.

BTDMTTF X = Y = S; HMTSF X = Se, Y = CH$_2$
HMTTF X = S, Y = CH$_2$

Table 17.13. Crystal data for three related structures. The first two crystals are isomorphous while the third is isostructural. $Z = 2$ for the three structures

	{[BTDMTTF][TCNQ]}		{[HMTSF][TCNQ]}	{[HMTTF][TCNQ]}
	298K	130K	RT	RT
Space group	C2/m	C2/m	C2/m	Pmna (no. 53)*
a	21.296	21.205	21.999	12.462
b	15.567	12.530	12.573	3.901
c	3.928	3.859	3.980	21.597
β	92.74	93.30	90.29	
Vol	1050	1023.6	1076	1050
Mol. Wt.	524.7	524.7	676.3	488.7
Refcode/ reference	Rovira et al. (1995).	Espinosa et al. (1997).	SEOTCQ; Philips et al. (1976).	HMTFCQ; Chasseau et al. (1978).

* The space group was misprinted as *Pmma* (no. 51) in the paper by Rovira *et al.* (1995).

We shall introduce this area by first considering three related structures for which crystal data are given in Table 17.13.

In the *C2/m* structures the cations (symmetry 2/*m* at 0, 0, 0) and anions (symmetry 2/*m* at 0, 1/2, 1/2) are located in segregated stacks along [001], with each donor stack surrounded by four acceptor stacks, and conversely; both types of molecule have 'ring-over-bond' overlap; the *Pmna* structure is analogous. The BTDMTTF cations are separated by 3.521 Å and the TCNQ anions by 3.194 Å at 130K. The structures are shown in projection down the short ≈ 4 Å axes (Fig. 17.45). There are distances shorter than the sum of van der Waals radii between 'internal' S and N (3.24 Å) and between 'external' S and C (of cyano) at 3.44 Å. There are similar short 'internal' S(Se)...N distances in the other two structures but the CH_2 that replaces 'external' S does not make short contacts. The similarities and differences between the two structure types are shown in the two parts of Fig. 17.45.

More detail emerges from the 130K structure analysis of {[BTDMTTF][TCNQ]}. Bond lengths in the two components, x-ray diffuse scattering and Raman and IR frequency measurements all agree that there is transfer of 0.56*e* from BTDMTTF to TCNQ. Multipolar refinement of the electron density showed a transfer of 0.34*e* from each of the two external S atoms (almost entirely) to the four N of TCNQ. The concentration of electron density was found to be appreciably higher in TCNQ than in BTDMTTF. Detailed analysis showed that 'external' S interacted not with cyano N or C but with the –CN triple bond itself, with the interaction having features similar to those of weak hydrogen bonds. It was concluded that this was the mechanism that controlled the electronic properties – the conductivity and the charge density waves. These very interesting results pose the question 'what happens when 'external' S is replaced by CH_2?' Study at low temperatures of {[HMTTF][TCNQ]} and of the isostructural pair {[1,6-dithiapyrene][TCNQ]} (DAKTIS) and {[1,6-diaminopyrene][TCNQ]} (WEMHEB) (see Table 17.10) should prove illuminating.

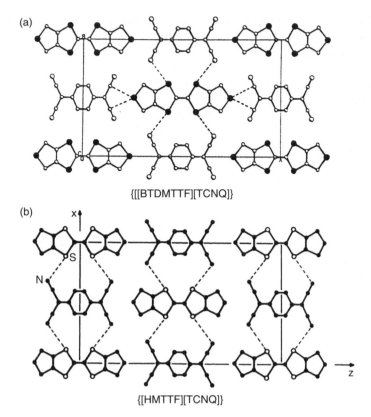

{{[BTDMTTF][TCNQ]}}

{{[HMTTF][TCNQ]}}

Fig. 17.45. Comparison of the short-axis projections of {{[BTDMTTF][TCNQ]}} (*C2/m*) and {{[HMTTF][TCNQ]}} (*Pmna*). S is denoted by filled circles in the upper diagram and by open circles in the lower diagram. (Adapted from Rovira *et al.* (1995) and Chasseau *et al.* (1978).)

17.8   Theoretical studies of some segregated stack complexes

Theoretical studies of the molecular complexes discussed in this chapter have been directed to two ends: the first is to provide an explanation for the overwhelming predominance of stacked structures; the second is to provide an explanation for the physical properties of the complexes, first and foremost the electrical conductivity. Structural ends are usually expressed in the standard terminology of quantum chemistry, physical properties in terms of band structure. In accord with the structural bias of this book, we shall not discuss the theoretical approaches to the physical properties but concentrate on the quantum chemistry, essentially restricting ourselves to TTF, TCNQ and the [TTF]·[TCNQ] system.

We start with some experimental facts. TCNQ$^{-\bullet}$ dimerizes in aqueous solution to give (TCNQ)$^{2-}$ (Boyd and Philips, 1965) with an equilibrium constant (at 298K) of 2.5×10^3 litre/mol and $\Delta H = -43.5$ kJ/mol dimer. The authors comment "The solution dimer is presumably related to the structure of some of the ion radical salts of TCNQ, where paired electrons are found in a singlet state with an ESR-detectable population of a triplet state."

Neutral TCNQ has a herringbone packing of planar molecules with an interplanar spacing of 3.45 Å at room temperature (Long, Sparks and Trueblood, 1965, see their Fig. 3; TCYQME); polymorphs have not been reported. There are two polymorphs of neutral TTF – monoclinic (Cooper *et al.*, 1974) and triclinic (Weidenborner *et al.*, 1977; Ellern *et al.*, 1994). Only the (apparently more stable) monoclinic polymorph is relevant in the present context. It also has a herringbone crystal structure, with a spacing of 3.63 Å between planar, parallel molecules. As we have seen, the interplanar spacings between adjacent TTF cations and between TCNQ anions can be 0.2–0.3 Å less than these values, suggesting attractive interactions additional to those due to van der Waals interactions. The tasks of theory are to provide an explanation for the reduced spacings and then to indicate the relative stabilities of various mutual moiety arrangements.

The first of these does not appear to have been met at all as authors generally fix the interplanar spacing at experimental values (3.17 Å for TCNQ, 3.47 Å for TTF), and then investigate interaction energy as a function of slip along the long molecular axis. Calculations up to 1980 by extended Hückel MO (EHMO) methods have been reviewed by Lowe (1980); minima are found for the appropriate degrees of slip for the neutral and singly and doubly charged entities; however, it would not be unfair to say that the point has been illustrated rather than demonstrated. Shifts other than along the long molecular axes do not appear to have been considered.

There have been a number of attempts to calculate the stable overlap modes for TCNQ π-dimers and the results have been summarized critically (Silverman, 1981). The available results do not seem to be easily comparable with experiment. The essential problem is that the molecular orbital programmes in current use, despite their wide range, were designed primarily for calculating interactions between atoms at or around covalent bonding distances and do not perform well when interactions at van der Waals distances are involved. Silverman (1981) applied a procedure due to Gordon and Kim (1972) to the calculation of interaction energies for TTF neutral dimers but this does not yet seem to have been done for TCNQ anions.

A bridge between the quantum chemical and energy band approaches is provided (Starikov, 1998) by a three-dimensional HF crystal orbital calculation using CRYSTALS92 (Dovesi, Saunders and Roetti, 1992); we discuss only the results for {[TTF][TCNQ]}. As the calculation was based on the experimental crystal structure, no insight is provided about stacking or mutual arrangement of moieties but the component charge distributions were obtained (for TTF positive charges on S and negative charges on C, for TCNQ negative charges on N) and there is charge transfer of 0.5–0.9 e (depending on details of the calculation) from TTF to TCNQ, which fills the conduction band in the [010]* direction. The mobile charge density is carried on the TCNQ stack while the rest of the charge density residing on the TTF stack is essentially localized and immobile. This corresponds to the experimental finding that the TCNQ stacks carry most of the conductivity and the TTF stacks most of the magnetic susceptibility.

17.9 Studies of {[TTF][TCNQ]} and some related materials

In the previous section we described the structural chemistry of cation radical–anion radical salts in general terms, with most of the results referring to room temperature. Here we give more detail about the most famous member of this class {[TTF][TCNQ]}, whose physical and structural properties have been studied over a wide range of temperature and pressure; some closely related materials are included in the discussion. Above 54K crystalline {[TTF][TCNQ]}, is disordered in subtle fashion, crucial for explanation of its unusual physical properties. Three phase transitions (of somewhat unusual type) occur around 50K, below which temperature ordering occurs. We start by considering physical properties and room temperature crystal structure and then discuss the dependence of structure on temperature, finally relating structural features to physical properties on the basis of current theoretical treatments. A number of comprehensive reviews have been published (Garito and Heeger, 1974; Berlinsky, 1976; Friend and Jerome, 1979; Schulz, 1980; Conwell, 1988a, b), and the structural physics of {[TTF][TCNQ]}, and related compounds has been discussed in particular detail by Pouget (1988).

As mentioned at the beginning of this chapter, interest in {[TTF][TCNQ]}, rocketed when it was found to have a large metallic-type conductivity along the [010] axis over the range 300–56K (cf. the conductivity of Pb shown in Fig. 17.1), followed by a drastic drop as the temperature was lowered further (Figs. 17.1 and 17.46). The three crystalline compounds {[TTF][TCNQ]}, {[DSeDTF][TCNQ]} and {[TSeF][TCNQ]} are iso-morphous (Table 17.10) and their conductivities along [010] show very similar types of temperature dependence (Fig. 17.46); such similarity may well be expected to extend to

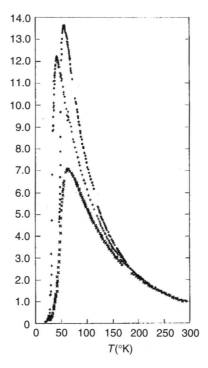

Fig. 17.46. The ordinate shows the normalized conductivity s/s_{295} along [010] as a function of temperature for {[TTF][TCNQ]} (highest values), {[DSeDTF][TCNQ]} (lowest) and {[TSeF] [TCNQ]} (intermediate). The conductivity of {[TTF][TCNQ]} at $\approx$5K is about 5×10^{-6} S/cm (Cohen and Heeger, 1977). (Reproduced from Etemad *et al.*, 1975.)

other physical properties. Microwave measurements (9400 Mhz ≈ 0.3 cm) (Miane *et al.*, 1986) give the following room temperature values (units of S/cm) along the principal axes of the (markedly anisotropic) conductivity ellipsoid : $\sigma_a = 1.00 \pm 0.16$, $\sigma_b = 325$, $\sigma_{c^*} = 0.0115$ (anisotropy $\sigma_b/\sigma_a \approx 330$; $\sigma_b/\sigma_{c^*} \approx 28000$; $\sigma_a/\sigma_{c^*} \approx 100$). These values are essentially DC conductivities.

The conductivity of {[TTF][TCNQ]} has been studied more intensively than that of any related material not only because of its intrinsic interest but also because comparatively large, high quality crystals can be grown. Results for the *b*-axis conductivity of some 600 crystals studied in 16 laboratories are summarized in Fig. 17.47 (Thomas *et al.*, 1976). There is some evidence for an increase in conductivity with time as crystal quality has improved. An average value of σ_{max}/σ_{RT} would be about 20. Values of $\sigma_{max}/\sigma_{RT} \approx 500$ were reported (Coleman *et al.*, 1973) for three crystals out of 70 studied by the University of Pennsylvania group but are not generally accepted (Thomas *et al.*, 1976; Cohen *et al.*, 1976). Clearly there was, some 25 years ago, no such thing as a standard and reproducible value for the conductivity of {[TTF][TCNQ]} at a given temperature comparable to the values available for copper and other substances. Despite these reservations, the general dependence of conductivity on temperature is well established as well as the overall anisotropy of the conductivity.

The crystal structure of {[TTF][TCNQ]} is shown in Fig. 17.48. Interplanar distances are 3.47 Å in the TTF stack (ring over double bond overlap) and 3.17 Å in the TCNQ

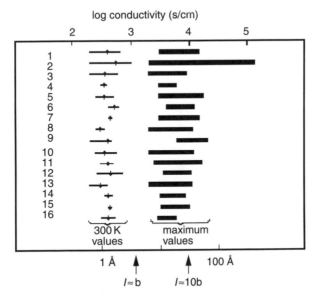

Fig. 17.47. Diagram summarizing the values of the room temperature and maximum conductivities along [010] (stack axis) for {[TTF][TCNQ]}, as reported by different laboratories: 1. Johns Hopkins U.; 2. U. Pennnsylvania; 3. Stanford U. and IBM, San Jose; 4. Bell Labs.; 5. Dupont; 6. IBM, Yorktown Heights; IBM, San Jose; 8. U. Illinois; 9. NBS; 10. Monsanto; 11. U. Chicago; 12. U. British Columbia; 13. Hughes; 14. Clemson; 15. Tennessee; 16. Tokyo. The effective mean free path is shown below the frame of the diagram. (Adapted from Thomas *et al.*, 1976.)

stack (R/EB overlap). The angles between molecular plane normal and [010] are not very different (24.5° for the TTF stack and 34.0° for the TCNQ stack) but have opposite signs, giving rise to the overall herringbone arrangement of the stacks. Each TCNQ stack has six nearest-neighbor stacks, four TTF and two TCNQ. Alternatively the overall arrangement

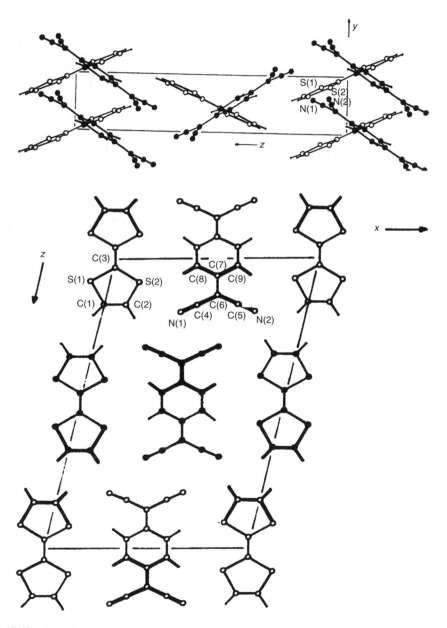

Fig. 17.48. Crystal structure of {[TTF][TCNQ]} viewed in projection down [100] (above) and [010] (below) axes. There is a herring-bone arrangement of segregated TTF and TCNQ stacks that are located in alternate (100) planes. (Reproduced from Kistenmacher *et al.*, 1974.)

can be described in terms of sheets of like stacks, with TTF and TCNQ moieties arranged in alternate (200) planes. Some approaches between S atoms of the cation and N atoms of the anion are closer than the sum of the van der Waals radii $1.85 + 1.75 = 3.60$ Å; for example, in {[TTF][TCNQ]} there are two S ... N distances of 3.20 and 3.25 Å directed roughly normal to the sheets of like stacks. Such approaches, which are roughly along [100], suggest that the intrastack interactions are supplemented by interactions between them, thus giving rise, in some of these materials, to what has been termed a quasi-two dimensional character. The molecular dimensions measured at room temperature by XRD (Kistenmacher et al., 1974) and neutron diffraction (Filhol, Bravic et al., 1981) are in good agreement, and were interpreted to imply a charge transfer Z of between 0.5 to 1.0 electron from TTF to TCNQ (Kistenmacher et al., 1974). We draw a similar conclusion about the imprecision of the determination of the amount of charge transfer from our discussion of the moiety dimensions in {[TTF][TCNQ]} over a range of temperature (#13.3.6). A much more precise value of Z is obtained from the diffuse scattering measurements to be discussed below.

There are similar moiety arrangements in {[TMTTF][TCNQ]}, {[TMTSeF] [DMTCNQ]}, and {[TMTTF][HCBD]} but the herring-bone pattern is less marked in {[TTF][DETCNQ]} and {[DBTTF][TCNQCl$_2$]}. Description in terms of 'sheets of stacks' can also be applied to the other examples in Table 17.10. The long axes of the moieties are parallel in projection in all these examples except for {[TMTSeF][DMTCNQ]} and {[TTF][DETCNQ]}.

Some fundamental thermodynamic quantities for {[TTF][TCNQ]} and its components are summarized in Table 17.14. The standard enthalpy of formation of the salt from its components ΔH_c is -37.4 kJ/mol, in good agreement with determination of the enthalpy of sublimation of the salt as 38 kJ/mol more endothermic than the sum of the values for the components. The entropies of salt and components do not appear to have been measured. The value of ΔV_c is -8.1%, which is considerably larger than, for example, the

Table 17.14. Fundamental thermodynamic quantities for {[TTF][TCNQ]} and its components at room temperature and pressure

Parameter	TTF	TCNQ	{[TTF][TCNQ]}	ΔX^*
m.pt. (K)	567–9	464	483(dec.)	–
Volume (Å^3)	202.1	254.9	419.9	-37.1
$\Delta H_{f^{\circ}}$ (kJ/mol)	290.8 ± 1.7	664.9 ± 0.7	918.4 ± 2.1	-37.4
(Metzger, 1977)				
ΔH_{subl}(kJ/mol)	97.1 ± 7.1	131.8 ± 5.4	266.9 ± 18.4	38
(Kruif and Govers, 1980)				

Notes:
1. $\Delta X = [X\{[TTF][TCNQ]\} - \{X(TTF) + X(TCNQ)\}]$. This follows from the definition of the measured standard thermodynamic quantities of formation as $\Delta X_c = X[DA]_c - \{X(D_c) + X(A_c)\}$, where the reaction is $D_c + A_c \Rightarrow [DA]_c$. The temperature is usually taken as 298K and the subscript c denotes that all substances are in the crystalline state. This definition applies to a 1 : 1 composition and has been specialised from the more general definition given in Appendix I.
2. The volume per moiety in the crystal is calculated from measured cell dimensions.
3. The measured values, at various temperatures, have been corrected to 298K by Govers (1978).

value of -4.8% found for the ionic mixed stack π-molecular compound {TMPD$\cdots$ TCNQ}.

The cohesion of the crystal attracted early interest. The crystal consists of segregated stacks of TTF^{Z+} and TCNQ^{Z-} moieties; the electrostatic energy is repulsive within the stacks and between like stacks but attractive between unlike stacks. The ionization energy of TTF is 672.5 kJ/mol and the electron affinity of TCNQ is 272.1 kJ/mol; if one assumes (as discussed below) that the charge transfer Z is 0.59, then the energy needed to stabilize the crystal is at least 236.2 kJ/mol. The Madelung energy, calculated on the basis of various quantum-mechanical models of the charge distributions in the TTF and TCNQ moieties, ranged from -71.4 to -91.7 kJ/mol (Metzger and Bloch, 1975). Later calculations, summarized by Govers (1978), give a mean value of -72.9 and a range of -57.5 to -83.7 kJ/mol for $Z = 0.59$. Thus the Madelung energy is not enough to stabilize the crystal. This was termed "the electrostatic binding energy defect" by Metzger and Bloch (1975), and much effort has been devoted to refinement of its calculation and finding routes to its elimination (Metzger, 1981). One possibility is to include polarization, dispersion and charge-dipole contributions in the cohesive energy; this has been done by Govers (1978), using the atom-atom approximation of Kitaigorodskii. The sum of the van der Waals and repulsive energies calculated for {[TTF][TCNQ]} amounts to ≈ -196 kJ/mol, thus giving an overall cohesive energy of -269 kJ/mol, which is enough to stabilize a crystal containing partially ionized moieties. These results suggest that it is essential to include both dispersion and Madelung contributions in any calculation of the cohesive energy of {[TTF][TCNQ]}, but there are so many uncertainties in the numerical values that it does not appear justified to give them in detail.

The elastic properties (thermal expansion and compressibility) are summarized in Fig. 17.49 and have been discussed in some detail (Filhol et al., 1981). The thermal expansion coefficient along [010] for {[TTF][TCNQ]} has also been measured by a microwave technique in the range 80–300K (Krause et al., 1983) and agrees well with the analogous diffraction results. The temperature region of the phase transitions was not covered in these sets of measurements.

The anisotropy of elastic properties is much less than the anisotropy of conductivities. The maximum compressibility κ_b and thermal expansion α_2 are both directed along [010] while the other two, much smaller values, have very similar directions with respect to the packing of the molecules; there is a minor dependence of the orientations of the two ellipsoids in the ac plane on pressure and temperature respectively. The "softest" direction is along the stack axis, roughly normal to the molecular planes, while the "harder" directions are between stacks. A similar situation is found in mixed stack molecular compounds (Chapter 15).

The specific heat of {[TTF][TCNQ]} has been measured in the region of the phase transitions (Craven et al., 1974; Djurek et al., 1977). The effects are very small (Fig. 17.50). It is now accepted that there are three transitions at ambient pressure (T_L at 38, T_M at 48 and T_H at 54K), in accordance with the results of Djurek et al. (1977). The entropy change in the transition around 52K has been estimated to be ≈ 0.03 R, with similar values for the other transitions. The low-temperature ($\approx$1.5–4K) specific heat can be fitted by a T^3 equation, without any linear term arising from electronic or magnetic excitations. The Debye temperature was estimated as 89.5K and the material characterized as a small band gap nonmagnetic semiconductor (Wei et al., 1973).

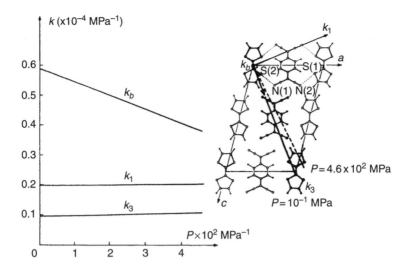

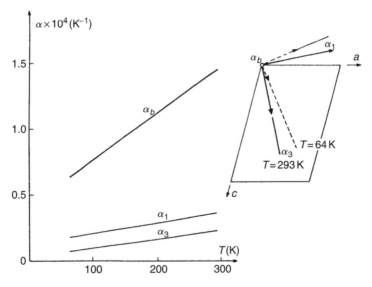

Fig. 17.49. (Top) The principal compressibilities ($\kappa_i = -(1/l_i)((dl_i/dp))$) at 293K of {[TTF][TCNQ]} and their directions at two representative pressures (10^{-1} MPa on left and 4.6×10^2 MPa on the right); (bottom) principal thermal expansion coefficients $\alpha_i = (1/l_i)((dl_i/dT))$ as functions of temperature and their directions at two representative temperatures. κ_b and α_2 are along [010] because of the monoclinic symmetry of the crystals. (Reproduced from Filhol, Bravic *et al.*, 1981.)

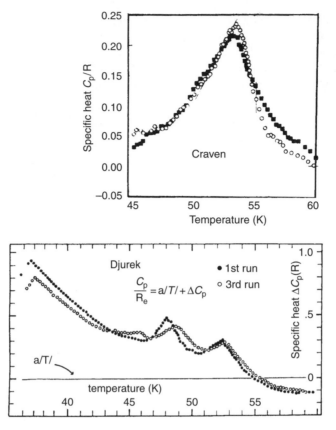

Fig. 17.50. Specific heat of {[TTF][TCNQ]} in the range 36–62K; the values of Craven *et al.* (1974) (background subtracted) and Djurek *et al.* (1977) (background not subtracted) are in qualitative agreement over part of the range but the latter show two (or perhaps three) additional transitions. (Adapted from Craven *et al.* (1974) and Djurek *et al.* (1977).)

The low temperature maximum in the conductivity naturally encouraged investigation of diffraction patterns over a range of temperatures and we shall distinguish here between evidence obtained from Bragg diffraction and diffuse scattering (x-rays and neutrons have been used in both types of study). Bragg diffraction gives information about the average structure while the diffuse scattering gives information about deviations from this average. We start with some of the results obtained from Bragg diffraction. Soon after the first studies of {[TTF][TCNQ]}, it was reported (Skelton *et al.*, 1974) that (x-ray) oscillation photographs at 294, 77 and 10K showed that there was no doubling of the *b* axis below the metal–insulator transition at about 50K. A full set of cell dimensions was measured somewhat later (single crystal diffractometer, Mo Kα radiation) over the range 40–295K and crystal structures determined at 60, 53 and 45K (Schultz *et al.*, 1976). The cell dimensions in the range 40–295K (the diagrams in Schultz *et al.* (1976) include data points at 5 and 25K of unstated provenance) showed some irregularities (of $\approx$0.02 Å and $\approx$0.2°) in the values of *a*, *c* and β in the temperature region of the transitions. These have not been confirmed in later work. One should note that cell dimensions, especially if measured using Mo Kα radiation, are susceptible to larger errors (perhaps by a factor of 5 for cell lengths and 2.5 for cell angles (Taylor and Kennard, 1986; Herbstein, 2000) than may appear from an internal statistical analysis. The lack of irregularity along the stacking direction b was confirmed by a capacitance dilatometric study (Schafer *et al.*,

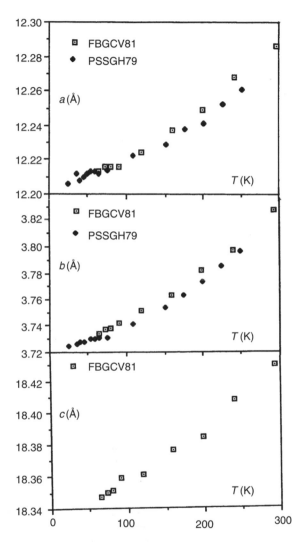

Fig. 17.51. Cell dimensions of {[TTF][TCNQ]} as a function of temperature. The values labelled FBGCV81 (Filhol *et al.*, 1981) were measured by neutron diffraction using 17–26 centered reflections and have been reported (perhaps optimistically) to have standard uncertainties of 0.002–0.004 Å. The value of β is 104.51(2)° at 293K and is essentially constant at 104.40(1)° in the region 90–60K. The values labelled PSSGH79 were measured by neutron diffraction (Pouget *et al.*, 1979) using the 300 and 020 reflections and have reported standard uncertainties of $\approx$0.008 Å (a was calculated here from $a \sin \beta$ by using the values of β noted above).

1975), while a later and more comprehensive study by the same method showed that there were no anomalous length changes ($\Delta L/L$) greater than about 2×10^{-5} in *all three* crystallographic directions (Ehrenfreund *et al.*, 1981). The available cell dimensions are summarized in Fig. 17.51. Lattice parameters and transition temperatures (see below) of hydrogenated (used mainly for x-ray diffraction studies) and deuterated (used mainly for neutron diffraction studies) samples do not differ appreciably, implying that the hydrogens do not have more than a minor role, if any, in the structural chemistry of {[TTF][TCNQ]}.

Parenthetically we note here two other crystals for which the variation of cell dimensions with temperature has been measured. The back-reflection Weissenberg method (Cu Kα $\theta_{Bragg} > 80°$; estimated precision of 3 in 10^4) has been applied to {[TMTSeF][DMTCNQ]} (Guy *et al.*, 1982) (triclinic, Table 17.10) over the range 300–10K; there is a metal-to-insulator transition at 40K. There were no indications of anomalies in lattice parameters at the Peierls transition within the precision of the experiment, in agreement with what was found for {[TTF][TCNQ]}. Single crystal diffractometer measurements (Mo Kα radiation) have been made for {[TTF][2,5-diethyl-TCNQ]}; Fig. 17.52; Schultz and Stucky, 1977). Irregularities in cell dimensions in the region of the transition are similar to those reported for (but not confirmed) in {[TTF][TCNQ]}. Only the irregularities in *a* would appear to be significant. This salt has a transition at 111K as inferred from the specific heat, which shows a fairly sharp anomaly (Fig. 17.53(a)) corresponding to $\Delta S \approx 0.17(2)$ R, about six times larger than that found for {[TTF][TCNQ]}. The conductivity of {[TTF][DETCNQ]} shows a quite different temperature dependence (Fig. 17.53(b)) from that of {[TTF][TCNQ]} and there is no peak in the σ-T curve.

We now return to {[TTF][TCNQ]} and its structure analysis using Bragg reflections. Intensities of the *fundamental* reflections have been used for crystal structure

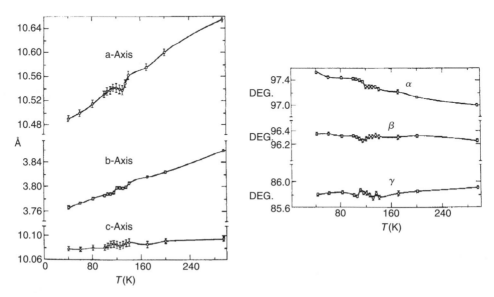

Fig. 17.52. Plot of unit cell dimensions vs. T for {[TTF][DETCNQ]}. (Reproduced from Schultz and Stucky, 1977.)

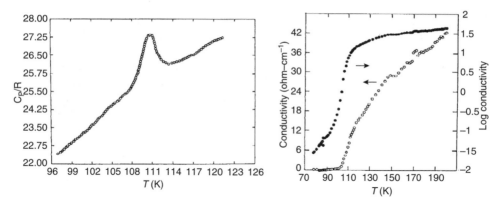

Fig. 17.53. (a) The specific heat {[TTF][DETCNQ]} in the vicinity of the transition; (b) variation of conductivity with temperature. (Reproduced from Schulz, Stucky, Craven *et al.*, 1976.)

determinations at 60, 53 and 45K (i.e. above and in the region of the phase transitions) and the differences found to be very small Schultz, Stucky, Blessing and Coppens, 1976). Thus the averaged structure (i.e. of the subcell, $a = 12.30$, $b = 3.82$, $c = 18.47$ Å, $\beta = 104.4°$, $P2_1/c$, $Z = 2$) (Table 17.10)) is essentially unchanged over the whole temperature range from 300K downwards. We conclude, from the evidence summarized above, that the drastic changes in conductivity are not accompanied by appreciable changes in *average* crystal structure. Whatever structural changes occur in {[TTF] [TCNQ]} during the phase transformations must be very different from the appreciable structural changes that occur in first-order transformations (for example, those illustrated for the urea-paraffin channel inclusion complexes discussed in Chapter 7) or even in second-order transitions like that in {pyrene···PMDA} at ≈160K (see Chapter 16). Furthermore, the transitions in {[TTF][TCNQ]}, and {[TTF][DETCNQ]} would appear to be different in nature, although it is difficult at this stage to define these differences.

We now consider the deviations from the average structure of {[TTF][TCNQ]} described above. x-ray and neutron diffraction patterns show, in addition to the Bragg reflections and thermal diffuse scattering, an additional type of diffuse scattering (found to sharpen to discrete satellite reflections at the very lowest temperatures), that provides vital evidence for the explanation of the electronic and other physical properties. The extreme weakness of this type of scattering, and the low temperatures involved, has greatly complicated the experimental investigations. We summarize some of the experimental results for {[TTF][TCNQ]}, (Kagoshima *et al.*, 1976; Khanna *et al.*, 1977; Kagoshima, 1982) before considering the physical implications of such scattering.

The diffuse scattering considered here is analogous to the "ghost" spectra obtained, in addition to the diffraction pattern of the averaged grating, from optical diffraction gratings with periodic errors of ruling (Wood, 1946). Similar effects are obtained in the diffraction of x-rays or slow neutrons from appropriately distorted crystals; the basic theory has been set out by James (1950). A periodic lattice distortion (PLD) [also sometimes called a periodic structural distortion (PSD)] will give satellite reflections displaced by wave vectors $\pm\mathbf{q}$ from the parent reciprocal lattice point $\mathbf{G}_i$. For small amplitude distortions the first order intensity of such satellites is given by

$$I(\mathbf{q}) = NI_c[F(\mathbf{G}_i)]^2 \times [\mathbf{S}_j\langle u_{\mathbf{q}j}^2\rangle(\mathbf{S}\cdot\mathbf{e}_{\mathbf{q}jk})(\mathbf{S}_{\mathbf{q}jk'}^*)/(m_r m_{r'})^{1/2}].$$

In this equation $F(\mathbf{G}_i)$ is the structure factor of Bragg reflection $\mathbf{G}_i$, $|\mathbf{u_q}|$ is the distortion amplitude with wave vector $\mathbf{q}$, $\mathbf{S}$ is the diffraction vector ($\mathbf{S} = \mathbf{G}_i + \mathbf{q}$), N is the number of unit cells in the crystal, I_c is the electron scattering intensity, m_r denotes the mass of the rth atom and j denotes the mode of lattice modulation. The mean square amplitude of modulation of the jth branch with wave vector $\mathbf{q}$ is given by $\langle u_{\mathbf{q}jk}^2 \rangle$ and $\mathbf{e}_{\mathbf{q}jk}$ is the corresponding polarization vector. When $\mathbf{S}$ and $\mathbf{e}_{\mathbf{q}jk}$ are parallel the scalar product ($\mathbf{S} \cdot \mathbf{e}_{\mathbf{q}jk}$) will be unity, while it will be zero when $\mathbf{S}$ and $\mathbf{e}_{\mathbf{q}jk}$ are mutually perpendicular. This allows distinction to be made between modulations of different polarizations. The modulation may be either static, when measurement of the intensities of the satellites will give, after suitable analysis, the modulation amplitudes in various directions, or dynamic. Dynamic lattice modulations are due to phonons, and $\langle u_{\mathbf{q}j}^2 \rangle$ is proportional to $\langle E_{\mathbf{q}j} \rangle / w_{\mathbf{q}j}^2$, where $\langle E_{\mathbf{q}j} \rangle$ and $w_{\mathbf{q}j}$ are the average energy and angular frequency of the ($\mathbf{q}j$) phonon.

The diffuseness of the ghosts from an optical grating depends on the degree of correlation of the errors of ruling; the greater the correlation, the sharper the ghosts. Similar considerations apply to distorted crystals. The first studies of the {[TTF][TCNQ]} satellites used photographic methods for surveying regions of reciprocal space, and the dependence of the scattering on temperature. Counter methods were introduced later for more quantitative investigation of limited regions of the reciprocal lattice; presumably the next step will be to employ area detector diffractometers. We shall attempt to put the available results in perspective, starting with the structure of the insulating phase as it is generally easier to proceed from ordered to disordered structures.

The sharp satellite reflections at the lowest temperatures show that the structure tends towards complete order in this region. Below 38K, the incommensurate superstructure has dimensions $a' = 4a$, $b' = 3.4b$, $c' = c$ (unprimed values refer to the subcell) and the unit cell has the five-dimensional superspace group $P : P2_1/c : cmm$. (Bak and Janssen, 1978) The situation in {[TTF][TCNQ]} is complicated by the fact that satellites due to static and dynamic modulations have similar $\mathbf{q}$ wave vectors. The following quotation (Bouveret and Metgert, 1989) is illuminating:

The structure of the low temperature modulated phases was only slightly touched even if the available data were of first importance for the community. This state came from the fact that there was no sample big enough to perform exhaustive inelastic scattering studies and because the weakness of the satellite intensities (typically 10^{-3}–10^{-4} of the main Bragg reflections) rendered x-ray data collection time prohibitive. Nowadays new intense light sources like rotating anodes or synchrotron radiation facilities became customary and make almost possible what was unrealistic before.

Two independent measurements have been made of the intensities of the $2k_F$ group of satellite reflections of the insulating phase. Coppens et al. (1987) used synchrotron radiation to measure the intensities of 437 satellite reflections from a crystal at 15K. Bouveret and Metgert (1989) measured the intensities of 137 unique satellite reflections from a crystal at 13K, using Cu Kα from a rotating anode tube, The occurrence of the satellites was ascribed to displacements of the molecules by translational and librational modulation waves, whose parameters were determined by least squares analysis (Petricek et al., 1985); the Coppens group treated the molecules as rigid bodies while Bouveret and Megtert allowed for their segmentation into a number of separate rigid parts (e.g. the fulvalene rings, the quinoid ring, the $C(CN)_2$ groups). The largest modulation was a slip of

TTF in the direction of the long axis of the molecule with an amplitude of ≈ 0.02 Å; the TTF molecules are rigid. The TCNQ translational modulations are of similar nature but of even smaller magnitude, and there are also rotations of 0.1–0.6° about its long axis; the TCNQ molecules undergo intramolecular distortions involving displacement of the quinoid ring normal to the mean molecular plane. While it is difficult to generalize about modulation amplitudes, the {[TTF][TCNQ]} values would appear to be at the lower end of the range; in α-bis(1,2-benzoquinone dioximato)Pd(II) at 19.5K, for example, the commensurate superlattice has a transverse modulation wave with a maximum displacement of 0.84 Å (Kistenmacher and Destro, 1983; BZQXPD04, 12, 13). The zero-point translational amplitude of thermal vibration in an organic crystal would be expected to be about 0.1 Å at these temperatures, while librational amplitudes could lie in the range 0.5 to 3°. Thus the structural effects in {[TTF][TCNQ]} are remarkably subtle. Even these exacting analyses are incomplete as the much weaker group of $4k_F$ satellites was not included.

We now consider the changes in the diffraction pattern of {[TTF][TCNQ]} on heating from ≈ 20K. The behavior of {[TTF][TCNQ]}, in the temperature range between 20 and 54K is intricate with changes in both positions and intensities of superlattice reflections (Fig. 17.54); it is clear that gradual disordering is taking place and the superlattice structure (incommensurate along [010]) changes from $a' = 4a$, $b' = 3.4b$, $c' = c$ below 38K to $a' = 2a$, $b' = 3.4b$, $c' = c$ just below 54K. We shall not attempt to discuss the theoretical treatments proposed (Bak and Emery, 1976).

The satellite reflections are diffuse above 54K and typical diffractometer scattering curves obtained (at various temperatures) in two of the many zones investigated are shown in Figs. 17.55. There are diffuse peaks at $\eta = 0.29$, 0.41 and $0.59b^*$. The latter pair are

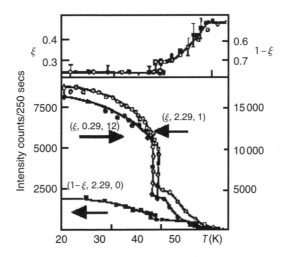

Fig. 17.54. In the upper portion of this figure the variation of the positions of the satellite reflections along the a^* axis with temperature is shown for the satellites at $0.29b^*$ in the (021), (00,12) and (020) zones. The corresponding intensity dependence is shown in the lower part of the figure. The index x represents the wave number of the lattice modulation along a^* in units of a^*. (Reproduced from Kagoshima *et al.*, 1976.)

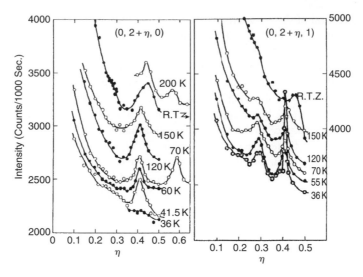

Fig. 17.55. Observed intensity in the (020) (left) and (021) (right) zones at several temperatures. The right hand scale is for the RT measurements and the left hand scale for all the other temperatures. The anomaly appearing at $0.41b^*$ below 150K moves toward $0.45b^*$ with increasing temperature. The sharp peak at $T = 36$K is an example of the satellites used in the determination of the modulation amplitudes. Note the absence of a diffuse peak at $0.29b^*$ in the left portion of the diagram, where the scattering vector **S** is parallel to **b**. (Reproduced from Kagoshima *et al.*, 1976.)

considered equivalent in the reduced zone scheme ($0.59 = 1 - 0.41$) and the two remaining peaks are considered to be related by a factor of 2; that at $\eta = 0.29b^*$ is the 2kF anomaly and that at $0.59b^*$ is the 4kF anomaly. The average charge density on the moieties is estimated from $Z = 2(2k_F/b^*) = 0.59$ (electrons for TCNQ and holes for TTF).[9] However, their different temperature dependences suggest that 4kF is not a simple second harmonic of 2kF, nor is it clear that the change in 4kF from 0.59 at ≈ 50K to 0.55 at 300K implies a parallel change in ρ. From the dependence of diffuse peak intensity on zone of measurement it appears that the 2kF anomaly is mainly polarized along c^* (transverse polarization) and the 4kF anomaly along b^* (longitudinal polarization). Kagoshima (1982) has noted that "Naively speaking the periodic lattice distortion should be a longitudinal mode." Thus the polarization of the 4kF anomaly fits in well with a modulation of the stacks of TTF and TCNQ moieties along b (**b** and **b*** are parallel in a monoclinic crystal) while that of the 2kF anomaly could be related to displacements of the stacks in the direction of the long molecular axes, which is close to c^*. However, the more generally accepted explanation is that tilting of the molecules in the stacks leads to perturbation of the longitudinal (along [010]) electron transport by the transverse (along **c***) mode and thus it is the 2kF anomaly that demonstrates the existence of charge density waves

---

[9] Other similar values come from XPS and Raman scattering measurements. Pouget (1988) gives a comprehensive collection of Z values for twelve radical cation : radical anion salts; all lie within the range 0.5 to 0.75.

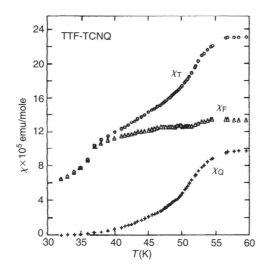

Fig. 17.56. The total susceptibility χ_T of {[TTF][TCNQ]} and the susceptibilities of the TTF (χ_F) and TCNQ (χ_Q) stacks as a function of temperature. (Reproduced from Tomkiewicz, 1980.)

along **b** and these, as described below, account for the high conductivity in this direction above 54K.

There is independent experimental evidence from ESR measurements indicating that the TTF and TCNQ stacks behave differently in this temperature range (Tomkiewicz, 1980). The susceptibility of the crystal can be decomposed into contributions from the separate stacks (Fig. 17.56). The metal–insulator transition at 54K is ascribed to ordering of the TCNQ stack, where the greatest change in susceptibility occurs in this temperature region, while the transition at 38K is ascribed to changes in the TTF stack; there do not appear to be clear magnetic effects of the 48K transition.

The complicated phase transition behavior of {[TTF][TCNQ]} suggests that there is much to be said for investigating an analogous but simpler material; {[TSeF][TCNQ]} is a good candidate because there appears to be only a single (metal-insulator) transition, at 29K, but it has been studied (Kagoshima *et al.*, 1978; Yamaji *et al.*, 1981) less extensively than {[TTF][TCNQ]}. Below 29K an incommensurate superlattice is formed with $a' = 2a$, $b' = 3.17b$, $c' = c$ (a, b, c are the lattice parameters of the subcell). A photographic study (Yamaji *et al.*, 1981) of a twinned crystal at 55K showed diffuse sheets of weak scattering normal to [010]* that derive from a $2k_F$ anomaly transversely polarized along **c***, as in {[TTF][TCNQ]}. The analysis was based solely on the TSeF molecules as, according to the authors, the contribution of the TCNQ molecules to the diffuse scattering is two orders of magnitude less and can thus be neglected. The TSeF molecules are found to slide along their long axes, as in {[TTF][TCNQ]}; translations of TSeF molecules are weakly correlated, being in-phase along [001] and out-of-phase along [010]. A $4k_F$ anomaly was not observed and thus does not appear to be required for the explanation of the high conductivity peak at low temperatures (Fig. 17.46).

The next stage is to relate the physical properties (we shall consider only the conductivity) of {[TTF][TCNQ]} and its analogs to the diffraction phenomena (Gill, 1986).

The electronic structure of {[TTF][TCNQ]} is derived from the one-dimensional stacking of TTF and TCNQ moieties along the b axis; there are two bands, an electron band on the TCNQ stack and a hole band on the TTF stack. These are filled to the $2k_F$ level; this value is measured as 0.59 electron (hole) from the diffuse scattering. Thus the stacks are composed of disordered arrays of molecules and ions (in the ratio 41 : 59) and a qualitative, chemically oriented explanation of the high conductivity along the stacks follows from the assumption of facile electron (hole) transfer between adjacent molecules and ions. A more physically oriented explanation is generally given in terms of charge density waves. Diffuse scattering shows there is a periodic structural distortion (PSD) along b. The PSD is accompanied by a corresponding redistribution of the conduction electrons in order to restore electrical neutrality along the stack (Fig. 17.57); the new distribution is the charge density wave. When the CDW is incommensurate with respect to the average lattice, it can slide along the stack in a coherent manner giving rise to a contribution to the b-axis conductivity which, according to Jérome (1980), can be as large as 80% of the total conductivity at $T \approx 60K$. This contribution drops to zero at 300K and also below the phase transitions where the CDW is pinned to the lattice.

A quantitative explanation of the temperature dependence of the conductivity along the stack axis is much more controversial, and perhaps a dozen models (or mechanisms for scattering of the electrons) have been proposed, of which only some will be noted here. The temperature dependence of the conductivity of {[TTF][TCNQ]} (above the region of the phase transitions) has been explained in terms of single-phonon scattering of electrons (Conwell, 1980), by two-phonon scattering (the libron model) (Weger, 1980), and by a combination of single-phonon scattering and charge-density wave (CDW) conductivity (Jérome, 1980; Jérome and Schultz, 1982). Weger (1980) has remarked: "Different materials display different mechanisms, and even the same material may change from one region to another when the temperature, pressure, doping or other conditions are varied. There is not yet complete consensus as to where {[TTF][TCNQ]} at 100K (say) is."

The resistivity $(\rho = \sigma^{-1})$ of normal three-dimensional metals is not zero because the electron waves are scattered by phonons; as the phonon amplitudes decrease with temperature the conductivity increases with decreasing temperature (e.g. Fig. 17.1). On

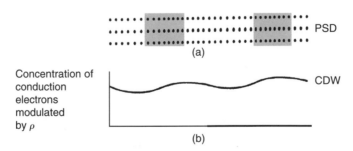

Fig. 17.57. The periodic structural distortion of a previously regular one-dimensional lattice is represented in (a); the regions of compression are shaded. The associated charge density wave (CDW) is shown in (b). Both diagrams are much exaggerated. (Reproduced from Gill, 1986.)

the simple Bloch model the resistivity is proportional to T, for $T > \Theta_D$ (the Debye temperature). For {[TTF][TCNQ]} it is found that $\rho = BT^n$, in the range 60–300 K, with $n \approx 2.3$ and B strongly pressure dependent; thus there appear to be differences between the behaviour of one and three dimensional metals (the basic physics of one-dimensional systems has been summarized by Schulz (1980).

A most comprehensive discussion has been given by Conwell (1988a, b). The room temperature conductivity of {[TTF][TCNQ]} is linearly dependent on pressure (Fig. 17.58; other TCNQ salts behave analogously) and this indicates that a significant portion of the temperature dependence of σ is actually due to volume dependence. When correction is made for this effect, then the resistivity at constant volume is found to have a nearly linear temperature dependence; thus the behaviour becomes more metal-like. Calculation, using a simplified model with one molecule per unit cell and neglect of mode mixing, shows that two-phonon scattering, and other sources of scattering, are negligible compared with acoustic one-phonon scattering, which dominates the resistivity at least above 200K, and that the TCNQ stacks contribute about four times as much to the conductivity as the TTF stacks. At lower temperatures there is a distinct dip in the σ–P curve at ≈ 20 kbar, the 80K behaviour being shown in Fig. 17.58. When this is coupled with diffraction evidence for the occurrence of commensurability at ≈ 20 kbar and 80K, then the conclusion is that charge density wave conductivity makes an appreciable contribution to the overall conductivity at lower temperatures; the charge density waves can slide at lower pressures and higher temperatures but are pinned to the lattice at higher pressures and lower temperatures. The transverse conductivity, lower than the longitudinal conductivity by a factor of 100 to 1000, is ascribed to hopping.

This brief account has neglected many important physical properties which can also be accounted for in terms of the above general picture; some references are given for completeness (Jacobsen, 1988; Scott, 1988).

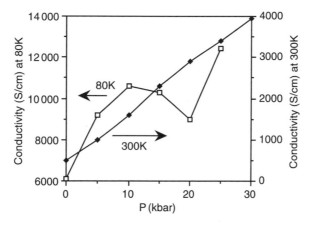

Fig. 17.58. Dependence of conductivity of {[TTF][TCNQ]} on pressure (Andrieux *et al.*, 1979). The room temperature dependence is approximately linear ($\sigma = 118P + 470$) but the 80K values show a distinct dip at $P \approx 20$ kbar.

17.10   Concluding summary

Such a tortuous path has been followed in this chapter that it is perhaps desirable to retrace the way in which we have developed the material; for convenience we include references to figures and chapter sections in which the various aspects are described in greater detail. Our major goal has been to describe the structures and physical properties of segregated stack π-molecular complexes, as we have done for other classes of molecular complexes and compounds in previous chapters. In order to keep the treatment within reasonable bounds, we concentrate on {[TTF][TCNQ]}, which is the most famous and widely studied example of such molecular complexes because of its unusually high electrical conductivity (Fig. 17.1). We begin in a wider context by introducing the current range of electron donors (Section 17.2.2) and acceptors (Section 17.2.3) used for formation of segregated stack π-molecular compounds. We then describe structures of cation and anion radical salts (i.e. those with closed shell counterions), emphasizing those of the TTF (Section 17.3.3) and (the much more numerous) TCNQ salts (Section 17.4). The principal structural feature common to these salts is the stacking of the cation (anion) radicals, and thus they are described as essentially *one-dimensional*. The TCNQ salts show considerable structural variety, with stacks extending in length from isolated π-dimers (Fig. 17.18) to unlimited. The structures of the stacks show considerable variety, with periodicities (the number of TCNQ moieties in a crystallographic repeat unit) ranging from diads through octads (Fig. 17.31). Indeed it seems reasonable to anticipate that the periodicity of TCNQ stacks will be limited only by the size of the counterions, with the fine structure within the periods determined by the mutual interaction between cation arrangement and anion stacking; for example, the octad stacks shown in Fig. 17.31 consist of two tetrads, each made up of two π-dimers. The stack axis conductivities of the crystals depend critically on the fine structure of the TCNQ stacks. The overall phase behaviour (and hence some of physical properties) of TCNQ salts are often strongly influenced by the cation arrangement, whether this is ordered or disordered and whether there is cation–cation hydrogen bonding and/or cation–anion hydrogen bonding; examples such as DMM $(TCNQ)_2$ and MEM $(TCNQ)_2$ are discussed in Section 17.4.9. The limited available evidence suggests that these conclusions can be generalised to other anions (e.g. chloranil and $M(dmit)_2$ (Section 17.4)) and cations (e.g. TTF and, to a lesser extent, TMPD (Section 17.3.4)). Thus stacking appears to be *the* essential structural feature of one dimensional cation and anion radical salts.

   The group of cation radical, anion radical salts discussed here all have their cations and anions arranged in separate homologous (hence "segregated") stacks, and the molecular complexes of greatest current interest, exemplified here by {[TTF][TCNQ]} (Section 17.7), have average structures with monad stacks. However, {[TTF][TCNQ]} is not entirely typical because it crystallizes in the monoclinic system whereas most of the segregated stack π-molecular compounds are triclinic, and a very few orthorhombic (Table 17.10). These crystallographic differences imply different mutual arrangements of symmetry-related stacks. The average structure of {[TTF][TCNQ]} does not change appreciably with temperature but the detailed structure, as revealed by the diffuse scattering in the diffraction pattern, is dependent on temperature (and also on pressure, although this aspect has not been considered at any length here). In thermodynamic terms the temperature dependence is shown by the occurrence of three phase changes at 54, 48 and 38K at ambient pressure (Fig. 17.50). However, these phase changes are very different in nature from the first or second order changes encountered in other crystalline molecular complexes and compounds; there are no appreciable changes of average structure and the changes in enthalpy ($\approx$5 J/mol) and entropy ($\approx$0.03 R) are very small. The diffuse scattering shows an incommensurate modulation of the structure along [010], the stack axis, and commensurate modulations in the other directions. The incommensurate modulation indicates that the average charges on the moieties are $Z = 0.59$ (holes for the TTF stack and electrons for the TCNQ stack); in chemical terms one can say that each stack has a disordered array of molecules and ions in ratio 0.41:0.59 and that the facile exchange of charge between the moieties leads to the high conductivity. The parallel description due to physicists is couched in terms of partially filled energy bands. The key to the conductivity is provided by the existence of a disordered array of molecules and ions or, alternatively, a partially filled band. {[TTF][TCNQ]} becomes ordered as the temperature is reduced below 38K (Fig. 17.47) and measurement (at $\approx$ 15K) and analysis of the intensities of the satellite reflections (no longer diffuse) suggests very small and subtle translational ($\approx$0.02 Å) and librational ($\approx$0.1–0.6°) modulations in the TTF and TCNQ arrangement. None of its analogs has been studied to the same extent as {[TTF][TCNQ]} itself, and so it is not yet clear what generalizations are permissible.

One other caveat is needed. The remarks above apply to "essentially one-dimensional" structures but it is clear that many donor-acceptor π-compounds have structures appreciably different from the mixed and segregated stack prototypes. Sometimes there are clear indications of two- (Fig. 17.42), and even three-dimensional interactions, while in other examples the structures are quite unexpected (Fig. 17.44). It seems inevitable that further work will present us with a landscape even more variegated than that which we have today.

References

Abashev, G. G., Vlasova, R. M., Kartenko, N. F., Kuzmin, A. M., Rozhdestvenskaya, I. V., Semkin, V. N., Usov, D. A. and Russkikh, V. S. (1987). *Acta Cryst.*, C43, 1108–1112.

Abrahams, S. C. (1955). *Acta Cryst.*, 8, 661–671.

Acker, D. S. and Hertler, W. R. (1962). *J. Am. Chem. Soc.*, 84, 3370–3374.

Aharon-Shalom, E., Becker, J. Y. and Agranat, I. (1979). *Nouv. J. Chim.*, 3, 643–645.

Alcácer, L. and Maki, A. H. (1974). *J. Phys. Chem.*, 78, 215–217.

Alcácer, L. (1985). *Mol. Cryst. Liq. Cryst.*, 120, 221–228.

Alcácer, L., Novais, H., Pedroso, F., Flandrois, S., Coulon, C., Chasseau, D. and Gaultier, J. (1980). *Solid State Comm.*, 35, 945–949.

Alizon, J., Galice, J., Robert, H., Delplanque, G., Weyl, C., Fabre, C. and Strzelecka, H. (1976). *Mol. Cryst. Liq. Cryst.*, 33, 91–100.

Amzil, J., Catel, J.-M., Costumer, J. le, Mollier, Y., Sauve, J.-P. and Flandrois, S. (1986). *Mol. Cryst. Liq. Cryst.*, 133, 333–353.

Andersen, J. R. and Jorgensen, O. (1979). *J. Chem. Soc., Perkin I*, pp. 3095–3098.

Andersen, J. R., Bechgaard, K., Jacobsen, C. S., Rindorf, G. Soling, H. and Thorup, N. (1978). *Acta Cryst.*, B34, 1901–1905.

Andersen, J. R., Bechgaard, K., Pedersen, H. J., Chasseau, D. and Gaultier, J. (1982). *Mol. Cryst. Liq. Cryst.*, **85**, 187–193.

Andersen, J. R., Engler, E. M. and Bechgaard, K. (1978). *Ann. N. Y. Acad. Sci.*, **313**, 293–300.

Andrieux, A., Schulz, H. J., Jerome, D. and Bechgaard, K. (1979). *Phys. Rev. Lett.*, **43**, 227–230.

Anzai, H., Toumoto, M. and Saito, G. (1982). *Mol. Cryst. Liq. Cryst.*, **125**, 385–392.

Aronson, S. and Mitelman, J. S. (1981). *J. Solid State Chem.*, **36**, 221–224.

Ashton, P. R., Balzani, V., Becher, J., Credi, A., Fyfe, M. C. T., Mattersteig, G., Menzer, S., Nielsen, M. B., Raymo, F. M., Stoddart, J. F., Venturi, M. and Williams, D. J. (1999). *J. Am. Chem. Soc.*, **121**, 3951–3957.

Ashwell, G. J., Allen, D. W., Kennedy, D. A. and Nowell, I. W. (1982). *Acta Cryst.*, B**38**, 2525–2528.

Ashwell, G. J., Bartlett, V. E., Davies, J. K., Eley, D. D., Wallwork, S. C., Willis, M. R., Harper, A. and Torrance, A. C. (1977). *Acta Cryst.*, B**33**, 2602–2607.

Ashwell, G. J., Cross, G. H., Kennedy, D. A., Nowell, I. W. and Allen, J. G. (1983). *J. Chem. Soc., Perkin II*, pp. 1787–1791.

Ashwell, G. J., Eley, D. D., Drew, N. J., Wallwork, S. C. and Willis, M. R. (1977). *Acta Cryst.*, B**33**, 2598–2602.

Ashwell, G. J., Eley, D. D., Drew, N. J., Wallwork, S. C. and Willis, M. R. (1978). *Acta Cryst.*, B**34**, 3608–3612.

Ashwell, G. J., Eley, D. D., Fleming, R. J., Wallwork, S. C. and Willis, M. R. (1976). *Acta Cryst.*, B**32**, 2948–2952.

Ashwell, G. J., Eley, D. D., Harper, A., Torrance, A. C., Wallwork, S. C. and Willis, M. R. (1977). *Acta Cryst.*, B**33**, 2258–2263.

Ashwell, G. J., Eley, D. D., Wallwork, S. C. and Willis, M. R. (1975). *Proc. Roy. Soc. Lond.*, A**343**, 461–475.

Ashwell, G. J., Eley, D. D., Wallwork, S. C., Willis, M. R., Peachey, G. F. and Wilkos, D. B. (1977). *Acta Cryst.*, B**33**, 843–848.

Ashwell, G. J., Eley, D. D., Wallwork, S. C., Willis, M. R., Welch, G. D. and Woodward, J. (1977). *Acta Cryst.*, B**33**, 2252–2257.

Ashwell, G. J., Wallwork, S. C., Baker, S. R. and Berthier, P. I. C. (1975). *Acta Cryst.*, B**31**, 1174–1178.

Aumüller, A. and Hünig, S. (1986a). *Liebigs Ann. Chem.*, **142**, 142–164.

Aumüller, A. and Hünig, S. (1986b). *Liebigs Ann. Chem.*, **142**, 165–176.

Aumüller, A., Erk, P., Meixner, H., Schütz, J.-U. von, Gross, H. J., Langohr, U., Werner, H.-P., Wolf, H. C., Burschka, C., Klebe, G., Peters, K. and Schnering, H. G. von, (1988). *Synth. Met.*, **27**, B 181–188.

Bailey, J. C. and Chesnut, D. B. (1969). *J. Chem. Phys.*, **51**, 5118–5128.

Bak, P. and Emery, V. J. (1976). *Phys. Rev. Lett.*, **36**, 978–982.

Bak, P. and Janssen, T. *Phys. Rev.*, B**17**, 436–439 (1978).

Bandrauk, A. D., Ishii, K., Truong, K. D., Aubin, M. and Hanson, A. W. (1985). *J. Phys. Chem.*, **89**, 1478–1485.

Bandrauk, A. D., Truong, K. D., Carlone, C., Jandl, S. and Ishii, K. (1985). *J. Phys. Chem.*, **89**, 434–442.

Batail, P., Ouhab, L., Halet, J.-F., Padiou, J., Lequon, M. and Lequon, R. M. (1985). *Synth. Met.*, **10**, 415–425.

Bechgaard, K. and Andersen, J. K. "Molecular properties of molecules used in conducting organic solids," in *The Physics and Chemistry of Low Dimensional Materials*, edited by L. Alcácer, NATO ASI Series C, Reidel: Dordrecht etc., pp. 247–263 (1980).

Bechgaard, K., Kistenmacher, T. J., Bloch, A. N. and Cowan, D. O. (1977). *Acta Cryst.*, B**33**, 417–422.

Bell, S. E., Field, J. S. and Haines, R. J. (1991). *J. Chem. Soc., Chem. Comm.*, pp. 489–491.

Berg, C., Bechgaard, K., Andersen, J. R. and Jacobsen, C. S. (1976). *Tetrahedron Letts.*, pp. 1719–1720.

Berger, P. A., Dahm, D. J., Johnson, G. R., Miles, M. G. and Wilson, J. D. (1975). *Phys. Rev.*, B**12**, 4085–4089.

Berlinsky, A. J. (1976). *Contemporary Physics*, **17**, 331–354.

Bodegom, B. van and Bosch, A. (1981). *Acta Cryst.*, B**37**, 863–868.

Bodegom, B. van, and Boer, J. L. de. (1981). *Acta Cryst.*, B**37**, 119–125.

Bodegom, B. van, Boer, J. L. de and Vos, A. (1977). *Acta Cryst.*, B**33**, 602–604.

Boer, J. L. de, Vos, A. and Huml, K. (1968). *Acta Cryst.*, B**24**, 542–549.

Boer, J. L. de, and Vos, A. (1972a). *Acta Cryst.*, B**28**, 835–839.

Boer, J. L. de, and Vos, A. (1972b). *Acta Cryst.*, B**28**, 839–848.

Bollinger, C. M., Darkwa, J., Gammie, G., Gammon, S. D., Lyding, J. W., Rauchfuss, T. B. and Wilson, S. R. (1986). *Organometallics*, **5**, 2386–2388.

Born, L. and Heywang, G. (1990). *Z. Kristallogr.*, **190**, 147–152.

Born, L. and Heywang, G. (1991). *Z. Kristallogr.*, **197**, 223–233.

Bosch, A. and Bodegom, B. van. (1977). *Acta Cryst.*, B**33**, 3013–3021.

Bousseau, M., Valde, L., Legros, J.-P., Cassoux, P., Garbauskas, M. and Interrante, L. V. (1986). *J. Am. Chem. Soc.*, **108**, 1908–1916.

Bouveret, Y. and Megtert, S. (1989). *J. Phys. France,* **50**, 1649–1671.

Boyd, R. H. and Phillips, W. D. (1965). *J. Chem. Phys.*, **43**, 2927–2929.

Brook, T. J. R. and Koch, T. H. (1997). *J. Mater. Chem.*, **7**, 2381–2388.

Browall, K. W., Bursh, T., Interrante, L. V. and Kasper, J. S. (1972). *Inorg. Chem.*, **11**, 1800–1806.

Bryce, M. R. and Howard, J. A. K. (1982). *Tetrahedron Letts.*, **23**, 4273–4276.

Bryce, M. R., Moore, A. J., Bates, P. A., Hursthouse, M. B., Liu, Z.-X. and Nowak, M. J. (1988). *J. Chem. Soc., Chem. Comm.*, pp. 1441–1442.

Bryce, M. R., Moore, A. J., Lorcy, D., Dhindsa, A. S. and Robert, A. (1990). *J. Chem. Soc., Chem. Comm.*, pp. 470–472.

Büchner, R., Field, J. S. and Haines, R. J. (1997). *J. Chem. Soc., Dalton Trans.*, pp. 2403–2408.

Bunzel, W., Vögtle, F., Franken, S. and Puff, H. (1984). *J. Chem. Soc., Chem. Comm.*, pp. 1035–1037.

Burggraf, M., Dragan, H., Gruner-Bauer, P., Helberg, H. W., Kuhs, W. F., Mattern, G., Muller, D., Wendl, W., Wolter, A. and Dormann, E. (1995). *Z. Phys. B Condensed Matter*, **96**, 439–450.

Chasseau, D., Gaultier, J., Hauw, C. and Jaud, J. (1973a). *Compt. Rend. Acad. Sci., Paris, Ser. C*, **276**, 661–664.

Chasseau, D., Gaultier, J., Hauw, C. and Jaud, J. (1973b). *Compt. Rend. Acad. Sci., Paris, Ser. C*, **276**, 751–753.

Chasseau, D., Comberton, G., Gaultier, J. and Hauw, C. (1978). *Acta Cryst.*, B**34**, 689–691.

Chasseau, D., Gaultier, J. and Hauw, C. (1972). *Compt. rend. Acad. Sci., Paris, Ser. C*, **274**, 1434–1437.

Chasseau, D., Gaultier, J. and Hauw, C. (1976). *Acta Cryst.*, B**32**, 3262–3266.

Chasseau, D., Gaultier, J., Hauw, C. and Schvoerer, M. (1972). *Compt. rend. Acad. Sci., Paris, Ser. C*, **275**, 1491–1493.

Chasseau, D., Gaultier, J., Hauw, C., Fabre, J. M., Giral, L., and Torreiles, E. (1978). *Acta Cryst.*, B**34**, 2811–2818.

Chen, W., Cava, M. P., Takassi, M. A. and Metzger, R. M. (1988). *J. Am. Chem. Soc.*, **110**, 7903–7904.

Chesnut, D. B. and Arthur, P., Jr. (1962). *J. Chem. Phys.*, **36**, 2969–2975.

Chiang, T. C., Reddoch, A. H. and Williams, D. F. (1971). *J. Chem. Phys.*, **54**, 2051–2055.

Cohen, M. J. and Heeger, A. J. (1977). *Phys. Rev.*, B**16**, 688–696.

Cohen, M. J., Coleman, L. B., Garito, A. F. and Heeger, A. J. (1976). *Phys. Rev.*, **B13**, 5111–5116.

Coleman, L. B., Cohen, M. J., Sandman, D. J., Yamagishi, F. G., Garito, A. J. and Heeger, A. J. (1973). *Sol. State Commun.*, **12**, 1125–1132.

Conwell, E. M. (1980). "Phonon scattering in quasi 1-d conductors," in *The Physics and Chemistry of Low Dimensional Solids*, edited by L. Alcácer, NATO ASI Series C, Reidel, Dordrecht etc., Vol. 56, pp. 213–222.

Conwell, E. M. (1988a). "Introduction to highly conducting quasi-one-dimensional organic crystals" in *Semiconductors and Semimetals*, Vol. 27 (Highly Conducting Quasi-One-Dimensional Organic Crystals, edited by E. M. Conwell), Academic Press, Boston, pp. 1–27.

Conwell, E. M. (1988b). "Transport in quasi-one-dimensional conductors," in *Semiconductors and Semimetals*, Vol. 27 (Highly Conducting Quasi-One-Dimensional Organic Crystals, edited by E. M. Conwell), Academic Press, Boston, pp. 215–292.

Cooper, W. F., Edmonds, J. W., Wudl, F. and Coppens, P. (1974). *Cryst. Struct. Comm.*, **3**, 23–26.

Coppens, P., Petricek, V., Levendis, D., Larsen, F. K., Yan, Y. and LeGrand, A. D. (1987). *Phys. Rev. Lett.*, **59**, 1695–1697.

Corfield, P. W. R. and LaPlaca, S. J. (1996). *Acta Cryst.*, **B52**, 384–387.

Cowan, D. O., Kini, A., Chiang, L.-Y., Lerstrup, K., Talham, D. R., Poehler, T. O. and Bloch, A. N. (1982). *Mol. Cryst. Liq. Cryst.*, **86**, 1–26.

Craven, R. A., Salamon, M. B., Depasquali, G., Herman, R. M., Stucky, G. D. and Schultz, A. J. (1974). *Phys. Rev. Lett.*, **32**, 769–772.

Daoben, Z., Ping, W., Minwie, Q. and Renyuan, Q. (1985). *Mol. Cryst. Liq. Cryst.*, **128**, 321–327.

Detty, M. R., Murray, B. J. and Perlstein, J. H. (1983). *Tetrahedron Letts.*, **24**, 539–542.

Deuchert, K. and Hünig, S. (1978). *Angew. Chem. Int. Ed. Engl.*, **17**, 875–886.

Diekmann, J., Hertler, W. R. and Benson, R. E. (1963). *J. Org. Chem.*, **28**, 2719–2724.

Dietz, K., Endres, H., Keller, H. J. and Moroni, W. (1981). *Z. Naturforsch.*, **36b**, 952–955.

Dietz, K., Keller, H. J. and Wehe, D. (1984). *Acta Cryst.*, **C40**, 257–260.

Djurek, D., Franulovi'c, K., Prester, M., Tomi'c, S., Giral, L. and Fabre, J. M. (1977). *Phys. Rev. Lett.*, **38**, 715–718.

Dovesi, R., Saunders, V. B. and Roetti, C. (1992). CRYSTAL92 Users Manual (Torino, Italy and Daresbury, U. K.)

Dunitz, J. D. (1985). *Science*, **228**, 353–354.

Dupuis, P. and Neel, J. (1969). *Compt. rend. Acad. Sci. Paris, Ser. C*, **268**, 557–560, 653–655.

Dupuis, P., Flandrois, S., Delhaes, P. and Coulon, C. (1978). *J. Chem. Soc. Chem. Comm.*, pp. 337–338.

Ehrenfreund, E., Steinitz, M. O. and Nigrey, P. J. (1981). *Mol. Cryst. Liq. Cryst.*, **69**, 173–176.

Elbl, K., Krieger, C. and Staab, H. A. (1986). *Angew. Chem. Int. Ed. Engl.*, **25**, 1023–1024.

Ellern, A., Bernstein, J., Becker, J. Y., Zamir, S., Shahal, L. and Cohen, S. (1994). *Chem. Mater.*, **6**, 1378–1385.

Emge, T. J., Bryde, W. A., Wiygul, F. M., Cowan, D. O. and Kistenmacher, T. J. (1982). *J. Chem. Phys.*, **77**, 3188–3197.

Emge, T. J., Wiygul, F. M., Ferraris, J. P. and Kistenmacher, T. J. (1981). *Mol. Cryst. Liq. Cryst.*, **78**, 295–310.

Endres, H. (1982). *Angew. Chem. Int. Ed. Engl.*, **21**, 524.

Endres, H. (1984). *Angew. Chem. Int. Ed. Engl.*, **23**, 999–1000.

Endres, H., Keller, H. J., Moroni, W., Nöthe, D. and Vu Dong. (1978a). *Acta Cryst.*, **B34**, 1703–1705.

Endres, H., Keller, H. J., Moroni, W., Nöthe, D. and Vu Dong. (1978b). *Acta Cryst.*, **B34**, 1823–1827.

Endres, H., Keller, H. J., Müller, B. and Schweitzer, D. (1985). *Acta Cryst.*, **C41**, 607–613.

Endres, H., Keller, H. J., Queckbörner, J., Schweitzer, D. and Veigel, J. (1982). *Acta Cryst.*, **B38**, 2855–2860.

Enkelmann, V., Göckelmann, K., Wieners, G. and Monkenbusch, M. (1985). *Mol. Cryst. Liq. Cryst.*, **120**, 195–204.

Enkelmann, V., Morra, B. S., Krönke, Ch., Wegner, G. and Heinze, J. (1982). *Chem. Phys.*, **66**, 303–313.

Espinosa, E., Molins, E. and Lecomte, C. (1997). *Phys. Rev.*, **B56**, 1820–1833.

Etemad, S., Penney, T., Engler, E. M., Scott, B. A. and Seiden, P. (1975). *Phys. Rev. Lett.*, **34**, 741–744.

Euler, W. B., Melton, M. E. and Hoffman, B. M. (1982). *J. Am. Chem. Soc.*, **104**, 5966–5971.

Fabre, J. M., Torreilles, E. and Giral, L. (1978). *Tetrahedron Letts.*, No. 39, pp. 3703–3706.

Fanghängel, E., Schukat, G., Schuetzenduebel, J. and Humsch, W. (1983). *J. Prakt. Chem.*, **325**, 976–980.

Farges, J.-P. (1980). "Electronic properties and new forms of instabilities in TCNQ-salts with intermediate conductivity," in *The Physics and Chemistry of Low Dimensional Solids*, edited by L. Alcácer, Reidel, Dordrecht, NATO ASI Ser. C, Vol. 56, pp. 223–232.

Farges, J.-P. (1985a). *J. Phys. (Les Ulis, Fr)*, **46**, 465–472.

Farges, J.-P. (1985b). *J. Phys. (Les Ulis, Fr)*, **46**, 1249–1254.

Ferraris, J. P. and Saito, G. (1979). *J. Chem. Soc., Chem. Comm.*, pp. 992–993.

Ferraris, J. P., Cowan, D. O., Walatka, V. V., Jr. and Perlstein, J. H. (1973). *J. Am. Chem. Soc.*, **95**, 948–949.

Figgis, B. N., Sobolev, A. N., Kepert, C. J. and Kurmoo, M. (2001). *Acta Cryst.*, **C51**, 991–993.

Filhol, A. (1994). In *Organic Conductors: Fundamentals and Applications*, edited by J. P. Farges, Marcel Dekker Inc., New York.

Filhol, A. and Thomas, M. (1984). *Acta Cryst.*, **B40**, 44–59.

Filhol, A., Bravic, G., Gaultier, J., Chasseau, D. and Bethier, C. (1981). *Acta Cryst.*, **B37**, 1225–1235.

Filhol, A., Zeyen, C. M. E., Chenavas, P., Gaultier, J. and Delhaes, P. (1980). *Acta Cryst.*, **B36**, 2719–2726.

Flandrois, S., Chasseau, D., Delhaes, P., Gaultier, J., Amiell, J. and Hauw, C. (1979). *Bull. Chem. Soc. Jpn.*, **52**, 3407–3414.

Friedel, J. and Jérome, D. (1982). *Contemporary Physics*, **23**, 583–624.

Friend, R. H. and Jérome, D. (1979). *J. Phys. C*, **12**, 1441–1477.

Fritchie, C. J. and Arthur, P. Jr. (1966). *Acta Cryst.*, **21**, 139–145.

Fritchie, C. J., Jr. (1966). *Acta Cryst.*, **20**, 892–898.

Fritz, H. P., Gebauer, H., Friedrich, P., Ecker, P., Artes, R. and Schubert, U. (1978). *Z. Naturforsch.*, **B33**, 498–506.

Gallois, B., Gaultier, J., Granier, T., Ayroles, R. and Filhol, A. (1985). *Acta Cryst.*, **B41**, 56–66.

Gama, V., Almeida, M., Henriques, R. T., Santos, I. C., Domingos, A., Ravy, S. and Pouget, J. P. (1991). *J. Phys. Chem.*, **95**, 4263–4267.

Gama, V., Henriques, R. T., Bonfait, G., Almeida, M., Meetsma, M., van Smaalen, S. and de Boer, J. L. (1992). *J. Am. Chem. Soc.*, **114**, 1986–1989.

Gama, V., Henriques, R. T., Bonfait, G., Pereira, L. C., Waerenborg, J. C., Santos, I. C., Duarte, M. T., Cabral, J. M. P. and Almeida, M. (1992). *Inorg. Chem.*, **31**, 2598–2604.

Gama, V., Henriques, R. T., Almeida, M., Veiros, L., Calhorda, M. J., Meetsma, A. and de Boer, J. L. (1993). *Inorg. Chem.*, **32**, 3705–3711.

Gammell, J. T. and Krumhansl, J. A. (1983). *Phys. Rev.*, **B27**, 1659–1668.

Garito, A. F. and Heeger, A. J. (1974). *Accts. Chem. Res.*, **7**, 232–240.

Gemmer, R. V., Cowan, D. O., Poehler, T. O., Bloch, A. N., Pyler, R. E. and Banks, R. H. (1975). *J. Org. Chem.*, **40**, 3544–3547.

Gill, J. C. (1986). *Contemp. Phys.*, **27**, 37–59.

Goldberg, S. Z., Eisenberg, R., Miller, J. S. and Epstein, A. J. (1976). *J. Am. Chem. Soc.*, **98**, 5173–5182.

Goldberg, S. Z., Spivack, B., Stanley, G., Eisenberg, R., Braitsch, D. M., Miller, J. S. and Abkowitz, M. (1977). *J. Am. Chem. Soc.*, **99**, 110–117.

Goldstein, P., Seff, K. and Trueblood, K. N. (1968). *Acta Cryst.*, **B24**, 778–791.

Gordon, R. G. and Kim, Y. S. (1972). *J. Chem. Phys.*, **56**, 3122–3133.

Govers, H. A. J. (1978). *Acta Cryst.*, **A34**, 960–965.

Graja, A., Przybylski, M., Swietlik, R., Wallwork, S. C., Willis, M. R. and Rajchel, A. (1983). *Mol. Cryst. Liq. Cryst.*, **100**, 373–387.

Günther, G., Hünig, S., Peters, K., Rieder, H., Schnering, H. G. von, Schütz, J.-U. von, Söderholm, S., Werner, H.-P. and Wolf, H. C. (1990). *Angew. Chem., Int Ed Engl.*, **29**, 204–205.

Guy, D. R. P., Marseglia, E. A., Parkin, S. S. P., Friend, R. H. and Bechgaard, K. (1982). *Mol. Cryst. Liq. Cryst.*, **79**, 337–341.

Halfpenny, J. (1985). *Acta Cryst.*, **C41**, 119–121.

Hanson, A. W. (1968). *Acta Cryst.*, **B24**, 768–778.

Harms, R. H., Keller, H. J., Nöthe, D. and Wehe, D. (1982). *Acta Cryst.*, **B38**, 2838–2841.

Hart, H. and Sasaoka, M. (1978). *J. Am. Chem. Soc.*, **100**, 4326–4327.

Henriques, R. T., Alcácer, L., Pouget, J. P. and Jerome, D. (1984). *J. Phys. C, Solid State Physics*, **17**, 5197–5208.

Henriques, R. T., Almeida, M., Matos, M. J., Alcácer, L., and Bourbonnais, C. (1987). *Synth. Met.*, **19**, 379–384.

Herbstein, F. H. (2000). *Acta Cryst.*, **B56**, 547–557.

Heywang, G., Born, L., Fitzky, H.-G., Hassel, T., Hocker, J., Müller, H.-K., Pittel, B. and Roth, S. (1989). *Angew. Chem. Int. Ed. Engl.*, **28**, 483–485.

Hoekstra, A., Spoelder, T. and Vos, A. (1972). *Acta Cryst.*, **B28**, 14–25.

Hove, M. J., Hoffman, B. M. and Ibers, J. A. (1972). *J. Chem. Phys.*, **56**, 3490–3502.

Howard, I. A. (1988). "A reference guide to the conducting organic quasi-one-dimensional molecular crystals," in *Semiconductors and Semimetals*, edited by E. Conwell, Academic Press, Boston Vol. 27, pp. 29–85,

Hurtley, W. R. H. and Smiles, S. (1926). *J. Chem. Soc.*, **129**, 2263–2270.

Ikemoto, I., Katagiri, G., Nishimura, S., Yakushi, K. and Kuroda, H. A. (1979). *Acta Cryst.*, **B35**, 2264–2265.

Ikemoto, I., Yamada, M., Sugano, T. and Kuroda, H. (1980). *Bull. Chem. Soc. Jpn.*, **53**, 1871–1876.

Inabe, T., Mitsuhashi, T. and Maruyama, Y. (1988). *Bull. Chem. Soc. Jpn.*, **61**, 4215–4224.

Inabe, T., Okinawa, K., Ogata, H., Okamoto, H., Mitani, T. and Maruyama, Y. (1993). *Acta Chim., Hung.*, MODELS IN CHEMISTRY **130**, 537–554.

Interrante, L. V., Browall, K. W., Hart, H. R., Jr., Jacobs, I. S., Watkins, G. D. and Wee, S. H. (1975). *J. Am. Chem. Soc.*, **97**, 889–890.

Ishii, K., Kanako, K. and Kuroda, H. (1976). *Bull. Chem. Soc. Jpn.*, **49**, 2077–2081.

Iwasawa, N., Saito, G., Imaeda, K., Mori, T. and Inokuchi, H. (1987). *Chem. Lett.*, pp. 2399–2402.

Jacobsen, C. S. (1988). "Optical properties," in *Semiconductors and Semimetals*, Vol. 27 (Highly Conducting Quasi-One-Dimensional Organic Crystals, edited by E. M. Conwell), Academic Press, Boston, pp. 293–384.

James, R. W. (1950). *The Optical Principles of the Diffraction of X-Rays*, see pp. 205–207, 563–571; Bell, London.

Jérome, D. and Schulz, H. J. (1982). "Quasi-one-dimensional conductors: the Peierls instability, pressure and fluctuation effects,, in *Extended Linear Chain Compounds*, edited by J. S. Miller, Plenum Press, London, Vol. 2, 159–204.

Jérome, D. (1980). "Fluctuating collective conductivity and single-particle conductivity in 1-D organic conductors," in *The Physics and Chemistry of Low Dimensional Solids*, edited by L. Alcácer, NATO ASI Series C, Reidel Dordrecht, Vol 56, pp. 123–142.

Johnson, C. K. and Watson, C. R., Jr. (1976). *J. Chem. Phys.*, **64**, 2271–2286.

Johnson, C. K., Watson, C. R., Jr. and Warmack, R. J. (1975). Abstracts of American Crystallographic Assocn. Meeting, **3**, 19.

Kagoshima, S. (1982). "x-Ray, neutron, and electron scattering studies of one-dimensional inorganic and organic conductors", in *Extended Linear Chain Compounds*, edited by J. S. Miller, Plenum Press, New York and London, **2**, 303–337.

Kagoshima, S., Ishiguro, T. and Anzai, H. (1976). *J. Phys. Soc. Jpn.*, **41**, 2061–2071.

Kagoshima, S., Ishiguro, T., Schulz, T. D. and Tomkiewicz, Y. (1978). *Solid State Commun.*, **28**, 485–490.

Kaminskii, V. F., Shibaeva, R. P. and Atovmyan, L. O. (1973a). *J. Struct. Chem. USSR*, **14**, 645–650.

Kaminskii, V. F., Shibaeva, R. P. and Atovmyan, L. O. (1973b). *J. Struct. Chem. USSR*, **14**, 1014–1019; *Zh. Strukt. Khim.*, **14**, 1082–1088. .

Kamminga, P. and van Bodegom, B. (1981). *Acta Cryst.*, B**37**, 114–119.

Kaplan, M. L. (1976). *J. Cryst. Growth*, **33**, 161–165.

Katayama, C. (1985). *Bull. Chem. Soc. Jpn.*, **58**, 2272–2278.

Katayama, C., Honda, M., Kumagai, H., Tanaka, J., Saito, G. and Inokuchi, H. (1985). *Bull. Chem. Soc. Jpn.*, **58**, 2272–2278.

Kathirgamanathan, P. and Rosseinsky, D. R. (1980). *J. Chem. Soc., Chem. Comm.*, pp. 356–357.

Kathirgamanathan, P., Mazid, M. A. and Rosseinsky, D. R. (1982). *J. Chem. Soc., Perkin II* pp. 593–596.

Kathirgamanathan, P., Mucklejohn, S. A. and Rosseinsky, D. R. (1979). *J. Chem. Soc., Chem. Comm.*, pp. 86–87.

Kato, R., Kobayashi, H., Kobayashi, A. and Sasaki, Y. (1985). *Chem. Letts.*, pp. 131–134.

Kato, R., Mori, T., Kobayashi, A., Sasaki, Y. and Kobayashi, H. (1984). *Chem. Lett.*, pp. 1–4.

Keller, H. J. Steiger, W. and Werner, M. (1981). *Z. Naturforsch.*, **36**B, 1187–1189.

Keller, H. J., Nöthe, D., Pritzkow, H., Wehe, D., Werner, M., Koch, P. and Schweitzer, D. (1980). *Mol. Cryst. Liq. Cryst.*, **62**, 181–200.

Khanna, S. K., Pouget, J. P., Comes, R., Garito, A. F. and Heeger, A. J. (1977). *Phys. Rev.*, B **16**, 1468–1479.

Khidekel, M. L. and Zhilyaeva, E. L. (1981). *Synth. Metals*, **4**, 1–34.

Kistenmacher, T. J. and Destro, R. (1983). *Inorg. Chem.*, **22**, 2104–2110.

Kistenmacher, T. J., Phillips, T. E. and Cowan, D. O. (1974). *Acta Cryst.*, B**30**, 763–768.

Kistenmacher, T. J., Phillips, T. E., Cowan, D. O., Ferraris, J. P., Bloch, A. N. and Poehler, T. O. (1976). *Acta Cryst.*, B**32**, 539–547.

Kistenmacher, T. J., Rossi, M., Chiang, C. C., Van Duyne, R. P., Cape, T. and Siedle, A. R. (1978). *J. Am. Chem. Soc.* **100**, 1958–1959.

Kistenmacher, T. J., Rossi, M., Chiang, C. C., Van Duyne, R. P. and Siedle, A. R. (1980). *Inorg. Chem.*, **19**, 3604–3608.

Kobayashi, A., Sasaki, Y., Kato, R. and Kobayashi, H. (1986). *Chem. Letts.*, pp. 387–390.

Kobayashi, H. and Kobayashi, K. (1977). *Bull. Chem. Soc. Jpn.*, **50**, 3127–3130.

Kobayashi, H. (1974). *Bull. Chem. Soc. Jpn.*, **47**, 1346–1352.

Kobayashi, H. (1975). *Bull. Chem. Soc. Jpn.*, **48**, 1373–1377.

Kobayashi, H. (1978). *Acta Cryst.*, B**34**, 2818–2825.

Kobayashi, H. (1982). *Bull. Chem. Soc. Jpn.*, **55**, 2693–2696.

Kobayashi, H., Danno, T. and Saito, Y. (1973). *Acta Cryst.*, B**29**, 2693–2699.

Kobayashi, H., Kato, R., Kobayashi, A. and Sasaki, Y. (1985). *Chem. Letts.*, pp. 535–538.

Kobayashi, H., Kato, R., Kobayashi, A., Nishio, Y. and Kajita, K. (1986). *Physica* B + C (Amsterdam), **143**, 512–514.

Kobayashi, H., Marumo, F. and Saito, G. (1971). *Acta Cryst.*, B**27**, 373–378.

Kommandeur, J. (1980). "ESR in alkali TCNQ salts" in *The Physics and Chemistry of Low Dimensional Solids*, edited by L. Alcácer, Reidel, Dordrecht, NATO ASI Ser. C, Vol. 56, pp. 197–212.

Kondo, K., Matsubayashi, G., Tanaka, T., Yashioka, H. and Nakatsu, K. (1984). *J. Chem. Soc., Dalton Trans.*, pp. 379–384.

Konno, M. and Saito, Y. (1973). *Acta Cryst.*, B**29**, 2815–2824.

Konno, M. and Saito, Y. (1974). *Acta Cryst.*, B**30**, 1294–1299.

Konno, M. and Saito, Y. (1975). *Acta Cryst.*, B**31**, 2007–2012.

Konno, M. and Saito, Y. (1981). *Acta Cryst.*, B**37**, 2034–2043.

Konno, M., Ishii, T. and Saito, Y. (1977). *Acta Cryst.*, B**33**, 763–770.

Konno, M., Kobayashi, H., Marumo, F. and Saito, Y. (1973). *Bull. Chem. Soc. Jpn.*, **46**, 1987–1990.

Kramer, G. J., Groeneveld, L. R., Joppe, J. L., Brom, H. B., De Jongh, L. J. and Reedijk, J. (1987). *Synth. Met.*, **19**, 745–750.

Krause, A., Schaefer, H. W. and Helberg, H. W. (1983). *J. Phys. Colloq.*, C**3**, 1429–1432.

Krief, A. (1986). *Tetrahedron*, **42**, 1209–1252.

Kröhnke, C., Enkelmann, V. and Wegner, G. (1980). *Angew. Chem., Int. Ed Engl.*, **19**, 912–919 [pp. 913–918 advertisement].

Kruif, C. G. de, and Govers, H. A. J. (1980). *J. Chem. Phys.*, **73**, 553–555.

LaPlaca, S. J., Corfield, P. W. R., Thomas, R. and Scott, B. A. (1975). *Solid State Commun.*, **17**, 635–638.

Lau, C.-P., Singh, P., Cline, S. J., Seiders, R., Brookhart, M., Marsh, W. E., Hodgson, D. J., and Hatfield, W. E. (1982). *Inorg. Chem.*, **21**, 208–212.

Lee, V. Y., Engler, E. M., Schumaker, R. R. and Parkin, S. S. P. (1983). *J. Chem. Soc., Chem. Comm.*, pp. 235–236.

Legros, J.-P., Bosseau, M., Valade, L. and Cassoux, P. (1983). *Mol. Cryst. Liq. Cryst.*, **100**, 181–192.

Lequon, R.-M., Lequon, M., Jaouen, G., Ouahab, L., Batail, P., Padioux, J. and Sutherland, J. G. (1985). *J. Chem. Soc., Chem. Comm.*, pp. 116–118.

Lindqvist, O., Andersen, L., Seiler, J., Steinmecke, G. and Hoyer, E. (1982). *Acta Chem. Scand.*, A**36**, 855–856.

Lindqvist, O., Sjölin, L., Seiler, J., Steinmecke, G. and Hoyer, E. (1979). *Acta Chem. Scand.*, A**33**, 445–448.

Long, R. E., Sparks, R. A. and Trueblood, K. N. (1965). *Acta Cryst.*, **18**, 932–939.

Lowe, J. P. (1980). *J. Am. Chem. Soc.*, **102**, 1262–1269.

Malatesta, V., Millini, R. and Montanari, L. (1995). *J. Am. Chem. Soc.*, **117**, 6258–6264.

Marks, T. J. (1978). *Ann. N. Y. Acad. Sci.*, **313**, 594–616.

Mathieu, F. (1986). *Acta Cryst.*, C**42**, 1169–1172.

Matsubayashi, G.-E., Ueyama, K. and Tanaka, T. (1985). *Mol. Cryst. Liq. Cryst.*, **125**, 215–222.

Maxfield, M., Cowan, D. O., Bloch, A. N. and Poehler, T. O. (1979). *Nouv. J. Chim.*, **3**, 647–648.

Mayerle, J. J. and Torrance, J. B. (1981a). *Acta Cryst.*, B**37**, 2030–2034.

Mayerle, J. J. and Torrance, J. B. (1981b). *Bull. Chem. Soc. Jpn.*, **54**, 3170–3172.

McCullough, R. D., Kok, G. B., Ledstrup, K. A. and Cowan, D. O. (1987). *J. Am. Chem. Soc.*, **109**, 4115–4116 (for correction see p. 7583).

McGhie, A. R., Garito, A. F. and Heeger, A. J. (1974). *J. Cryst. Growth*, **2**, 295–297. McGhie, A. R., Nigrey, J. F., Yamagishi, F. G. and Garito, A. F. (1978). *Ann. N. Y. Acad. Sci.*, **313**, 293–300.

McPhail, A. T., Semeniuk, G. M. and Chesnut, D. B. (1971). *J. Chem. Soc.*, (A), pp. 2174–2180.

Melby, L. R. (1965). *Can. J. Chem.*, **43**, 1448–1453.

Melby, L. R., Harder, R. J., Hertler, W. R., Mahler, W., Benson, R. E. and Mochel, W. E. (1962). *J. Am. Chem. Soc.*, **84**, 3374–3387.

Metzger, R. M. and Bloch, A. N. (1975). *J. Chem. Phys.*, **63**, 5098–5107.

Metzger, R. M. (1981). "Cohesion and ionicity in organic semiconductors and metals," in *Crystal Cohesion and Conformational Energies, Topics in Current Physics,* No. 26, edited by R. M. Metzger, Springer Verlag, pp. 80–107.

Metzger, R. M., Thiessen, W. E., Hopkins, T. E. and Simpson, P. G. (1974). *American Crystallographic Association*, Spring Meeting, p. 135.

Miane, J. L., Filhol, A., Almeida, M. and Johannsen, B. (1986). *Mol. Cryst. Liq. Cryst.*, **136**, 317–333.

Michaud, M., Carlone, C., Hota, N. K. and Zauhar, J. (1979). *Chem. Phys.*, **36**, 79–84.

Middeldorp, J. A. M., Visser, R. J. J. and de Boer, J. L. (1985). *Acta Cryst.*, B**41**, 369–374.

Miles, M. G., Wilson, J. D. and Cohen, Morell H. (1972). U. S. Patent No. 3779814.

Miller, J. S. and Epstein, A. J. (1976). *Prog. Inorg. Chem.*, **20**, 1–151.

Miller, J. S., Zhang, J. H., Reiff, W. M., Dixon, D. A., Preston, L. D., Reis, A. H., Jr., Gebert, E., Extine, M., Troup, J., Epstein, A. J. and Ward, M. D. (1987). *J. Phys. Chem.*, **91**, 4344–4360.

Miller, J. S., Zhang, Z. H. and Reiff, W. M. (1987). *Inorg. Chem.*, **26**, 600–608.

Mitsuhasi, T., Goto, M., Honda, K., Maruyama, Y., Inabe, T., Sugawara, T. and Watanabe, T. (1988). *Bull. Chem. Soc. Jpn.*, **61**, 261–269.

Mizuno, M., Garito, A. F. and Cava, M. P. (1978). *J. Chem. Soc.*, pp. 18–19.

Morgado, J., Santos, I. C., Veiros, L. F., Henriques, R. T., Duarte, M. T., Almeida, M. and Alcacer, L. (1997). *J. Mater. Chem.*, **7**, 2387–2392.

Mori, T. and Inokuchi, H. (1986). *Solid State Comm.*, **59**, 355–359.

Morssink, A. and Bodegom, B. van. (1981). *Acta Cryst.*, B**37**, 107–114.

Murakami, M. and Yoshimura, S. (1980). *Bull. Chem. Soc. Jpn.*, **53**, 3504–3509.

Nakano, H., Miyawaki, K., Nogami, T., Shirota, Y., Harada, S. and Kasai, N. (1989). *Bull. Chem. Soc. Jpn.*, **62**, 2604–2607.

Nakasuji, K., Kubota, H., Kotani, T., Murata, I., Saito, G., Enoki, T., Imaeda, K., Inokuchi, H., Honda, M., Katayama, C. and Tanaka, J. (1986). *J. Am. Chem. Soc.*, **108**, 3460–3466.

Nakasuji, K., Nakatsuka, M., Yamochi, H., Murata, I., Harada, S., Kasai, N., Yamamura, K., Tanaka, J., Saito, G., Enoki, T. and Inikuchi, H. (1986). *Bull. Chem. Soc., Jpn.*, **59**, 207–214.

Nakasuji, K., Sasaki, M., Kotani, T., Murata, I., Enoki, T., Imaeda, K., Inokuchi, H., Kawamoto, A. and Tanaka, J. (1987). *J. Am. Chem. Soc.*, **109**, 6970–6975.

Narita, M. and Pittman, C. U., Jr. (1976). *Synthesis*, pp. 489–514.

Nielsen, M. B., Lomholt, C. and Becher, J. (2000). *Chem. Soc. Rev.*, **29**, 153–164.

Onnes, H. Kammerlingh and Tuyn, W. (1929). *Int. Crit. Tables*, **6**, p. 128, McGraw-Hill, New York.

Painelli, A., Pecile, C. and Girlando, A. (1986). *Mol. Cryst., Liq. Cryst.*, **134**, 1–19.

Petricek, V., Coppens, P. and Becker, P. (1985). *Acta Cryst.*, A**41**, 478–483.

Phillips, T. E., Kistenmacher, T. J., Bloch, A. N. and Cowan, D. O. (1976). *J. Chem. Soc., Chem. Comm.*, pp. 334–335.

Phillips, T. E., Kistenmacher, T. J., Bloch, A. N., Ferraris, J. P. and Cowan, D. O. (1977). *Acta Cryst.*, B**33**, 422–428.

Pouget, J. P. (1988). "Structural instabilities," in *Semiconductors and Semimetals*, Vol. 27 (Highly Conducting Quasi-One-Dimensional Organic Crystals, edited by E. M. Conwell), Academic Press, Boston, pp. 88–214.

Pouget, J. P., Shapiro, S. M., Shirane, G., Garito, A. F. and Heeger, A. J. (1979). *Phys. Rev.*, B **19**, 1792–1799.

Prinzbach, H., Berger, H. and Lutringhaus, A. *Angew. Chem. Int. Ed. Engl.*, **4**, 435 (1965).

Rasmussen, P. G., Hough, R. L., Anderson, J. E., Bailey, O. H. and Bayon, J. C. (1982). *J. Am. Chem. Soc.*, **104**, 6155–6156.

Reis, A. H., Jr., Preston, L. D., Williams, J. M., Peterson, S. W., Candela, G. A., Swartzendruber, L. J. and Miller, J. S. (1979). *J. Am. Chem. Soc.*, **101**, 2756–2758.

Richard, P., Zanghi, J.-C., Guedon, J.-F. and Hota, N. (1978). *Acta Cryst.*, B**34**, 788–792.

Ristagro, C. V. and Shine, H. T. (1971). *J. Org. Chem.*, **36**, 4050–4055.

Rizkallah, P. J., Wallwork, S. C. and Graja, A. (1983). *Acta Cryst.*, C**39**, 128–131.

Rosseinsky, D. R. and Kathirgamanathan, P. (1982). *Mol. Cryst. Liq. Cryst.*, **86**, 43–55.

Rovira, C., Tarres, J., Llorca, J., Molins, E., Veciana, J., Yang, S., Cowan, D. O., Garrigou-Lagrange, C., Amiel, J., Delhaes, P., Canadell, E. and Pouget, J. P. (1995). *Phys. Rev.*, B**52**, 8747–8758.

Saakyan, G. P., Shibaeva, R. P. and Atovmyan, L. O. (1972). *Dokl. Akad. Nauk. SSSR*, **207**, 1343–1346.

Saeva, T. D., Reynolds, G. A. and Kaszczuk, L. (1982). *J. Am. Chem. Soc.*, **104**, 3524–3525.

Saito, G. and Ferraris, J. P. (1979). *J. Chem. Soc., Chem. Comm.*, pp. 1027–1029.

Saito, G., Enoki, T., Inokuchi, H., Kumagai, H. and Tanaka, J. (1983). *Chem. Letts.*, pp. 503–506.

Saito, G., Enoki, T., Toriumi, K. and Inokuchi, H. (1982). *Sol. State Commun.*, **42**, 557–560.

Sakata, Y., Tatemitsu, H., Yamaguchi, S. and Misumi, S. (1984). *Mem. Inst. Sci. Ind. Res., Osaka Univ.*, **41**, 81–90.

Sandman, D. J., Epstein, A. J., Holmes, T. J. and Fisher, A. P., III. (1977). *J. Chem. Soc., Chem. Comm.*, pp. 177–178.

Sandman, D. J., Epstein, A. J., Holmes, T. J., Lee, J.-S. and Titus, D. D. (1980). *J. Chem. Soc., Perkin II*, pp. 1578–1585.

Sandman, D. J., Fisher, A. P., III, Holmes, T. J. and Epstein, A. J. (1977). *J. Chem. Soc., Chem. Comm.*, pp. 687–688.

Sandman, D. J., Stark, J. C. and Foxman, B. M. (1982). *Organometallics*, **1**, 739–742.

Sanz, F. and Daly, J. J. (1975). *J. Chem. Soc., Perkin II*, pp. 1146–1150.

Sato, M., Lakshmikanatham, M. V., Cava, M. P. and Garito, A. P. (1978). *J. Org. Chem.*, **43**, 2084–2085.

Schafer, D. E., Thomas, G. A. and Wudl, F. (1975). *Phys. Rev.*, B**12**, 5532–5535.

Schukat, G., Richter, A. M. and Fanghängel, E. (1987). *Sulfur Rep.*, **7**, 155–231.

Schultz, A. J. and Stucky, G. D. (1977). *J. Phys. Chem. Solids*, **38**, 269–273.

Schultz, A. J., Stucky, G. D., Craven, R., Schaffman, M. J. and Salamon, M. B. (1976). *J. Am. Chem. Soc.*, **98**, 5191–5197.

Schultz, A. J., Stucky, G. D., Blessing, R. H. and Coppens, P. (1976). *J. Am. Chem. Soc.*, **98**, 3194–3201.

Schulz, T. D. (1980). "Physics in one dimension", in *The Physics and Chemistry of Low Dimensional Solids*, edited by L. Alcácer, NATO ASI Series C, Vol 56, Reidel, Dordrecht, pp. 1–30.

Scott, B. A., La Placa, S. J., Torrance, J. B., Silverman, B. D. and Welber, B. (1977). *J. Am. Chem. Soc.*, **99**, 6631–6639.

Scott, H., Kronick, H., Chairge, P. and Labes, M. M. (1965). *J. Phys. Chem.*, **69**, 1740–1742.

Scott, J. C. (1988). "Magnetic properties," in *Semiconductors and Semimetals*, Vol. 27 (Highly Conducting Quasi-One-Dimensional Organic Crystals, edited by E. M. Conwell), Academic Press, Boston, pp. 293–384; pp. 386–436.

Shchegolev, I. F. and Yagubskii, E. B. (1982). "Cation-radical salts of tetrathiotetracene and tetraselenotetracene: synthetic aspects and physical properties." In *Extended Linear Chain Compounds*, edited by J. S. Miller, Plenum Press, London, Vol. 2, pp. 385–434.

Shibaeva, R. P. (1982). "Structural aspects of one-dimensional conductors based on tetrathiotetracene and tetraselenotetracene," In *Extended Linear Chain Compounds*, edited by J. S. Miller, Plenum, New York; Vol. 2, 435–467.

Shibaeva, R. P. (1983). *Sov. Phys. Cryst.* **28**, 644–646; *Kristallgrafiya*, **28**, 1094–1096.

Shibaeva, R. P. and Atovmyan, L. O. (1972). *J. Struct. Chem. U.S.S.R*, **4**, 514–531.

Shibaeva, R. P. and Ponomarev, V. I. (1975). *Sov. Phys. Cryst.*, **20**, 183–186.

Shibaeva, R. P. and Yarochkina, O. Y. (1975). *Sov. Phys. Dokl.*, **20**, 304–305.

Shibaeva, R. P., Atovmyan, L. O. and Ponomarev, V. I. (1975). *J. Struct. Chem. USSR*, **16**, 792–795; *Zh. Strukt. Khim.*, **16**, 860–865.

Shibaeva, R. P., Atovmyan, L. O. and Rozenberg, L. P. (1971). *Tetrahedron Letts.*, pp. 3303–3306.

Shibaeva, R. P., Atovmyan, L. O. and Rozenberg, L. P. (1975). *J. Struct. Chem. USSR*, **16**, 135–137; *Zh. Strukt. Khim.*, **16**, 147–149.

Shibaeva, R. P., Atovmyan, L. O., Ponomarev, V. L., Filipenko, O. S. and Rozenberg, L. P. (1974). *Sov. Phys. Crystallogr.*, **19**, 54–57.

Shibaeva, R. P., Kaminskii, V. F. and Simonov, M. A. (1980). *Cryst. Struct. Comm.*, **9**, 655–661.

Shibaeva, R. P., Rozenberg, L. P. and Atovmyan, L. O. (1973). *Sov. Phys. Cryst.*, **18**, 326–329.

Shibaeva, R. P., Rozenberg, L. P., Aldoshina, M. Z., Lyubovskaya, R. N. and Khidekel, M. L. (1979). *J. Struct. Chem. USSR*, **20**, 409–414.

Shibaeva, R. P., Shvets, A. E. and Atovmyan, L. O. (1971). *Dokl. Akad. Nauk SSSR*, **199**, 334–337.

Shirotani, L. and Kobayashi, H. (1973). *Bull. Chem. Soc. Jpn.*, **46**, 2595–2596.

Silverman, B. D. and LaPlaca, S. J. (1978). *J. Chem. Phys.*, **69**, 2585–2589.

Silverman, B. D. (1981). "Slipped versus eclipsed stacking of tetrathiafulvalene (TTF) and tetracyanoquinodimethane (TCNQ) Dimers", in *Crystal Cohesion and Conformational Energies, Topics in Current Physics*, No. 26, pp. 108–136, edited by R. M. Metzger, Springer Verlag, Berlin.

Skelton, E. F., Bloch, A. N., Ferraris, J. P. and Cowan, D. O. (1974). *Phys. Lett.*, **47**A, 313–314.

Soling, H., Rindorf, G. and Thorup, N. (1981). *Acta Cryst.*, B**37**, 1716–1719.

Somoano, R., Hadek, V., Yen, S. P. S., Rembaum, A. and Deck, R. (1975). *J. Chem. Phys.*, **62**, 1061–1067.

Soos, Z. G. (1965). *J. Chem. Phys.*, **43**, 1121–1140.

Spencer, H. K., Cava, M. P. and Garito, A. F. (1977). *J. Chem. Soc., Chem. Comm.*, pp. 966–967.

Starikov, E. B. (1998). *Int. J. Quant. Chem.*, **66**, 47–68.

Steurer, W., Visser, R. J. J., van Smaalen, S. and de Boer, J. L. (1987). *Acta Cryst.*, B**43**, 567–574.

Stokes, J. P., Emge, T. J., Bryden, W. A., Chappell, J. S., Cowan, D. O., Poehler, T. O., Bloch, A. N. and Kistenmacher, T. J. (1982). *Mol. Cryst. Liq. Cryst.*, **79**, 327–336.

Su, W. P and Schrieffer, J. R. (1981). *Phys. Rev. Letts.*, **46**, 738–741.

Suguira, K., Mikami, S., Iwasaki, K., Hino, S., Asato, E. and Sakata, Y. (2000). *J. Mater. Chem.*, **10**, 315–319.

Sundaresan, T. and Wallwork, S. C. (1972a). *Acta Cryst.*, B**28**, 491–497.

Sundaresan, T. and Wallwork, S. C. (1972b). *Acta Cryst.*, B**28**, 1163–1169.

Sundaresan, T. and Wallwork, S. C. (1972c). *Acta Cryst.*, B**28**, 2474–2480.

Sundaresan, T. and Wallwork, S. C. (1972d). *Acta Cryst.*, B**28**, 3065–3074.

Sundaresan, T. and Wallwork, S. C. (1972e). *Acta Cryst.*, B**28**, 3507–3511.

Suzuki, T., Kabuto, C., Yamashita, Y., Saito, G., Mukai, T. and Miyoshi, T. (1987). *Chem. Lett.*, pp. 2285–2288.

Tanaka, C., Tanaka, J., Dietz, K., Katayama, C. Tanaka, M. (1983). *Bull. Chem. Soc. Jpn.*, **56**, 405–411.

Taylor, R. and Kennard, O. (1986). *Acta Cryst.*, B**42**, 112–120.

Teitelbaum, R. C., Marks, T. J. and Johnson, C. K. (1980). *J. Am. Chem. Soc.*, **102**, 2986–2989.

Thomas, D. D., Keller, H. and McConnell, H. M. (1963). *J. Chem. Phys.*, **39**, 2321–2329.

Thomas, G. A., Schafer, D. A., Wudl, F., Horn, P. M., Rimai, D., Cook, J. W., Glocker, D. A., Skove, M. J., Chu, C. W., Groff, R. P., Gillson, J. L., Wheland, R. C., Melby, R. L., Salamon, M. B., Craven, R. A., De Pasquali, G., Bloch, A. N., Cowan, D. O., Walatka, V. V.,

Pyle, R. E., Gemmer, R., Poehler, T. O., Johnson, G. R., Miles. M. G., Wilson, J. D., Ferraris, J. P., Finnegan, T. F., Warmack, R. J., Raaen, V. F. and Jerome, D. (1976). *Phys. Rev.*, **B13**, 5105–5110.

Thorup, N., Rindorf, G., Jacobsen, C. S., Bechgaard, K., Johannsen, I. and Mortensen, K. (1985). *Mol. Cryst. Liq. Cryst.*, **120**, 349–352.

Tilak, B. D. (1951). *Proc. Indian Acad. Sci.*, **33A**, 71–77.

Tomkiewicz, Y. (1980). "EPR of organic conductors", in *The Physics and Chemistry of Low Dimensional Solids*, edited by L. Alcácer, NATO ASI Series C, Vol 56, Reidel: Dordrecht., pp.187–195.

Tormos, G. V., Bakker, M. G., Wang. P., Lakshmikanathan, M. V., Cava, M. P. and Metzger, R. M. (1995). *J. Am. Chem. Soc.*, **117**, 8528–8535.

Torrance, J. B. (1978). *Ann. N. Y. Acad. Sci.*, **313**, 210–233.

Torrance, J. B., Mayerle, J. J., Bechgaard, K., Silverman, B. D. and Tomkiewicz, Y. (1980). *Phys. Rev.*, **B22**, 4960–4965.

Torrance, J. B., Mayerle, J. J., Lee, V. J. and Bechgaard, K. (1979). *J. Am. Chem. Soc.*, **101**, 4747–4748.

Truong, K. D., Bandrauk, A. D., Ishii, K., Carlone, C. and Jandl, S. (1985). *Mol. Cryst. Liq. Cryst.*, **120**, 105–110.

Uchida, T., Ito, M. and Kozawa, K. (1983). *Bull. Chem. Soc. Jpn.*, **56**, 577–582.

Ueda, N., Natsume, B., Yanagiuchi, K., Sakata, Y., Enoki, T. Saito, G., Inokuchi, H. and Misumi, S. (1983). *Bull. Chem., Soc. Jpn.*, **56**, 775–779.

Ueno, Y., Nakayama, A. and Okawara, M. (1978). *J. Chem. Soc., Chem. Comm.*, pp. 74–75.

Valade, L., Legros, J.-P., Bousseau, M., Cassoux, P., Garbauskas, M. and Interrante, L. V. (1985). *J. Chem. Soc., Dalton Trans.*, pp. 783–794.

Valade, L., Legros, J.-P., Cassoux, P. and Kubel, F. (1986). *Mol. Cryst. Liq. Cryst.*, **140**, 335–351.

Visser, R. J. J., Bouwmeester, H. J. M., de Boer, J. L. and Vos, A. (1990a). *Acta Cryst.*, **C46**, 852–856.

Visser, R. J. J., Bouwmeester, H. J. M., de Boer, J. L. and Vos, A. (1990b). *Acta Cryst.*, **C46**, 860–864.

Visser, R. J. J., de Boer, J. L. and Vos, A. (1990a). *Acta Cryst.*, **C46**, 857–860.

Visser, R. J. J., de Boer, J. L. and Vos, A. (1990b). *Acta Cryst.*, **C46**, 864–866.

Visser, R. J. J., de Boer, J. L. and Vos, A. (1990c). *Acta Cryst.*, **C46**, 869–871.

Visser, R. J. J., de Boer, J. L. and Vos, A. (1993). *Acta Cryst.*, **B49**, 859–867.

Visser, R. J. J., de Boer, J. L. and Vos, A. (1994). *Acta Cryst.*, **C50**, 1139–1141.

Visser, R. J. J., de Boer, J. L. Smaalen, S., van, and Vos, A. (1990). *Acta Cryst.*, **C46**, 867–869.

Visser, R. J. J., Oostra, S., Vettier, C. and Voiron, J. (1983). *Phys. Rev.*, B **24**, 2074–2077.

Waclawek, W., Wallwork, S. C., Ashwell, G. J. and Chaplain, E. J. (1983). *Acta Cryst.*, **C39**, 131–134.

Wal, R. J. van der and Bodegom, B. van. (1978). *Acta Cryst.*, **B34**, 1700–1702.

Wal, R. J. van der and Bodegom, B. van. (1979). *Acta Cryst.*, **B35**, 2003–2008.

Warmack, R. J., Callcot, T. A. and Watson, C. R., Jr. (1975). *Phys. Rev.*, **B12**, 3336–3338.

Webster, O. W. (1964). *J. Am. Chem. Soc.*, **86**, 2898–2902.

Weger, M. (1980). "Theory of the electrical conductivity of organic metals," in *The Physics and Chemistry of Low Dimensional Solids*, edited by L. Alcácer, NATO ASI Series C, Vol 56, Reidel, Dordrecht etc., pp. 77–100.

Wei, T., Etemad, S., Garito, A. F. and Heeger, A. J. (1973). *Phys. Lett.*, **45**A, 269–270.

Weidenborner, J. E., La Placa, S. J. and Engler, E. M. (1977). *American Crystallographic Association*, Ser. 2, p. 74.

Wheland, R. C. and Martin, R. L. (1975). *J. Org. Chem.*, **40**, 3101–3109.

Willi, C., Reis, A. H., Jr., Gebert, E. and Miller, J. S. (1980). *Inorg. Chem.*, **20**, 313–318.

Williams, J. M., Beno, M. A., Wang, H. H., Leung, P. C. W., Emge, T. J., Geiser, U. and Carson, K. D. (1985). *Accts. Chem. Res.*, **18**, 261–267.

Williams, R., Lowe Ma, C., Samson, S., Khanna, S. K. and Somoano, R. B. (1980). *J. Chem. Phys.*, **72**, 3781–3788.

Wolmerhäuser, G., Schnauber, M. and Wilhelm, T. (1984). *J. Chem. Soc., Chem. Comm.*, pp. 2693–2696.

Wood, R. W. (1946). *Physical Optics*, pp. 254–260, MacMillan, New York, 3rd Edition.

Wudl, F. and Southwick, E. W. (1974). *J. Chem. Soc., Chem. Comm.*, pp. 254–255.

Wudl, F., Nalawajek, D., Troup, J. M. and Extine, M. W. (1983). *Science*, **222**, 415–417.

Wudl, F. (1984). *Accts. Chem. Res.*, **17**, 227–232.

Wudl, F., Schafer, D. E. and Miller, B. (1976). *J. Am. Chem. Soc.*, **98**, 252–254.

Wudl, F., Smith, G. M. and Hufnagel, E. J. (1970). *J. Chem. Soc. Chem. Comm.*, pp. 1453–1454.

Wudl, F., Wobschall, D. and Hufnagel, E. J. (1972). *J. Am. Chem. Soc.*, **94**, 670–672.

Wudl, F., Zellers, E. T. and Bramwell, F. B. (1978). *J. Org. Chem.*, **44**, 2491–2493.

Yakushi, K., Nishimura, S., Sugano, T., Kuroda, H. and Ikemoto, I. (1980). *Acta Cryst.*, **B36**, 358–363.

Yamaguchi, S., Miyamoto, K. and Hanafusa, T. (1989). *Bull. Chem. Soc. Jpn.*, **62**, 3036–3037.

Yamaji. K., Megtert, S. and Comes, R. (1981). *J. Physique,* **42**, 1327–1343.

Yamashita, Y., Saito, K., Suzuki, T., Kabuto, C., Mukai, T. and Miyashi, T. (1988). *Angew. Chem. Int. Ed. Engl.*, **27**, 434–435.

Yamashita, Y., Suzuki, T., Mukai, T. and Saito, K. (1985). *J. Chem. Soc., Chem. Comm.*, pp. 1044–1045.

Yamashita, Y., and Tomura, M. (1998). *J. Mater. Chem.*, **8**, 1933–1944.

Yan, G. and Coppens, P. (1987). *Acta Cryst.*, **C43**, 1610–1613.

Yasui, M., Hirota, M., Endo, Y., Iwasaki, F. and Kobayashi, K. (1992). *Bull. Chem. Soc. Jpn.*, **65**, 2187–2191.

Yui, K., Ishida, H., Aso, Y., Otsubo, T., Ogura, F., Kawamoto, A. and Tanaka, J. (1989). *Bull. Chem. Soc. Jpn.*, **62**, 1547–1555.

Appendix 1

Thermodynamic measurements on binary adducts

A.1   Introduction

Treatments of the thermodynamics of solids (Westrum and McCullough, 1963; Swalin, 1987; Gaskell, 1995) will not be duplicated here. Instead, we intend only to summarise the specific features of the various methods that have been used to determine thermodynamic parameters for crystalline binary adducts (molecular compounds and complexes). Quantitative results for the different types of binary adduct, and consequent structural implications, are discussed in the body of the text.

We principally consider the thermodynamic parameters of the reaction in which a crystalline binary adduct is formed from its crystalline components, i.e.

$$m\mathrm{D_c} + n\mathrm{A_c} \Leftrightarrow [\mathrm{D}_m\mathrm{A}_n]_c$$

$$\Delta X_c = X\,(\mathrm{D}_m\mathrm{A}_n)_c - \{\,mX(\mathrm{D_c}) + nX(\mathrm{A_c})\}$$

Here ΔX_c represents Gibbs free energy (G), enthalpy (H), entropy (S) or volume (V) of formation of the crystalline molecular compound, $[\mathrm{D}_m\mathrm{A}_n]_c$, from its components, $\mathrm{D_c}$ and $\mathrm{A_c}$; the subscripts denote that the substances are in the crystalline state. In most of the following we shall set $m = n = 1$ for simplicity. If the components and/or the binary adduct are polymorphic, then care must be taken to define the polymorphs involved in the reaction. The ΔX_c values refer to the temperatures at which the measurements were made. We shall, in general, restrict ourselves to atmospheric pressure, although pressure provides another variable that has hardly been explored.

Use of thermodynamic functions for the *crystalline* substances implies that we hope to extract from the values of ΔX_c information about the factors which lead to the formation of the binary adduct; in general such hopes will be best realized when the nature of the components changes minimally on formation of the binary adduct (cf. our original definition in Chapter 1). Thus the thermodynamic functions of packing complexes are expected to be easier to interpret than those of, say, molecular compounds of metal salts. Formation of a binary adduct means that ΔG_c must be negative, its magnitude giving a measure of the stability of the binary adduct. However, the individual values of ΔH_c and ΔS_c are more informative from a structural point of view. As holds in general, three different situations are possible at a particular temperature:

(i)   when ΔH_c is negative and ΔS_c positive, then the binary adduct will be both enthalpy and entropy stabilized;

(ii)   when ΔH_c and ΔS_c are both positive, then ΔG_c will be negative if $T\Delta S_c > \Delta H_c$ and the binary adduct will be entropy stabilized;

(iii) when ΔH_c and ΔS_c are both negative, then ΔG_c will be negative if $T|\Delta S_c| < |\Delta H_c|$ and the binary adduct will be enthalpy stabilized.

All three situations are realised in practice and a detailed illustration is given in Chapter 16 where values (at 298K) of ΔH_c are plotted against ΔS_c for some π-molecular compounds (see Fig. 16.1).

A.2   Experimental methods

A.2.1   *Methods that permit determination of all three thermodynamic parameters*

A.2.1.1   *Electrochemical method*

This method was first used by Brønsted (1911), extended by Bell and Fendley (1949) and described in detail by Abdel-Rehiem *et al.* (1974). The EMF difference is measured between the two cells shown, a picric acid molecular compound being used for illustration.

$$\mathbf{E}_1: \text{ glass electrode } \left|\begin{array}{l}\text{saturated picric}\\\text{acid solution}\end{array}\right. \Big/\!\Big/ \text{ Hg}^{\text{I}} \text{ picrate}\,\Big|\,\text{Hg(I)}$$

$$\mathbf{E}_2: \text{ glass electrode } \left|\begin{array}{l}\text{solution saturated with}\\\text{both picric acid molecular}\\\text{compound and donor}\end{array}\right. \Big/\!\Big/ \text{ Hg}^{\text{I}} \text{ picrate}\,\Big|\,\text{Hg(I)}$$

It is assumed that the complexed picric acid does not ionize; the results do not depend on the solvent (water or mixed solvents such as $9:1$ CH_3CN-H_2O have been used). The free energies are obtained from the EMF measurements as follows:

$$\Delta G_c = -F(E_2 - E_1)$$

where F is the Faraday. The entropy of formation ΔS_c is obtained from the temperature dependence of the EMF difference, i.e.

$$\Delta S_c = (\partial[\Delta G_c] / \partial T)_p$$

The enthalpy of formation values $\Delta H_c(298)$ are then obtained from the Gibbs-Helmholtz equation $\{\Delta H = \Delta G - T(\partial[\Delta G] / \partial T)_p\}$. A typical set of experimental measurements is shown in Fig. A1.1. This method appears to have been applied only to π-molecular compounds and the results are discussed in Chapter 16.2.

A.2.1.2   *Solubility method*

If c_1 is the solubility of component A in a suitable solvent and c_2 its solubility in the solvent saturated with respect to both the other component and the binary adduct, then $\Delta G_c = RT \ln(c_2/c_1)$. The assumption is made that the solutions are sufficiently dilute for activities to be proportional to concentrations. Repetition of the measurements at a number of temperatures allows calculation of ΔS_c and ΔH_c. The method was used to determine the ΔG_c values at 298K for a number of TNB and TNT π-molecular compounds of various aromatic donors (Hammick and Hutchison, 1955), water being used as solvent for these systems; values of ΔS_c and ΔH_c were not obtained.

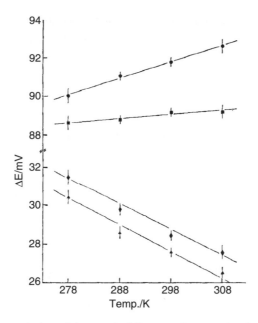

Fig. A1.1. Temperature variation of the e.m.f. differences for π molecular compound formation. The curves are (in order, from the top) for {naphthalene···picric acid}; {[^{2}H$_8$]-naphthalene···picric acid}; {naphthalene···styphnic acid}; {[^{2}H$_8$]-naphthalene···styphnic acid}. (Reproduced from Shahidi and Farrell (1976).)

A.2.1.3 *Dissociation pressure method*

We illustrate using results obtained (Choi and Brown, 1966) for the thermal decomposition of {C$_6$H$_6$ ... Al$_2$Br$_6$} according to the equation:

$$\{C_6H_6 \ldots Al_2Br_6\}(s) \Rightarrow Al_2Br_6(s) + C_6H_6(g).$$

The equilibrium constant $K_p = P$ and the free energy of the dissociation reaction is given by $\Delta G = -RT \ln K_p$. If the dissociation pressure is measured at a number of temperatures, then the enthalpy of dissociation is obtained from the van't Hoff equation

$$d(\ln K_p)/dT = \Delta H/RT^2.$$

The dissociation pressures were measured as 11.5 mm at 273K, 0.650 mm at 237.3K and 0.223 mm at 229.5K; we calculate $\Delta H = 42.5$ kJ/mol, with $\Delta G = 9.5$ kJ/mol and $\Delta S = 121$ J/mol K (all at 273K), in agreement with the Choi–Brown values ($\Delta H = 44.4$ kJ/mol; $\Delta S = 125.5$ J/mol K). However, the reaction relevant in the present context is

$$Al_2Br_6 \text{ (s)} + C_6H_6 \text{ (s)} \Rightarrow \{C_6H_6\cdots Al_2Br_6\}(s).$$

and so the above values must be corrected for the enthalpy and entropy of sublimation of benzene, where, as an approximation, we use the values appropriate to the melting point of benzene (278K), which are ΔH(sublimation) $= 44.0$ kJ/mol and ΔS(sublimation) $= 44.0/278 = 158$ J/mol K. Thus $\Delta H_c = -42.5 - (-44.0) = +1.5$ kJ/mol and $\Delta S_c = -121 - (-158) = +37$ J/mol K. The molecular compound is entropy stabilized with respect to its crystalline components.

A.2.1.4   *Specific heat measurements*

The thermodynamic functions of a crystal over a range of temperatures can be obtained from measurements of the specific heat c_p as a function of temperature:

Enthalpy funtion $\qquad\qquad H(\mathrm{T}) - H(0) = \int_0^T c_p \, \mathrm{d}T$

entropy $\qquad\qquad\qquad S(\mathrm{T}) - S(0) = \int_0^T c_p \, \mathrm{d}(\ln T)$

free energy function $\quad [G(T) - H(0)]/T = [G(T) - H(0)]/T - [S(T) - S(0)].$

$H(0)$ is a constant characteristic of the substance, while $S(0)$ is set equal to zero for perfect crystalline materials, in accordance with the third law of thermodynamics; thus care must be taken when there is reason to believe that a material has residual zero-point entropy. Specific heat measurements on a binary adduct and its crystalline components will give as functions of temperature the enthalpy of formation of the adduct ΔH_c, (apart from the contribution of the three $H(0)$ values), the entropy of formation ΔS_c (subject to the applicability of the third law), and the free energy of formation function, (apart from the contribution of the three $H(0)$ values). These functions all represent integrals (of different kinds) over the specific heats. This is the only method that will give values for the thermodynamic parameters at low temperatures. It has been suggested that $\Delta c_p = \{c_p$ (binary adduct)$- [c_p(\text{component 1}) + c_p(\text{component 2})]\}$ could give useful information (Dunn, Rahman and Staveley, 1978) but it seems preferable to work with the enthalpy function and the entropy.

We illustrate the calculations for the yellow polymorph of {naphthalene$\cdots$PMDA}, where the results are simplified by the absence of phase transformations in molecular compound and components; the sources of the thermodynamic data are: naphthalene (McCullogh, Finke, Messerly, Todd, Kincheloe and Waddington 1955); PMDA (Dunn, Rahman and Staveley, 1978); {naphthalene$\cdots$PMDA} (Boerio-Goates and Westrum, 1980). For convenience in plotting we show values for ΔH and $T\Delta S$.

The values of ΔH in Fig. A1.2 do not include the contributions of the three H(0) values and thus the position of the curve along the ordinate is not known; this could be fixed by an independent measurement of ΔH_c but such is not available for (naphthalene$\cdots$PMDA) nor, indeed, for any of the other π-molecular compounds for which specific heats have been measured. The requirement of negative ΔG_c means that {H(0)(naphthalene$\cdots$PMDA) $-$ [H(0)(naphthalene) $+$ H(0)(PMDA)]} must be less than $\approx$1100 J/mol at 300K. Application of the third law gives ΔS equal to ΔS_c. The temperature dependence of ΔC_p is qualitatively similar to that of ΔS_c. At this stage we cannot say whether {naphthalene$\cdots$PMDA} is only entropy stabilized or entropy and enthalpy stabilised. Both ΔH and $T\Delta S$ are strongly temperature dependent.

A.2.2   *Determination of ΔG_c only*

Measurement of the depression of the freezing point gives ΔG_c at a particular temperature (Brown, 1925); thus ΔS_c and ΔH_c cannot be determined.

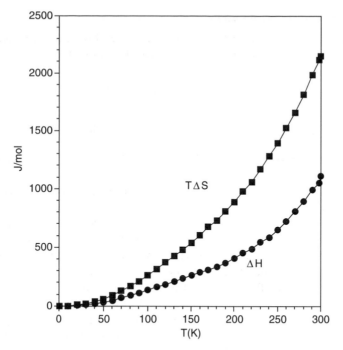

Fig. A1.2. The temperature dependence of ΔH and $T\Delta S$ (see text for definitions) for {naphthalene···PMDA}.

A.2.3   *Determination of ΔH_c only*

A.2.3.1   *Combustion method*

This method will give absolute enthalpies and is illustrated here for quinhydrone (Suzuki and Seki, 1953). The enthalpies of combustion (kJ/mol) of crystalline quinhydrone and its components were measured and their difference gives ΔH_c(quinhydrone).

A.2.3.2   *Dissolution method*

This method was introduced by Suzuki and Seki (1953) for the study of quinhydrone; further applications have been made to other quinhydrones (Artiga, Gaultier, Haget and Chanh, 1978) and to some π-molecular compounds (Suzuki and Seki, 1955). The difference is measured between the enthalpies of solution (under the same conditions) of the same quantities of binary adduct and a mechanical mixture of the two components.

The results obtained for quinhydrone from measurement of enthalpies of combustion are compared in Table A1.1 with those obtained by the enthalpy of dissolution method at 24°, using acetone as solvent.

A.2.3.3   *Phase diagram method*

The enthalpy of formation of a binary adduct at its melting point (T_m) from the *liquid* components (reaction below) can be calculated from the detailed form of the

Table A1.1. Measurement of enthalpy of formation of quinhydrone by various methods. All values are in kJ/mol

Method	Hydroquinone	p-Benzoquinone	Quinhydrone	ΔH_c(Quinhydrone)
Enthalpy of combustion	2863.0 ± 1.7	2746.3 ± 2.5	5575.4 ± 4.6	-33.9 ± 5.5
Enthalpy of dissolution	1.54	-17.27	-38.25	-22.52 ± 0.18

Table A1.2. Calculation of enthalpy of formation for a binary adduct by the method of Boerio-Goates, Goates, Ott, and Goates, (1987)

Reactants (state, T(K))	Products (state, T(K))	ΔH (kJ/mol)
C_6H_6(l, 239.12) + CCl_4(l, 239.12)	{$C_6H_6 \ldots CCl_4$}(s, 239.12)	-15.05
C_6H_6(s, 278.66)	C_6H_6(l, 278.66)	$+9.87$
C_6H_6(l, 278.66)	C_6H_6(l, 239.12)	-0.50^*
CCl_4(s, 250.41)	CCl_4(l, 250.41)	$+2.52$
CCl_4(l, 250.41)	CCl_4(l, 239.12)	-0.08^*
C_6H_6(s, 239.12) + CCl_4(s, 239.12)	{$C_6H_6 \ldots CCl_4$}(s, 239.12)	-3.24

liquidus (solid–liquid equilibrium) curves (Boerio–Goates, Goates, Ott, and Goates, 1985).

$$m\text{A}(l) + n\text{B}(l) = [\text{A}_m\text{B}_n](s) \text{ at } T_m.$$

A thermodynamic cycle was then used to allow for the enthalpies of fusion of A and B and their specific heats in order to calculate $\Delta H_c(T_m)$ for the reaction

$$m\text{A}_c(T_m) + n\text{B}_c(T_m) = [\text{A}_m\text{B}_n]_c(T_m).$$

We illustrate for {$C_6H_6 \cdots CCl_4$} ($T_m = 239.12$K), using data from Boerio-Goates, Goates, Ott, and Goates, (1987) (Table A1.2).

Enthalpies of fusion are more generally available than specific heats and thus approximate values of ΔH_c can be obtained if the generally small specific heat contributions (asterisked) can be neglected. These methods have been applied to a number of binary adducts (Goates, Boerio-Goates, Goates and Ott, 1987); the uncertainties in the enthalpies were conservatively estimated to be ± 0.50 kJ/mol.

References

Abdel-Rehiem, A.G., Farrell, P. G. and Westwood, J. V. (1974). *J. Chem. Soc. Faraday I*, **70**, 1762–1771.

Artiga, A., Gaultier, J., Haget, Y. and Chanh, N. B. (1978). *J. Chim. Phys.*, **75**, 378–383.

Bell, R. P. and Fendley, J. A. (1949). *Trans. Farad. Soc.*, **45**, 121–122.

Boerio-Goates, J. and Westrum, E. F., Jr. (1980). *Mol. Cryst. Liq. Cryst.*, **60**, 249–266.

Boerio-Goates, J., Goates, S. R., Ott, J. B. and Goates, J. R. (1985). *J. Chem. Thermodynam.*, **17**, 665–670.

Brønsted, J. H. (1911). *Z. phys. Chem.*, **78**, 284–292.

Brown, F. S. (1925). *J. Chem. Soc.*, pp. 345–348.

Choi, S. U. and Brown, H. C. (1966). *J. Amer. Chem. Soc.*, **88**, 903–909.

Dunn, A. G., Rahman, A. and Staveley, L. A. K. (1978). *J. Chem. Thermodynamics*, **10**, 787–796.

Gaskell, D. R. (1995). "Introduction to the Thermodynamics of Materials," Third Edition, Taylor and Francis, Washington, D. C.

Goates, J. R., Boerio-Goates, J., Goates, S. R. and Ott, J. B. (1987). *J. Chem. Thermodynamics*, **19**, 103–107.

Hammick, D. Ll. and Hutchison, H. P. (1955). *J. Chem. Soc.*, pp. 89–91.

McCullogh, J. P., Finke, H. L., Messerly, J. F., Todd, S. S., Kincheloe, T. C. and Waddington, G. (1957). *J. Phys. Chem.*, **61**, 1105–1161.

Shahidi, F. and Farrell, P. G. (1978). *J. Chem. Soc. Chem. Comm.*, pp. 455–456.

Suzuki, K. and Seki, S. (1953). *Bull. Chem. Soc. Jpn.*, **26**, 372–380.

Suzuki, K. and Seki, S. (1955). *Bull. Chem. Soc. Jpn.*, **28**, 417–421.

Swalin, R. A. (1987). "Thermodynamics of Solids." 2nd edition. 400 pp. Wiley–VCH, Berlin.

Westrum, E.F.,Jr. and McCullough, J.D. (1963). "Thermodynamics of crystals" in *Physics and Chemistry of the Organic Solid State*, edited by D. Fox, M.M. Labes, and A. Weissberger, Interscience, New York and London.

Book Index

Some notes:

1. Although there is no separate Author Index, names of prominent personalities are indexed together with some leading references.
2. Many, but not all, crystal structures mentioned in the text are indexed. CSD Refcodes are given where possible.
3. Information given in the overall and chapter Tables of Contents is only partially duplicated here.